JORDAN L.

SOURCE BOOK ON FORMING OF STEEL SHEET

A discriminative selection
of outstanding articles
from the periodical and
reference literature

Compiled by
Consulting Editor:

John R. Newby
Senior Staff Metallurgist
Research and Technology Division
Armco Steel Corporation

AMERICAN SOCIETY FOR METALS
Metals Park, Ohio 44073

First printing, October 1976

Library of Congress Cataloging in Publication Data
Source book on forming of steel sheet

Includes Index.
1. Sheet-steel—Addresses, essays, lectures.
2. Steelwork—Addresses, essays, lectures.
I. Newby, John R. II. American Society for Metals.
TS360.S68 672'.8'23 76-28176

Library of Congress Catalog Card Number: 76-28176

PRINTED IN THE UNITED STATES OF AMERICA

Preface

The volume of parts produced by the forming of metal sheet is a measure of the industrial importance of the metal-forming processes. The annual total of formed parts produced by the automotive industry alone is staggering, as is the tonnage of low-carbon steel sheet required to fulfill the burgeoning demand.

Because of its versatility, strength and low cost, low-carbon steel sheet is the material most widely used in the production of formed parts. Quite logically, therefore, most of the articles in this book deal with the forming of low-carbon steel sheet and with the innumerable improvements—both metallurgical and mechanical—that have contributed to parts-forming technology. A significant portion of that technology arises from, and depends on, the evaluation of a critical and complex property of sheet materials—formability.

SOURCE BOOK ON FORMING OF STEEL SHEET addresses all aspects of forming technology as it applies to steel sheet. The contents of the book have been arranged in eight principal sections, reflecting specific areas of interest that the reader may wish to pursue. There is also an index to help guide the reader to specific subject matter. A brief summary of the contents of the book by section suggests its breadth of coverage.

Classification of Sheet Forming Operations. This 83-page section contains five articles that reflect the experience and practice of forming a wide variety of components in several major industries, including automotive and appliance manufacturing. The section concludes with the comprehensive article on deep drawing that appeared originally in Vol. 4 of the *Metals Handbook* (8th Edition).

Sheet Materials for Forming. This section contains five articles that, in effect, serve as a materials guide to the selection of carbon and high-strength low-alloy steel sheet for forming applications. Unlike the conventional materials guide in tabular form, these articles provide in-depth treatment of key variables that affect formability, including mill processing controls, mechanical properties, aging characteristics and surface condition.

Metallurgical Properties of Steel Sheet. This section, which contains four articles, opens with the classic ASM Campbell Memorial Lecture, "Sheet Steel—Micrometallurgy by the Millions," by D. J. Blickwede. This article provides a comprehensive and fascinating treatment of the subjects of the mechanics of forming, anisotropy, and the metallurgical and processing characteristics of steel sheet. It is supplemented by three outstanding articles of more recent vintage.

Formability Tests for Steel Sheet. In common with the previous section, this section begins with a classic technical paper—a paper whose impact on criteria for predicting sheet-metal formability (and consequently on test methods and theory) has probably never been exceeded. It is followed by R.L. Whiteley's fundamental treatment of the importance of directionality in drawing-quality sheet steel and two other definitive articles concerned with formability testing.

Blanking. This section brings together Harding Hugo's practical treatise on improving metal stamping die performance and the fundamental coverage of blanking provided by the *Metals Handbook* articles "Blanking of Low-Carbon Steel" and "The Selection of Material for Blanking and Piercing Dies."

Forming and Drawing Dies and *Lubricants.* The articles "The Selection of Material for Press Forming Dies" and "The Selection of Material for Deep Drawing Dies," both taken from Vol. 1 of *Metals Handbook,* and the article "Selection and Use of Lubricants in Forming of Sheet Metal," taken from Vol. 4 of *Metals Handbook,* comprise the contents of these sections.

Presses and Auxiliary Equipment. The concluding section of the book provides the reader with two basic references on presses and auxiliary equipment—the article "Selection of Power Presses," by V. J. Mankowsky and "Presses and Auxiliary Equipment for Forming of Sheet Metal," an article from Vol. 4 of *Metals Handbook.*

The American Society for Metals is especially grateful to Mr. John R. Newby of Armco Steel Corporation who served as consulting editor of this book. A recognized authority of long standing in the metallurgy of low-carbon steel sheet and in all aspects of its fabrication, Mr. Newby guided the selection and organization of articles that comprise the book. His professional experience and his familiarity with the technical literature, to which he has been a frequent contributor, are among the many assets he brought to this assignment. Mrs. Mary Ann Podboy, a member of ASM staff, supervised all aspects of the production of the book. Finally, the contribution of the authors whose articles appear in this book is gratefully acknowledged; their professional experience and technical competence account in largest part for the value of this book to industry and the engineering community.

Cover Photo Credit: Inland Steel Company.

Paul M. Unterweiser
Staff Editor
Manager, Publications Development
American Society for Metals

Allen G. Gray
Technical Director
American Society for Metals

Contributors to This Source Book*

ITARU AOKI
Nippon Steel Corporation

M. V. BALAKRISHNAN
Inland Steel Co.

J. A. BAUSCHER
Carnegie-Illinois Steel Corp.

HELMUT BEISSWANGER
Research Association for
Sheet Metal Fabrication

D. J. BLICKWEDE
Bethlehem Steel Corp.

R. D. BUTLER
British Leyland Motor Corp.

G. M. DAVIES
British Leyland Motor Corp.

HARRY S. ERISMAN
Bethlehem Steel Corp.

SATOSHI HANAI
Nippon Steel Corporation

R. M. HOBBS
Melbourne Research Laboratories
The Broken Hill Proprietary Co. Ltd.

W. HOLYMAN
Rubery Owen & Co. Ltd.

TAKASHI HORITA
Nippon Steel Corporation

HARDING R. HUGO
Danly Machine Corp.

W. E. JOHNSON
General Electric Co.

O. KINPARA
Nippon Kokan K. K.

HARUO KUBOTERA
Nippon Kokan K. K.

W. T. LANKFORD
Carnegie-Illinois Steel Corp.

JOHN P. MALONEY
General Motors Corp.

V. J. MANKOWSKY
USI Clearing

KAZUO MATSUDO
Nippon Kokan K. K.

ROBERT L. MEYERS
Armco Steel Corp.

M. NAKANO
Nippon Kokan K. K.

KAZUHIDE NAKAOKA
Nippon Kokan K. K.

JOHN R. NEWBY
Armco Steel Corp.

D. NORTH
British Leyland Motor Corp.

R. O'NEIL
Inland Steel Co.

SHIGEO SAKAI
Nippon Steel Corporation

TAIICHI SATOH
Nippon Steel Corporation

S. C. SNYDER
Carnegie-Illinois Steel Corp.

A. STANWAY
Rubery Owen & Co. Ltd.

R. L. WHITELEY
Bethlehem Steel Corp.

KOZO YAMADA
Nippon Kokan K. K.

*List does not include names of contributors to Metals Handbook

Table of Contents

Section V: Blanking

Section VI: Forming and Drawing Dies

Section VII: Lubricants

Section VIII: Presses and Auxiliary Equipment

Section I: Classification of Sheet Forming Operations

Prediction and analysis of press performance for sheet steels

R.M. Hobbs
Senior Research Officer,
Melbourne Research Laboratories.
The Broken Hill
Proprietary Co. Ltd.

Abstract:
In the production of a press formed component in large volumes from sheet steel a number of critical steps must be made for optimum results. First the correct design for the press tools must be established and the optimum material grade selected. Secondly, the dies must be arranged correctly for production, and the performance of the material accurately monitored during setup and production.

This paper reviews the rules for predicting die design and material performance. A description is then given of the procedure for setting up dies using the grid strain analysis technique for assessing the material press performance and obtaining the optimum combination of die settings, material and lubricant.

Reference: Hobbs, R.M. "Prediction and Analysis of the Press Performance of Sheet Steels", BHP Tech. Bull., 18 (2), Nov. 1974, p2.

Key Words: press forming procedures, lubricants, mechanical properties, sheet steels.

Introduction

Today, after fifty years of practice, the designing of dies, selection of sheet materials and lubricants for forming a component is still more of an art than a science. The reason lies with the complex interaction of the large number of variables in the process. Table 1 lists the most important variables and their effects on the process. No practical theory has been developed which can cope with all these variables — however, by analysing the way material flows in simplified practical cases, a number of "rules of thumb" have been developed. These provide a starting point for die design which is generally refined by a period of trial and error development of the dies. The end result of the development depends on the skill of the die setter and in general the most cost effective combination of material, lubrication and die setting is not achieved, but rather an arrangement which is "adequate to do the job" is developed. Recently the use of strain analysis in conjunction with the "Forming Limit Curve" of Keeler and Goodwin[1] has introduced a yardstick into die development which enables the die setter to achieve a better combination of material, lubricant and dies.

The object of this paper is to summarise the rules of thumb used to design dies and predict press performance, and then to describe the use of strain analysis during die development to obtain the optimum combination of die settings, and material and lubricant properties.

DESIGN GUIDES

A number of standard texts have been written on press forming, the most comprehensive being Metals Handbook vol. 4[2] which has been compiled by a number of committees. It gives copious examples of methods of forming different parts and includes many rules of thumb for die design.

Other useful texts are the series of articles published by F. Stanley Cope in the American Machinist, 1953-1955[3], and the book by Sachs and Voegeli[4]. For austenitic stainless steels, there is the volume published by International Nickel[5]. It is usual to divide the types of shells into categories:

1. Singly curved parts.
2. Contour flanged parts.
3. Double curvature sections.
4. Deep recessed parts with vertical walls
   (a) shallow circular shells
   (b) deep circular shells
   (c) wide flanged circular shells
   (d) rectangular and non-circular shells
5. Deep recessed parts with tapered walls
   (a) conical shells
   (b) spherical shells
   (c) rectangular or non-circular tapered shells
6. A combination of the above, particularly categories 4 and 5.

Singly Curved Parts

These are normally produced by folding, by a press brake or by roll forming (see Figure 1). Rules for designing tooling for press brakes have been described recently by Barrack[6] and are relatively simple. The force F(N) required for forming the bend can be determined by the empirical equation.

$$F \approx k\frac{T^2 L}{d} \qquad (1)$$

where T is the material thickness (mm), L the length of bend (mm) and d the die opening (mm). The constant k depends on material and the type of bend, and the typical values are given in Table 2 for die openings of $d < 10T$. For die openings of about 20T the values can be reduced by 10%.

Prediction of springback angle, or the degree of overbending required is more difficult and depends on the yield stress of the material, on the radius of bend and on the bottoming force in the die. An equation for estimating springback angle, θ, for a bend of angle, α around a radius R (mm), is

$$\theta \approx \alpha . \frac{R}{T} \cdot \frac{S_y}{70\,000} . N \qquad (2)$$

where S_y (MPa) is the yield stress of the material, and N is a constant, $N \sim 1$ for die opening of $\leqslant 3T$ to $N = 2$ for a die opening of 12T. This equation is applicable both for press brake forming and roll forming.

The minimum bend radius for a given material can be found[7] from knowledge of the uniform elongation, e_l, or reduction of area, A_r, in the tensile test.

$$R\ min \approx \left(\frac{50}{e_l} - \frac{1}{2}\right).T \qquad (3)$$

or

$$R\ min \approx \left(\frac{50}{A_r} - 1\right).T \qquad (4)$$

In calculating the width of blank, allowance must be made for material in the bends. Table 3 gives values for mild steel of various thicknesses. First the perimeter of the part is calculated assuming the bends are sharp and then the allowance for each bend radius is subtracted to find the width of the blank.

Table 1. Variables in the press forming process

MAJOR VARIABLES	
Sheet Material	— n-value (ability to strain harden - a measure of the stretchability of the material)
	— r-value (resistance to thinning — a measure of the deep drawability of the material)
	— anisotropy in the plane of the sheet (r_0, r_{45}, r_{90} values — measures the tendency to earing)
	— uniformity of thickness
Lubricant	— pressure sensitivity
	— temperature sensitivity
	— stability
	— thickness and position of application
Blank	— size, shape
Tooling	— stiffness of die and blankholder plates (use of shims to flex the blankholder plate)
	— surface roughness
	— die radius (may sometimes be alterable)
MINOR VARIABLES	
Sheet Material	— strain rate sensitivity of the yield stress
	— surface roughness (affects lubrication)
Blank	— edge condition (burred, heavily worked)
	— location on the die plate
Press	— ram speed
	— method of blankholding
	— stiffness of the frame, accuracy of movement in guides.

Table 2. Values of k in Equation 1 for determining the force for press brake bending.

k	Bend Type	Material	Thickness
600	V-bend	Mild steel with yield point of 270-310 MPa	> 2 mm
800	V-bend	High strength, hot rolled yield point 400 MPa	> 2 mm
2000-3000*	V-bend	Sheet steels	< 1.5 mm
2400	Channel with corner radii r > 8T	Mild steel with yield point 300 MPa	> 2 mm
3000	Channel with corner radii r > 8T	High strength hot rolled steel, yield point~ 400 MPa	> 2 mm

* For sheet with thickness < 2 mm it is usually necessary to bottom the dies to obtain a sharp bend. The additional force required for bottoming is included here.

Table 3. Bend allowances for 90° bends in steel[(2)].

	Material Thickness (mm)					
Bend Radius	0.8	1	2	3	6	10
0.8	1.4	1.7	3.1	5.0	9.2	15.1
1	1.5	1.8	3.3	5.2	9.5	15.3
2	2.0	2.3	3.9	5.7	10.0	15.7
3	2.5	2.8	4.3	6.0	10.5	16.1
6	4.0	4.3	5.5	7.5	11.5	17.5
10	5.5	5.8	7.0	8.5	13.5	19.5

Table 4. Limiting radii for contour flanging[(2)]

Material	Thickness (mm)	Maximum R_2/R_1	
		Stretch Flanging	Shrink Flanging
Deep drawing grade	1	1.44	1.10
Aluminium stabilised steel	2	1.44	1.20
Drawing grade rimmed steel	1	1.38	1.10
Cold rolled mild steel	1	1.22	1.10
Hot rolled mild steel	2	1.18	1.20

Contour Flanged Parts

These are generally produced in a press brake using tooling which incorporates a blankholder, or as part of a press forming operation. Figure 2 shows the two types of contour flange: shrink flange and stretch flange, when the curvature in the flange is convex and concave, respectively. The two types of failure are wrinkling of the free edge in the shrink flange and splitting at the edge of the stretch flange. The limits are determined by the condition of the cut edge, thickness of the material, ductility and cleanliness (freedom from stringer inclusions) of the material and the method of forming. The limits are expressed in terms of the radius of the free edge before and after flanging, R_1 and R_2 (see Figure 2) and are given in Table 4. These limits can be improved considerably if, in the case of stretch flanging, the blank edge is ground and, in the case of shrink flanging, urethane or rubber pad forming tools are used.

Doubly Contoured Shells

In forming doubly contoured shells the major problem is in controlling wrinkling in the outer portion of the blank. In thin materials, which are prone to wrinkling, it is usually necessary to incorporate a blankholder to constrain the outer edge. Alternatively, for lower volume production the use of rubber pad, urethane or fluid forming techniques may be used. When a blankholder is used the material is stretch formed over the punch and stretching limits can be determined by measuring the initial blank length, R_1 and the final perimeter length of several cross-sections through the centre of the part, R_2, and using the maximum R_2/R_1 values given in Table 4 for stretch flanging.

Deep Recessed Shells with Vertical Walls

In these shells the walls are produced by drawing in the outer portion of the original blank then bending the material through 90° and restraightening it. Due to the reduction in diameter of the blank, compressive circumferential stresses are developed and wrinkles will tend to form in the outer flange (analogous to the buckling of a thin plate under compression) — the thinner the material the more prone to wrinkling. To prevent this, pressure is applied by means of the blankholding plate — for mild steel sheet the interfacial pressure is typically 2-3.5 MPa and for austenitic stainless steel 10-14 MPa*.

In the initial stage of the draw the blank is stretched across the punch (see Figure 3(a)). As the punch is pushed through the draw ring the metal is bent over the die radius and straightened into the cup wall (see Figure 3(b)). The outer portion of the blank is reduced in

**NOTE:* These interfacial pressures are for controlling wrinkling only. Higher pressures are used when it is necessary to hold back the blank, as in tapered shells — see later.

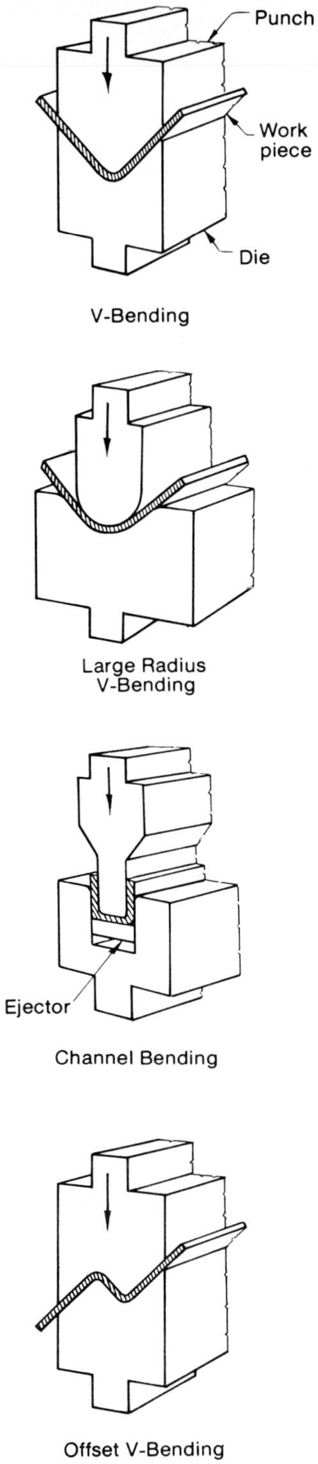

Figure 1. Examples of press brake forming

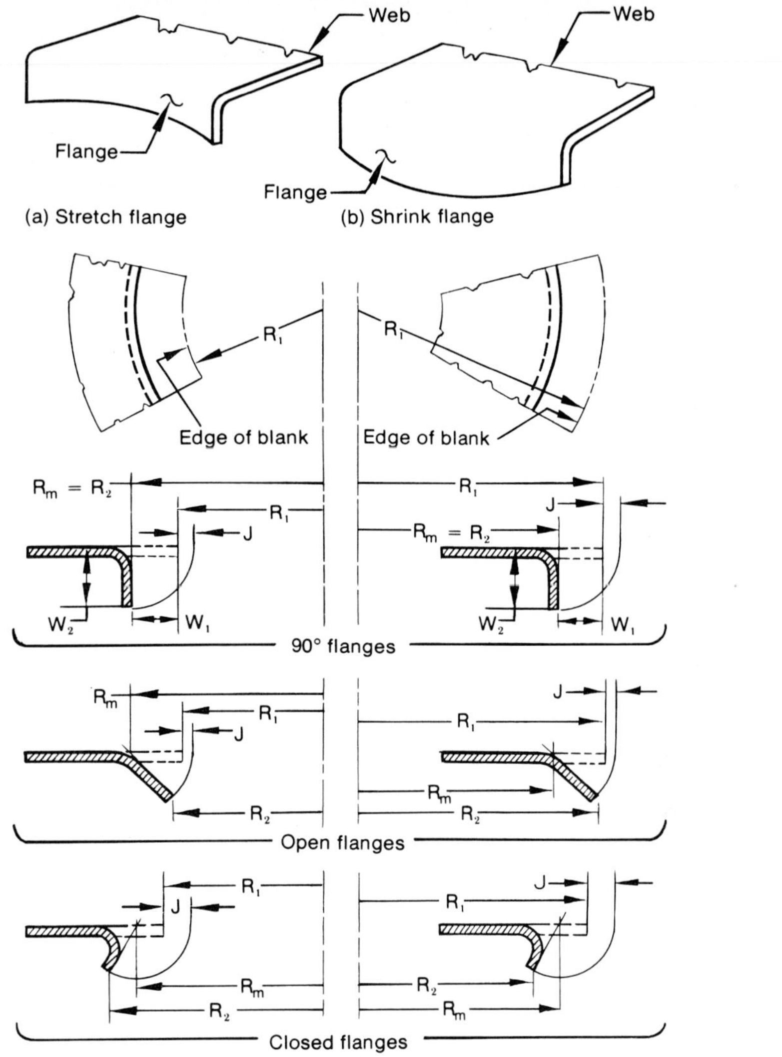

R_1 = edge radius before forming
R_2 = edge radius after forming
R_m = mould line radius
W_1 = width of flange before forming
W_2 = width of formed flange

J = bend allowance (difference between flange width measured from the mould line after bending and the length of the neutral axis of the metal before bending).

Figure 2. Three types of stretch and shrink flanges

Table 5. Maximum height and draw ratio of the single drawn shallow flat bottomed circular shell

Material Grade	Maximum h/d	Maximum D/d
Drawing quality, rimmed steel	0.75	2.0
Deep drawing quality aluminium stabilised steel	0.85	2.1
Low carbon, niobium stabilised cold rolled steel	1.1	2.25
Hot rolled drawing quality steel	0.7	1.9
Austenitic stainless steel grade 302	0.75	2.0

diameter as the metal is drawn in (see Figure 3(c)). The load required to shrink the blank and to bend and unbend the metal over the die radius is carried by the wall. Therefore the strength of the wall controls the maximum size of shell which can be drawn. An analysis of wall thickness shows that there are two thin areas at which fracture may occur. The most common fracture is at the "nose" of the punch, where the material was stretched between punch and die in the initial stage of the draw — this is related to the punch radius[8] as shown in Figure 4. The punch radius should be above 6T for drawing grade mild steel and 4T for austenitic stainless steel (where T is the thickness of the material) to avoid severe thinning.

The second region occurs further up the wall, above the die impact line. i.e. in the drawn region where the strain history is complex — the material has been reduced in diameter, bent and unbent under tension while passing over the draw radius and then carries the high tensile load to draw in the remainder of the blank. This thin area can be related to the draw radius[8] (see Figure 5) which should be greater than 8T for drawing grade mild steel and 10T for austenitic stainless steels to prevent severe thinning. However the die radius should not be too great as the conventional flat blankholder does not contact the blank over the draw radius and control of wrinkling may be difficult especially near the end of the draw.

A third form of fracture may be

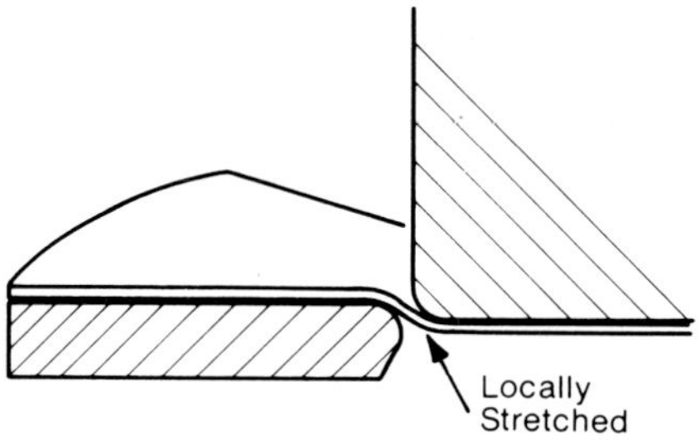

Figure 3(a). Stretching as punch contacts blank.

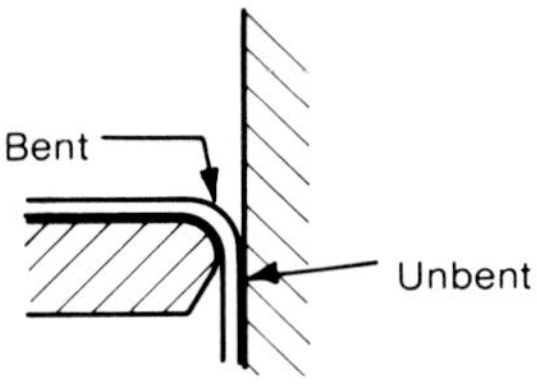

Figure 3(b). Bending and unbending over die radius.

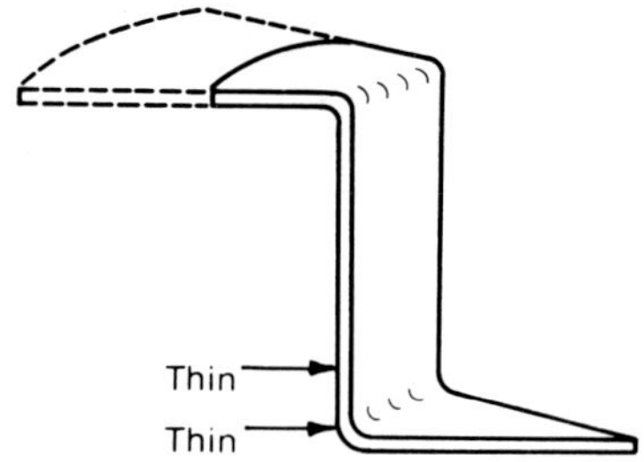

Figure 3(c). Local thin areas produced by stretching and bending.

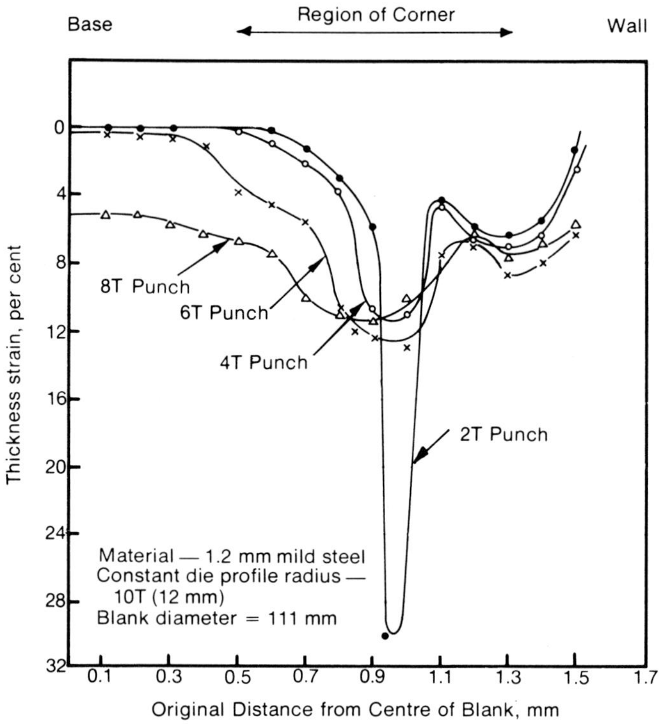

Figure 4. Effect on thickness strain of increase in punch profile radius — using a constant blank diameter.

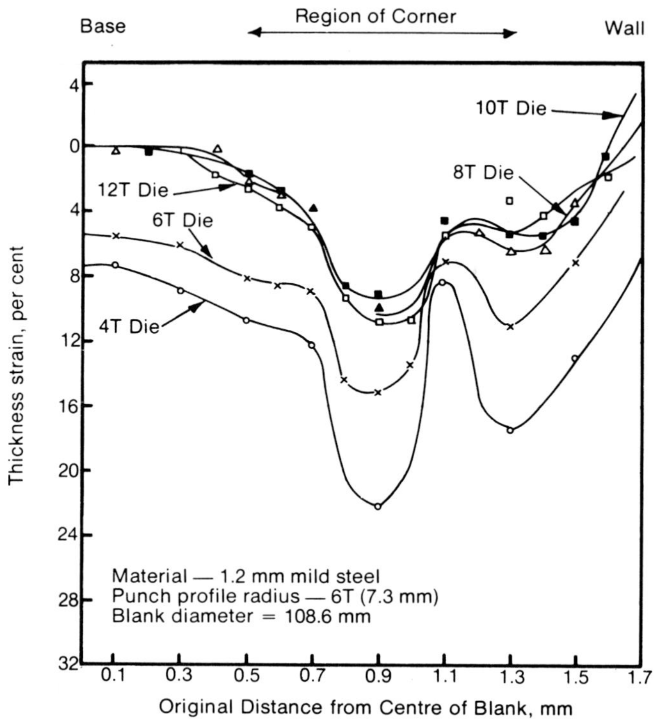

Figure 5. Effect on thickness strain of increase in die profile radius — using a constant blank diameter.

encountered in deep draws when the die profile radius is too small. This takes the form of a vertical crack in the wall and is found most often in thick walled cups when T/d >0.1 (see Figure 6).

As mentioned previously it is usual to subdivide this category into shallow circular shells, deep circular shells, wide flanged circular shells, and rectangular and non-circular shells.

Shallow Circular Shells with Vertical Walls

These shells are produced in a one stage draw. The maximum height of flat bottomed shell which can be achieved in typical commercial practice for drawing quality rimmed steel is generally assumed to be h/d∼0.75. This assumes that the optimum die profile radius and punch profile radius, a moderate punch speed (200 mm/s) and a good drawing compound-type lubricant are used. Deeper shells are produced with improvement in steel grade and the effect is summarised in Table 5.

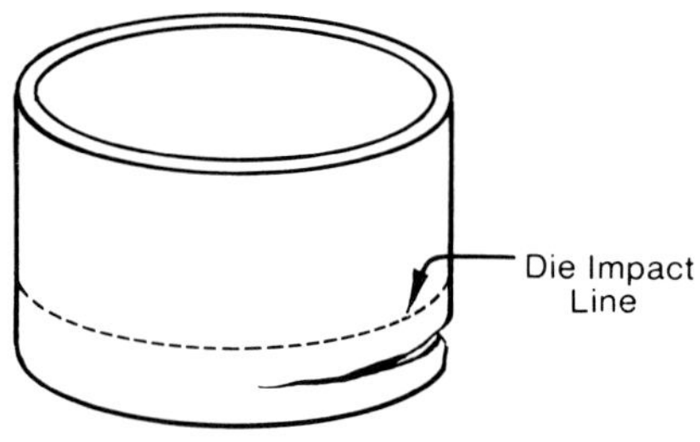

(a)

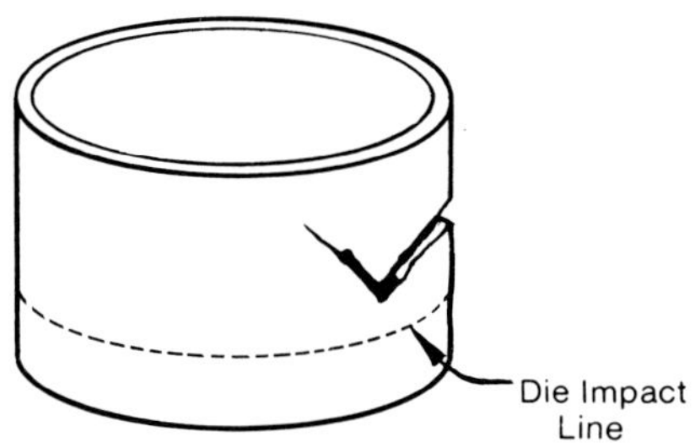

(b)

(a) Fracture over punch nose — punch nose radius is too sharp

(b) Chevron fracture in wall — die profile radius is too sharp.

(c) Vertical crack in thick walled cups — die profile radius may be too sharp and blank edge condition may be poor.

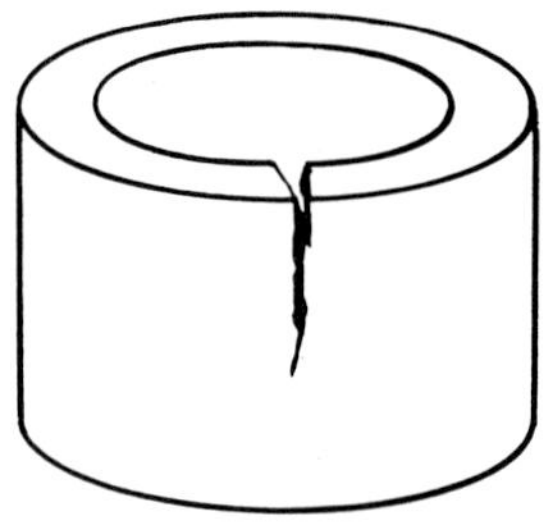

(c)

Figure 6. Three types of failure in deep drawing.

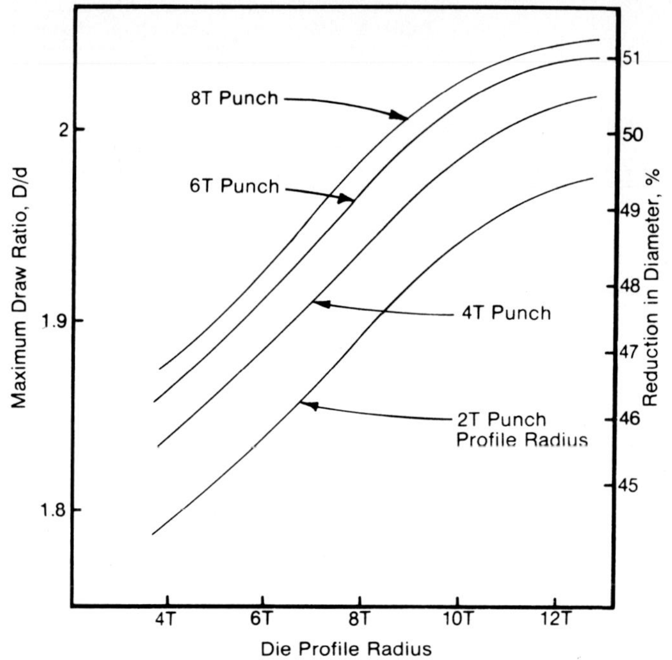

Figure 7. Effect of die profile radius and punch profile radius on maximum draw ratio of a single stage cup drawn in DQ mild steel.

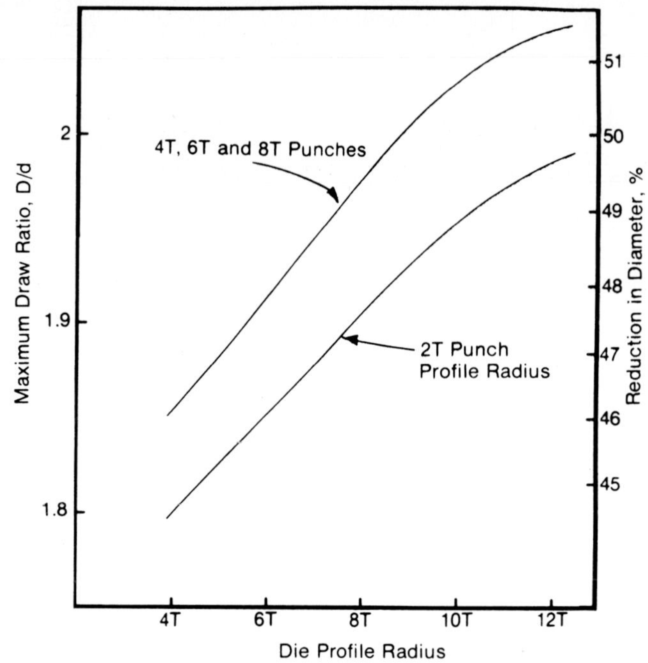

Figure 8. Effect of die profile radius and punch profile radius on the maximum draw ratio of a single stage cup drawn in austenitic stainless steel.

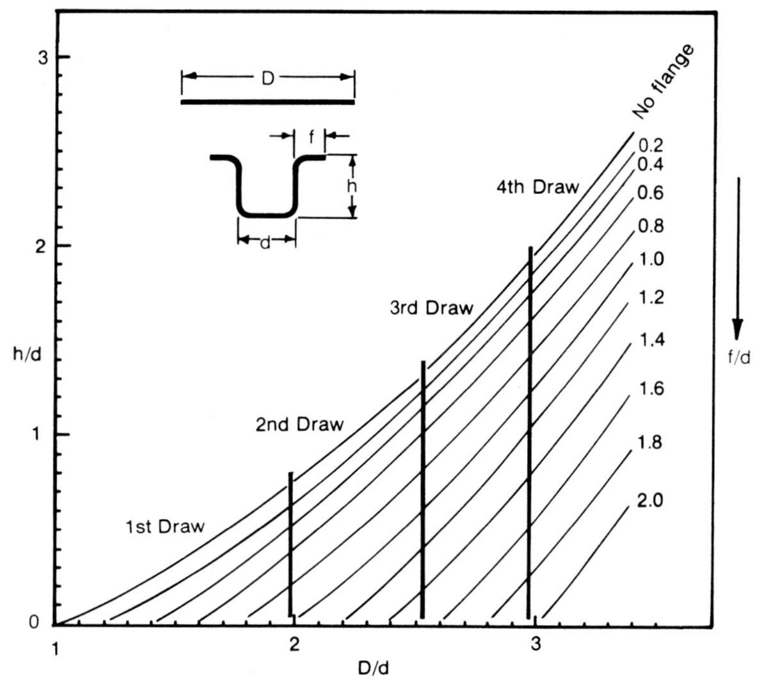

Figure 9. Relationship of blank diameter to cup height and flange width.

The effect of altering punch profile radius and die profile radius is shown in Figure 7 and Figure 8 for drawing grade rimmed steel and austenitic stainless steel, respectively.

The use of better lubricant and slower press speeds should improve the limits for rimmed or capped steels. In the case of continuously cast aluminium stabilised steel there is evidence to suggest that higher press speeds (5000 mm/s) will improve the drawability[9].

Much more significant improvements are obtained by the use of fluid forming in which the female die is a rubber diaphragm backed by hydraulic fluid at high pressure. A simple version of this has been developed by Granzon[10] in which the female die is filled with water. During the draw the water becomes pressurized and is squeezed out between the die and the shell providing lubrication over the die radius and more importantly forcing the material against the punch which reduces thinning over the punch nose. The limiting draw ratio D/d may be improved from 2.0 to 2.5 and the height to diameter ratio, h/d, from 0.75 to 1.3 for drawing quality rimmed steel by this process.

The limits given above have been for flat bottomed shells, but also apply to contoured bottom shells where h is the height of the vertical portion of the wall, provided the material is not too thin, $T > d/100$.

To determine the blank diameter, D, for a given shell diameter, d, and height, h, the following formula is used[2].

$$D = \sqrt{d^2 + 4d.h} \quad (5)$$

This is shown in terms of D/d vs h/d in Figure 9.

The punch force to draw the shell is given by

$$F = \pi dTS(D/d-k) \quad (6)$$

where S (MPa) is the ultimate tensile strength of the material and k is usually 0.6 to 0.7[2], F in newtons provided d, D, and T are in mm.

Deep Circular Shells

For deep cups, multistage drawing must be used as shown in Figure 10. The reduction of diameter in each stage is lower with succeeding stages. Table 6 gives practical maximum values for reduction in diameter and height to diameter ratios achieved in four successive stages using the optimum die and punch profile radii, a good drawing compound, and drawing quality rimmed steel sheet. Figure 9 also shows the number of stages necessary as a function of height to diameter ratio.

Substantially deeper draws, better control and less likelihood of wrinkling can be obtained by reverse redrawing

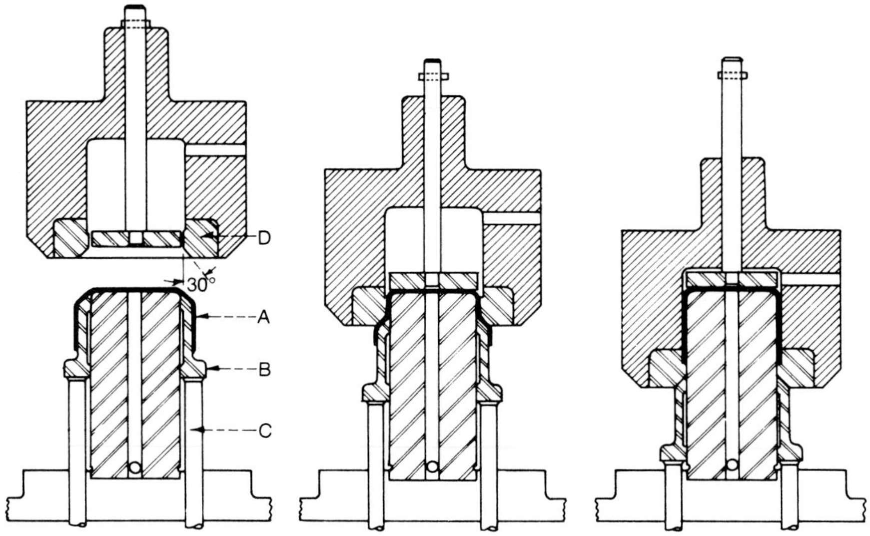

When a drawn shell A is to be redrawn, it is placed over the draw collar B (with dies open). Partially redrawn shell illustrates how metal is caused to flow into a shell of smaller diameter and greater length. Metal flow is caused by descent of die ring D against the resistance of the draw collar B, supported by the pressure pins C. The draw collar will strip the shell from the punch during the press upstroke.

Figure 10. The redrawing operation. The optimum die throat angle for redrawing thin sheet steel is 30°.

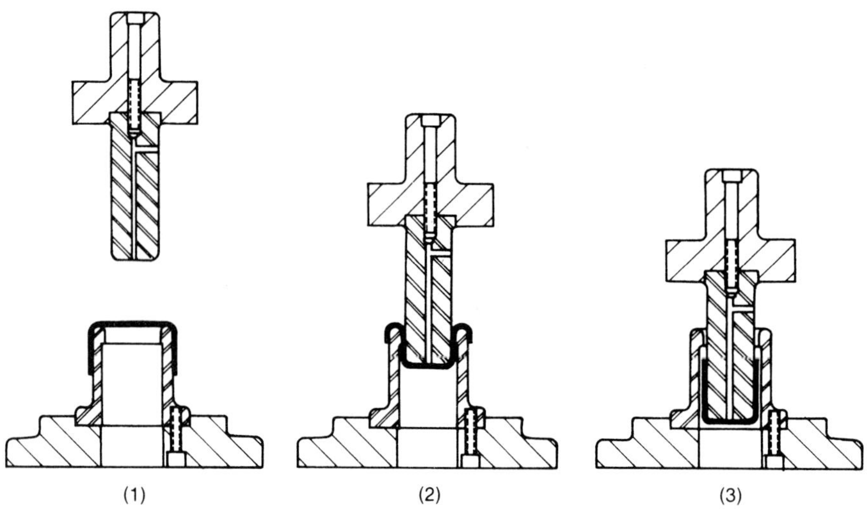

Three steps in reverse drawing a small thick-walled shell:
(1) The shell is placed over the die member;
(2) The punch turns the shell inside out;
(3) The reduced shell will be stripped by the sharp edge in the die cavity, when the punch ascends.

Figure 11. The reverse redrawing operation.

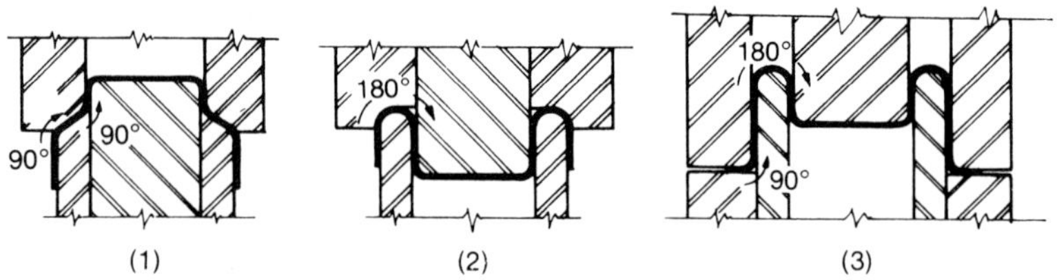

(1) Reducing a shell supported by a collar, causes the metal to flow inward 90° then upward 90°.
(2) Reverse drawing also causes 180° flexure, but in one direction.
(3) Overstressing can occur if the metal is pulled through more than 180°.

Figure 12. Flexure in redrawing.

Table 6. Maximum reductions in diameter and cup height/diameter for successive redraws — drawing quality rimmed steel

	$\frac{d_i}{d_{i-l}}$	$\frac{D_{blank}}{d_i}$	$\frac{h_i}{d_i}$
First draw	0.53	1.90	0.75
Second draw	0.75	2.50	1.30
Third draw	0.84	3.00	~2.0
Fourth draw	0.86	3.50	~2.8

Table 7. Maximum reductions in diameter for successive reverse redraws

	$\frac{d_i}{d_{i-l}}$	$\frac{D_{blank}}{d_i}$	$\frac{h_i}{d_i}$
First (conventional) draw	0.53	1.90	0.75
Second (reverse) redraw	0.70	2.70	1.50
Third (reverse) redraw	0.75	3.60	~2.9

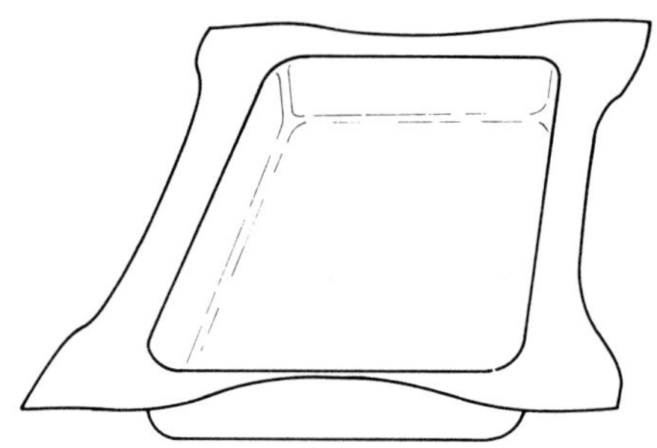

Figure 13. Rectangular pressing with earing at corners.

instead of conventional redrawing. In this process the shell is turned inside out while reducing its diameter (see Figure 11). The flexure of the metal as it passes over the die radius is then all in one direction rather than first flexing one way then the other as in conventional redrawing (see Figure 12). Reverse redrawing also has the advantage that it can be accomplished in one stroke of a triple action press or double action press with die cushion. Also greater interstage reductions are possible (see Table 7).

Wide Flange Shells

The additional flange to the shell is of course drawn in by the wall and therefore raises the load carried by the wall. Thus the maximum height which can be drawn in one stage is reduced.

Figure 9 gives a plot of the blank size necessary and the number of stages required to draw a flanged shell as a function of height to diameter and flange width, f. Again, the optimum die and punch profile radii and a good lubricant are assumed.

Rectangular Shell

The depth of rectangular shells drawn in one operation is basically dictated by corner radii. Table 8 gives the maximum height to width ratio for square or rectangular shells as a function of corner radius to shell width. Die profile radius and punch profile radius are assumed to

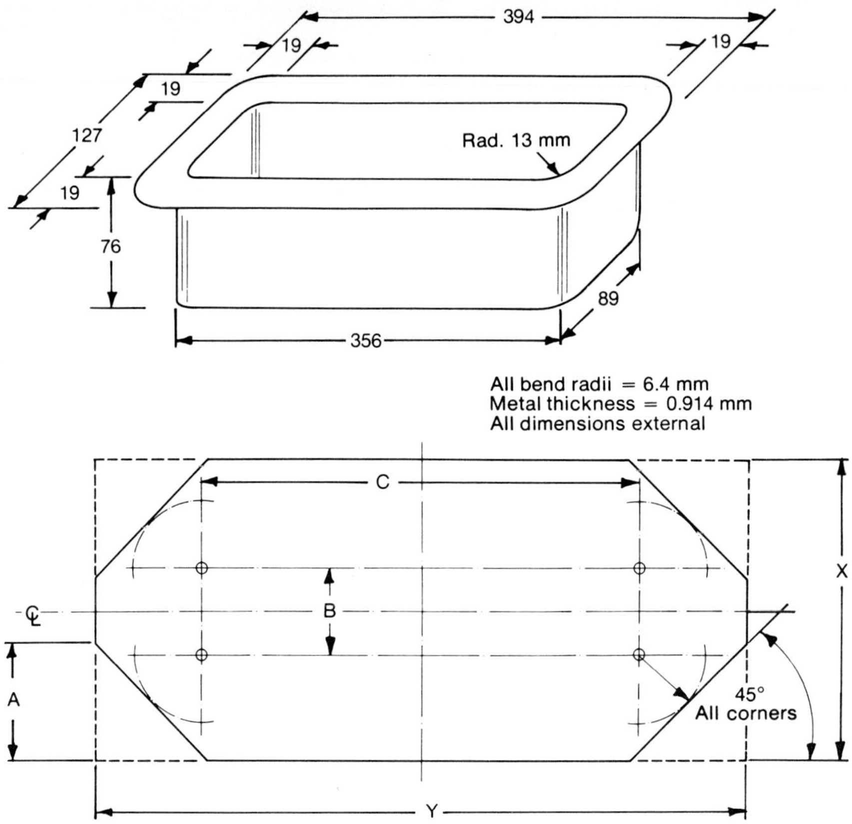

Laying out the Blank

The dimensions of the rectangular blank are established as X = 264 mm and Y = 531 mm by measuring the perimeter along the major and minor centre lines. Mark out the positions of the centre lines and then the positions of the centres of the corner radii. This establishes B = 64 mm and C = 330 mm. From the four centre points scribe a radius selected from figures for the corner radii of 13 mm and depth of 76 mm. This gives R = 47 mm with the note that this will be a difficult pressing and will probably require further blank development. Add ⩽ 17 mm to R to allow for the flange to give R = 64 mm.

Figure 14. Laying out the blank.

Table 8. Maximum* height to width ratios (h/w) for square or rectangular shells drawn in one operation as a function of corner radius to width ratio (r/w)

r/w	h/w
0.4	1.0
0.3	0.9
0.2	0.76
0.15	0.65
0.10	0.50
0.05	0.30
0.02	0.12

* These figures are conservative especially for r/w <0.15, however considerable blank and die development will be necessary if they are exceeded.

be the optimum (10T and 8T respectively) for drawing quality rimmed sheet steel.

As in the case of all deep drawn shells it is important to use the correct blank size and shape in order to decrease the stresses in the walls and reduce the probability of fracture. If a rectangular blank is used for pressing a rectangular shell then large "ears" will remain at the corners after pressing (see Figure 13).

It is general practice to cut the corners of the blank in order to avoid the ears and "even up" the flow. Figure 14 shows a worked example for a rectangular pressing 355 mm x 89 mm x 76 mm deep with a 19 mm flange[11]. First the rectangular blank dimensions are calculated by measuring the total perimeter along the major and minor centre lines. Then the amount of corner cutting is determined by reference to Figure 15. In cases which may be critical the blank is "developed" further by rounding the cut corners as shown in Figure 16. (there are three examples depending on whether R = 0.65, R >0.65 or R<0.65).

Recently Okamoto and Hayashi[12] have challenged the validity of this traditional approach and have shown that by leaving the corners of the blank uncut, fracture high in the corners is eliminated and deeper shells can be obtained. Tests by the author have confirmed this but difficulty in controlling the flow of material in the side walls was encountered leading to slackness and "oil canning" effects.

Blank orientation can play a major part because commercial capped and stabilised drawing quality steels develop the highest r values (i.e. best resistance to thinning) parallel and at right angles to the rolling direction. If the blank can be oriented with these directions running into the corners then deeper shells will be obtained. However, unless the developed blank has substantial corner cutting this will lead to extra scrap in cutting blanks from coil (see Figure 17).

The new low carbon, niobium stabilised cold rolled steels behave considerably better than commercial capped, rimmed, and stabilised steels in rectangular pressings, because they develop high r-values at 45° to the rolling direction. If the blank is cut across the rolling direction the good drawing properties then coincide with the corners. Tests[13] indicate that improvements in shell depth of 30-40% can be achieved using the low carbon, niobium steels, especially for shells with tight corners, r/w<0.2.

A further variable often used to control blank flow is the flexing of the blankholder plate by inserting shims between the plate and blankholder bolster to provide higher local pressures on the blank. This can be very effective and completely alter the flow pattern. For a rectangular shell the shims are usually placed to hold the sides back and encourage the corner to flow in. Draw beads can be used in a similar manner and provide more effective control, however they tend to leave galling marks on the part.

Very deep rectangular pans should be drawn in two operations following the procedure[3] shown in Figure 18. The width and length of the first stage shell should be larger than the final shell by about 3 times the final corner radius. The corner radius in the first stage should be 4 to 5 times the final corner radius and the punch profile radius should be about 1.5 times the final corner radius.

Deep Recessed Shells with Sloping Walls

Shells with sloping walls are more difficult to produce than shells with vertical sides because of the unsupported material in the side walls which will tend to buckle or "pucker". As explained previously the blank will attempt to buckle as it is reduced in diameter. The blankholder prevents this, but after the material has passed the die orifice and moves into the shell wall it is not constrained by a blank holder and with further reduction in diameter it will tend to pucker. Puckering can only be prevented by stretching the material. Failure modes are therefore two. Failure 1 occurs in the stretch region over or near the punch nose and is related to the work hardening ability of the material, punch profile radius and coefficient of friction between punch and blank. Failure 2 is puckering in the sloping wall and is related to the angle of taper and the material thickness to shell diameter ratio. Obviously the thicker the material the greater will be its resistance to puckering. For thickness, T>d/100 where d is the die orifice diameter, puckering does not occur.

Since puckering is prevented by stretching the material in the shell wall, the drawing process is controlled by the blankholder force. At low blankholder force the walls will pucker at low shell depths, as the blankholder force is increased deeper shells will be drawn before puckering occurs. At high blankholder force the metal will fracture over the nose of the punch and as blankholder force is increased further the shell height will reduce. A plot of shell height against blankholder force (see Figure 19), illustrates this effect. There is an optimum blankholder force which produced the deepest shell — at this force both wall puckering and splitting over the punch nose occur simultaneously at the failure point. The difficulty of drawing a shell of a given height can be determined from this plot — e.g. a shell drawn to half maximum height can be produced over a wider range of blankholder force than a shell drawn to ¾ maximum height, and therefore it will be a less critical draw.

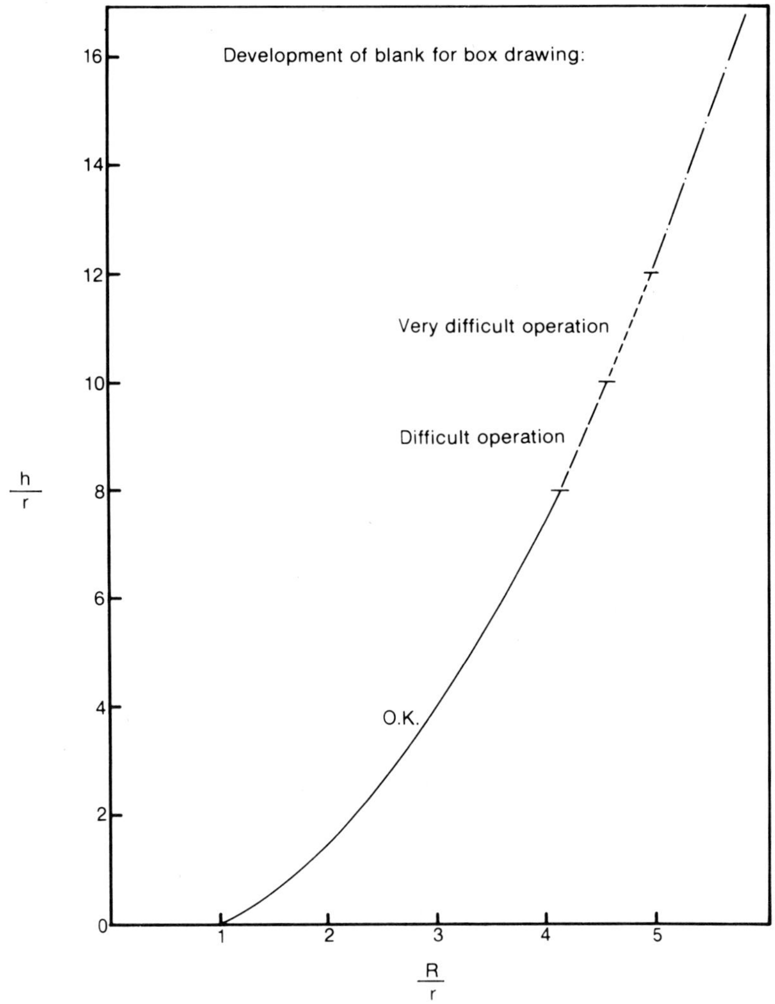

Figure 15. Plot for determining amount of corner cutting R, as a function of corner radius, r and height, h.

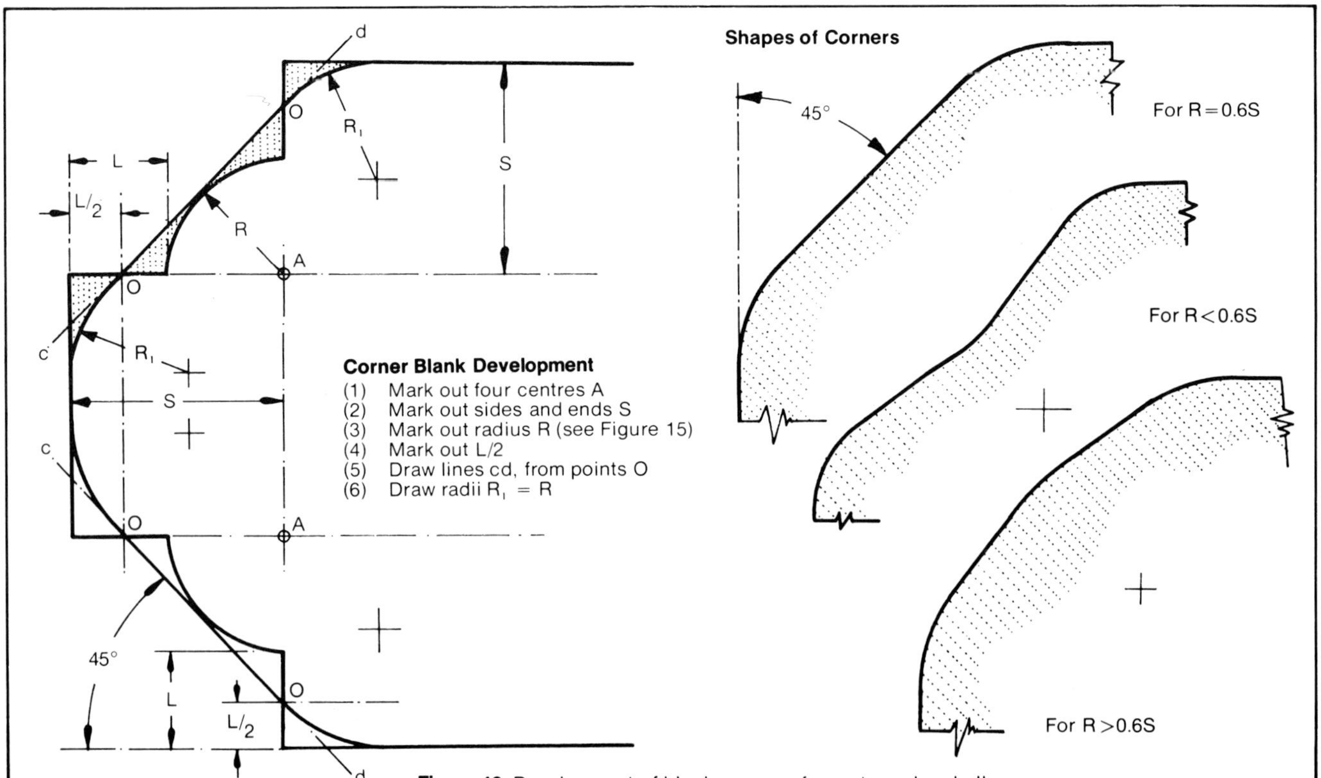

Figure 16. Development of blank corners for rectangular shells.

The curves in Figure 20 illustrate the importance of metal thickness in drawing tapered shells[14]; doubling the metal thickness allows the shell to be drawn 50% deeper through a die throat diameter of 100 mm with a punch diameter of 75 mm. The effect of sheet steel grade is seen[15] in Figure 21; changing from rimmed drawing quality to aluminium stabilised deep drawing quality steel improved the depth of shell by 20% through a 203 mm die throat diameter. Work is in progress to determine these curves for a wide range of tapered shells[14,15]. When these are available the prediction of tapered shell depth will be possible — at present there are no firm rules for tapered shells.

General rules for tapered shells are:

1. shallow tapered shells with height less than one quarter the diameter at the open end, d, and with thickness $T > d/200$ can be drawn in one operation.
2. tapered shells with $1/4 < h/d < 1/2$ should be drawn into a similar shape having vertical or a small taper in a preliminary operation.

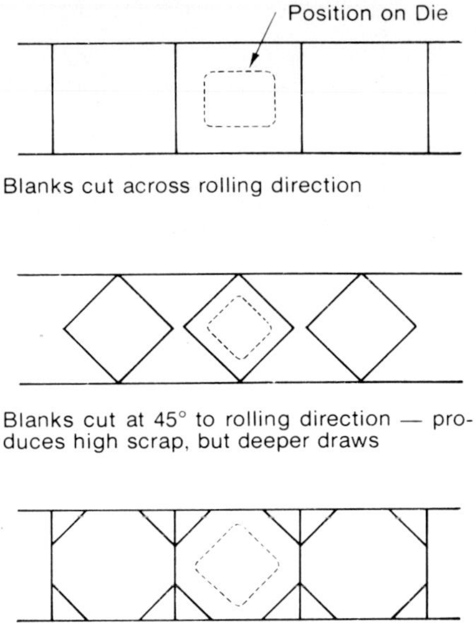

Figure 17. Methods of cutting blank from coil for drawing quality rimmed and aluminium stabilised steels.

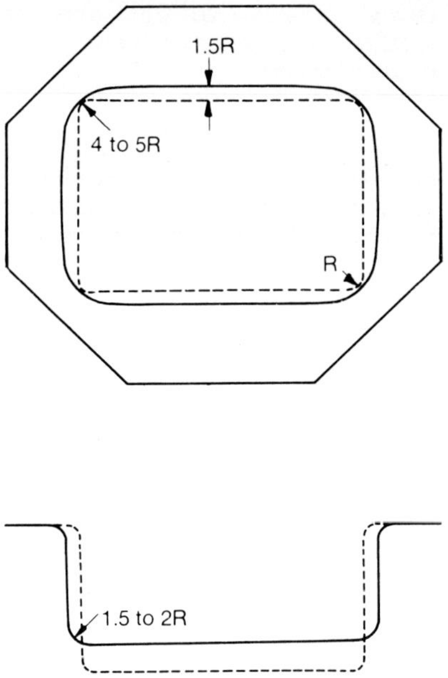

Figure 18. Profiles of first and second stages in drawing a very deep rectangular shell with tight corner radius, R[3].

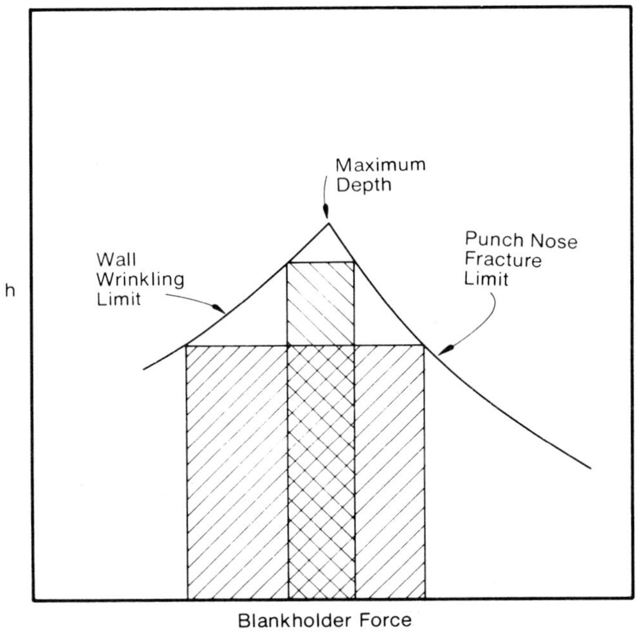

Figure 19. Drawing of a tapered shell — a typical plot of depth of shell against blankholder force. The shallower the shell the greater the latitude in blankholder force to produce a successful shell.

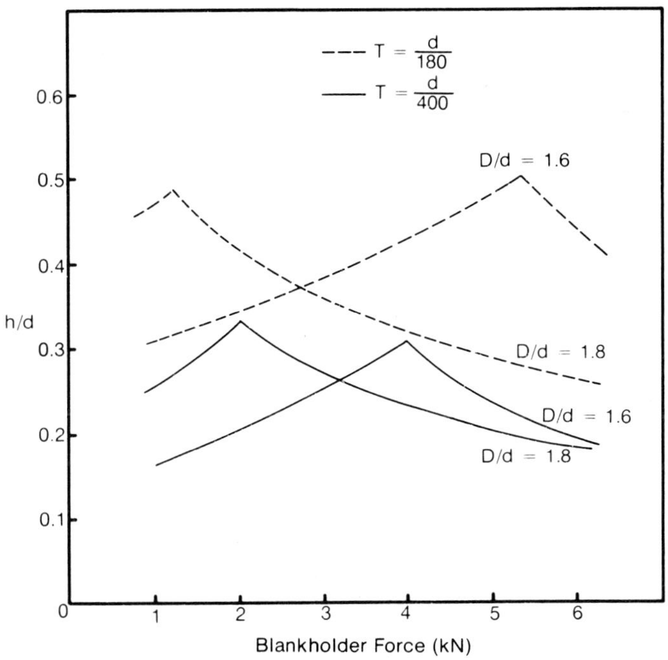

Figure 20. Depth of a tapered shell as a function of blankholder depth. The effect of metal thickness and blank size is shown. Die orifice diameter is 100 mm and punch orifice diameter is 75 mm.

3. deep tapered shells with $h/d > 1/2$ must be drawn first into stepped shells having vertical walls and then stretched to shape in a final operation.
4. where $T < d/100$ tapered shells can be drawn to between 0.4d and 0.5d provided the punch profile radius is 10T or greater. However, the sides will tend to bow out at the top and control of the shape will be difficult unless matching male and female dies are used and the part is coined in the dies at the bottom of the stroke.

Hemi-Spherical Shells

As with tapered shells the main problem encountered in drawing spherical shells is puckering. To avoid this, blankholder forces are higher than in drawing vertical sided shells in order to control flow and promote stretching.

The important parameters in the drawing of spherical shells are:
height of crown, h.
diameter of blank, d.
thickness of material, T.
flange width, f.

A full hemisphere can be drawn to a diameter equal to 100T and this diameter can be proportionately increased as the height of the shell is decreased[3].

Shallow shells give problems with springback and to counteract this the flanges must be drawn to a larger diameter, and the shell to a greater depth.

When controlling flow it is sometimes found that the blankholder force is insufficient and recourse is then made to:

1. Oversize blanks—giving an increase in the area acted upon by the blankholder

2. Draw beads—which provide localised resistance to flow in the required areas.

When a shell proves impossible to produce in a single operation, either

1. A two piece punch is used — this is a method of reducing the pull exerted by a domed punch by dividing the punch into two parts which act separately, or
2. The shell is drawn in two operations —where the final shape is either stretched or redrawn from a straight walled shell.

Special Shells

A special shell is one in which the body is not concentric with other parts of the shell, or the flange at the open end is not parallel with the bottom, or the shell is made up of a combination of straight, tapered and spherical surfaces. Most special shells have sections shaped like the common shells covered previously and it is possible to make judicious use of the rules for the parts in combination to produce the die design for the special shell.

DIE DEVELOPMENT AND PRESS SETTING

After the dies have been manufactured, finished to a high polish and mounted in the press the first job is to "bed-in" the blankholder with the die plate. It is most important that neither the blankholder nor die plate have any high spots which might locally restrain the blank and prevent even flow of material during the draw. This is checked by coating "engineers' marking paste" on one face and then closing the blankholder to the force expected in practice. Any high spots will be marked on the opposite face and should be polished away.

Secondly, all the radii should be checked to ensure they blend in smoothly, then a blank cut to the shape as predicted by the rules for blank size and shape should be tried in the press. A lubricant should be selected to suit the operation. For deep drawing a good quality drawing compound or for tapered shells where restraint by the blankholder is needed, a lubricant which prevents galling of the sheet steel to the press tools but which has a relatively high coefficient of friction will be more suitable.

After a few blanks have been drawn and the trouble spots identified, a series of blanks which have been electromarked* [16] with grids of small circles, usually 2 to 5 mm diameter, are drawn to varying depths say $^1/_2$, $^3/_4$, $^7/_8$ and full depth. Note that it is important to maintain the same press speed as is used in practice, so in a mechanical press the slide should be adjusted so that at the bottom of the press stroke the tool is only closed to the required depth. As the blank is drawn into the required shape the circles on the steel surface will be distorted into ellipses with the major and minor axes providing a measure of the direction and magnitude of the principal strains.

* See Appendix

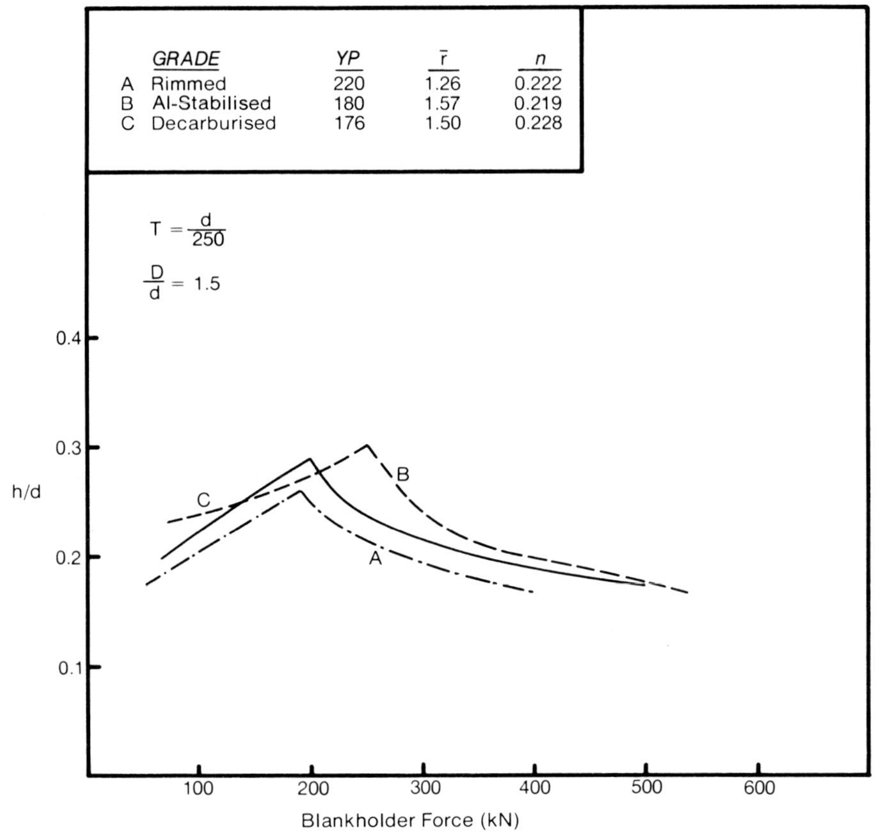

Figure 21. Drawing of tapered shell, showing the effect of steel grade on depth of shell. Die orifice diameter is 203 mm and punch diameter is 150 mm.

$$e\ \text{major} = \frac{\text{major axis} - d_0}{d_0}$$

$$e\ \text{minor} = \frac{\text{minor axis} - d_0}{d_0}$$

where d_0 is the diameter of the initial circles.

Knowledge of the local distribution of strains in the part can provide a measure of the criticality of the pressing by reference to the Keeler—Goodwin forming limit curve[1,17], (see Figures 22, 23 and 24). Essentially this is a map of the necking and fracture strains of the material under varying strain conditions, from equi-biaxial strain (as in hydraulic bulging) through stretching failures (as in most spherical or tapered shells) to drawing failures. The tensile test condition lies on the diagram in the tension—compression region at a small minor strain (between 0 and −20% depending on the width of the tensile specimen).

If the strains measured in the part are close to the critical strain level then the part can be expected to give trouble during the production run, and development of the dies, or blank or change of material or lubricant should be made to reduce the critical strains.

Three typical examples of the use of the technique are given below for illustration.

Example 1: A deep drawn rectangular pan first fractured over the punch nose in the corner at half depth.

Examination of the strain distribution just prior to fracture showed that the strains were low either side of the fracture and rose very quickly to a peak on the punch profile radius. The use of a more expensive steel grade with higher n-value (work hardening rate) caused the strains to be spread more evenly over the radius (see Figure 25) and prevented fracture. A cheaper solution which produced the same effect was to increase the punch profile radius, however, with this done the part was then drawn to full depth but fractured in the corner high in the wall. Using a better and more expensive lubricant prevented fracture, however, the strains were still close to the critical strain level. A tendency to wrinkling over the die profile radius in the corners prevented enlarging of this radius but further improvement was obtained by cutting the blank at 45° to the rolling direction and accepting a slightly increased offal scrap rate. Alternatively a thicker material could have been used (see Figure 22).

Example 2: A stainless steel sink bowl to be produced in a single draw showed critical strains in the corners. The strain condition was close to plane strain, indicating a high degree of stretching across the corners. When the blank shape was altered, to lower the amount of stretching across the corner, unacceptable wrinkles developed in the side walls. Improving the lubrication also introduced wrinkles into the walls. The only solution which would satisfy these constraints was to specify a grade of steel with sufficiently high n-value to

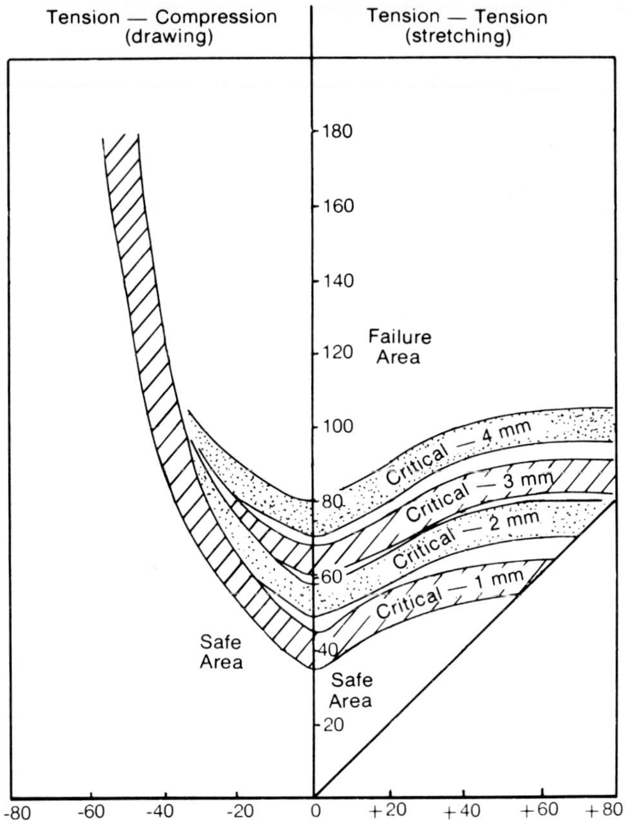

Figure 22. Forming limit curves for drawing quality and mild sheet steels of various thicknesses[17].

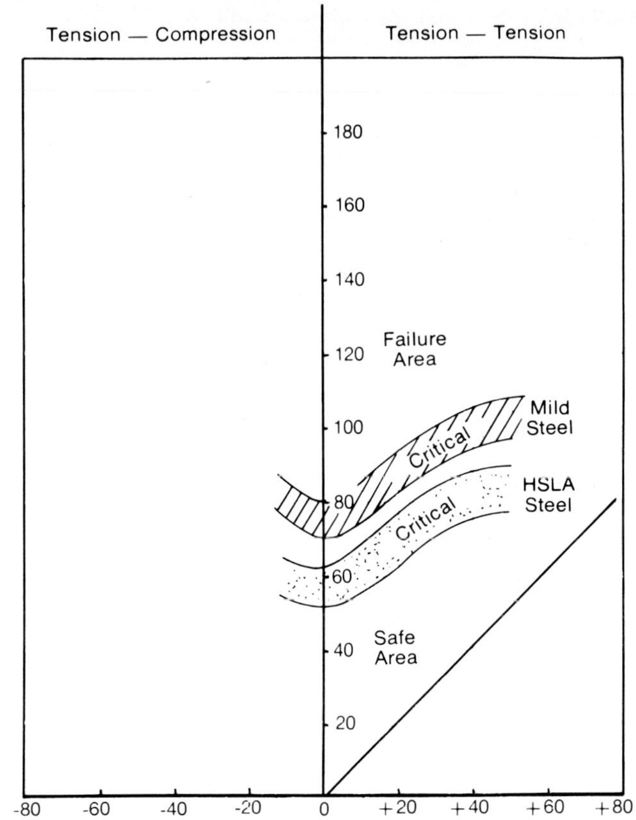

Figure 23. Forming limit curves for high strength low alloy sheet steel compared with mild sheet steel both 4 mm thick[17].

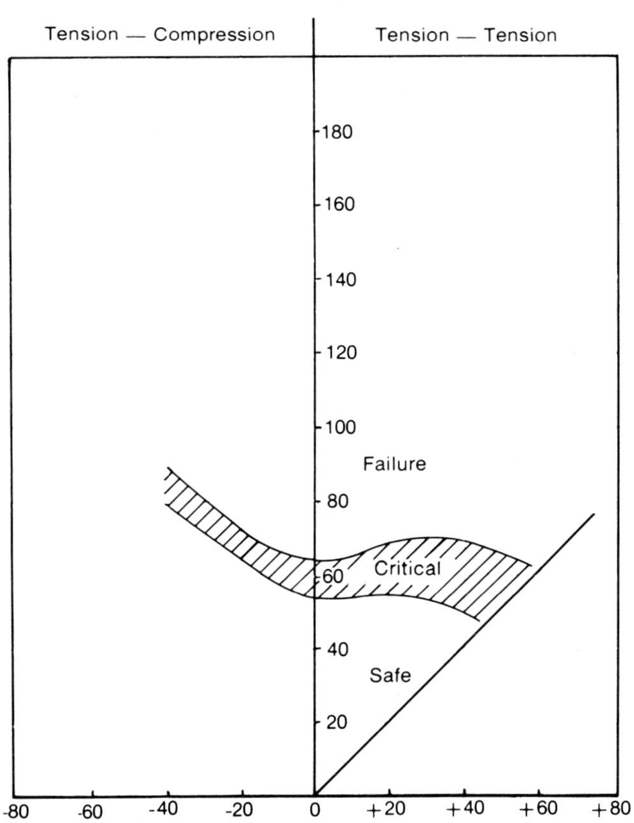

Figure 24. Forming limit curve for 301 austenitic stainless steel[18] ($M_d = 13°C$).

spread the strains more uniformly over the punch profile radius.

Example 3: The nose of a domestic electric iron contained critical strains close to the die profile radius above the die impact line. From the information cited for doubly contoured shells the thin area in a drawn shell which lies above the die impact line is generally associated with a die profile radius which is too sharp. The solution was provided by locally increasing the die radius at the nose which lowered the strains. Using a better lubricant or changing the blank shape only introduced unacceptable wrinkling.

ASSESSING MATERIAL PERFORMANCE DURING PRODUCTION

At the beginning of the production run a grid-marked master pressing should be taken using a standard steel grade (preferably stabilised) with a standard lubricant. At the end of each production run another pressing should be made using the same steel and lubricant and the strain patterns compared. Areas of die wear which may lead to trouble in future runs will be detected, and these can be repaired during the normal service periods.

In some cases as the dies wear the strains in the pressings are lowered, particularly over punch radii, and in some cases a cheaper lubricant or steel grade may be substituted. Measurement of the strains produced in a cheaper material or by a cheaper lubricant will indicate whether the substitute will perform satisfactorily during a production run.

In the case of a consistent fault occurring during a production run it is worthwhile making a pressing with the standard steel and lubricant and checking against the master pattern. This will quickly establish whether the dies are worn or out of alignment, or whether the current batch of material or lubricant is at fault.

If there is a change in the strain pattern the dies are at fault and the standard material can be used for correction. If there is no change in the strain pattern the material is at fault and analysis of the strain patterns may indicate whether simple changes could be made to, say, blank shape or shims to reduce the criticality.

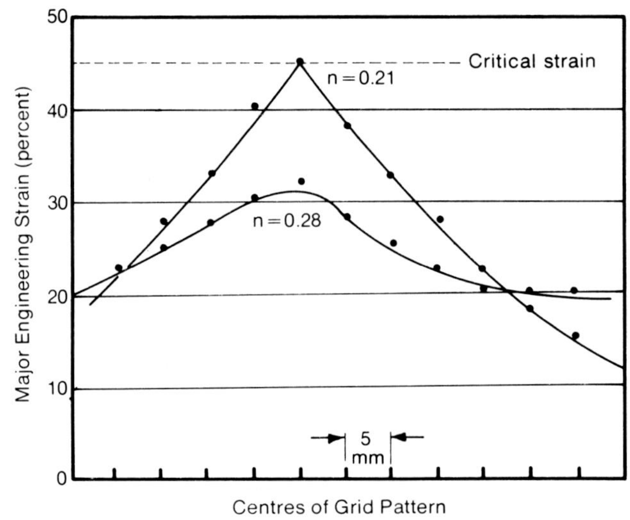

Figure 25. Showing the effect of n-value on strain distribution in a critical region of a drawn part. The strains are spread more evenly by a steel with high n-value.

CONCLUSIONS

This paper has covered some of the important steps in achieving optimum results in the press forming of sheet steels.

The metal performance in a variety of standard shell types has been described. The performance in single bends, stretch and shrink flanges and in deep recessed shells are reasonably well understood, however, knowledge of the press forming of shells with tapered sides is not precise, at present. Most commercial pressings can be described by a combination of the standard shell types so the design of dies for commercial pressings should be aided by the information given in the paper.

The procedure for development of dies and metal blank and selection of material and lubricant for optimum results has also been described, and the use of the grid-circle strain analysis for aiding this procedure and locating critical areas which may become troublesome during the subsequent production run has been illustrated.

Finally the use of the grid-strain analysis to monitor the metal and lubricant performance during the production run, and to assist in trouble shooting has been descirbed.

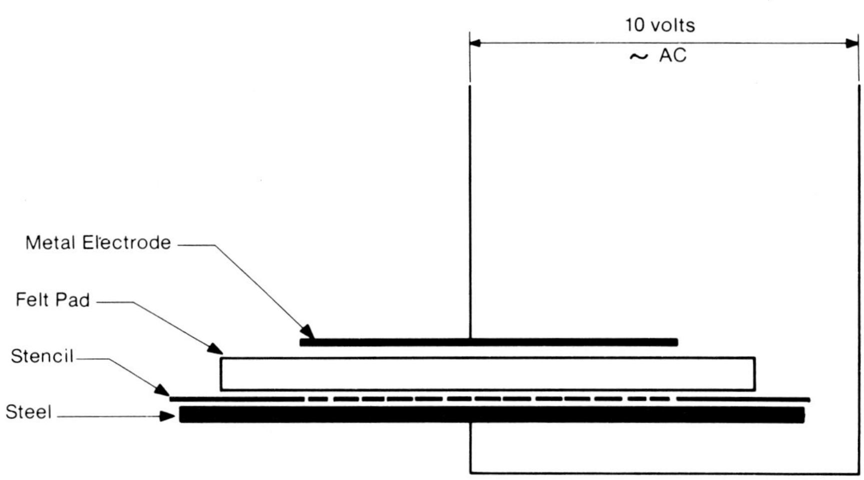

Figure 26. Electro-etching Arrangement.

ACKNOWLEDGEMENTS

The author would like to acknowledge the useful discussions with his colleagues T. Thomson, Dr. G. Glover, J. Havranek and T. Gibson during the preparation of this paper.

APPENDIX

Electromarking of Circle Grids on Steel

In order to measure the strain distribution in the component, it is necessary to mark the starting blank with a grid of small circles (2 or 2.5 mm diameter).

A number of kits for electromarking are available on the local market. However, a simple technique used by the author which does not require an elaborate etching rig and produces reasonably clear patterns is shown in Figure 26. The sheet to be marked is degreased, placed on the ground and connected to one terminal of the power pack. A stencil is dampened with water then placed on the sheet and all wrinkles smoothed out. Next a felt pad, about 10 mm thick and slightly smaller than the stencil, is dunked in a tray of electrolyte, allowed to drain so that it is damp rather than wet and placed over the stencil. Finally, a piece of scrap sheet steel (about 2 mm thick) with the same dimensions as the stencil is placed on the pad and connected to the other terminal of a power pack, pressure is applied (most easily by standing on the "sandwich" or by the use of a rocking pad or roller) and the power pack switched on. After etching, the electromarked area is neutralised by washing with a weakly alkaline solution (0.5N NaOH solution containing 0.018% $NaHCO_3$ buffer) and coated with rust preventing oil.

The power pack provides 10 volts AC at 200 amperes and etches a pattern in 10-15 seconds in low carbon steel sheet and 45-60 seconds on stainless steel sheet.

The electrolyte used is

KCl	80 g
NaCl	90 g
HNO_3	100 ml
HCl	100 ml
Water	4.5 litre

The stencil contains the pattern of circles and can be purchased from Electromark (UK) through its Australian Agent, Westeels.

References

1. Keeler, S.P. Sheet Metal Industries, **48** October 1971, p739;
   Chatfield, D.A. and Keeler, S.P. Metal Progress, **99** May 1971, p60.
2. Metals Handbook, vol.4., "Forming". American Society for Metals, Metals Park, Ohio, 1969.
3. Stanley Cope, F. American Machinist 1953-55.
   1953 August 3rd, p110; August 17th, p106; August 31st, p96; Sept 14th, p164; Sept. 28th, p126; October 26th, p120; Dec. 7th, p156;
   1954 Jan. 4th, p100; Feb. 15th, p148; Mar. 15th, p134; April 26th, p114; May 10th, p168; June 7th, p134; July 5th, p126; July 19th, p130; Aug. 2nd, p134; Aug. 16th, p122; Aug. 30th, p109; Sept. 13th, p140; Sept. 27th, p120; Dec. 6th, p166; Dec. 20th, p94;
   1955 Jan. 7th, p112; Feb. 28th, p106; Mar. 14th, p145; April 11th, p110; April 15th, p122; May 23rd, p130; June 6th, p130.
4. Sachs, G. and Voegeli, H.E. "Principles and Methods of Sheet Metal Fabricating", Reinhold Publishing Corp, N.Y., 1966.
5. "Forming of Austenitic Chromium-Nickel Stainless Steels" International Nickel Co., New York.
6. Barrack, F.E. Sheet Metal Industries, **51** April 1974, p174.
7. Yang, C.T. Metal Progress, **98** November 1970, p170.
8. PERA Report No. 115, 1963, "Deep Drawing. Influence of Punch Profile Radius, Die Profile Radius and Material Thickness".
9. Klein A.J. Metal Progress, **103** June, 1973, p69.
10. Granzon, W. Manufacturing Engineering and Management, Jan 1974, p38.
11. Grainger, J.A. Institute of Sheet Metal Engineering Year Book, 1972/1973. London.
12. Akamoto, T. and Hayashi, Y. Sumitomo Search. **10** 1973.
13. Thomson, T., Brammar, I.S. and Hobbs, R.M.: BHP/MRL Report No. Presented at the 1972 Conference of I.D.D.R.G Amsterdam.
14. Havranek, J. and Hobbs, R.H. to be published.
15. Kubotera, H. Nakana, M. Ueno, Y. and Uchida, Y. Nippon Kokan Technical Report (Overseas) June, 1973, No. 16, p1.
16. Gibson, T.J. and Hobbs, R.H. Proceedings of the International Conference on Production Technology, Melbourne, 1974, p210.
17. Lee, P. and Hiam, J.R., Dofasco, Canada. "Factors Influencing the Forming Limit Curves used for Press Forming". 1973.
18. Jackson, R., BHP Melbourne Research Laboratories. Private Communication. 1974.

How GE Optimizes Steel Usage

By W. E. Johnson

WITH THE RADICAL change in the economic climate, emphasis has shifted from output at virtually any cost to extreme conservation of all costs — particularly those related to materials.

In the manufacture of major appliances, material costs may run in the neighborhood of 40% of net sales billed.

The Major Appliance Business Group at GE has organized many programs to look at conserving our resources. One of them, our Steel Optimization Program, concentrates on sheet metal analysis techniques and specification practices.

Three Major Considerations

We buy a substantial quantity of steel annually. So you can see why there is a continuous concerted effort to optimize our purchasing specifications and our subsequent usage of this valuable material.

An analysis of our utilization of the sheet steel purchased showed that for every 100 tons bought, 22 tons of scrap were generated, mostly as offal from blanking operations. Seems amazing, doesn't it? But check your own fabricating shop. You may be surprised too.

On top of the cost of the scrap material itself, of course, must be added the labor and energy involved in generating it and transporting it. A small percentage of this scrap is used as input to our gray iron foundry in Louisville; the rest is sold back to scrap brokers at 25 cents on the purchase dollar. This is a multimillion dollar area for concentrated effort.

Second, the designing, specifying, and purchasing functions need to work hard to reduce the cost of premiums for size, grade, surface finish, packaging, minimum coil size, and special requirements of all types. Over 10% of the cost of cold-rolled sheets in many fabricating shops goes for these extras. This is another multimillion dollar area for concentration.

Third, a good deal of loss can occur when splitting, galling, wrinkling, and all the other causes of shop scrap become excessive. Using the latest sheet metal analyses

Keeler-Goodwin Forming Limit Diagram

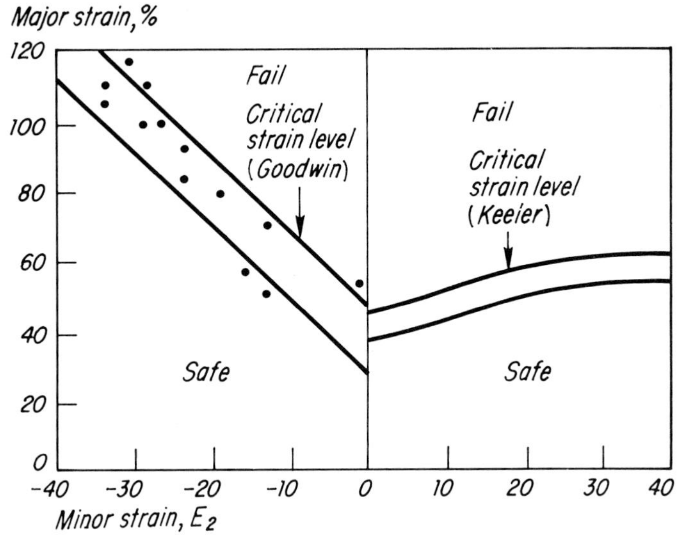

Fig. 1 — Goodwin diagram (left) shows tension-compression strain rate; Keeler diagram (right), tension-tension strain.

Modified Stretch-Draw Chart

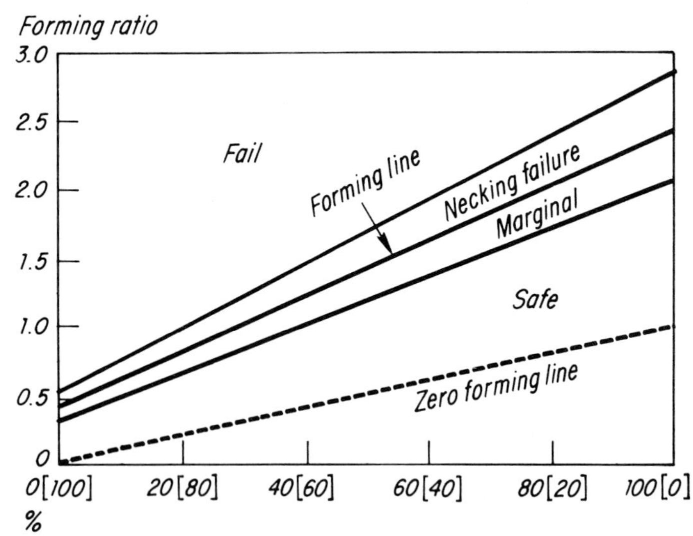

Shape analysis diagram for determination of severity of design. Read % draw from left to right; % stretch (in brackets) from right to left.

in Sheet Metal Forming

techniques, we can determine the best steel to use on a sheet metal part. Where problems arise, we can usually determine what corrective action to take.

Optimizing Metal Utilization

Analyses start with an electroetched gridded circle pattern on the blank. As the part is formed, the circles are elongated; by plotting maximum strains, we can gather much useful information.

Forming limits have been established by Keeler and Goodwin. The major strain in a circle is plotted vertically on the Forming Limit Diagram (FLD) and the strain at 90° is plotted horizontally (Fig. 1, top). Stretching occurs when both strains are positive and drawing occurs when the minor strain is negative.

Our metallurgy laboratory has adapted these curves to a Combined FLD with bands of strain depicting necking, failure, and safer zones. Say we grid a part, find the most severely strained circle, and plot its strains. With our experience, we can determine if the steel can be downgraded, or if it is correct for the job; and we can pinpoint metal flow to show the tooling man exactly where he should work. After he has made his alterations, another gridded part can be run to show the severity change, if any.

The next technique calls for a more elaborate measurement of the grid. It involves the changes in longer lengths of the blank and shows the design severity. Called Shape Analysis, it was originated by Kasper at Chrysler Corp. We have modified it slightly to show similar levels of design severity (Fig. 1, bottom left). Using this technique plus the FLD, we can usually determine whether tools, design, or poor steel is causing the problem.

One other technique is related to splitting along an edge (Fig. 1, bottom right). As burr height increases, the amount of stretching an edge can take is reduced.

Material Edge Condition Analysis

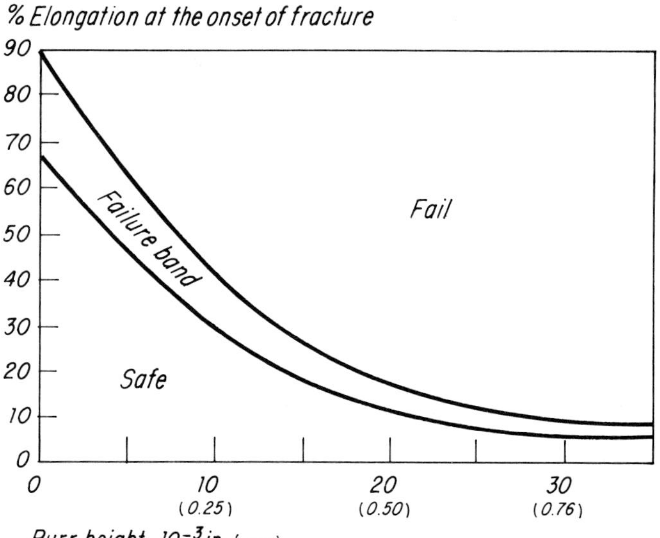

Edge stretchability is directly related to burr height.

Several Case Histories

Let's go through the operations as they occur in normal sequence and see where we can optimize our steel usage.

First, we have design of the part. Here we need to keep the word "forgiving" in mind. This is another way of saying "easily producible" because we all realize that perfect conditions are sometimes not present in the shop.

If for example the recommended draw radius is 6 to 10× thickness, try to stay in the middle of the range. Resist marketing or stylist "feelings" that push designs to unreasonable limits. The cost of having to live with a poor design from the manufacturing standpoint can be overwhelming. Two examples with different results point this out.

A new design for a dryer base presented a splitting edge problem. Parts made from prototype tools gave some splits. Shape analysis revealed several areas for improvement (Fig. 2) where strains could easily be made less critical by minor design variations. In general, the idea is to increase the flow of metal from neighboring, less critical segments into the critical points, thus lowering the maximum strain. Designs which lock up areas of metal from free flow are troublesome.

Sometimes the design simply cannot be compromised,

Shape Analysis Pinpoints Most

as was the case with a dryer inner door. A new design was compared with the present part rather empirically. The present part was shown to be marginal by Shape Analysis. A comparison of critical dimensions revealed that the new design would need to have radii opened up and other unacceptable changes. So the part was redesigned for plastic. Perhaps a costly problem was avoided.

In another example, a progressive die sequence was retooled and high scrap rates were encountered in forming a motor end shield (Fig. 3). Close comparison between present production parts and the newly tooled stages of this end shield showed clearly the higher strains in the new tools. They resulted from a poorly designed first station which did not gather enough metal for later forming. Sharp radii called for on the part print were found to be unnecessary. These also prevented easy metal flow.

Tool maintenance can sometimes cause as much trouble as it eliminates. After a shutdown for buildup of an expander plug, splits were excessive in a dishwasher tub. A grid analysis showed the tool repairman exactly where the metal was not flowing in. He immediately grasped the situation and knew exactly where to grind.

In 20 min another gridded part revealed the reduction in maximum strain. The total required strain had been distributed over the neighboring area, and splits were virtually eliminated.

Prior to the tooling change, the major strain in circle 1 was +65% and the minor strain was +17%. After the tooling change, the major strain was reduced to +20%, the minor strain to +10%.

In another instance, two forming lines were running the same refrigerator main frame cup. One line produced around 9% scrap due to cracks. The other had a reject rate of about 1%. The shop had the material tested, and the lab could determine no difference.

Further inquiry revealed that on the line with low scrap tooling was comparatively new. Cracking was occurring around the periphery of a hole which was expanded after punching (Fig. 4). Analysis of the edge condition revealed a heavy burr around the punched holes on the line with high scrap. Sharpening of the punches on an improved preventive maintenance basis eliminated the burr and the cracking problem.

A refrigerator liner approximately 25 ft² (2.3 m²) used drawing quality steel at a premium of over 7%. Strain analysis showed less than 0.2% of the over-all surface actually required this grade. Recommendations were made to alter the radii on the tools in the critical areas. Significant savings have been realized by eliminating the need for drawing quality steel.

Another Application: An advantage of using strain analysis is to determine whether a part is primarily stretched or primarily drawn as it is formed. The deep drawability of killed steel is effectively utilized most naturally on a drawn part.

Figure 2

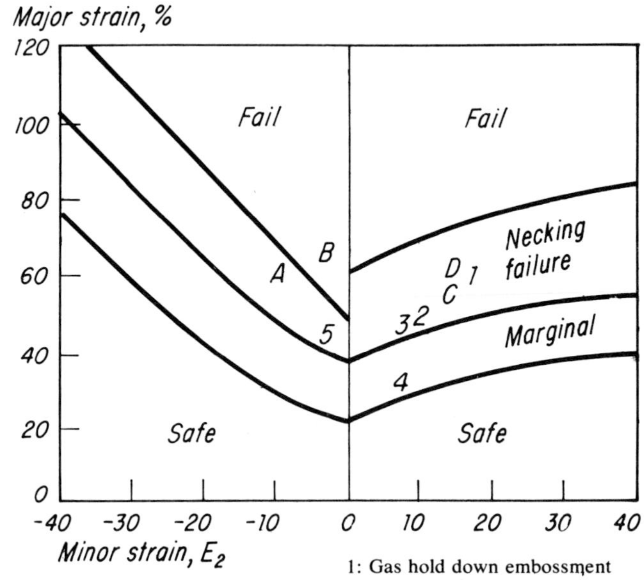

A: Right front corner
B: Left front corner
C: Left rear corner
D: Right rear corner

1: Gas hold down embossment
2: Blower hold down embossment
3: Blower hold down embossment
4: Blower hold down embossment
5: Idler bar embossment

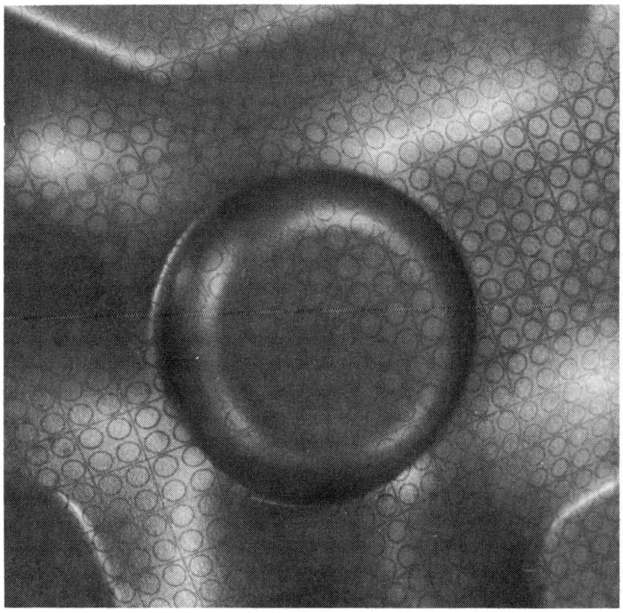

Blower hold down embossment (area 2)

An air conditioner shroud was made from killed steel for some time before a strain analysis showed that the primary mode of deformation was by stretch. The tearing problem of the punched-out bottom was attributed to the punch trimming being done with out-of-line tooling which essentially tore rather than cut out the bottom. The maximum strain levels were rather low and no draw — or negative minor strains — were observed. A trial of commercial quality steel was recommended, as well as a correction of the punch alignment.

Severe Dryer Base Strains

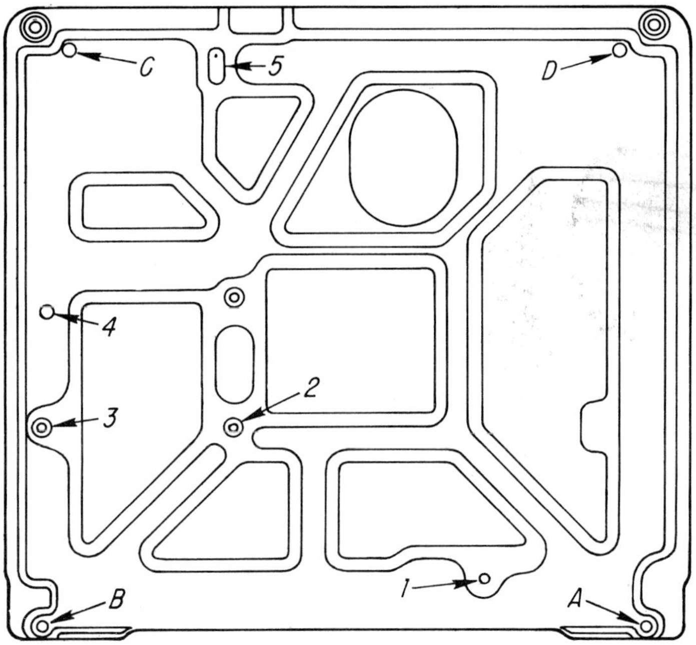

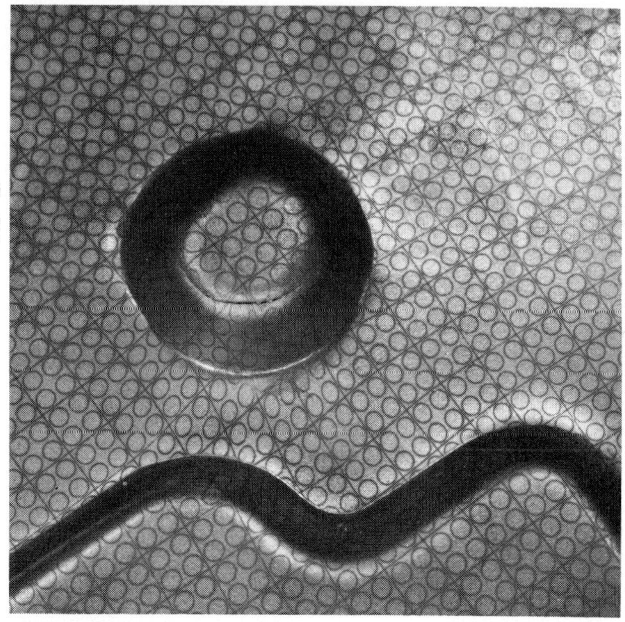

Right rear corner (area D)

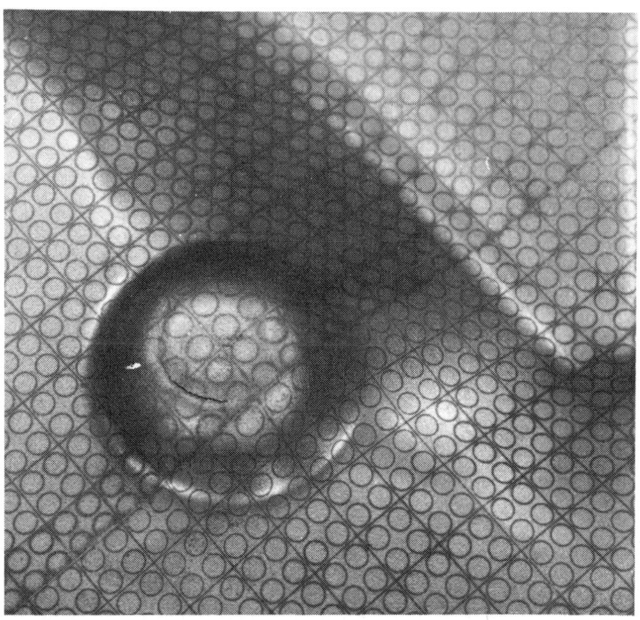

Gas hold down embossment (area 1)

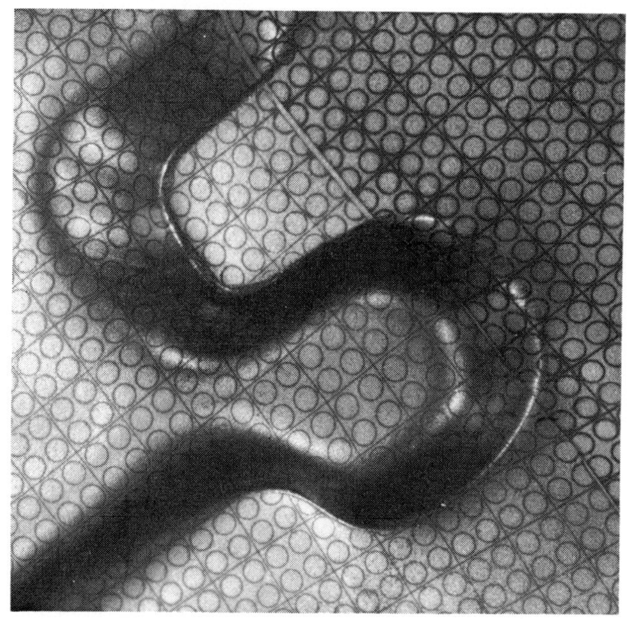

Right front corner (area A)

Checking New Designs

Use of strain analysis on a new design may be useful in the die design concept. An integral shelf support was desired in a refrigerator wall. Using the total strain allowable of 15% as a basis, the support was designed to be made in two steps.

First, the 15% "extra" metal was generated by drawing a uniform depression over a wider area of the wall than ultimately would be used. Then it was a relatively easy matter to precisely push the metal back in the final shape with only a little further stretching. This technique has the advantage of uniformly distributing the strains needed to make this part. A one-step draw would concentrate the critical strains in the shelf top due to the restriction caused by the sharp radius needed between the shelf and the liner. Splitting would no doubt occur.

On each part analyzed, basic formability data are obtained, such as material specifications; draw, punch nose, and corner radii; punch to die clearances; drawing force,

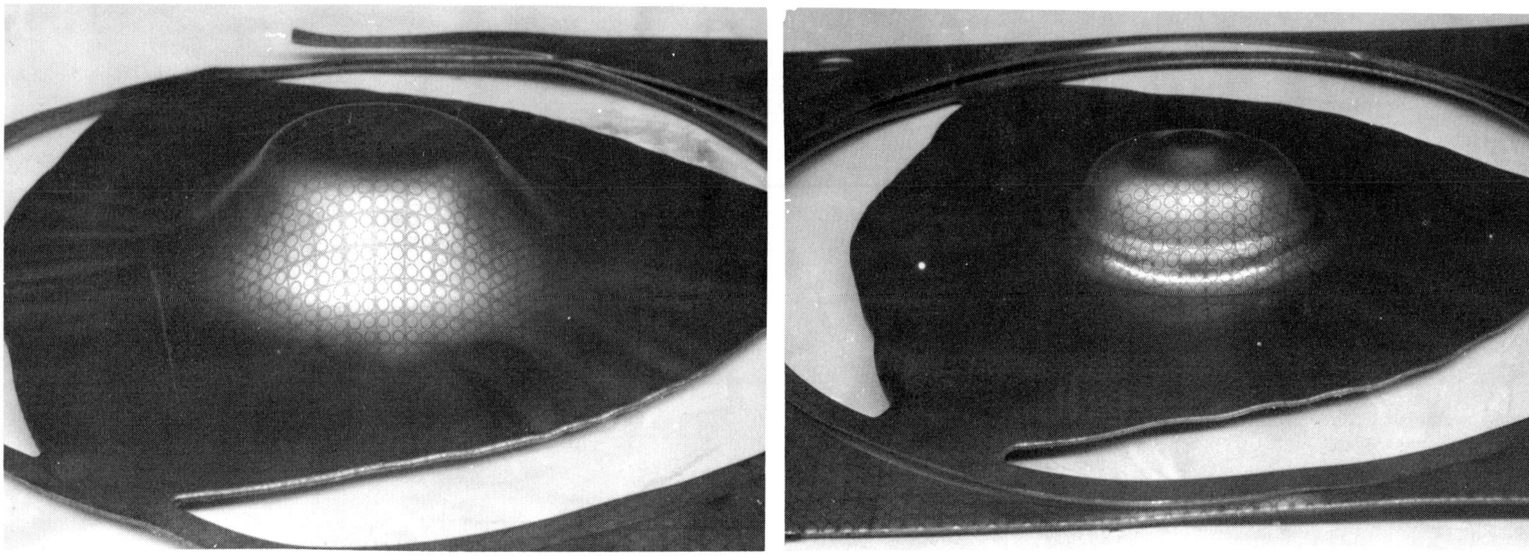

Fig. 3 — In first draw on motor end shield, new dies (right) generated high scrap for two reasons: 40% less material was prestretched, and the shape restricted metal flow. End shield made on old die is shown at left for comparison.

type of blankholder; location and dimension of draw beads; press speed; lubricant; drawing against a blanked edge; and various aspects of part geometry, such as round shell without flange, round shell with flange, square or rectangular shells.

By building up this bank of information, we eventually will be able to provide design guidelines to the engineers and industrial designers for sheet steel parts. We have used some literature data as a start on draw radii and verified it with experience as shown in Fig. 5. Note that too large a draw radius induces wrinkling and that the standard recommendation of 6 to 10× thickness decreases with increased thickness. Another design limitation is encountered in drawing a rectangular boxlike part. The part depth-to-corner radius ratio should be between three and six to be drawn in one operation.

Above six takes multiple draws: two draws for a part depth-to-corner radius ratio of 6 to 12; three for a ratio of 12 to 17; and four for a ratio of 17 to 22. In designing a part try to stay away from the borderline conditions because they either create extra scrap or cost extra tooling dollars.

Other Routes to Saving

Other cost saving approaches to optimizing your steel purchasing include the following:

1. Truncating the fourth decimal — Since the advent of the theoretical minimum weight method of buying steel from mills, the buyer pays according to the specified thickness. Thus by dropping the fourth decimal from the ordered thickness — by ordering 0.035 in. (0.88 mm) instead of 0.0359 in. (0.91 mm) — you will save 9/359, or about 3%, and the same product is received from the mill.

We have yet to detect any difference in the thickness of the received sheets in several years of experience. This may not sound like a significant amount of money, but if you are buying a million dollars worth of sheet steel (3125 tons), you can pay for the salary of a top engineer with the savings.

2. Using hot-rolled steel as a substitute for cold-rolled on heavier gage parts — Since the development of improved pickling systems and computerized hot mills, the quality of hot-rolled sheets has improved dramatically. It is often difficult for the layman to detect the difference between hot and cold-rolled samples.

For many applications where cold-rolled is now specified on thicknesses over approximately 0.060 in. (1.5 mm) hot-rolled may be used without any design or appearance penalty. The saving is on the order of 15%. A slight increase in strength on the average also accompanies the hot-rolled usage.

One drawback must be investigated. The thickness tolerance on the hot-rolled is about twice that of cold-rolled steel, so die clearances must be known to insure they will take heavier stock. This can conveniently be done by placing lead shims between the punch and cavity and cycling the press. In general, if the dies are at all worn, no clear-

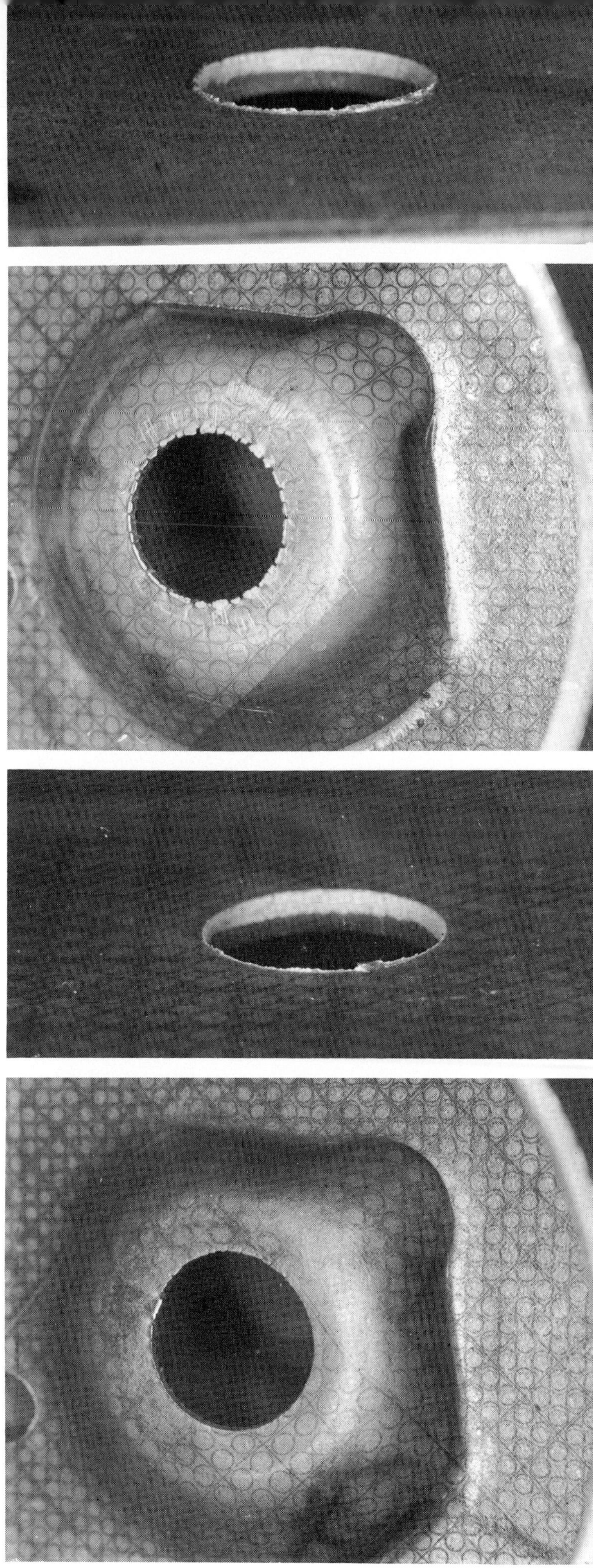

A B C D

Fig. 4 — Burrs on blank (A) for refrigerator main cup caused splits in formed parts (B). Sharpening punches eliminated the problem as shown by blanked (C) and finished part (D). Scrap rate fell to 1%.

ance problems will be encountered. We even use this steel for heavy-gage, bright-plated parts such as hinges with no adverse appearance problems.

3. Utilizing Class 2 surface on unexposed parts — To reduce costs on all but the most stringent appearance requirements, specify the Class 2 surface on cold-rolled sheets at a saving of $8 per ton. Because this material is often not run through the temper mill after annealing, coil breaks and lack of flatness may have to be contended with; but more often than not, little or no difference can be seen between Class 1 and Class 2 cold-rolled steel in the shop.

4. Minimizing size extras — With the advent of computerized cost accounting methods, steel mills have developed price extras which more nearly reflect true costs. Thus premiums up to 15% of base price are being charged for ordering narrower coils and sheets. The point to watch here is to try to avoid paying any width extra by ordering multiple widths where feasible.

The optimum width for 0.035 in. (0.88 mm) thick material is, for example, 45 to 60 in. (1143 to 1524 mm). If your parts are 22½ to 30 in. (572 to 762 mm) wide, order double width and slit. If they are narrower, you can order several widths and slit to size. Ordinarily, service center slitting or even in-house slitting costs are less than the mill width extra.

If your widths do not fit any of these categories, consider rotating the blank direction 90°; possibly the new width will eliminate the width extra. Ordinarily, directionality is not that critical as far as formability is concerned.

Using coils instead of cut length blanks, of course, is an obvious saving if your operation can handle coils. Flatness and close inspection for surface defects are sacrificed, however. The saving by eliminating the blank length extra is about 7%. To realize a net saving, your shop must be able to shear either on the press line or on a cut-up line and absorb more loss due to rejectable surface defects, all at less than the 7% material saving.

Another device which we have found to be helpful with older coil feed tooling is a low-cost system for accurately controlling the offal scrap between blanks. Essentially, it consists of a shaft encoder with a preset counter which precisely initiates the blanking shear. Blank lengths which previously had varied as much as 1½ in. (38 mm) are now controlled to 1/16 in. (1.5 mm) tolerance.

5. Reducing weight — Much has been written about the usage of higher-yield-strength, low-alloy steels. They have particular usefulness where increased elastic deflection can be tolerated without permanent set. In general, they have not been widely applied in the appliance industry. Possibly we are missing a good bet, but compression buckling and die problems have prevented their acceptance.

Draw Radius vs Material Thickness

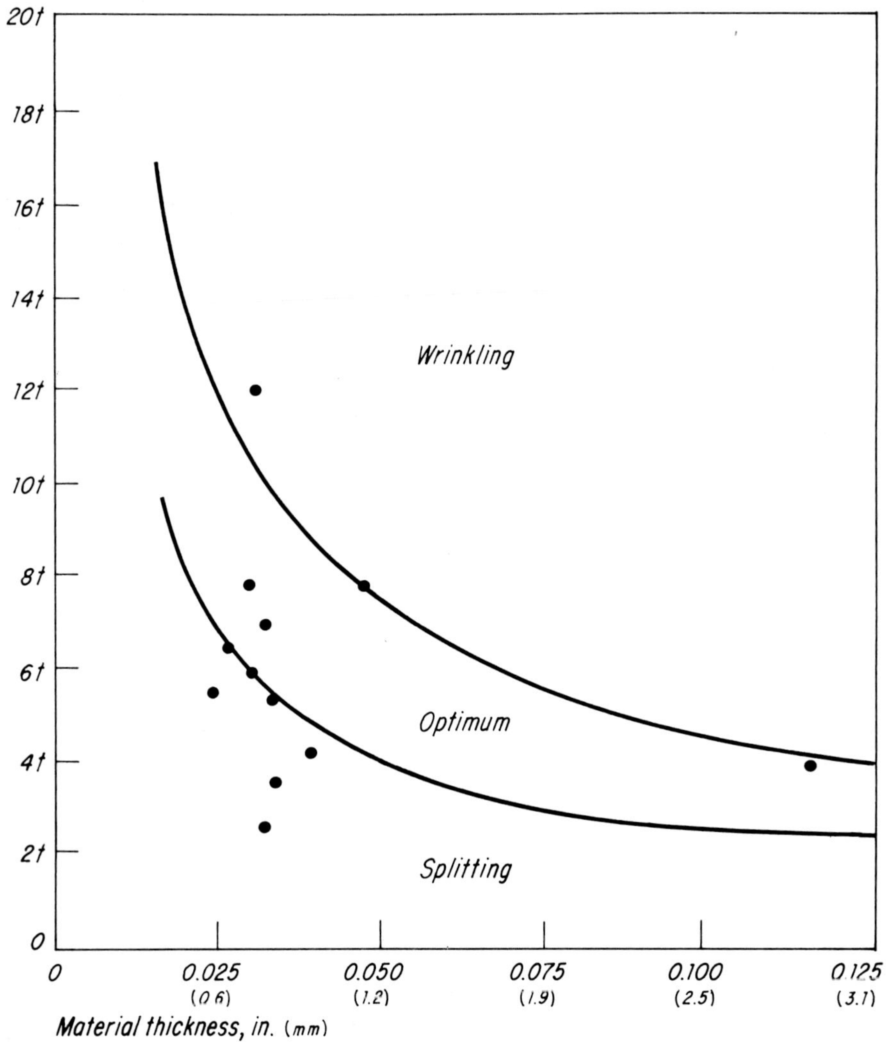

Fig. 5 — Splitting was encountered in using a draw radius that was too sharp; wrinkling was problem with too large radius.

In another approach to weight reduction (using low-carbon, cold-rolled sheets), deflection or rigidity is changed by geometry. Usage of embossments, ribs, corrugations, or textures should be investigated to optimize the usage of every pound of steel purchased.

6. Putting a good shop man, say a tooling specialist or foreman, or shop planner, in the "front office" for a few weeks — Ask him to go through each part and look for size and grade opportunities. Have him ask why one part needs a 2 in. (51 mm) trim after forming while another only needs 1 in. (25 mm) all around.

Ask why similar parts are made from different grades. Why does one line run consistently higher scrap rate than its duplicate? Why is die maintenance better on the south side of the shop than on the north? Is the optimum lubricant used on every job?

A conscientious man intimately familiar with the shop can often spot inconsistencies and can also effectively rebut shop opposition to change.

A Summing Up

There are all kinds of approaches to steel optimization, some technical, some managerial, and some psychological. Those of you who have tried to change a well-running but materials-wasteful job in the shop (and encountered a hundred reasons why it couldn't be done) will appreciate the need for many techniques for getting the job done. But with the tremendous leverage offered in high volume production, the rewards are extremely high.

We find this work challenging and oftentimes frustrating, but when an idea pays off big, it's hard to beat the excitement and sense of fulfillment; and this is the part of the job we remember. Best of all, this is the part that top management remembers. We are thankful for that.

THE ASSESSMENT OF SHEET STEEL FORMING SEVERITY IN AUTOBODY STAMPINGS

By Satoshi Hanai*, Shigeo Sakai*, Itaru Aoki*,
Taiichi Satoh* and Takashi Horita*

SYNOPSIS

As for the severity assessment, the conventional methods using the forming limit strain or critical press working conditions are not so practical from the predictability point of view.

On account of the greatest difficulties of establishing the unified assessment system of severity in the press forming of a large-sized and geometrically complicated parts such as autobody parts, this paper deals with the possibilities of each assessment system of forming severity for the limited number of large-sized geometrically basic models, paying an attention to the importance of size effect, for the purpose of designing the suitable material properties of sheet steels.

The obtained systematic understanding of forming severity corresponding to each basic model based on the fundamental difference in forming pattern is found effective for improving the forming conditions and selecting the suitable material properties of sheet steels in its application to some actual cases in high severity forming of autobody parts.

1. Introduction

There have been published many works concerning the evaluation of forming severity in press forming of sheet metals. However, no generalized standard has so far been established with respect to the assessment of press forming severity of a large-sized and geometrically complicated part such as autobody parts. As it is difficult to complete the synthetic assessment system of forming severity including various forming defects which contain breakage, wrinkling, puckering, surface damage and low dimensional accuracy, it may be more effective for the solution of this problem to employ a practical way for investigating the selection system of a suitable method corresponding to each case of different character in many press forming operations.

Steel makers have been interested in the assessment system of sheet steel forming severity in relation with the following problems:

---

* Technical Research Center, Nagoya Works, Nippon Steel Corporation, Aichi Pref., JAPAN

1) In the production process of sheet steels, the design and application of suitable material properties conforming to the variety, shape and size of pressed parts and press working conditions. a) Establishment of the changing system of sheet steel producing method according to the model change in car body and grade down in material properties. b) Necessity to know the suitability of a selected material for the expected forming operation by means of a little information such as the name and shape of pressed parts under inquiry. c) Understanding of the mutuality of factors for lowering the forming severity in terms of the required press performance and improvement in the formability of sheet steels.

2) Investigation of the formability and applicability of sheet steels produced by a newly developed production process such as continuous casting and continuous annealing line, for various pressed parts.

3) Decision of the aim or direction of development in the producing conditions and the quality of sheet steels, as in super deep drawing quality or high strength low alloy sheet steel.

The present paper describes the effects of press working conditions and sheet material properties which play an important role in the forming operations on the forming limit determined by breakages or geometrical defects. The main investigations concentrate in the above-mentioned item (1) through the way of examining the press performance of large-sized parts in comparison with the past proposed assessment system of forming severity.

There is very limited an information from sheet steel users in the case of assessing the press formability from the above-mentioned point of view. As it is usually a rare case to know the deformation condition and the strain distribution of the pressed part, it is necessary to judge the suitability of a material with the limited knowledge of simple geometrical factors in the required part. The assessing method for the severity of simple shape forming should be appreciated again in consideration of the conflict between the practical usability and possible accuracy in the severity assessment system.

2. Basic considerations on forming severity

2-1 Forming limit diagram

The forming limit diagram introduced by Keeler[1] and Goodwin[2] has widely been utilized in the analysis of sheet metal formability. This method is understood to be useful only in cases that fractures occur in the stretch deformation region. Quantitative data and information are required to be accumulated with respect to the shape and strain distribution for various kinds of autobody stampings.

The effect of material properties on sheet formability is investigated by press forming more than ten practical models, which produces various types of deformation successfully in our laboratory. Fig. 1 shows the effect of n and r values on the strain distribution in the fracture-expected region. In these tests are introduced three fender type models which include the following three different types of deformation and fracture, that is to say, type A has the whole deformation of stretching and stretch fracture, in type B the whole deformation is of deep drawing while the fracture occurs under stretch deformation, and type C is characterized by the whole deformation of deep drawing and the fracture located in the drawn side wall.

Weak points of the forming limit diagram are as follows. 1) The limit is extremely varied depending on the forming condition. 2) This method is of no help to the analysis of limiting dimensional accuracy. 3) It is impossible to predict the forming severity before the examination of an actual forming operation. As it is obviously important to know the suitability of a material for a practical forming operation prior to the running stage of production, it may be said in these respects that the success of the forming limit diagram is limited to some extent.

2-2 Limiting conditions in forming operations

The forming limit representation specified by fracture is useful in the actual operations. Example of this method is shown in Fig. 1 for the possible forming region relating to the blankholder force. This method has no general standard for different dies, though it is of use in the evaluation of material formability.

2-3 Limit of whole deformation

Many attempts have been made for the expression of severity by means of profile length or surface area increase in a pressed part. In the analysis of practical forming information, an increase in the profile length occurs usually in the case of high severity forming, but the large profile length in a die cavity does not always bring the forming operation to higher severity, because the high increasing rate of profile length is sometimes caused mainly by the draw-in of a shrink flange. In general, the severity becomes comparatively lower, when the geometrical defects can easily be contolled even in the presence of large draw-in. It is important to study the effect of profile length change between the adjacent cross sections of a pressed part on the severity of forming, because a sharp change in the profile geometry is apt to cause the dimensional defects.

Another effective method for the severity assessment uses the depth of a pressed part. In the type of stretch forming, there is present a rather simple relationship between the depth and the growth of strain, where it is easy to correlate the limiting depth to the forming limit strain as follows:

$$\text{Severity} = \varepsilon/\varepsilon_{max} \sim H/H_{max} .$$

But in the type of deep drawing, the above-mentioned relation becomes more complicated and therefore should be refined with regard to the tool shapes and dimensions by more detailed experiments. It should be noted that the maximum height is not always determined only by the occurrence of fracture but also by wrinkling.

3. Experimental on typical forming patterns

The methods described in the previous chapter are not entirely satisfactory to the assessment of material suitability for a particular forming operation, from the following reasons:

1) Limited usefulness where the deep drawing component is predominant.

2) Little help to the evaluation of geometrical accuracy.

Therefore, it is necessary to make the study of sheet forming tests relating to the basic simple shapes, in order to acquire the more detailed information as to the severity assessment of sheet metal forming. The conventional testing procedures are not so effective, probably because of their limited dimensions. A variety of failures are not completely realized by the small size simulation.

A 400 ton hydraulic double action press was used to perform various types of forming which were well simulated to the forming patterns of actual autobody parts. The critical forming limit, the technique for avoiding the geometrical defects and the effect of material properties on the above factors were experimentally investigated through a variety of

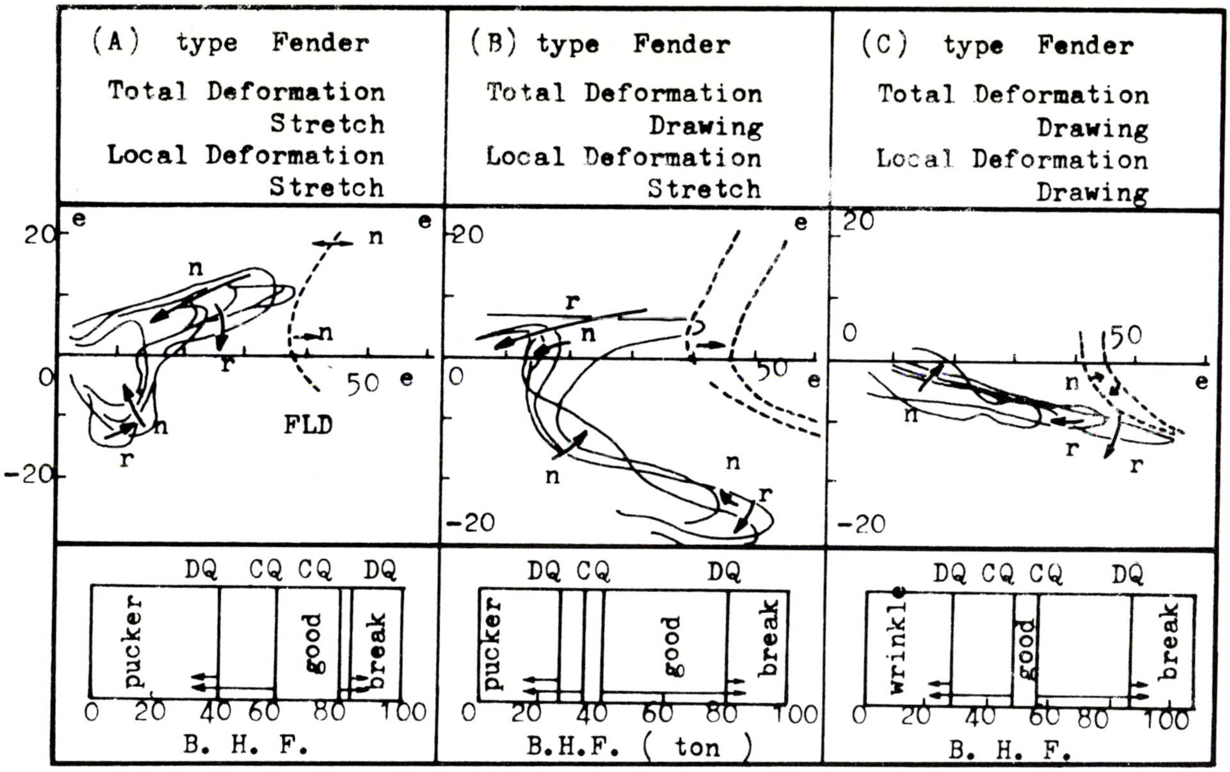

Fig. 1 Deformation constitutional diagrams and operational forming limit in actual parts stamping.

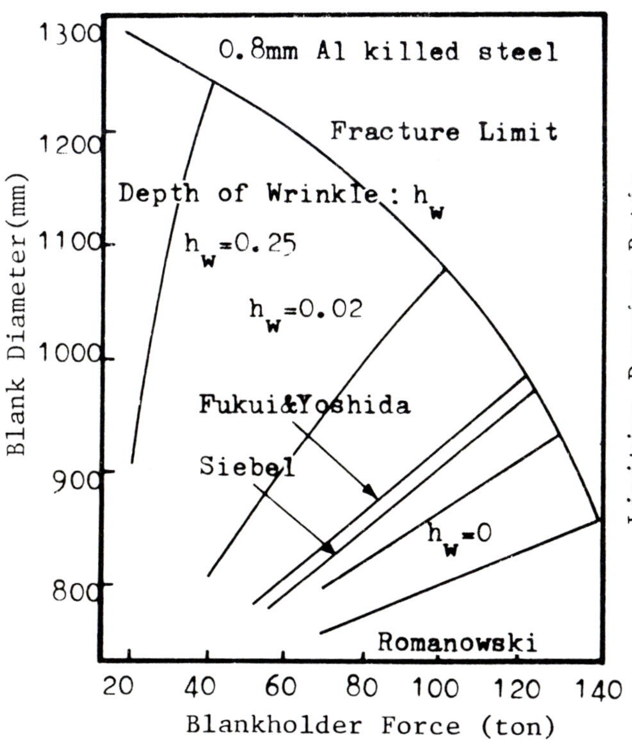

Fig. 2 Forming limit in 600 ϕ cylindrical cup drawing.

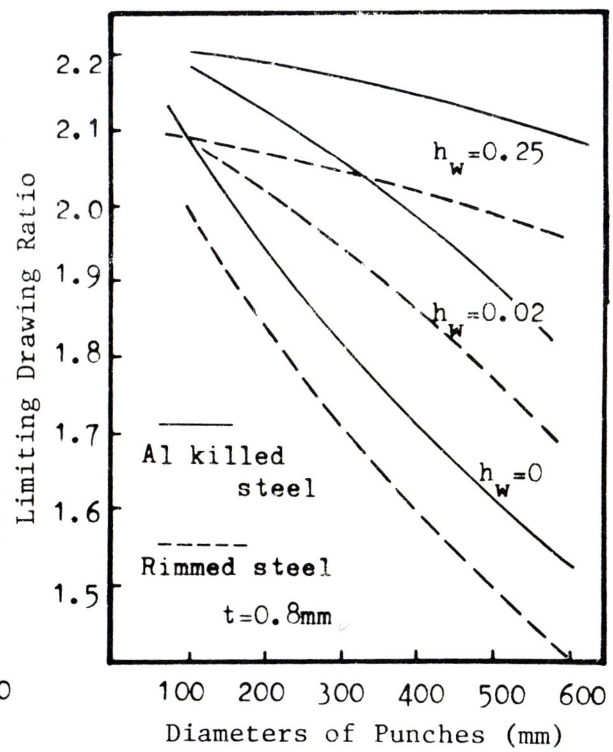

Fig. 3 Size effect in L.D.R.

fundamental model forming.

3-1 Deep drawing of cylindrical shells

Effect of tool geometry on sheet formability was examined for various sized cylindrical shells in deep drawing operation. The diameter of punches ranging from 60 to 600 mm was employed for various punch profile radii of 4 to 40 mm. Die profile radius was 4 to 10 mm, and then the tests were carried out in the cases of the ratio (punch diameter/sheet thickness) being 50 to 1000.

The results are summarized below.

1) The criterion of wrinkling has a greater influence on the forming limit in the deep drawing of larger sized parts. The empirical assessing formulas of optimum blankholder force proposed by Fukui and others [3~5] are available even for the large sized cup drawing as shown in Fig. 2.

2) An increase in the punch diameter results in the reduction of LDR as indicated in Fig. 3, provided that the blankholder force is constant. However, if the criterion of wrinkling is eased, LDR increases to a great extent.

3) The influence of profile radius on LDR seems rather small.

4) Effect of r value on LDR is strong, as an increase in r value by 0.5 improves LDR almost by 0.1. This effect results mainly from the improvement of fracture limit. An increase in r value is of little help to the improvement of wrinkling limit.

3-2 Deep drawing of square shells

Another typical deep drawing is the square shell forming which is similar to many autobody stampings. The relationship between the forming limit and various forming factors is investigated for the square shell forming of different shapes and sizes. Tool geometry is shown in Table 1.

The following results are obtained.

1) Three typical patterns are found through examining the mode of failures, critical forming depth and critical forming conditions as indicated in Fig. 4. Smaller tool size region ($L_p/t < 200$) where the forming limit is specified by the fracture at the profile radius (α fracture). Medium tool size region ($L_p/t = 200 \sim 600$) where the forming limit is specified by the fracture occurring at the punch profile or in the side wall (wall breakage). Larger tool size region ($L_p/t > 600$) where the forming limit is specified by the wall breakage

due to localized deformation in the square shell corner.

2) Deep drawing performance is shown in Fig. 4 examining the criterion of wrinkling and fractured blank sizes.

3) Critical forming depth is shown in Fig. 5 for the individual tool dimension.

4) In the smaller and medium size regions, an increase in radius of corner results in an increase in the critical forming depth. In the larger sized region, the critical forming depth has the maximum value in the vicinity of 50 mm corner radius in Fig. 6.

5) Effects of material properties on the forming limit are: a) In the medium and larger sized regions, an increase in r value results in an increase in the critical forming depth. As the limiting curves of wrinkling are slightly improved with increasing r value, the critical forming depth is in proportion to r value. b) The influence of r value on the forming limit seems to be greater in smaller sized forming than in larger sized one. c) Effect of r value on the forming limit is greater in the case of larger corner radius and is negligible for the condition of $r_c/L_p < 1/20$. Little effect of material properties on the forming limit is found in the case of smaller punch profile radius. d) Influence of workhardening characteristic (n value) on the forming limit is little except for the severe tool geometry.

6) It is possible to improve the critical forming depth with the blank prepared by appropriate corner cut. In the medium and larger sized regions, the critical forming depth is improved at most by 20 %.

3-3 Deep drawing of conical square shell

A technique for setting the slope of side wall is employed in the majority of autobody pressings in order to ease the severity of deep drawing. It is difficult to control the geometrical defect like puckering. In this case the trouble arises from the puckering in the side wall due to the presence of free surface from the direct control of blankholder force and the fracture may occur in the case of eliminating the puckering with higher blankholder force. Tests are carried out by the tool condition indicated in Fig. 7. The following results are obtained as to the corner radius and the clearance.

1) A reduction in the clearance results in the improvement of the puckering limit but lowers the fracture limit and also the effect of

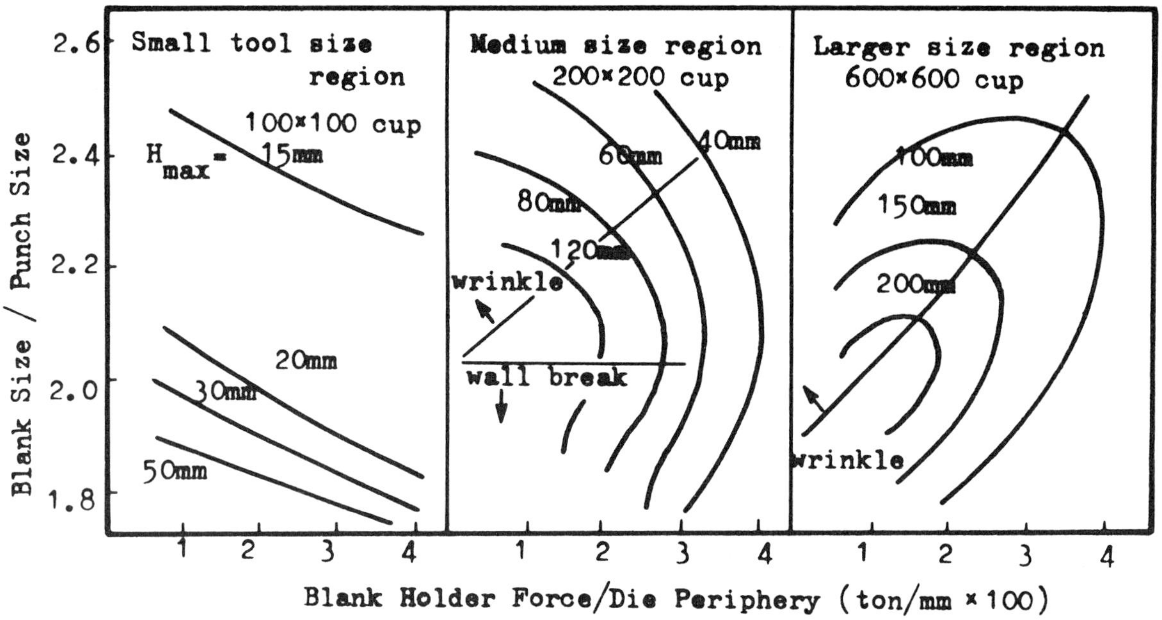

Fig. 4 Forming limit heights in various sized square shell drawing.

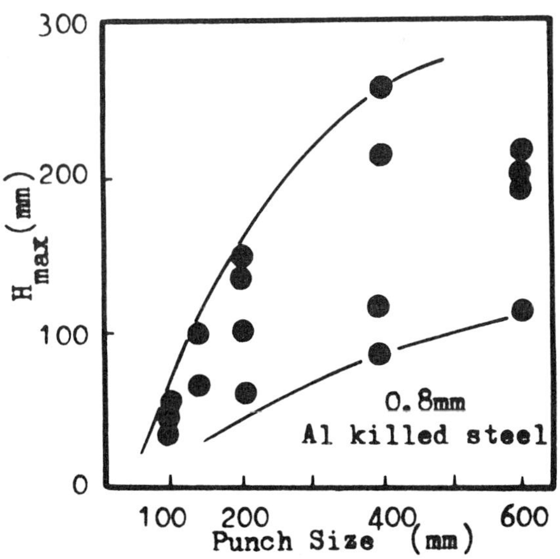

Fig. 5 Possible height in square shell drawing

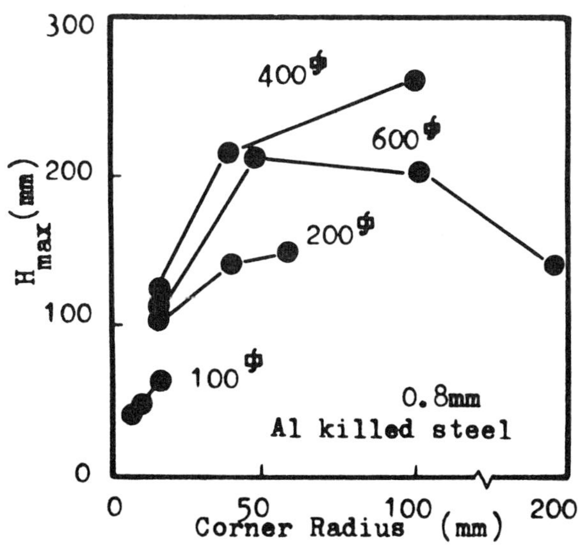

Fig. 6 Influence of corner radius on maximum height of square shell

	Punch	Die	r_p	rd	rc
Small tool size region	99 × 99	101.5Φ	5	5	2,5,10,20
	149×149	151.2Φ	5,10	5	10,20
Medium tool size region	200×200	204×204	10,20	10,20	20,40,60
	400×400	404×404	20,40,80	10,20	20,40,60,100
Larger size "	600×600	604×604	20,40	10	20, 50,100

Table 1 Tools in square shell drawing

blankholder force on the control of puckering.

2) The puckering becomes larger in the approximately same radius of corner for the punch and die, provided that the clearance is constant.

3) The smaller the r_{pc}, the larger the puckering in the case of $r_{pc} = r_{dc}$.

4) The relationship between the critical forming depth and the slope of the side wall is shown in Fig. 8. A decrease in the clearance results in a reduction in the critical forming depth because of a decrease in stretch deformation. When the smaller and proper clearance avoides the puckering, the critical forming depth increases abruptly. The forming with further smaller clearance approaches to that of square shells.

5) For example, the critical forming depth H_{max} for 5 mm clearance is as follows.

$$H_{max}\ (mm) = 35.2 \cdot t\ (mm) + 5.8 \cdot \bar{r} - 2.1 \cdot YP\ (kg/mm^2) + 58.0$$

3-4 Cylindrical step drawing

This technique is frequently employed for the avoidance of geometrical defects and the improvement of dimensional accuracy in autobody stampings. It is effective in a case that the shape of a panel can not easily be controlled by changing the press working conditions. Results of some tests on cylindrical step drawing are shown in Fig. 9. The possible forming depth H_1 without fracture and puckering is determined by geometrical factors such as step width l, depth H_2 and step profile radius r_{id}. Two types of forming are observed in relation with changes in the step width and depth. Type A ($H_2 < H_1$) is a severe forming where the inner shell is freely formed under insufficient control of puckering. Type B ($H_2 > H_1$) is characterized by a kind of stretch forming localized in the conical shell. As for the effect of material properties on the forming limit in this type of forming, higher r value is effective in type B and the strength and/or elongation has a great influence as shown in Fig. 12.

3-5 Forming with draw beads

Draw beads technique is also available for the general geometrical defects of pressed parts. According to the experimental results obtained in the rectangular model forming, the press forming conditions can easily be controlled by using draw beads, but this technique weakens the effect of material properties on formability.

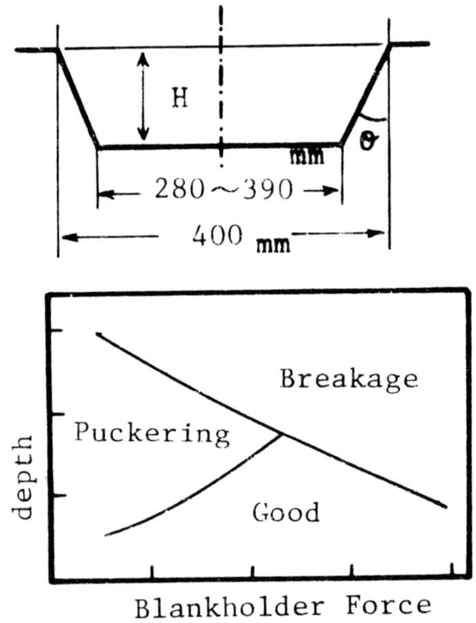

Fig. 7 Conical square shell. drawing

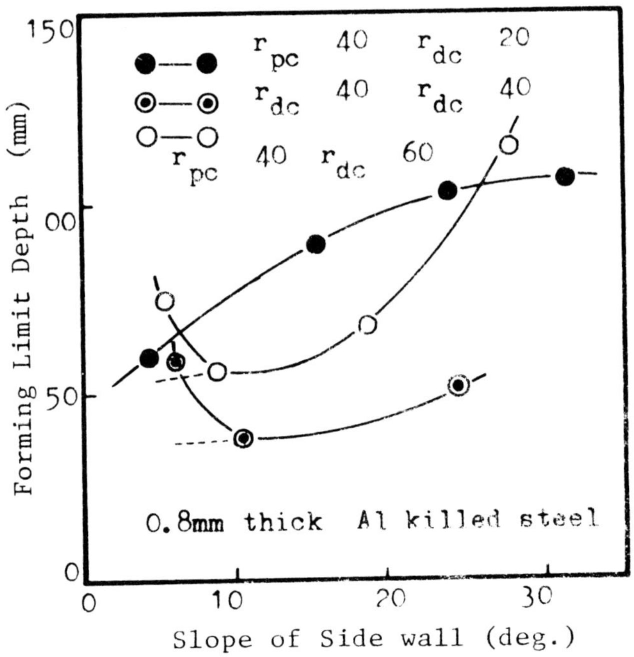

Fig. 8 Effect of corner radius and slope in side wall on forming limit of conical square shell drawing.

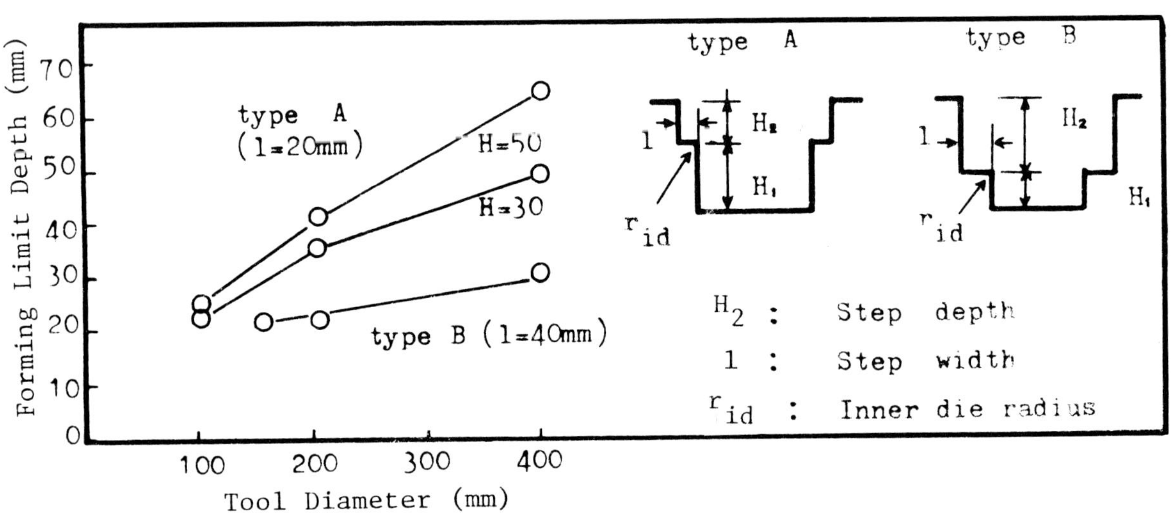

Fig. 9 Forming limit depth in cylindrical step drawing..

	Y. P	T. S.	E 1	n value	r value
Al-killed steel	17.0kg/mm^2	31.1kg/mm^2	46.4 %	0.218	1.64
Rimmed steel	20.4	31.0	46.5	0.218	1.23

Table 2 Mechanical properties of sheet steels used in the experiments which are described in chapter 3.

4. Applications for practical pressings

4-1 Production selection of suitable material

There is generally little trouble in the majority of autobody parts where the normal commercial grade sheet steels are employed. consequently, the assessment of sheet steel severity necessary for assessing the suitability of a material is especially effective in more severe forming operations. The criteria required for the above purpose are as follows: (1) To determine merely the selection of sheet steel grade, for example, rimmed or aluminium killed sheet steel, (2) To produce sheet steels with some specially designed material properties, (3) To supply the special quality sheet steel to more severe forming. Previous information concerning the increasing rate of profile length and strain distribution proves its usefulness as first order approximation. Actual applications are as follows.

1) Fender forming

Two types of deformation take place in this forming. One is the type (a) forming where the stretching deformation is predominant, and the other the type (b) forming which is of the drawing deformation. Deep drawing quality steel sheet may be applicable in the case of the depth of a formed part being over 50 mm in the type (a). The type (b) forming needs the drawing quality material when the maximum drawing depth is over 150 mm. It may be difficult to control puckerings in the forming of a part with the deep drawing depth, the small profile radius of corner and the large angle of the overhang part. As for the material properties, the suitable selection of the yield point and r value is required, as indicated in the formula of the critical forming depth. The decarburized sheet steel is available for the forming of severe lateral embossing in the side wall of type (b).

2) Quarter panel forming

Deep drawing quality sheet steel with lower yield point may be indispensable for this type of forming. Possible classification of this forming is made by means of the depth of the pressed part, where the sheet steel with the highest quality is available for the depth being over 200 mm. The sheet steel having higher tensile strength and r value is effective in the forming restricted by fracture in the side wall where the tool geometry has small corner radius and large punch

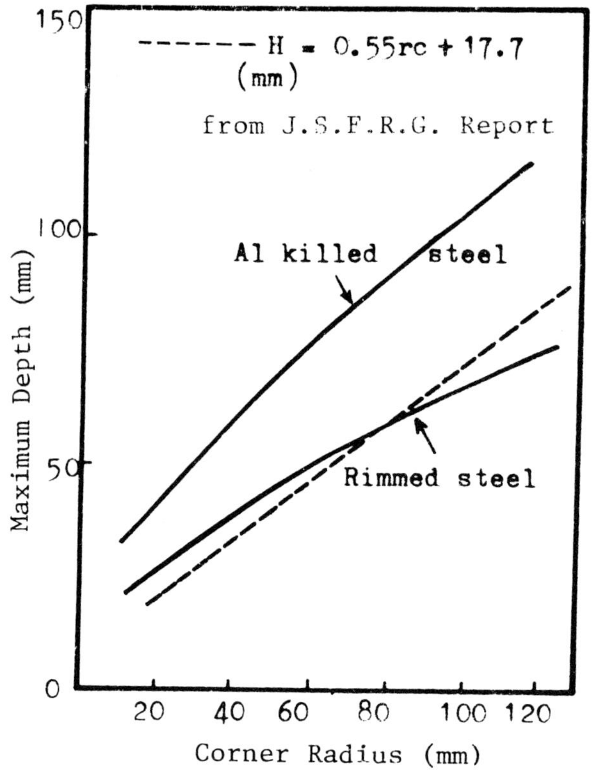

Fig. 10 Relationship between corner radius and maximum height in square cone forming.

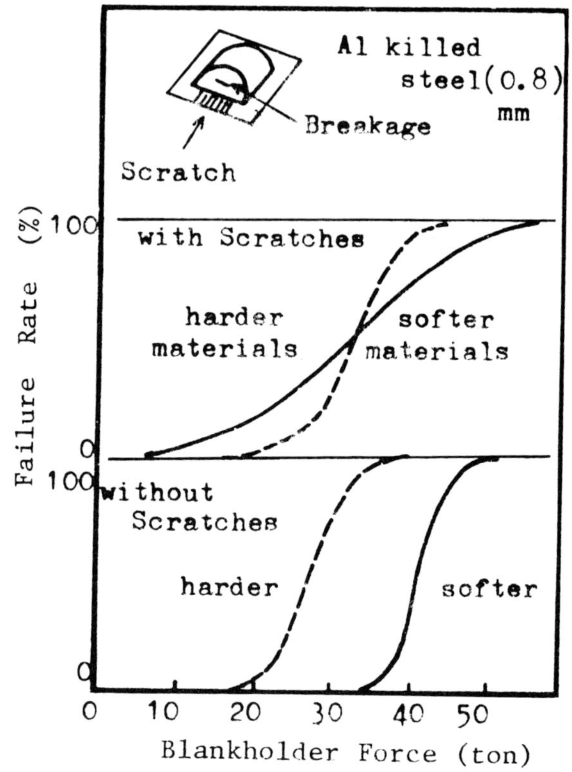

Fig. 11 Effect of die face condition to press performance.

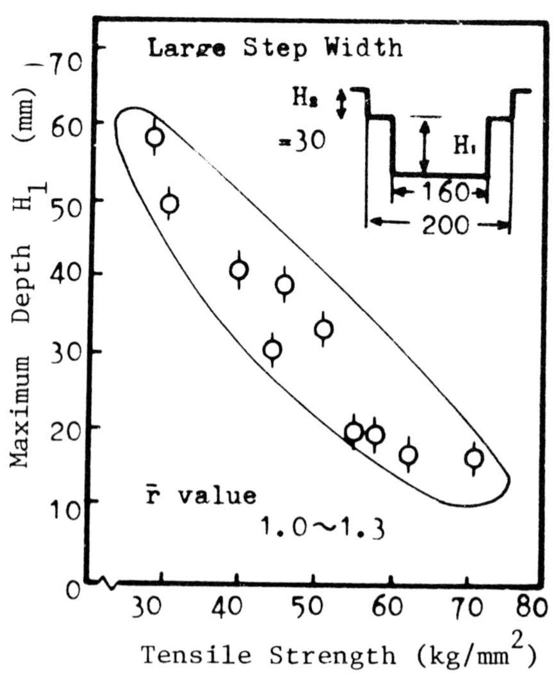

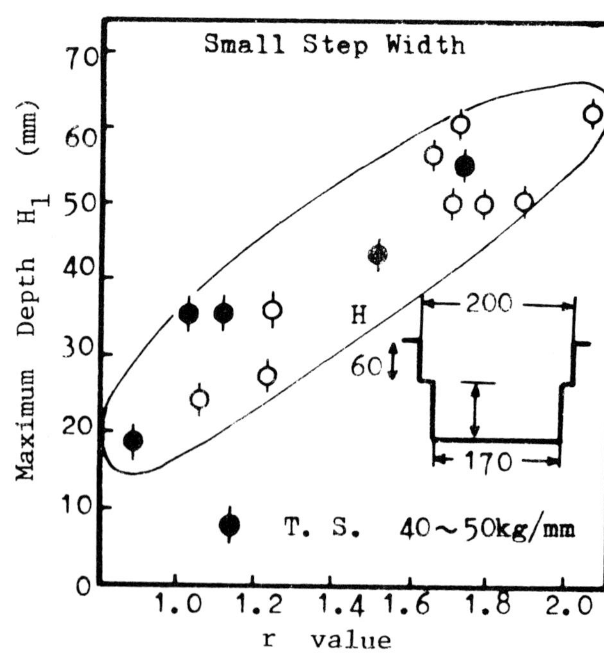

Fig. 12 Influence of mechanical properties on step drawing height.

profile radius. The aluminium killed sheet steel specified by lower manganese content or titanium killed sheet steel may be of use in the forming which requires the controlling of slight puckering on the panel surface.

3) Door inner panel forming

The forming limit depends mainly on the corner radius as illustrated in Fig. 10. The suitable selection in designing the material properties is made with reference to this diagram. The dotted line in this figure shows the experimental formula obtained from the cooperative investigations in the Japan Sheet Metal Forming Research Group(6).

4-2 Other aspects in press performance

An effective achievement is not obtained by improving material properties in order to lower the rejected rate in some cases of press forming. It is important to produce the uniform and stable material properties. The technique of producing sheet steels recently has been improved remarkably as for the uniformity of properties, and consequently the rejected rate has become lower. An attention should be paid for the phenomena caused by the interaction factors between press working conditions and material properties. Fig. 11 shows that the press performance changes noticeably due to a slight difference in material hardness detected by means of the growth of scratches on the die face in the simulative forming test.

5. Conclusions

In the press forming of large-sized and geometrically complicated autobody parts characterized by various kinds of defects, the assessment system of forming severity should be selected and applied under some appropriate limitations according to its suitability for the characters of the required press forming.

The limited number of basic models which are almost the same in size as actual autobody parts may be sufficient to obtain the more direct and effective information of the forming limit depending on the fracture and geometrical defects as usually seen in the autobody parts. The practicability and predictability of severity assessment are very important for engineers designing and manufacturing sheet steels.

The relative contribution of sheet thickness and material properties to the forming limit relating especially to fracture and wrinkling is estimated quatitatively.

The basic model experiments employed in this investigation have very often been applied to the forming tests of a large-scaled repetition, which are available for finding the most practical and effective way of reducing the rejected rate.

Acknowledgement

The authors wish to express their gratitudes for the guidance and encouragement received from Dr. K. Yoshida, and also their thanks to the press forming engineers of Toyota Motor Co., Ltd. and to the members of press forming research group of Nippon Steel Corporation.

References

1) S.P. Keeler: S.A.E. Report, No. 65o535 (1965).
2) G.M. Goodwin: S.A.E. Report, No. 680093 (1968).
3) S. Fukui, K. Yoshida, et al.: Aero. Res. Inst. Report, Univ. Tokyo, Vol. 8, No. 1 (1954) 23.
4) E. Siebel: Stahl und Eisen: Vol. 74 (1954), 155.
5) W.P. Romanowski: Handbuch der Stanzereitechnik, (1959) VEB Verlag Technik
6) Report of Japan Sheet Metal Forming Research Group, No. 44 (1973).

Some Practical Aspects of Pressing and Deep Drawing Using High Strength Hot Rolled Steels

By: A. Stanway B.Sc (Eng) Met., A.C.T. * and
W. Holyman C.Eng., M.I. Mech.E. **

INTRODUCTION

Many engineering components are cold formed from hot rolled strip or plate steel, the choice of this material usually being governed by economic reasons or by reason of material availability in the size and thickness range concerned. Material formability relative to strength level is a vital criterion, if sufficient advantage is to be taken of the mechanical properties exhibited by steels currently available and under development.

In many, if not most instances, the components remain in the cold formed condition without further treatment, when the changes in properties resultant from the forming operations are not deleterious to the application or ultimate use of the component. Often designers take advantage of enhanced mechanical properties produced by cold forming operations, but where the nature of the subsequent use of the component precludes this, an appropriate heat treatment must be given.

In this paper, some practical aspects of pressing and deep drawing various types of component comprising small pressure vessels, automobile wheels, light axle cases and commercial vehicle sidemembers are discussed together with appropriate testing. This includes simple conformity to specification, certain formability tests and actual or simulated service performance tests on finished components or assemblies. The steels considered are available from many of the principal wide and narrow continuous strip mills and non-continuous plate mills in the U.K. and abroad, and exhibit yield stress levels ranging from approximately 200 to 800 N/mm^2. The changes in mechanical properties of the steels brought about by cold forming, and the response to heat treatment are also discussed.

SMALL PRESSURE VESSELS

(a) Compressor Housings

These components are manufactured from extra deep drawing quality mild steel to BS.1449: Part 1: 1972 HR2P in a series of pressing operations with a minimum inter-stage delay.

Conventional draw tooling is used on double action presses. The severity of the operations necessitate the use of chlorinated extreme pressure lubricants. Since it is important that the pressings are clean and free from rust, it is essential that the lubricant contains a stain inhibitor. The lubricant viscosity is kept as low as possible consistent with satisfactory performance, in order to ease application.

Figure 1 shows yield stress, ultimate tensile strength and elongation values exhibited by test pieces cut from

* Manager, Technical Services, Rubery Owen & Co. Ltd., Darlaston, U.K.

** Development Supervisor, Rubery Owen & Co. Ltd.

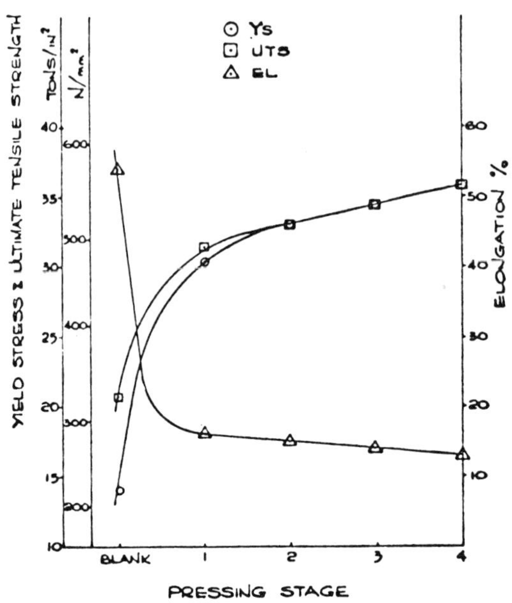

Fig 1. Mechanical properties of compressor housing pressings.

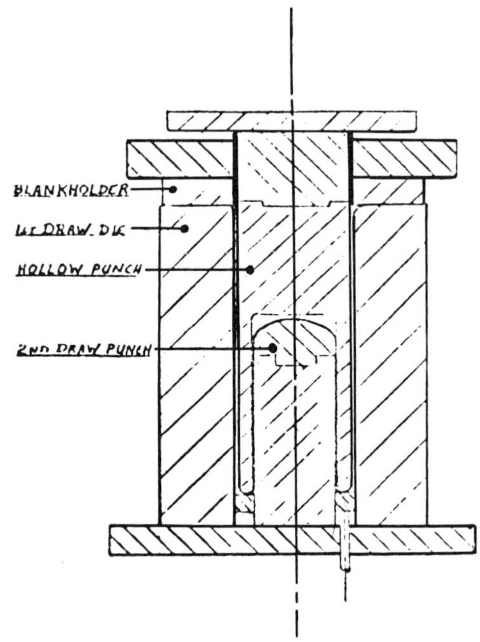

Fig 2. Reverse redraw tool

Fig 3. Appearance of L.P.G.cylinder pressing in the as drawn condition following impact.

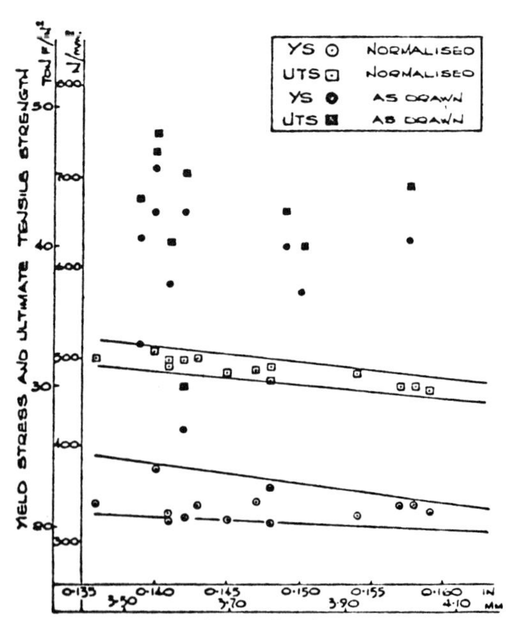

Fig 4. Relationship between mechanical properties and wall thickness of cylinder pressings in type B steel.

(a) Compressor Housings Cont'd

the blank and various stage pressings. Although the yield stress and the ultimate tensile strength become identical at the second pressing operation, there is still a significant percentage elongation which indicates that the as drawn pressing is not excessively brittle. Thus as the component is subjected only to low internal pressures and is not abused by frequent handling, it is possible to place it in service without a stress relieving heat treatment.

Test requirements are minimal for this component and only in the case of manufacturing problems is any such work carried out.

(b) Gas Cylinders

These are chiefly used for liquefied petroleum gases and for dissolved acetylene. They are manufactured either as two or three piece fabrications from three principal types of steel, the compositions and specified mechanical properties of which are detailed in Table 1. Only two-piece construction will be considered here. Such cylinders consist of two deep drawn shells joined by a circumferential weld. Various attachments such as a valve boss and a footring are also welded to the cylinder bodies.

The less exacting shells are produced in a single draw using conventional double action tooling with lubricants similar to that used for the compressor housing. For the deeper pressings it is necessary to use two draws. All blanks for these are lubricated with a soap and borax coating which is applied hot. When conventional redrawing is employed, the two operations are carried out on separate presses. Otherwise the reverse re-draw technique developed by Weatherhead in U.S.A. (1) is employed. The two operations are carried out in tandem (figure 2) on a press with a stroke in excess of 1700 mm.

Originally a 1:2.4 scale tool was manufactured to check the press specification, the method of lubrication and the cylinder steel types (2). The scale ratio was chosen to suit the press and the steel thickness available at the time.

The "as drawn" shells are excessively hard and brittle and totally unsuitable for the ultimate application, where the cylinder may be subjected to wide temperature variations, fluctuating pressures and considerable abuse during handling. Figure 3 illustrates rather dramatically the effect of dropping an "as drawn" shell onto a hard floor. Thus it is essential that the shells following welding to each other and to the various attachments are normalised from 900/920°C prior to testing and finishing. A detailed

TABLE 1

Specification for types A, B and 2 steels

Steel	*A* per cent.	*B* per cent.	*2* per cent.
C	0·20	0·18	0·20
Mn	0·60	0·40–1·20	0·70–1·50
Si	0·30	0·25	0·15
S	0·050	0·040	0·050
P	0·050	0·040	0·050
Nb	—	—	0·01–0·08
Yield stress, N/mm.^{2}	220 min.	275 min.	310 min.
Ult. ten. str., N/mm.^{2}	340–430	400–490	430 min.
Elong. on 200 mm., per cent.	22 min.	17 min.	15 min.
Elong. on 50 mm., per cent.	30 min.	24 min.	21 min.
Bend 180°	1T	1½T	1T

Values are maxima unless otherwise stated.

Mechanical properties quoted must be obtained from normalized samples.

TABLE 2

Chemical analysis and mechanical properties of carbon-manganese and pearlite-reduced steels used in fatigue tests

Steel	*C-Mn* per cent.	*P.R.S. (higher carbon)* per cent.	*P.R.S. (lower carbon)* per cent.
C	0·21	0·10	0·07
Mn	1·39	0·96	0·98
Si	0·21	0·21	0·02
S	0·032	0·031	0·015
P	0·021	0·017	0·026
Nb	—	0·021	0·026
Yield stress, N/mm.^{2}	330	395	400
Ult. ten. str., N/mm.^{2}	540	495	495
Y.S.: U.T.S.	0·61	0·80	0·81
Elong. on 50 mm., per cent.	37	39	36
Elong. on 100 mm., per cent.	28	29	26

TABLE 3

Fatigue test results on full-scale chassis side-members

Speciman	*Cycles to failure*	*Remarks*
C-Mn steel	368,550	Cracked 6 in. behind position of maximum stress
	427,875	Cracked 15 in. behind position of maximum stress
	553,800	Cracked at point of maximum stress
P.R.S. (higher carbon)	748,250	Cracked 30 in. behind position of maximum stress
	964,375	Cracked over front axle position
	975,600	Cracked 15 in. behind position of maximum stress
P.R.S. (lower carbon)	1125,000	—
	1320,000	—
	1440,000	—

(b) Gas Cylinders Cont'd

comparison of the mechanical properties exhibited by "as drawn" cylinder shells and "as normalised" finished cylinders manufactured from B type steel, expressed as a function of wall thickness is given in figure 4. The "as drawn" properties are widely scattered and bear no relationship to wall thickness, but the normalising removes most of the scatter, and although this is not shown in the figure, it restores the ductility of the steel. A slight but definite relationship is evident, the lower wall thicknesses being associated with higher mechanical properties and vice-versa. This is clearly a function of cylinder cooling rate following the heat treatment.

The "as rolled" mechanical properties of type A steel are compared with the corresponding properties exhibited by test pieces cut from finished 104 lb. propane cylinders in figure 5. It will be seen that there is generally a small reduction in both yield stress and ultimate tensile strength consequent from forming and heat treatment. Similar comparisons can be made for types B and 2 steels, but whereas there is a slight gain in strength with the former, the latter shows a considerable drop in strength. With type 2 steel the precipitation hardening effect of the niobium addition is virtually completely destroyed by the normalising operation, which clearly over-ages the steel.

With gas cylinders of minimum designed wall thickness of 6 mm or less, it is not usual to specify impact criteria at room or sub-normal temperatures. However, on occasions it is necessary to carry out appropriate impact testing. Figure 6 gives the impact properties of a number of batches of type B steel at various temperatures in the "as rolled" "as drawn" and normalised conditions. The lower impact properties of the cold worked steel as compared with the unworked steel are evident as is the significant improvement brought about by normalising.

Acceptance tests for gas cylinder steel are generally limited to conventional chemical analysis and mechanical testing. Very roughly, if the yield stress value is limited, the production performance can be predicted providing the steel is clean and free from significant surface defects. Some formability tests have been carried out on steel for dissolved acetylene cylinders (type 2). There was very little variation in r and n values and less in the L.D.R. At first there appeared to be some correlation between the results of the biaxial stretch test and press shop performance, but the situation had been confused by material thickness variations and changes in shop conditions. The mode of failure in the simulative

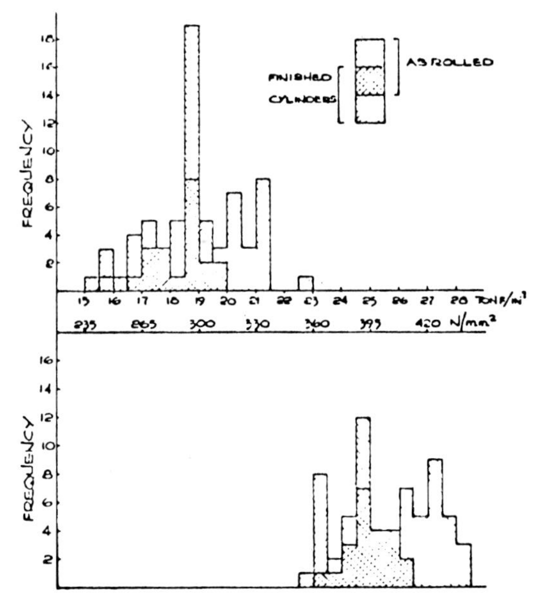

Fig 5. Comparison of yield stress and Ultimate tensile strength of type A steel

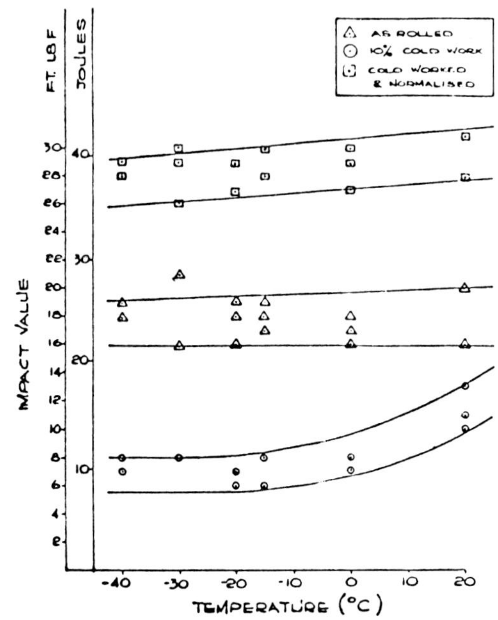

Fig 6. Impact properties of type B steel. Charpy test pieces 10 x 2 mm.

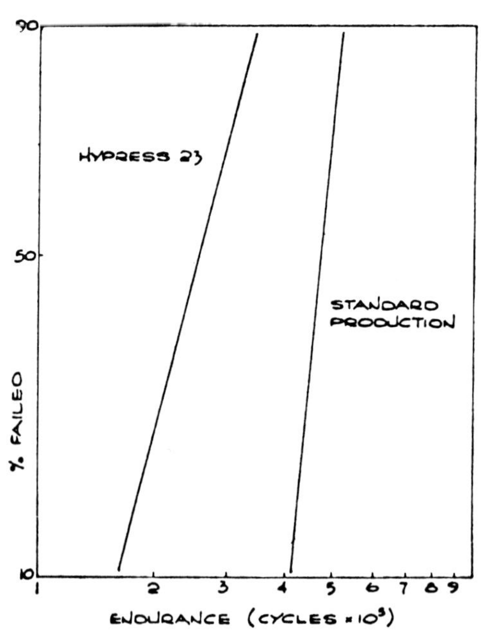

Fig 7. Fatigue test results for wheels manufactured from Hypress 23, 90% thickness of standard.

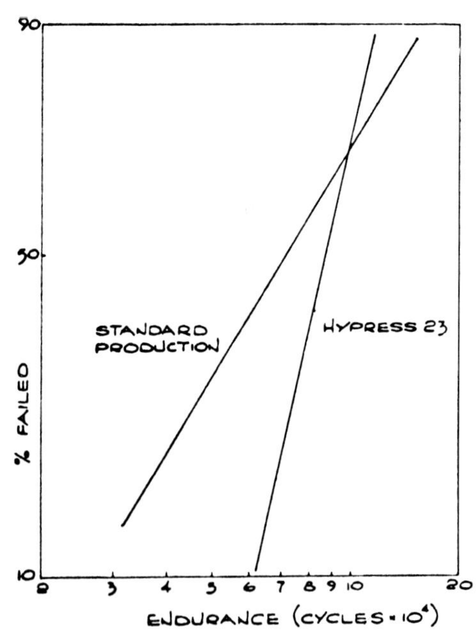

Fig 8. Fatigue test results for 13 inch diameter wheels.

(b) Gas Cylinders Cont'd

tests proved to be different from the vertical splitting usually experienced in the press shop. It was concluded that conventional formability tests do not seem to be appropriate for type 2 steel.

In practice very low scrap values can be achieved providing that the acceptance tests are satisfactory and the material thickness is within tolerance.

Final testing of completed cylinders is well described in (3, 4 & 5).

AUTOMOBILE WHEELS

Steel passenger vehicle roadwheels are fabricated by joining a rim section to a centre pressing usually by spot welding, but occasionally by rivetting. The rim is made by rolling a cut length of flat strip into a hoop or band and joining the ends by flash butt welding. The resulting cylinder is rolled into the desired shape and size. The steel is ordered to BS.1449: Part 1: 1972 HR4P with limitations to the yield stress value. Generally the quality normally supplied by the leading mills is adequate for the job, thus wheel rims will not be considered further here.

Wheel centre pressings vary in complexity. In some cases it is sufficient to specify a deep drawing quality steel to BS.1449: Part 1: 1972 HR3P to make the job without excessive scrap. However for the difficult ROSTYLE wheels it is essential to use an extra deep drawing quality viz HR2P to the same Standard.

Centre pressings are first drawn on double action presses followed by a series of forming operations on single action presses. All steps are relatively small, thus problems with steel formability are rare. The lubricant applied when the steel is pickled and oiled is usually adequate for the forming operations. Consequently acceptance tests are limited to chemical analysis and tensile testing.

Considering the shapes of wheel centre pressings, and to a slightly lesser extent wheelrims, it is difficult for a Designer to calculate working stresses accurately.

Consequently, wheels are designed on the basis of past satisfactory experience and extensive fatigue and road testing. The work hardening of the metal during pressing is an essential feature and is relied upon by the Designer to add strength in appropriate areas, and for the above steels does not have any deleterious effects. The maximum hardness values found are usually of the order of 140 to 160 HV corresponding to actual ultimate tensile strength values of 370 to 420 N/mm^2 (24 to 27 $tonf/in^2$). The original steel rarely has ultimate

AUTOMOBILE WHEELS CONT'D

tensile strength values exceeding 340 N/mm^2 (22 tonf/in^2).

Attempts have been made to use higher strength steels for wheel centre pressings with differing degrees of success. Work has been concentrated on pearlite reduced steels by reason of their outstanding ductility at the strength levels concerned. To be commercially viable, such steels could only be used at a lower thickness to compensate for their higher cost, and to some extent for higher scrap percentages resulting from their reduced formability as compared with the normal steel. Work hardening is greater than with the extra deep drawing quality steels and hardness values of 170 to 200 HV have been experienced corresponding to actual ultimate tensile strength values of 450 to 525 N/mm^2 (29 to 34 tonf/in^2) with low elongation values. The original steels used had ultimate tensile strength values in the range 400 to 430 N/mm^2 (26 to 28 tonf/in^2).

Figure 7 compares the fatigue endurance of standard production wheels with that of wheels made with centres pressed from a proprietary brand of pearlite reduced steel some 10% thinner than standard. It can be seen that the high strength wheels have much the poorer performance. Figure 8 compares results for the two types of steel at standard thickness. In this case, there is clearly a benefit to be gained by using the pearlite reduced steel. However, since standard wheels are perfectly satisfactory and service failures under normal conditions of use are almost unknown, the commercial penalty of using the higher strength steel is unacceptable.

It is difficult to account for the relatively poor performance of the high strength steel. Possibly with the very high degree of work hardening involved, differences in notch sensitivity of the respective steels are important and failure can take place from tooling damage, but as yet this behaviour cannot be satisfactorily explained.

LIGHT AXLE CASES

Axle cases were traditionally hot formed. However, with the development of pearlite reduced steels, cold pressing is being increasingly used for passenger car and light commercial vehicle casings.

Half pressings are produced from pickled and oiled steel in four press operations, blanking, prebending, forming and coining the edges. The latter operation allows the two half pressings to be welded together without further preparation. A viscous chlorinated mineral oil is applied to minimise scoring. Although internal bend radii are sometimes as small as $1\frac{1}{2}t$ (where t = metal thickness) few problems have been experienced with cracking on axle cases.

LIGHT AXLE CASES CONT'D

Acceptance tests for the steel are confined to chemical analysis and mechanical testing. The latter reveals the main problem experienced with pearlite reduced steels viz variable yield stress values. Figure 9 gives an indication of such variability. The production difficulties manifest themselves as shape problems consequent from unpredictable springback.

Simulated service tests are carried out on a hydraulic fatigue rig using strain information obtained during road track testing of vehicles. Such testing has resulted in the production of improved axle casings. For example cold formed casings failed at 222,000 and 104,000 cycles according to material thickness, while the previous heavier design hot formed case, with welded on stiffeners, failed after only 22,000 cycles at the same load.

CHASSIS SIDEMEMBERS

These components vary in length having a current maximum of over 11,000 mm (36 feet) and may be simple channels in shape or may be quite complex involving tapers, bends and inwards and outwards joggles. They are formed on large specially designed presses, the latest of which has a capacity of 40 MN (4000 tonf) and has been described in the literature (6).

Blanks are prepared either from narrow strip using a crop and pierce tool or from wider material using a blank and pierce tool. Although blanking uses more material, there are problems with edge camber on narrow strip. The steel is in the hot rolled (i.e. black) condition.

A section through a typical forming tool is shown in figure 10. Marquette pins act on the pad to hold the blanks against the punch face to prevent movement during forming. Setting of the cushion pressure is critical to control springback of flanges, web flatness and twist in the finished pressing. Springback of the flanges is also controlled by coining at the bottom of the stroke.

Lubrication is necessary but not critical for forming sidemembers. A light mineral oil or a water soluble paste is used.

As the same tool is used for cropping or blanking both left and right hand sidemembers, it is inevitable that on one hand, the burr from the shearing action is on the outside of the bend at the ends of the pressing. For this reason the ends of such sidemember blanks are locally annealed using an acetylene torch prior to forming.

While the forming operation cannot be described as simple bending, it approximates to this in the less complex areas of any sidemember. However, in the region of bends and joggles there may be significant amounts of cold working, hence steel

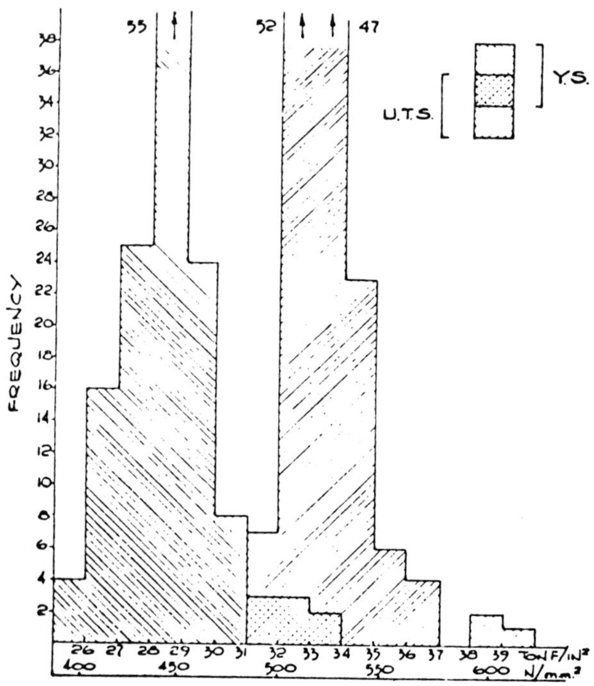

Fig 9. (left) Yield stress and ultimate tensile strength values of pearlite reduced steel plate 0.280 inch (7.1mm) thickness in as rolled condition.

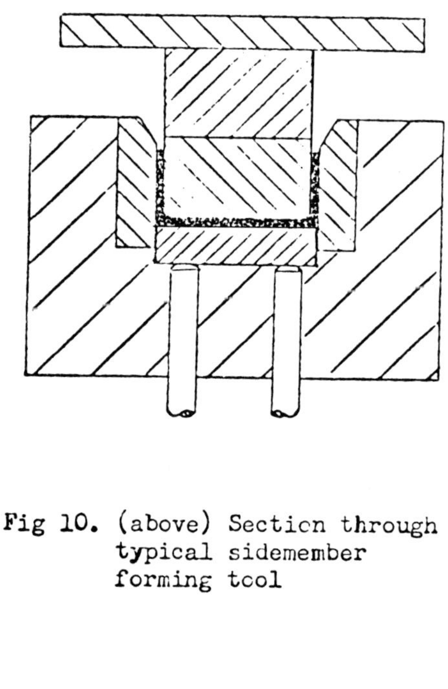

Fig 10. (above) Section through typical sidemember forming tool

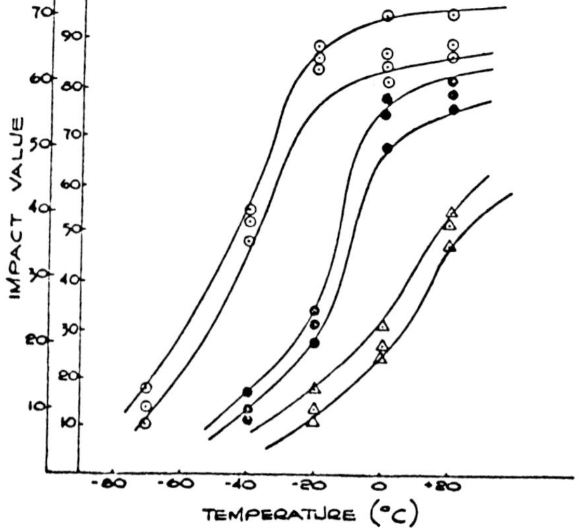

Fig 11. (left) Comparison of impact properties of pearlite reduced steel and conventional carbon-manganese steel plate 0.25-0.375 inches (6.3-9.5 mm) in thickness.
Charpy test pieces 10 x 5 mm.

ductility is an important criterion.

The components are designed on the basis of minimum specified yield stress for the steel concerned, with appropriate consideration for fatigue behaviour and expected service conditions. Some years ago, the only high yield stress steels available for this type of application were conventional carbon-manganese steels such as B.S.968:1962, with a minimum specified yield stress of 355 N/mm^2 (23 tonf/in^2). Dependent upon the source, such steels tended to have both carbon and manganese contents near the top of the specified range and had relatively poor ductility. The use of these steels resulted in a higher than desirable degree of defective components involving re-working, repair, or even in the worst instances total scrapping. The development of pearlite reduced steels and their availability in quantity at no extra cost since 1966 has offered an almost complete solution to this problem. These low carbon, relatively low manganese, niobium-treated steels have much improved ductility and are usually much cleaner than conventional analysis steels of similar strength levels. For these reasons their use in sidemember manufacture is to be preferred.

Originally, there was some resistance to the introduction of these steels by Production Staff and by Customers because of their somewhat unconventional properties, and in particular their high yield stress to ultimate tensile strength ratio. It was thought that this might result in inferior impact resistance, particularly at low temperatures, and lowered fatigue properties. Thus, an extensive programme of work was undertaken by Rubery Owen in collaboration with the two United Kingdom suppliers of these steels, in order to prove their suitability for the intended application (7). Some of the results obtained are discussed below.

Conventional carbon-maganese steels supplied to B.S.968:1962 or similar specifications have yield stress values generally between 355 and 390 N/mm^2 (23 and 25 tonf/in^2) in the "as rolled" condition. The yield stress to ultimate tensile strength ratio rarely exceeds 0.75. Pearlite reduced steels supplied to the same specification have rather higher yield stress values, and the above ratio is of the order of 0.85 to 0.90. The actual value of the yield stress is very variable both between and within the hot rolled coils.

Figure 9 above gives an indication of the spread of yield stress and ultimate tensile strength values exhibited by samples of this type of steel. This clearly illustrates the wide yield stress range which is reflected in various production difficulties. These are mainly confined to the shape of the pressing, as it is not always possible to allow for the variable springback of the pressings resulting from the range of yield stress values encountered even in quite a small batch of steel.

CHASSIS SIDEMEMBERS CONT'D

Figure 11 gives the results of a number of impact tests carried out on samples of two types of pearlite reduced steel and a conventional carbon-manganese steel. From these results the superiority of the low carbon pearlite reduced steel over the higher carbon variety and also the conventional steel is clearly evident. As shown above for the liquid petroleum gas cylinder steel type B in figure 6, cold working lowers the impact values of pearlite reduced steel at a given temperature. Results obtained to date indicate that such reduction is not serious.

Fatigue tests were carried out on a conventional carbon-manganese steel and two batches of pearlite reduced steel, initially in the form of specially prepared test pieces retaining the original "as rolled" surface of the steel plate. The work was carried out on a 200KN Avery-Schenk Pulsator using an alternating load of zero mean value. Failure was defined as complete fracture of the specimen. The results of the tests are presented in figure 12 and chemical analysis and mechanical properties of the steels used are given in table 2.

Subsequently steels from the same batches from which the plain fatigue test specimens had been prepared, were pressed into commercial vehicle chassis sidemembers. They were made up into complete chassis frames and were loaded statically by means of coil springs and a dynamic load was applied by a mechanical vibrator. Failure of each frame was detected by a change in amplitude of vibration and was usually manifested by cracking through the flanges, the cracks almost always starting from a punched hole. The results of these tests are given in table 3.

The results of the tests on the prepared specimens showed that the pearlite reduced steel had slightly lower fatigue properties than the conventional carbon-manganese steel. However, tests on the full-scale components clearly demonstrated the superiority of the frames constructed from the pearlite reduced steel.

One possible explanation for the better performance of the pearlite reduced steel in component form as opposed to prepared test piece form, when compared to the performance of the conventional carbon-manganese steel, is that it could be less notch sensitive than the carbon-manganese steel at the same strength level. Work carried out by British Steel Corporation South Wales Group, and reported in a private communication, indicated that if there was any difference in notch sensitivity between the two types of steel, then the pearlite reduced steel was marginally the poorer. However, other variables may have influenced their results. It has been postulated that pearlite reduced steels punch, blank and shear more cleanly than conventional carbon-manganese steels of similar strength levels. This is confirmed by simple observation on the press shop floor quite frequently. Since this would result in far fewer stress raisers in the full scale component, the difference in performance could be explained wholly in these terms:

Acceptance tests for the steel are generally limited to conventional chemical analysis and tensile and bend testing. When production troubles other than pressing shape have been encountered, these have been a result of sub-cutaneous non-metallic inclusions in the steel. These are related to higher than normal sulphur and aluminium contents and are more harmful at the higher yield stress levels.

Looking into the future, a variety of steels with yield stress values up to approximately 800 N/mm^2 (table 4) have been formed on the experimental channel pressing tool which was used for the previous work. Results are given in figure 13.

Material D2 is known to be a poor sample of a pearlite reduced steel. This highlights one of the problems when testing steel samples; the difficulty of ensuring that samples are truly representative of their type. The other pearlite reduced steels, A1 and D1 were fair samples.

The steels from mill C were outstanding. These were carbon-manganese steels containing titanium.

Since this work was carried out, mills A and D have also supplied lower carbon steels containing titanium. Samples were used for channel pressing tests with an inside bend radius of 1.4t. The largest crack was 19 mm. This was from the burr on the outside of the bend on the 660 N/mm^2 yield stress steel. There were no cracks at 1.75t. As a result of the tests it was considered that it should be possible to:

1. Reduce scrap and rectification on existing sidemembers. (Over 100 Tonnes of this steel have been processed into sidemembers with negligible rejects.)

2. Produce sidemembers at present strength levels but with smaller inside bend radii. (About 80 sidemembers with $1\frac{1}{4}$t inside bend radius have been produced to date using titanium treated steel with a yield stress of approximately 450 N/mm^2).

3. Produce sidemembers at higher strength levels using present inside bend radii (2t). The limitations discussed below with regard to quenched and tempered steels also apply to any high strength steel.

Returning to figure 13 steel F2 and the steels from mill E were quenched and tempered. E3 and F2 formed around a radius of 2t without cracking.

TABLE 4 - DETAILS OF STEELS FOR CHANNEL PRESSING EXPERIMENTS

Mill	Matl.	Thick-ness mm	0.5% P.S. N/mm²	U.T.S. N/mm²	Elongation on 50 mm %	Chemical Analysis %						Others
						C	Mn	P	S	Si	Mo	
A	A1	6.35	469	561	38	.155	1.31	.018	.018			Nb .019
B	B1	6.35	412	599	18 on 200 mm	.18	1.5	.025	.022	.03		
	B2	8										
	B3	6.35	471	616	27.5							
	B4	6.35	500	623	30							
C	C1	6.35	L 470	594	39	.17	1.20	.012	.021	.37		Ti .174
			T 477	610	32							
	C2	8	L 470	613	39	.18	1.49	.014	.012	.55		Ti .165
			T 474	617	39							
D	D1	6.35	522	582	32	.08	.98	.016	.018			Nb .045
	D2	8	448	533	33	.09	.99	.011	.029			Nb .05
E	E1	6.35	L 515	616	32	.17	1.3	.03	.033	.035	.023	Cu .39
			T 564	632	34							Ni .014
	E2	6.35	L 579	639	27							Cr .17
			T 531	650	24							Va .70
	E3	6.35	L 774	828	24	.19	.99	.029	.016	.54	.37	B .0033
			T 849	896								Ti .035
												Cr .73
												Va .015
F	F1	8	L 436	565	40	.175	1.33	.014	.013	.43		Nb .024
			T 440	565	42							
	F2	8	L 803	819	24	.145	1.48	.014	.015	.22	.525	
			T 798	811	27							

KEY: A = B.S.C. Rotherham; B = B.S.C. Scunthorpe; C = Hoesch A.G. Dortmund; D = B.S.C. Llanwern; E = B.S.C. Clydebridge; F = Oxelösunds Järnverk

CHASSIS SIDEMEMBERS CONT'D

One of the problems anticipated with high strength steel is springback. This is related to a number of factors and when the various measurements were studied the results were confusing. The extent of the problem became clear when springback measurements from channels produced at one press setting were compared with the yield stress value of the steels (figure 14).

A further sample of steel F2 was obtained in 2134 mm (8 ft) lengths and formed on sidemember tools over a 2t radius. No cracks were observed but there was about 6° springback. A number of methods of overcoming springback without a separate setting operation have been proposed (8) but none of these are practical for sidemembers pressed from quenched and tempered steel.

The available force of 40 MN (4000 Tonf) and energy of 1,329,000 Joules (976,000 ft lbs) on our largest press also limit the use of high strength steels.

If there is sufficient demand for components in high strength steels these problems will undoubtedly be overcome.

CONCLUSIONS

One of the main objectives of the use of high strength steel for components is material economy, but this cannot be realised if production methods are not or cannot be modified to take account of the more difficult forming characteristics of such materials. In this paper certain practical aspects of the pressing and deep drawing of differing types of engineering component from a number of medium to high yield stress hot rolled steels has been discussed in some detail. It has been indicated that such steels can be used in a viable manner at yield stress levels approaching 550 N/mm^2. The use of even higher yield stress steels for similar applications can be anticipated from experimental work in progress.

ACKNOWLEDGEMENTS

The authors would like to thank the Directors of Rubery Owen & Company Limited for permission to publish this paper. They also wish to thank several of their colleagues, in particular Mr. B.J. Allen and Mr. J.F.R. Amoore for providing certain material and for most helpful discussion. Without their co-operation, the preparation of this paper would not have been possible. Mr. A.J. Wynn prepared many of the figures.

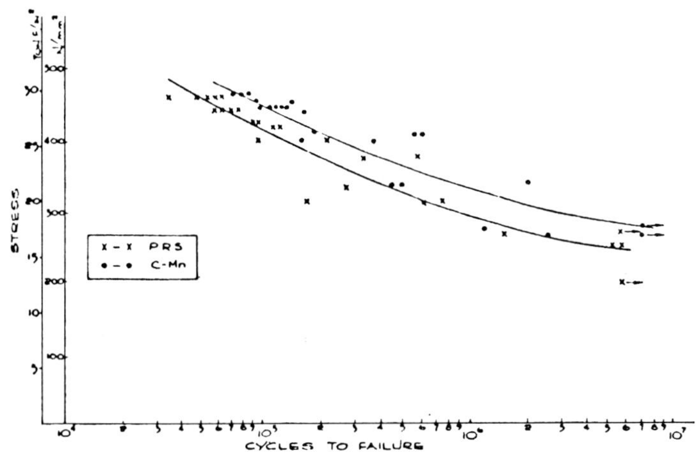

Fig 12. Stress endurance results for pearlite reduced steel and for conventional carbon-manganese steel.

Fig 13. Results of experimental forming tests on various high yield stress steels.

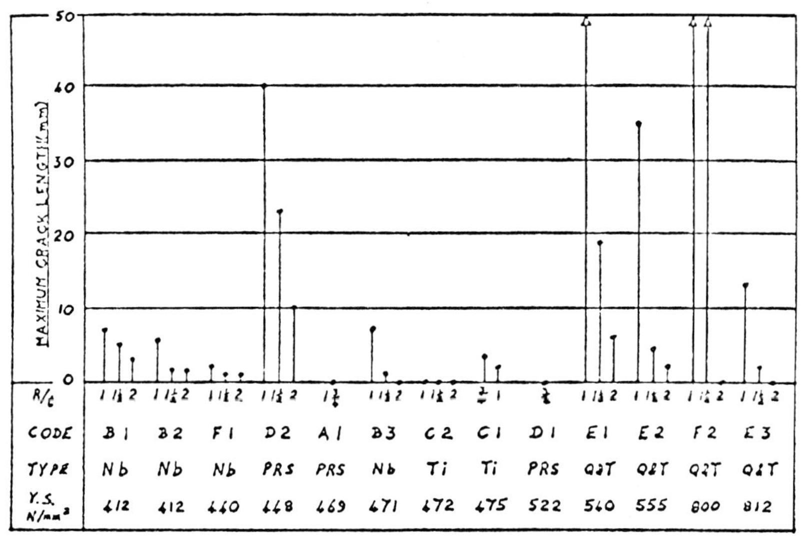

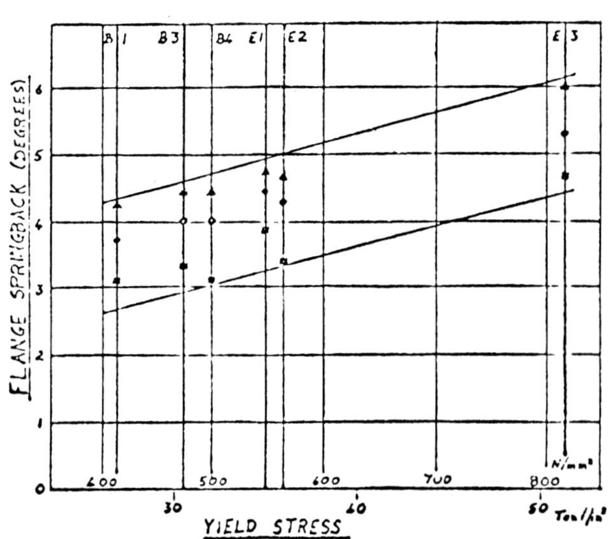

Fig 14. Comparison of flange springback for certain high yield stress steels as a function of the yield stress.

REFERENCES

1. ELWERS G. One draw reduces 40 inch diameter blank 66% Iron Age January 18th 1951 Vol. 167 pp 55-58.

2. COUPLAND H.T. and HOLYMAN W. Laboratory and press shop examination of some dry film lubricants. SHEET METAL INDUSTRIES January 1965 Vol. 42 No. 453 pp 7-14.

3. PROPOSED BRITISH STANDARD 5045 Part 2 "Specification for transportable gas containers up to 130 litres water capacity with automatically welded circumferential seams" Section 5 (Document 74/71057).

4. THIRD PROPOSAL FOR A DRAFT EUROPEAN STANDARD. FOR THE DESIGN, CONSTRUCTION AND ACCEPTANCE OF WELDED GAS CYLINDERS - Section 8 Document CEN/WG23 (Secretariat 49) 68E.

5. PROPOSAL FOR A DRAFT I.S.O. RECOMMENDATION FOR THE DESIGN, CONSTRUCTION AND ACCEPTANCE OF WELDED GAS CYLINDERS - Section 7 Document ISO/TC58/SC3 (Secretariat 94) 312E.

6. Midlands Company install 4000 ton chassis sidemember press. SHEET METAL INDUSTRIES 1966 Vol. 43 p 641.

7. SPEAR P., COUPLAND H.T., AND STANWAY A. Some experience in pressed and formed components. Proceedings of the joint conference of B.I.S.R.A. and I.S.I. Scarborough 4-6 April 1967 on Strong, tough structural steels.

8. Controlling material springback. TOOLING May 1968 Vol. 22 No. 5 pp 37 & 38.

Deep Drawing

*By the ASM Committee on Forming of Sheet Metal in Presses
and the ASM Committee on Press Forming of Steel*

DRAWING is a process for forming sheet metal between an edge-opposing punch and a die (draw ring) to produce a cup, cone, box or shell-like part. The work metal is bent over and wrapped around the punch nose. At the same time, the outer portions of the blank move radially toward the center of the blank until they flow over the die radius as the blank is drawn into the die cavity by the punch. The radial movement of the metal increases the blank thickness as the metal moves toward the die radius; as the metal flows over the die radius, this thickness decreases, because of the tension in the shell wall between the punch nose and the die radius, and (in some instances) because of the clearance between the punch and the die.

The distinction between shallow drawing and deep drawing is arbitrary, although shallow drawing generally refers to the forming of a cup no deeper than half its diameter, with little thinning of the metal. In deep drawing, the cup is deeper than half its diameter, and wall thinning, although not necessarily intentional, may be more than in shallow drawing. Wall thinning is more pronounced in stainless steel parts than in those made of low-carbon steel. One or more redraws, with or without intermediate annealing, are often required in order to complete a deep drawn part.

The fundamentals of drawing discussed in this article are generally applicable to all ductile metals, although drawing practice varies for different work metals. Most of the examples of practice presented here are concerned with the processing of flat rolled low-carbon steels such as 1006, 1008 and 1010. For drawing other metals, the reader may consult the articles in this volume that deal with forming of specific metals.

Mild drawing of low-carbon steel, as one stage of a forming sequence, is dealt with in the article "Press Forming of Low-Carbon Steel", which begins on page 112. Many of the practices discussed in that article can be applied also to the forming of other ductile metals.

Additional press operations performed on drawn workpieces are redrawing, reducing, expanding, bulging, sizing and ironing.

Redrawing is used to reduce the diameter and increase the height of a previously drawn cup, either for the full length or for a specified distance from the bottom. A punch pulls the cup into the die and reduces the diameter by either direct or reverse redrawing.

Reducing differs from redrawing in that the diameter is reduced at the mouth of the shell rather than at the bottom or for the full length. The top of the shell is pushed or compressed into the die rather than pulled through the die by the punch. Reducing is often referred to as necking, tapering, nosing or closing.

Expanding or bulging is the forming of irregular contours or surfaces of revolution on shells or rings by expanding or reducing the diameter for a portion of its cylindrical length. Such shapes cannot be removed from a regular punch or die. The change in shape is produced by a wedge-action punch or die, or by the use of a fluid or a rubber punch.

Sizing is used for forming workpieces to specific dimensions. The final forming of sharp radii should be combined with a small reduction in the cross-sectional area of the part. The metal should be crowded and not stretched into sharp corners. When the sizing operation is designed to eliminate oil-canning in the sidewall of a box or to establish a depth, the metal is stretched.

Ironing is the intentional thinning of the wall of a drawn shell. The operation provides uniform wall thickness or produces a thin-wall, thick-bottom part. The clearance between the punch and the die is less than metal thickness; this reduces wall thickness and increases the height as the shell is drawn through the die by the punch. Some slight redrawing takes place as the wall is thinned by ironing. The small reduction in diameter is employed mainly to permit free entry of the punch into the shell. Because the ironed shell clings to the punch, the stresses in the metal are compressive. A part with an ironed sidewall is less susceptible to cracking and distortion in machining than is a drawn part that has not been ironed.

Product Shape. Drawing is used to produce both shallow and deep straight-wall shells (or cups) from flat stock. These shells can be used as formed, like some pots and pans, or they may need attachments such as handles and pouring spouts. Drawn shells can also be thread rolled and edge curled to form a variety of products (such as screw caps). Drawing is used to develop shell walls that slope in or out, or to produce simple or complex square or rectangular boxlike shapes. It is also possible to produce drawn shells that have walls of nonuniform thickness; cartridge cases are typical.

When shell-derived products are small and are required in large numbers, they are usually produced in a high-production press, such as an eyelet-type transfer machine.

By reducing, expanding and bulging operations, shells can have contoured walls, which may be noncylindrical and embossed for decorative purposes (typical products are metal tumblers and cooking utensils). The screw ends on electric light bulbs are drawn, as are many metal slip cases and tubular products (shells with the bottoms trimmed). Seamless tubular products, ranging in diameter from a fraction of an inch to several feet, with a closed or partly closed end, may be made from drawn shells.

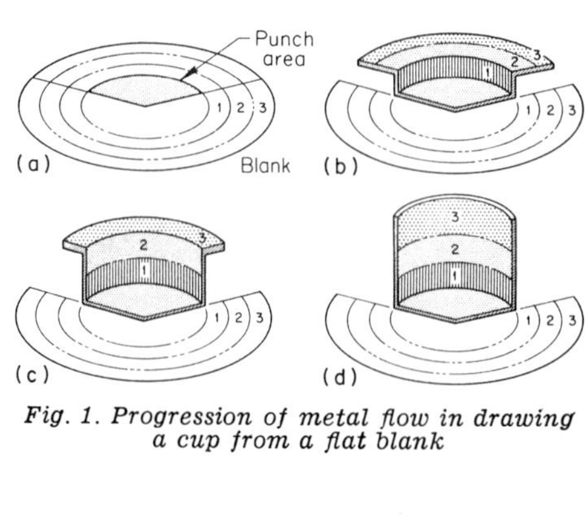

Fig. 1. Progression of metal flow in drawing a cup from a flat blank

Shallow shells can be formed by combined drawing and stretching. For some shells, the process may be similar to embossing, which requires that the metal being formed remain under tensile load. For a panel that must have compound curvatures, this requirement frequently cannot be met, and such a panel may be difficult to draw. However, by use of a process such as stretch forming, the required curvatures can be developed with minimum wrinkling; wrinkles in the scrap area can be removed by trimming.

Sheet-metal shapes such as used in aerospáce structures are usually assembled from drawn and stretch formed components. An automobile body is assembled from shallow drawn, deep drawn, and stretch formed components. Smaller structural shapes, such as home-appliance components, are also made from drawn shells.

Fundamentals of Drawing

A flat blank is formed into a cup by forcing a punch against the center portion of a blank that rests on the die ring. Progressive stages of metal flow in drawing a cup from a flat blank are shown schematically in Fig. 1. During the first stage, the punch contacts the blank (Fig. 1a), and the metal section denoted as 1 in Fig. 1 is bent and wrapped around the punch nose (Fig. 1b). Simultaneously and in sequence, the outer sections of the blank, denoted as 2 and 3 in Fig. 1, move radially toward the center of the blank until the remainder of the blank has bent around the punch nose and a straight-wall cup is formed (Fig. 1c and d). During drawing, the center of the blank (denoted as punch area in Fig. 1a) is essentially unchanged as it forms the bottom of the drawn cup. The areas that become the sidewall of the cup (1, 2 and 3 in Fig. 1) change from the shape of annular segments to longer parallel-side cylindrical elements as they are drawn over the die radius. Metal flow occurs until all the metal has been drawn over the die radius.

A blankholder is used in a draw die to prevent the formation of wrinkles as compressive action rearranges the metal from flange to sidewall. Wrinkling starts because of some lack of uniformity in the movement or the resistance to movement in the cross section of the metal. A blankholder force sufficient to resist or compensate for this nonuniform movement prevents wrinkling. Once a wrinkle starts, the blankholder is raised from the surface of the metal so that others can form easily. The force to hold the blank flat during drawing of cylindrical shells varies from practically zero for relatively thick blanks to about one-third of the drawing load for a blank 0.030 in. thick. Thinner blanks often require proportionally greater blankholder force.

Conditions for drawing without a blankholder depend on the ratio of the supported length of the blank to its thickness (see page 173 and Fig. 20), the amount of reduction from blank diameter to cup diameter, and the ratio of blank diameter to stock thickness. For thick sheets, the maximum reduction of blank diameter to cup diameter in drawing without a blankholder is about 25%. This ratio approaches zero for thin foil-like sheet. If a blankholder is employed, the maximum reduction is increased to about 50% for metals of maximum drawability and 25 to 30% for metals of marginal drawability in the same equipment.

Draw Ratios

The reduction in drawing cylindrical shells is generally expressed in terms of the diameters of the blank and the cup. Strain depends only slightly on the blank thickness. (See Appendix 2 at the end of this article.) The drawability of a metal is expressed either as the ratio of the blank diameter to the punch diameter, D/d, or as the percentage reduction from the blank diameter to the cup diameter, $100(1 - d/D)$.

The ratio of the height of the cup to the diameter of the cup sometimes is used as an expression of drawing ratio. However, this ratio varies with the drawability of the metal and the amount of thickening or thinning of the sidewall. Draw-reduction values for cupping and redrawing cylindrical shells in a double-action press or a die provided with a blankholder are given in Table 1.

It is emphasized that the values in Table 1 are approximate; they vary with work-metal composition, thickness and hardness, and with shell contour. For instance, a metal with maximum drawability, such as copper, can be reduced initially as much as 55%, but a refractory metal may allow an initial reduction of only 25%.

In drawing a shell with enough punch-to-die clearance to prevent ironing, the sidewall is thicker than the original blank for a short distance from the top; near the bottom, the wall is a little thinner than blank thickness. The amount of thickening or thinning depends on the metal, the percentage reduction, and the tensile stress in the sidewall.

The first, or cupping, reduction percentage is given in Table 1 as 40%. In general, flaws introduced into the cup during the first draw cannot be removed in subsequent processing. Rather, they are aggravated, especially if they have resulted from strain instability, which can lead to wrinkles, puckers or metal thinning. Intermediate annealing is frequently not needed, and may be undesirable. For example, metal processed in an eyelet machine, a transfer die, or a progressive die cannot be annealed during forming, yet the shells processed in such equipment frequently undergo several redraws. Metal fed into these machines must be selected to allow redrawing without annealing.

Table 1. Typical Relations of Shell Depth to Draw Reduction for Ductile Metal(a)

Depth of shell, in diameters	Number of draws	Reduction, %(b) Cupping	First redraw	Second redraw	Third redraw
½	1	40	..	..	..
1	2	40	25	..	..
1½	3	40	25	15	..
2	4	40	25	15	11

(a) Values are based on deep drawing of steel in a double-action press or a die provided with a blankholder. These values can serve as a guide for drawing of most ductile metals. Values are based on annealing between operations. (b) Percentage reduction $= 100(1 - d/D)$.

Table 2. Nominal Height of Cup Calculated From Blank Diameter, Cup Diameter, and Percentage Reduction

Reduction, %(a)	Cup diam (d), in.	Cup height (h), in.(b)	Ratio h/d
Blank Diameter, 2 In.			
20	1.6	0.23	0.14
30	1.4	0.37	0.26
40	1.2	0.54	0.45
50	1.0	0.75	0.75
Blank Diameter, 5 In.			
20	4.0	0.55	0.14
30	3.5	0.92	0.26
40	3.0	1.35	0.45
50	2.5	1.87	0.75
Blank Diameter, 10 In.			
20	8	1.15	0.14
30	7	1.87	0.26
40	6	2.7	0.45
50	5	3.75	0.75
Blank Diameter, 20 In.			
20	16	2.2	0.14
30	14	3.6	0.26
40	12	5.3	0.44
50	10	7.4	0.74

(a) Reduction $= 100(1 - d/D)$.
(b) $h = (D^2 - d^2)/4d$.

Developed from Charts IV and V, "Computations for Metal Working Presses", published by the E. W. Bliss Co.

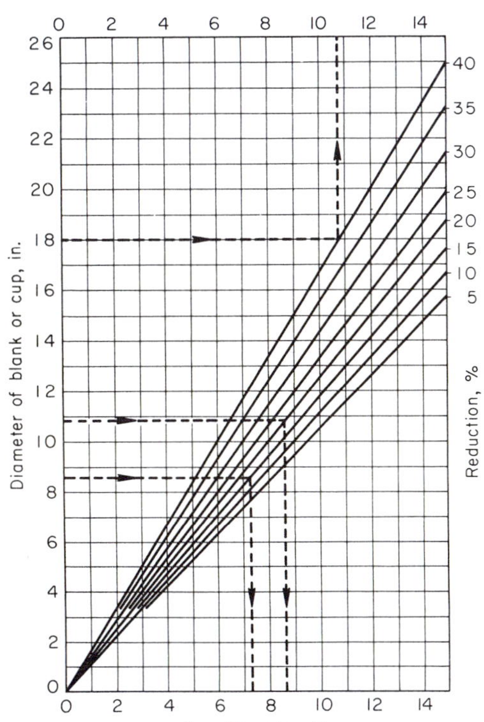

Fig. 2. Chart for checking percentage reduction in the drawing of cups. See text for sample calculation. The inside diameter is ordinarily used for the cup diameter.

Some metals cannot be redrawn without first being annealed. This is especially true of metals (such as stainless steels) that initially possess high strength or that cold work or strain harden rapidly to high strength levels. High work-metal strength causes excessive force on the drawing tools, which can cause tool breakage. In addition, excessive force can cause lubrication difficulties. Breakdown of lubricant causes galling of the workpiece on the tools and leads to shell breakage.

Annealing for redrawing, if not closely controlled, can impair drawability. This is especially true if grain coarsening results from the annealing. This characteristic varies widely among different work metals.

Blank Size vs Cup Diameter. Reduction of drawn cups is usually expressed as the percentage reduction from the diameter of the blank to the inside diameter of the cup, or from the inside diameter of one shell to that of the next. Percentage reduction is calculated using the formula $100(1 - d/D)$. Thus, the drawing of a 6-in.-diam cup from a 10-in.-diam blank results in a reduction of 40%. Nominal cup heights resulting from diameter reductions of 20, 30, 40 and 50% for blanks 2, 5, 10 and 20 in. in diameter are given in Table 2.

A chart that can be used to estimate cup size from blank size or cup size from an earlier draw is shown in Fig. 2.

Sample Calculation Using Fig. 2. To illustrate the use of Fig. 2, assume the problem of determining whether a cup of 7½-in. ID can be made in three draws of 40, 20, and 15% reduction, respectively, from an 18-in.-diam blank. To find the diameter of the cup after the first draw, trace the line for 18-in. blank diameter horizontally to its intersection with the diagonal for 40% reduction. From this intersection, draw a vertical line to the top of the chart and read 10.8 in. for the inside diameter of the cup after the first draw. Next, trace a horizontal line from 10.8 on the vertical axis until it intersects the diagonal for 20% reduction. From this point draw a vertical line to the bottom of the chart and read 8.6 in., which is the inside diameter of the cup after the second draw. The inside diameter of the cup after the third draw, assumed to be 15% reduction, is found using the same procedure, by drawing a line horizontally from 8.6 to its intersection with the diagonal for 15% reduction and from there to the bottom of the chart, which gives a reading of 7¼ in. Accordingly, it is concluded that a 7½-in.-ID cup can be drawn from an 18-in.-diam blank in the three assumed reductions.

Rectangular Shells. The drawing of oval, square or rectangular shells involves true drawing at the corners only; at the sides and ends, metal movement is more accurately described as bending. The stresses at the corner of the shell are compressive in the metal moving toward the die radius and are tensile in the metal that has already moved over the die radius. The metal between the corners is in tension only, in both the sidewall and flange areas. Stresses in the metal in the corner areas are severe, compared with the sidewall areas where only bending occurs. A portion of the surplus metal in the corner areas tends to move into the sidewall areas. If the punch-to-die clearance at the corners or in the sidewalls adjacent to the corners is too great, the metal may wrinkle rather than shrink or be compressed.

For square and rectangular boxes, except those with a width slightly more than twice the corner radius, the depth that the box can be drawn in one operation depends more on the size of the corner radius and on the stock thickness than on the width of the box.

When the ratio of the depth to the corner radius is 6 or less, the shell can usually be drawn in one operation. The basic ratio is 12 for two draws, 17 for three draws, and 22 for four draws. Square, or nearly square, cups and nearly cylindrical oval cups can be drawn in one operation if the area of the blank does not exceed the cross-sectional area of the punch by four times. This ratio is 4½ for boxes with a length-to-width ratio of 3, and decreases as the length-to-width ratio increases. Square boxes with a small corner radius (about 3% of box width) can be drawn to a maximum depth of about 80% of the width from low-carbon steel or some stainless steels, and to 70 to 75% of the width from aluminum alloys 3003-O and 5052-O.

Other factors that influence the producibility of a rectangular shell are work-metal composition, hardness and thickness, and radius at the bottom of the shell. Soft brass can easily be drawn to a depth of six times the corner radius in one operation, but some stainless steels can be drawn to a depth of only three times the corner radius. Likewise, thin material and small corner radii at the bottom of the shell reduce the maximum depth of draw. Shells that have a depth of more than six times the corner radius must be drawn in two or more operations. If two or more draws are necessary, the first-draw die contour can be determined from Fig. 3 as follows:

Dimension a is half the length or width of the finished box; b is the decrease in box size on a side; c is the thickness of an internal blankholder at the corner; r is the finished-shell radius; and R is the first-draw radius.

A rectangular shell can be reduced on a side by an amount b that depends on corner radius r and is determined by multiplying r by 1½. If the corner radius is greater than ½ in., the reduction should remain the same as for a ½-in. radius. The first-draw radius R should be four or five times larger than r when laid out, so that c is not less than ⅛ in. The R/r ratio will vary according to other dimensions.

If more than one redraw is necessary, the proportions of additional redraws can be determined by the above method, unless the end of the cup is semicircular after the first draw, under which condition reductions for round cups apply.

Defects in Drawn Parts

Most drawing problems can be averted by precautionary measures based on a knowledge of their causes. Table 3 contains some frequently encountered evidences of trouble, with the possible causes. Some of these defects and the means employed for correcting them are discussed in this article.

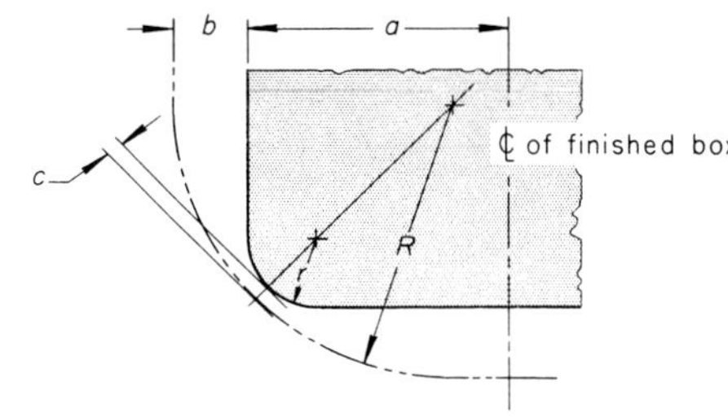

Fig. 3. Relation of draw and redraw dies for drawing a rectangular box. See text for explanation.

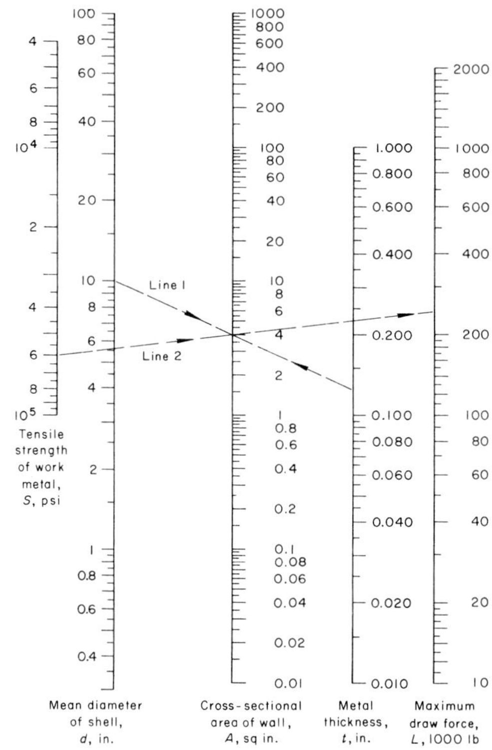

Fig. 4. Nomograph for estimating drawing force for a round shell. See text for sample calculation.

Presses

Sheet metal is drawn in either hydraulic or mechanical presses. Descriptions of these machines are given in the article "Presses and Auxiliary Equipment for Forming of Sheet Metal", which begins on page 1 in this volume. Also see the article "Selection of Presses for Forming of Sheet Metal", which begins on page 18.

Double-action presses are required for most deep drawing, because a more uniform blankholding force can be maintained for the entire stroke than is possible with a spring-loaded blankholder. Double-action hydraulic presses with a die cushion are often preferred for deep drawing, because of their constant drawing speed, stroke adjustment, and uniformity of clamping pressure. Regardless of the source of power for the slides, double-action straight-side presses with die cushions are best for deep drawing. Straight-side presses provide a wide choice of tonnage capacity, bed size, stroke and shut height.

Factors in Selection. Tonnage requirements, die space, and the length of stroke are the most important considerations in selecting a press for deep drawing. The condition of the crankshaft and connection bearings and gibs is also a factor in press selection.

Tonnage required for producing round cups is equal to the product of the cross-sectional area and the tensile strength of the work metal.

The force (load) required for drawing a round cup is expressed by the following empirical equation:

$$L = \pi d t S\left(\frac{D}{d} - k\right)$$

where L is press load, in pounds; d is cup diameter, in inches; D is blank diameter, in inches; t is work-metal thickness, in inches; S is tensile strength, in pounds per square inch; and k, a constant that takes into account frictional and bending forces, is usually 0.6 to 0.7.

The force (load) required for drawing a rectangular cup can be calculated from the following equation:

$$L = tS(2\pi R k_A + l k_B)$$

where L is press load, in pounds; t is work-metal thickness, in inches; S is tensile strength, in pounds per square inch; R is corner radius of the cup, in inches; l is the sum of the lengths of straight sections of the sides, in inches; and k_A and k_B are constants. Values for k_A range from 0.5 (for a shallow cup) to 2.0 (for a cup of depth five to six times the corner radius). Values for k_B range from 0.2 (for easy draw radius, ample clearance, and no blankholding force) and 0.3 (for similar free flow and normal blankholding force of about $L/3$) to a maximum of 1.0 (for metal clamped too tightly to flow).

Figure 4 can be used as a general guide for computing maximum drawing load for a round shell. These relations are based on a free draw with sufficient clearance so that there is no ironing, using a maximum reduction of 50%. The nomograph gives the load required to fracture the cup. A typical example of its use is as follows:

Sample Calculation Using Fig. 4. Assuming that 0.125-in.-thick metal with a tensile strength of 60,000 psi will be drawn into a 10-in.-diam cup, the maximum drawing load L is determined from Fig. 4 as follows: Draw line 1 to connect point 10 on the d scale to point 0.125 on the t scale. Line 1 intersects the A scale at 4.0, which is the approximate cross-sectional area of the wall. Then draw line 2 to connect point 4.0 on the A scale to 60,000 on the S scale. When line 2 is extended to the right, it intersects the L scale at 240,000 lb, or approximately the maximum drawing load for this cup.

When blankholder cylinders are mounted on the main slide of the press, the blankholder tonnage must be added to the calculated main-slide or drawing tonnage. When a die cushion is used to eject workpieces, the main slide works against this force; therefore, tonnage is required in addition to the tonnage calculated for the deep drawing load.

In toggle draw presses, the blankholding force is taken on the rocker shaft bearings in the press frame, so that the crankshaft bearings sustain only the drawing load. In other types

Table 3. Causes of Common Problems in Deep Drawing

Metal Fracture During Cupping

1 Blank thickness wrong or out-of-tolerance
2 Metal hardness, grain size or quality not suitable for part being drawn. Correct by proper annealing or upgrading quality of work metal.
3 Scratches, blemishes, dirt, etched-in stains or other defects on surface of blank, with cupping causing additional stress in the metal at surface defects
4 Draw radii, draw beads, die ring and blankholder surfaces not smooth. These surfaces should be smooth, sometimes highly polished, with no radial lines, scratches, galling or welding.
5 Draw radius too small, creating resistance to metal flow sufficient to cause undue thinning or fracture. Correct by increasing die radius or by changing die entrance to conical or elliptical shape.
6 Clearance between punch and die too little, too great, or uneven
7 Blankholding pressure too great
8 Percentage reduction of blank diameter to cup diameter too great
9 Punch nose radius too small. Correct by drawing cup with a larger bottom radius and restriking to smaller radius.
10 Lubrication inadequate or unsuitable

Wrinkles on Top Edge or Flange of Cup

1 Blankholding pressure too light
2 Draw radius too large
3 Punch nose radius too large. Dome-shape cups frequently have this defect because of the amount of free metal between the punch and draw radius.
4 Wrinkles on one side of cup or flange, caused by burr on blank or by an unbalanced blankholding pressure

Uneven Top Rim or Flange in Drawn Cup

1 Earing or formation of scallops around the edge, caused by differences in directional properties of the sheet metal
2 Nicks or burrs along the periphery of the blank
3 Punch, die or blank locators not concentric. Too much material pulled into the die by off-center forming punch or from off-center blank prevents forming of an even cup depth or flange width.
4 Scratches, blemishes, dirt, etched-in stains, or other types of defects present on the surface of the blank
5 Blankholder exerting an unbalanced force on the blank, permitting metal to be drawn unevenly into the die

Flange Wrinkled, Puckered or Split During Redrawing

1 Blank thickness wrong or out-of-tolerance
2 Draw radius too large
3 Depth of previously drawn cup too short or redraw operation too deep. Flange of previously drawn cup is drawn into the die, or sidewall of shell is thinned.
4 Clearance between punch and die too great for the stock thickness

Severe or Localized Wall Thinning Near Shell Bottom

1 Quality of steel unsuitable for redrawing
2 Nose radii of redraw punch not properly blended to bottom and side of punch
3 Draw radii too small, thus increasing tension in sidewall and stretching the metal
4 Diameter reductions too severe
5 Lubricant inadequate or unsuitable

Fracture at Bottom of Redrawn Shell

1 Surface defects that were present in original blank but that did not cause rejection after cupping
2 Draw radius and blankholding surfaces scratched, nicked or galled, restricting metal flow into the die
3 Punch nose and draw radii too small to permit proper metal flow
4 Diameter reduction too great or previously drawn shell too short for redrawing operation
5 Metal in previously drawn shell wall adjacent to bottom radius too thin to withstand redrawing
6 Lubrication inadequate or unsuitable

Fracture in or at Flange of Redrawn Cup

1 Surface defects or improper grain size in original blank that did not cause rejection after cupping
2 Punch and die clearance too small, and excess metal crowded into flange

Excess Material at Top of Redrawn Shell

1 Metal thickness too great or punch-to-die clearance too small or uneven. Shell wall is ironed and excess metal is crowded into flange or shell-mouth radius.
2 Punch nose radius larger than cup-bottom radius, causing excess metal to flow up sidewall
3 Draw radius in preceding operation too large, permitting wrinkles to start that enlarge during redrawing of cup
4 Cup is too long for the redrawing operation, which causes the redraw punch to bottom before the sidewall of the cup is drawn into the die. The same thing happens if the redraw punch is too short.

SEE NOTES BELOW FOR DISCUSSION OF ILLUSTRATIONS.

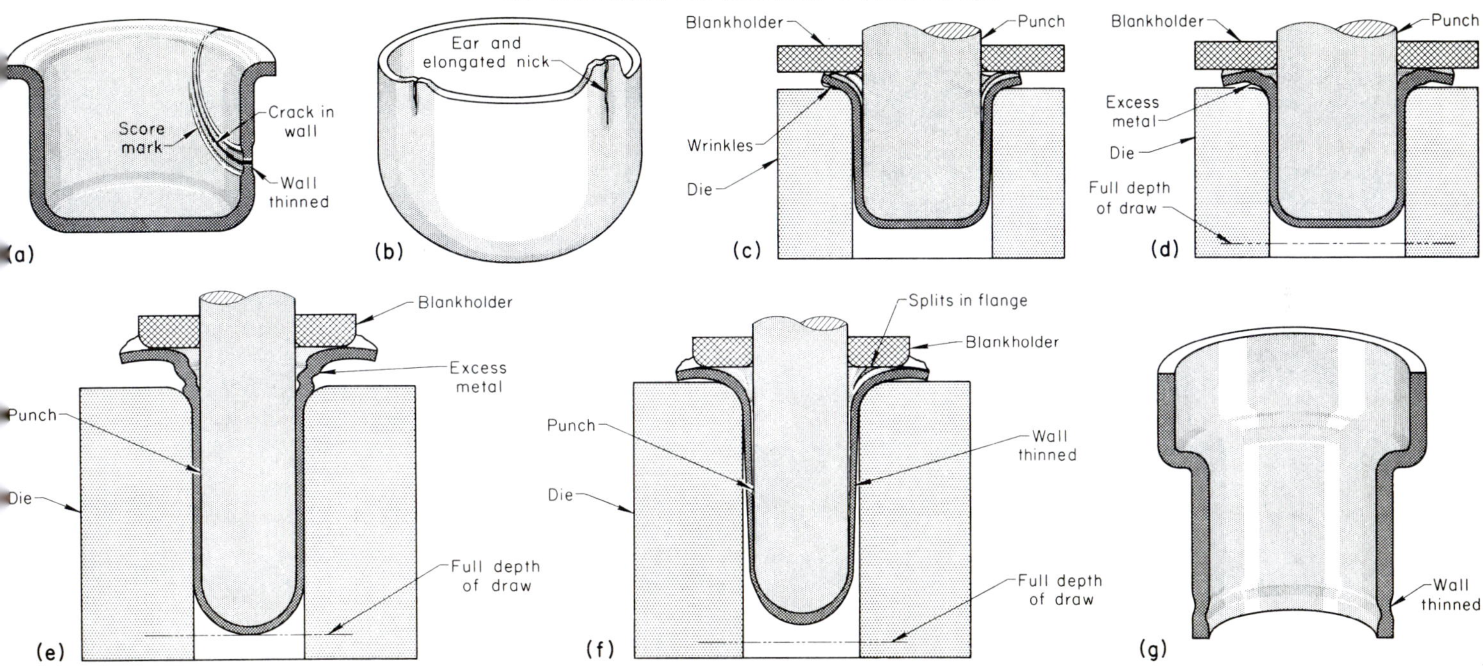

Notes Regarding the Illustrations

(a) Material flaw affects cupping. Original flaw widens near bottom of cup, reducing wall thickness. Marked cracking appears along the flaw and in the flange. (b) Earing, and elongation of nicks during cupping. Nicks were in periphery of blank before cupping.

(c) Wrinkles in flange from insufficient blankholding pressure or thin metal. (d) Excess metal in shell-mouth radius from thick metal or insufficient punch-to-die clearance. (e) Cup too long for redraw operation. Excess metal is forced up under flange, distorting shell-mouth radius. (f) Cup too short for redraw operation. Shell wall thinned, possible fracture at shell bottom, or flange splitting. (g) Localized wall thinning resulting from punch nose radius that is too small or not well blended with side or end of punch.

of presses, both the drawing and blankholding loads are on the crankshaft, and allowances are made when computing press capacity. For round work, the allowance for blankholding should be 30 to 40% of the drawing force. For large rectangular work, the drawing force is relatively lower than for round work, but the blankholding force may be equal to the drawing force. Where stretching is involved and the blank must be gripped tightly around the edge (and a draw bead is not permissible), the blankholding force may be two or three times the drawing force.

Blank size governs the size of the blankholder surfaces. Some presses with sufficient force cannot be considered for deep drawing because the bed size and shut height are inadequate.

Depth of Draw. Both length of stroke and the force required at the beginning of the working portion of the stroke are important considerations. When drawing parts that have straight walls, often the part can be drawn through the die cavity, and then stripped from the punch and ejected from the bottom of the press. Even under these ideal conditions, the minimum stroke will be equal to the sum of the length of the drawn part, the radius of the draw die, the stock thickness, and the depth of the die to the stripping point, plus some clearance for placing the blank in the die.

Workpieces with flanges or tapered walls must be removed from the top of the die. In drawing these workpieces, the minimum press stroke is twice the length of the drawn workpiece, plus clearance for loading the die. In an automatic operation using progressive dies or transfer mechanisms, at least half the stroke must be reserved for stock feed, because the tooling must clear the part before feeding starts for the next stroke. For automatic operation, it is common practice to allow a press stroke of four times the length of the drawn workpiece. For this reason, some equipment is not suited to automatic operation, or it is necessary to use manual feed with an automatic unloader, or vice versa, because of shortage of suitable presses.

Slide Velocity. When selecting a press, it is also necessary to check slide velocity through the working portion of the stroke (see the section in this article on Effect of Press Speed, page 175).

Means of Holding the Blank. Double-action presses with a punch slide and a blankholder slide are preferred for deep drawing. Single-action presses with die cushions (pneumatic or hydraulic) can be used, but are less suitable for drawing complex parts.

Draw beads are incorporated in the blankholder for drawing parts requiring greater restraint of metal flow than can be obtained by using a plain blankholder or for diverting metal flow into or away from specific areas of the part (see Restraint of Metal Flow, page 172).

Selection vs Availability. Often the ideal press equipment for a specific job is not available. This makes it necessary to design tools and choose product forms of work metal in accordance with available presses and supplementary equipment. For instance, if available presses are not adequate for drawing large workpieces, the manufacturing sequence must be completely changed. It may be necessary to draw two sections and weld them together. Also, operations otherwise combinable, such as blanking, piercing, drawing and trimming, may have to be performed singly in separate presses.

On the other hand, some manufacturers have placed more than one die in a single press, because of the availability of a large press and the shortage of smaller presses. Usually this procedure causes lower production, because all blanks must be positioned before the press can be operated. However, storage of partly formed workpieces and additional handling between press operations are eliminated. Where several small dies are used to reduce over-all tool cost, there is economic justification for the use of small-capacity presses. If small presses are not available, frequently it is more economical to use compound dies. This is particularly true if over-all part production is likely to exceed original estimates.

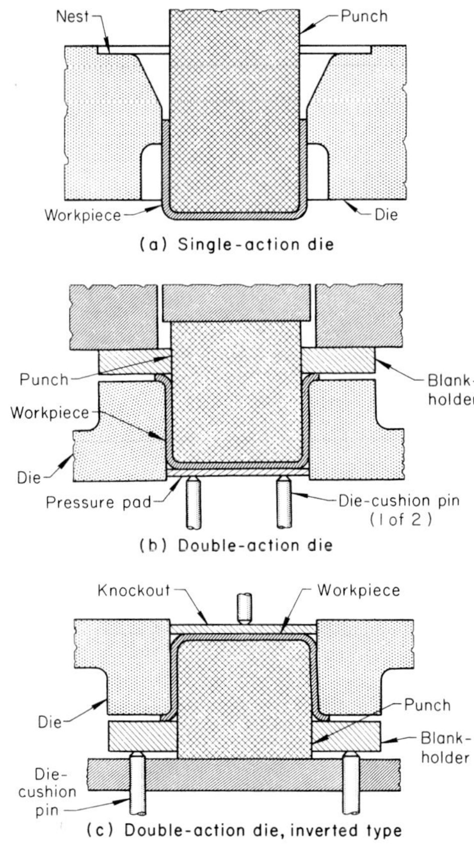

Fig. 5. Components of three types of simple dies, shown in setup for drawing a round cup. See text for discussion.

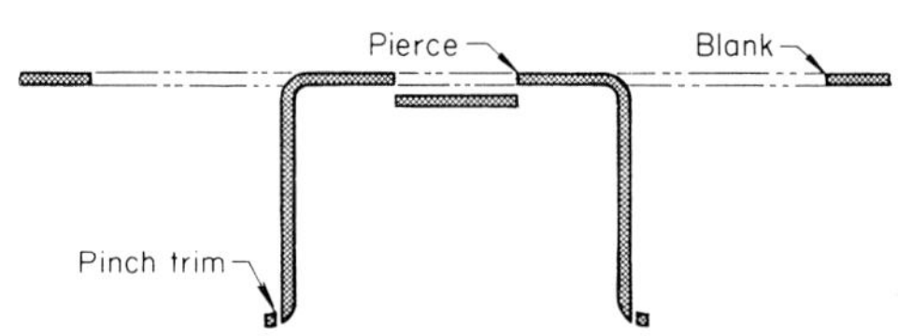

Fig. 6. Progressive blanking, drawing, piercing and pinch trimming of a cup in a compound die

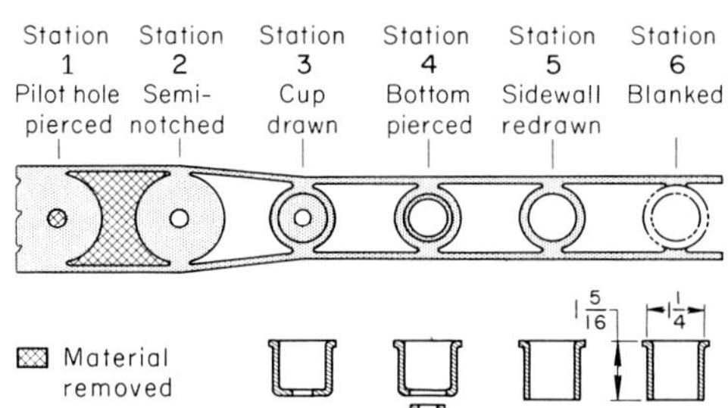

Fig. 7. Production of a small ferrule in a six-station progressive die

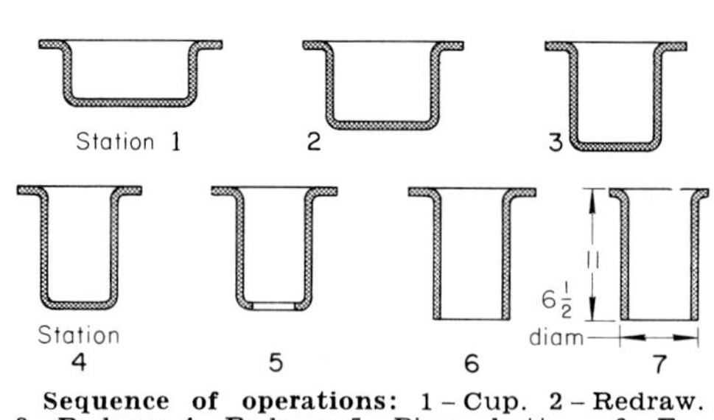

Sequence of operations: 1 – Cup. 2 – Redraw. 3 – Redraw. 4 – Redraw. 5 – Pierce bottom. 6 – Extrude bottom. 7 – Trim flange.

Fig. 8. Seven-station drawing and piercing of a cylinder in a multiple die and transfer mechanism

Availability of auxiliary equipment may also influence the type of press and tooling used. For instance, if equipment is available for handling coils, plans will be made accordingly. However, if coil-handling equipment is not available and straight lengths of sheet or strip are to be processed, a compatible tooling procedure must be used, even though it might not be the most economical procedure.

Dies

Dies used for drawing sheet metal are usually one of the following basic types or some modification thereof: (*a*) single-action dies; (*b*) double-action dies; (*c*) compound dies; (*d*) progressive dies; or (*e*) multiple dies with transfer mechanism. Selection of type depends largely on part size, severity of draw, and quantity of parts to be produced.

Single-action dies (Fig. 5a) are the simplest of all drawing dies and have only a punch and a die. A nest or locator is provided to position the blank. The drawn part is pushed through the die and is stripped from the punch by the counterbore in the bottom of the die. The rim of the cup expands slightly to make this possible. Single-action dies can be used only when the forming limit permits cupping without the use of a blankholder.

Double-action dies have a blankholder. This permits greater reductions and the drawing of flanged parts. Figure 5(b) shows a double-action die of the type used in a double-action press. In this design, the die is mounted on the lower shoe; the punch is attached to the inner, or punch, slide; and the blankholder is attached to the outer slide. The pressure pad is used to hold the blank firmly against the punch nose during the drawing operation, and to lift the drawn cup from the die. If a die cushion is not available, springs or air or hydraulic cylinders can be used; however, they are less effective than a die cushion, especially for deep draws.

Figure 5(c) shows an inverted type of double-action die, which is used in single-action presses. In this design, the punch is mounted on the lower shoe; the die on the upper shoe. A die cushion can supply the blankholding force, or springs or air or hydraulic cylinders are incorporated in the die to supply the necessary blankholding force. The drawn cup is removed from the die on the upstroke of the ram, when the pin-like extension of the knockout strikes a stationary knockout bar attached to the press frame.

Compound Dies. When the initial cost is warranted by production demands, it is practical to combine several operations in a single die. Blanking and drawing are two operations commonly placed in compound dies. Figure 6 shows how a cup is blanked, drawn, pierced and pinch trimmed in a single operation using a compound die. By this means, workpieces can be produced up to four times as fast as by the simple dies shown in Fig. 5.

Progressive Dies. The initial cost and length of bed needed for progressive dies usually limit their application to relatively small workpieces. Figure 7 hows a typical six-station progression or making small shell-like workpieces n a mass-production basis (see also xamples 229 and 230). However, larger arts, such as liners for automobile eadlights, have been drawn in progressive dies.

Total number of parts to be produced nd production rate often determine vhether or not a progressive die will be sed when two or more operations are equired. There are, however, some ractical considerations that may rule gainst a progressive die, regardless of uantity. These considerations are: (*a*) he workpiece must remain attached to he scrap skeleton until the final staion, without hindering the drawing perations; (*b*) drawing operations iust be completed before the final staion is reached; (*c*) sometimes in deep rawing it is difficult to move the workiece to the next station; (*d*) if the raw is relatively deep, stripping is ofen a problem; and (*e*) the length of ress stroke must be more than twice he depth of draw.

Assuming that a progressive die can e used to make acceptable drawn arts, cost per piece usually is the final onsideration. Progressive-die drawing usually considered economical if savng in material and labor can pay for he die in one year. Ordinarily, the savng effected by the use of a progressive ie results from decreased labor.

Multiple dies, in conjunction with ransfer mechanisms, are often used nstead of progressive dies for mass roduction of larger parts. Multiple dies nd transfer mechanisms are practical or a wider range of workpiece sizes han are progressive dies. Although the yelet-type transfer method is most videly used for making parts less than in. in diameter, transfer dies are ractical for much larger workpieces. he seven-station operation for making 6½-in.-OD cylindrical shell shown in ig. 8 represents a typical sequence for he transfer-die method. The workpiece s mechanically transferred from one ie to the next. One advantage of the ransfer-die method, as opposed to the rogressive-die method, is the greater exibility permitted in processing proedure, mainly because in transfer dies he workpiece does not remain attached o the scrap skeleton during forming. ecause of this, precut blanks can be rawn by the transfer method.

Preforms can also be used as blanks. or instance, oil pans for automobiles re blanked and partly drawn in a ompound die, then finish formed, ierced and trimmed by the transfer nethod.

Dies for producing a given part isually cost more for the transfer-die nethod than for a separate-die operaion, but about the same as for a proressive-die operation. The cost of dapting the transfer unit to the part s not included in the die cost. Likewise, he production rate for the transfer nethod is usually greater than for sinle-die operation, but 10 to 25% less han for drawing in a progressive die. Aany parts can be produced equally vell by all these methods. Under these onditions, tool cost, rate of production, nd total quantity of parts to be drawn etermine the choice of procedure.

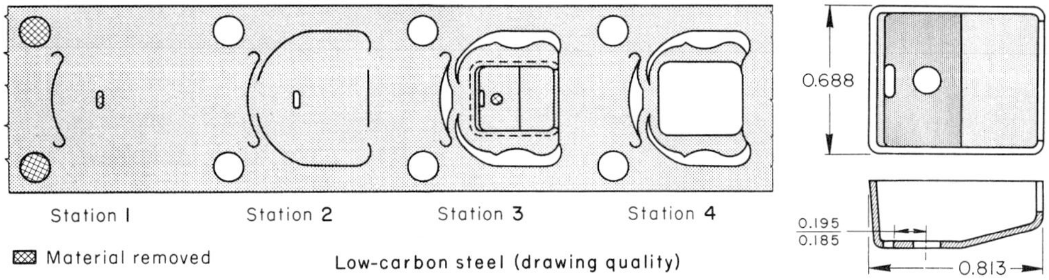

Station 1: Pilot holes and oval hole are pierced; arc is lanced. **Station 2:** Outline of blank is lanced preparatory to drawing. **Station 3:** Cup is drawn; as punch bottoms, mounting hole is pierced upward. **Station 4:** Part is pinched off; scrap skeleton is trimmed off.

Fig. 9. Producing a small drawn part in a four-station progressive die (Example 229)

Table 4. Comparison of Costs for Drawing Ball Races in Separate vs Progressive Dies (Example 230)

Item	Separate dies	Progressive die
Tool cost (total)	$3190	$4575
Cost of labor and burden:		
Per piece	$0.0173	$0.0039
For 750,000 pcs/year..	$12,975	$2925
First-year saving(a) ...	...	$8665

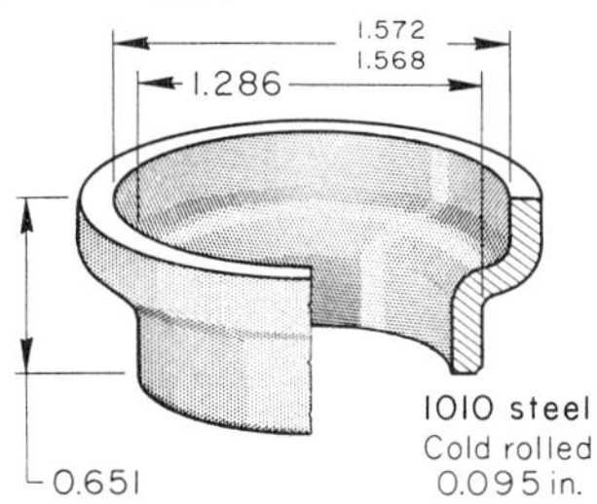

(a) $10,050 saving in cost of labor and burden, minus $1385 additional tool cost. Tools were amortized over one year.

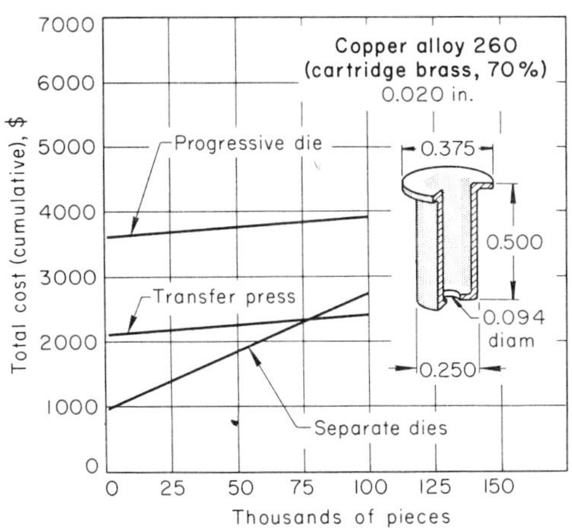

Item	Separate dies	Transfer press	Progressive die
Production, pcs/hr ...	1200(a)	5400(b)	7800(b)
Tool cost	$960	$2100	$3600
Cost per 1000 Pieces			
Labor, production(c) ..	$15.71	$0.56	$0.38
Labor, maintenance(d)	...	0.48	0.48
Material	2.16	1.76	2.16
Total	$17.87	$2.80	$3.02

(a) Number of pieces drawn (four draws), pierced and trimmed; blanks were produced in an automatic press at the rate of 4200 per hour. (b) Rate for all operations, including blanking. (c) Direct labor, at $3 per hour. Cost for separate-die method represents $2.50 (5/6 hr, for 1000 pieces, times $3 hourly rate) for each of the six operations (four draws, pierce and trim) plus $0.71 for blanking (5/21 hr, for 1000 pieces at 4200 pieces per hour, times $3). (d) At $6 per hour.

Fig. 10. Effect of quantity on cost of producing a flanged cup by three different methods (Example 231)

Effect of Tooling Procedure on Cost

Parts made from blanks less than about 3½ in. across can be drawn by various methods, ranging from a hand press for making a few pieces to a progressive die for continuous high production. For these parts, quantity is usually the major influence on choice of procedure. Usually, the initial cost for individual dies is less than for a progressive die, and so individual dies are used for producing small quantities. However, when several operations must be performed, the initial cost of a single progressive die is often less for small parts than the cost of several individual dies, and so a progressive die is used regardless of production quantity.

As workpiece size increases, the cost of a progressive die is likely to become prohibitive or its use impractical. Thus, for larger workpieces, the choice of tooling is more often a choice between separate dies and multiple dies with a transfer mechanism, based principally on the quantity of workpieces to be drawn. The four examples that follow compare costs for producing the same parts by two or more types of tooling. The first two of these examples compare production by separate versus progressive dies; the third example compares the cost of production by three types of tooling; the fourth example compares the cost of tooling for two different die setups. (See also Examples 233 and 234, which show the effect of production quantity on cost.)

Example 229. Saving of 37% in Die Cost by Use of a Progressive Die Instead of Four Individual Dies (Fig. 9)

The small cup-shape part shown in Fig. 9 was produced from low-carbon steel in a four-station progressive die, in the steps shown. The only critical dimension was the location of the mounting hole, which had a tolerance of ±0.005 in.; this was held without difficulty. The progressive die cost $1975, compared to an estimated $3150 total for four individual dies to make the same part.

The progressive die was made from an air-hardening tool steel (Rockwell C 60 to 62) and operated in a 50-ton dieing machine at 75 strokes per minute. Lubricant was die oil. Average production run was 75,000 pieces.

Example 230. Cost Savings by Change From Separate-Die to Progressive-Die Production (Table 4)

Ring-shape ball races with two diameters (see drawing in Table 4) were originally produced in three operations by separate dies. In the first operation, the workpiece was blanked and the center hole was pierced. Drawing, partly done in the second die, was completed in the third die, where the inside radius was coined.

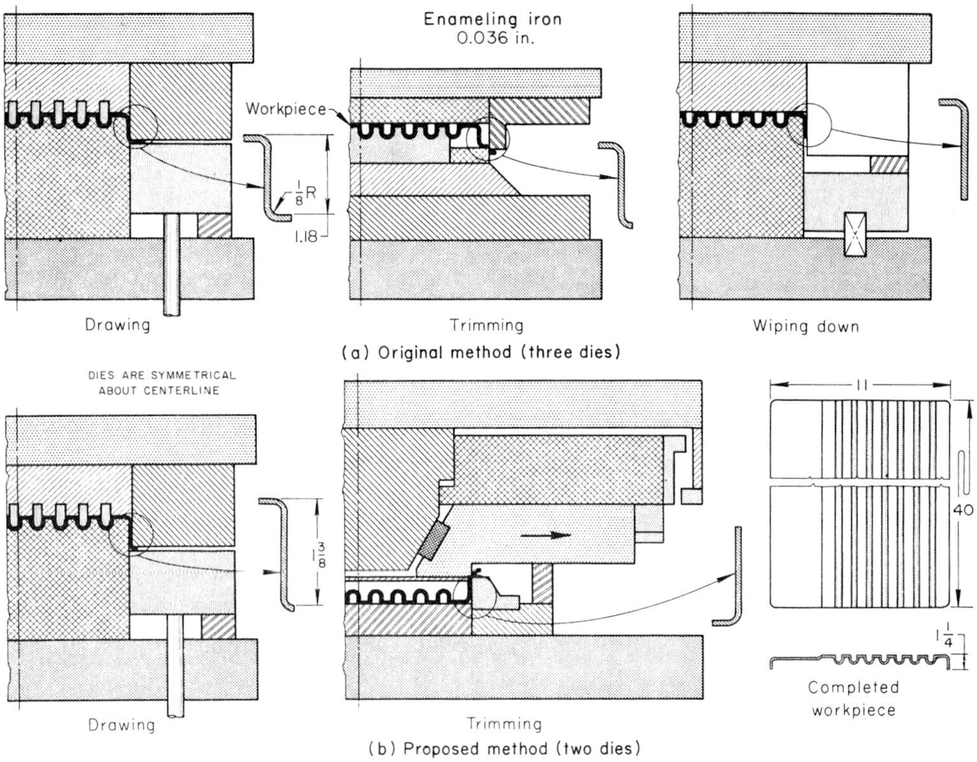

Operation	Tool cost	Setup cost(a)	Labor cost(a)
Original Method (Three Dies)			
Draw	$ 4,400	$200	$298.50
Trim	2,800	150	298.50
Wipe down	3,200	150	298.50
Total	$10,400	$500	$895.50

Operation	Tool cost	Setup cost(a)	Labor cost(a)
Proposed Method (Two Dies)			
Draw	$ 4,400	$200	$298.50
Trim	9,000	150	298.50
Total	$13,400	$350	$597.00

(a) Based on $5 per hour, and on ten setups of 1000 pieces per run (total of 10,000 pieces)

Fig. 11. Two-die and three-die methods for producing range-drawer panels. Higher tool cost for the two-die method made it impractical for 10,000 pieces per year. (Example 232)

When annual production increased to 750,000 parts, use of a progressive die became advisable. A six-station (two idle stations) progressive die produced the parts from strip by piercing a center hole in station 1, shaving the center hole in station 3, blanking out and pushing back into strip (as a means of feeding the blank to the next station) in station 4, and extruding (flanging) and drawing in station 6. The last operation (at station 6) produced a smooth surface on the inside radius without coining.

As shown by the cost comparison in Table 4, 750,000 parts were produced in a progressive die for about one-fourth the cost of producing the same quantity in separate dies, although the die itself was more costly.

Example 231. Relation of Quantity to Cost of Producing Flanged Cups in Separate Dies, a Transfer Press, or a Progressive Die (Fig. 10)

The cost of producing a flanged cup in various quantities by three different methods is shown graphically in Fig. 10. At zero production, tool costs only are shown: $960 for separate dies, $2100 for dies used in a transfer press, and $3600 for a progressive die.

As shown in the table that accompanies Fig. 10, production rate was lowest for separate dies and highest for the progressive die. Although labor cost for production was highest for separate-die operation and lowest for processing in a progressive die, cost per piece for labor and material was lowest for the transfer method, because blanks for both the separate and the progressive dies were taken from a single-width strip, whereas the transfer method permitted use of wider strip. By staggering the blank layout, it was possible to obtain twice the number of workpieces using a strip less than twice the width required for a single blank. The wider strip was then run through the dies twice. This technique resulted in a material saving of 18%.

Figure 10 shows that for fewer than 77,000 pieces the separate-die method was the least costly (considering initial cost of tools), and that for quantities of more than 77,000 the transfer method was the least costly. A progressive die would not be economical for any quantity of this part.

Example 232. Cost Comparison for Two Methods of Producing Drawer Panels (Fig. 11)

In the original method of drawing panels for electric-range drawers (Fig. 11a), a wipe-down operation was required because, with the trim line beyond the ⅛-in. draw radius, a recoil line remained on the completed workpiece. The recoil line was caused by the metal at the draw radius taking a set; thinning of the metal usually occurred in this area.

In a proposed method (Fig. 11b), the depth of draw was 1⅜ in. (slightly deeper than in the original method), to allow for relocation of the trim line ahead of the radius, and a cam trimming die eliminated the recoil line and hence the need for wiping down. Tool cost for the proposed method, however, was $3000 more than for the original method (see cost comparison with Fig. 11), and although there was a saving of $298.50 in labor and $150 in setup, the proposed method could not be justified for an annual production of only 10,000 panels since the break-even point was 67,000 pieces per year. Each operation produced 335 pieces per hour, at a press speed of 22 strokes per minute. A dry-coat lubricant was applied with heat and an over-and-under roller coater.

Material for Dies and Punches

The selection of material for dies and punches for drawing sheet metal depends on work-metal composition, workpiece size, severity of the draw, quantity of parts to be drawn, and tolerances and surface finish specified for the drawn workpieces.

To meet the wide range of requirements, punch and die materials ranging from epoxy resin to highly alloyed tool steels with nitrided surfaces, and even carbide, are used. On rare occasions, for drawing only a few pieces, masonite or hard wood has been used as a tool material.

For detailed information on tool materials, see the articles "Selection of Material for Press Forming Dies" (which begins on page 145) and "Selection of Material for Deep Drawing Dies" (which begins on page 194). Selection tables in those articles consider the variables enumerated above.

Tooling for Low, Medium and High Production

The term "short run" may refer to quantities ranging from prototype manufacturing up to 10,000 pieces, depending on the type of product being considered.

In addition to low demand for a specific drawn shape, the possibility of a change in part design is often a reason for minimizing the tooling investment.

For producing any given shape, data such as shown in Fig. 10 should be obtained; that is, manufacturing costs should be calculated, or at least estimated, on the basis of quantity requirements. Usually, short-run tooling is relatively low in cost, but production labor costs for using it are high. Conversely, more elaborate tooling is initially expensive, but labor and other manufacturing costs are invariably less than when using short-run tooling. Therefore, when comparative costs are plotted graphically as in Fig. 10, there is usually a break-even point at some specific quantity. For quantities lower than the break-even point, short-run tooling is cheaper on a cost-per-piece basis, whereas greater quantities justify the cost of more elaborate tooling.

Total quantity to be drawn is not always the most practical basis for a choice between short-run and long-run tooling. Production demands may be such that the quantity of parts to be produced within a given time must be considered, and this can rule out short-run tooling because of low productivity. On the other hand, short-run tooling is sometimes used because of an anticipated change in workpiece design, even though total quantity would warrant the cost and use of long-run tooling.

Short-run tooling differs from long-run in one or more of three ways: (*a*) lower-cost materials are used for short-run tools; (*b*) short-run tools are made for individual operations, which often are performed in different presses; and (*c*) short-run tooling may need handwork to finish the part and involves a minimum of automation.

On the basis of material and processing costs alone, a large die or punch often can be made at much less cost by using a low-cost material instead of a high-cost tool steel that is difficult to machine and heat treat (for example, zinc alloy versus hardened D2 tool steel). However, factors other than tool materials are usually more influential in determining the cost of tooling.

One means of minimizing tool cost is to produce the workpiece in a series of simple operations (like blanking, drawing and piercing) without attempting to combine them. In this way, it is often possible to use standard tools already on hand.

In short-run tooling, the accuracy of the contour of the finished part edge determines the need for, and means of, trimming. A hand-cut developed blank is often used to eliminate the need for a trimming die. Numerous manual operations can be substituted for machine operations, often eliminating the need for a die or at least simplifying the construction of a die. Individual die sets are often eliminated for short runs, the punch and die being installed directly into a master die set, which is then mounted in the press. Standardization of tools can result in low-cost short-run tooling, because required tooling components are more likely to be on hand.

One approach to low-cost tools for short runs is the use of interchangeable inserts. Figure 12 shows how three different inserts can be used in the same draw ring, so that the ring can produce parts of different radii. The use of inserts is especially appropriate when a design change or a change in work-metal composition is expected. Insert practice can also be an important element in a standardization program.

Medium-production tooling usually has the following features:

1. Punches and dies with adjustable holders for use after regrinding
2. Jig-ground locating holes
3. Carbide or high-alloy die steel on wearing surfaces
4. Ejector pins
5. Engineered mechanisms for stripping and lifting out formed workpieces
6. Heeled trim punches
7. Mechanized equipment for either loading or unloading of the dies.

High-production tooling would include the above items, but would be built so as to reduce the downtime needed for maintenance. Holes in the die shoes would provide accessibility to lifters and air cylinders, and for the removal of drawing compound that has accumulated beneath the pressure pad.

Automatic equipment would be used to load the blank and to unload finished workpieces. The upper surface of the lifters would be covered with soft brass to prevent marking the panel. Steel guides would control movement of the incoming blank and the outgoing part.

The two examples that follow show cost-quantity relations for different tooling setups.

Example 233. Comparative Costs for Three Methods of Producing Bearing Separators (Fig. 13)

Separators for tapered roller bearings were made by three different tooling methods. Relations of cost to quantity for the three methods are shown in Fig. 13. Costs of tools, setup and labor are itemized in the table with Fig. 13.

The higher tooling cost for the intermediate-volume method balanced the higher labor cost for the low-volume method at approximately 37,000 pieces, where cost per piece was 19¢. The break-even point for the high-volume method occurred at 345,000 pieces, where cost per piece was 5¢. Each 1000 parts required 230 lb of steel, which was purchased as auto-body sheet and slit to width.

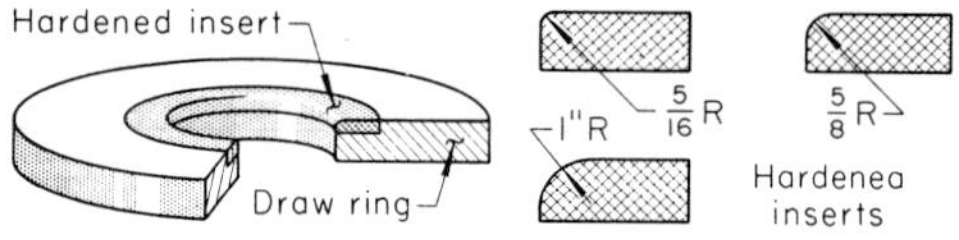

Fig. 12. Interchangeable inserts for use in a gray cast iron draw ring

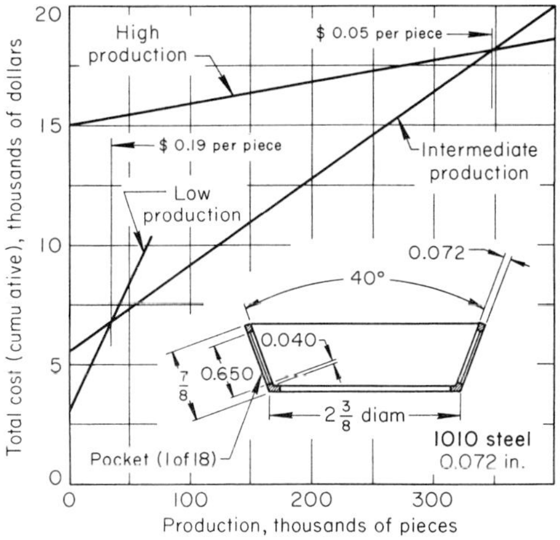

Operation or tool	Tool cost: Fixed	Tool cost: Maintenance(a)	Setup cost(b)	Labor cost(c)
Low-Volume Production				
Blank; draw ..	$2200	$1.80	$15.63	$ 1.76
Index; pierce..	300	2.60	18.75	35.20
Index; form ..	175	1.48	18.75	35.20
Bottom blank.	300	0.60	9.37	8.80
Spread	70	0.30	9.37	8.80
Total	$3045	$6.78	$71.87(d)	$89.76
Intermediate-Volume Production				
Blank; draw ..	$2200	$1.80	$15.62	$ 1.76
Bottom blank.	300	0.60	9.37	8.80
Pierce; form..	2800	7.02	12.50	8.80
Spread	140	0.40	12.50	5.28
Total	$5440	$9.82	$49.99(e)	$24.64
High-Volume Production (One operation in a transfer press)				
Dies	$13,333	$7.27	$0.25(f)	$2.31
Transfer fingers and cams ...	1,600	...	...	...
Total	$14,933	$7.27	$0.25	$2.31

(a) Cost per 1000 pieces. (b) At $6.25 per hour. (c) Per 1000 pieces; at $5.50 per hour. (d) Setup cost per 1000 pieces was $4.10. (e) Setup cost per 1000 pieces was $1.18. (f) Setup cost per 1000 pieces.

Fig. 13. Effect of quantity on cost of producing bearing separators by three different tooling methods (Example 233)

Example 234. Use of High-Production Tooling to Increase Output of Wheel Housings (Fig. 14)

Wheel housings (Fig. 14) were produced in pairs to reduce scrap losses and to minimize the number of dies required. At first, these housings were made with medium-production tooling. Tool cost was $88,000, and production rate was 3200 pairs of housings per week (two 8-hr shifts for five days).

Changing to high-production tooling, at a cost of $120,000, increased production from 3200 to 18,000 pairs of housings for the same time. Relation of cost per pair (in arbitrary units) to total production by the two different tooling methods is shown in Fig. 14. The break-even point occurred at approximately 45,000 pairs. Operations involved in the two methods are listed in the table with Fig. 14. Automation in the high-production procedure was the major difference. Medium-production tools were designed for a total production of 50,000 to 200,000 pairs, whereas high-production tooling was designed for a total production of 500,000 to 2 million pairs.

The work metal was commercial quality, cold rolled low-carbon steel that was roller coated with a water-soluble drawing oil before being drawn.

Effect of Punch and Die Radii

As the blank is struck by the punch at the start of drawing, it is wrapped around the punch and die radii; the stress and strain that develop in the workpiece are similar to those in bending, with an added stretching component. The bends, once formed, have the radii of the punch and die corners. The bend over the punch is stationary with reference to both punch and shell wall. The bend over the die radius, however, is continuously displaced with reference to both the punch radius and the blank, and it also undergoes a gradual thickening as the shell is drawn. The force required to draw the shell at the intermediate position has a minimum of three components: (*a*) the force required for bending and unbending the metal flowing from the flange into the sidewall; (*b*) the force required for overcoming frictional resistance of the metal passing under the blankholder and over the die radius; and (*c*) the force required for circumferential compression and radial stretching of the metal in the flange.

Because of the variation in metal volume and in resistance to metal flow, the punch force increases rapidly, passes through a maximum, and gradually decreases to zero as the edges of the flange approach and enter the die opening and pass into the shell wall. With the cup diameter remaining constant, the maximum press load and the length of stroke required to draw the cup depend on the size of the blank.

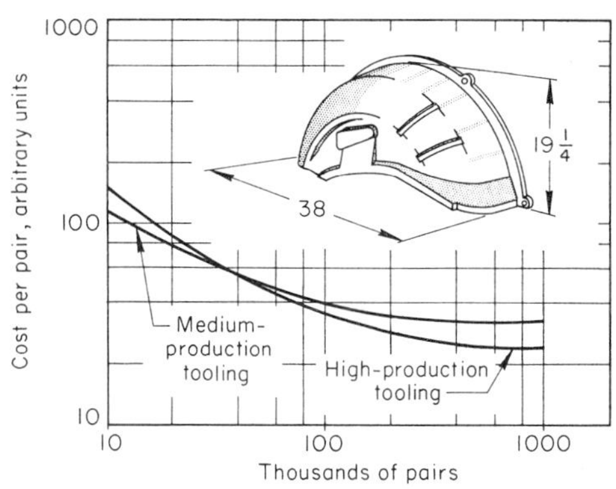

Medium-Production Tooling	High-Production Tooling
First Set of Operations: Pair of Wheel Housings Handled Together	
Draw and cut off corners; lift up	Draw and cut off corners; scrap drops through; lift up
Trim	Trim; lift up and eject to conveyor; scrap drops through
Pierce and part	Pierce and part; lift up and eject to conveyor; scrap drops through
Second Set of Operations: Housings Formed Separately, but in Two Dies on Same Press	
Flange and restrike	Flange and restrike; lift up and eject to conveyor
Final form	Final form; lift and eject to conveyor

Fig. 14. Relation of quantity and cost for deep drawing of pairs of wheel housings by medium-production vs high-production tooling. Cost units include amortization of tools. (Example 234)

The punch force-stroke relations for drawing blanks of various diameters from brass sheet 0.060 in. thick, using a 2-in.-diam punch, are shown in Fig. 15.

Under the conditions shown in Fig. 15, during cupping, the shell bottom is subjected to tensile stress in all directions, while the lower portions of the shell wall, particularly the radiused portion connecting the bottom with the wall, are subjected primarily to longitudinal tension. The stress in the metal being drawn into the shell wall consists of combined compressive and tensile stresses. Separation of the shell bottom from the wall is likely if a reduction is made that requires a force greater than the strength of the shell wall near the bottom (see curve for 4.40-in.-diam blank in Fig. 15).

The punch and die radii and percentage of reduction determine the load at which the bottom of the shell is torn out. Drawing is promoted by increasing punch and die radii. For a given drawing condition, the punch force needed to move the metal into the die decreases as the die radius increases, as shown in Fig. 16.

The reduction of drawing force in a double-action die by modification of the effective die radius can be accomplished in two convenient ways, which are shown in Fig. 17 and 18. In the conic lead-in die (Fig. 17), the cutout is effective in reducing frictional loads by removal of the portions of the die surface that are usually heavily loaded and increase friction. In Fig. 18, the sheet metal is formed into a conic shape before appreciable drawing begins. This has the effect of reducing the area of contact over the die radius by an amount proportional to $\alpha/90$ degrees (where α is the angle to declination of the hold-down surface to the horizontal, as shown in Fig. 18).

If the punch nose radius can be increased from one to five times metal thickness, the breaking load in the sidewall of the shell will decrease so that the reduction in blank diameter will increase from 35% to about 50% (for steel), and the shell can be drawn deeper before the sidewall tears.

If the shell bottom radius is less than four times the sheet thickness, it is usually desirable to form it with a larger-radius punch and then to restrike to develop the specified radius. This will minimize bottom failures. However, the bottom corner radius usually cannot be increased beyond ten times the sheet thickness without likelihood of wrinkling. The metal in dome-shape parts is likely to pucker in the unconfined area between the punch nose and die radius. Frequently, high blankholding forces or draw beads are used to induce combined stretching and drawing of the metal when forming dome shapes (see Example 255).

Deep drawing of boxes in stainless steel and other high-strength alloys with sides longer than fifty times stock thickness may result in a stability problem called "oil canning". The deflection of the sides by snap action can be eliminated by drawing the part in two operations with slightly different punches and an intermediate anneal. The first draw punch will have a larger nose radius than the second so that in the second drawing operation the metal can be stretched to eliminate the oil-canning effect. Stretching of the metal in parts with long sidewalls can be improved by gradually increasing the punch nose radius from the corner toward the center. A constant nose radius is used on the second-draw punch.

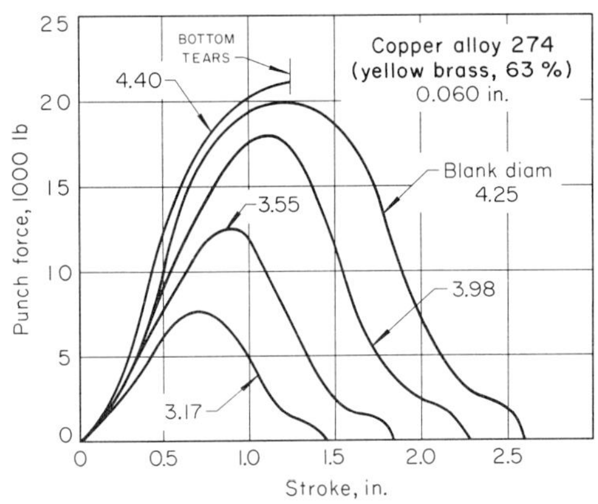

Fig. 15. Force-stroke relations for drawing blanks of various diameters from 0.060-in. brass sheet, using a 2-in.-diam punch

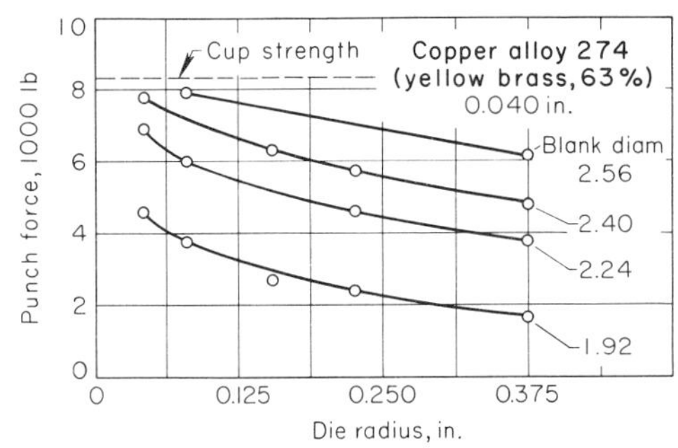

Fig. 16. Effect of die radius on punch force required for cupping various diameters of 0.040-in.-thick brass blanks, using a 1.2-in.-diam punch with a nose radius of 0.24 in.

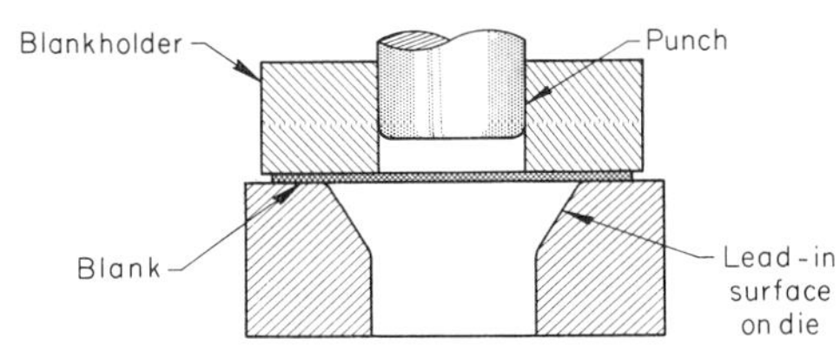

Fig. 17. Conic lead-in on die, which reduces drawing load

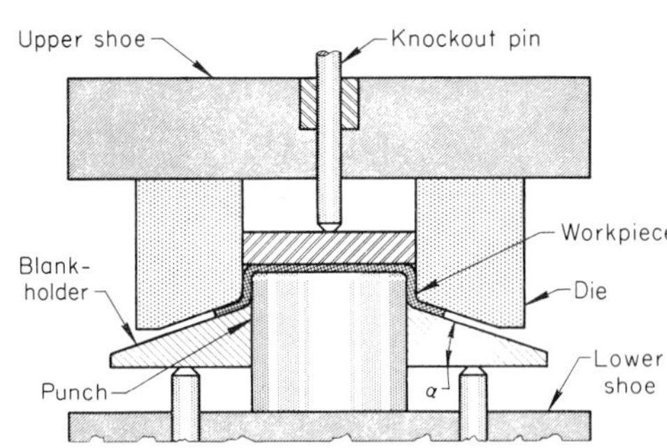

Fig. 18. Conic blankholder for reducing the severity of draw

Effect of Punch-to-Die Clearance

The selection of punch-to-die clearance depends on the requirements of the drawn part and on the work metal. Because there is a decrease, then a gradual increase, in the thickness of the metal as it is drawn over the die radius, clearance per side of 7 to 15% greater than stock thickness (1.07 to $1.15t$) helps prevent burnishing the sidewall and punching out the bottom of the cup.

The drawing force is minimum when the clearance per side is 15 to 20% greater than stock thickness (1.15 to $1.20t$) and the cupped portions of the part are not in contact with the walls of the punch and die. The force increases as the clearance decreases, and a secondary peak occurs on the force-stroke curve where the metal thickness is slightly greater than the clearance and ironing starts.

Redrawing operations require greater clearance, in relation to blank thickness, than the first draw, to compensate for the increase in metal thickness during cupping. A sizing redraw is used where the diameter or wall thickness is important or where it is necessary to improve surface finish to reduce finishing costs. The clearance used is less than that for the first draw.

Table 5 lists clearances for cupping, redrawing, and sizing draws of cylindrical parts from metal of various thicknesses. As the tensile strength of the stock decreases, the clearance must be increased.

Clearance between the punch and die for a rectangular shell, at the sidewalls and ends, is about the same as, or a little less than, for a circular shell. Clearance at the corners may be as much as 50% greater than stock thickness, to avoid ironing in these areas and to increase drawability.

Sheet Steel for Drawing

The principal factors affecting the selection of grade and quality of low-carbon steel sheet for deep drawing are (*a*) severity of draw (as determined by the amount of reduction and the punch nose radius), (*b*) thickness of sheet, (*c*) shape of the part (round, rectangular or conical), (*d*) flange requirements, (*e*) ironing requirements, (*f*) desired finish, (*g*) grain size, (*h*) press speed, (*i*) availability of material, and (*j*) cost.

The three principal classes of available low-carbon sheet steel are commercial quality, drawing quality, and aluminum killed ("special killed"), in order of increasing drawability.

For hot rolled sheets, the mill oxide on the surface is detrimental for deep drawing, because it impairs the flow of

Table 5. Punch-to-Die Clearance for Drawing Operations

Metal thickness(t), in.	Clearance-to-metal-thickness relationship for: Cupping	Redrawing	Sizing draws
Up to 0.015	1.07 to $1.09t$	1.08 to $1.10t$	1.04 to $1.05t$
0.016 to 0.050	1.08 to $1.10t$	1.09 to $1.12t$	1.05 to $1.06t$
0.051 to 0.125	1.10 to $1.12t$	1.12 to $1.14t$	1.07 to $1.09t$
0.126 and up	1.12 to $1.14t$	1.15 to $1.20t$	1.08 to $1.10t$

Table 6. Selection Table for Low-Carbon Steel Sheet for Deep Drawing of Cylindrical Cups

Punch nose radius	Reduction in drawing(a) — 0.036 in. thick (20 gage) 20%	30%	40%	50%	0.105 in. thick (12 gage) 20%	30%	40%	50%	0.250 in. thick 20%	30%	40%	50%
No Flange, No Ironing												
2t	CR,DQ	....	....	....	....	....	....	....	....	....	....	....
4t	CR,DQ	CR,DQ	CR,DQ,SK	....	HR,DQ	HR,DQ CR,CQ	HR,DQ CR,DQ	....	HR,CQ	HR,CQ	HR,DQ	....
8t	CR,CQ	CR,CQ	CR,DQ	CR,DQ,SK	HR,CQ	HR,CQ	HR,DQ CR,CQ	HR,DQ CR,DQ	HR,CQ	HR,CQ	HR,DQ	HR,DQ
16t	CR,CQ	CR,CQ	CR,CQ	CR,DQ	HR,CQ	HR,CQ	HR,CQ	HR,DQ CR,CQ	HR,CQ	HR,CQ	HR,CQ	HR,DQ
32t	CR,CQ	CR,CQ	CR,CQ	CR,DQ	HR,CQ	HR,CQ	HR,CQ	HR,CQ CR,CQ	HR,CQ	HR,CQ	HR,CQ	HR,DQ
No Flange, Ironed(b)												
2t	CR,DQ,SK	....	....	....	....	....	....	....	....	....	....	....
4t	CR,DQ	CR,DQ,SK	....	....	HR,DQ	HR,DQ	HR,DQ CR,DQ	....	HR,DQ	HR,DQ	....	....
8t	CR,DQ	CR,DQ,SK	CR,DQ,SK	....	HR,CQ	HR,CQ	HR,DQ	HR,DQ CR,DQ	HR,CQ	HR,DQ	HR,DQ	....
16t	CR,DQ	CR,DQ	CR,DQ,SK	CR,DQ,SK	HR,CQ	HR,CQ	HR,DQ	HR,DQ	HR,CQ	HR,CQ	HR,DQ	HR,DQ
32t	CR,DQ	CR,DQ	CR,DQ	CR,DQ,SK	HR,CQ	HR,CQ	HR,CQ	HR,DQ	HR,CQ	HR,CQ	HR,CQ	HR,DQ
With Flange, No Ironing												
2t	....	....	....	....	....	....	....	....	....	....	....	....
4t	CR,DQ	CR,DQ,SK	CR,DQ,SK	....	CR,DQ	CR,DQ,SK	CR,DQ,SK	....	HR,DQ	HR,DQ	....	....
8t	CR,DQ	CR,DQ	CR,DQ,SK	CR,DQ,SK	HR,DQ	CR,DQ	CR,DQ	CR,DQ,SK	HR,DQ	HR,DQ	HR,DQ	HR,DQ
16t	CR,CQ	CR,CQ	CR,DQ	CR,DQ	HR,CQ	HR,CQ CR,CQ	HR,DQ CR,DQ	CR,DQ	HR,DQ	HR,DQ	HR,DQ	HR,DQ
32t	CR,CQ	CR,CQ	CR,CQ	CR,DQ	HR,CQ	HR,CQ	HR,DQ CR,CQ	CR,CQ	HR,CQ	HR,DQ	HR,DQ	HR,DQ
With Flange, Ironed(b)												
2t	....	....	....	....	....	....	....	....	....	....	....	....
4t	CR,DQ,SK	CR,DQ,SK	....	....	CR,DQ,SK	CR,DQ,SK	CR,DQ,SK	....	HR,DQ	....	....	....
8t	CR,DQ,SK	CR,DQ	....	....	HR,DQ	CR,DQ	CR,DQ,SK	....	HR,DQ	HR,DQ	....	....
16t	CR,DQ	CR,DQ	CR,DQ,SK	CR,DQ,SK	HR,DQ	HR,DQ CR,DQ	HR,DQ CR,DQ	CR,DQ,SK	HR,DQ	HR,DQ	HR,DQ	HR,DQ
32t	CR,CQ	CR,CQ	CR,DQ	CR,DQ,SK	HR,DQ	HR,DQ	HR,DQ	HR,DQ CR,DQ	HR,DQ	HR,DQ	HR,DQ	HR,DQ

CR = cold rolled; CQ = commercial quality; DQ = drawing quality; HR = hot rolled; SK = special killed.

NOTES: (1) Two entries in one space indicate that the choice is borderline and will be influenced by uniformity, price, quantity, and availability. (2) Omission of an entry (....) in this table

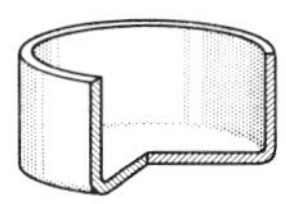

indicates that the draw is usually too severe to be completed successfully.

(a) In terms of reduction from blank diameter to cup diameter. (b) Operation that will bring the sidewall to a uniform thickness with the minimum amount of reduction.

metal and increases tool wear. "Special surface" can be obtained for any of the three qualities at extra cost, or oxide can be removed by pickling, which also adds to the cost. Hot rolled sheets for deep drawing are usually purchased pickled and oiled.

The primary consideration in selection between hot and cold rolled steel for deep drawing is availability of the required sheet thickness. Where superior surface finish is required, cold rolled sheet or strip is specified. The choice between commercial or drawing quality in either the hot or cold rolled sheets depends on the severity of draw.

Drawing quality is generally specified for severe draws because, with a nominal cost extra, required drawability is guaranteed. The problem is to recognize when it is expedient to pay the extra price for drawing performance guaranteed by the supplier.

Rimmed steels are satisfactory for most draws of mild to moderate severity. The use of special killed steels is usually limited to cold rolled steel for draws requiring freedom from strain aging, maximum uniformity within a shipment and among successive shipments, and maximum drawability.

The first approach to selection can be made as follows: (*a*) When surface conditions of the finished piece are most important, a cold rolled, rimmed steel will be specified; (*b*) when drawability requires freedom from aging or if the design specifies severe forming, such as flanges, ironing, sharp corner radii, extreme cup depth, or lack of symmetry, special killed quality is indicated. Assuming correct tooling and normal reductions, any scrap losses with special killed steel will probably result from the opening of small surface disturbances during drawing. This occurs more often in coil stock than in sheet.

"Special soundness" refers to the internal condition and is not a prerequisite for deep drawing. It is specified only when subsequent operations such as bright plating, highly critical welding, or restrictive carburizing require maximum uniformity and homogeneity.

Severity Classification. Table 6 shows the grades of low-carbon steel recommended for drawing round cups. Severity is based on cup reduction and punch nose radius for specific sheet thicknesses. Optimum tooling and lubrication are assumed for the most difficult draws. Within the four sections of Table 6, the effects of flanges and ironing are also considered. Similarly, Table 7 shows grades recommended for drawing rectangular cups or boxes having corner radii equal to the punch nose radius. In both tables, severity of drawing increases toward the upper right corner of each of the subdivisions that define the various conditions. The sequence of steels recommended to meet the increasing severity conditions is: commercial quality, drawing quality, and drawing quality special killed.

Availability is considered in Tables 6 and 7 as follows: Steel recommendations in the 20-gage thickness column (0.036 in.) are confined to cold rolled. Recommendations for the 12-gage thickness (0.105 in.) are predominantly hot rolled, although some cold rolled steels are included for extreme severities. Only hot rolled grades are available for the 0.250-in. thickness (borderline between sheet and plate).

The portions of Tables 6 and 7 relating to 0.105-in. sheet include both cold and hot finished grades; this sheet thickness is within the range where a cold rolled sheet is competitive with the hot rolled material. When all the requirements of one particular part are considered, the price differential between the two grades for surface finish only is often a negligible fraction of the cost of either. When the usually specified extra for pickling and oiling is added to the base price of hot rolled sheet, the price of hot finished steel is approximately the same as the cold finished grade of the same quality level.

The cold finished grade provides the advantage of greater uniformity of properties throughout the shipment, a result of the closer process control of annealing temperature, compared with finishing temperature in hot rolling. Another influential factor is that the cold rolled sheet has a more nearly uniform thickness extending to the edges of the sheet. Roll wear on the hot mill sometimes makes close thickness tolerances at the edges difficult to maintain. In some applications, this "feathered" edge is unusable, and the outside portion of the sheet must be discarded; thus, in blanking, the blank must be confined within the area of more uniform thickness. The alternative is to specify the hot rolled sheet with a cut edge at extra cost.

Table 7. Selection Table for Low-Carbon Steel Sheet for Deep Drawing of Rectangular Box-Shape Cups Without a Flange

Punch nose radius	Reduction in drawing(a): 0.036 in. thick (20 gage) 20%	30%	40%	50%	0.105 in. thick (12 gage) 20%	30%	40%	50%	0.250 in. thick 20%	30%	40%	50%
No Flange, No Ironing												
4t	....	....	....	....	....	....	....	....	....	....	....	....
8t	CR,DQ,SK	....	....	....	HR,DQ	HR,DQ CR,DQ	....	....	HR,CQ	HR,DQ	....	....
16t	CR,DQ,SK	CR,DQ,SK	CR,DQ,SK	....	HR,DQ	HR,DQ	HR,DQ CR,DQ	....	HR,CQ	HR,DQ	HR,DQ	....
32t	CR,DQ	CR,DQ,SK	CR,DQ,SK	....	HR,DQ	HR,DQ	HR,DQ CR,DQ	CR,DQ,SK	HR,CQ	HR,CQ	HR,CQ	HR,DQ
No Flange, Ironed(b)												
4t	....	....	....	....	....	....	....	....	....	....	....	....
8t	CR,DQ,SK	....	....	....	HR,DQ	CR,DQ	....	....	HR,DQ	....	....	....
16t	CR,DQ,SK	CR,DQ,SK	....	....	HR,DQ	HR,DQ	CR,DQ,SK	....	HR,CQ	HR,DQ	HR,DQ	....
32t	CR,DQ	CR,DQ,SK	CR,DQ,SK	....	HR,DQ	HR,DQ	HR,DQ CR,DQ	CR,DQ,SK	HR,CQ	HR,CQ	HR,DQ	HR,DQ

CR = cold rolled; CQ = commercial quality; DQ = drawing quality; HR = hot rolled; SK = special killed.

NOTES: (1) Two entries in one space indicate that the choice will be influenced by uniformity, price, quantity, and availability. (2) Omission of an entry (....) in this table indicates that the draw is too severe to be completed successfully.

(a) In terms of reduction from blank width to cup width. Corner radii are the same as punch nose radius. (b) Operation that will bring the sidewall to a uniform thickness with the minimum amount of reduction.

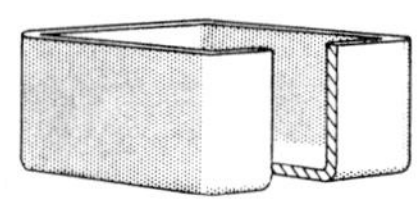

Between 12 and 18 gage (0.105 and 0.048 in.), both hot and cold rolled grades are widely available and offer a greater range of selection. The hot and cold rolled steels become most closely competitive at about 16 gage (0.060 in.). Finish requirements on the parts will usually determine the final choice between the two grades indicated for each specific severity.

Either normalizing or annealing may be necessary for the most severe drawing requirements of Tables 6 and 7 — for instance, in Table 7, for the 0.250-in. thickness. In such borderline applications, quality of tooling and the experience of operators will often determine whether even the highest grade of sheet will make the part.

Grain size and surface finish affect drawability and may influence the selection of a grade. Grain sizes of 5 and coarser may result in objectionable surface roughness as well as reduced drawability. The dull finish normally supplied on drawing sheets is designed to hold lubricants and improve drawability. Brighter finishes are required if parts are to be electroplated.

Higher-Strength Grades. Higher strength can be obtained in drawn steel products at some sacrifice in drawability by using "physical quality" carbon steel (available with tensile strength as high as 80,000 psi min), high-manganese steel (up to about 0.20% C and 1.60% Mn), high-strength low-alloy steel conforming to ASTM A242 or SAE 950, or higher-carbon steels such as 1020 to 1060.

An example of a deep drawn part where high strength is required and heat treatment is impractical is a pressure tank for propane gas, which is being made from a 1.5% manganese steel by deep drawing the two halves and joining them by welding. For additional discussion and examples of deep drawing higher-strength grades, see the section on Deep Drawing of Pressure Vessels, on page 188.

Ironing. In cup drawing, the lip of the cup is usually thicker than the sidewall adjacent to the base. When uniformity of both wall thickness and strength is required in the drawn article and the steel is not subsequently heat treated, a certain maximum amount of ironing is allowable during the drawing operation. Permissible sidewall reductions for maintaining uniform strength in steels of various carbon contents are given in Table 8. The allowable reduction decreases as the carbon content and the amount of work hardening increase.

Table 8. Comparative Allowable Ironing to Maintain Uniform Strength of Cold Rolled Carbon Steel During Cup Drawing

Steel	Permissible reduction in wall, %
1008	
Drawing quality, special killed	12
Drawing quality	8
Commercial quality	7
1020	
Drawing quality, special killed, electric furnace	10
Drawing quality, annealed	8
Commercial quality, annealed	6
1030	
Special killed, electric furnace, annealed	8
Commercial quality, annealed	5
1040	
Special killed, electric furnace, annealed	7
Commercial quality, annealed	4

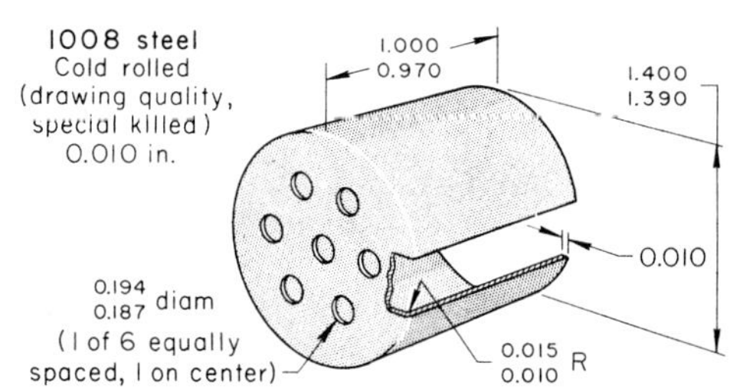

Fig. 19. Thin-wall cup that was made in a compound die, using a special lubricant containing graphite (Example 235)

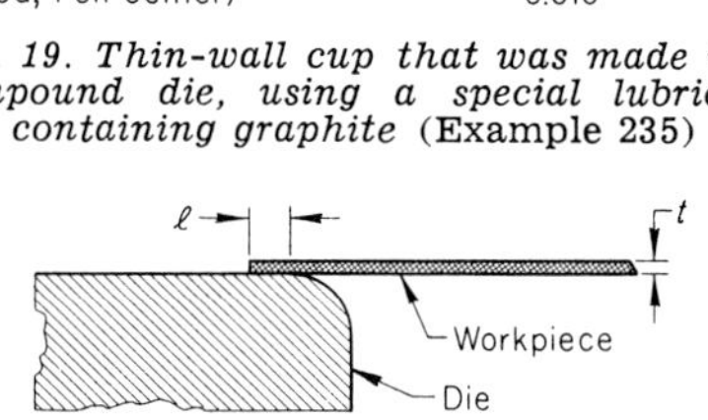

Fig. 20. Ratio of sheet thickness(t) *to supported length*(l) *determines need for use of a blankholder when drawing.*

Effect of Sheet Thickness

The development of wrinkles in drawing is caused by mechanical instability in the plane of the blank. This instability accompanies the circumferential compression to which the blank is subjected as it is drawn to the die radius. When dome shapes are drawn, the same compressive stress can develop in the shell wall or bottom, and a pucker will result.

The probability of wrinkling increases as the sheet thickness decreases to 0.032 in. or less. For a given drawing application, the sheet should be as thick as possible to suppress wrinkle formation. However, as thickness increases, press load rises, thereby increasing the probability of scoring, and production rate decreases.

Parts made of thin metal may be difficult to remove from the punch because the lubricant fills the space between the part and the punch, causing a vacuum. An air vent in the punch eliminates this problem.

In the following example, the thin-wall cup and the narrow scrap ring had to be removed from the drawing die carefully to prevent damage to the die and the cup.

Example 235. Thin-Wall Cup That Was Susceptible to Damage During Removal From the Die (Fig. 19)

The thin-wall cup shown in Fig. 19 had to be carefully removed from the drawing die to prevent denting the sides and distorting the bottom. Also, the narrow pinch-trimmed scrap ring had to be removed by the operator to prevent damage to the strip during feeding and damage to the die because of more than one stock thickness.

The cup was drawn from 0.010-in.-thick strip 3 in. wide (1008 steel, cold rolled, drawing quality, special killed). The steel was pickled, had a No. 4 or 5 temper (Rockwell B 65 max), and a No. 2 surface finish. A specially compounded lubricant containing graphite was used to blank, draw, pinch trim, and pierce seven 0.194/0.187-in.-diam holes, in a compound die. The die was made of D5 tool steel.

A 45-ton mechanical press with a die cushion, operating at 50 strokes per minute, was used. An air-operated hitch-type stock feeder moved the strip through the die.

Restraint of Metal Flow

Even in the simplest drawing operation, as shown in Fig. 5(a), the thickness of the work metal and the die radius offer some restraint to the flow of metal into the die. For drawing all but the simplest of shapes, some added restraint is usually required to control the flow of metal. This additional restraint is usually obtained by the use of a blankholder, as illustrated in Fig. (5b) and (c). The purpose of the blankholder is to suppress wrinkling and puckering, and to control the flow of the work metal into the die.

Drawing Without a Blankholder. A blank is not susceptible to wrinkling, and a blankholder need not be used, if the ratio of supported width to sheet thickness is within certain limits. In Fig. 20, the supported length (l) is the length from the edge of the blank to the die cavity (point of tangency). The sheet thickness is denoted as t. The l/t ratio is influenced little by other geometrical conditions, and differs little for the various metals commonly drawn. When the l/t ratio does not exceed 3 to 1, a cup can be drawn from annealed brass, aluminum (to half-hard), and low-carbon steel without a blankholder. For slightly harder work metals, such as hard copper or half-hard brass, this ratio should not exceed 2.5 to 1.

An elliptical or conical die opening, such as in Fig. 17, can be used where the die radius required to draw the part reduces the length of the blank-supporting surface to less than three times stock thickness. The distance between the die opening and the punch should not exceed ten times stock thickness.

A 30° elliptical radius derived from a circle created by a given draw radius increases the strain on the metal being drawn by 4.2%, but it decreases the metal out of control by 47% of the length of the original draw radius. This shape has been helpful in the drawing of tapered shells from a flat blank. For these draws, it is desirable to increase the strain slightly to prevent puckers and to reduce the metal out of control for the same reason.

A 45° elliptical radius derived as above reduces the strain on the metal being drawn by 1.03% and also reduces the metal out of control by 33% of the length of the original draw radius. The 45° ellipse is useful only when a large radius will draw the part, but produces wrinkles, and a smaller radius will not permit the draw.

A 60° elliptical radius does not measurably reduce drawing strain and accounts for only a 9% reduction of metal out of control. Its use on draw dies is not economically feasible when the small gains derived are considered in relation to the cost of producing the contour.

Drawing thick metal without a blankholder is frequently done when the blank diameter is no greater than 20 times stock thickness.

The part described in the following example was drawn in a die having a conical opening and a ratio of blank diameter to thickness of 19 to 1.

Example 236. Drawing a Thick Blank Without Use of a Blankholder (Fig. 21)

The cup (Fig. 21) for a disk-brake piston was drawn and sized in four operations from a drawing quality, hot rolled 1008 steel blank 0.250 in. thick and 4¾ in. in diameter without the use of a blankholder to prevent wrinkles. The workpieces were carried from station to station in a hydraulic-actuated dial feed. Spring-loaded fingers stripped the workpiece from the punch on the upstroke and held it in an indexing table to carry it to the next station. A die cushion provided the force for a pressure pad during the first three forming operations and also ejected the parts from the die cavities.

Blanks were stacked in a magazine with the cut (burnished) edge up and were fed into the first station of the die by an air-actuated slide. As the blank moved forward, it was roller coated with molybdenum disulfide lubricant.

In the first operation, the cup was drawn to 3-in. OD, 2¹⁄₃₂ in. high, with a ½-in. inside radius. The top inside corner of the cup was chamfered by the draw punch. A blankholder was not necessary, because of the stock thickness and the relatively low ratio of blank diameter to stock thickness (19 to 1). The die opening was conical (30° from top of die), and a 1-in. radius blended the entry angle with the sidewall.

In the second operation, the cup was reduced to 2.800-in. OD, 2⅛ in. deep, with a ¼-in. inside radius. The third station compressed the flange to a height of 2 in. to form sharp inside and outside corners.

The last station sized the cup to 2.770-in. OD, formed the concave end, and provided the elliptical form on the outside corner and the flat on the bottom. The cup height was 2¹⁄₃₂ in. The sizing ring was originally made of tungsten carbide, which required considerable maintenance to produce the required surface finish on the sidewalls. Changing to an aluminum bronze ring increased ring life and reduced maintenance.

Clearances per side between the punch and die for each of the operations performed in the dial index setup were: first operation, 0.305 in.; second operation, 0.239 in.; third operation, 0.253 in.; and fourth operation, 0.250 in.

The die operated in an 800-ton mechanical press at 14 strokes per minute. Yearly production was over 3 million pieces.

In subsequent operations, the part was machined to 2.753/2.751-in. OD, and height to 1.937 in. The groove was machined in and the bottom corner was burnished to 15-micro-in. finish.

Another method of making the same part competitively was by cold extrusion from a steel slug.

A similar setup was used to make a cup 2.376 in. in diameter from a 4½-in.-diam blank.

Blankholders. The purpose of a blankholder is to prevent wrinkles from forming in the flange of a part during drawing. The formation of wrinkles interferes with, or prevents, the compressive action that rearranges the metal from flange to sidewall. Much greater reductions are possible when a blankholder is used.

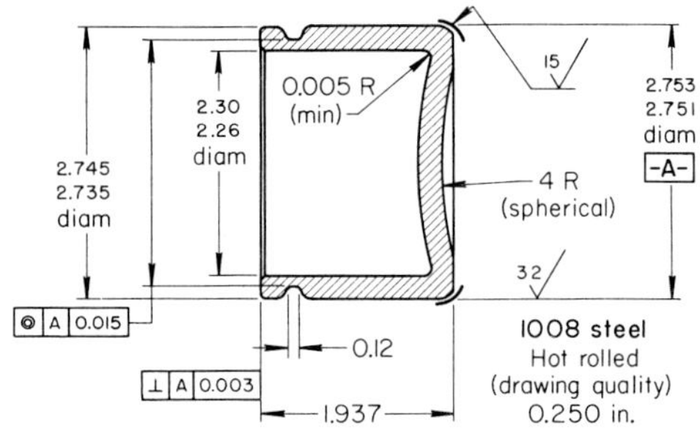

Fig. 21. Thick-wall cup that was drawn without a blankholder, then machined to size and shape shown (Example 236)

Blankholders can be used in both double-action and single-action presses. In a double-action press, the blankholder advances slightly ahead of the punch, and dwells at the bottom of its stroke throughout the drawing phase of the punch cycle. The blankholder dwell usually extends to a point on the punch upstroke at which positive stripping of the shell is assured. By using a die cushion and an inverted die, similar action can be obtained in a single-action press. A die cushion in a double-action press supports the blank and holds it against the punch during the drawing operation; it then lifts the finished part out of the die.

A blankholder must allow the work metal to thicken as the edge of the blank moves inward toward the working edge of the die. The amount of thickening is expressed by:

$$t_1/t = \sqrt{D/D_1}$$

where t is the thickness of the blank; t_1 is the thickness of the flange at any instant during the drawing operation; D is the diameter of the blank; and D_1 is the diameter of the flange at any instant during the drawing operation, or the mean diameter of the workpiece without the flange. As the metal flows, paths of least resistance are taken; therefore, the actual value of t_1 will be less than that calculated from the formula.

Types of Blankholders. The simplest type of blankholder is fixed to the die block and has a flat hold-down surface, as shown in Fig. 22(a). A disadvantage of this type is that maintenance of the optimum gap between the die surface and the flat hold-down surface requires careful adjustment. As shown in the illustration, the blankholder does not quite touch the work metal as drawing begins; restraint begins and increases as the flange portion thickens. A gap that is either too small or too large increases force and reduces drawability. For best results, the gap should be slightly smaller than the flange thickness, allowing 50 to 75% of the final thickening before the work metal touches the blankholder.

Some of the difficulty in maintaining adjustment of a fixed flat blankholder can be avoided by tapering the hold-down surface, as shown in Fig. 22(b). A tapered fixed blankholder can tightly clamp the blank edge at the beginning of the operation. The amount of taper is not critical and is usually ½ to 1½ times the theoretical taper (thickness of the blank to maximum thickness of the flange).

The flat controlled-pressure blankholder shown in Fig. 22(c) is generally preferred in production operations, because it can be adjusted to a predeter-

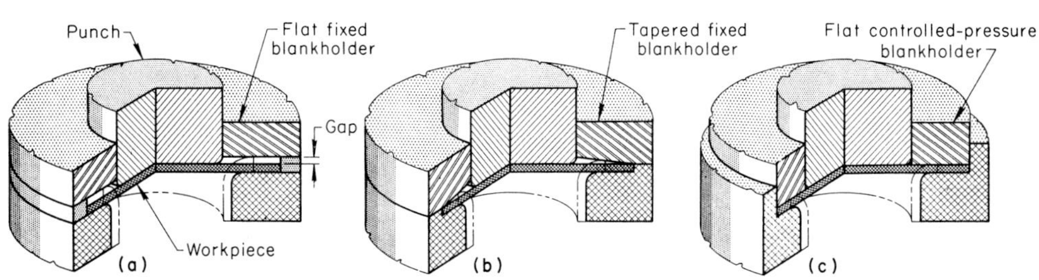

Fig. 22. Setups showing use of three types of blankholders. See text for discussion.

mined and closely controlled value by hydraulic or pneumatic pressure. Springs, unless extremely long, are not suitable for supplying pressure to a blankholder during deep drawing, because the force exerted by a spring increases rapidly as it is compressed. The force on hydraulic or pneumatic die cushions will increase about 20% when compressed the full stroke length. Some hydraulic systems have pressure control valves that supply a more nearly constant pressure during the entire stroke.

The fixed-type blankholder (Fig. 22a and b) draws a cup without a flange and ejects it through the bottom of the die. The blankholder shown in Fig. 5(b) and (c) and Fig. 22(c) can be used for drawing a cup with or without a flange. The cup is stripped from the punch by the blankholder and ejected from the die by a knockout or pressure pad. Cups without a flange can be pushed through the die if a pressure pad is not needed to support the blank.

Thin-face shimmed blankholders, like the one shown in Fig. 23, are used when the flange must be restrained with a minimum of marking or damage to the material, and where draw beads normally would be needed. (Draw beads are used to restrain metal flow at points found by trial or by experience to give the needed control.) The shim is not continuous, but exerts pressure at selected points to restrict flow. This kind of blankholder is inexpensively produced and maintained. Costs for setup and production are about the same as for a conventional blankholder.

Blankholder Pressure. The restraint needed in drawing varies during the press stroke as the draw progresses. Ideally, the pressure of the blankholder should be light at first, increasing as the draw progresses, and then should become lighter again as the flange thickens. However, it is not practical to vary the blankholder pressure, and an average value is used.

Optimum blankholder pressure depends largely on the amount of reduction and the die contour. However, assuming a generous radius on the die, hold-down force varies little for a wide range of work-metal thicknesses, and is approximately proportional to the area of the portion of the blank initially held by the blankholder.

Approximate blankholder pressures, based on mechanical properties of the metal, are given for several different metals in Table 9. Blankholder forces for drawing a wide range of thicknesses of low-carbon steel, as a percentage of drawing forces, are given in Table 10.

The pressures in Tables 9 and 10 are used only to prevent wrinkling at the edges of the blank and flange. Wrinkling can occur in other areas of the workpiece that are not confined between the tools. When the punch radius is large in proportion to work-metal thickness, wrinkling is likely to occur (see Fig. 49). Wrinkling can often be prevented by superimposing a high radial tension on the normal stress developed in drawing. This is normally accomplished by means of draw beads. Another means of eliminating wrinkling is to use oversize blanks, but this is wasteful of work metal. Also, an

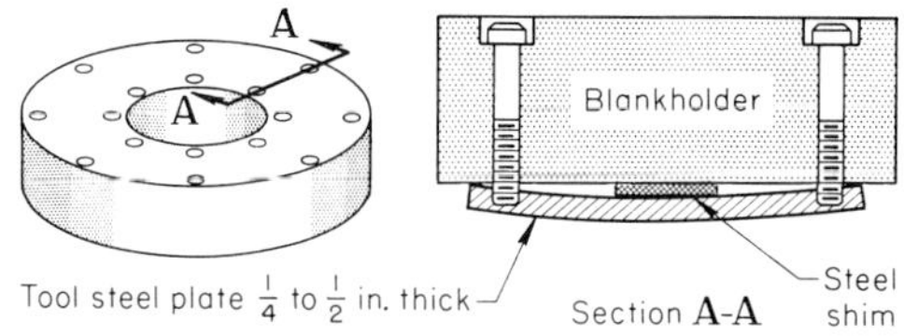

Fig. 23. Thin-face shimmed blankholder

Table 9. Blankholder Pressures for Several Annealed Metals

Metal	Pressure, psi(a)
Low-carbon steel	500
Austenitic stainless steel	1000
Aluminum	100
Aluminum alloys	200 to 500
Copper	200
Alpha brasses	250 to 300
Phosphor bronze	400

(a) Values are approximately 1/150 to 1/200 of the sum of the yield strength and tensile strength of the specific metal.

Table 10. Blankholder Forces for Deep Drawing of Low-Carbon Steel(a)

Work-metal thickness, in.	Blankholder force, % of drawing force	Work-metal thickness, in.	Blankholder force, % of drawing force
0.005	85	0.050	23
0.010	67	0.070	14
0.015	57	0.100	9
0.020	50	0.125	8½
0.025	44	0.187	8¼
0.030	39	0.250	8

(a) Values given should not be used for shallow drawing, particularly of elliptical or hemispherical shapes.

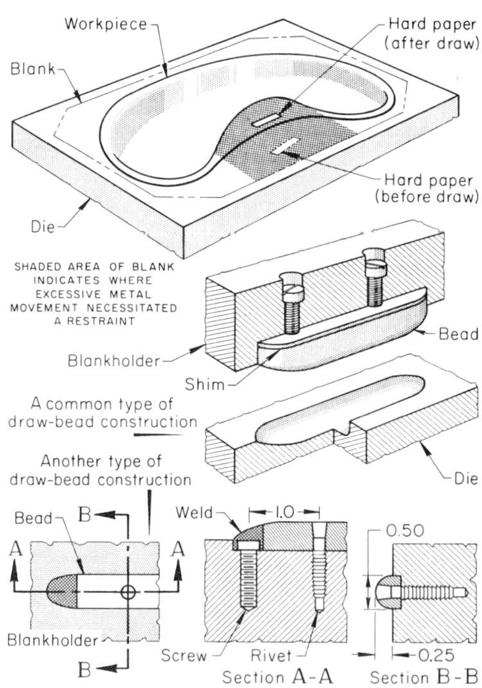

Fig. 24. Method of determining optimum bead location, and construction of typical draw beads. See also Fig. 49.

oversize blank may restrict metal flow and cause excessive thinning or even fracture.

Draw beads (Fig. 24) help prevent wrinkles and control the flow of metal in the drawing of shells. The use of draw beads increases the cost of tools, product development, and tool maintenance. However, they often are the only means of controlling metal flow in drawing odd shapes. Draw beads are ordinarily used for the first draw only; thus, production rates are the same as when conventional blankholders are used.

Optimum location of the draw bead can be determined by placing a piece of hard paper on the flange of the workpiece, as shown in Fig. 24. The paper will move as the workpiece is drawn, indicating where restraint is needed.

Two methods of constructing draw beads are illustrated in Fig. 24 (lower views). Beads are usually made from water-hardening tool steel, such as SAE W108, to resist wear from the sliding work metal. The height of a bead fastened with screws from the top can be adjusted by shimming.

The second bead design shown in Fig. 24 uses a drive rivet that has hardened serrations and a soft tapered head. Rivets are driven into holes in the blankholder, and the heads are peened and ground flush with the bead. Additional anchorage is provided by welding the ends of the bead to a low-carbon steel screw after the bead is flame hardened. The weld is ground to blend with the bead and blankholder surface. The bead is seated in a close-fitting groove to prevent tipping during the draw operation.

For low production, draw beads often are made by laying a weld bead on the die after the optimum location has been determined.

Restraint of the metal flow, to the extreme of locking the flange of the blank to prevent motion, is needed for some draws. A deep shell with sloping walls may be made by drawing followed by several redraws. This results in a stepped workpiece. The final sizing draw is a stretching operation, made with the flange secured by a locking bead in the blankholder, as shown in Fig. 25. This kind of blankholder is also used in making shallow drawn panels.

For additional information on the design and use of draw beads, see page 125 of "Press Forming of Low-Carbon Steel", in this volume.

Sometimes draw beads need not be used for the full depth of draw, or need be used only in certain locations, such as corners of regular or irregular polygon-shape parts. In such applications, some of the work metal may be allowed to slip through the draw bead, to be retained at the end of the stroke only by blankholder pressure, as in the following example.

Example 237. Forming a Large Irregular Shape With the Aid of Draw Beads (Fig. 26)

When parted, the drawn shape shown in Fig. 26 made a right-hand and left-hand wheel housing for an automobile. Forming the two housings together reduced scrap loss and minimized the number of dies required. A 1010 steel blank 48 in. wide by 54 in. long and 0.042 in. thick was used. Operations were:

1 Shear blank to size and trim corners 4 in. by 45°.
2 Draw in a toggle press with 400-ton outer slide and 440-ton inner slide.
3 Separate shell into two pieces in a 300-ton mechanical press.
4 Restrike to sharpen radii and flare bottom edge in a 300-ton mechanical press.
5 Trim periphery of part in a 300-ton mechanical press.

Draw beads similar to those shown in Fig. 24 were placed in the blankholder about 6 in.

from the cavity, surrounding it. This arrangement provided additional control of the blank during the initial drawing stage.

After drawing to about half the full depth, the edge of the blank had moved inside the draw bead at certain places along the sides and ends, and it was held only by the blankholder surfaces. The amount of metal controlled by the draw beads decreased as the depth of draw increased. At the end of the operation, four to eight points of the blank were still confined by the bead. The amount of metal still outside the draw bead depended on the earing characteristics of the metal, the centering of the blank under the punch, and the uniformity of metal movement between blankholder faces.

A pigmented emulsified drawing compound slightly thinned with water was used during drawing.

The drawing die for operation 2 was made of ductile iron. The dies for operations 3 and 5 were gray iron faced with O1 tool steel cutting edges. The restrike die for operation 4 was ductile iron.

Production rate was 150 draws per hour. Die life was three model years.

Effect of Press Speed

Drawing speed is usually expressed in linear feet per minute. Under ideal conditions, press speeds as high as 75 ft per min are used for deep drawing of low-carbon steel. However, 20 to 55 ft per min is the usual range — up to 55 ft per min for single-action presses and 35 to 50 ft per min for double-action presses (Table 11). Ideal conditions include: (*a*) use of a drawing quality of work metal; (*b*) symmetrical workpieces of relatively mild severity; (*c*) adequate lubrication; (*d*) precision carbide tools; (*e*) carefully controlled blankholding pressure; and (*f*) presses that are maintained to a high level of accuracy. When one or more of the above conditions is less than ideal, some reduction in press speed is required. If all, or nearly all, are substantially less than ideal, press speed may have to be reduced to 20 ft per min. When the operation includes ironing, the drawing speed is usually reduced to about 25 ft per min, regardless of other factors.

The punch speed in hydraulic presses is relatively constant throughout the stroke. In mechanical presses, "punch speed" is that at mid-stroke, since the velocity changes in a characteristic manner throughout the drawing stroke, from maximum velocity to zero. The only adjustment in speed that can be made is to decrease flywheel speed or to use a press with a shorter stroke that operates at the same number of strokes per minute. This proportionately decreases maximum punch speed.

Speed is of greater significance in drawing stainless steels and heat-resisting alloys than in drawing softer, more ductile metals. Excessive press speeds have caused cracking and excessive wall thinning in drawing these stronger, less ductile metals. Nominal speeds for drawing various metals are given in Table 11. Effect of speed on drawability of low-carbon steel is shown in Table 12.

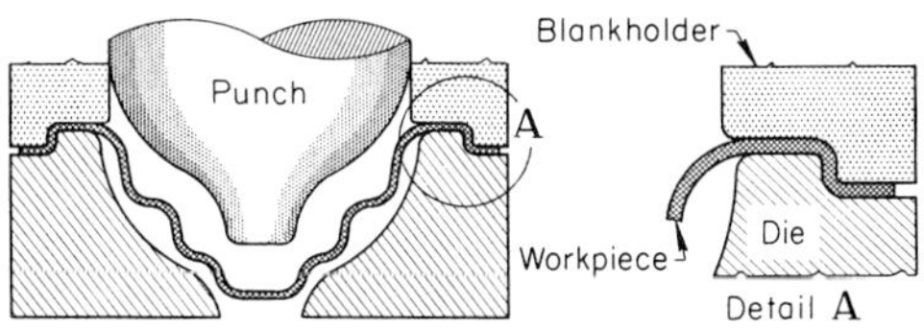

Fig. 25. Locking blankholder used for final sizing draw

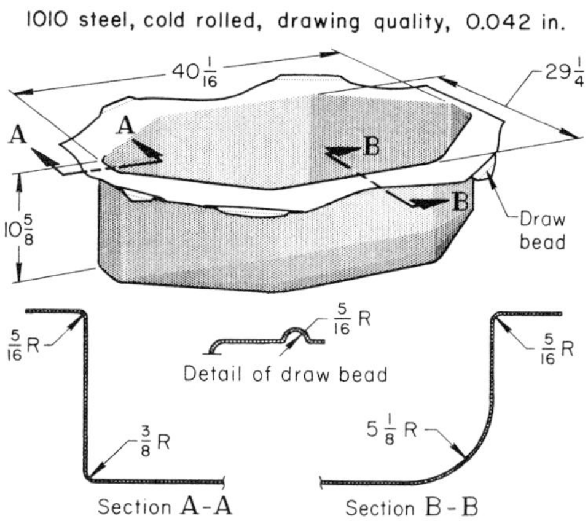

Fig. 26. Large irregular shape that was drawn with the aid of draw beads. The shell was later parted to make two wheel housings. (Example 237)

Table 11. Nominal Speeds for Drawing of Various Metals

Metal	Speed, ft per min: Single-action press	Double-action press
Aluminum	175	100
High-strength aluminum	...	30 to 40
Brass	200	100
Copper	150	85
Steel	55	35 to 50
Steel (with carbide dies)	...	60
Stainless steel	...	20 to 30
Zinc	150	40

Effect of Lubrication

When two metals are in sliding contact under pressure, as with the dies and the work metal in drawing, galling (pressure welding) of the tools and the work metal is likely. When extreme galling occurs, drawing force increases and becomes unevenly distributed, causing fracture of the workpiece.

The likelihood of pressure welding depends on the amount of force and the work-metal composition. Some work metals are more "sticky" than others. For instance, austenitic stainless steel is more likely to adhere to steel tools than is low-carbon steel.

Lubricants are used in most drawing operations. They range from ordinary machine oil to pigmented compounds.

Selection of lubricant is based primarily on ability to prevent galling, wrinkling or tearing during deep drawing. It is also influenced by ease of application and removal, corrosivity and other factors, as described on page 25 in the article on Lubrication.

If a lubricant cannot be applied uniformly by ordinary shop methods, its purpose is defeated, regardless of its ability to prevent pressure welding. In general, as the effectiveness of a lubricant increases, the difficulty of removing it also increases. For example, grease or oil can be easily removed, whereas special procedures (frequently including some hand scrubbing) are required for removing lubricants that contain zinc oxide, lithopone, white lead, molybdenum disulfide, or graphite.

Sometimes a lubricant is too corrosive for use on certain metals. For example, copper alloys are susceptible to staining by lubricants that contain large amounts of sulfur or chlorine compounds. Lubricants containing lead or zinc compounds are not recommended for drawing stainless steel or heat-resisting alloys, because if not thoroughly removed, the compounds can cause intergranular attack when the workpieces are heat treated or placed in high-temperature service.

Table 12. Effect of Press Speed on Drawability of Low-Carbon Steel Sheet, Cold Rolled, Drawing Quality, Aluminum Killed

Punch nose radius	Reduction in blank diameter: 20%	30%	40%	50%
10 Strokes per Minute				
2*t*	Satisfactory	Some thinning at radius	Fracture at radius	Complete fracture at radius
4*t*	Satisfactory	Satisfactory	Some thinning at radius	Slight fracture at radius
8*t*	Satisfactory	Satisfactory	Satisfactory	Excessive thinning at radius
16*t*	Satisfactory	Satisfactory	Satisfactory	Satisfactory
32*t*	Satisfactory	Satisfactory	Satisfactory	Satisfactory
60 Strokes per Minute				
2*t*	Some thinning at radius	Some thinning at radius	Fracture at radius	....
4*t*	Satisfactory	Satisfactory	Some thinning at radius	Fracture at bottom
8*t*	Satisfactory	Satisfactory	Satisfactory	Excessive thinning at radius
16*t*	Satisfactory	Satisfactory	Satisfactory	Excessive thinning at radius
32*t*	Satisfactory	Satisfactory	Satisfactory	Satisfactory
150 Strokes per Minute				
2*t*	Excessive thinning at radius	Fracture at radius	....	....
4*t*	Some thinning at radius	Some thinning at radius	Fractured at radius	....
8*t*	Satisfactory	Some thinning at radius	Excessive thinning at radius	....
16*t*	Satisfactory	Satisfactory	Satisfactory	Almost complete fracture at radius
32*t*	Satisfactory	Satisfactory	Satisfactory	Partial fracture at radius

Simple round cups, no flange, not ironed, 20-gage sheet, 0.036 in. thick. Blank spaces indicate bottom pushed out.

Suitable safety precautions are necessary with toxic or flammable lubricants.

Some metals, such as magnesium and titanium, are drawn at elevated temperature, which complicates selection of the lubricant. Most oil-base and soap-base lubricants can be used successfully up to 250 F, but above this temperature the choice narrows rapidly. Some special soap-base lubricants can be used on work metals up to 450 F. Molybdenum disulfide and graphite can be used at higher temperatures.

Any lubricant must remain stable, without becoming rancid, when stored for a period of several months at various temperatures.

The cost of application and removal of the lubricant, as well as its initial cost, must be considered, because all these items can add substantially to the cost of the drawn workpieces.

In some plants, when a new application is started, a heavily pigmented drawing lubricant is used, regardless of the difficulty of applying and removing it. Lubricant is then downgraded as much as possible, to simplify the operation and reduce cost. In other plants, the reverse of this practice is used — that is, a simple lubricant, such as machine oil, is used at first, and lubricant is then upgraded when necessary.

In the following example, two similar parts were drawn from heavy stock, using extreme-pressure lubricants. However, a higher-grade lubricant was needed for one part because of embossing in the bottom.

Example 238. Variations in Lubrication for Two Similar Caps (Fig. 27)

Figure 27 shows the top and bottom caps for a tubular part that were made in similar production lines. The top cap, of cold rolled 1008 or 1010 steel 0.125 in. thick, had a 9⁄16-in. hole pierced in its center. The bottom cap also was made of cold rolled 1008 or 1010 steel, but 0.140 in. thick. It had a stud welded to the base and three locating lugs embossed into the bottom. The top cap was lubricated with water-soluble oil-base drawing compound, and the bottom cap with a mixture of wax and a drawing compound.

For each cap, the sequence of operations was about the same:

1 Blank in a 200-ton mechanical press. (Strip for the top cap was lubricated with water-soluble oil-base drawing compound.)
2 Lubricate with drawing compound (dripped on while blanks were moving on a conveyor).
3 Draw in a 190-ton inclined, straight-side press.
4 Pierce hole in top cap (dial index table mounted in a 75-ton press).
5 Coin chamfer.
6 Size in a 75-ton press by pushing through a sizing die.

The example that follows describes an application in which the lubricant was changed from a pigmented drawing compound to a water-soluble chlorinated oil to prevent excessive rise in workpiece temperature during three-stage drawing:

Example 239. Change of Lubricant to Facilitate Handling of Workpiece (Fig. 28)

The oil-filter shell shown in Fig. 28 was drawn in three stages from a cold rolled rimmed steel blank 0.089 in. thick and 13.19-in. diam. The three dies were installed in a single 550-ton mechanical press. Between strokes, the shells were transferred from die to die by hand. Direct redrawing was used for the last two stages, and the bottom was embossed in the last stage. The shell was trimmed to height in a separate operation by a trimming machine.

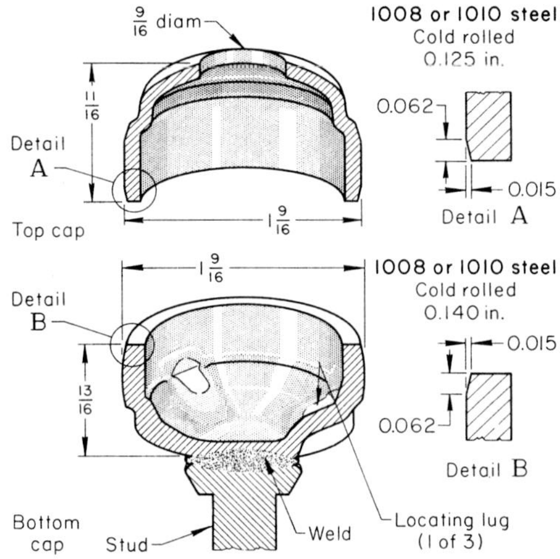

Fig. 27. Caps that required different lubricants for forming (Example 238)

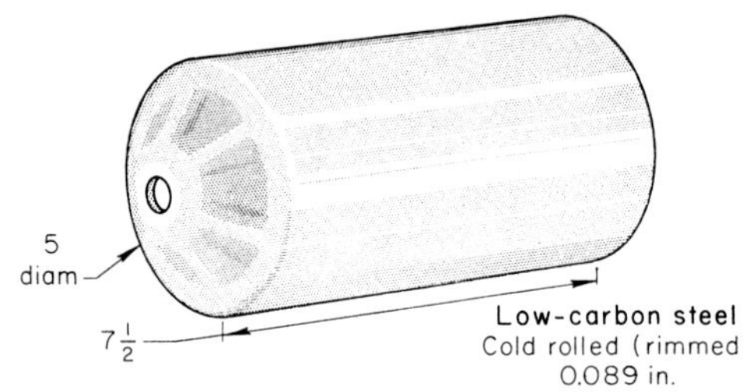

Stock size ....0.0885 ± 0.0035 in. by 39⅝ by 68 in.
Sequence of operations:
Cup to 7½-in. diam by 4 in. deep
First redraw to 6-in. diam by 6 in. deep
Final redraw to 5-in. diam by 7½ in. deep
Tool material:
Draw dies ....7140 alloy steel
Draw punches ....O2 tool steel
Lubricant finally selected ....Water-soluble chlorinated drawing oil
Production lot (average) ....2500 pieces
Production rate ....120 pieces per hour

Fig. 28. Oil-filter shell for which lubricant was changed to reduce workpiece temperature during drawing (Example 239)

Table 13. Typical Lubricants Used in Drawing Steel
(See page 25 for more complete list.)

Mild Severity (10% or Less)(a)

1 Mineral oil, at a viscosity of 250 to 350 SUS
2 5% soap solutions

Average Severity (About 25%)(a)

1 Heavy-duty emulsions
2 Dissimilar metals deposited on steel, using heavy-duty emulsions

Maximum Severity (50% or More)(a)

1 Phosphate or dissimilar metal coatings, with dried soap or wax film
2 Phosphate or dissimilar metal coatings, with compounds containing one or more of materials such as whiting, lithopone, mica, white lead, talc, china clay, or molybdenum disulfide.

(a) Percentage reduction from blank diameter to cup diameter

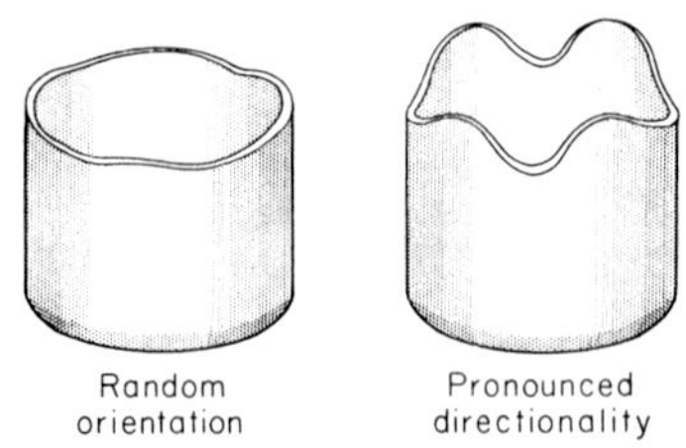

Fig. 29. Effect of grain orientation on earing of drawn 80 Cu – 20 Ni sheet

Two lubricants were tried. With a pigmented drawing compound, results were satisfactory but the temperature of the workpieces rose to 400 F during drawing — hot enough so that the operator had to use asbestos gloves for handling the workpieces. The lubricant was then changed to a water-soluble chlorinated oil, which would keep the parts cool enough so they could be transferred without the use of special gloves. This lubricant was also easier to remove; workpieces were cleaned in a warm alkaline solution before the next operation (brazing).

Tools lasted for three years (about 200,000 pieces). Draw dies were of nitrided 7140 alloy steel and punches of O2 tool steel. The dies did not have to be reground, but occasionally buildup of work metal on the die radius had to be stoned away.

Radii and clearance on the die elements were as follows:

Item	Cupping	First redraw	Second redraw
Die radius, in.	11⁄32	¼	¼
Punch radius, in. ..	⅜	15⁄32	¼
Punch-to-die clearance per side, in. ..	0.090	0.122	0.114
Pressure-pad radius, in.	Flat	13⁄32	¼

The difficulty of removing drawing lubricants is an important consideration in production operations. In a number of applications, changes in drawing techniques (such as increasing the number of draws) or in workpiece design (larger radii, for example) have been made solely to permit use of an easier-to-remove drawing lubricant. (Cleaning methods are discussed in the section of this article on Cleaning of Workpieces, page 187.)

Typical lubricants used in drawing steel are given in Table 13, according to severity of draw, or the percentage reduction from blank to cup diameter. The entries in Table 13 are arranged in the order of increasing effectiveness, but there are exceptions. For example, in some applications, dried soap and wax perform better than pigmented compounds for more severe drawing operations.

For more detailed information on lubricants for drawing of steel and other metals, see the article on page 23 of this volume, and articles dealing with the forming of specific metals.

Zinc phosphate coating of the steel to be drawn is helpful for any drawing operation, and the importance of phosphate coating increases as severity of draw increases. Although the phosphate coating alone has some value as a lubricant, its principal function is to serve as a wick to retain lubricant in the areas where the need is greatest. For methods of application and other details on the use of phosphate coating, see the article that begins on page 531 in Volume 2 of this Handbook.

Effect of Rolling Direction

The mechanical properties and drawability of a sheet or plate depend considerably on its processing history and to a certain extent on the quality of the ingot from which it was rolled. Elongation of the metal during rolling introduces a phenomenon called directionality, anisotropy or, sometimes, fibering. Any wrought product exhibits different properties in different directions. The directionality may be significant for certain applications and immaterial for others.

Gross mechanical directionality seldom hampers the forming of sheet and strip. However, if it is associated with an agglomeration of hard constituents or segregations, mechanical directionality may make thick sheet and plate susceptible to separation along a plane parallel to the surfaces. Such laminations occasionally occur in drawing and heading of cartridge cases.

Crystallographic directionality is common in both annealed and cold worked sheet and strip. The best-known result from directionality is "earing" — the development of ears, or scallops, on a deep drawn cup or tubular part (Fig. 29).

Abnormal earing requires additional trimming, causes scrap loss, and may require special stripping procedures to remove the workpiece from the punch. Sheet with pronounced directionality is also more susceptible to cracking during drawing than sheet with random orientation.

The drawing of a small cylindrical cup is a common test method for evaluating the directionality of sheet metal. Another method is the tear-length test. Slots are cut in various directions on the edges of a piece of sheet to be tested, forming short tabs. Each of these tabs is gripped with pliers or a slotted key, and the metal is torn until a triangular tongue is separated from it. Variation in length of the tongues indicates varying directionality.

The directionality of grain orientation produced by rolling is sometimes a factor in the selection of low-carbon steel sheet for deep drawing. Although it can affect the deep drawing characteristics of heavier-gage material, it is particularly important in deep drawing sheet 0.062 in. thick or less.

Direct Redrawing

Shells that are too deep to be drawn in a single operation are completed by one or more redraws.

In direct redrawing in a single-action die, the drawn cup is slipped over the punch and is loaded in the die, as shown in Fig. 30. At first, the bottom of the cup is wrapped around the punch nose without reducing the diameter of the cylindrical section. Then the sidewall section enters the die and is gradually reduced to its final diameter. Metal flow takes place as the cup is drawn into the die, so that the wall of the redrawn shell is parallel to and deeper than the wall of the cup at the start of the redraw. At the beginning of redrawing, the cup must be supported and guided by a recess in the die, or by a blankholder, to prevent it from tipping, because tipping would result in an uneven shell.

In a single-action redraw, the metal must be thick enough to withstand the compressive forces set up in reducing the cup diameter without wrinkling. Wrinkling can be prevented by the use of an internal blankholder and a double-action press (upper right view in Fig. 30), which usually permits a shell to be formed in fewer operations than by single-action drawing without the use of a blankholder.

Internal blankholders (Fig. 31) are slip-fitted into drawn shells to provide

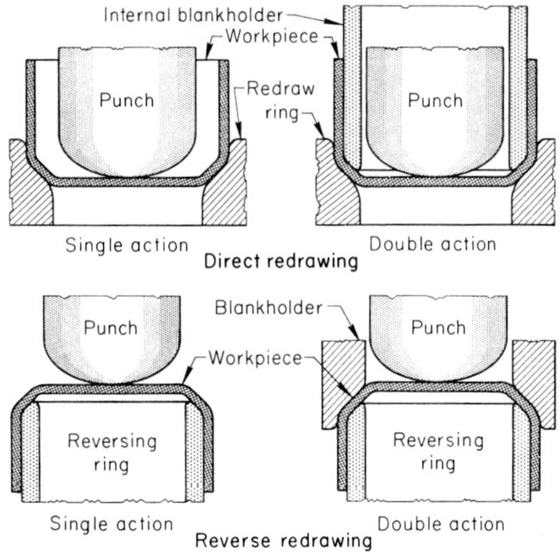

Fig. 30. Direct and reverse redrawing in single-action and double-action dies

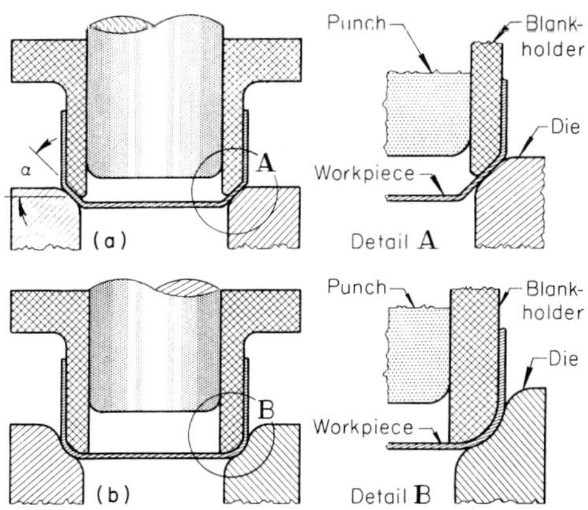

Fig. 31. Setups employing internal blankholders for restraint of metal in redrawing shells. See text for discussion.

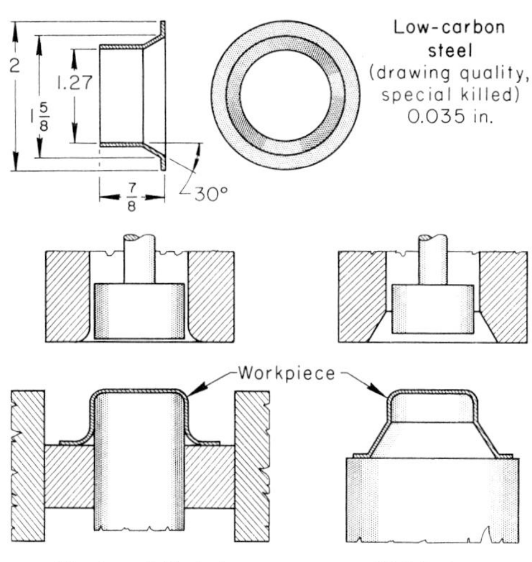

Fig. 32. Ferrule drawn in two operations to minimize wall thinning (Example 240)

support and prevent wrinkling during direct redrawing. The blankholder presses on the drawn shell at the working edge of the die before the punch contacts the bottom of the shell and begins the redraw. It dwells against the shell as the metal is drawn into the die by the punch, preventing wrinkles.

The bottom of the cup to be redrawn can be tapered (Fig. 31a) or radiused (Fig. 31b), with the tip of the blankholder and the mouth of the die designed accordingly. An angle of 30° is used for metal thinner than 1/32 in. and 45° for thicker work metal. A modification of the above is a blankholder fitted against an S-curve die (Fig. 31b). The main disadvantage of an S-curve die is that it is more expensive to make and maintain.

Near the bottom of a redrawn shell, there is usually a narrow ring, caused by the bottom radius of the preceding shell, that is thinner and harder than the adjacent metal. (See Example 290 in the article on Rubber-Pad Forming.)

Redrawing may be required for reasons other than the severity of the drawn shape. The two examples that follow describe redrawing that was necessary to prevent thinning and bulging.

Example 240. Use of Redrawing to Prevent Wall Thinning (Fig. 32)

Figure 32 shows a ferrule and the setups for drawing it in two operations from coiled special-killed steel 2¾ in. wide by 0.035 in. thick. After being drawn, the ferrules were pierced, finish formed, and trimmed in two additional operations.

The ferrules could have been produced in a single draw, but excessive thinning of the walls would have resulted. The additional operation permitted larger radii on tools used for the first draw. The first operation (blanking and drawing) was done in a 50-ton, hand-fed mechanical press with a 3-in. stroke. A 35-ton press with a 3-in. stroke was used for subsequent operations. Forming components of the punch and die were made of O1 tool steel and were hardened to Rockwell C 58, and cutting components were made from the same tool steel and were hardened to Rockwell C 61. Each operation produced 2000 pieces per hour. The lubricant applied to the strip stock was a commercial nonpigmented drawing compound.

Example 241. Change From One to Two Draws to Prevent Bulging (Fig. 33)

A 36-in.-diam cup was originally made in one draw from a circular blank in a double-action mechanical press with a capacity of 700 tons on the ram and 500 tons on the blankholder. Ram stroke was 42 in. and blankholder stroke was 27 in. The cup was produced successfully, but with bulged sides (see "Original method" in Fig. 33).

Bulging was prevented by first drawing to the shape shown in operation 1 of the improved method in Fig. 33, and then restriking to the final shape (operation 2). Punches and dies were made from gray iron with tool steel inserts; average life was 100,000 pieces. Production rate was about 150 pieces per hour. Lubricant was dry soap.

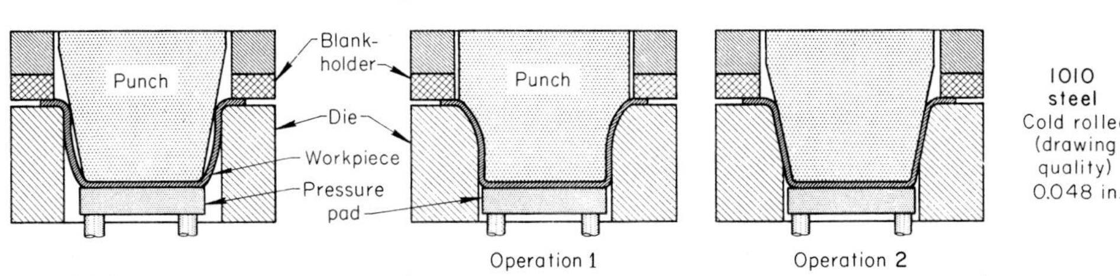

Fig. 33. Change from single draw to two-operation draw that prevented bulging of the sidewalls of a cup (Example 241)

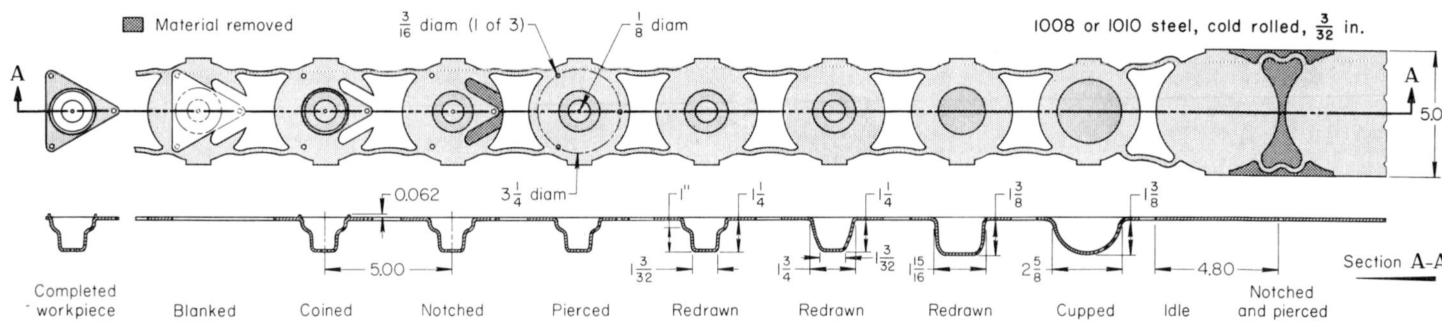

Fig. 34. Drawing a mass-produced automotive spring seat in a progressive die (Example 242)

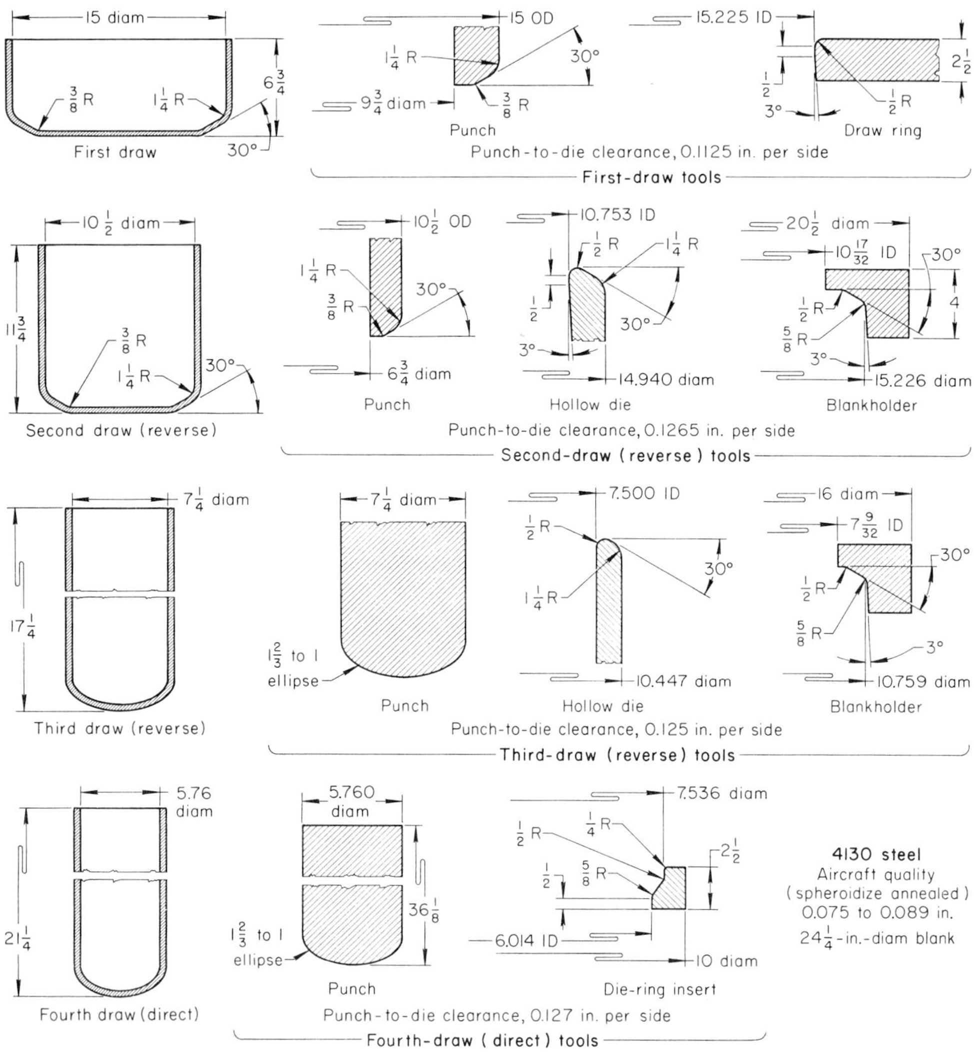

Fig. 35. Progression of shapes produced, and details of tooling used, in forming a deep shell in four draws, including reverse and direct redrawing (Example 243)

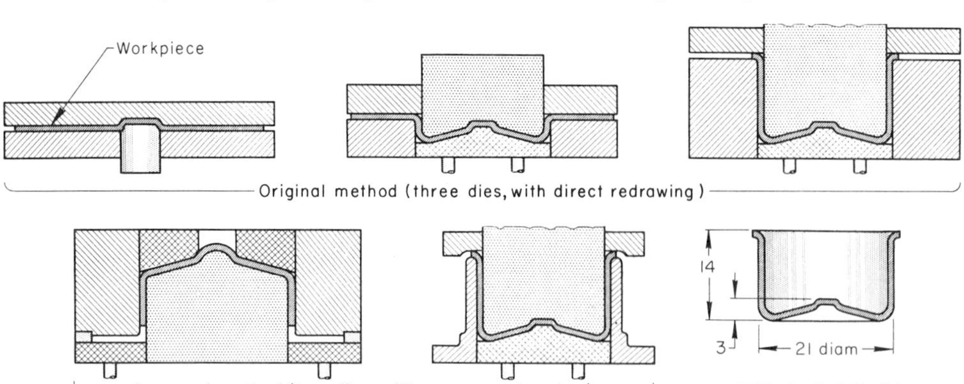

Fig. 36. Original and improved methods of drawing a washing-machine tub (Example 244)

The draw radius for operation 1 of the improved method was 2 in., with a 10-in. radius blending with the draw radius and the straight sides. The depth of draw was 11.4[illegible] in. For operation 2, the draw radius was 0.125 in., and the depth of draw was 11.50 in. The punches for both operations 1 and 2 had a nose radius of 1.195 in.

Redrawing can be done in a progressive die while the part is still attached to the strip. Where space permits the extra stations, the amount of work done in each station will be less than that done in a single die. This reduces the severity of draw and promotes high-speed operation. In the following example, parts were direct redrawn without using an internal blankholder to assist in forming the cup, although a flat blankholder was used to control the metal in the surrounding flange.

Example 242. Redrawing a Spring Seat in a Progressive Die (Fig. 34)

The spring seat shown in Fig. 34 was drawn in a progressive die in quantities of more than a million pieces per year. After cupping, three redraws were used to form the cup to shape, depth and diameter. The die was operated in a 100-ton mechanical press at 60 strokes per minute.

Stock was cold rolled 1008 or 1010 steel 5 in. wide by $\frac{3}{32}$ in. thick. A sulfurized oil was originally used as the lubricant, but it was found that a water-soluble oil did the job and was easier to remove. The lubricant was applied by intermittent spraying.

The sequence of operations in the progressive die was as follows:

1. Pierce and notch blank outline, leaving two narrow deformable carrier strips.
2. Cup.
3. Redraw to reduce diameter.
4. Redraw to reduce diameter and form taper.
5. Redraw to reduce diameter for part of depth only.
6. Pierce four holes (one in bottom of cup).
7. Partially trim outline.
8. Coin ridge around cup.
9. Cut off.

Tool material was D2 tool steel. Tool life was three years or more, but design changes made the tool obsolete before it wore out.

Reverse Redrawing

In reverse redrawing, the cupped workpiece is placed over a reversing ring and redrawn in the direction opposite to that used for drawing the initial cup. As shown in the two lower views in Fig. 30, reverse redrawing can be done with or without a blankholder. The blankholder serves the same purposes as in direct redrawing.

Advantages of reverse redrawing as compared with direct redrawing include: (*a*) drawing and redrawing can be accomplished in one stroke of a triple-action hydraulic press, or of a double-action mechanical press with a die cushion, which can eliminate the need for a second press (Example 246); (*b*) greater reductions per redraw are possible with reverse redrawing (Examples 243 and 244); (*c*) one or more intermediate annealing operations can often be eliminated by using the reverse technique; and (*d*) better distribution of metal can be obtained in a complex shape (Example 245).

In borderline applications, annealing is required between redraws in direct redrawing, but is not needed in reverse redrawing.

Disadvantages of reverse redrawing are: (*a*) the technique is not practical for work metal thicker than ¼ in., and (*b*) reverse redrawing requires a longer stroke than does direct redrawing.

Usually, metals that can be direct redrawn can be reverse redrawn. All of the carbon and low-alloy steels, austenitic and ferritic stainless steels, aluminum alloys, and copper alloys can be reverse redrawn.

Reverse redrawing requires more closely controlled processing than does direct redrawing. This control must begin with the blanks, which should be free from nicks and scratches, especially at the edges.

The restraint in reverse redrawing must be uniform and low. For low friction, polished dies and effective lubrication of the work are needed. Friction is also affected by hold-down pressure and by the shape of the reversing ring. Radii of tools should be as large as practical — ten times the thickness of the work metal if possible.

The examples that follow describe specific conditions where it was advantageous to use reverse redrawing.

Example 243. Use of Reverse Redrawing to Obtain Greater Reduction Than Possible by Direct Redrawing (Fig. 35)

The 5.76-in.-ID by 21¼-in.-deep shell shown in Fig. 35 was drawn from an aircraft-quality spheroidize-annealed 4130 steel blank, 0.075 to 0.089 in. thick by 24¼ in. in diameter, in four drawing operations. The shells were annealed after the first and second redraws. The total reduction in diameter of 76% would have needed six direct redrawing operations for a part four diameters deep. By using the reverse redrawing technique for two operations, acceptable workpieces were made in four operations, as illustrated in Fig. 35. The reduction in diameter for each operation was 38, 30, 31 and 21%, respectively.

The first three operations were done in a double-action hydraulic press at a drawing speed of 200 in. per min, using a blankholder. The direct redraw (operation 4) was done in a single-action press at the same speed but without a blankholder. The forces used for the operations were as follows:

	Force, tons, for operation: 1	2	3	4
Drawing	150	100	100	50
Die cushion	6	25	20	..
Blankholder	60	56	32	..

A phosphate coating was applied to the steel strip, and then the stock was immersion coated with a soap solution and dried before being fed into the blanking die. The shells were relubricated after each anneal.

The shapes of the die elements for the various operations and of the shells after each operation are shown in Fig. 35.

Example 244. Direct vs Reverse Redrawing of Tubs (Fig. 36)

A 21-in.-diam tub for a washing machine was originally produced in three direct draws, as shown in the sequence at the top of Fig. 36. This procedure required three presses, three dies, and three operators. Changing to reverse redrawing permitted the tub to be produced in two draws (sequence shown at the bottom of Fig. 36). By this procedure, cost was reduced, because one press, one die, and one operator were eliminated. Reverse redrawing also eliminated thinning and rings on the diameter, which had occurred in the radiused area at the junction of the flange and the sidewall (formed in the second draw), as the flange was drawn into the sidewall during the third draw.

The material was cold rolled, drawing quality 1010 steel, coated with a dry soap lubricant. Production rate was 130 parts per hour.

The first drawing operation was done in a single-action press. The punch-nose radius was 1.75 in., the die radius was 0.187 in., and punch-to-die clearance was 0.053 in. per side. Depth of draw was 10 in.

Reverse redrawing was done in a double-action press. The punch-nose radius was 2 in., die radius 0.250 in., and punch-to-die clearance 0.085 in. per side.

Tubs having depths of 14 and 16 in. were drawn in the same dies.

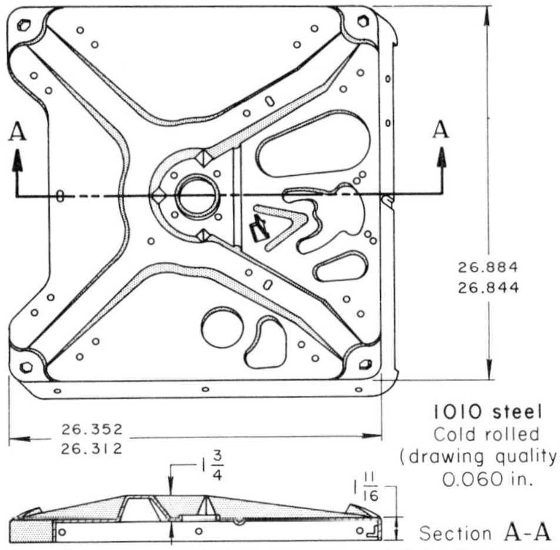

Sequence of Operations

1 Draw to form the center area in a 400-ton hydraulic press.
2 Reverse redraw to form spider in a 700-ton hydraulic press.
3 Wipe down sides and restrike in a 250-ton mechanical press.
4 Trim and pierce sides in a 175-ton trimming machine.
5 Pierce top holes and cam-form front.
6 Form flanges around openings in top.

Operations 5 and 6 were done simultaneously in separate dies in a 200-ton mechanical press.

Fig. 37. Washing-machine base that was produced by drawing and reverse redrawing to provide better distribution of metal (Example 245)

Example 245. Reverse Redrawing for Better Distribution of Metal in a Complex Shape (Fig. 37)

The base of an automatic washer (Fig. 37) required drawing and reverse redrawing for adequate distribution of the metal. Bases were produced from 0.060-in.-thick cold rolled, drawing quality 1010 steel. The first drawing operation provided sufficient metal to form the center of the base before metal was trapped by the spider (leg) sections. The workpiece was then reverse redrawn, which allowed the metal needed for forming the spiders to be drawn from the outside. Drawing was followed by restriking, trimming and piercing the sides, piercing holes in the top and cam-forming the front edge, and flanging the holes in the top.

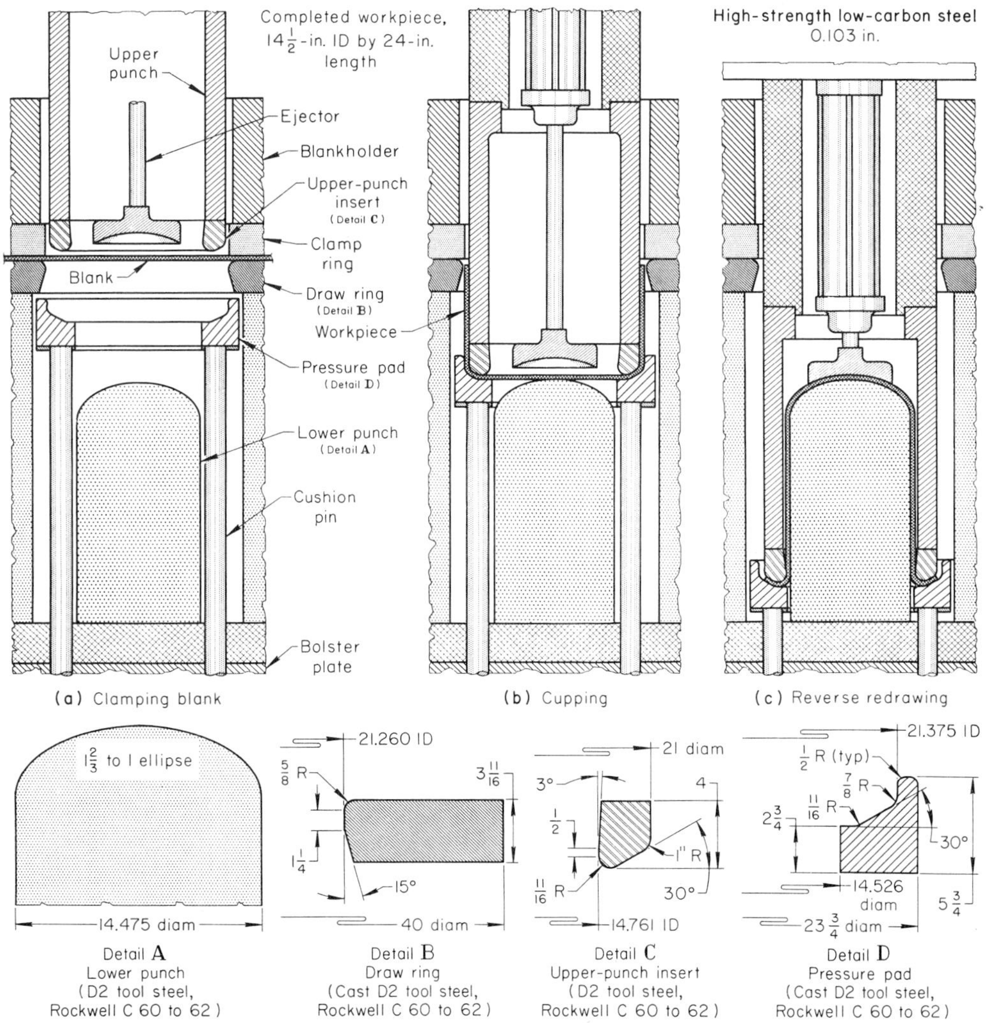

Fig. 38. Drawing a shell in a double-action press with a die cushion (Example 246)

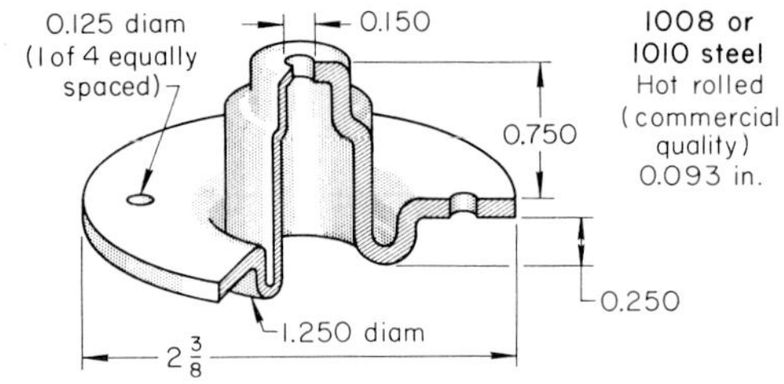

Sequence of operations: 1 – Pierce and notch to outline blank. 2 – Draw. 3 – Redraw. 4 to 8 – Reverse redraw (five stages). 9 – Pierce center hole. 10 – Blank. 11 – Cut scrap.

Fig. 39. Retainer cap that was made using five reverse redraws in an 11-station progressive die (Example 247)

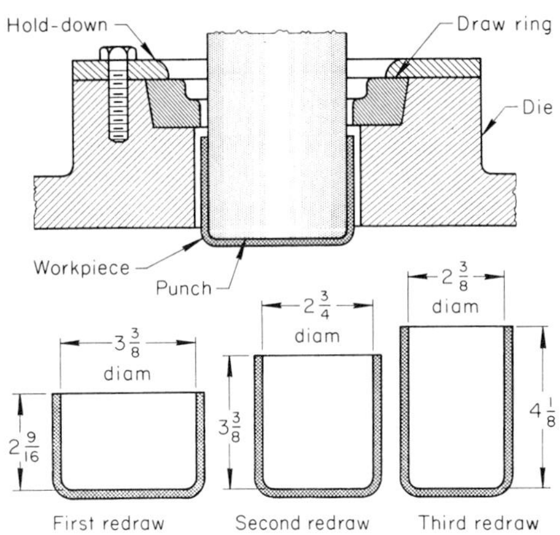

Fig. 40. Die in which draw rings and punches can be replaced for producing three successive redraws

The dies were made of cast iron with tool steel inserts. A dry soap lubricant was used on the blanks. Production rate was 220 bases per hour.

In the following example, reverse redrawing followed cupping to produce a deep drawn shell in one stroke of a hydraulic press. A description of a similar setup is given in Example 4, on page 22 in the article on Selection of Presses. Drawing of a similar part in two operations is described in Example 264 in the present article.

Example 246. Reverse Redrawing in a Double-Action Press for One-Stroke Forming of a Flanged Shell (Fig. 38)

By using a double-action hydraulic press with a die cushion and reverse redrawing, it was possible to produce flanged shells with a 14½-in. ID and a length of 24 in. from 38½-in.-diam flat blanks.

As shown in Fig. 38(a), the lubricated blank was first clamped by the downward stroke of the blankholder slide against the draw ring. The hollow upper punch then advanced to draw the cup, as shown in Fig. 38(b). A pressure pad, supported by die-cushion pins, held the shell against the upper punch during the final stages of cupping and during reverse redrawing.

In a third action, as the upper punch continued its downward motion, the cup was turned inside out over the lower punch by reverse redrawing to produce the flanged part, as shown in Fig. 38(c). Finally, after the upper slides had returned to their starting position, the pressure pad lifted the workpiece from the lower punch. A cylinder on the punch slide pushed the part out of the hollow upper punch.

The material was commercial quality, box-annealed, high-strength low-carbon steel. A dry soap coating was used as a lubricant.

The shapes of the draw ring, pressure pad, and upper punch nose are shown in Fig. 38.

Clearance per side between the draw punch and die for the first operation was 130% of stock thickness, and for the redraw operation, 143% of stock thickness. The force on the upper punch was 350 tons, and on the blankholder and die cushion, 100 tons.

The die surfaces subject to wear were made of D2 tool steel and hardened to Rockwell C 60 to 62.

Reverse redrawing can be done in a progressive die as well as in single-stage dies, provided the operations are divided to distribute the work and reduce the severity of each stage, as in the following example.

Example 247. Progressive-Die Forming of Retainer Cap, Using Reverse Redrawing (Fig. 39)

Reverse redrawing was used in five of the 11 stations of a progressive die used to form the part shown in Fig. 39. Although the completed workpiece shows severe forming, the work was divided among stations so that at no station did the forming present a serious problem. A water-soluble oil-base drawing compound was sprayed on the strip as it entered the die.

The work material was 3¼-in.-wide by 0.093-in.-thick commercial quality hot rolled 1008 or 1010 steel that had been pickled and oiled. The blank was outlined by a combined pierce-and-notch stage, but carrier tabs kept the blank attached to the skeleton until the last stage of the die. A scrap cutter broke up the skeleton.

The drawing tools were mounted in a 100-ton mechanical press that operated at 60 strokes per minute. Considering setup time, coil-loading time, and maintenance, production efficiency was estimated at 75%.

The dies were made of D2 or D5 tool steel. Total tool life was about 8 million pieces.

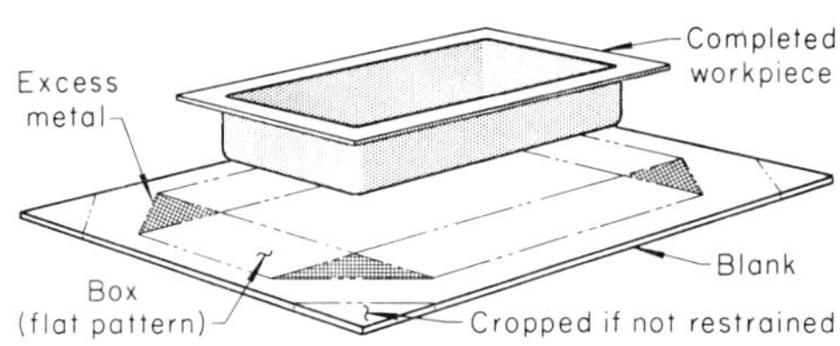

Fig. 41. Flanged rectangular box drawn from a blank with restraint at the corners. See text for discussion.

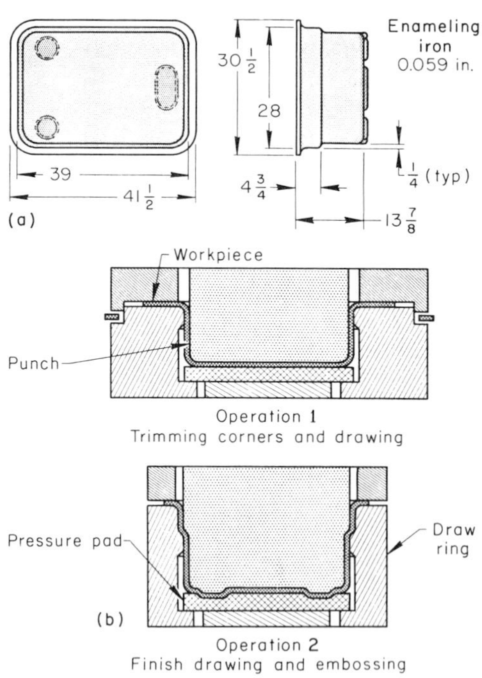

Fig. 42. Large rectangular container produced by two drawing operations in the setups shown (Example 248)

Tooling for Redrawing

Tooling for redrawing depends mainly on the number of parts to be redrawn and on available equipment. In continuous high production, a complete die is used for each redraw; the workpieces are conveyed from press to press until completed. In low or medium production, it is common practice to use a die with replaceable draw rings and punches. A die of this type used for three redrawing operations is shown in Fig. 40; the three redraws were made by changing to successively smaller draw rings and punches. The cup was drawn in a compound blank-and-draw die from a blank 0.067 in. thick and 6¾ in. in diameter.

Boxlike Shells

Square or rectangular shells can be formed by redrawing circular shells when there is no flange. When flanges are required, the difficulty of producing acceptable boxlike shapes by drawing is increased. For deep drawn square or rectangular shells (for example, where the depth is greater than either length or width), the best approach for forming a narrow flange is to allow sufficient stock, and to form the flange after redrawing from a cylindrical shell.

Shallower boxlike shapes (for instance, with proportions similar to the box illustrated in Fig. 41) can be drawn with a flange, which is then trimmed to the desired width. Calculations for the area of a blank used for a circular workpiece cannot be used for a square or rectangular box. These require metal in the bottom, ends, sides and flange, as shown when a box is unfolded ("flat pattern" in Fig. 41). The excess metal at the corners (shaded areas in Fig. 41) is a problem.

A seamless square or rectangular shell is made by drawing metal into the corners. The metal not needed for the corners is pushed into the walls adjacent to the corner radius, and into earlike extensions of the corners. The compressive stresses set up when the metal in the corners is rearranged cause the metal to be thicker in the corners than in the sidewall, or in the original blank. Drawing limits and blank development for square or rectangular shells are given in Appendix 2 at the end of this article.

The more difficult draws are made more easily by using a blank that is carefully developed. There are methods of developing the shape at the corners of a blank for a square or rectangular shell so that there is a minimum of excess metal. However, by cropping the corners as shown in Fig. 41 and by using a blankholder, satisfactory parts generally can be made. Draw beads in the blankholding surface surrounding the die, as in Example 237, are frequently used.

In the following example, the corners of the blank were trimmed and a blankholder was used to draw a large rectangular shell.

Example 248. Stepped Container Produced by Two Drawing Operations (Fig. 42)

The part shown in Fig. 42(a) was drawn in two operations from an enameling iron blank 60 by 70 in. by 0.059 in. thick. The shape after each drawing operation is shown in Fig. 42(b). A double-action press was used to trim the corners and make the first draw. The same press was used for drawing the $4\frac{3}{4}$-in.-deep step and embossing the bottom. The end of the second-operation punch fitted the previously drawn shell and drew metal from the flange to finish the part. A pressure pad held the part against the punch during the second operation and ejected the finished part from the die. It also served as a die to emboss three pads in the bottom of the part. Embossing the bottom eliminated an oil-canning condition. After the two drawing operations, workpieces were inverted for piercing 24 holes in the flange and trimming excess metal from the flange.

The double-action mechanical presses used had 700-ton capacity and a 42-in. stroke on the ram, and 500-ton capacity and a 27-in. stroke on the blankholder. Dies were made from gray iron, and drawing surfaces were flame hardened. Setup time was 4 hr. Dies produced approximately 100,000 parts before reconditioning (not including polishing of radii). Production rate was 80 to 100 parts per hour.

The draw radius for operation 1 was 0.62 in. along the sides, but increased to 0.75 in. at the corners. Punch-to-die clearance was 0.090 in. per side. For operation 2, the draw radius was 0.50 in. and the punch-to-die clearance 0.075 in. per side. The punch nose radius for both operations was 1 in., and the corner radius 3.25 in.

Workpieces With Flanges

Regardless of whether the drawn workpiece is circular, rectangular or unsymmetrical, producing acceptable small-width flanges on workpieces is seldom a problem. Flanged workpieces are usually drawn in two or more operations, frequently with restriking as a final operation. The two examples that follow demonstrate procedures for producing small parts with flanges.

Example 249. Separate Dies for Drawing a Small Flanged Cup (Fig. 43)

The flanged cup shown in Fig. 43 was produced from $3\frac{1}{2}$-in.-wide enameling iron sheet in three consecutive operations: blanking and drawing; redrawing; and restriking. The changes in shape from blank to restrike are shown in Fig. 43. Three open-back inclinable mechanical presses with capacities of 90 to 100 tons and 5-in. strokes were used. A die cushion was used for the first operation. Each operation could produce approximately 1500 pieces per hour, using conventional hardened tool steel dies. Lubricant was a compound of zinc phosphate and dry soap.

Example 250. Use of a Progressive Die for Drawing a Small Flanged Ferrule (Fig. 44)

The production of a small flanged ferrule in a five-station progressive die is shown in Fig. 44. The cost per piece by this method was 40% of that estimated for producing the part in separate dies, three of which would have been required. This did not include the cost of setup, which would be greater for three separate dies.

The progressive die was made of D2 tool steel and was hardened to Rockwell C 60 to 61. A 35-ton press with a 1-in. stroke and hand feed produced 1500 pieces per hour from $1\frac{1}{2}$-in.-wide coil stock. The die produced 10,000 parts before reconditioning was required. Commercial drawing lubricant was supplied by an automatic oiler.

Cylindrical workpieces with wide flanges are troublesome to draw because of excessive wrinkling or fracturing in the sidewall due to lack of metal flow. Even though the metal is restrained by a blankholder, it is difficult to obtain acceptable flatness without special procedures.

Wide flanges on relatively large workpieces can be made flat by coining after drawing. Another means of dealing with wrinkling, when design permits, is to provide ribs in the flange. (This controls the wrinkling by allowing space for the excess metal.) Ribs are usually spaced radially around the flange (as in the example that follows), although circular, concentric ribs also are effective.

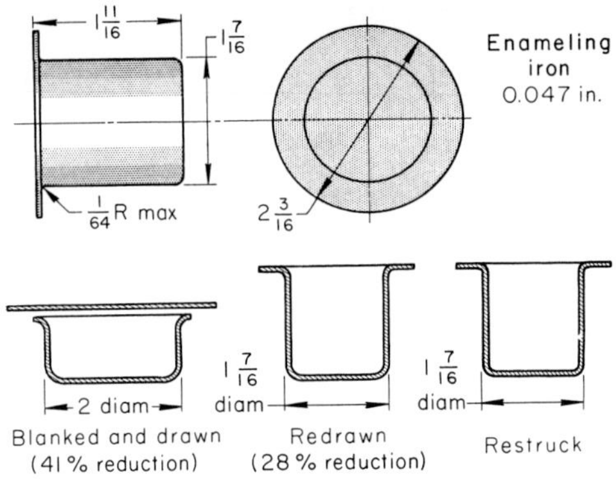

Fig. 43. Flanged cup drawn in three operations in separate dies (Example 249)

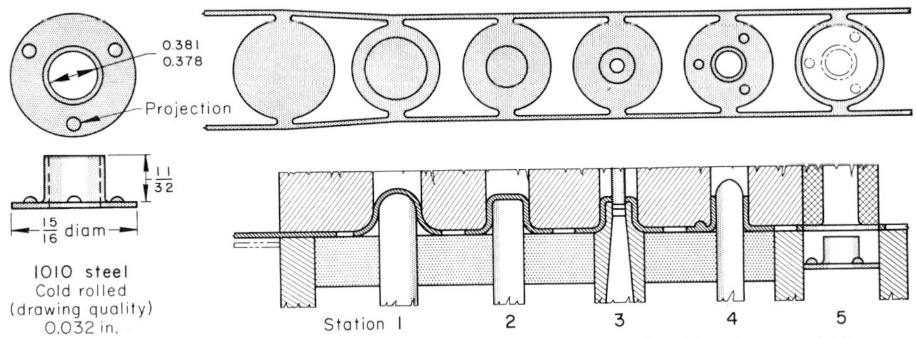

Fig. 44. Production of a small flanged ferrule in a progressive die (Example 250)

Example 251. Addition of Ribs to a Wide Flange for Attaining Specified Flatness (Fig. 45)

The wide-flange workpiece shown in Fig. 45 was originally designed with a flat flange and was drawn using a conventional blankholder. By this method, flatness of the flange could not be held within 0.030 in. TIR as specified. However, by adding the eight ribs shown in Fig. 45, which were formed by the die, it was possible to obtain acceptable flatness. Twenty-five ribbed workpieces were measured for flange flatness; the pieces were supported on three equally spaced points, and measurements were taken on a 7.75-in.-diam circle. Twenty-four had flatness within 0.026 in.; the twenty-fifth was only slightly out of the specified range (0.034 in.).

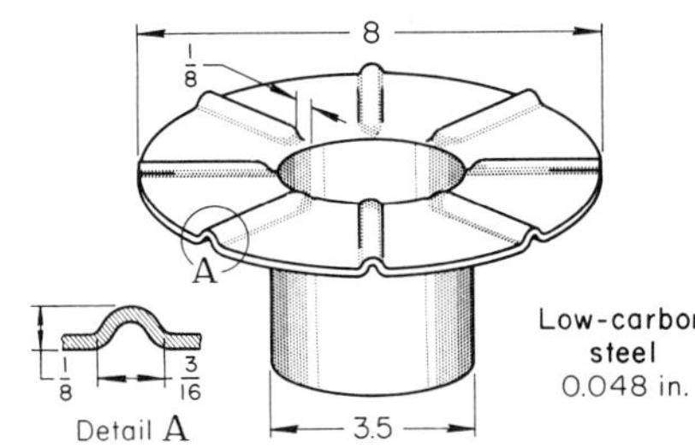

Fig. 45. Part on which addition of ribs to the wide flange enabled it to be drawn to specified flatness (Example 251)

Rectangular, boxlike workpieces that have flanges are difficult to redraw in such a manner that the flange is unaffected in redrawing operations. For this reason, it is common practice to draw the part first to a shallower depth and with larger bottom radii than needed for shaping the final contour. The part is then re-formed in a final operation. The following example demonstrates one procedure for maintaining flatness in a flanged workpiece.

Example 252. Maintaining Flatness Within 0.010 In. on a Flanged Rectangular Box (Fig. 46)

One of the requirements for the workpiece shown in Fig. 46 was that the bottom and end should be flat within 0.010 in. The depth of draw was established by trial, and was assured in each setup by the position of the knockout pad that formed the bottom of the die. The cavity usually was about $\frac{1}{32}$ in. deeper than the workpiece. The shape of the blank used was determined by trial. The blank size and shape, in conjunction with the mating of the blankholder surfaces, were important in minimizing wrinkling. After the blank size and its location on the blankholder surface were determined, four spring-loaded locating pins were placed in the die. The blankholder surfaces were spotted-in from a drawn part.

Along three sides, the bottom radius of the finished part was $\frac{3}{64}$ in., or equal to stock thickness; therefore, two draw punches were used. The first punch had a nose radius four times stock thickness ($\frac{3}{16}$ in.). The box was drawn to approximately $\frac{3}{64}$ in. from the bottom, and was then annealed and restruck with the second punch, which had a nose radius to suit the finished-part dimensions. The flange was trimmed as shown in the view at the lower left of Fig. 46.

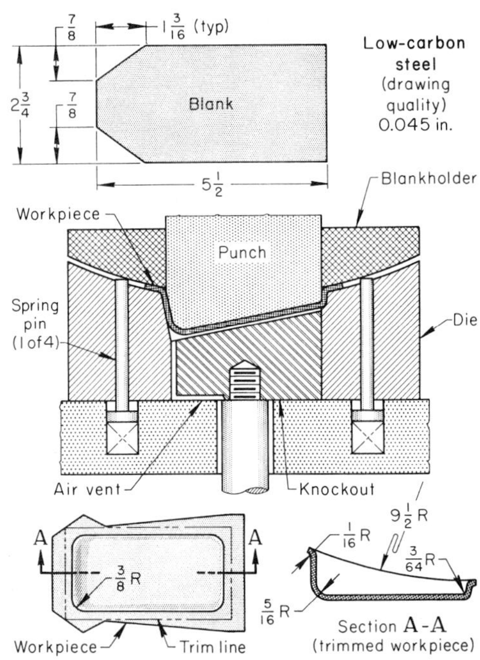

Fig. 46. Setup used for drawing a flanged rectangular box to flatness within 0.010 in. for bottom and end (Example 252)

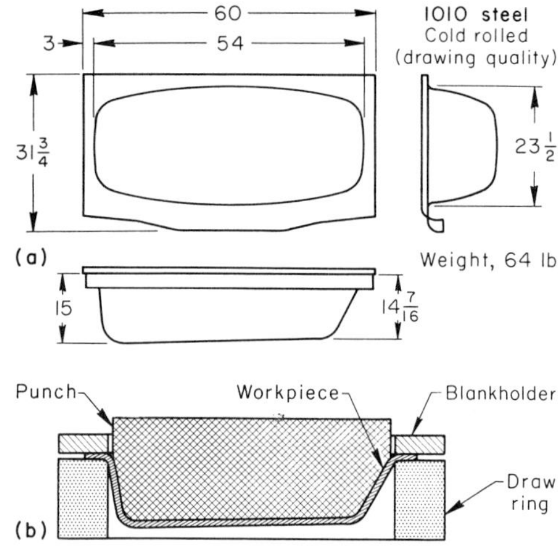

Fig. 47. Bathtub completed in one draw except for forming flange and trimming (Example 253)

The part was drawn in a double-action press. Hardened tool steels were used for the punch, die, blankholder and knockout.

Unsymmetrical workpieces that have flanges are often difficult to draw, particularly when neither draw beads in the die nor ribs in the workpiece can be permitted. Under these conditions, considerable development is usually required to determine the blankholder pressure that will result in the desired metal flow without using a larger blank than necessary. Drawing of bathtubs, for example, requires careful control of the die surfaces and blankholding pressure, even though the tubs can be drawn in one operation, as in the following example.

Example 253. Forming a Bathtub in One Operation (Fig. 47)

The bathtub shown in Fig. 47(a) was made from cold rolled, drawing quality 1010 steel in a single drawing operation. The flange was formed and trimmed in a subsequent operation. Because the bathtub was unsymmetrical and neither draw beads in the die nor ribs in the workpiece were permitted, metal flow had to be controlled otherwise. Design of the tub permitted no straight lines or flat surfaces, and constantly changing curves kept the metal stretching, so that no "loose" areas or oil canning would develop. In addition, extreme care was used in spotting-in the draw ring and the blankholder, and in balancing them with blankholder pressure. Blanks were no larger than necessary, and corners were trimmed before drawing. Bathtubs were produced in quantities of 100,000 to one million at a rate of 100 per hour.

Complex Shapes

Several operations are usually required in drawing complex shapes. However, for mass production, it is usually feasible to combine press operations, so that extremely complex shapes can be produced in no more than six or seven separate operations. For small complex parts, progressive dies are generally used, but for parts as large as the automobile wheel disk described in the following example, progressive dies would not be practical.

Example 254. Drawing a Complex Shape From 0.136-In. Steel (Fig. 48)

Automobile wheels were produced by welding a drawn outer disk to a roll formed inner rim (Fig. 48). Outer disks for 15-in. wheels were completed in six operations (including one reverse redraw), as shown in Fig. 48.

The first operation (blanking and drawing) was done in a 1200-ton press with a 20-in. stroke. The center portion of the disk was reverse redrawn in an 800-ton press with a 14-in. stroke. A 500-ton press with a 14-in. stroke was used for the first reduction, and an 800-ton press with a 16-in. stroke for the second reduction. A 1500-ton press with a 30-in. stroke was used for the fifth operation; then the disks were transferred to a 500-ton, 14-in.-stroke press for piercing and coining.

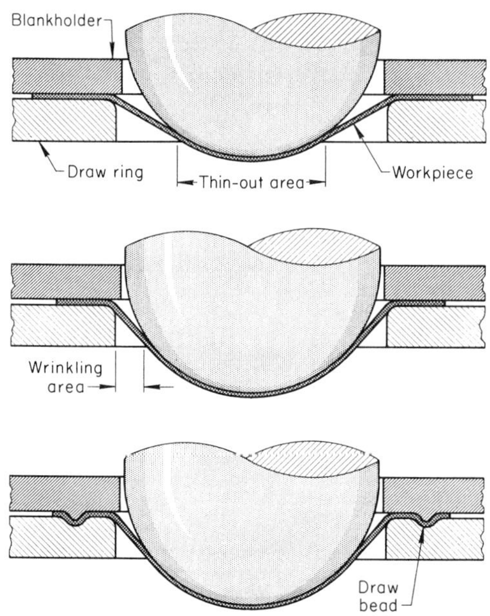

Fig. 49. Drawing a hemisphere with and without a draw bead. Text discusses effect of restraint on thin-out and wrinkling.

Hemispheres

In the drawing of a hemisphere, metal flow must be closely controlled for balance between excessive thinning in one area and wrinkling in another. In the upper sketch of Fig. 49, the punch has begun to stretch the round blank, which is restrained by the blankholder, and the crown section of the hemisphere is being formed. At this stage, the crown section is subjected to biaxial tension, which results in metal thinning. With correct pressure on the blankholder, thinning is in the range of 10 to 15%. More than 15% thinning is likely to result in fracture of the crown section. In the upper sketch of Fig. 49, the portion of the blank under the blankholder has not begun to move.

As the drawing operation continues, the metal begins to move from the blankholder and a different problem develops (center sketch of Fig. 49). Here the metal has been drawn into a partial hemisphere with unsupported metal in a tangential slope between the punch and the clamped surface. Unlike the drawing of straight-sided shapes, having the wide gap (indicated in Fig. 49 as wrinkling area) prevents the use of the draw-ring bore as a means of forcing the metal against the punch surface, and thus the probability of wrinkling increases. Because the metal cannot be confined between the punch and die, wrinkling is likely to occur in this area. To prevent wrinkles, the metal must flow from the flange area and, at the same time, must be securely held in tension. This requires an additional stretching force, derived from the portion of the blank that remains clamped. The area of metal between the clamping surfaces is gradually reduced as the punch advances, but the draw radius offers some resistance because the metal must follow a sharper bend as it moves into the die. One means of controlling wrinkling is by the use of draw beads, as shown in the lower sketch of Fig. 49, and another means is by a sharp draw radius. Small radii are susceptible to metal pickup and, depending on sharpness, can produce undesirable circumferential grooves in the hemisphere if the punch does not move at a steady rate.

In the following example, metal flow was controlled by a draw bead that was outside the trim line.

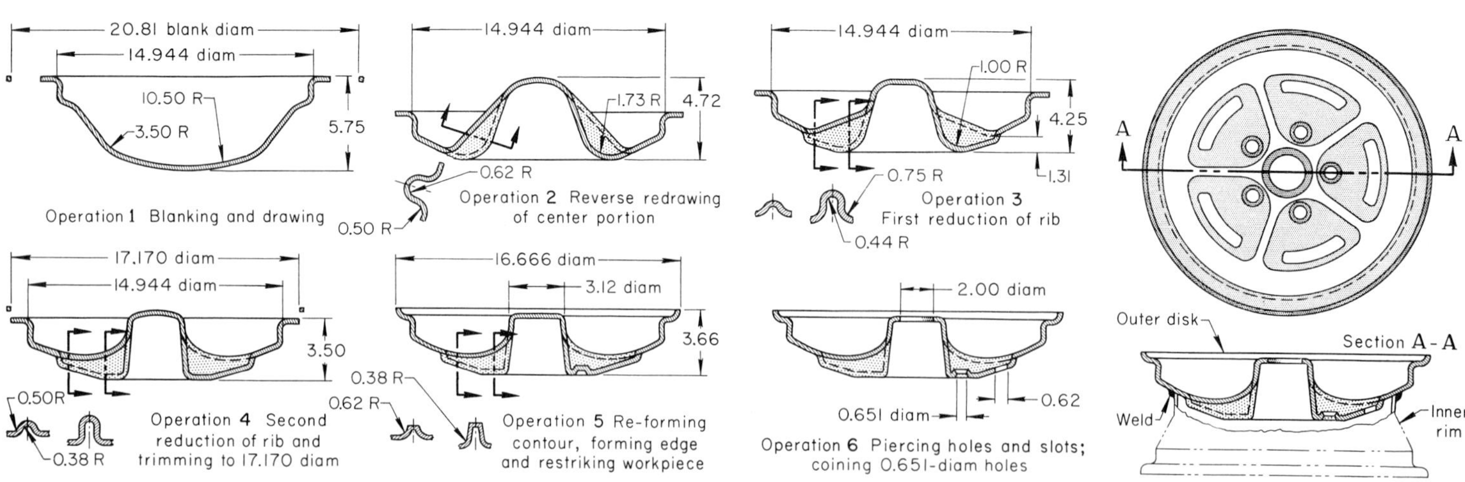

Fig. 48. Production of the outer disk of an automobile wheel in six operations (Example 254)

Example 255. Maximum Thinning of 13% Maintained by a Draw Bead (Fig. 50)

The 37-in.-ID hemisphere shown in Fig. 50 was originally drawn by simply clamping the blank between the blankholder and the draw ring. By this procedure, it was impossible to produce hemispheres free of wrinkles without having areas that were thinner than 0.211 in. (the minimum allowable thickness).

By using a draw bead in the draw ring and blankholder like that illustrated in the bottom view in Fig. 49, it was possible to produce hemispheres that were free of wrinkles and thicker in all areas than the allowable minimum. Figure 50 shows 31 thickness measurements on a typical hemisphere produced with a draw bead. Note that the minimum thickness reading (0.221 in.) was only about 13% less than the 0.255-in. nominal thickness of the blank.

The material was ASTM A285, grade C steel (hot rolled, flange quality). The lubricant used was a mixture of equal parts of chlorinated and sulfurized oils.

The press had a maximum blankholder pressure of 600 tons. Although hemispheres free of wrinkles could not be produced with flat draw rings, with draw beads the parts were successfully produced free of wrinkles or draw-ring marks using 500 to 600 tons of blankholding force.

The blankholder and draw ring were made of an abrasion-resistant cast steel and were hardened to Rockwell C 58 to 63. Figure 50 shows the draw bead and draw radius that were machined into the die.

In the following example, special killed, drawing quality steel was used to minimize thinning in a dome-shape deep drawn housing.

Example 256. Drawing a Compressor Housing in a Compound Die (Fig. 51)

The lower compressor housing for a refrigeration unit shown in Fig. 51 was made in a compound blank-and-draw die. The part was deep drawn from 1008 steel (hot rolled, drawing quality, special killed), 0.120 in. thick. Coil stock 15⅞ in. wide was fed into a 300-ton press, which produced 350 parts per hour. The special killed steel permitted drawing the part to a depth of 7¼ in. in one operation, while also forming several embossments (not shown) on the bottom of the housing without excessive loss because of cracking or thinning. The amount of thinning was important because the operating pressure for the housing assembly was 650 psi. Tests on the assembly were conducted at 1300 psi.

The blanking and drawing components of the die were made of A2 tool steel and were hardened to Rockwell C 56 to 58. Clearance between the blanking punch and die was 0.008 in. (about 7% of stock thickness) per side.

Clearance between the draw punch and draw ring was 0.125 in. per side, and the punch had a back taper of 0.001 in. per inch to compensate for thickening of the metal in the sidewalls. Draw radius was 0.50 in. A pigmented drawing compound was used as the lubricant.

After drawing, the shell was trimmed to a height of 6⅝ in. in a trimming machine. Annual production was one million housings.

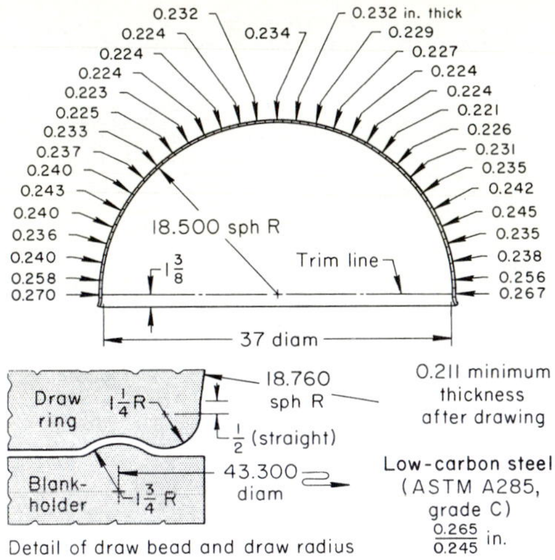

Fig. 50. Variation in wall thickness of a hemisphere drawn with the use of a draw bead (Example 255)

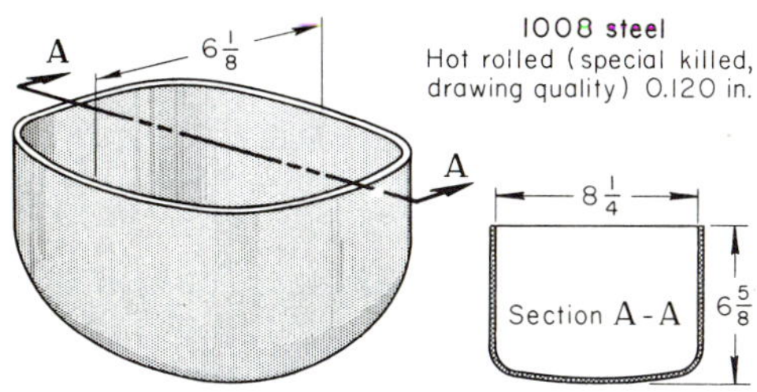

Fig. 51. Compressor housing that was blanked and drawn in a compound die in one operation (Example 256)

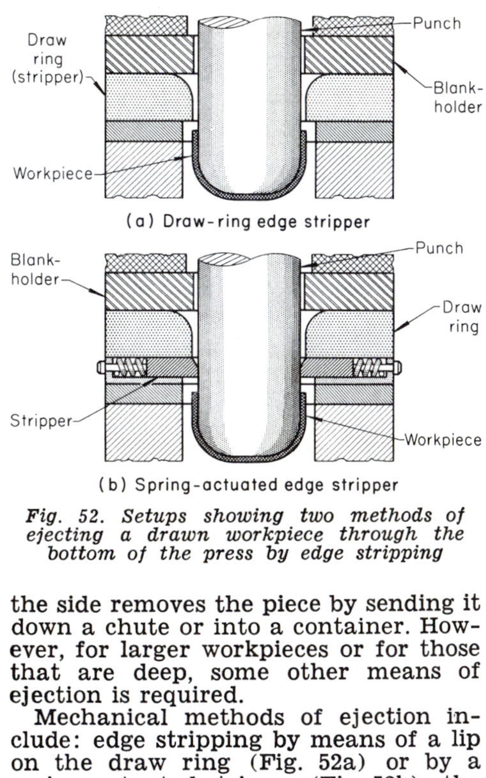

Fig. 52. Setups showing two methods of ejecting a drawn workpiece through the bottom of the press by edge stripping

Ejection of Workpieces

In drawing operations, the drawn workpiece may adhere to either the punch or the die. Adherence is increased by depth of draw, straightness of workpiece walls, and viscosity of lubricant.

The simplest means of ejecting a small workpiece is by compressed air through jets in the punch or the die. Timed air blast is widely used for ejecting relatively small workpieces — for example, where cup diameter is no greater than 4 or 5 in. In some production drawing operations, not only is the workpiece ejected by compressed air, but another timed blast of air from the side removes the piece by sending it down a chute or into a container. However, for larger workpieces or for those that are deep, some other means of ejection is required.

Mechanical methods of ejection include: edge stripping by means of a lip on the draw ring (Fig. 52a) or by a spring-actuated stripper (Fig. 52b); the use of a blankholder in combination with an upper ejector (Fig. 53); and the use of a lower ejector in combination with an upper stripper ring (Fig. 54).

Numerous other mechanical methods using cams or links have been devised to meet specific requirements. These methods are usually modifications of those described above. For instance, thin shells are sometimes stripped from punches near the top of the press stroke by a cam-actuated rod that extends through the punch. This method often is used to avoid damage to the open end of the shell, which can occur when the piece is ejected by other methods.

Major factors influencing the method of ejection are workpiece design (especially the presence or absence of a flange), work-metal composition and thickness, and the type of equipment available.

Ejection by Edge Stripping. The simplest means of mechanical ejection is illustrated in Fig. 52(a). Here, the cup is drawn in a double-action press using a blankholder, and with relief between the bore of the draw ring and its support. As the edge of the workpiece passes below this relief, slight springback is sufficient for the workpiece to catch on the lower edge of the draw ring and be stripped from the punch on its return stroke. The workpiece is removed from the bottom of the press. However, there are four requirements for success with this method: (*a*) there must be an opening in the press bed and the bolster plate, or the die must be raised above the bolster plate; (*b*) the workpiece must permit straight-through ejection (no flange); (*c*) the work metal must have some springback; and (*d*) work metal must be sufficiently thick.

Minimum work-metal thickness that permits use of the above procedure depends on work-metal characteristics and on punch-to-draw-ring clearance. There must be enough springback and sufficient work-metal thickness to catch the lower edge of the draw ring. In the drawing of steel, which always has some springback, the minimum shell wall thickness that allows simple edge stripping is usually about 18 gage (0.0478 in.). For the drawing of soft nonferrous metals, the thickness should be greater than 18 gage.

A modification of the draw-ring-edge method is shown in Fig. 52(b). The principle is the same, except that a spring-loaded stripper is used. This refinement in method allows stripping of thin-wall workpieces. However, requirements for workpiece design and type of press are the same as for the simple edge stripper. When this procedure is used, two strippers mounted 180° apart are usually sufficient; for stripping large workpieces, three or four strippers equally spaced around the draw-ring periphery are ordinarily used.

Ejection From Inverted Dies. The edge-stripping principle can also be used as an aid in ejecting from an inverted die if the workpiece does not have a flange. The top view in Fig. 53 shows how the sheet metal is drawn over a stationary lower punch. In the reverse stroke (bottom view in Fig. 53), the blankholder moves upward, lifting the workpiece off the punch. Movement of the blankholder is usually delayed to avoid pushing the part into the draw ring. If the part remains inside the draw ring, an ejector actuated by a cam or by a knockout bar ejects the workpiece from the draw ring (middle view in Fig. 53). This design functions equally well on parts with a flange.

Upward Ejection. Presses with a die cushion or without an opening in the bed do not permit ejection through the press bed even though the workpiece shape permits straight-through ejection. Also, a flange on the workpiece precludes straight-through ejection, regardless of the press design. Under these conditions, the workpiece can be ejected by the method shown in Fig. 54. Here, the workpiece is drawn downward through the draw ring, as indicated in the upper view in Fig. 54. On the return stroke (Fig. 54, lower view), the workpiece is forced upward by the ejector; if the workpiece adheres to the punch, it is stripped off the punch by the stripper ring, as shown. The stripper ring is held in position by three or four support legs. The procedure illustrated in Fig. 54 is adaptable only to single-action drawing or redrawing. The support legs must be positioned so that the die can be loaded and unloaded.

Developed Blanks vs Final Trimming

The most important factor that influences a choice between using a developed blank or a final trimming operation is whether or not the shape of the drawn edge is acceptable. Sometimes a semideveloped blank is necessary to draw an acceptable part, and the edge must be trimmed to meet dimensional tolerances.

The next consideration is the cost of blanking versus final trimming. This would include the adaptability of the process to the available equipment, based on expected production requirements. The principal advantage of a developed blank is that strip or coiled work metal can be used. The use of strip eliminates the need for shearing the work metal to a rough-blank shape, as is sometimes required when final trimming is used. The developed-blank approach is usually more economical than final trimming, because a blanking die is frequently less expensive than a final-trimming die.

When using developed blanks, the draw dies are made and several blanks are drawn to select the optimum developed shape before the blanking die is made. This delays putting the draw die into production. However, with proper planning and scheduling, this should not be a problem.

Another disadvantage of developed blanks occurs when variations in work-metal properties and thickness are sufficient to affect the uniformity of the drawn workpiece. Under these conditions, closer tolerances are obtained by final trimming.

It is possible to develop blank contours accurately enough so that the outline of the drawn part is within tolerance, thus avoiding a final-trimming operation, as in this example:

Example 257. Developed Blank vs Final Trimming for Table Bases (Fig. 55)

The table base shown in Fig. 55 was producible either from a developed blank or by final trimming; a study was made to determine the more economical procedure. Because of work-metal thickness (3/16 in.) and part shape, both roll and pinch trimming were ruled out. Therefore, the study was made on the basis of trimming a partly formed workpiece, followed by a wipe-down operation in the die.

For either method, three separate sets of dies were required: dies for blanking, drawing and piercing for the developed-blank method; and dies for partial drawing, combination trimming and piercing, and final drawing (wiping down) for the final-trimming method. For both methods, available space and equipment permitted the use of dies in progression, so that a completed workpiece was obtained with each stroke of the press. It was estimated that the three dies used for the trimming method would cost 30% more than those for the developed-blank method. In addition, shearing would increase production cost of the workpiece by 500% over the cost of the developed-blank method. This large difference was caused by the necessity of preparing rectangular blanks with trimmed corners for the trimming method, whereas strip metal was used for the developed-blank method. To minimize cost, it was decided to use the developed-blank method, despite a delay in the start of production. Workpieces were less accurate than if they had been finally trimmed, but the variations obtained were acceptable.

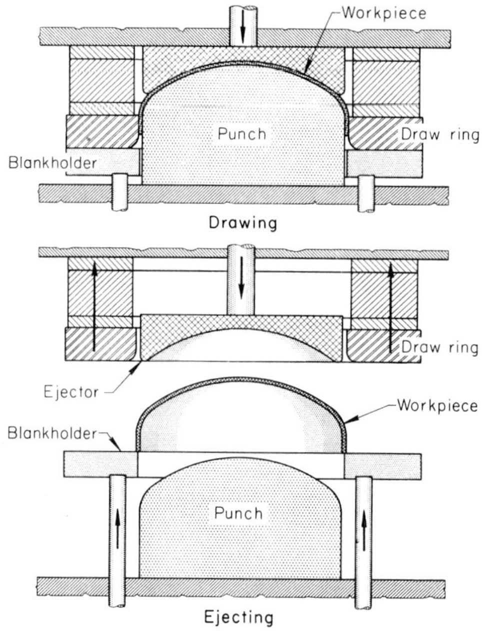

Fig. 53. Use of a blankholder and ejector for stripping of a drawn workpiece from inverted dies

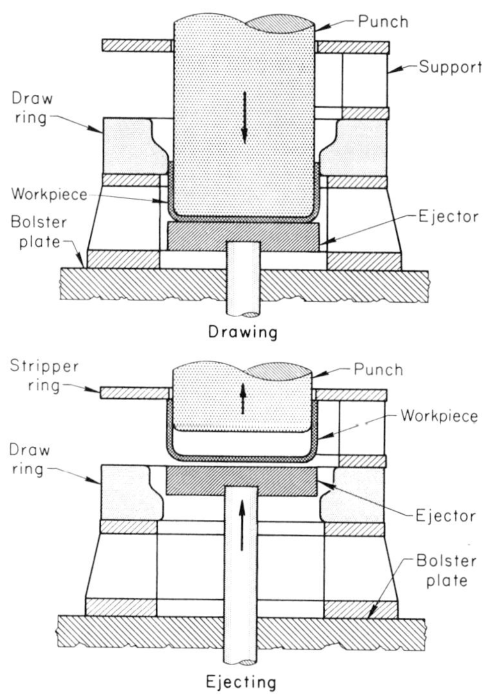

Fig. 54. Setup in which a drawn workpiece is lifted from the die by an ejector and is stripped off punch by a fixed stripper ring

Trimming

Trimming in a lathe (using a cutting tool), roll trimming in a lathe, rotary shearing, die trimming (regular and pinch), and trimming on special machines are the methods most commonly used for trimming drawn workpieces. In one plant, the costs of trimming methods (not including tool cost) are rated according to the following scale of percentages:

Lathe trimming (cutting tool)	100%
Lathe trimming (roll)	30
Rotary shearing	30
Die trimming	5 to 20
Special-machine trimming	10

Methods for Specific Shapes. Cylindrical workpieces like the one in Fig. 56(a) can be trimmed by at least four different methods:

1 In a lathe, with a cutting tool, but production cost is high.
2 By roll trimming in a lathe or in a rotary shear. Production cost is lower than for trimming with a cutting tool, but the finish of a rolled edge is poor and maintenance cost of the rolls is high.
3 By pinch trimming in the press at the bottom of the drawing stroke. This entails almost no increase in production cost, but requires a more expensive die. This method produces a thinned edge at the trim line, which may be unacceptable.
4 In a shimmy die or trimming machine, but production quantities must be high to warrant the investment.

Cylindrical workpieces with flanges, like the one in Fig. 56(b), can also be trimmed in a lathe, although shapes like this are ideal for trimming in a die and can be die trimmed for about 5% of the cost of trimming in a lathe. Rotary shearing can also be used for trimming circular drawn parts with flanges, provided the dimensional tolerance is 0.030 in. or more.

Drawn workpieces with an irregular trim line, as in Fig. 56(c), can be trimmed in a die, for low-production requirements, or with a shimmy die or trimming machine, for high-production requirements. The cost of a trimming die is about half that for a special trimmer (excluding the cost of the original machine). However, the trimming cost per piece with the special trimmer will be only about half the per-piece cost with multiple dies, and the trimmed edges will be better.

Flanged workpieces like the one in Fig. 56(d) can be trimmed in a die for 5% of the cost of trimming in a rotary shear. In low production, drawn workpieces of such a shape are frequently trimmed in a nibbler and filed to conform to a template. This means of trimming costs up to 60 times as much as die trimming.

Reducing Drawn Shells

Necking and nosing are used for reducing the diameter of a drawn cup or shell for a part of its height.

Necking. By the die-reduction method, the work metal is forced into compression, resulting in an increase in ength and wall thickness.

The thicknesses of a shell before and after necking are related by the following formula:

$$t_2 = t_1 \sqrt{d_1/d_2}$$

and heights before and after necking, by the formula:

$$h_2 = h_1 \sqrt{d_1/d_2}$$

where t_1 is shell thickness before necking, t_2 is shell thickness in the necked area after necking, d_1 is mean diameter of the shell before necking, d_2 is mean diameter after necking, h_1 is unit of height before necking, and h_2 is unit of height after necking.

Figure 57 shows the flow of metal in a necking operation. As the metal flows, paths of least resistance are taken; therefore, the actual values for t_2 will be less, and for h_2 greater, than those calculated from the formulas.

Necking results are uniformly better if the workpiece has been slightly cold worked. This provides added strength to resist bulging in the column section and buckling in the section being reduced. The entry angle on the necking die is important, because the probability that the metal will collapse is decreased as the angle with the vertical becomes smaller. This angle should be less than 45° (Fig. 57). If the angle is greater than 45°, a series of reductions may be necessary with localized annealing between reducing operations. With a die entry angle less than 45°, thin-wall tubes can be reduced as much as 15% in diameter; thick-wall tubes can be reduced as much as 20%.

The following example describes how a missile component was produced by reducing a drawn shell.

Example 258. Necking a Shell for a Missile Case (Fig. 58)

The drawn shell shown at the left in Fig. 58 had been slightly cold worked by previous ironing of the wall. The setup for necking the shell is illustrated in the center sketch of Fig. 58. The split bottom die, which had a hinged door with a cam latch, supported the shell during necking, and also provided the holding friction needed when the workpiece was stripped from the necking die. The necking die, by use of the entry configuration, wedged the compressed metal into the desired shape and size. The sizing plug controlled the inside diameter and kept it within tolerance, preventing bellmouthing.

The drawn shell was necked as follows:

1. Shell was loaded into the split bottom die, which was closed by locking with a cam.
2. Downstroke of the press formed the neck.
3. Return stroke of the press lifted the necking die and sizing plug.
4. Split bottom die was unclamped and the workpiece was removed.

The necking die and sizing plug were made of D2 tool steel and were hardened to Rockwell C 58 to 60. Tallow was applied to the shell for lubrication.

A 750-ton hydraulic press was used (because of availability, not tonnage requirements).

Nosing reduces the open end of a shell by tapering or rounding the end (usually by cold reduction), and is used chiefly in making ammunition. Shells often are machined before, instead of after, nosing. Shells often are cold reduced as much as 30% of their original diameter by nosing. Typical nosing practice is described in Example 401, in the article "Bending and Forming of Tubing", in this volume.

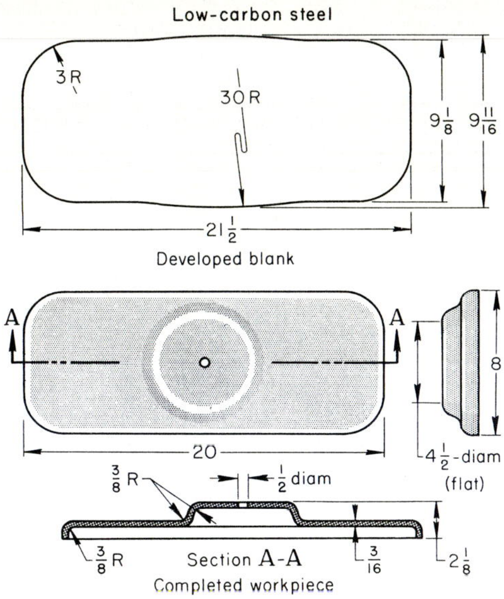

Fig. 55. Table base drawn from a developed blank (Example 257)

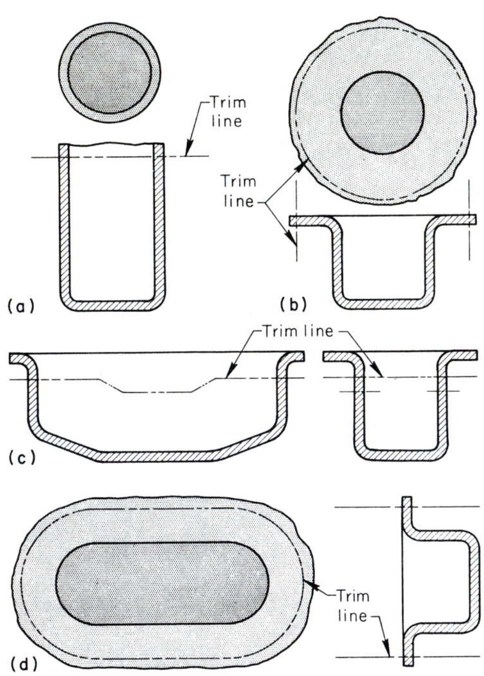

Fig. 56. Typical trim lines on drawn parts. (Text discusses trimming methods applicable to these representative shapes.)

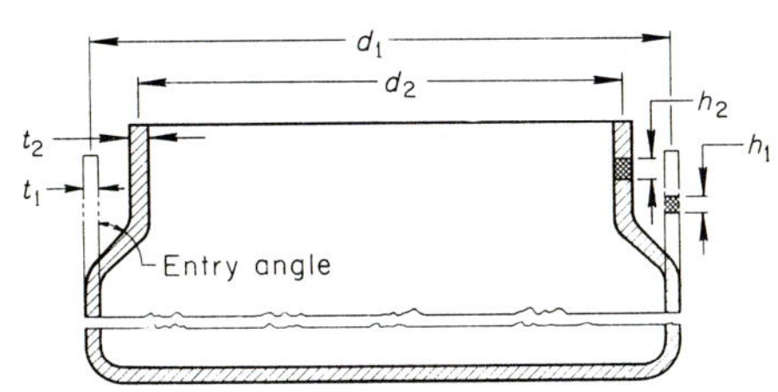

Fig. 57. Flow of metal in the reduction of a drawn shell by necking. See text for formulas and discussion.

Ironing is the intentional reduction in wall thickness of a shell by confining the metal between the punch and the die wall. When ironing occurs, the force to displace the punch often increases to a secondary maximum in the force-displacement curve. The second force maximum can be of such magnitude that the shell will break. However, after ironing has started and metal has been wrapped around the punch, the force is uniform and frequently less than that for redrawing operations.

Ironing is seldom used with redrawing operations unless the amount of wall thinning is relatively small, because it results in excessive die wear, causes workpiece breakage, and increases press tonnage requirements. If a shell with constant wall thickness is needed, however, it can be obtained only by ironing.

In the following example, the wall thickness was reduced during the cupping operation, as well as in two ironing operations.

Example 259. Ironing Used to Make a Thin-Wall, Thick-Bottom Shell (Fig. 59)

The part shown in Fig. 59 was produced by reducing the sidewalls during the cupping operation and two ironing operations. The 12-in.-diam blanks were cut on a 400-ton press from hot rolled, drawing quality 1008 steel 0.188 in. thick. The blanks were phosphate coated, and a pigmented drawing compound was applied before cupping.

A 400-ton press was used to draw a 9-in.-diam cup 1.50 in. deep. The wall was ironed during cupping by using a punch-to-die clearance of 0.150 in. per side. The punch nose radius was 0.25 in. and draw radius was 0.50 in. for this operation. The punch, draw ring and blankholder were made of A2 tool steel and hardened to Rockwell C 56 to 58.

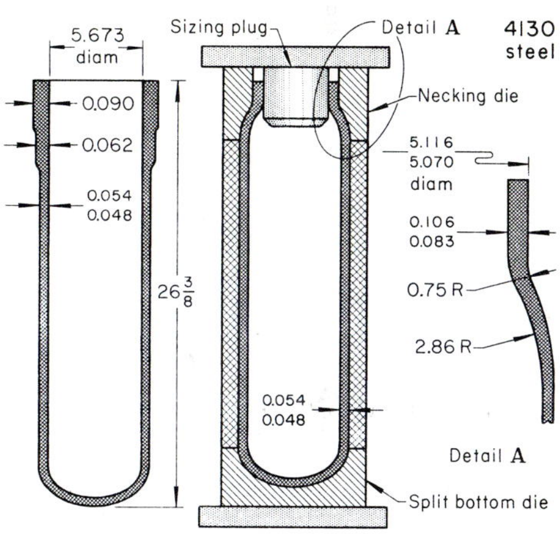

Fig. 58. Reduction of a shell for a missile case by necking (Example 258)

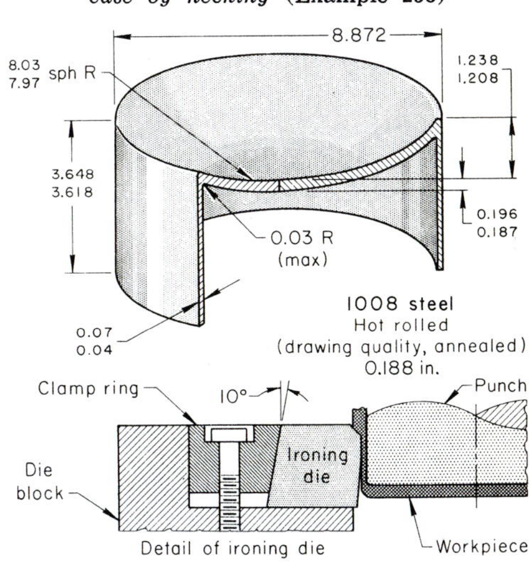

Fig. 59. Cup with a reversed spherical bottom, which was drawn in a 400-ton press, using phosphate-carried drawing compound (Example 259)

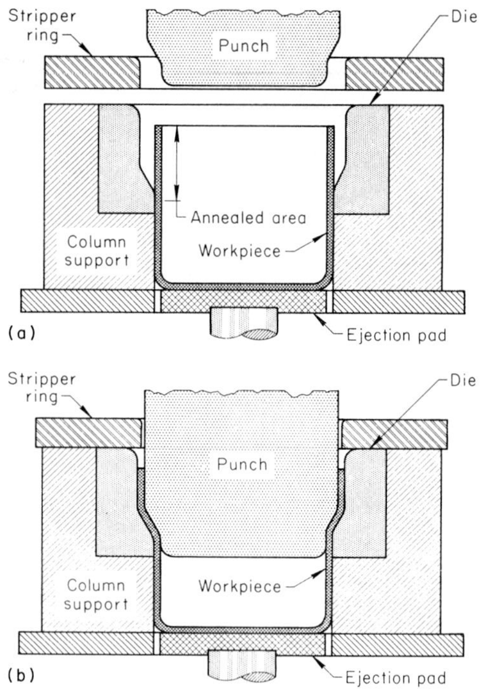

Fig. 60. Expanding the mouth of a drawn shell with a punch

The first ironing operation reduced the outside diameter of the shell to 8.952 in., and the wall thickness to 0.110 in. The bottom radius was 0.125 in. The bottom radius after the second ironing operation was 0.06 in.; the final diameter and wall thickness were as shown in Fig. 59. Figure 59 also shows a detail of the ironing die.

The ironing dies were made of tungsten carbide. They were held in a low-carbon steel die block with an O1 tool steel clamp ring. The ironing punches were made from A2 tool steel. Ironing was done in a 350-ton press at 300 pieces per hour. Annual production quantity was 750,000 pieces.

After a rotary shear trimmed the part to height, the spherical bottom was reverse redrawn and the corner radius reduced to 0.030 in. in a 350-ton press.

Expanding Drawn Workpieces

There are several methods for expanding portions of drawn workpieces in a press. Because the wall thickness is reduced during expansion, it is not advisable to increase the diameter for ductile metal shells (such as low-carbon steel or copper) more than 30%. If a diameter increase of more than 30% is required, the operation should be done in two or more stages, with annealing between stages.

Expanding With a Punch. In expanding with a punch, as in Fig. 60, the portion to be expanded is first annealed. Localized annealing, instead of annealing the entire cup, helps retain strength in the remainder of the cup. Regardless of whether or not the strength is required in the finished part, maximum column strength is desirable to prevent buckling as the punch enters the cup.

After the cup has been placed in the die (Fig. 60a), the punch moves downward and expands the top of the cup (Fig. 60b). During the return stroke, the workpiece is stripped from the punch by the stripper ring and is ejected from the die by the ejection pad.

In an expanding operation of this kind, die dimensions are predetermined within reasonably close limits during the design stage. However, the possibility of later design changes must always be considered. Depending on the shape and location of the expanded section, a height reduction of the cup may occur that will require some modification of the die and punch after tryout.

Expanding with segmented dies is often used for forming sidewalls of drawn shells or sections of tubing. The forming segments are contracted by compression springs and expanded radially by a tapered punch. The die is made of two or more segments held apart by compression springs. As the press ram descends, cams move the die segments together. The punch then moves the inner segments outward, thus forming the contours in the sidewall. This method is illustrated by Fig. 16 in the article "Bending and Forming of Tubing", in this volume. The presence of gaps between the forming segments is one of the drawbacks of this method and is the reason that an alternative method, such as rubber-pad forming, is sometimes selected.

Edge Curling

Curling the edge of a curved sheet, tube or shell is essentially a stretch or shrink flanging operation.

A spring-loaded segmented punch can support a drawn workpiece during edge-curling operations in which the punch must contract to free the workpiece, as in the following example.

Example 260. Supporting a Dome-Shape Workpiece During Edge Curling (Fig. 61)

The workpiece shown in Fig. 61 required support by a spring-loaded segmented punch during curling of the flange section. Without this support, the work-hardened dome wall would have resisted curling, so that the flange section would have pulled in and sections of the head would have buckled.

The design and sequential operation of the tooling used for this edge-curling operation are also shown in Fig. 61. Eight expanding segments moved radially in and out on the taper of the punch. A garter spring held the segments tightly against the punch, and compression springs moved the segments to their contracted position. Stop pins were used to control the up position of the segments. The punch was supported by a die cushion that had greater force than the springs, thus assuring that the punch would expand before curling took place.

In operation, the drawn workpiece was positioned on the contracted segments of the punch, as shown in Fig. 61. As the press ram moved downward with the blankholder, the workpiece was contacted and in turn pushed the segments downward and into an expanded position, as shown. Continued movement of the ram forced the die cushion downward, carrying the part into the curling die. On the return stroke of the press ram, the die-cushion pressure stripped the part from the curling die, and the springs returned the segments to their original (contracted) position. With the segments contracted, the workpiece was removed without difficulty.

Dimensional Accuracy

Dimensional accuracy in deep drawing is affected by the variation in work-metal thickness, variation in work-metal condition (chiefly hardness), drawing technique (particularly the number of operations), accuracy of the tools, rate of tool wear, and press condition.

Control of dimensions begins with the purchase of sheet to closer-than-commercial thickness tolerance, which adds substantially to the cost. Close control of sheet hardness also costs more. In-process annealing may be required to minimize springback or warpage; it will not be needed if tolerances are more liberal. Annealing, handling and cleaning operations are costly.

As tolerances become closer, it is often necessary to add more die stations to minimize the amount of drawing in any one station. Close tolerances may demand restriking operations that would not be necessary for parts with more liberal tolerances. Additional operations increase tool costs and decrease productivity, thereby increasing the cost per piece.

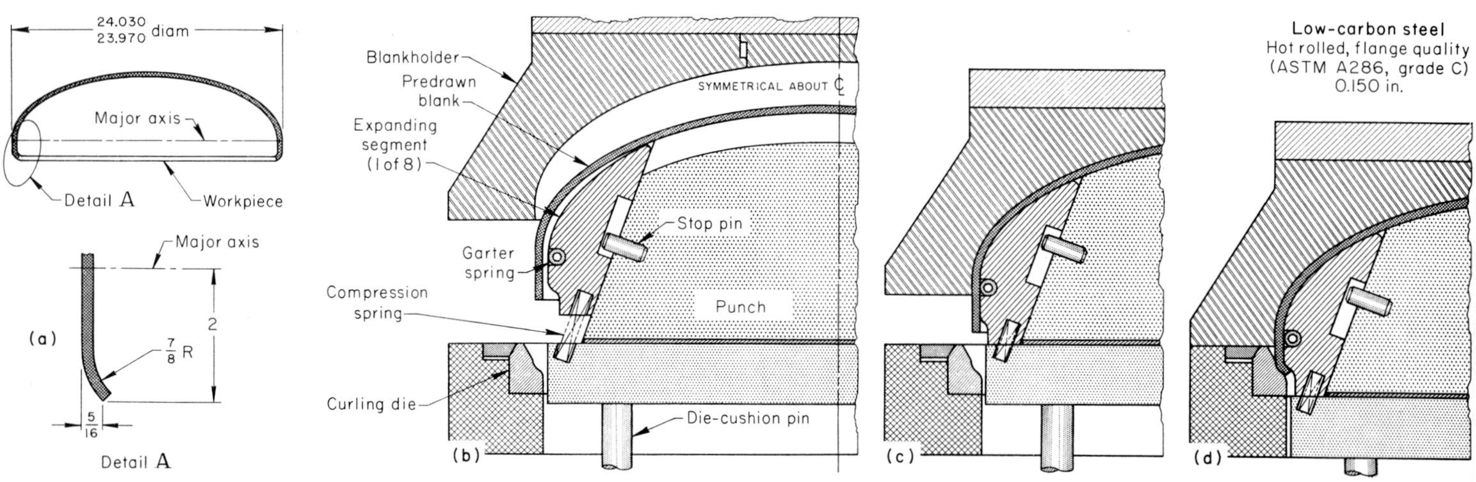

Fig. 61. Use of a spring-loaded segmented punch for support of a dome-shape workpiece during edge curling (Example 260)

The initial cost of tools increases as tolerances become closer, because of greater cost for precision machining and grinding, or more costly tool materials. In addition, tool life before reconditioning and total tool life decrease as tolerances become closer. Maintenance costs and downtime of presses are also greater.

When required, extremely close tolerances can be maintained on some parts by precise control, as shown in the following example:

Example 261. Inside Diameter of Cups Maintained Within 0.0005 In. (Fig. 62)

The needle-bearing cup shown in cross section in the upper part of Fig. 62 had to be produced to a tolerance of 0.0005 in. on the smallest inside diameter (bearing pathway), and had to be free from scratches, nicks or pits, both inside and outside. Originally, a high speed steel die was used, but about 45 pieces were under the specified diameter until the die warmed up, after which tolerances could be maintained. Changing to a carbide die eliminated this problem.

For cups produced by both types of die material, the bearing-pathway diameter was measured on every fifth piece for the first 100 pieces, and then at intervals of 2000 pieces for the cups produced in the high speed steel die and 5000 pieces for the cups produced in the carbide die. Results, plotted in the graph in Fig. 62, show that (after warm-up of the tool steel die) all pieces measured were within the 0.0005-in. tolerance.

To obtain this high degree of accuracy, process variables were closely controlled: cups were drawn from a special grade of steel strip (copper washed); the press was in first-class condition (plungers were square with the bed within 0.001 in. and were carried in a slide that was guided on 3½-in.-diam posts with 0.0015-in. maximum clearance in the bushings); and only nine stations of an 11-station transfer press were used. The 60-ton multiple-pillar press operated at 55 strokes per minute. A flowing water-soluble die lubricant was used.

Cups were gaged before curling to allow the entire pathway to be engaged by an air spindle. Cups had to pass through a 1.1911-in.-diam gage, and then were pressed into confinement rings with an inside diameter of 1.1875 ± 0.0001 in. — at which time all three inside diameters were measured.

In most deep drawing, the accuracy shown in Fig. 62 is either impossible or impractical. The more usual practice when dimensional accuracy is important is to check critical dimensions at specified intervals during a production run and plot the variation. Data from this method of quality control show the capabilities of the process under shop conditions and the magnitude of drift during a production run. When results (either initially or during a run) are unacceptable, one or more of the controls discussed at the beginning of this section can be applied.

Typical variation in diameter of deep drawn parts during a production run is shown in the following example:

Example 262. Variation in Diameter of Deep Drawn Parts (Fig. 63)

Figure 63 shows the variation in the inside diameter of 25 washer tubs produced by deep drawing. The 25 tubs were selected at random, and dimensions were obtained by measuring the circumference with a tape and calculating diameter from circumference. As shown in Fig. 63, diameters of only two of the 25 tubs were out-of-tolerance. Out-of-roundness, measured by inside micrometers, averaged 0.105 in. for the 25 pieces. Height of the 25 tubs varied from 13.303 to 13.326 in.

The press in which the tubs were drawn was in good condition; gib clearance was 0.005 to 0.008 in., and bolster and ram faces were parallel within 0.005 to 0.010 in.

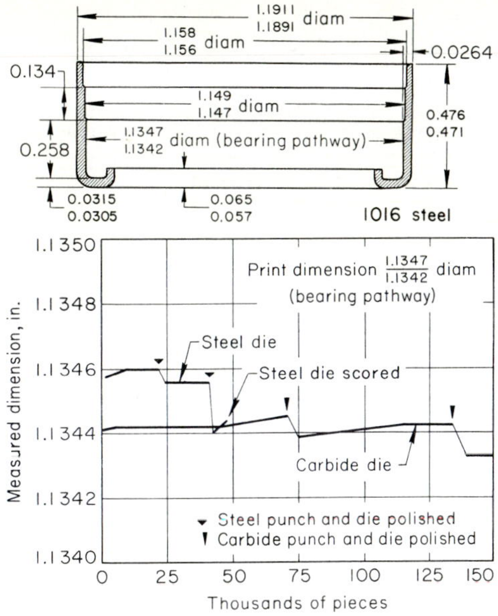

Fig. 62. Variation in bearing-pathway diameter of a needle-bearing cup drawn with high speed steel and carbide dies (Example 261)

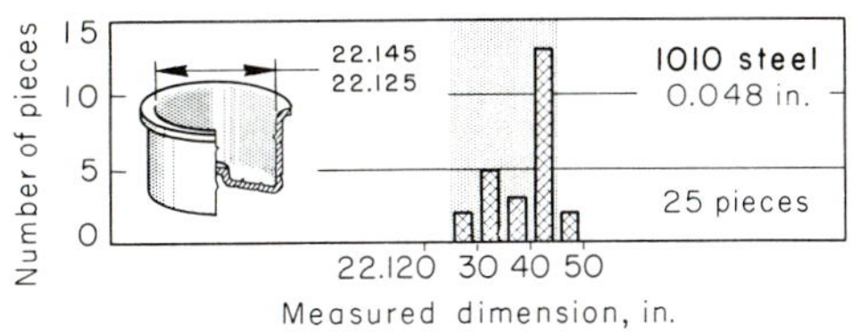

Fig. 63. Variation in inside diameter of 25 deep drawn washer tubs (Example 262)

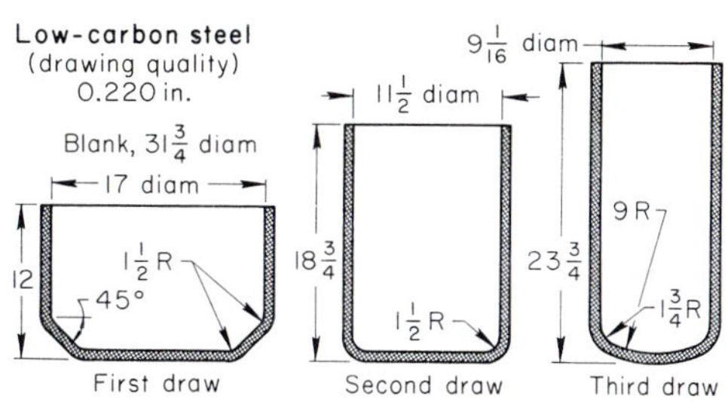

Fig. 64. Progression of shapes in the production of a shell in three drawing operations with intermediate annealing (Example 263)

In-Process Annealing

For severe drawing (more than 50% reduction), it is often helpful, and sometimes necessary, to anneal the workpiece one or more times. Although annealing between draws greatly extends the total amount of reduction that can be done, it is expensive. Handling and cleaning (before annealing and sometimes after annealing) are usually major factors in the total cost of in-process annealing. The following discussion and the production example relate strictly to processing of low-carbon steel. For in-process annealing of other metals, see the articles in this volume that deal with forming of specific metals.

Practice for Steel. Pack annealing requires the workpieces to be packed in an inert material, heated above the upper critical temperature (usually 1700 F for low-carbon steel), and then cooled slowly. This procedure is effective, but it is harmful if the packing material is not absolutely inert, because low-carbon steels carburize readily if heated in a carbonaceous atmosphere. In addition, some cleaning is usually required after annealing.

A preferred method of annealing is to heat the workpiece to 1700 F in an exothermic atmosphere, and then to cool it in a water-jacketed controlled-atmosphere chamber; in this way, scaling of the workpiece is prevented during the entire annealing cycle. Exothermic atmospheres are decarburizing (not harmful and sometimes helpful to deep drawing steels) rather than carburizing. Exothermic atmospheres are relatively low in cost, and if the furnace has been properly purged, cleaning of the workpiece after annealing is not necessary.

Sometimes, subcritical annealing at 1250 to 1300 F (process annealing) is used between draws, but this treatment is seldom recommended, because: (*a*) grain growth may occur, which will impair rather than enhance further drawing, even though the steel is softer; and (*b*) low-cost atmospheres are not suited to the temperature range involved, and therefore scaling will occur and cleaning after annealing will probably be required.

Although in-process annealing is costly, it often makes deeper draws possible and so eliminates some redrawing operations, and its cost is often justified. The benefits of annealing are more often realized when drawing thick-wall parts, as in this example:

Example 263. Number of Draws Reduced by Annealing (Fig. 64)

Figure 64 shows the evolution of a deep shell that was made in three drawing operations from a blank 31¾ in. in diameter by 0.220 in. thick. The shells were annealed in a controlled atmosphere at 1700 F after the first and second operations. Without annealing, at least four redrawing operations would have been needed. For this part, annealing cost less than two more redraws.

Cleaning of Workpieces

In general, the more effective the lubricant, the more difficult it is to remove. For this reason, an overly effective drawing lubricant should be avoided.

The cleaning method depends on the work-metal composition, the lubricant, the degree of cleanness required, workpiece shape, and sometimes on the length of time between application of lubricant and its removal.

Some metals will be attacked by cleaners that are not harmful to others. For example, strong alkaline cleaners are suitable for cleaning steel and many other metals, whereas they are likely to attack aluminum alloys. For detailed information, refer to articles on the cleaning and finishing of the specific metals in Volume 2 of this Handbook.

Unpigmented oils and greases can be removed from steel workpieces by several simple shop methods, which include alkaline dipping, emulsion cleaning, and cold solvent dipping. These methods are usually sufficient for in-process cleaning. However, if the workpieces are to be painted, a more thorough cleaning by emulsion spray or

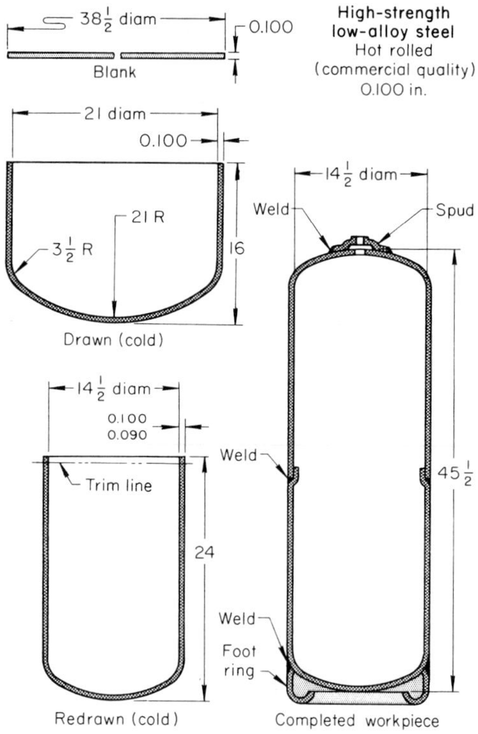

Draw press	500-ton hydraulic
Redraw press	300-ton hydraulic
Lubricant	Dry soap
Die materials:	
Draw	Graphitic tungsten tool steel(a)
Redraw	D2 tool steel
Die hardness	Rockwell C 60
Tool life, pieces per grind	1000
Lot size, pieces	10,000
Annual production, pieces	60,000

(a) 1.50% C, 0.40% Mn, 0.65% Si, 2.80% W

Fig. 65. Cylinder (for liquefied petroleum gas, LPG*) that was made of two shells, each of which was deep drawn in two draws, as shown at left* (Example 264)

vapor degreasing is required. For plating, electrolytic cleaning plus etching in acid (immediately prior to plating) is required. These latter methods usually follow a rough cleaning operation.

Pigmented drawing lubricants and waxes greatly increase cleaning problems. At a minimum, in-process cleaning usually requires slushing in a hot emulsion or vapor degreasing. If the lubricant is not removed for several days after application, soaking in a hot alkaline cleaner or an emulsion cleaner may be required. Particularly for complex workpiece shapes, some hand or power brush scrubbing may be needed. If the workpieces are to be painted or plated, additional cleaning will be required, as described above.

For detailed information on choice of method, see the article "Selection of Cleaning Process", pages 307 to 316 in Volume 2 of this Handbook.

Deep Drawing of Pressure Vessels

Various grades of steel — many of them high-strength alloys — are deep drawn to make cylinders for compressed gases. Joints (when they are made) are around the girth of the vessel, rather than longitudinal. Integrity of the vessel and enough strength to hold pressure are important features and are regulated by Interstate Commerce Commission (ICC) specifications. Specification number and other important information are coined on each cylinder head after a test that proves that the cylinder is able to meet the specification. Commercial quality, hot rolled steels in the as-rolled condition are generally used. Work metal is usually induction heated or induction annealed, because this results in a minimum of scale. The following three examples indicate the range of practice in producing low, medium and high-strength compressed gas cylinders.

Example 264. Deep Drawing Two Pieces for a 100-Lb LPG Cylinder (Fig. 65)

The cylinder shown in Fig. 65 was drawn in two parts from commercial quality, hot rolled low-alloy high-strength steel with a minimum yield strength of 50,000 psi. The 38½-in.-diam blank, 0.100 in. thick, was lubricated with dry soap and drawn to a 21-in.-ID cup in a 500-ton, double-action hydraulic press. After drawing, the cup was induction annealed and relubricated with dry soap. It was redrawn to 14½ in. in diameter in a 300-ton, double-action hydraulic press. The shell was trimmed to length with a cutoff wheel. One shell of each pair had an offset formed so that it would fit inside the other shell for a strong joggle-butt weld. The top shell was pierced and fitted with a spud for later insertion of a valve. Before welding, the shells were washed.

After welding, the cylinder (for LPG) was heat treated to relieve welding and drawing stresses, and to develop strength sufficient to meet ICC specification 4BA240. It was pressure tested, stamped with its rating, and painted with a prime coat.

In blanking, clearance between the punch and die was 6% of stock thickness per side. The radius for the draw die was 0.800 in., or eight times stock thickness. Clearance between the draw die and punch was 0.125 in. The redraw die had a 40°-per-side conical entrance that was 3 in. deep, and a ⅝-in.-wide land connected to the conical entrance by a 1¾-in. radius. There was a 10° relief below the land. The clearance between the punch and redraw die was 0.110 in. The sidewall thickened enough to be slightly ironed for a distance of about 2 in. at the top of the shell.

The drawing of a similar shell is described in Example 246 in this article.

Example 265. Deep Drawing a Single Shell for a 100-Lb Chlorine Cylinder (Fig. 66)

The chlorine-gas cylinder shown in Fig. 66 was made in one piece from a 39-in.-diam blank, 0.220 in. thick, by drawing, redrawing and ironing. The end was reduced by spinning to produce the spherical top and spud for tapping.

The blank was lubricated with sulfonated oil, and cold drawn to a 20-in.-diam cup in a 1500-ton, double-action hydraulic press. To withstand a draw of this severity (almost 50% reduction), the material selected was commercial quality, hot rolled 1020 steel.

The cup was next heated to 1700 F, and three redraws were made while it was still hot. In these operations, the cup was redrawn successively to inside diameters of 14 in., 10½ in., and 8¼ in. Redrawing was done in 500-ton presses — one double-action press and two single-action presses.

The 8¼-in.-diam cup was annealed to relieve stresses left by hot redrawing, pickled, and lubricated with dry soap. Then it was reduced to 8$\frac{1}{16}$ in. in inside diameter, while the sidewalls were ironed to an average thickness of 0.210 in. After another anneal, pickle and dry-soap lubrication, ironing was repeated, reducing the inside diameter to 8 in. and wall thickness to 0.150 in. Ironing gave the sidewalls an excellent finish, both inside and outside.

The shell was cut to length. Part of the scrap was used for a foot ring. Sidewalls were inspected for thickness and finish. Then the open end of the shell was heated and spun down to a small neck. The cylinder was heat treated to the strength required by ICC specification 3A480. Finishing the cylinder included drilling and tapping threads in the neck for a valve, proof testing, stamping, painting and baking.

In blanking, punch-to-die clearance was 0.012 in. per side, or about 6% of stock thickness. For the drawing and redrawing operations, punch-to-die clearance was 0.270 in. per side. The ironing operation (6) used a punch-to-die clearance of 0.148 to 0.150 in.

In the example that follows, a different procedure was used, because the cylinders were much heavier than those described in Examples 264 and 265, and

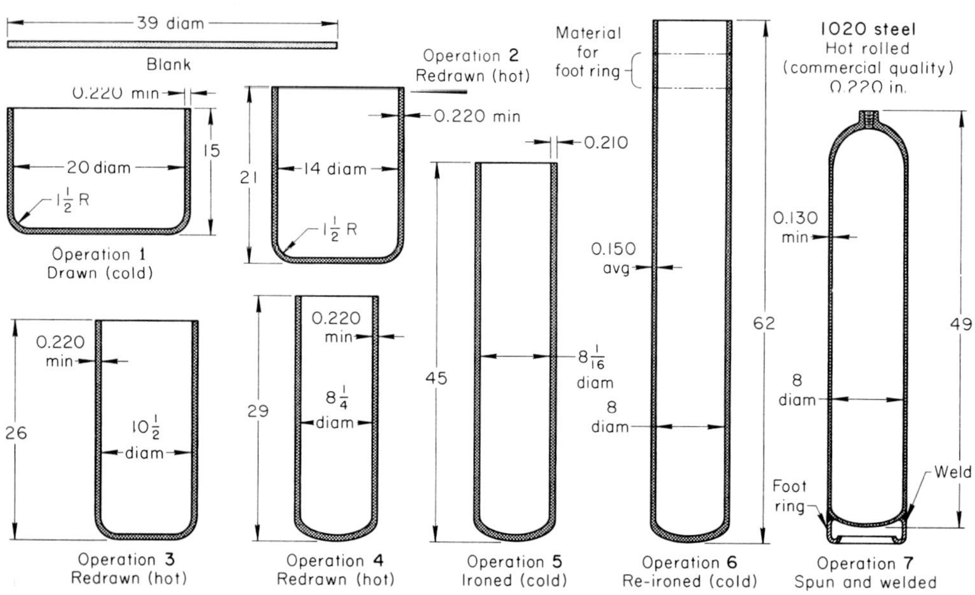

Presses:	
Draw	1500-ton double-action hydraulic
First redraw	500-ton double-action hydraulic
Other redraws	500-ton single-action hydraulic
Lubricants:	
Draw	Sulfonated oil
Fourth and fifth redraws(a)	Dry soap

Die materials:	
Hot operations	Graphitic tungsten tool steel(b)
Cold operations	D2 tool steel
Die hardness	Rockwell C 60
Tool life, pieces per grind(c)	1000
Lot size, pieces	1000
Annual production, pieces	1000

(a) Including ironing operations, which needed additional lubrication. (b) 1.50% C, 0.40% Mn, 0.65% Si, 2.80% W. (c) Tools were reconditioned after annual run.

Fig. 66. Cylinder (for chlorine gas) that was produced from one blank (Example 265) the service pressure rating was much higher — 2015 psi, compared to 240 psi and 480 psi for the LPG and chlorine-gas cylinders, respectively.

Example 266. Drawing a Compressed-Oxygen Cylinder From an Extruded Shell (Fig. 67)

The first step in the production of the compressed-oxygen cylinder shown in Fig. 67 was the hot extrusion of a shell from a slug. The slug was cut from a billet of hot rolled 4130 steel, induction heated to 2300 ± 50 F, and backward extruded into a shell with an 8.7-in. inside diameter and an 0.83-in.-thick wall in a double-action, 1750-ton hydraulic press. While still hot, the shell was reduced to 8.56 in. in inside diameter and 0.230 in. in wall thickness in a draw bench by pressing it through seven dies. The force on the hydraulic ram was 250 tons. Closing the end and tapping threads for the valve were done in much the same manner as described in the previous example. The cylinder was heat treated to ICC specification 3AA2015.

The dies were made of nickel-molybdenum-silicon gray iron and were heat treated to Rockwell C 40. Tool life was 1200 pieces per grind. Lubricant was graphite in oil. Annual production was 150,000 cylinders, made in lots of 5000.

For propane gas, pressure tanks must have high strength at minimum weight. In one application, the weight of such a tank was reduced from 130 to 70 lb by changing from 1025 steel to a high-manganese, deep drawing steel (0.20% C, 1.60% Mn, 0.025% P and 0.030% S — all percentages maximums). Before drawing, the high-manganese steel had a minimum yield strength of 50,000 psi and a minimum tensile strength of 70,000 psi.

Bottles for dispensing small quantities of liquefied gases or gases under high pressure are commonly made of drawing quality low-carbon steel to take advantage of the improved mechanical properties produced by deep drawing. The bottles range in size from ½-in. diameter by 1¼ in. long to 1½-in. diameter by 6 in. long. The following example describes the drawing of a medium-size bottle.

Example 267. Deep Drawing of a Carbon Dioxide Bottle (Fig. 68)

The bottle shown in Fig. 68(a) was drawn and necked from aluminum-killed, acid-etched 1008 steel 0.037 to 0.039 in. thick. The permissible hardness range was Rockwell B 38 to 52, but the lower part of the range was preferred. The strip stock was 5½ in. wide and had a No. 2 finish, with a rolled edge.

In the first operation (Fig. 68b), two pieces per press stroke were blanked and cupped, using a compound die in a 70-ton mechanical press operating at 45 strokes per minute. Blank diameter was 2.825 in. and cup diameter was 1.590 in., for a reduction of 43.7%. The progression was 4.375 in. and the blank center distance across the strip was 2.500 in.

The cup was direct redrawn in five operations to the dimensions and shape shown in Fig. 68(c), pinch trimmed to 2½ in. long, and sized to 0.735/0.730 in. in diameter in a transfer die. The first redraw reduced the shell diameter 20%. The percentage reduction was reduced with each redraw, the last being 11%. The die operated in a 75-ton straight-side mechanical press at 44 strokes per minute. Draw rings were tungsten carbide with highly polished working surfaces. Punches and blankholders were D3 tool steel, hardened to Rockwell C 61 to 63.

The open end of the shell was annealed to permit necking in five steps to the dimensions shown in Fig. 68(d). The bottle was moved to each of the five tungsten carbide necking dies by a rotary-feed table. Necking was done in a 50-ton mechanical press at 3200 pieces per hour.

An automatic lathe was used to counterbore the neck, form internal serrations that retained the seal, chamfer the end, and face the bottle to length. A caution notice and date were imprinted by a roll-type stamping device. Some bottles had a circumferential indentation, which was roll formed. This indentation was used as a locator in the application and, because it was a depressed surface, provided a safe place for a color band indicating the type of gas in the container.

A water-soluble mineral oil was used as a lubricant for drawing, reducing and necking.

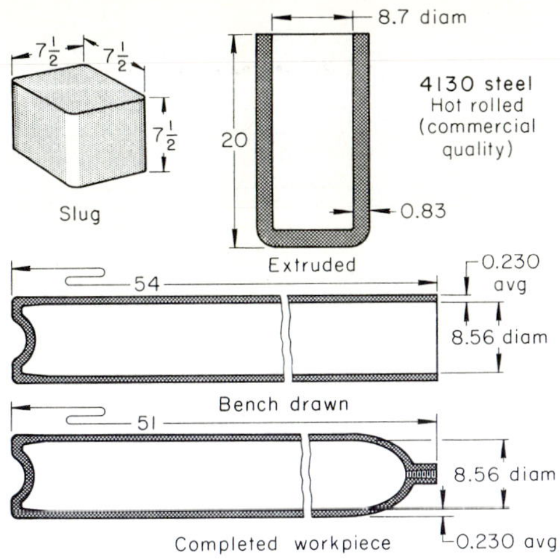

Fig. 67. High-strength cylinder (for compressed oxygen) that was drawn from a hot extruded shell (Example 266)

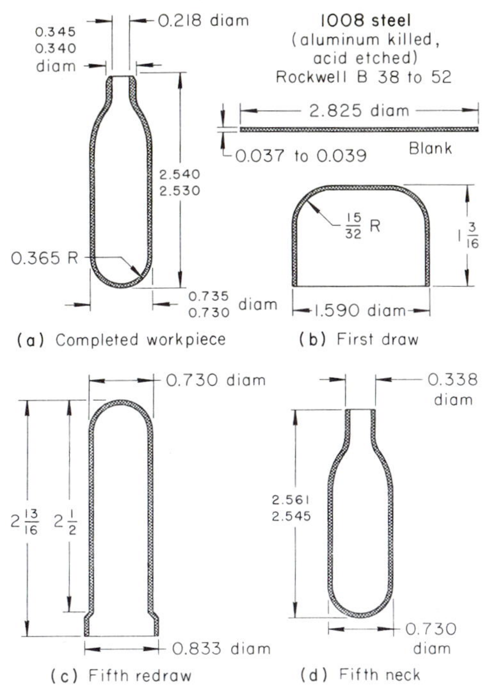

Fig. 68. Drawing and necking a bottle for carbon dioxide gas (Example 267)

Safety

Deep drawing, like other press operations, involves potential hazards to operators, maintenance men and other personnel in the work area. No press, die, or auxiliary equipment can be considered ready for operation until these hazards are eliminated by the installation of necessary safety devices. Operators should be properly instructed in safe operation of equipment. The article that begins on page 1 in this volume contains information and literature references on safe operation.

APPENDIX 1

Deep Drawing With Ultra-High Pressure

By F. J. Fuchs and J. W. Archer*

In conventional deep drawing, the maximum depth to which a shell can be drawn in a single operation is limited by the tensile strength and ductility of the work metal and by friction between workpiece and tooling. Expressed as the ratio of blank diameter to shell diameter, the accepted limit of depth of draw for the most drawable metals is only slightly greater than 2 to 1. The high-pressure techniques described here make possible a blank-to-shell diameter ratio of at least 3.3 to 1, or a reduction of nearly 70% in a single draw.

The important differences between high-pressure drawing and conventional drawing are that the ductility of the work metal is actually increased during high-pressure drawing and stress patterns are controlled so that the tensile strength of the work metal is supplemented by applied compressive stress, which effectively cancels out the strength limitation. The high pressure, properly controlled, can be applied to the surface of the blank either by a fluid or by mechanical means.

Fluid Techniques

The high-pressure-fluid technique is the most efficient way of overcoming work-metal strength limitations, low ductility, and friction. In conventional drawing, forces to overcome friction on the surfaces of the blank and to compress the blank circumferentially to reduce its size can come only from the force generated by the punch pressing on the bottom of the cup. This force creates tensile stress in the cup wall and will pull the cup into two parts when the tensile strength of the work metal is exceeded. Also, in the conventional operation, ductility is constant or decreases as the work metal deforms during drawing. Because of the high blankholding force needed to prevent flange wrinkling, friction is high and is distributed over the top and bottom surfaces of the blank in contact with the draw ring.

In the high-pressure-fluid method, the fluid pressure acting on the edge of the blank produces radial compressive forces through the blank and the radial stress terminates as an axial compressive force in the cup, partially or completely balancing the tensile force normally found. The balance can be controlled by altering the pressure of the fluid.

Improved ductility arises from the effect demonstrated by P. W. Bridgman that metals under high hydrostatic pressure become more ductile. In the high-pressure-fluid method of drawing, the portion of the blank that extends beyond the die radius is completely enveloped in a high-pressure fluid; therefore, it has increased ductility, and is the portion of the metal un-

*Engineering Research Center, Western Electric Co. This Appendix is condensed from ASM Technical Report C7-13.2 (October 1967).

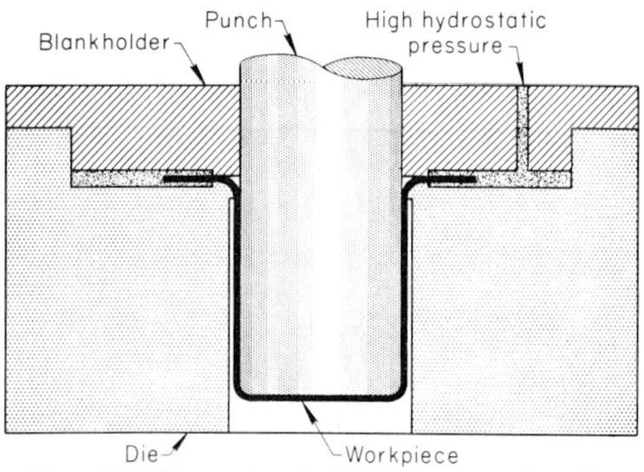

Fig. 69. Elements of deep drawing by the high-pressure-fluid technique

dergoing the most deformation. The remaining metal under deformation is a small annular section at the die radius. The compressive forces between punch and die wall in conjunction with the radial compressive force contributed by the fluid pressure mechanically subject this small annular section to high environmental compression so it, too, has increased ductility. Referring to Fig. 69, it is evident that since the only mechanical clamping force on the blank is a small seal area outside the die radius, friction forces are minimized. The fluid will flow inward in a restricted way both above and below the blank to create positive lubrication, further decreasing friction.

Fluid-Pressure Equipment. The equipment shown in Fig. 70 is designed for use in a hydraulic press having a 1000-ton upper ram and a 150-ton lower ram. The die is made to draw $\frac{3}{32}$-in.-thick blanks having a maximum diameter of 3.5 in. The blanks are placed on the lower half of the die, which is secured to the bottom platen of the press. A locating ring centers the blank. The assembly mounted to the upper ram of the press consists of three telescopically related parts. As the upper ram is lowered, the high-pressure chamber seats on the lower die assembly. As the ram continues down, the upper die moves within the chamber to seat on the lower die and to clamp the blank. The ram continues down to pressurize the fluid (castor oil), which is above the die.

The pressurized fluid flows through the channels in the die to surround the blank, except for the circular areas on the blank described by the punch and the throat of the die. The fluid is prevented from entering the punch and die areas by annular protrusions on both halves of the tool. The protrusions pinch the blank, and effect a metal-to-metal seal. The punch and die are vented to prevent any buildup of back pressure caused by leakage past the blank. To form the shell, the punch actuation is coordinated with the generation of high pressure.

The two die halves are made of modified H11 tool steel. The contact surfaces of the die halves are polished to a mirror finish and hard chromium plated. The punch is of high speed steel, polished and hard chromium plated. The chamber is a pressed-ring vessel with a 4-in.-diam bore, fabricated of maraging steel. The chamber is designed for use to fluid pressures of 125,000 psi. The maximum working pressure is generated by using approximately 640 tons of the available capacity of the press.

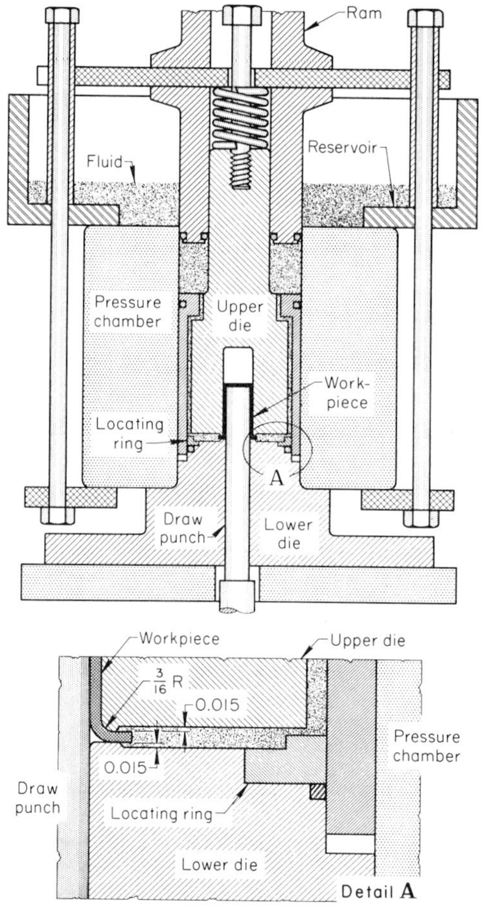

Fig. 70. Equipment for fluid-pressure deep drawing in a hydraulic press

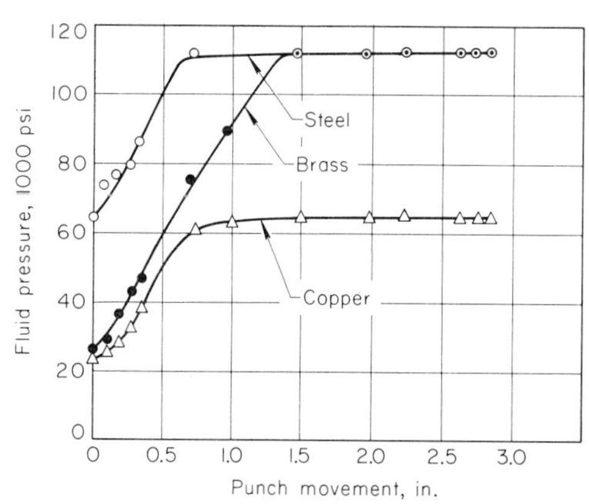

Fig. 71. Fluid pressures used to form shells by high-pressure drawing from copper, brass and steel blanks

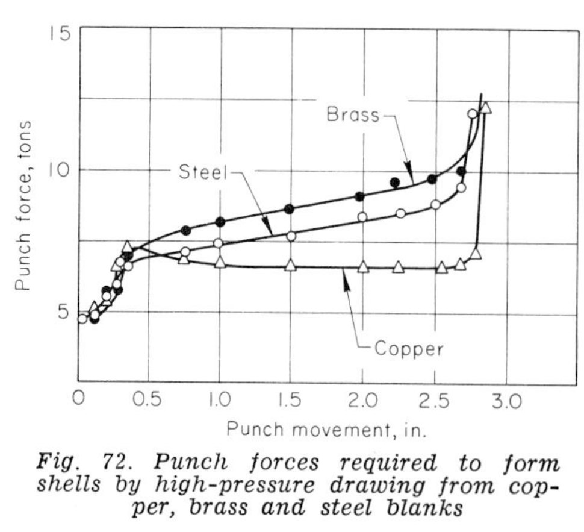

Fig. 72. Punch forces required to form shells by high-pressure drawing from copper, brass and steel blanks

From the 3.5-in. blank, a flanged shell is formed having a depth of 2.78 in., an outside diameter of 1.054 in., and an inside diameter of 0.875 in. The blank-to-cup diameter ratio is approximately 3.3 to 1, or a reduction of nearly 70%. The die has a $\frac{3}{16}$-in. radius, and the punch a $\frac{1}{16}$-in. radius. These radii are considerably less than those that could be used in conventional deep drawing of a $\frac{3}{32}$-in.-thick blank. Punch-to-die clearance is 0.0035 in. less than nominal stock thickness so that the sidewall of the shell is slightly ironed to give a uniform wall thickness and improved surface finish.

The sealing surface on the upper die extends 0.015 in. beyond the surrounding surface. The $\frac{3}{16}$-in. die radius blends into the surrounding die surface without a flat at the tangent point to make a 0.015-in.-high sealing ring. The sealing surface on the lower die also is 0.015 in. high, but it has a short flat beyond the tangent point of the upper die radius before blending into the surrounding die surface. The size and shape of the sealing surfaces assure a good metal-to-metal seal. The space between the upper and lower die surfaces is greater than the amount of stock thickening that occurs in the blank; thus, the blank is always surrounded by the pressurized fluid and is in a state of increased ductility.

The high-pressure tool operation is completely automatic. When the cycle is started, the tool closes and the high pressure is generated. The bottom ram or draw punch is then actuated by a pressure switch mounted on the input port to the upper ram, which generates the high pressure. At the end of the punch movement, the punch is hydraulically locked in position and the upper ram is retracted. After the upper ram pulls the die off the shell, the punch retracts. The higher friction force between shell and punch, and the opening sequence used, prevent the shell from remaining in the die. Unloading the shell and loading the next blank are done manually before the next cycle is initiated.

Results. A series of tests was conducted on each of three commercial sheet metals: (*a*) 1010 steel (commercial quality, annealed, Rockwell B 51 to 53); (*b*) oxygen-free copper (annealed); and (*c*) cartridge brass (annealed 1½ hr at 1000 F from the half-hard condition).

The steel blanks were phosphate coated and lubricated with soap. The copper and brass were lubricated with a drawing wax. Some blanks of each material were lightly sand blasted and coated with 0.002-in.-thick Teflon. The steel and brass blanks had a nominal thickness of 0.089 in.; copper blanks of 0.089 and 0.093-in. thicknesses were used to establish semiquantitatively the effect of ironing on the forces required to form a shell.

The first phase of testing each material consisted of establishing the maximum fluid pressure to be used in the tool and the pressure at which the punch should be actuated. The pressure curve was purely a function of time and took 4 to 5 sec to reach its maximum value, because of the slow speed of the upper ram. It was found that the punch must move during buildup of the high pressure. If the punch started to move when the pressure was too low, a hole was pierced in the blank. If the hydrostatic pressure was too high, extrusion of the blank occurred without the aid of the punch, and a shell that had a bulged end was produced.

Each material tested was formed to the full depth of the die. Figures 71 and 72 show the converted values for the fluid pressure and the punch force required, as a function of punch movement. These curves were obtained while forming acceptable shells.

All shells showed some earing, but the anisotropy was not severe for shells of this wall thickness. All shells were structurally sound with no shear cracks in the walls. For very deep draws, beyond the limits of the present tools, it is conceivable that anisotropy of the sheet metal may be the limiting factor. Cross-rolled or drawing quality materials may be required.

The quality of the parts made by the high-pressure single-draw method was far better than obtainable by conventional multiple-draw techniques. The results show that expensive redrawing and annealing operations can be eliminated, significantly decreasing the cost of making shells. Because of ironing in the high-pressure process, the shells had smooth surfaces, approaching the mirror finish of the punch and die. The ironing also resulted in uniform shell thicknesses. The axial variation found for copper and steel shells was never more than 0.0007 in. on the outside diameter. The inside diameters varied no more than the axial tolerance of the punch and were about 0.0005 in. smaller than the punch. The variation between outside diameters of copper shells was less than 0.0007 in., for steel shells less than 0.0012 in., and for brass shells less than 0.0006 in.

The equipment shown in Fig. 70 can be used for production cycles of less than 1 min. However, the high-pressure piston has to be the same in area as the blank being formed, and therefore when a sizable blank is used, the press requirement can be very large. For instance, if a 6-in.-diam blank is to be formed with a pressure of 300,000 psi, a 4500-ton press would be needed. Again referring to Fig. 70, large radial forces are created in the chamber because the fluid under pressure operates over a considerable height in the chamber. So a massive chamber would be required for large work.

In order to improve the mechanical relations, a new type of deep drawing equipment has been evolved, as shown in Fig. 73. Here the vertical forces due to pressure are contained by a quick-opening breech mechanism that mounts into the walls of the pressure vessel. The breech can be closed and opened by direct in-and-out motion of a small ram. The pressurizing fluid occupies a space only slightly larger than the blank itself, and therefore exerts a low radial force. The high-pressure fluid is supplied by small built-in intensifiers mounted in the base block of the tool. These are powered by a relatively small lower ram on the press. With this type of equipment, a 6-in.-diam blank in fluid at 300,000-psi pressure can be formed with a 100-ton press.

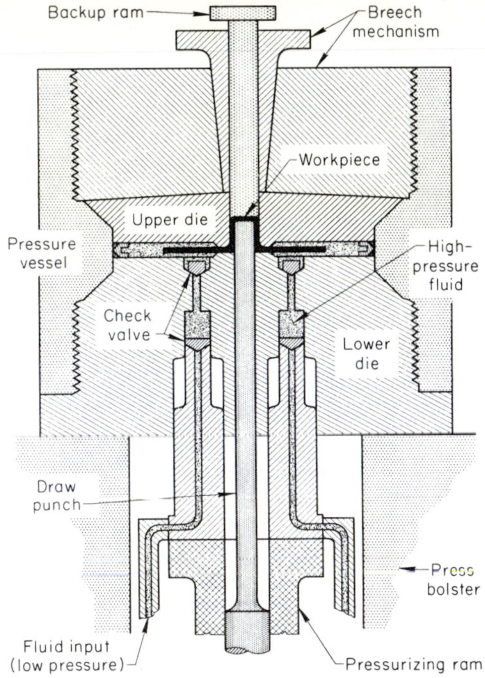

Fig. 73. Improved equipment for fluid-pressure deep drawing

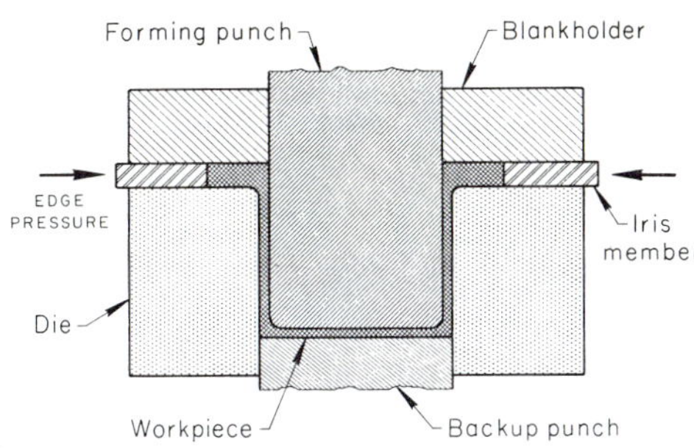

Fig. 74. High-pressure mechanical deep drawing process

Mechanical Edge-Pressure Technique

Another approach to high-pressure drawing involves the use of mechanical members to apply the high compressive forces to attain high work-metal ductility. Figures 74 and 75 show the process and the equipment. The pressure on the edge of the blank is created by an iris assembly that closes radially. The iris members are driven by high-intensity hydraulic rams. The vertical forces applied to the blank result from the top and bottom plate closure activated by the press ram. In this respect, the mechanical technique suffers the same limitation of press capacity as the fluid technique, although the press stroke is much shorter because it is a simple clamping action. Another disadvantage of the mechanical technique is the considerably higher friction on the surfaces of the blank, making excellent lubrication desirable.

In spite of these problems, the mechanical tool has desirable capabilities. The ease of operation and decreased stress on tools afforded by not using the high-pressure fluid are helpful. The mechanical iris members make possible shells without the anisotropic distortion of flange edges. Because of this, parts can be made with finished flanges or scrap-free ends. The pressures that can be generated in this type of tool are limited only by the compressive strength of the tool steels used.

The dies and punches are made of 18% nickel maraged steel. The iris members are driven by high-intensity rams of 150 tons capacity, each fed by 35,000-psi intensifiers. The backup ram

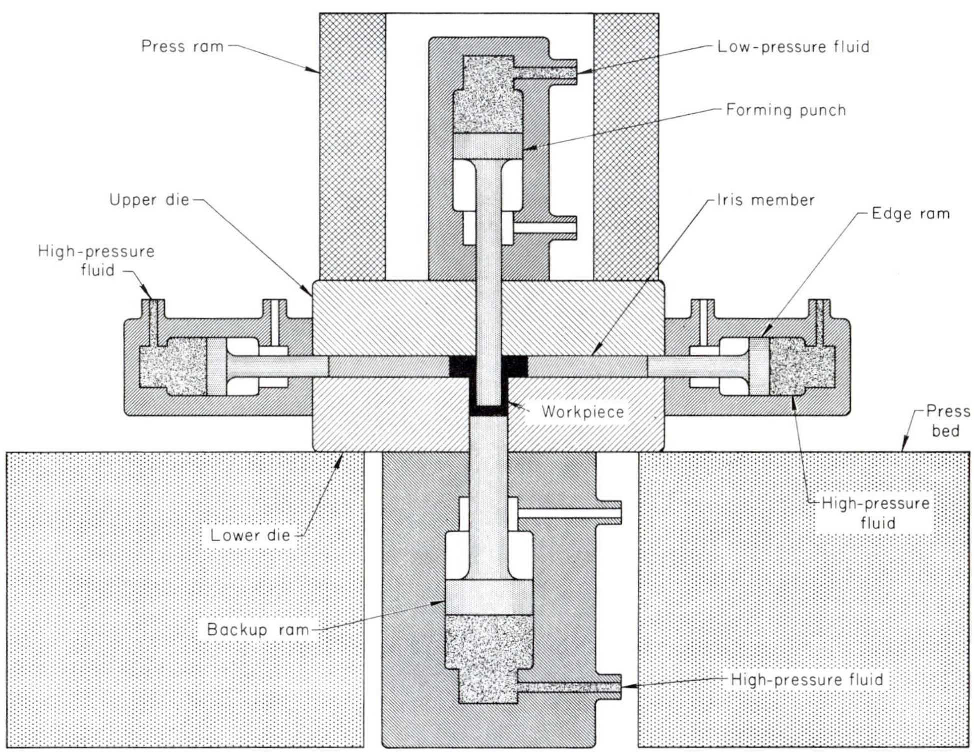

Fig. 75. Tools for deep drawing by the high-pressure mechanical process

is also a 150-ton subpress. The draw punch is actuated by a 150-ton ram built into the 1000-ton press, which itself does the clamping. The equipment is designed to make a rectangular shell of 1/16-in. wall up to 5 in. long from a blank 1/4 in. thick.

In an example of the application of this equipment, a fully annealed 4 1/4-by-3 1/8-by-1/4-in. blank of oxygen-free copper was used. Blanks were coated with either wax or 0.002-in.-thick Teflon. During initial punch movement, the mechanical pressure acting on the edge of the blank built up to about 140,000 psi. This pressure extruded the 1/4-in. blank into the 1/16-in. walls of the rectangular shell. Metal flow was satisfactory, even with the small punch and die radii used. The shell walls were uniform as a result of the compressive forces being applied in all directions.

APPENDIX 2

Effect of Metal Thickness on Deep Drawing

By H. E. Ihle

The stock thickness and its relation to the diameter or width of the blank affects the formability of round, square and rectangular shells. Stock thickness also affects the depth of shell that can be drawn in one operation.

Table 14. Formability Factors (d/D Ratios) for Drawing Round Shells Without Flanges From 1010 Steel(a)

Stock thickness, % of blank diameter	Formability factor (d/D ratio): First draw	Second draw	Third draw	Fourth draw	Fifth draw
2.0	0.48	0.73	0.76	0.78	0.80
1.5	0.50	0.75	0.78	0.80	0.82
1.0	0.53	0.76	0.79	0.81	0.84
0.6	0.55	0.78	0.80	0.82	0.85
0.3	0.58	0.79	0.81	0.83	0.86
0.15	0.60	0.80	0.82	0.85	0.87
0.08	0.63	0.82	0.84	0.86	0.88

(a) For draw-die radii of $8t$ to $15t$ (t = stock thickness); factors may increase about 2% for draw-die radii of $4t$ to $8t$.

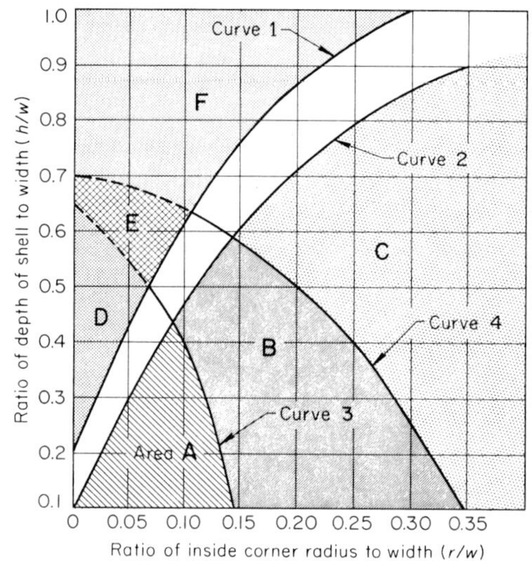

Fig. 76. Dimensional relations defining six areas of formability of square or rectangular drawn shells of 1010 steel. See text.

Round Shells

The number of drawing operations needed to produce a finished round shell depends on blank diameter, depth and diameter of the shell, properties and thickness of the work metal, and draw-die and punch-nose radii.

The ratio of the punch diameter to the blank diameter (d/D) is a simple and practical measure for use in determining formability. The maximum d/D ratio depends markedly on the metal thickness and the draw-die radius. The ratio of metal thickness to blank diameter (t/D) reflects the principle of volume constancy between the blank and the shell and, as shown in Tables 14 and 17, affects the formability factor d/D. The formability factors in Table 14 are multiplied by the blank diameter, or by the previous shell diameter, to obtain the shell diameter. For each redraw, the factors increase, indicating a decrease in formability. The size of the draw-die radius also has an effect on the formability factors, as footnoted in Table 14. When the factors given in Table 14 are not exceeded, three or four draws can be made in 1010 steel without annealing if the punch-nose radius is at least 75% of the draw-die radius.

The ratio of draw depth to shell diameter, and the size of the draw-die radius, depend on the t/D ratio, as shown in Table 15.

Table 15. Maximum Depth-to-Diameter (h/d) Ratios for Drawing Round Shells Without Flanges From 1010 Steel(a)

Drawing operation	Maximum h/d ratio for steel with thickness of the following percentages of blank diameter: 2.0 to 1.5%	1.5 to 1.0%	1.0 to 0.6%	0.6 to 0.3%	0.3 to 0.15%	0.15 to 0.08%
First	0.94 to 0.77	0.84 to 0.65	0.70 to 0.57	0.62 to 0.50	0.52 to 0.45	0.46 to 0.38
Second ..	1.88 to 1.54	1.60 to 1.32	1.36 to 1.1	1.13 to 0.94	0.96 to 0.83	0.90 to 0.70
Third ...	3.5 to 2.7	2.8 to 2.2	2.3 to 1.8	1.9 to 1.5	1.6 to 1.3	1.3 to 1.1
Fourth ..	5.6 to 4.3	4.3 to 3.5	3.6 to 2.9	2.9 to 2.4	2.4 to 2.0	2.0 to 1.5
Fifth	8.9 to 6.6	6.6 to 5.1	5.2 to 4.1	4.1 to 3.3	3.3 to 2.7	2.7 to 2.0

(a) The larger h/d ratios in each range are for draw-die radii of $4t$ to $8t$ (t = stock thickness); the smaller ratios are for draw-die radii of $8t$ to $15t$. Also, draw-die radii increase as ratios of stock thickness to blank diameter (t/D ratios) decrease.

Table 16. Maximum Depth-to-Diameter (h/d) Ratios for Drawing Round Shells With Flanges in One Operation From 1010 Steel(a)

D_f/d ratio(b)	Maximum h/d ratio for steel with thickness of the following percentages of blank diameter: 2.0 to 1.5%	1.5 to 1.0%	1.0 to 0.6%	0.6 to 0.3%	0.3 to 0.15%
1.1	0.90 to 0.75	0.82 to 0.65	0.70 to 0.57	0.62 to 0.50	0.52 to 0.45
1.3	0.80 to 0.65	0.72 to 0.56	0.60 to 0.50	0.53 to 0.45	0.47 to 0.40
1.5	0.70 to 0.58	0.63 to 0.50	0.53 to 0.45	0.48 to 0.40	0.42 to 0.35
1.8	0.58 to 0.48	0.53 to 0.42	0.44 to 0.37	0.39 to 0.34	0.35 to 0.29
2.0	0.51 to 0.42	0.46 to 0.36	0.38 to 0.32	0.34 to 0.29	0.30 to 0.25
2.2	0.45 to 0.35	0.40 to 0.31	0.33 to 0.27	0.29 to 0.25	0.26 to 0.22
2.5	0.35 to 0.28	0.32 to 0.25	0.27 to 0.22	0.23 to 0.20	0.21 to 0.17
2.8	0.27 to 0.22	0.24 to 0.19	0.21 to 0.17	0.18 to 0.15	0.16 to 0.13
3.0	0.22 to 0.18	0.20 to 0.16	0.17 to 0.14	0.15 to 0.12	0.13 to 0.10

(a) The larger h/d ratios in each range are for draw-die radii of $4t$ to $8t$ (t = stock thickness); the smaller ratios are for draw-die radii of $8t$ to $15t$. Also, draw-die radii increase as t/D ratios decrease. (b) Ratio of flange diameter (D_f) to shell diameter (d).

When drawing flanged round shells, the ratio of the flange diameter to the shell diameter (D_f/d) and the t/D ratio both affect the maximum depth-to-diameter ratio of the shell (see Table 16). The formability factors d/D shown in Table 17 should not be exceeded during the first drawing operation for a flanged round shell.

Square or Rectangular Shells

Drawing of square or rectangular shells subjects the metal in the corners to compressive stresses. Metal in the sidewalls is subject to tensile stresses after the metal has been drawn over the die radius. Compressive stresses in the sidewalls are caused by the flow of metal from the corner areas.

Compression of the metal from the corner areas into the sidewalls depends on the ratio of the inside vertical corner radius of the shell to the width of the shell (r/w). Resistance of the metal to bending and wrinkling depends on the ratio of metal thickness to blank width (t/W). Formability factors ($0.71\sqrt{r/h}$) for square or rectangular shells drawn in one operation from 1010 steel are listed in Table 18; formability factors for less ductile metals are up to 10% larger.

Maximum ratios of depth to width for square or rectangular shells drawn in one operation are given in Table 19; these values are based on the ratio of shell corner radius to shell width (r/w), and on the ratio of metal thickness to blank width (t/W).

The number of drawing operations needed is influenced by the ratio of shell depth to shell width (h/w), and the ratio of the inside vertical corner radius of the shell to the width of the shell (r/w), as shown in Fig. 76. Areas A, B and C below curve 2 in Fig. 76 represent the limits for drawing square or rectangular shells in one operation. Shells in areas D, E and F above curve 1 need two or more operations. The space between the curves 1 and 2 represents shells that are borderline for drawing in one operation. Shells in areas A, B and C have a stock thickness of no more than 0.6% of the blank width; in areas D, E and F, the stock thickness is not less than 2% of blank width.

Each area in Fig. 76 represents shells with certain characteristics. Area A represents shallow shells with small inside corner radii and an $r/(w-h)$ ratio of no more than 0.17. With these shells, only a small amount of metal moves into the sidewalls from the corner areas; thus, the sidewall height is constant. Typical blank development for area-A shells is shown in Fig. 77.

Area B in Fig. 76 represents shallow shells with medium-size corner radii and an $r/(w-h)$ ratio of 0.17 to 0.4. The movement of metal into the sidewalls from the corner areas causes some earing in this type of shell. Figures 77(c) and 78 show a typical corner of a blank for an area-B shell.

Area C in Fig. 76 represents shells with medium to large corner radii and with an $r/(w-h)$ ratio of more than 0.4. The compression of metal into the sidewalls of these shells causes considerable earing. One method of developing the corner of the blank for wide or shallow area-C shells is illustrated in Fig. 78.

Workpieces in area D in Fig. 76 have an h/w ratio no greater than 0.65, and a blank shaped as shown in Fig. 77(d). Area F represents shells with an h/w ratio greater than 0.7. The ranges of areas D and F overlap into area E.

The procedure for laying out the corners of blanks for shells in areas A, B and D in Fig. 76 is described below, with reference to Fig. 77(a):

1. Lay out a square or rectangle by drawing lines through the loci of the workpiece corner radii. These lines are the primary datum lines.
2. The width of metal needed for the sidewall and bottom radius is $h + 0.57r_b$, where h is the shell depth, in inches, and r_b is the bottom inside radius, in inches. This width is added to each side of the layout of step 1, from the primary datum lines.
3. The corner radius R of the blank, which provides the metal needed for each corner of the shell, is:

$$R = \sqrt{r^2 + 2hr - 0.86r_b(r + 0.16r_b)}$$

where r is the inside corner radius of the

workpiece, in inches. For parts in area D, radius R is increased by 10 to 20%.

4 Lay out corner radius R, using the loci of the workpiece corner radii as center points, so as to intersect the lines drawn in step 1 at points b and c.

5 Bisect ab and cd. Through the dividing points e and f, draw gh and ij tangent to arc of corner radius R drawn in step 4. [These tangents may coincide (Fig. 77b), they may cross each other outwardly (Fig. 77c), or they may cross each other inwardly (Fig. 77d).]

6 Blend the intersections of the tangents and the edges of the blank with an arc of radius R. When the tangents intersect inwardly, blend the intersections with an outward arc of radius R (detail A in Fig. 77a).

Shells in area B in Fig. 76 and shallow or wide shells in area C have a larger amount of metal compressed into the sidewalls than in area A. The blank is adjusted for this by increasing the metal for the corners and decreasing the metal for the sidewalls, as described below, with reference to Fig. 78:

1 Start the blank layout using steps 1, 2, 3 and 4 in the procedure detailed above. If the corner radius is not the same as the bottom radius, the loci of the bottom radii establish secondary datum lines from which the metal for the sidewalls found in step 2 in the six-step procedure previously listed is projected.

2 The corrected corner radius R_c is calculated by the formula:

$$R_c = R[0.074(R/2r)^2 + 0.982]$$

3 The width of metal for the sidewalls is reduced by h_w and h_l, calculated by the formulas:

$$h_w = yR^2/(w - 2r) \quad \text{and} \quad h_l = yR^2/(l - 2r)$$

Values for factor y are given in Table 20.

4 Blend the intersections of the width, length and corner radii of the blank by making R_w tangent to R_c and the blank width; and R_l tangent to R_c and the blank length.

Square shells in area F, and in area C with a large h/w ratio, can be drawn from round blanks. The blank diameter is determined according to the following formula:

$$D = 1.13\sqrt{w^2 + 4w(h - 0.43r) - 1.72r(h + 0.33r)}$$

Rectangular shells in areas C and F with large h/w ratios usually are drawn from blanks with semicircular ends connected by two parallel sides; their length is the difference between the length and width of the shell.

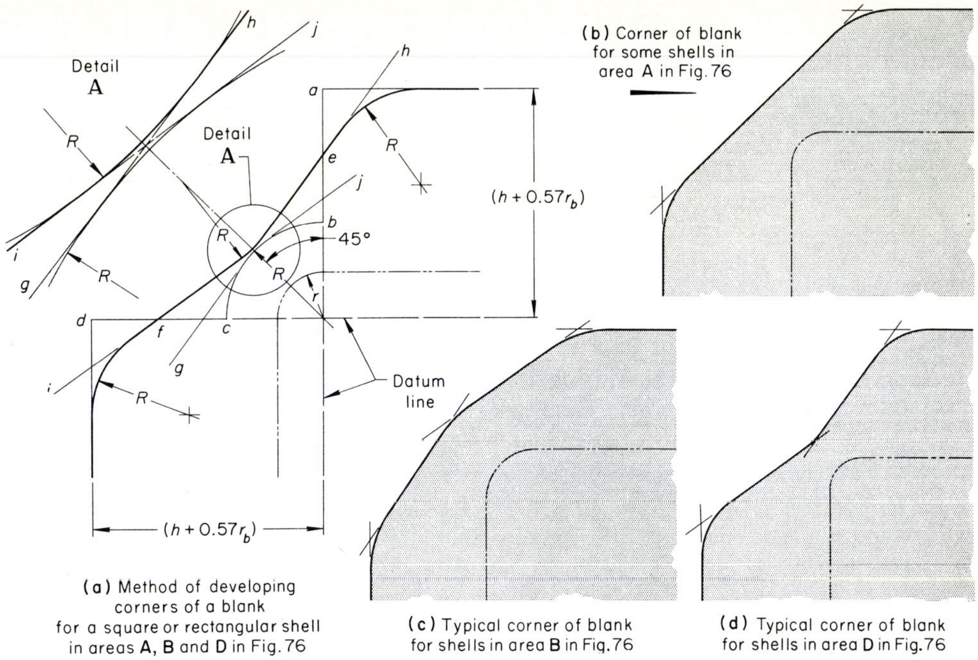

Fig. 77. Corner-layout method, and typical corners, for blanks for square or rectangular shells in areas A, B and D in Fig. 76. Corner shape varies with shell dimensions, and the shape typical of area B and that typical of area D extend into adjacent areas.

Table 17. Formability Factors (d/D Ratios) for Drawing Round Shells With Flanges in One Operation From 1010 Steel

D_f/d ratio(a)	Formability factor (d/D ratio) for steel with thickness of the following percentages of blank diameter: 2.0 to 1.5%	1.5 to 1.0%	1.0 to 0.6%	0.6 to 0.3%	0.3 to 0.15%
1.1	0.51	0.53	0.55	0.57	0.59
1.3	0.49	0.51	0.53	0.54	0.55
1.5	0.47	0.49	0.50	0.51	0.52
1.8	0.45	0.46	0.47	0.48	0.48
2.0	0.42	0.43	0.44	0.45	0.45
2.2	0.40	0.41	0.42	0.42	0.42
2.5	0.37	0.38	0.38	0.38	0.38
2.8	0.34	0.35	0.35	0.35	0.35
3.0	0.32	0.33	0.33	0.33	0.33

(a) Ratio of flange diameter to shell diameter

Table 18. Formability Factors for Square or Rectangular Shells Drawn in One Operation From 1010 Steel

r/w ratio(a)	Formability factor ($0.71\sqrt{r/h}$) for steel with thickness of the following percentages of blank width: 2.0%	1.0%	0.6%	0.3%
0.4	0.40	0.42	0.44	0.48
0.3	0.36	0.38	0.40	0.42
0.2	0.33	0.34	0.36	0.38
0.1	0.25	0.25	0.25	0.25
0.05	0.15	0.15	0.15	0.15

(a) Ratio of inside vertical corner radius of shell to width of shell

Table 19. Maximum Depth-to-Width (h/w) Ratios for Square or Rectangular Shells Drawn in One Operation From 1010 Steel

r/w ratio(a)	Maximum h/w ratio for steel with thickness of the following percentages of blank width: 2.0 to 1.5%	1.5 to 1.0%	1.0 to 0.6%	0.6 to 0.3%
0.3	1.0	0.95	0.90	0.85
0.2	0.90	0.82	0.76	0.70
0.15	0.75	0.70	0.65	0.60
0.10	0.60	0.55	0.50	0.45
0.05	0.40	0.35	0.30	0.25

(a) Ratio of inside vertical corner radius of shell to width of shell

Table 20. Factor y for Corner Radius of Blank for a Square or Rectangular Shell in Area C in Fig. 76

r/w ratio(a)	Factor y for shell with h/w ratio(b) of: 0.3	0.4	0.5	0.6
0.10	...	0.15	0.20	0.27
0.15	0.08	0.11	0.17	0.20
0.20	0.06	0.10	0.12	0.17
0.25	0.05	0.08	0.10	0.12
0.30	0.04	0.06	0.08	...

(a) Ratio of inside vertical corner radius of shell to width of shell. (b) Ratio of shell depth to shell width.

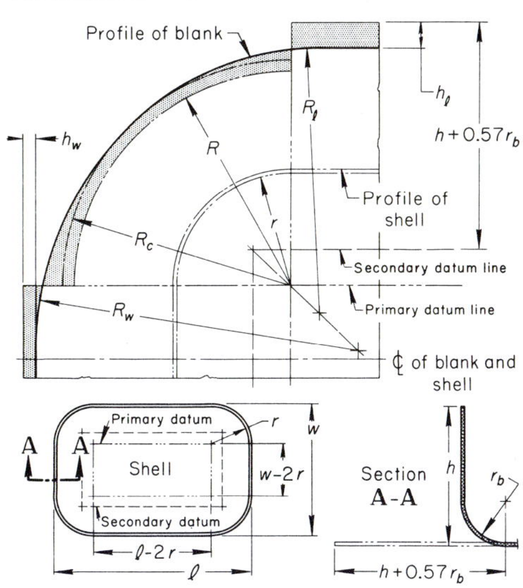

Fig. 78. Corner-layout method for wide shells with large corner radii in areas B and C in Fig. 76

Selected References

George Sachs, "Principles and Methods of Sheet Metal Fabricating", revised and enlarged by Henry E. Voegeli, Second Edition, Reinhold Publishing Co., New York, 1966. [A textbook on theory and practice. Chapters 11 through 15 (109 p) in particular analyze the processes of drawing thin-wall and thick-wall cylindrical cups and box-shape parts, and the redrawing and ironing of tubular parts. Extensive bibliographies are included.]

J. Dudley Jevons, "The Metallurgy of Deep Drawing and Pressing", Second Edition, Chapman and Hall, Ltd., London (England), 1941; 735 p. [A fundamental textbook. Comprehensive up to the date of its publication.]

"Die Design Handbook", prepared for the American Society of Tool and Manufacturing Engineers, Second Edition, McGraw-Hill Book Co., New York, 1965. [Section 10 (28 p), by Anton E. Mohrnheim, examines the mode of displacement of metal in drawing, and details algebraic and graphic methods for developing the blanks for cylindrical and rectangular draws. Section 11 (37 p) gives examples of the draw reductions used in forming many typical parts and illustrates the dies employed.]

Gerhard Oehler, "Schnitt-, Stanz- und Ziehwerkzeuge", Fifth Edition, Springer-Verlag, Berlin (West Germany), 1966; 643 p. [A standard textbook on cutting, stamping and deep drawing of metals. In German.]

J. M. Alexander, An Appraisal of the Theory of Deep Drawing, *Metallurgical Reviews*, 5(19), 1960, 349-411. [A critical review of the mathematical approach to the determination of the stresses in a material during deep drawing. The theory can assist in specifying punch and die profile radii and clearances. 38 references.]

W. P. Romanowski, "Handbuch der Stanzereitechnik", Veb Verlag Technik, Berlin (East Germany), 1959; 569 p. [A comprehensive textbook. 232 references. In German.]

C. W. Hinman, "Press Working of Metals", Second Edition, McGraw-Hill Book Co., New York, 1950; 551 p. [Chapter 16 (46 p) illustrates the die sequences used in some typical deep drawing operations, and Chapter 18 (57 p) contains formulas and reference tables for developing shell blanks.]

John A. Grainger, The Deep Drawing and Spinning of Sheet Metal, with Particular Reference to Non-Ferrous Materials, *Journal of the Institute of Metals*, 84, 1955-1956, 133-146. [Discusses deep drawing techniques, with special reference to the choice of lubricant.]

G. S. A. Shawki, Assessing Deep Drawing Qualities of Sheet, *Sheet Metal Industries*, 42, May, June and July 1965, 363-368, 417-424, 524-532. [A critical review, concluding that no single test is adequate. 124 references.]

J. C. Wright, Phenomenon of "Earing" in Deep Drawing, *Sheet Metal Industries*, 42, Nov 1965, 814-831. [Reviews the causes and control of earing. 32 references.]

Section II: Sheet Materials for Forming

Characteristics of Commonly Available Hot-Rolled and Cold-Rolled Steel Sheets

by Henry S. Erisman
Staff of Vice President-Operations
Bethlehem Steel Corporation

ABSTRACT

Steel sheets can be produced in a variety of types, grades, qualities and mechanical properties depending on end part requirements. This report discusses the commonly available low carbon varieties, with an explanation of how and why they are produced. Included is a simplified presentation of the aging phenomenon and how to cope with its effects. Surface finish and surface inspection are also discussed. A knowledge of the information contained in this report can be helpful in the purchase, the selection and proper use of the product--low carbon sheets.

INTRODUCTION

At the present time most steel for the production of carbon sheets is made by the basic oxygen, or open hearth or electric furnace process. The basic oxygen process is rapdily replacing the open hearth process as the dominant method of producing steel. In the United States, the 62-year reign of the open hearth ended in 1970.

This presentation deals primarily with low carbon steel having carbon content less than 0.15 per cent by heat analysis. TYPE of steel refers to the deoxidation practice used in making the steel, such as rimmed, capped or special killed. GRADE of steel refers to chemical composition only, whereas QUALITY refers to properties imparted to the product by manufacturing operations such as processing controls.

The correct steel for the end part application depends on the proper selection of:

TYPE, GRADE, and QUALITY

It is a misconception that all steels are the same, except for chemical composition. Characteristics can be widely dissimilar. For example, commercial quality can be produced for a moderate drawing operation, for a non-fluting requirement or for a moderately harder steel requirement, by proper selection of type and grade, and by use of proper controls in processing.

(Note. Strand cast steel, a relative newcomer in the field, has been developed and proved to be a worthy equivalent to steel made by the ingot casting process. Because the strand casting process does not lend itself readily to the rimming-type steel, there is strong evidence that killed steel (using silicon and aluminum as the deoxidizers) or fully non-aging special killed steel (with aluminum as the deoxidizer) will dominate as the production types. In some instances, vacuum degassing is used in making killed and special killed steel for strand casting. Hot-rolled sheets and cold-rolled sheets of strand cast steel can be produced with equally good surface, mechanical properties and non-aging characteristics as is produced with ingot cast steel. This paper deals with ingot cast steel and is not intended to include strand cast steels, developments of which have been reported by others).

There are many instances where a higher quality product is not necessary for the intended purpose. Also, the reverse is true, the characteristics of higher qualities might be required and are a sound, economical insurance. A knowledge of the available choices can be a valuable asset. It is the purpose of this paper to present those choices. After a look at a classification of products by size, we will explore:

TYPES, GRADES, QUALITIES
PROCESSING CONTROLS
MECHANICAL PROPERTIES
DIMENSIONAL CHARACTERISTICS
AGING
SURFACE

CLASSIFICATION BY SIZE

Hot-Rolled Sheet and Cold-Rolled Sheet products are described in ASTM A-568 and in AISI Product Manual entitled Carbon Sheet Steel. The following is a commonly used product category, by size:

Hot-Rolled Sheet
Coils and Cut Lengths

Width, In.	Thickness, In.
over 12 to 48 incl	0.0449 to 0.2299 incl
over 48	0.0449 to 0.1799 incl

Coils Only

over 12 to 48 incl	0.2300 to 0.5000 incl
over 48	0.1800 to 0.5000 incl

less than 12 is either Strip, Bar or Plate.

Cold-Rolled Sheet
Coils and Cut Lengths

Width, In.	Thickness, In.
2 to 12 incl(1)	0.0142 to 0.0821 incl
over 12 to 23-15/16 incl(2)	0.0142 and thicker
over 23-15/16	0.0142 and thicker

(1)Cold-rolled sheet coils and cut lengths, slit from wider coils with cut edge (only) and in thicknesses 0.0142 to 0.0821 incl., carbon 0.25 per cent maximum by heat analysis.

(2)When no special edge or finish (other than matte or luster) is specified, and/or single strand rolling of widths under 24 is not specified or required.

In widths 1/2 to 23-15/16 incl., and carbon over 0.25 per cent maximum by heat analysis, the product is known as Cold-Rolled Carbon Spring Steel Strip.

TYPES

The selection of TYPE of steel is an important decision. A knowledge of the end part application determines the choice. For the production of low carbon sheet steel, three major types are available, as follows:

A. RIMMED
B. CAPPED
C. SPECIAL KILLED.

When poured into ingot molds the steel will:

A. Rim or effervesce unless the gaseous activity is stopped by
B. Capping, using a heavy metal cap or by chemical additions; or by
C. Killing, using special deoxidizers. In the case of steel for sheets, aluminum is generally used as the deoxidizing agent and the product is known as SPECIAL KILLED steel.

Impurities and chemical elements (primarily carbon, phosphorus and sulphur) tend to rise and segregate to the center and to the top of the ingot. This effect is more pronounced in the case of RIMMED steel, intermediate in the case of CAPPED steel and least in the case of SPECIAL KILLED steel.

As illustrated, the ingot pattern of each type of steel is retained throughout the slabbing, the hot rolling and the cold rolling process. The final sheet product inherits its TYPE characteristics. Selection of best type is predicated on end part requirements and economics.

Refer to sketches entitled TYPES OF STEEL.

SURFACE: As a result of careful preparation of slabs, all three types result in sheet surfaces with a minimum of defects and imperfections. Because of its deep, protective skin, RIMMED steel gives best results, generally with the least amount of surface or sub-surface defects.

FORMABILITY. In spite of the non-uniformity that is a characteristic of RIMMED steel, most tonnage is made in this type, such as for exposed applications involving both drawing quality and commercial quality intended for moderate drawing applications. Due to its segregation pattern, most sheets made from RIMMED steels have softer edges with harder centers. RIMMED type sheets are generally used in the full-size dimensional form, such as for large, automotive panels. When sheared by consumers to smaller sizes, non-uniformity could be an undesirable characteristic.

CAPPED steel is not the type made for critical formability requirements.

SPECIAL KILLED steel is the best choice for formability by virtue of its superior mechanical properties, its uniformity and its non-aging characteristics.

UNIFORMITY. CAPPED steel is the economical choice for non-drawing applications where better uniformity is a desired characteristic. Multiples sheared or slit from CAPPED steel will give better uniformity than can be obtained from RIMMED steel with its variations from edge to center to edge. For example, CAPPED steel, slit to narrow widths, is used in the production of small diameter tubing. Also, for non-drawing applications, CAPPED steel is often used to obtain higher hardness levels, such as closer to the Rockwell B 60 maximum hardness for commercial quality, cold-rolled sheet. Cold-rolled sheets requiring hardness over Rockwell B 60 are generally produced from CAPPED steel using carbon over 0.15 per cent, or in combination with higher phosphorus and additional temper rolling. In some welding applications, CAPPED steel is more suitable because of the uniformity of the product, as opposed to the dissimilarity of rimmed steel with its center

TYPES OF STEEL

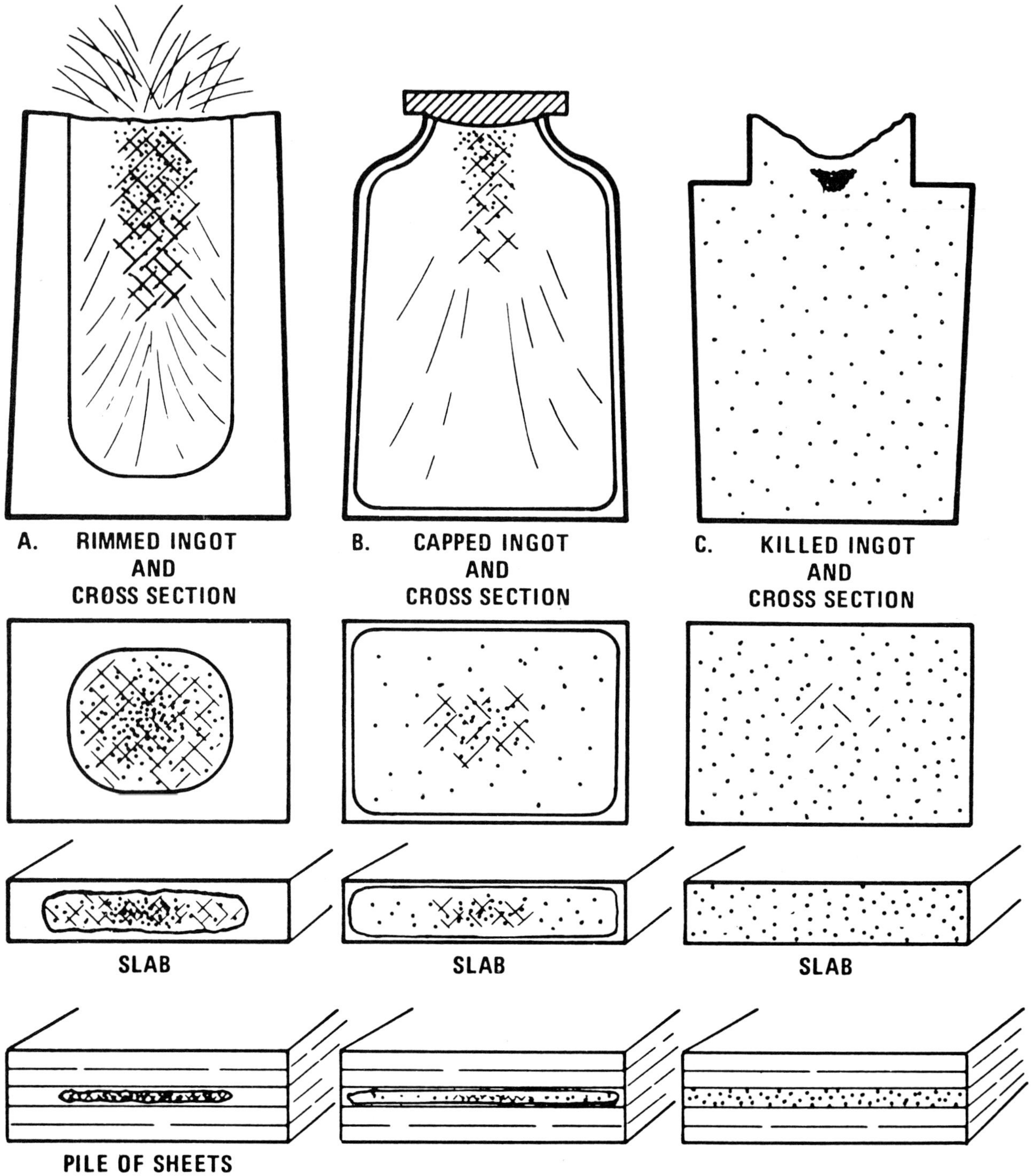

SURFACE	EXCELLENT	VERY GOOD	VERY GOOD
FORMABILITY	VERY GOOD	GOOD	EXCELLENT
UNIFORMITY	GOOD	VERY GOOD	EXCELLENT
AGING	YES	YES	NO
PRICE	BASE	BASE	PREMIUM

and edge differences. For the best in uniformity, as it would affect drawing performance as well as some sophisticated welding applications, SPECIAL KILLED is the choice. Not only is SPECIAL KILLED a more uniform product from edge to center to edge it has the best uniformity from sheet to sheet, regardless of position in ingot location from which the sheet was made.

QUALITIES AND GRADES

Hot-rolled sheet and cold-rolled sheet are furnished to three principal qualities, and usually to produce an identified part.

Commercial quality is ordinarily produced from rimmed or capped steel with a carbon maximum of 0.15 per cent by heat analysis (ASTM A-569 for hot-rolled and A-366 for cold-rolled can be used). The minimum requirement is that the sheet will bend flat on itself in any direction. In cold-rolled sheet the hardness is not expected to exceed Rockwell B 60 at time of shipment. Commercial quality can be furnished to produce identified parts within properly established limits (SAE J424).

The suitability of commercial quality can be assessed by use of the SAE Recommended Practice J863, Scribed Square Test. Experience has shown that commercial quality should be suitable if measurements are less than the following values:

Hot-Rolled Sheet

25% or less for thicknesses 0.0822" to 0.187" incl.
30% or less for thicknesses 0.188" and thicker.

Cold-Rolled Sheet

25% or less for all thicknesses.

For percentages of 25% or more, drawing quality or drawing quality special killed may be required.

Commercial quality can also be furnished to chemical composition, using a grade from SAE J403, Table 1 B or an equivalent composition:

Ladle Chemical Composition Limits, %

*SAE No.	C	Mn	P**	S**	Si
1006	0.08 max	0.25-0.45	0.040	0.050	0.10 max
1008	0.10 max	0.25-0.50	0.040	0.050	0.10 max
1009	0.15 max	0.60 max	0.040	0.050	0.10 max
1010	0.08-0.13	0.30-0.60	0.040	0.050	0.10 max
1012	0.10-0.15	0.30-0.60	0.040	0.050	0.10 max

*Commercial quality is also available in higher compositions.

**Phosphorus and Sulfur limits are generally lower, e.g. 0.035 maximum and 0.040 maximum, respectively, with typical analyses running less than 0.010 and 0.030.

Incidental elements, although not specified, are held to relatively low amounts. Such elements are copper, nickel, molybdenum and chromium.

Drawing quality is ordinarily produced from rimmed steel using SAE Grade 1006 or to a chemical composition as shown in Specification ASTM A-621 for hot-rolled or A-619 for cold-rolled. By the use of specially produced or selected steels in combination with special processing, drawing quality is furnished with better and more uniform drawing properties than commercial quality. It is used for more severe drawing applications to produce identified parts within properly established limits (SAE J424). See above, under commercial quality description, for assessment of quality required (refer SAE J863).

Drawing quality special killed is produced to make parts involving the most difficult draws or where a minimum of aging (hot-rolled) or non-aging characteristics (cold-rolled) are desired. ASTM Specification A-622 for hot-rolled and A-620 for cold-rolled can be used.

Physical quality can be furnished when mechanical property values are specified or required other than bend tests of commercial quality. (ASTM A-570 for hot-rolled and A-611 for cold-rolled can be used). Such values include those indicated by tension, hardness or other commonly accepted mechanical tests. Requirements to meet both mechanical tests and drawing performance are sometimes necessary. Care should be taken to ensure compatibility of such requirements.

Although mechanical properties are not guaranteed when commercial quality or drawing quality or drawing quality special killed steel is specified, certain tests are performed and property values are used as guidelines for producers and consumers. Processing controls are used which ensure ability of sheets to make identified parts within the product, the grade and quality specified.

PROCESSING CONTROLS

Following the selection of a TYPE and a GRADE of steel, certain processing controls are used to determine the sheet product characteristics. (Refer Steps in Production from Slab to Sheet). Included are:

Hot Rolling

1. Uniform heating of conditioned slabs to approximately 2200°F.

2. Proper finishing temperature at exiting from last stand of the hot rolling operation (approximately 1600°F min. corresponding to the upper critical temperature of low carbon steel).

3. Proper coiling temperature. Water sprays on the hot mill run-out table determine the coiling temperature, which is predicated on the next stages of processing, but pre-determined by the thickness to some extent. For the production of hot-rolled sheets, higher coiling temperatures produce softer sheets and lower coiling temperatures produce harder and stronger sheets. On hot-rolled coils intended for the production of cold-rolled sheets, a moderately high coiling temperature (over 1200°F) is used, except in the case of special killed steel when a relatively cold coiling temperature (less than 1100°F) is an absolute necessity for the production of best drawability with a unique "pan-cake" or elongated-type of ferritic grain in the cold-rolled sheet.

4. When required, hot-rolled sheets are temper-rolled (skin rolled) to improve flatness, to minimize surface disturbances such as coil breaks and fluting, and to alter mechanical properties.

Pickling (Scale Removal)

1. For the production of hot-rolled pickled or cold-rolled sheet products, full scale removal with a minimum of cross breaks is important.

2. In the production of hot-rolled sheets for formability, it is important to use a minimum of flex-rolling in the pickler processor in order to minimize the effects of aging which results from cold-work stress followed by the heat of pickling.

STEPS IN PRODUCTION FROM SLAB TO SHEET

EXAMPLE: 0.08 max. Carbon (1006) to Produce Drawing Quality Sheet

Rimmed Type

MICRO SECTION AT 100X MAGNIFICATION (SIMULATION)

Typical Hardness

Slab at 2200° F is Hot Rolled .100" thick, coil cooled, pickled

.100"

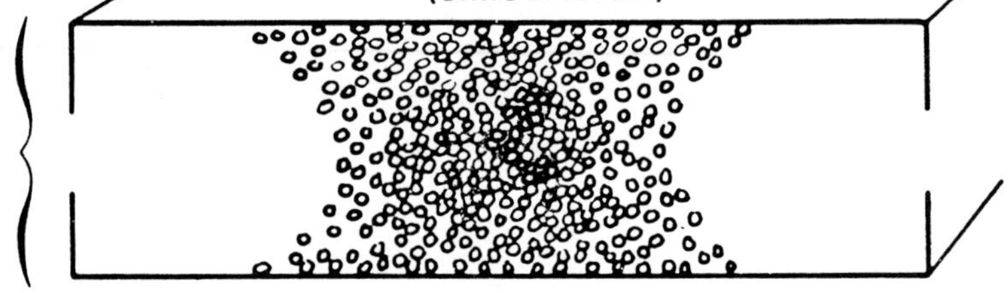

Ferrite Grains Plus Small Areas of Pearlite

B55

Tandem Reduced to .040" 60% Reduction

.040"

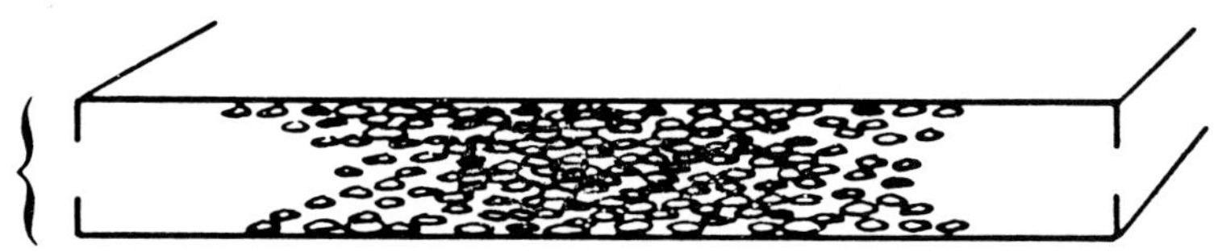

Elongated Ferrite and Pearlite

B100

Box Annealed Dead Soft

.040"

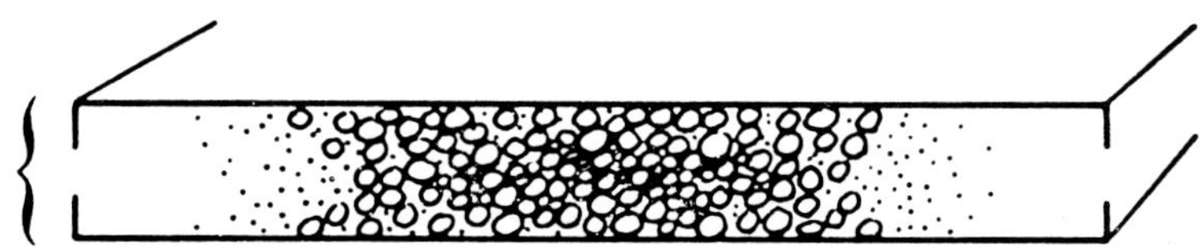

Recrystallized Ferrite and Spheroidized Carbides

B45

Skin Passed Approx.1% Reduction or Stretch

.040"

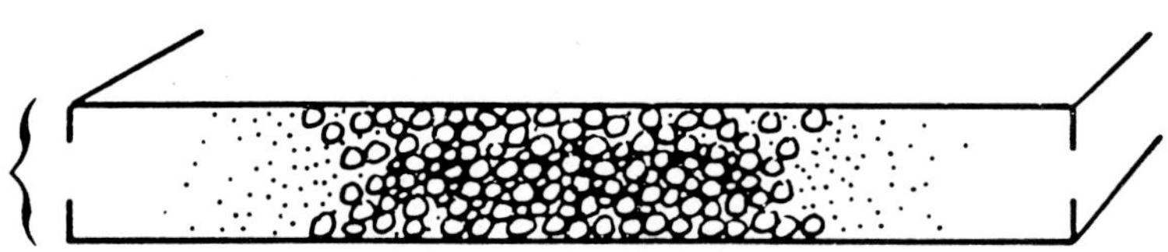

Recrystallized Ferrite and Spheroidized Carbides

B45+

The greater the percentage of reduction from hot band to cold rolled sheet the finer the grain size and the harder the sheet after annealing; the less the reduction, the coarser the grain size and the softer the sheet. The optimum grain size is a combination of correct chemistry, correct reduction and proper annealing practice.

Cold Reduction, Annealing & Temper Rolling

1. The percentage of cold reduction to final thickness is important. A greater reduction results in a finer, harder grain; a lesser reduction results in a coarser, softer grain. Too fine a grain size can cause breakage; too coarse can cause surface-granulation (roughening) on forming. The amount of reduction is approximately 50% - 70% to produce an optimum ferritic grain size of ASTM numbers 6 to 8 (approx.). For example, the cold reduction of 0.100" thick hot rolled coil to 0.040" is 60%. Reduction to 0.020" is 80% and results in a finer grain, harder sheet; to 0.060" is 40% and results in a softer, but coarser grain sheet. Likewise, reductions from thicker or from thinner hot bands to the same final thickness (e.g. to 0.040") results in finer and harder, or coarser and softer cold-rolled sheets.

2. Proper annealing temperature (approximately 1200° - 1300°F for drawing quality) along with proper heating, hold and cooling rate determines the optimum in properties, grain size and carbon spheroidization.

3. Although cold-rolled sheets are sometimes used in the annealed last or dead soft condition (for unexposed applications), most cold-rolled sheets are temper rolled (skin rolled), especially when used for exposed applications. The amount of temper rolling is important and is determined by the end part application. Where formability is not important but non-fluting is required, cold-rolled sheets are temper rolled as much as 2.0% which decreases drawability. Where deep drawing is required, the amount of temper rolling is approximately 1.0% or less but sufficient for sheets to be free from stretcher strains when drawn.

The actual percentages of temper rolling can be regulated very closely, for 0.50% or 0.75% or 1.25%. The amount is usually a compromise between drawability and non-straining propensity. With rimmed and capped steel the effects of temper rolling are temporary, therefore, for optimum results, the sheets should be used as promptly as possible.

Mechanical Properties

Hot-rolled and cold-rolled sheets, when specified as commercial quality, drawing quality or drawing quality special killed, are generally produced to make identified parts without guarantees on mechanical properties as rejection limits. However, after years of experience, producers and consumers have established guide lines of inspection, testing and acceptance which are usually negotiated and mutually agreed upon, based upon knowledge of end part applications.

Typical mechanical properties of the products are shown in the table that follows. For the same chemical composition, the properties are different for hot-rolled sheet and cold-rolled sheet, as well as from coil to coil or lot to lot of each product and each dimension, especially thickness. The variations in properties are greater for hot-rolled than for cold-rolled and the variations are greater for commercial quality than for drawing quality with variations at a minimum for special killed steel. An important point to bear in mind is that commercial quality does not connote a common set value for all sheets in a shipment since some sheets would be as good as drawing quality. There is an overlap of property values for all qualities. The higher the quality the better and more uniform the properties.

Typical Mechanical Properties of Low Carbon Sheet Steel

	Hardness Rockwell B	Yield Point 1000 psi	Tensile Str. 1000 psi	Tot. Elong. in 2"
Hot-Rolled (0.0598" to 0.2299")				
CQ	50/70	28/40	45/56	28/36
DQ	44/60	26/36	42/52	34/42
DQSK	48/65	28/38	42/54	36/44
Cold-Rolled (0.0239" to 0.1345")				
CQ	40/60	28/36	40/48	36/41
DQ	38/48	25/32	40/46	38/43
DQSK	36/45	23/28	40/46	40/45

CQ: Commercial Quality
DQ: Drawing Quality
DQSK: Drawing Quality Special Killed

Other test values are sometimes determined to better assess the formability of steel sheets, such as uniform elongation (prior to local neck down), plastic strain ratio "r", work hardening exponent, along with test values obtained by various cup-drawing tools and procedures, such as the Olsen, Erichsen, Swift and Fukui tests.

DIMENSIONAL CHARACTERISTICS

The tolerances for thickness, width, length, flatness, etc. are listed in various publications, such as ASTM A-568, AISI Carbon Sheet Manual (available in small pocket size) and in commercial brochures. In some cases, restricted tolerances are available, subject to negotiation.

Some interesting observations are these:

1. Thickness tolerances have been reduced in recent years as a benefit derived from improved automatic equipment. Also, a higher percentage of the material will generally be closer to the desired thickness than heretofore. Thickness tolerance can be applied equally or unequally, over and under but a recent trend has been to specify a minimum thickness with zero tolerance on the minus side.

2. Flatness is a standard rather than a process. Generally, there are two standards of flatness (commercial and stretcher leveled) which can be obtained by the use of various flattening methods such as roller leveling, tension leveling, special roller back-up leveling or by gripping and stretching. Flatness usually is not as critical for forming or drawing operations. In an effort to obtain better flatness, ductility must sometimes be sacrificed.

In the case of sheet steel, the thickness tolerance includes crown which is the amount that the sheet is thicker at the center than at the edge. Thickness is measured at any point across the width not less than 3/8 in. from a side, cut edge and not less than 3/4 in. from a mill edge (hot-rolled). The crown on hot-rolled sheets is generally not in excess of 0.004". The crown on cold-rolled sheets is generally not in excess of 0.001".

AGING

Aging is a phenomenon in low carbon steel sheets that progressively changes their as rolled characteristics over a period of time.

Aging can be expected with rimmed or capped steel. It can be minimized (in hot-rolled sheet) or nullified (in cold-rolled sheet) by the use of special killed steel, properly processed.

In rimmed or capped steels, its effect can be minimized by:

1. Prompt use of sheets, including proper rotation of stock so that older material is used first.

2. Effective roller leveling immediately before use. Roller leveling can not restore ductility lost by aging but it can minimize fluting and stretcher straining, although not indefinitely.

3. Reduction of delay between draws. Insofar as possible, progressive forming and drawing should be completed in prompt, successive stages. The frictional heat of the draw coupled with the delay between draws accelerates the aging.

4. By avoiding exposure of sheets to over-heated storage conditions.

Aging progresses with TIME and with TEMPERATURE. It results in two major changes:

1. Loss of ductility accompanied by increase in hardness, yield point and tensile strength, with decrease in elongation.

2. Stretcher straining and fluting when formed.

Although aging occurs to some extent in hot-rolled sheets it is more critical in the case of cold-rolled sheets because of the thinness of the material and because of the importance of surface appearance when used for exposed applications.

The following is a simplified example of the phenomenon, illustrating the elimination of the yield point elongation (Y.P.E.) by skin-passing, followed in time by increase in yield point (Y.P.) and by the return of the yield point elongation (fluting and stretcher straining) in the case of rimmed or capped but never changing in the case of special killed steel. Special killed steel retains the same yield point, (Y.P.) as well as the same hardness and ductility; and when skin-passed as strain-free, retains its original non-fluting, non-straining characteristics. Curves B, E and F remain the same, therefore, formability and freedom from surface disturbances remain the same.

AGING

CARBON ROLLED SHEET SHEET (CARBON 0.08 MAX)

Phenomenon of aging can be illustrated by plotting the results of tension testing as shown below.

The Yield Point Elongation (Y.P.E.), its occurrence and intensity, correspond to fluting or stretcher straining on the surface of sheets when formed.

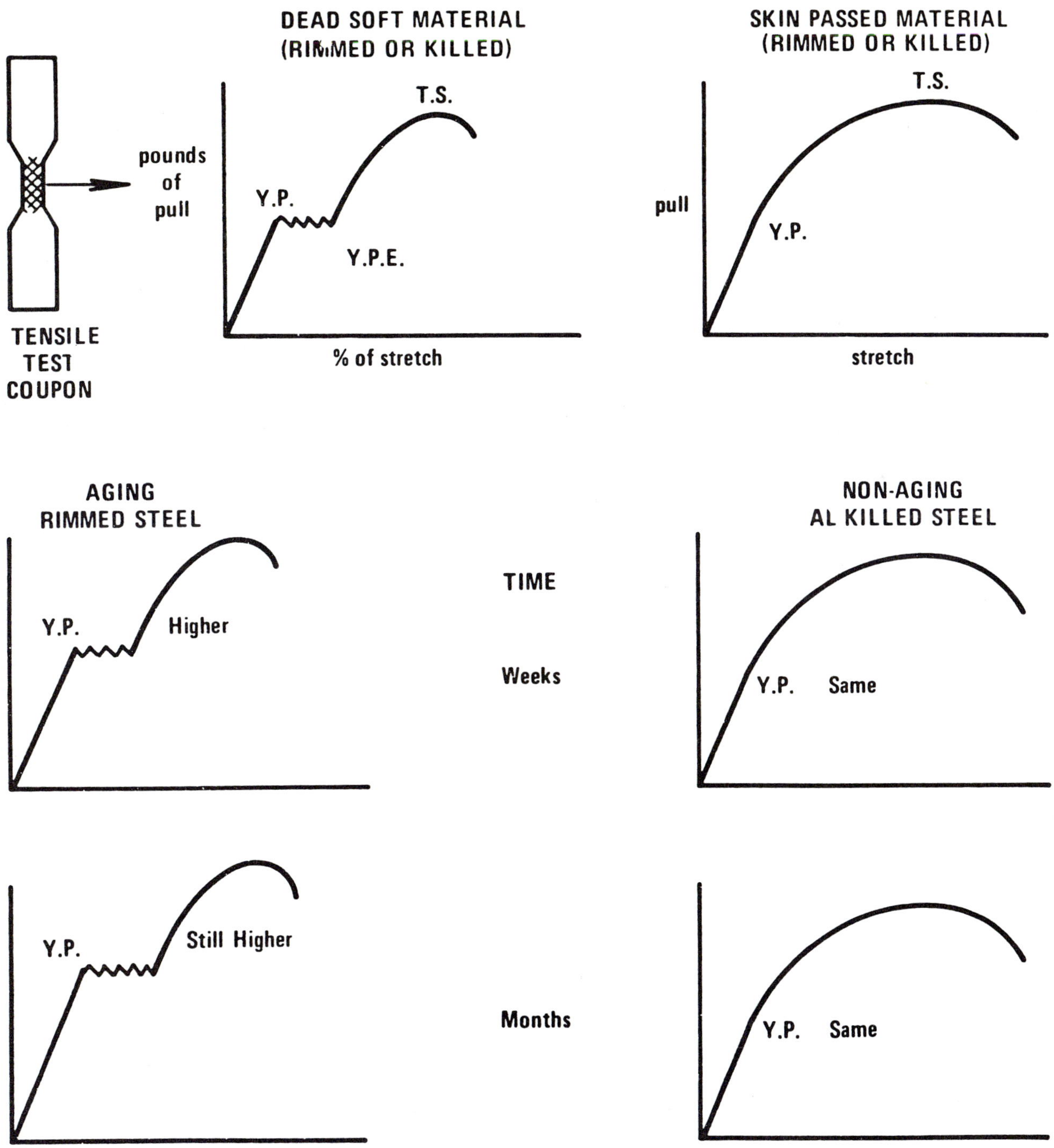

SURFACE

SURFACE refers to both surface finish and surface imperfections.

SURFACE FINISH (texture) refers to the degree of roughness, smoothness or luster of the surface as distinct from surface imperfections. The degree of surface roughness or surface smoothness is a factor in the drawing, painting, or plating of parts.

Hot-rolled sheets are categorized as either hot-rolled or hot-rolled pickled (scale removed), known as the CONDITION. The surface of hot-rolled products is generally of secondary importance. Hot-rolled sheets are not recommended for exposed parts where finish is a primary requirement. Hot-rolled sheets are subject to coil breaks, stretcher strains and fluting in the as rolled condition but can be furnished non-fluting when required. For bright plating, hot-rolled sheets like cold-rolled sheets, must be prepared by grinding and buffing.

Cold-rolled sheets are generally used for exposed applications where surface finish (texture) and a minimum of defects are of primary importance. Surface finish of cold-rolled sheets are sometimes classified as follows:

	Approximate Peak Height Microinch
Matte	20 - 80
Commercial Bright	10 max.
Luster Finish	5 max.

Microinch measurements are expressed as AA (Arithmetic Average) or as RMS (Root Mean Square, approximately 11% higher than the AA figures). A microinch equals one millionth of an inch and relates to the height of the peaks on the surface "topography" of a sheet. The number of peaks per inch is also an important consideration but no guide line numbers have been adopted, subject to further development.

Matte finish, which is produced from blasted rolls, is sometimes sub-divided into what is known as light matte or very light matte (approximately 20 to 40 microinches, AA). With some surface preparation matte finish is suitable for decorative painting, but not suitable for bright plating.

Luster finish is produced from ground rolls and is suitable for decorative painting and for plating with additional special surface preparation by the consumer.

Commercial bright finish is sometimes used for plating having less demanding requirements than luster finish.

SURFACE INSPECTION. It is common practice to furnish the inspected side on the outside of the coil or top side of cut lengths.

Sheet steel in coils or cut lengths may contain surface imperfections that can be removed with a reasonable amount of metal finishing.

Coils contain more surface imperfections than cut lengths because the producer does not have the same opportunity to sort portions containing such imperfections as is possible in the case of cut lengths.

METALLURGICAL FACTORS WHICH INFLUENCE THE FORMABILITY OF SHEET STEELS

M. V. Balakrishnan, Staff Metallurgist

R. O'Neil, Senior Metallurgist

Inland Steel Company, East Chicago, Indiana

Introduction

Only one-fifth of all metals produced are cast directly into shape; the remainder requires some kind of mechanical working operation to obtain the final part. Various types of sheet metals, in particular, sheet steels, are produced commercially in large tonnage for application into automotive, appliance, and other areas. These final parts are manufactured by various forming techniques such as drawing, stretching, roll forming, etc.; thus, sheet steel formability can be defined as the ability of the sheet to be successfully formed to a given shape.

In order to successfully form a part from a given piece of sheet steel, it has to possess certain properties that are conducive to the forming operation. These properties are greatly influenced by the way the sheet steels are produced. Beginning with the initial steelmaking process to the final finish rolling operation, each processing step must be carefully controlled in order to produce a sheet steel product possessing the required formability to suit the customer's needs.

In the following presentation, the authors have analyzed the metallurgical factors which influence the forming characteristics of sheet steels. The dependence of these factors on the various mill processing steps and their overall effect on the final sheet product is also discussed.

Quality Levels

Sheet steels are usually purchased as one of three product types: hot rolled, cold rolled or coated (galvanized, ALUMINIZED, or prepainted). Typical processing steps for the above products are shown in Plate I. Typical quality levels that are common to each product type are: Commercial Quality, Drawing Quality, Drawing Quality Special Killed, Physical Quality, and various High Strength Low Alloy steels. The level of formability within a product type will improve as the quality level shifts from Commercial Quality to Drawing Quality to Drawing Quality Special Killed. However, a Drawing Quality hot rolled product will not have the same formability as a Drawing Quality cold rolled product.

Commercial quality sheet products are intended to produce parts which involve simple bending or moderate forming. The commercial guarantees made for this quality level are Rb 60 maximum at time of shipment and adherence to standard bend tests. In the galvanized product, there is a major quality level which falls between the Commercial Quality and Drawing Quality levels. This product designation is "lock forming quality". This product is intended for end uses which require machine lock forming such as heating ducts, muffler shells, etc. Steel ordered to this quality level will be processed in such a manner that the ability to form a lockseam will be guaranteed.

Drawing quality steels are produced and processed with more stringent quality restrictions to produce a more formable end product. Established breakage limits are recognized by the supplier and claims for excessive breakage honored.

A Drawing Quality Special Killed steel will provide the best formability of the common quality levels. Also, the Drawing Quality Special Killed cold rolled product normally will be processed so that it is non-strain aging, whereas Drawing Quality or Commercial Quality cold rolled material will strain age. This quality level is intended for deep drawing applications and, as with Drawing Quality, properly established breakage limits are recognized.

Traditionally, physical quality and high strength steels are not regarded as being processed to any formability level. These materials are sold to a minimum physical strength level

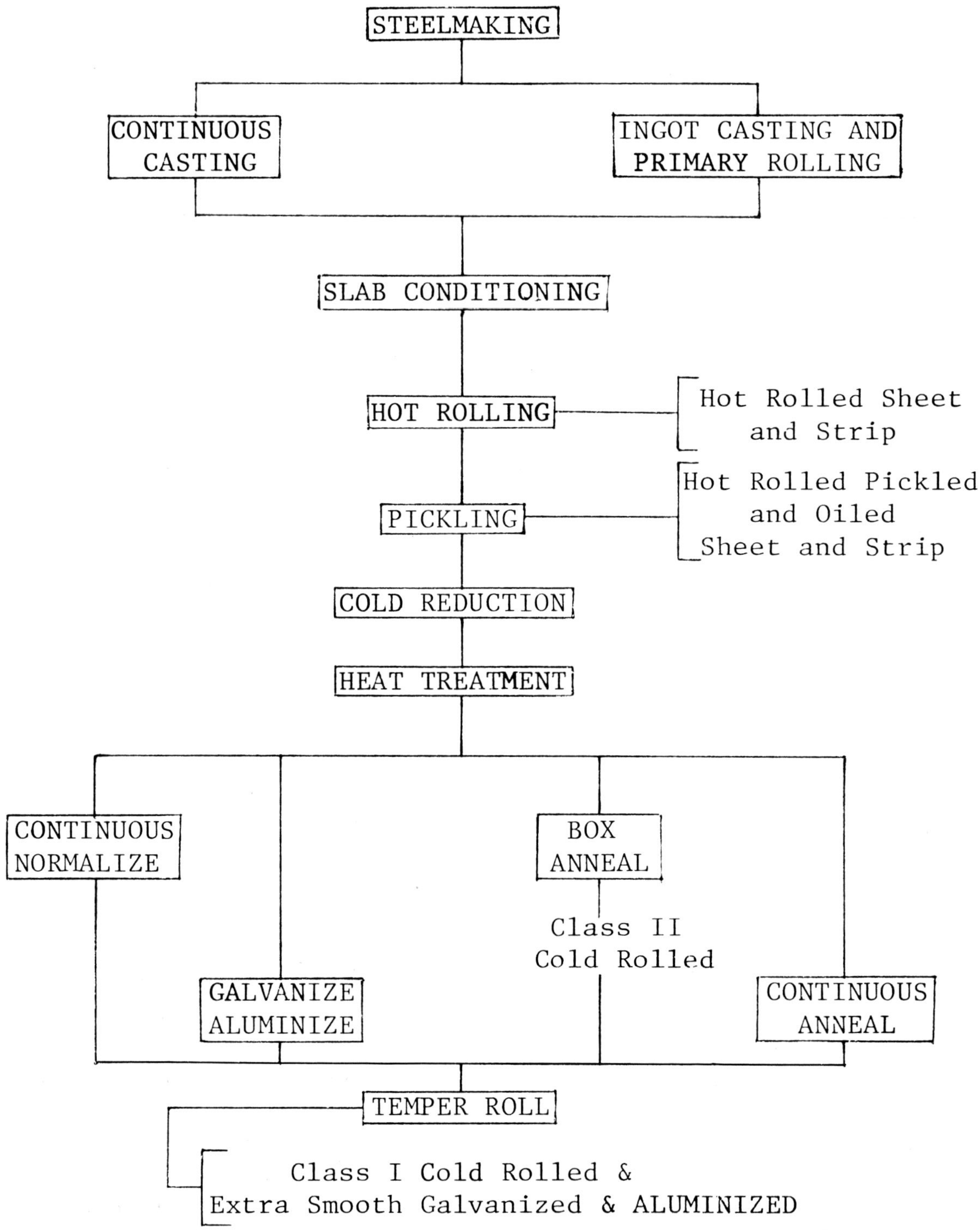

Plate I. **Flow** Sheet Showing Mill Processing Sequence for Production of Sheet Steels.

or in some cases a minimum hardness range with the implication that the customer has taken the reduced formability into account when designing the part. As a rule, when the minimum strength level ordered increases, the ductility decreases. Today, a new family of physical quality and high strength steels is being marketed in which a large improvement in formability is obtainable at a given strength level and the improvement is such that these steels will form similarly to Commercial Quality steels.

Factors Which Influence Formability

Regardless of the quality level, there are a variety of factors which influence the formability of a given type of steel. The common factors to be considered are: Steel chemical composition, preferred crystal orientation, macro and microstructures, surface texture and cleanliness, cold work and aging, and final shape and cross section of the strip. We shall look into the details of each of these variables and examine their influences separately and collectively.

Steel Chemistry

Base steel chemistry is probably one of the most important factors that affects formability. Steels used for forming as a rule are low in carbon content. Typical steels usually contain less than 0.15% carbon and less than one percent alloy plus residual elements. Manganese, the principal addition, normally ranges from 0.20% to 0.60%. Controlled amounts of aluminum, silicon, columbium, or titanium may be added as deoxidants or to develop certain properties. Residual elements such as phosphorus, sulfur, copper, chromium, nickel, molybdenum, nitrogen, etc., are usually held as low as possible. Normally, application limits at the steelmaking shops are based on the quality level being ordered.

Carbon is a significant element in steels used for complex forming applications. It is advisable to keep carbon in the 0.04 - 0.10% range. Increased amounts of carbon result in an increased amount of carbides (referred to as pearlite* and/or cementite) and finer final grain structures. The resultant effects increase the strength of the product and are detrimental to formability.

*Pearlite is actually a mechanical mixture of ferrite and cementite.

Manganese in the range of 0.20 - 0.60% provides good hot working qualities and promotes the development of desired grain size, especially in cold rolled grades during box annealing. A certain amount of manganese is necessary to neutralize the detrimental effect of sulfur upon hot workability. The manganese to sulfur ratio must be controlled to insure good surface quality. Sheet steels for tough forming applications usually contain manganese in the range of .25 - .50%. At higher levels manganese has a strengthening effect.

Phosphorus and sulfur are undesirable in high quality sheet steels used for formability applications and are kept to levels as low as economically feasible. Phosphorus is a solid solution strengthener and has an embrittling effect. Sulfur is present in the form of manganese sulfide inclusions. These inclusions retard the grain growth in the annealing processes and produce fine grained areas. In certain types of ingot poured steels, most notably rimmed steels, sulfur segregates positively towards the upper portion of the ingot. The allowable level of both phosphorus and sulfur depends on the final quality.

Residual elements such as copper, chromium, nickel, molybdenum, tin, and others should be kept to a minimum because of their detrimental effect upon the mechanical properties of sheet steel; the higher the level of these elements, the higher the strength. These elements are kept low by proper scrap selection and other steelmaking practices.

Oxygen and nitrogen also play an important role in sheet steels. The amount of oxygen in liquid steel determines the steel type and its solidification characteristics in the ingot. The level of oxygen is controlled by use of deoxidizers such as aluminum, silicon, titanium, etc. When oxygen is tied up with these deoxidants, complex non-metallics, eg, aluminates, silicates, oxysulfides, etc., are formed. Most non-metallics float out and combine with the slag; however, some remain entrapped and form inclusions that hamper uniform grain growth and may initiate failure during forming. If the inclusions are entrapped too near the surface of the sheet, they can initiate breakage.

Nitrogen is also an interstitial element and causes aging in sheet steels. It is introduced naturally in steels through contact with air. Nitrogen when artificially added can also be used as a strengthening element. In certain steel grades (fully aluminum killed) nitrogen combines with aluminum to produce, when processed properly, a finished product with superior forming characteristics. Cold rolled steels of this type are also non-aging.

Macrostructure

A strong interrelationship exists between carbon, oxygen and deoxidizers (such as Si, Al, Ti) and the relative amounts of these elements in liquid steel determine the steel type and macrostructure. The four basic steel types are: 1) Rimmed, 2) Mechanically capped, 3) Semi-killed, and 4) Fully killed. The details of these individual types are discussed below and typical slab macrostructures are exhibited in Plate II.

Rimmed Steel. Molten steel contains dissolved oxygen, the quantity being determined by the amount of carbon present, steel temperature and deoxidants. The higher the carbon content the lower the oxygen content and vice versa. Since rimmed steels are normally low in carbon and contain only small amounts of deoxidizers, the dissolved oxygen is at a high level. When the steel is poured into ingot molds, carbon monoxide gas will evolve and boil off as the molten metal solidifies. The evolution of this gas produces a characteristic metal action called rimming, and the resultant steel type is called rimmed steel. This rimming action takes place next to the ingot wall and produces a zone of almost pure iron. The core of the rimmed ingot, especially the upper portion, is chemically segregated, i.e., it contains higher amounts of most elements than the nominal steel analysis. The rim zone produces a very clean surface which in the final product form is especially useful for exposed applications.

Mechanically Capped. This steel type is a modification of rimmed steel where the rimming action is halted after a short time by the use of a metal plug in the ingot top. This interruption of the rimming process produces an ingot with a thinner rim zone and more uniform segregation characteristics than rimmed steel. The main application for this steel type is unexposed flat rolled products which require only moderate forming.

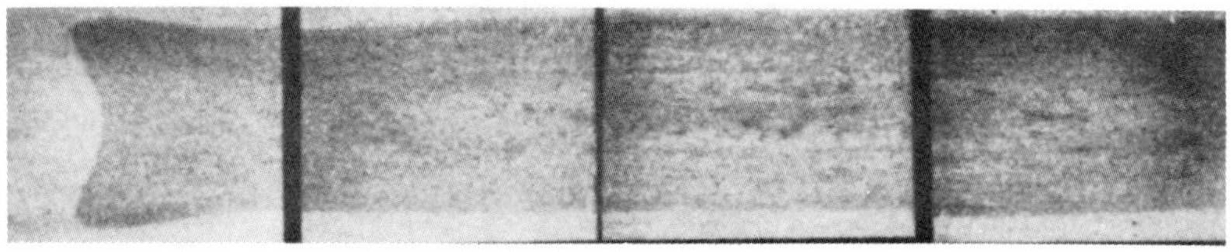

Rimmed Steel

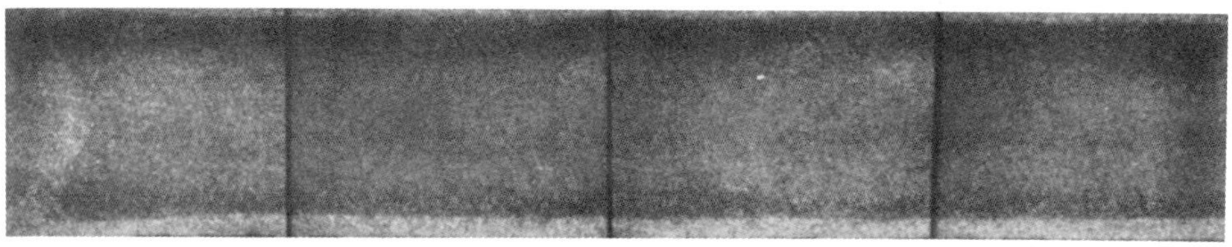

Mechanically Capped Steel

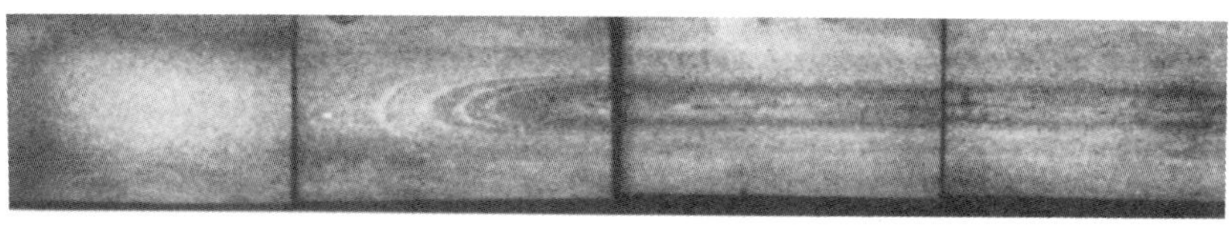

Semi-Killed Steel

Fully Aluminum Killed Ingot Cast

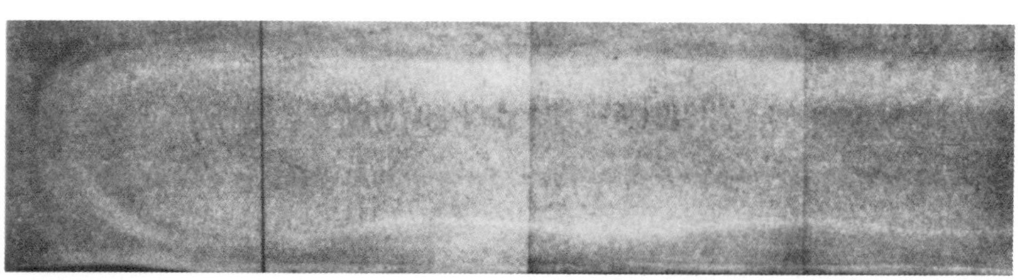

Fully Aluminum Killed Strand Cast

Plate II. Macro-etches of Slabs from the Various Steel Types.

Semi-Killed. This steel type contains controlled amounts of deoxidants (usually aluminum or silicon). The resultant steel has no rimming action; however, there is gas formation to prevent the formation of large shrinkage cavities or pipe. Such steel are usually produced to a higher carbon level (.10 - .15%) or where strengtheners are added (phosphorus, nitrogen, columbium, vanadium) to produce a final product with higher mechanical properties.

Killed Steel. This steel type contains a strong deoxidant to tie up the dissolved oxygen. The most commonly used element for deoxidation in sheet steels is aluminum. Other less commonly used elements are titanium, silicon, columbium or zirconium. When killed steel is poured into ingot molds there is no gas evolution. Hence, these ingots have large shrinkage cavities or pipe in the extreme upper portion of the ingot when solidified. To prevent the occurrence of pipe, these steels are usually poured into ingots equipped with "hot tops". These "hot tops" keep the steel molten at the top of the ingot and feed the molten steel into the shrinkage cavity. Aluminum killed steels have a relatively high degree of uniformity in chemical composition except near the vicinity of the shrinkage cavity. The shrinkage cavity or "hot top" portion of the ingot is removed by cropping during primary rolling. Aluminum killed sheet steel is primarily used for end uses that require a high degree of forming such as in deep drawing.

The proper selection of steel chemistry and deoxidation is dependent upon the intended end application. Stampings that are difficult will require special aluminum killed steels. For end uses that call for a superior surface such as an automotive hood, a rimming grade of steel is an excellent selection. Mechanically capped or semi-killed steels would be appropriate for unexposed end uses requiring moderate forming.

Microstructure

The microstructure of sheet steel is an extremely important factor in formability. The important microstructural features are grain size, grain shape, carbide morphology, and inclusion content. These micro-characteristics are controlled by steelmaking and subsequent processing.

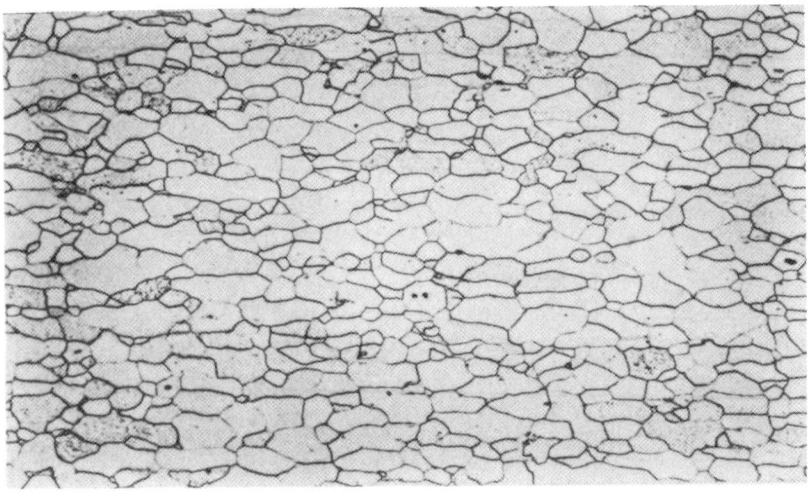

Cold Rolled
Box Annealed

G.S.: ASTM 7 - 8
100X

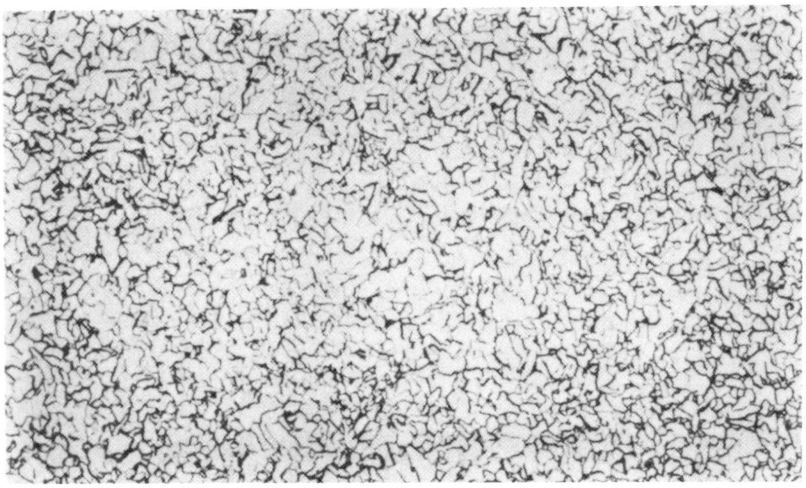

As Hot Rolled

G.S.: ASTM 9 - 10
100X

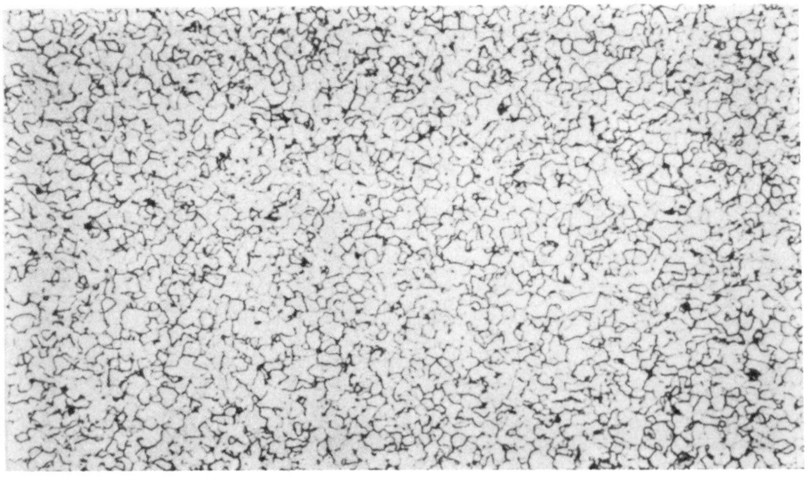

Galvanized

G.S.: ASTM 9 - 10
100X

Plate III. Photomicrographs of Cold Rolled Box Annealed, Hot Rolled, and Galvanized Strip from Fully Aluminum Killed Steel.

Fully recrystallized grains with properly developed texture and optimum size and shape are necessary for good formability.

Ferritic grain size has a strong relationship to the strength and hardness of the final product. Typical ASTM grain size range for optimum formability of cold rolled steels is between 7 and 9. Hot rolled and galvanized products tend to exhibit a slightly finer grain size. A fine grained product (ASTM 10 or larger), is harder, less ductile and has a high resistance to forming which can result in breakage. On the contrary, a coarse grained product (ASTM 5 or smaller) will have lower hardness but will exhibit a rough surface when formed. This rough surface often times is objectionable on exposed parts.

The typical grain shape for sheet steels is equiaxed. Equiaxed grains are found in rimmed and mechanically capped steels. Hot rolled and galvanized aluminum killed steel also have equiaxed grains. Cold rolled aluminum killed steels, when properly processed, exhibit pancake shaped ferritic grains. This grain shape which is associated with the proper texture is most desirable for deep drawing applications.

A random distribution and proper size of the carbide phase is also desirable. The amount of carbide phase present is proportional to the carbon content. The carbides are present in the form of cementite (Fe_3C) and/or pearlite (a lamellar mixture of ferrite and cementite).

In most hot rolled and galvanized material the carbides are pearlitic. In box annealed cold rolled steels the carbides are in the form of cementite. Hot rolling practice and subsequent heat treatments are controlled to produce the optimum carbide structure.

As mentioned earlier, inclusion content can affect the grain growth. If large inclusions are entrapped near the sheet surface, they can initiate breakage. The inclusion content is minimized by controlling steelmaking practices.

Typical microstructures of fully aluminum killed hot rolled, galvanized and cold rolled are shown in Plate III.

Surface Texture and Cleanliness

Both surface texture and cleanliness are also of importance in the formability of cold rolled sheet steels. In hot rolled or coated products these surface characteristics are not controlled to affect formability, per se.

In cold rolled, too smooth or too rough a surface texture can cause problems during forming. If the surface is too smooth, proper lubrication is difficult since the surface has too few sites to hold the lubricant. If the surface is too rough, the peaks can act as stress raisers and cause breakage. The traditional parameters measured to describe surface texture are peak count and the peak to valley depth. The optimum surface texture for best formability is conjectural. This is partially due to the lack of standardized measuring equipment to describe the parameters. It is also partially due to the disagreement among steel producers and users as to the range within which these parameters should fall; or whether or not these are the parameters that should be measured.

In cold rolled sheet steels surface cleanliness is controlled by proper selection of rolling oils and annealing atmospheres. Generally, the surface should be free of oxides and other residues. Highly clean surfaces are most important for steels intended for lithographing. A certain amount of carbon residue can be tolerated and sometimes desired for steels intended for deep drawing.

Aging and Cold Work

Aging is a phenomenon in steel which causes steel to harden. The two types of aging that commonly occur are quench aging and strain aging. Quench aging is the precipitation of carbides in the matrix after a rapid cooling process. The phenomenon is usually associated with a continuous heat treating operation such as Sendzimir Galvanizing or normalizing. Quench aging is controlled by either slowing the cooling rate or annealing after cooling.

The more commonly occurring and critical type of aging to sheet steel users is strain aging. The modern theory of strain aging as given by Cottrel states that with time, carbon and nitrogen atoms dissolved in the ferrite will segregate to dislocations and restrict their motion. This causes an increase in hardness and mechanical properties along with a return of the yield point. Sheet steels having yield point elongation would exhibit Luders bands on the formed parts. This effect is more typically associated with box annealed and temper rolled, cold rolled, rimmed, and mechanically capped steels. However, in properly processed box annealed cold rolled, fully aluminum killed steels, the strain aging effect does not occur. This is due to the tying up of the oxygen and nitrogen atoms by the aluminum. Hot rolled and galvanized sheet steels will all exhibit strain aging to various extents. The aging process is accelerated at higher temperatures (above ambient), which may occur in coil coating lines.

Surface strain lines can be minimized on cold rolled sheets by cold working after box annealing. The cold work is usually imparted by temper rolling. Temper rolling of the proper amount retards straining in fully aluminum killed steels. The effect will return after some period of time in rimmed and capped steels. Temper rolling just retards the return of the yield point. Therefore, it is important that the customer use temper rolled rimmed steels within a short time after delivery. If this is not possible, the effect can be minimized by flex rolling or flex levelling the sheets prior to forming.

Final Shape and Cross Section

The shape of the sheet, as delivered, can be as important as the previously mentioned factors. Good shape is associated with uniform longitudinal deformation across the strip width during forming. Poor shape can also cause difficulties in blanking or lubricating the sheet. Temper rolling and roller levelling are common methods to impart the proper shape to sheet steels.

Non-uniform cross section, either across the strip width or throughout the coil length can cause forming problems. These factors are controlled during both hot rolling and tandem reduction by the use of automatic gauge controls.

Processing

Steelmaking

The most commonly used method for producing sheet steels is the basic oxygen process. In this process, scrap and hot metal (blast furnace pig iron) are charged into a pear shaped, tiltable furnace lined with basic refractory material. The scrap is approximately 30 - 35% of the total charge. High purity oxygen is blown through a lance at supersonic velocities into the molten charge. The oxygen combines with the carbon and other elements in the hot metal to generate large amounts of heat needed for the refining process. Flux additions such as powdered lime and fluorspar are added to remove impurities. After 18 - 20 minutes of oxygen blowing, a 250-ton heat of steel will be at the desired carbon level and ready for tapping. A chemical analysis is performed at this time and further refining is done, if necessary. Ferro-alloy additions, as required, are made to the ladle as the steel is tapped from the furnace to the ladle.

The next process is either to pour ingots or continuously cast slabs. In ingot pouring, the molten steel is teemed into cast iron molds. After the steel has solidified, the molds are removed and the ingots are placed in a reheat furnace to prepare them for conversion to slabs. The four basic steel types: rimmed, mechanically capped, semi-killed, and fully killed can be produced as ingots.

All steels produced commercially by continuous slab casting are of the killed or semi-killed type. In continuous casting, molten steel is poured into a water cooled, oscillating, copper mold. A solid shell which contains liquid steel is formed in the mold. After the shell leaves the mold, it is further cooled by water sprays until the liquid core is solidified. The strand of continuous solidified metal is then cut to length. Strand cast steel will usually have more uniform segregation characteristics than ingot poured steel of the same grade, thus creating more uniform formability.

Primary Rolling

The next process step for steel in ingot form is conversion to slabs in a primary mill (commonly designated as a slabber or bloomer). This mill consists of soaking pits, which reheat the ingots up to rolling temperature, approximately 2500°F, a single stand, two high reversible rolling mill, a machine scarfer which conditions the slab surface, and a shear. While this process has little effect on formability as related to mechanical properties, it is of great importance in determining final product surface quality and internal soundness.

The reheating and rolling practices are varied depending upon the steel type (both deoxidation practice and steel chemistry) and ingot size. To insure that the cast structure of the ingot is eliminated, the rolling practice usually incorporates a minimum amount of edge reduction, the maximum reduction being based on mill limitations. The shearing step of this process is designated not only to cut the slabs to specified length but also to remove the pipe cavity.

If the final end use of the steel is an exposed application, both the cast and rolled slabs are sent to a hand conditioning dock. The slabs are inspected and all visible surface defects are removed using a hand torch. The slabs are then ready for hot rolling.

Hot Rolling

Hot rolling is the next processing step for sheet steels. In hot rolling, the slab is reheated to approximately 2400°F in a reheat furnace, reduced to an intermediate thickness in a series of roughing mills, rolled to final hot band gauge in a series of closely spaced finishing mills and then coiled. The most important aspect of this process as far as formability is concerned is temperature control throughout.

In reheating the slab, not only is the aim reheat temperature important, but also the temperature profile within the slab. An unevenly heated slab can cause non-uniform mechanical properties throughout the hot rolled coil. Too low a reheat temperature can prevent the solutioning of certain compounds such as AlN in the slab. This solutioning is necessary to produce the required final properties in both High Strength Low Alloy and Drawing Quality Special Killed steels. Control of reheat temperature and furnace atmosphere is also of importance in preventing the occurrence of rolled-in hot mill scale on the hot rolled strip. The roughing sequence breaks the slab down to an intermediate thickness (1" to 1-1/2") before hot rolling to the final hot rolled gauge. The scale (iron oxide) which forms on the slab surface during re-heating reforms during the roughing operations and is removed by the use of high pressure water sprays at each roughing stand. The usual aim during this process is to keep the bar temperature as hot as possible. A low temperature during roughing may cause problems in maintaining proper finishing and coiling practices. In addition, low temperatures may cause an oxide to form on the strip surface which cannot be removed by the descale sprays. The oxide will be rolled into the strip surface. This "rolled-in scale" defect can carry through to the final cold rolled sheet surface. It is objectionable for exposed applications and can initiate breakage.

The temperatures most closely monitored during hot rolling are the temperature of the strip after it has left the finishing train and the temperature before the strip enters the coiler. The finishing temperature is controlled by the use of descaling sprays at the entry end of the mill, and mill acceleration while the bar is in the finishing stand. The coiling temperature is controlled by water sprays which are located between the finishing stands and the coilers.

The usual aim for sheet steels is to maintain finishing temperatures above 1550°F. Low finishing temperatures tend to adversely affect the formability by producing a non-uniform grain structure in the hot rolled product. This structure can carry through subsequent processing and affect the cold rolled formability. A low finishing temperature in conjunction with a high coiling temperature can produce coarse grain on both hot rolled and cold rolled which could cause an objectionable surface condition when the product is formed. With too high a finishing temperature, the cooling capacity of the metallurgical sprays may be exceeded and it will be impossible to meet the coiling temperature restrictions.

Coiling temperatures and cooling rates are varied widely depending upon the product type. For example, a drawing quality special killed steel will receive an entirely different coiling practice depending upon whether it is intended for hot rolled, cold rolled, or galvanized. Coiling temperature and cooling rate are especially important in producing the desired properties in cold rolled killed steel. It is necessary to coil this grade cold (less than 1120°F) to keep aluminum nitride in solution in the hot rolled coil. During the annealing of the cold reduced coil the aluminum nitrides precipiate, causing the formation of the desired texture and grain shape for good formability. Cooling rate is also important in determining the final properties of hot rolled physical quality or High Strength Low Alloy steels. It can be seen that temperature control throughout the entire hot rolling process is of extreme importance. The formability of all product types in the sheet area can be adversely affected by non-optimum hot mill practices. As the quality level increases, the allowable variability from the optimum practice decreases.

The hot rolling step can be the final process for the hot rolled product, eg., hot rolled "bands". Further processing steps for hot rolled such as temper rolling, slitting and/or shearing will decrease the formability as some amount of mechanical cold work will be done on the strip during processing. This cold work may be necessary, however, if strain is a hazard or shape is critical.

Pickling

All cold rolled, galvanized and hot rolled sold as "hot rolled pickled and oiled" is pickled to remove the scale from the surface of the strip. This process usually takes place on a continuous line, which by design will include some mechanical working of the strip. The mechanical working is intended to break up the surface oxide to enhance the pickling. The pickling medium is acid (HCl or H_2SO_4) maintained at approximately 180°F. The pickled strip is then rinsed with water, air dried, and oiled.

The mechanical work that is imparted to the surface will have no effect on the formability on the cold rolled or galvanized product as they are further cold reduced. However, this mechanical work will decrease forming characteristics of the hot rolled pickled product and can cause poor press performance. One way to avoid the loss in formability is to batch pickle rather than continuous pickle. This process may be required if the end uses require superior formability. Mill applied dry lubricants which improve the forming characteristics of hot rolled pickled steel are also available.

Cold Reduction

The cold rolled and coated products are cold reduced after pickling. The amount of cold reduction can vary from 40 to 70%. A minimum amount of cold reduction is necessary to insure recrystallization of the cold worked grain structure during subcritical ($<1330°F$) and/or intracritical (1330 - 1650°F) annealing. A definite relationship has been established between formability and percent tandem reduction for Drawing Quality Special Killed cold rolled. Studies have shown that drawability as measured by a parameter known as $\bar{r}$ which is a measure of plastic anisotropy, increases as cold reduction increases to approximately 65%. The $\bar{r}$, which predicts the resistance to thinning during drawing, decreases as the amount of cold reduction increases beyond 70%. In rimmed and mechanically capped steels, cold reductions in excess of 70% will result in the formation of very fine recrystallized grains after box annealing.

The new tandem reducing mills being built in the steel industry at this time are more powerful and have more mill stands than the older generation of tandem mills. This has led to increased tandem reductions being used, both for metallurgical and productivity reasons. Both of these factors along with the trend to computerization had led to improved shape and gauge control. The improvement in gauge and shape control produces a more uniform product which should lead to better press performance. The final surface texture of the cold rolled sheet is determined in part by the type of surface texture produced at the tandem mill. Surface cleanliness of cold rolled steel can be adversely affected by the tandem rolling process. The lubricants used in rolling can break down on the coil surface during annealing and cause carbon deposits or "smutty steel". A heavy concentration of these deposits along the coil edges produce a defect commonly termed "snakes". Both of these defects can be avoided by close monitoring of the lubricating oils and proper control of anneal atmospheres.

Heat Treatment

Heat treating is an essential part of processing sheet steel, once it is cold reduced. The common methods of heat treatment are box annealing, continuous annealing and normalizing.

The most commonly used method for heat treating cold reduced coils is box annealing. In this batch process, coils are stacked up three or four high, placed under a cover, and heated to temperature in a special reducing atmosphere. The atmosphere prevents the decarburization and/or oxidation of the steel surface. The operating parameters of annealing are the temperature to which the steel is heated, the rate at which it is heated, and the time the steel is held at this temperature. In sub-critical annealing the steel is heated to just below the lower critical temperature (A_1 - 1330°F), at a rate of 50°F/hr., held for a predetermined time, and cooled very slowly to approximately 200°F. If the aim soak temperature is above 1330°F but below the A_3 temperature, the anneal is referred to as an intra-critical anneal.

During annealing of cold worked steel sheet, transformation of cold worked grains takes place in three stages, namely recovery, recrystallization, and grain growth. During the recovery period, the steel softens slightly and begins to attain the mechanical and physical properties of undeformed steel. During recrystallization, the final preferred orientation of the grain is obtained. The grain size increases during grain growth. Factors affecting the transformation of cold worked grains are alloy and impurity content and their distribution, hot rolling and cold rolling history, prior grain size, size and distribution of carbides and annealing variables.

The annealing variables that are normally controlled are heating rate, soak temperature, time at temperature, and annealing atmosphere. Heating rate plays an important role, in that, rapid heating causes smaller amounts of recovery. This results in a finer final grain size and higher mechanical properties. In Drawing Quality Special Killed cold rolled, a rapid heating rate prevents the development of the proper crystallographic texture that is necessary for the deep drawing characteristics normally desired.

The soak temperature and time at temperature is varied to match the quality level of the steel being annealed. As the quality level increases, the aim soak temperature and time at temperature also increase. Drawing Quality Special Killed steels are normally intra-critically annealed with longer soak times. Drawing Quality rimmed and mechanically capped steels will be sub-critically annealed just under the A_1 (1330°F) temperature for soak times comparable to those for Drawing Quality Special Killed. Commercial Quality steels will have lower soak temperatures and shorter soak times than Drawing Quality or Drawing Quality Special Killed cold rolled.

The hot rolled product can also be box annealed to improve formability. However, the hot mill practice must be carefully controlled to avoid surface cold work which will cause coarse surface grains during the stress relief box anneal.

Continuous annealing consists of rapidly heating the cold reduced strip to slightly below the A_1 (1330°F) and then rapidly cooling the strip to ambient temperatures in a continuous line under a reducing atmosphere. The lack of time for grain growth produces a finer grain structure than would be present in the same steel if box annealed and consequently the continously annealed steel would be less formable. This type of heat treatment is normally used for commercial quality steels requiring only moderate forming.

Normalizing is a form of heat treatment in which the steel is heated to a temperature above the A_3 (1650°F for steels with less than 0.1% carbon) and then cooled to ambient temperature in still air. Normalizing of cold reduced sheet steels is done to promote uniformity of structure and not to produce deep drawing characteristics. This uniformity of structure produces a sheet which has isotropic properties and is usually designated as non-earing. Normalized cold rolled sheet steels are generally intended for commercial quality applications or where a non-earing steel is desired. Cold rolled steel is usually normalized on a continuous heat treating line. Hot rolled steel is usually normalized as sheets in a semi-continuous unit. Normalized hot rolled sheets possess slightly better formability than the as-rolled hot rolled sheet.

Coated Products

Two different heat treatments are used on tandem reduced cold rolled intended for a galvanized or aluminized product. In a Wheeling line, a pre-annealed coil is hot dip galvanized on a continuous line, where the maximum strip temperature reached is approximately 1000°F

In the more common Sendzimir process, the full hard product is in-line heat treated by heating the strip above the upper critical temperature (normalizing) and hot dip galvanized. The modern Sendzimir line normally consists of payoff reels, either a chemical cleaning section or a burnoff section, a high temperature reducing section, a controlled cooling section, zinc pot, a leveller, and coilers. The strip surface should be free of any oils or residue so that the heat treated surface will be clean to promote good zinc adherence. The heating zones with a reducing atmosphere are maintained well above the A_3 temperature (Austenite zone) in order to normalize the strip. The strip is then control cooled to a temperature equal to the molten zinc temperature. The zinc adheres tightly to the steel surface with the formation of a thin alloy layer. Poor adherence will result if the line conditions are abnormal, unclean strip surface, etc. The galvanized strip produced this way will have properties conducive to lock forming quality applications. To obtain superior formability, galvanized steels are usually post annealed at a temperature of about 560°F for several hours. This will cause the steel to actually overage and soften by about 4 to 5 points of Rockwell B hardness. The strip can later be temper rolled to obtain a surface suitable for exposed applications.

ALUMINIZING lines are very similar to galvanizing lines, except the pot temperature is higher due to the melting point of aluminum.

Another product gaining commercial importance is prepainted steels. This product is either a box annealed cold rolled material or galvanized steel strip coated in a continuous paint line. Prepainted products are widely used for wall panels, appliance manufacture and certain automotive panels.

After annealing, the majority of the cold rolled product is temper rolled. This light cold rolling operation (1%) retards the strain aging affect in Drawing Quality Special Killed steels and temporarily in rimmed and mechanically capped steels. Though increasing the amount of temper rolling will further delay the return of strain aging, the formability of the sheet decreases as temper rolling increases.

A cold rolled commodity known as Class II, is sold to take advantage of the increased formability obtained by not temper rolling. This commodity should only be used where freedom from strain aging is not a requirement.

The final surface texture results from the temper mill operation as does the final product shape. Adjustments can be made using the proper combination of mill roll surface and roll contour.

While some cold rolled is shipped to the customer directly after temper rolling, the major part of this commodity receives one more processing step. If the material is sold as sheets the processing step is shearing to length. If sold as coils, the steel is rewound for surface inspection, edge slitting, and or mult slitting. Most shearing or rewind units have a levelling device as part of the unit. The leveller is used to improve shape and also retards the return of the strain aging effect. Although only a modicum of mechanical work is imparted to the sheet in levelling, the amount of work done can decrease the formability of the sheet.

Summary

It can be seen that each processing step during the production of sheet steels can have an effect on their formability. The factors which affect formability are closely controlled during processing. From the melt shop, through hot rolling, cold rolling, and heat treating, proper processing must be maintained to insure that the final product will meet the customer's requirements.

It should be recognized that, with the wide array of processing steps available to steel producers, each order is tailor made to meet the intended end use of the customer. A steel user should provide the producer with enough information about the the end use so that the producer can intelligently process the steel. It should be stressed that the user recognize the formability requirements of the part and order the appropriate steel type and grade. How to select the type and grade necessary will be discussed in the subsequent papers.

REFERENCES

1. Donald J. Blickwede, "New Knowledge About Sheet Steel", American Society for Metals, 1970.

2. American Iron and Steel Institute, "The Making of Steel", Second Edition.

3. United States Steel Corporation, "The Making, Shaping and Treating of Steel", Seventh Edition.

HIGH STRENGTH LOW ALLOY STEELS FOR AUTO PARTS

General Motors has completed a comprehensive evaluation of HSLA steels for automotive stampings. Aimed at finding out if the steels offer an alternative to conventional mild steels, this investigation shows that they offer 2 to 3 times the strength of regular mild steels, fabrication using traditional automotive techniques, weldability, and weight and potential cost savings.

by **John P. Maloney,** Manufacturing Development, **Roger A. Heimbuch,** Manufacturing Development and **Leon J. Rose,** Fisher Body Div., General Motors Corp.

High strength low alloy steels offer several advantages to the automotive designer. In simple tension a conventional mild steel member has to be 2½ times as thick to support the same load as an HSLA member (Fig. 1).

Furthermore, for a box section, which is more typical of structural panels, the HSLA member shown on the right in Fig. 2 will support 84% more buckling load than a mild steel part of the same geometry and material thickness. Thus, high strength low alloy steels can be used to improve product integrity or, more importantly, reduce weight and cost in ideal situations.

The main reason for interest in the steels is that they can provide a 2½ fold increase in yield strength for approximately a 40% cost premium over base hot rolled material. It is possible to design automotive structures that are stronger, lighter, and potentially less expensive, using HSLA steels than those using conventional 1010 carbon steel.

One of the first questions that comes to mind with HSLA steels is: What makes them so strong while retaining ductility? Fig. 3 shows the stress-strain curves of a typical HSLA steel and low carbon mild steel. Note the 2½ times increase in yield strength with about a 50% loss in ductility.

Composition varies

There are six basic ways of increasing the yield strength of a material. HSLA steels use four of these as follows:

- Grain size control — typically ASTM 12.
- Solid solution strengthening — almost all the elements added.
- Multiple phase-pearlitic structure.
- Precipitation strengthening — carbides and/or nitrides of columbium (niobium), titanium, vanadium and zirconium.

Metallurgically, HSLA steels can best be described as fine grained, precipitation hardened, fully-killed, low carbon steels with sulfide shape control. As shown in Table 1, their basic chemistry is that of an aluminum-killed, low carbon steel with small additions of columbium, titanium, vanadium or zirconium.

Also, each of the HSLA steels supplied by different metals producers has a different choice of elements. Since these elements form the nitride and/or carbide precipitate particles, their choice is very fundamental to the properties of the material. Note that material 'A' has a titanium additive, 'B' a vanadium additive, 'C' a columbium additive, 'D' a balance of columbium, vanadium

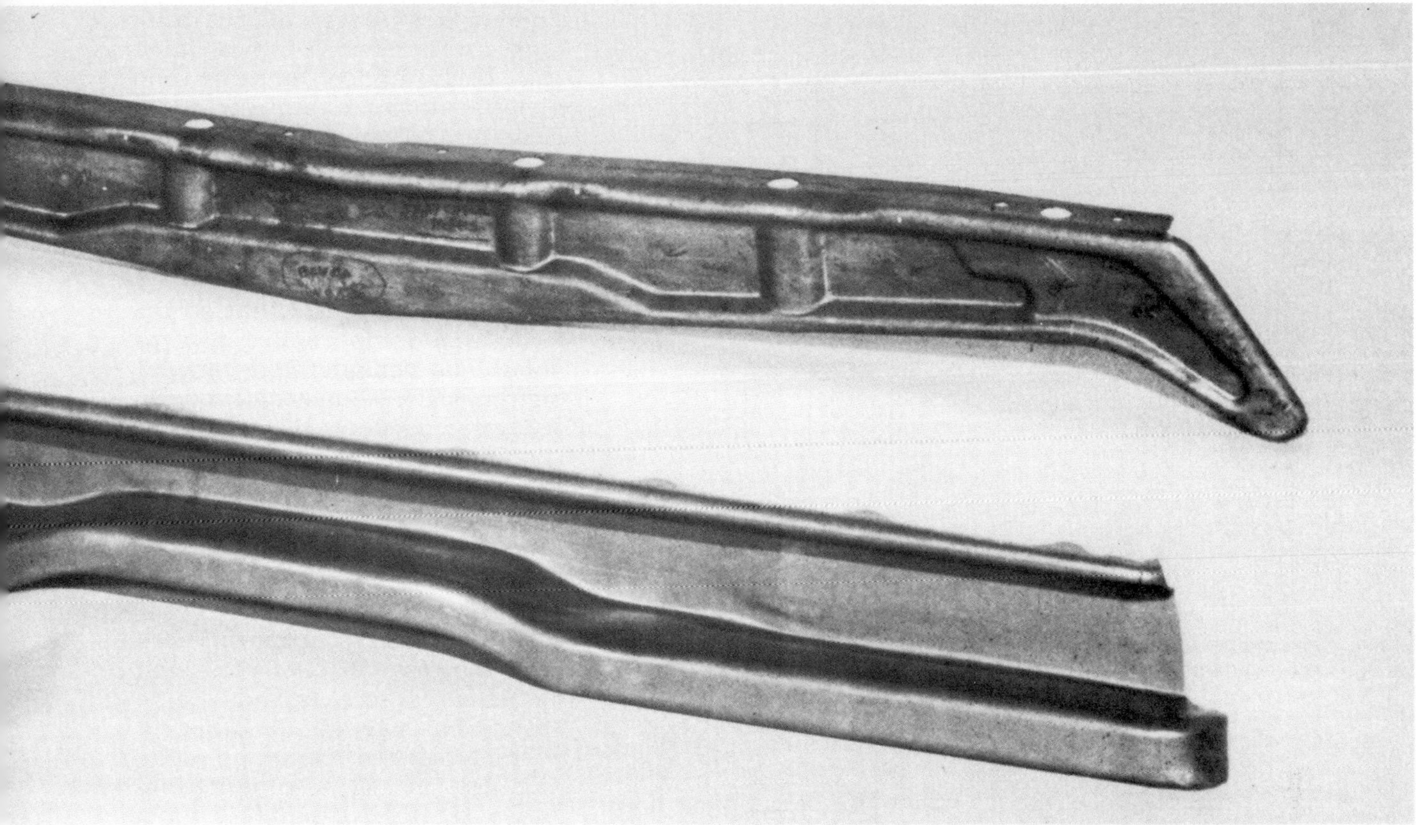

and zirconium, and 'E' a titanium additive. Since there are advantages and disadvantages to each, a particular type may not have a significant long range advantage.

Control of sulfide inclusions in HSLA steels contributes substantially to their transverse properties. Sulfur is controlled in two ways: by controlling the total amount of sulfur present, and by adding rare earths such as cerium or elements like zirconium for sulfide shape control. Control of other alloying elements is also important to overall performance; carbon and manganese, for example, must be controlled for good welding performance.

Proper mill processing of HSLA steel is as important as chemistry in determining engineering properties. In general, the basic mill practice is similar for each of the materials. For example, the strip must remain austenitic out of the last finishing stand — say 1650 F for a light gauge material. The strip is then quenched to about 1100 F in 15 to 20 sec, coiled, and then allowed to slow cool. Control of chemistry and mill practice results in a fine grained precipitation strengthened material.

Typical properties

With good chemistry control and mill practice, HSLA steels can have beneficial mechanical properties, as shown in Table 2. The first five columns give average data for 80,000 psi minimum yield strength steels with total elongations of 17% or better. Note that the directional properties of HSLA steels are very uniform, with less than 20% difference between longitudinal and transverse total elongations. This is directly attributable to their sulfide shape control.

Bending fatigue life. The S-N curve for HSLA steels, shown as a scatter band in Fig. 4, lies below the equivalent curve for mild steel on a percent of ultimate strength basis. The endurance limit occurs at a lower number of cycles, but at the same percentage of ultimate tensile stress. Therefore, high cycle base metal fatigue does not appear to be a problem with HSLA steels.

With spot welded HSLA samples the data are more difficult to interpret, but it appears that the load carrying capacity at infinite life for HSLA welds and mild steel welds is about the same. If this is not accounted for when designing fatigue loaded spot welded HSLA structures, spot weld fatigue could be a problem.

Strain rates. Many of the structures where HSLA steels are being considered for vehicles will be subject to impact type damage. Thus, their performance under high strain rates and low temperatures is important. Fig. 5 shows room temperature data for low carbon mild steel and a typical HSLA steel at high (10 in./in./sec) and low (0.01 in./in./sec) strain rates. The low rate would be considered quasi-static. The high rate is more realistic for impact conditions. On a percentage basis the mild steel is clearly more rate sensitive than the HSLA steel.

The change in strain rate characteristics with low temperature is show in Fig. 6. The data for HSLA steel are representative of a poor performing high strength steel, that is, other materials show greater ductility when tested under similar conditions.

Note that the HSLA steel shows a 75% loss in total elonga-

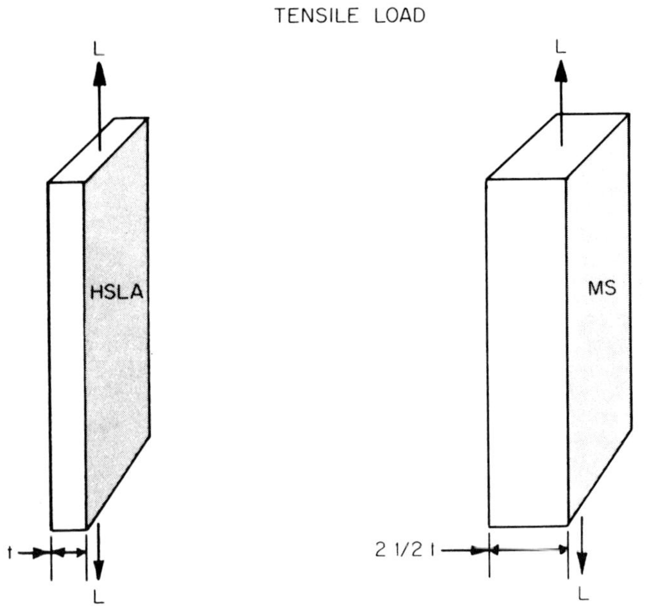

1. Comparative thickness of HSLA steel and mild steel under constant load and width.

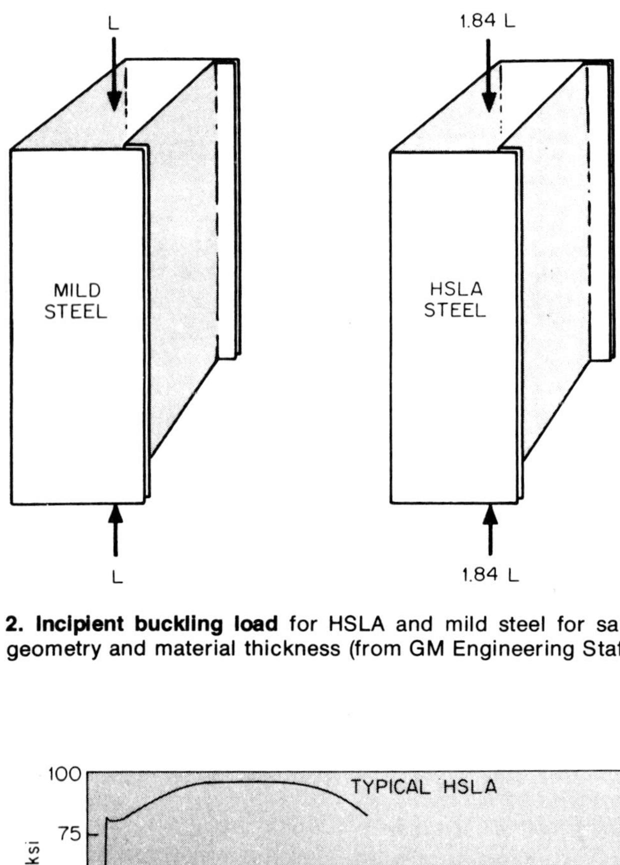

2. Incipient buckling load for HSLA and mild steel for same geometry and material thickness (from GM Engineering Staff.)

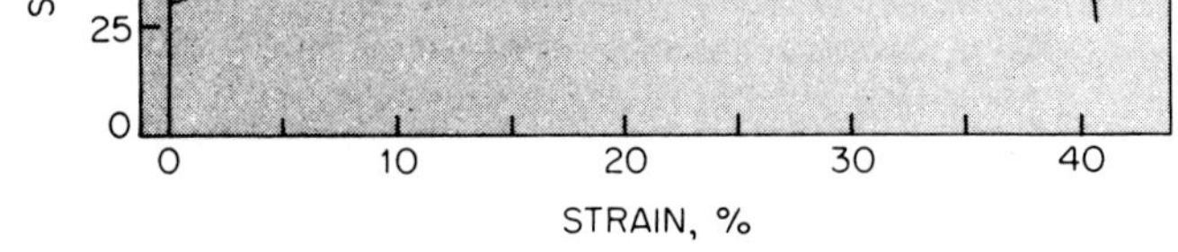

3. Stress-strain comparison of HSLA and SAE 1010 hot rolled steel sheet.

tion at low temperatures and high strain rates. A typical mild steel tested under the same conditions shows a similar loss. However, not enough testing has been done to generalize these data in terms of the chemistry of HSLA steels. Interpreting these data in terms of real world structures and environments isn't easy, but it suggests that strain rate and temperature effects should be considered when designing critical impact structures using HSLA steels.

Corrosion resistance. Compared to traditional automotive sheet steels, HSLA steels are no better or no worse in corrosion resistance (Fig. 7; results from GMR Metallurgy Department). In addition, current coating methods offer the same level of protection to HSLA steels as they do to traditional body steels. Based on current test results, stress corrosion does not seem to be a problem with HSLA steels. It should be emphasized, however, that HSLA steels will probably be used in lighter gauges than the mild steels they replace. Because of this gauge reduction, HSLA steels will have a lower factor of safety in corrosion than the mild steels they replace. So, good design practices and protective coatings, such as galvanizing and painting, should be used as required.

Forming the steels

Six high strength steels tested by GM, labeled A thru E, including one galvanized product are shown in Table 3. Commercial hot rolled mild steel, used as the basis of comparison, is listed at the bottom. Note that the yield strengths of HSLA steels are all about 80,000 psi. Elongations for the high strength steels are generally about half that of the mild steel. R and N values, — measures of drawability and stretch formability, respectively, — are obtained from tensile test data.

The most useful information that can be obtained in a laboratory regarding formability is the forming limit curve. Fig. 8 (from GMR Physics Dept.) compares mild steel to four HSLA steels — the higher the curve — the better the formability. Drawing performance is indicated to the left of the vertical major strain axis, the stretch formability is indicated to the right of this axis.

The area of most concern is from about −15% to +20% on the horizontal base line. Although the performance levels are obviously lower for all the high strength materials in both forming modes, the results indicate surprisingly good formability for steels with yield strengths in the 80,000 psi range.

A general data sheet format for recording forming test results is shown in Fig. 9. In the first test, frame members were formed and pierced complete by simply substituting high strength steel blanks in the production setup. Despite a minor split, which occurred in a severe stretch flange location, the results were better than expected.

In the seat back reinforcement tests (Fig. 9), HSLA steels are compared directly to mild steels at various punch depths. Because this shape was too demanding for high strength steels, it was difficult to make this part free of fractures and wrinkles with the punch at the bottom. By forming parts with the punch ¼ in. off bottom, however, it was possible to obtain practical data.

The rear rail die shown in Fig. 9 was first tuned-in to form OK mild steel parts. However, in attempting to form high strength steels with the same blank size and press settings, the high strength panel split badly. However, by using a series of hits with progressively smaller blanks, reducing binder pressures, and making other routine tryout adjustments, the transition from mild steel to high strength steel, in this particular application, was surprisingly simple.

A comparison of punch forces to form 0.075 in. mild steel with 0.075 in. high strength steel is shown in Fig. 10. The rear rail die was used for this test and parts were checked at various punch forces with a cross section template — the deviation from this template being the horizontal axis on the graph.

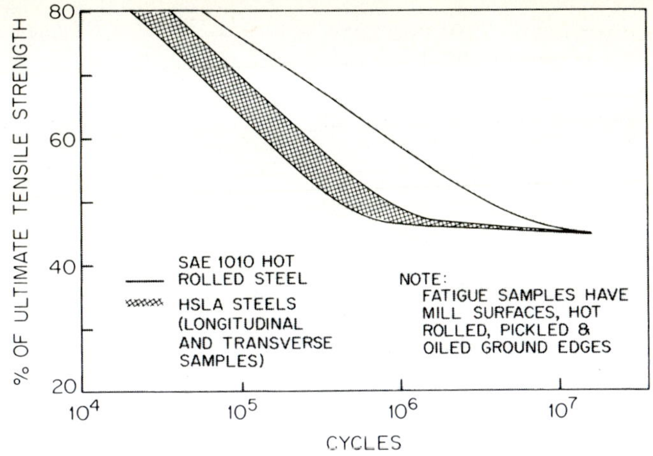

4. Fatigue life of HSLA steel compared to SAE 1010 hot rolled steel.

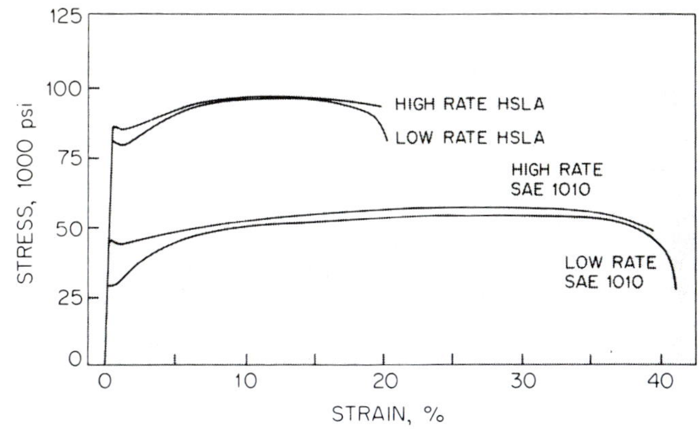

5. Typical high and low strain rate response of HSLA and low carbon mild steel.

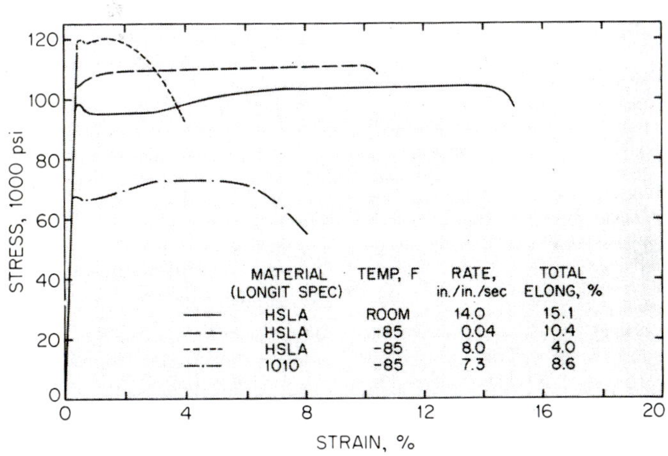

6. Change in strain of HSLA and 1010 steel with decreasing temperature.

As shown, the curves are nearly parallel, with the high strength steel requiring approximately double the force needed for similar mild steel shapes. This is not necessarily a representative test, however, since the normal tendency will be to use lighter gage high strength steel than mild steel for comparable parts.

Note also the forming force required for 0.054 in. high strength steel with 0.107 in. mild steel in Fig. 11. The graph shows that when two materials are compared with this 2 to 1 thickness ratio, the punch forces to get OK parts are about the same.

To sum up the formability properties of HSLA steels:

- Their performance is comparable to mild steels in simple bending as well as in straight line areas of draw dies.
- In stretch forming, performance is similar to mild steels except for reduced elongation to fracture and poor strain distribution. Stretch formed shapes must, therefore, be less demanding for the high strength steels.
- Solid forming is limited only by these same stretch formed areas.
- Regarding drawing, areas of radial cup shaped draws are borderline. Abnormal tonnages are required to avoid wrinkles and the tendency is toward fractures and laminations.
- With regard to springback, although HSLA steels generally spring back to a greater degree than mild steels, problems appear to be manageable within normal part design and die practices.
- Regarding forming loads punch forces are generally proportional to the tensile strength of the workpiece. To form parts as strong as mild steel parts, however, abnormal tonnages will not be required for HSLA steels.
- Indications are that press lines will require no more than the normal number of dies. Material utilization, as a function of blank size, is expected to be a standoff. And since the high strength materials carry a cost premium of about 40%, a thickness reduction of 30% is required to yield a material cost wash.

Table 1 Typical Chemical Analyses of HSLA Steels (Wt., %)

Steel Code	C	Mn	Si	Al	Ti	V	Cb	Zr
Low Carbon	.06	.39	.01	.005	.002	.005	.005	.005
Producer 'A'	.10	.34	.03	.070	.17	.005	.005	.005
Producer 'B'	.13	1.25	.60	.080	.002	.12	.005	.005
Producer 'C'	.10	1.20	.02	.065	.002	.005	.10	.005
Producer 'D'	.12	1.24	.22	.020	.002	.036	.037	.072
Producer 'E'	.12	.52	.03	.035	.31	.015	.005	.005

Table 2 Typical Mechanical Properties of HSLA Steels

Steel Code	Avg Yield Str, 1000 psi	Avg Ult Str, 1000 psi	Total Elong, %	Long Tot El, %	Trans Tot El, %
SAE 1010	33.6	45.6	40.5	42.4	38.7
Producer 'A'	85.2	98.1	17.8	18.5	15.8
Producer 'B'	87.6	105.7	18.1	20.2	17.8
Producer 'C'	80.3	88.7	21.6	22.3	19.2
Producer 'D'	84.7	96.0	21.5	20.7	19.3
Producer 'E'	85.2	93.7	19.2	19.3	17.7

- Since some formability variables can be identified only through volume usage, production applications are needed to gain quantitative data in areas such as material consistency and part quality, die lubrication and die life.

Welding characteristics

Resistance welding of HSLA steel is more difficult than mild steel. Fisher Body Manufacturing Development has made thousands of welds on standard test coupons.

Shown in Fig. 12 is a typical illustration of the lobe for welding mild steel to mild steel, in this case 0.072 cold rolled steel (CRS) to 0.072 CRS. A welding lobe is a graph of weld current and weld time combinations which will produce satisfactory welds with a specific set of conditions. For 0.072 CRS, the weld force was 1300 lb., tip geometry of ⅜ in. dia. contact surface on a ¾ in. electrode cap, ½ gpm water flow, and 100 conditioning welds.

A weldability lobe consists of 2 primary curves. The first curve on the left defines the combinations of weld current and times which will produce minimum acceptable (0.250 in.) nugget diameters. Points to the left of this curve will produce weldments which fail in shear, or produce nugget diameters less than the 0.250 in. minimum specified. The last curve on the right defines the flashing point. Points to the right of this curve result in weldments with excessive heat energy, as evidenced by heavy flash, electrode indentation, and/or electrode stocking.

Points within these two curves result in welds of satisfactory quality. The distance between these two curves at a given weld time is the *weld range*. A wide range is very desirable, indicating that the material would be tolerant of manufacturing process variables, such as ever-changing tip geometry and voltage fluctuations. As a convenient reference, each of the graphs has a third curve (shown as a dotted line) indicating the points which will produce a nominal 0.310 in. diameter weld nugget. A point on this line will be selected for a nominal weld schedule.

Shown in Fig. 13 is the weldability lobe for welding mild steel to an HSLA steel produced by producer 'A' (see Tables 1 and 2). The conditions are the same as the previous lobe shown in Fig. 12 for mild steel. The lobe is a slightly smaller and less current is required at any given weld time to produce the same size nugget, as evidenced by the lobe's position to the left of the previous lobe position. Thus, HSLA steels can be welded to mild steel with same equipment as for welding mild steel to mild steel. However, checking of weld quality would be slightly more difficult because of the stiffness of the HSLA steel.

A similar weldability lobe for

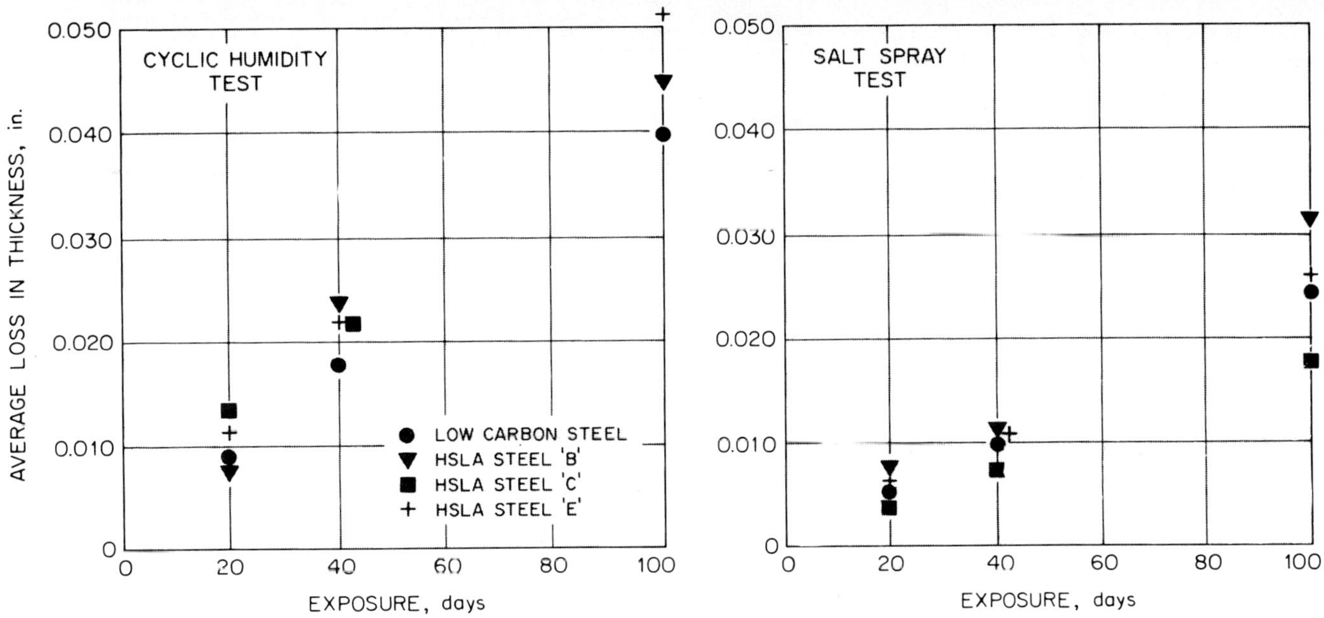

7. Accelerated environmental corrosion resistance of HSLA and low carbon steel (from GM Research).

welding HSLA steel from metal producer 'B' (see Tables 1 and 2) is shown in Fig. 14. The conditions are the same except for the weld force — 1600 lb. Because of the higher yield strength and stiffness of HSLA steel, the additional weld force was deemed necessary to pull the metal together under normal misfit production conditions. This lobe is not much different in size than the one in Fig. 13. The welds in this case will be much more difficult to check because both pieces are higher in strength and stiffness.

Although the results here are pretty good, not all HSLA steel is this good. In fact, even different heats or batches from the same metal producer will show other results. The steel producers are still learning about, and developing the materials and variations can be expected.

The weldability lobe for HSLA steel from producer 'E' (see Tables 1 and 2) is shown in Fig. 15. Note the very narrow range and the higher weld times. Here it was even necessary to go to pulsation welding to get reasonable electrode tip life. This particular HSLA steel would not be recommended for production use.

A composite of weldability lobes for four different producers of HSLA steel is shown in Fig. 16. It includes the lobes for HSLA steels from producers 'A', 'B', and 'E' previously shown, plus one more from producer 'C' (see Tables 1 and 2). A point shown at 36 cycles weld time

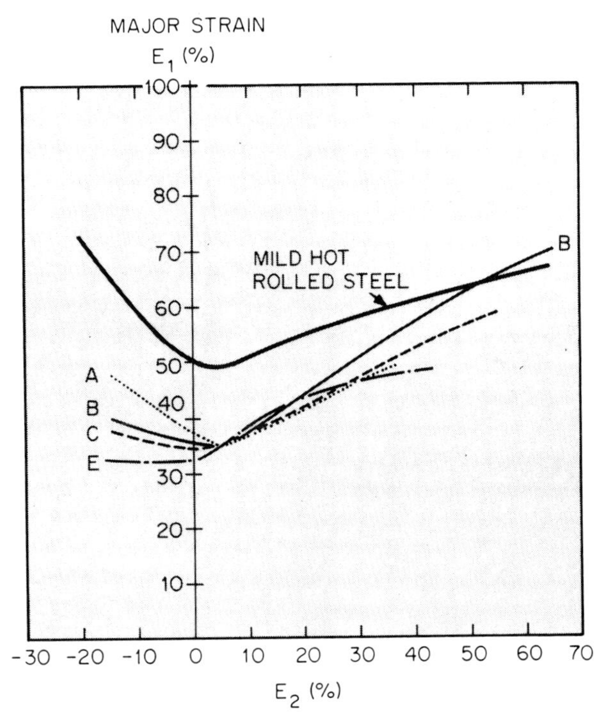

8. Forming limit curves of five HSLA steels vs mild hot rolled steel (from GM Research).

Table 3 Properties of Six HSLA Steels Evaluated for Formability

Steel Producer	Avg Yld Str, psi	Avg Ten Str, psi	%Elongation Uniform	Total	R	N
'A'	89,421	100,725	12.0	15.6	1.17	0.131
'B'	89,420	107,310		22.0		
	84,316	104,186	12.7	20.0	1.15	0.142
'C'	78,574	87,500		24.6		
	80,357	90,344	13.0	22.0	0.98	0.125
'C' Galv.	78,131	84,946	—	23.3	—	—
'D'	84,778	96,573	—	21.0	—	—
'E'	89,574	97,800	—	18.6	—	—
Comm. Mild Steel	33,677	45,896	24.0	40.0	1.05	0.184

PRODUCER	THICK, in.	YLD STR, psi	TEN STR, psi	% ELONGATION		R	N
'A'	0.075	89,421	100,725	12.0 UNIF	15.6 TOTAL	1.17	0.131

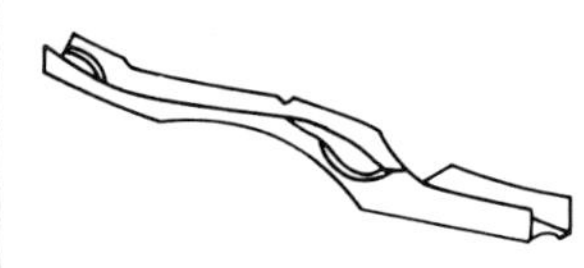	CHEVROLET FRAME MEMBER FORMED AND PIERCED COMPLETE. OK EXCEPT 1.88 in. LONG SPLIT IN FLANGE DIE. (TOOLS SET FOR 0.100 in. MIN GAGE).
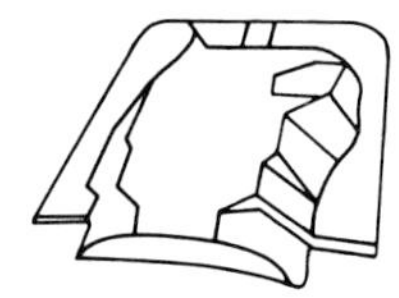	SEAT BACK REINFORCEMENT PUNCH 0.25 in. OFF BOTTOM-NECK IN 24% STRAIN AREA. WRINKLES IN FLAT AREA. PUNCH AT BOTTOM—FRACTURE 3 in. LONG. WRINKLES REMAIN.
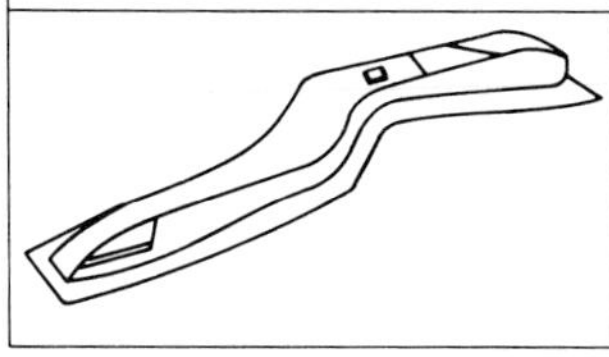	REAR RAIL BLANK SIZE AND PRESS SETTINGS ADJUSTED FOR HSLA STEEL. PART FORMED COMPLETE—NO DEFECTS.

9. Forming results with three test applications for HSLA steels.

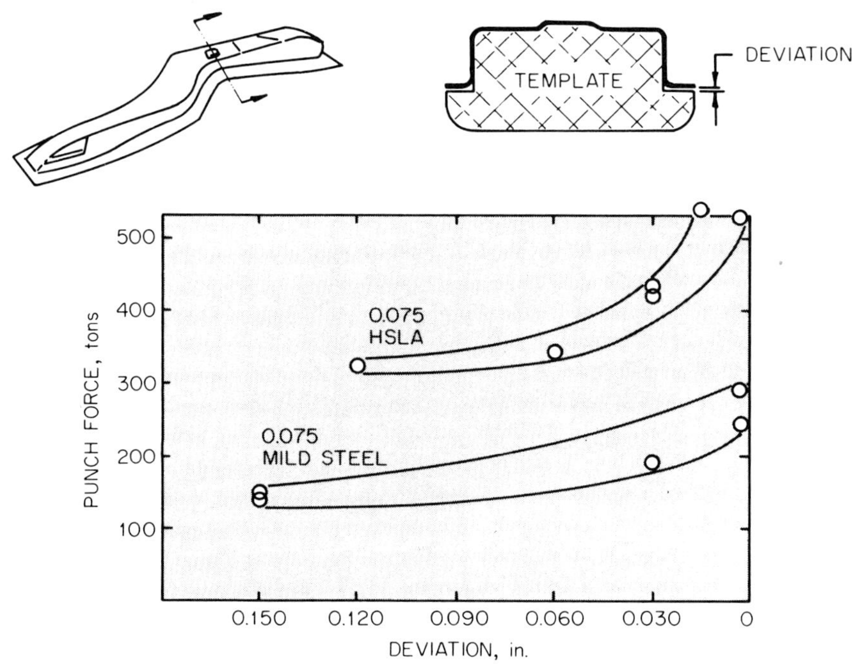

10. Comparison of forming pressure required for 0.075 in. thick HSLA and mild steel parts.

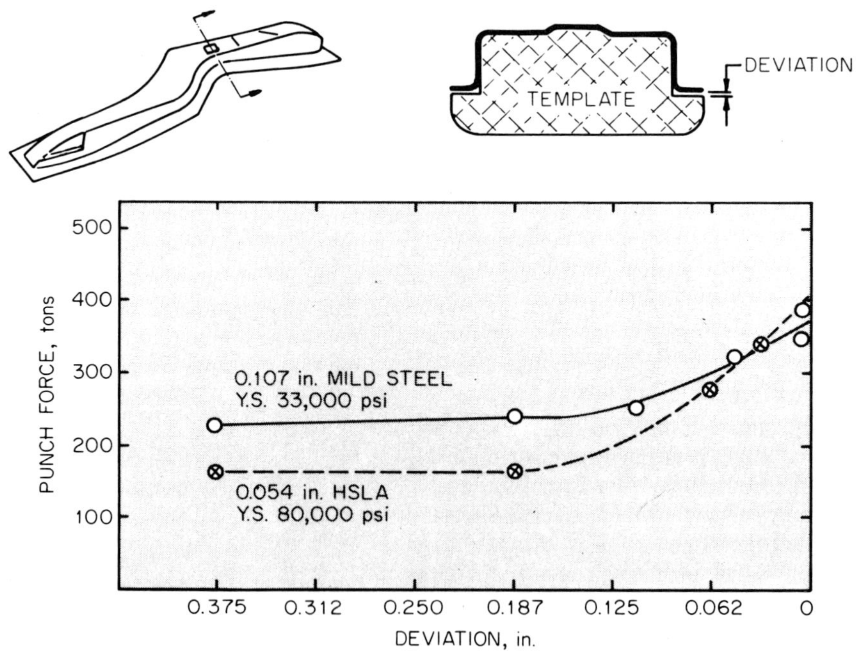

11. Forming pressure required for 0.054 in HSLA vs 0.107 in mild steel.

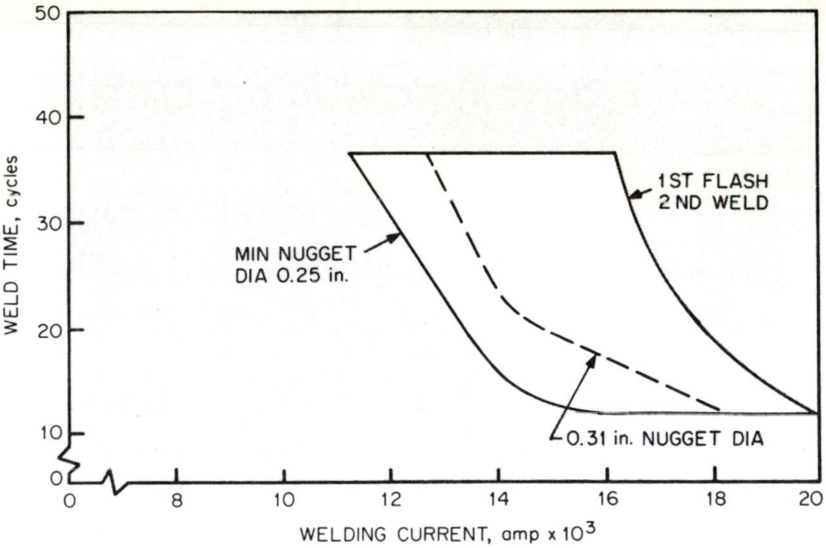

12. Lobe for resistance welding 0.072 in. cold rolled steel to itself.

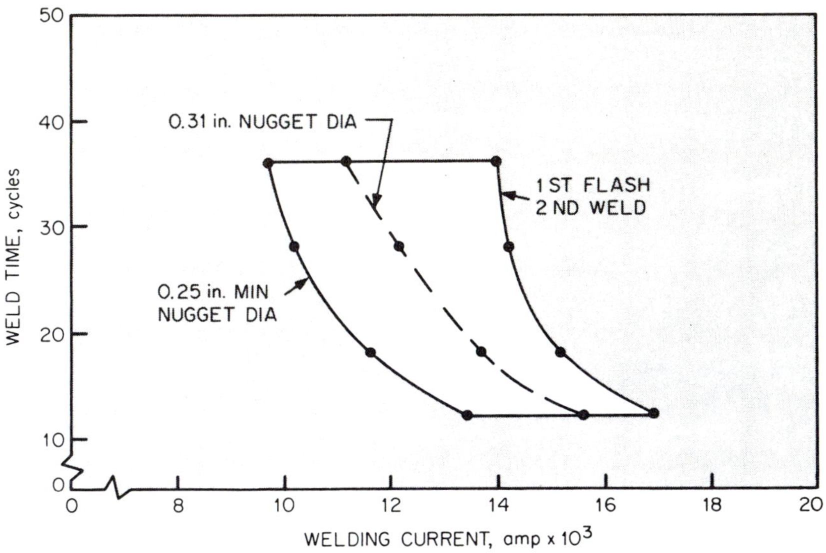

13. Same conditions produce this weldability lobe for welding 0.071 in. thick mild steel to 0.071 in. thick HSLA steel from producer 'A.'

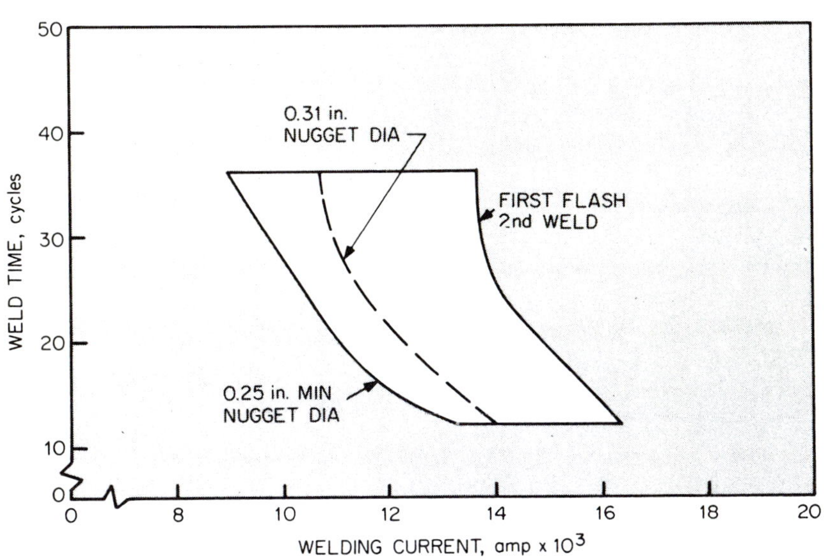

14. Welding results obtained in joining 0.082 in. HSLA steel 'B' to itself.

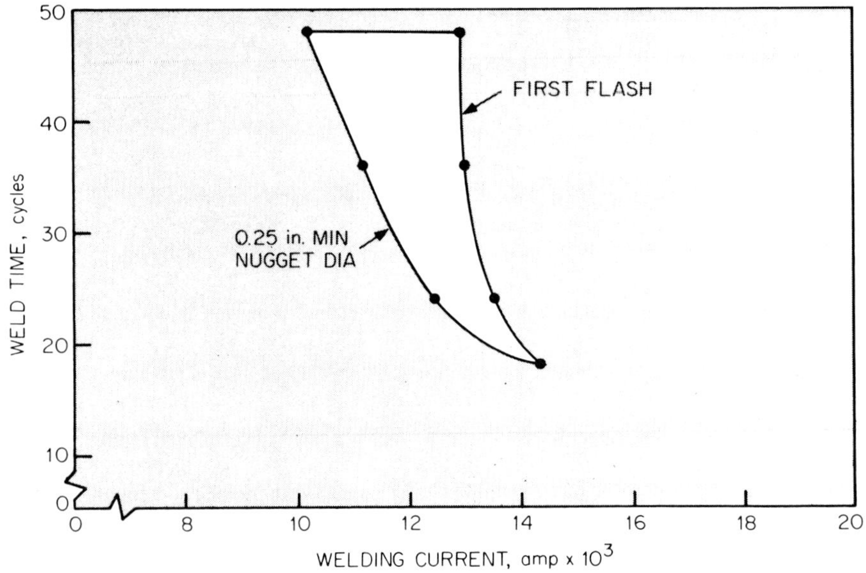

15. Welding results obtained in joining 0.075 in. thick HSLA steel 'E' to itself.

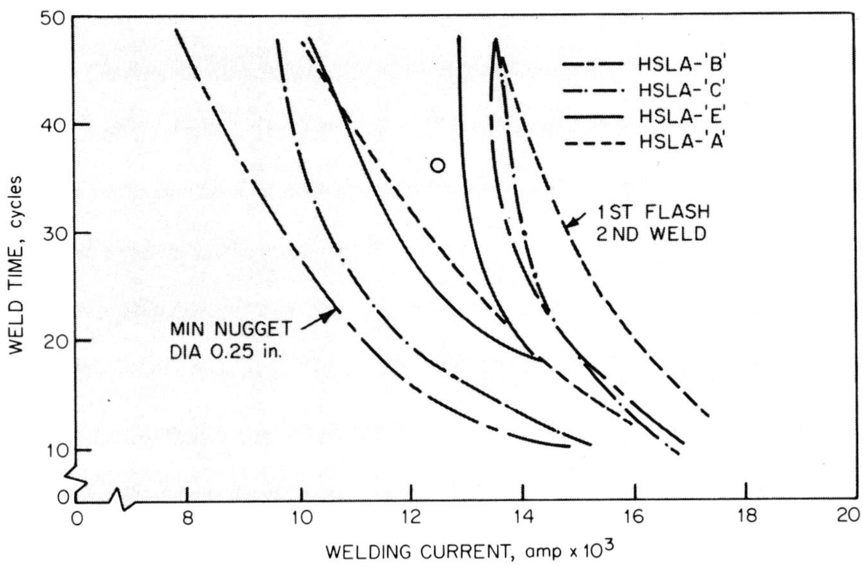

16. Comparison of welding characteristics of four different HSLA steels (0.075 in. thick steel to 0.075 in. thick steel.)

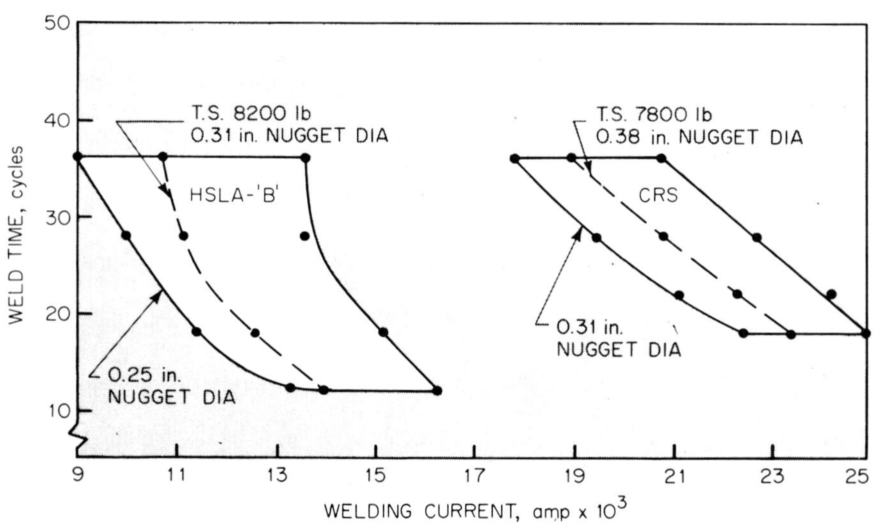

17. Comparison of welds in 0.120 in. cold rolled steel welds with welds in 0.082 in. HSLA steel from producer 'B.'

and 12,500 amp indicates that all four materials could be welded at a common schedule — but, and this is important, the welding range is very narrow.

Up to this point we have been comparing HSLA steel weldability to mild steel weldability on the basis of equal thickness of steel and, although there are some limitations, the results are encouraging. However, if we compare HSLA steel to mild steel on the basis of approximate equal strength, the weldability lobes are even more encouraging.

A comparison of 0.120 in. CRS lobes to a 0.082 in. HSLA-'B' lobe is shown in Fig. 17. The tensile shear strengths of single welds in each are approximately equal — 8200 lb. for the HSLA-'B' steel and 7800 lb. for the CRS. The weldability lobe for the HSLA steel is much larger than the mild steel lobe. The weld currents for equal weld time are noticeably lower for the HSLA steel and the other weld variables are also in favor of the HSLA steel. Weld force was 1600 lb. for HSLA and 2000 lb. for mild steel.

Electrode tip diameter and nugget diameter were both 1/16 in. larger for the mild steel than for the HSLA steel — this being necessary because of the 0.120 in. thickness of the mild steel, as compared to the 0.082 in. thickness of the HSLA steel. Overall, the results indicate a promising future for HSLA steel as far as welding is concerned.

In summary, the resistance welding limitations of HSLA steel are:

- HSLA steels are new materials and are still in the development stage.
- Production variations can be expected.
- More effort and time will be required to check weld quality.
- Poor fit-up of panels will reduce weldability.
- Costs will be higher because of increased equipment size, capacity, time, etc.
- Little has been done on galvanized or coated HSLA steels and corrosion can be a more significant factor where thinner gages are involved.

A guide to steel properties and applications

by Robert L. Myers • *Armco Steel Corp.*

Cold-rolled steel can be formed, welded, punched and finished as required for refrigerator door application.

This paper is intended to provide appliance engineers with a general guide to sheet steel, its properties and applications. The author discusses the three basic qualities of both hot- and cold-rolled steel sheet, the various finishes, and also covers enameling iron and stainless steel. Recent developments are treated in the context of suitable appliance applications.

Sheet steel has many characteristics that are used to the advantage of appliance engineers in the manufacture of major household appliances. Some of the more important are low cost, speed and ease of fabrication, speed and ease in joining, and high strength over the wide range of operating temperatures encountered.

Steel's high modulus of elasticity of 30,000,000 psi imparts good rigidity to appliance parts. Yield strengths of 25-35,000 psi resist permanent set or deformation of parts. On the other hand, the ductility of steel provides good fracture resistance and formability.

Typical elevated temperature mechanical properties of low-carbon flat-rolled steel are shown in Fig. 1. These data show that the yield strength decreases with an increase in temperature, but the tensile strength increases up to 500° F. This is due to aging and carbon precipitation between 300° and 500° F. These properties at elevated temperatures are responsible for the long-life performance of self-cleaning ranges and clothes dryers where these temperatures are attained in operation.

Sheet steel parts can be rapidly spot and seam welded together, which makes possible the high production rate required in the appliance industry. Zinc- and aluminum-coated steel sheets of the type used by appliance makers can also be readily spot and seam welded. However welding speeds are somewhat less than those achieved with uncoated sheets and more frequent electrode dressing is required. Other commonly used joining techniques include gas and arc welding, lock seam forming, mechanical fastening and metal stitching. The latter method is used mostly on pre-painted sheets that can't be resistance welded without marring the surface.

Basically, hot-rolled, cold-rolled, zinc-coated, and aluminum-coated steel sheets for the appliance industry are SAE 1006 or 1008 grade, with a low carbon content from 0.02 to 0.08% and 0.30% manganese. However, virtually an infinite variety of chemical compositions of steel is available, depending on the exact requirements of the specific application.

Three basic qualities of sheet steel are produced, available in either cut lengths or coils. Commercial quality (CQ) sheets are satisfactory where good flatness is essential and only moderate forming or shallow draws are required, such as for refrigerator cabinets. Drawing quality (DQ) is used on deep drawn parts such as compressor housings and washer spinner baskets. Drawing quality special killed (DQSK) steel sheets contain 0.03 to 0.05% aluminum which combines with nitrogen and prevents

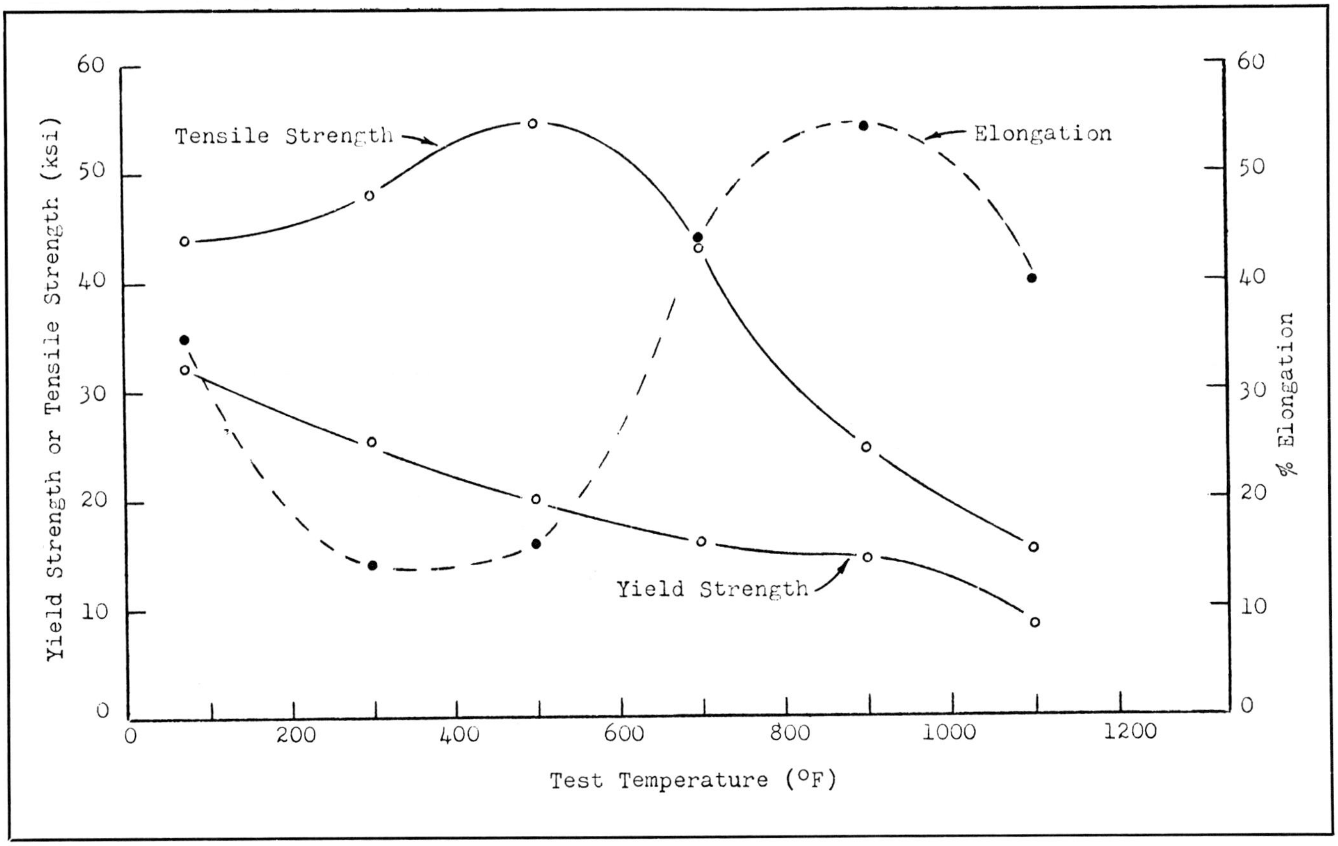

Fig. 1—Typical mechanical properties are shown for low carbon steel sheets at elevated temperatures. Yield strength decreases with rise in temperature, and tensile strength increases up to 500° F due to aging and carbon precipitation.

age hardening and stretcher strains in the steel. This grade is also superior to drawing quality in drawability.

Hot-Rolled Steel Sheets

Hot-rolled steel sheets are used mainly for structural strength in thicknesses of 0.060 in. and greater. Much of this tonnage goes into deeply drawn refrigerator and air conditioning compressor housings. These are deep draws and require DQ or DQSK. Other uses for hot-rolled sheets are base plates, angles, formed channels, brackets and a variety of other parts where high strength is required but good surface is not a critical consideration.

Typical mechanical properties of the three quality grades of hot-rolled sheets are shown in Table I. These data show the increase in ductility as measured in percent elongation from CQ to DQSK. Increased ductility and decreased hardness of drawing quality and drawing quality special killed provides better drawability than commercial quality. These property differences are achieved through chemistry selection and metallurgical control during hot-rolling, as they are not usually further heat treated.

One recent development in hot-rolled steel coils is the application of a dry lubricant on both surfaces at the mills. Pre-lubricated steel saves the customer the operation of applying drawing compounds at the presses or the need for operating and maintaining dry soap machines. These mill-applied lubricants also provide some corrosion resistance during transit and storage.

At least one steelmaker is supplying pre-lubricated enameling iron and decarburized enameling steel in coils. It appears that this usage will grow in the future.

Cold-Rolled Steel Sheets

Cold-rolled steel sheets are the most widely used of all steels in appliance production. Costing slightly more than hot-rolled, cold-rolled steel sheets are available in thicknesses as low as 0.014 in. and are therefore the most economical steel material on the basis of cost per square foot. Cold-rolled steel is specified in all major household appliances because of its excellent surface texture combined with other advantages of steel. Cold-rolled sheets may be specified U grade for unexposed parts, or E grade for exposed parts requiring the best surface finish. Surface textures can even be varied to conform to specific surface roughness requirements.

Cold-rolled sheets are produced by cold rolling hot-rolled pickled coils 50 to 70% and box annealing between 1100° and 1350° F. They are then temper rolled from 0.3 to 1.5% depending on end use. Cold reduction improves the surface and combined with the annealing, provides better drawability than hot-rolled sheets. Not only is ductility greater than hot-rolled (see Table I), but in addition the average plastic strain ratio (r_m) is increased from 1.0 on hot-rolled to 1.2 on cold-rolled and up to 1.6 on cold rolled drawing quality special killed steel sheets. The higher the average plastic strain ratio, the more the sheets resist thinning and breaking during

Robert L. Myers, manager of Research Materials Applications for Armco Steel Corp., was graduated from Purdue University in 1939 with a Bachelor of Science degree in chemical engineering. His professional affiliations include the American Ceramic Society; American Society for Metals, Materials System and Design Div.; Porcelain Enamel Institute Standards Committee; and the Society of Automotive Engineers. He has authored numerous papers on the application of steel to appliances.

drawing.

Some drawn parts currently produced from cold-rolled steels would not have been feasible previously. Recent technology in the area of steels with higher plastic strain ratios enables manufacturers to increase substantially their yield of prime parts on difficult draws, or to design parts that were previously impractical to draw. If any area seems to hold outstanding promise for continued development, this is it. Future refinements of cold-rolled steel should make it an even further improved material in years to come.

At present, the highest plastic strain ratio, or r_m value, of any commercially available steel is about 2.0. Besides having superior drawing qualities, this particular composition, called I-F Steel, has demonstrated it can often eliminate some drawing stages and/or annealing between draws. In some cases it may also make possible a reduction in gage because of its better metal-flow properties in critical areas of draws. Its typical mechanical properties appear in Table I.

Finishes on Cold Rolled

To resist rust, cold-rolled steel sheets must be coated. Organic coatings are used most widely, but other finishes include porcelain enamel, plastics, chrome, zinc and aluminum. The purpose and diverse advantages of these various finishes merit further explanation.

Zinc-coated steel sheets with various coating weights are used in appliances where corrosion is severe and extra protection is needed. Two typical applications are room air conditioners and clothes washer cabinets, but there are many others.

Galvanized steel sheets are made by continuously hot-dip coating cold-rolled steel. A recent development in making galvanized is jet finishing, where the coating thickness is controlled by jets of air or steam impinging on the strip surface just above the molten zinc bath. This produces better uniformity of coating thickness and broadens the range of coating weight that can be applied. There are many types, distinguished by specialized coatings and surfaces, and different thicknesses or weights of coating.

Standard commercial coating weight is G90 class in accordance with ASTM Specificaiton A525. It provides a minimum of 0.90 oz of zinc per square foot of sheet.

Painted dryer drum illustrates versatility of cold-rolled. Deeply drawn ribs assist tumbling action in the dryer.

Lighter or heavier coating weights up to G235 oz class, can be obtained if desired.

Regular spangled galvanized, because of the spangle outline, is best used for unexposed parts to replace chemically coated cold-rolled sheets when better corrosion protection is required. Minimized spangle surfaces have no visible spangle, and the best quality surface for exposed high gloss painted parts is known as extra smooth. This finish has both minimized spangle and the smoothest surface.

Although zinc-coated steel is seldom painted in appliance applications, the material can be painted when properly prepared or can be obtained with a mill phosphate treatment which requires only normal cleaning prior to painting.

Alloyed iron-zinc coatings are produced by special processing. They have no spangle and usually have coating weights of *0.40 to 0.60* oz per square foot (A40 and A60 coating designations). Lighter coating weights are also available. Spot welding characteristics of these alloyed coatings are better than other hot dip zinc-coated steels. Since the iron-zinc alloy does not have the ductility of a free zinc coating, some powdering of the coating may occur in drawing.

The unalloyed or free zinc coatings have good ductility and severity of fabrication is limited only by the properties of the base metal. These are shown in Table I. Even when deep drawn, no peeling or flaking of the coating occurs, assuring complete unbroken zinc protection on

Zinc-treated steel is welded to plain carbon steel for rigidity and corrosion resistance.

This washer tub is porcelain enameled on enameling steel. Note the exceptionally deep draw that is accomplished in only two operations.

finished products. As with the uncoated products, galvanized steel can be specified as DQ or DQSK. I-F Steel is also available in the zinc-coated product.

Benefits of Aluminum Coating

Aluminum-coated steel sheets are made much like galvanized steel sheets but offer some additional benefits. Cold-rolled strip is continuously coated by dipping in molten aluminum. The material is produced as either Type I or Type II.

Type I has a weight of coating of approximately 0.5 oz total per square foot both sides or 0.001 in. of aluminum on each side. It is intended for use where heat and corrosion resistance are required. Type I is the most commonly used grade in the appliance industry in kitchen ranges, clothes dryers, element retainers and baffles and reflector plates. Excellent heat resistance and reflectivity up to 900° F is one of its unusual properties. This is 300° to 400° F above the usable temperature range for galvanized steel.

Type II is used primarily for protection against atmospheric corrosion. Generally used unpainted, its coating life is four to five times that of the zinc coating on G90 Class galvanized steel.

Alunimum-coated steel combines the corrosion protection of aluminum with the strength and economy of steel. Type I is particularly advantageous at high operating temperatures up to 1250° F—though it loses its heat reflective property above 900° F.

Aluminum-coated steel Type I withstands moderate forming, drawing and spinning operations without flaking or peeling of the coating or damage to the base metal. It can be bent over a radius equal to twice its thickness without damage to the coating. The mechanical properties are listed in Table I. These steels show good ductility in drawing quality and drawing quality special killed for increased formability.

In cleaning aluminum-coated steel parts, strong acid and alkaline cleaners must be avoided as they are corrosive to aluminum. Although aluminum-coated steel accepts and holds paint well, some form of surface treatment is recommended for the very best paint adherence and life.

Steel for Porcelain Enameling

Enameling iron is another specialty grade produced for two-coat porcelain enameling by the appliance producer. An extra low carbon enameling iron is produced for one-coat porcelain enameling. Enameling iron is widely used in all major household appliances.

The typical composition of enameling iron is 0.03% carbon and 0.05% manganese, but the manganese content may be as high as 0.20% to meet special drawing or fabricating requirements. Processing in hot-rolling, pickling, and cold-rolling is similar to that of cold-rolled sheets, but metallurgical controls are different. Cold re-

TYPICAL MECHANICAL PROPERTIES OF VARIOUS SHEET STEELS

Steel Grade	Yield Strength psi	Tensile Strength psi	% Elongation 2″	Hardness RB
Hot-Rolled Commercial Quality	38,000	52,000	30	55
Hot-Rolled Drawing Quality	35,000	50,000	36	50
Hot-Rolled Drawing Quality Special Killed	35,000	50,000	40	50
Cold-Rolled Commercial Quality	34,000	46,000	35	45
Cold-Rolled Drawing Quality	30,000	45,000	39	40
Cold-Rolled Drawing Quality Special Killed	25,000	43,000	41	40
I-F Steel	22,000	46,000	46	46
Enameling Iron Commercial Quality	35,000	50,000	28	50
Enameling Iron Drawing Quality	25,000	40,000	38	37
Decarburized Enameling Sheets, Commercial Quality	33,000	43,000	37	40
Decarburized Enameling Sheets, Drawing Quality	25,000	40,000	45	35
Zinc-Coated Commercial Quality	32,000	50,500	35	57
Zinc-Coated Drawing Quality	29,000	48,500	40	52
Zinc-Coated Drawing Quality Special Killed	32,000	48,500	41	49
Aluminum-Coated Commercial Quality	35,000	51,000	32	58
Aluminum-Coated Drawing Quality	30,000	49,000	38	52
Aluminum-Coated DQ Special Killed	29,000	47,000	41	48
Stainless Type 301	40,000	90,000	60	80
Stainless Type 304	35,000	85,000	55	80
Stainless Type 430	45,000	70,000	25	85
Stainless (Armco 400)	32,000	62,000	32	68
Stainless Type 409	40,000	68,000	30	73

Table I—Listing shows yield strength, tensile strength, percent elongation and hardness for various sheet steel materials.

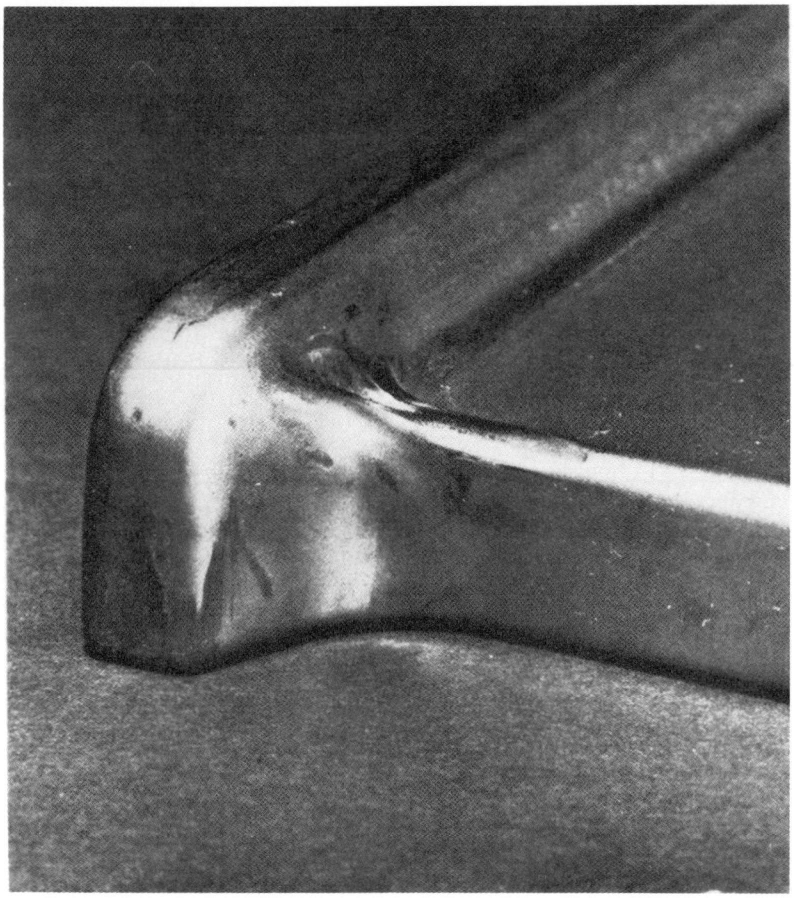

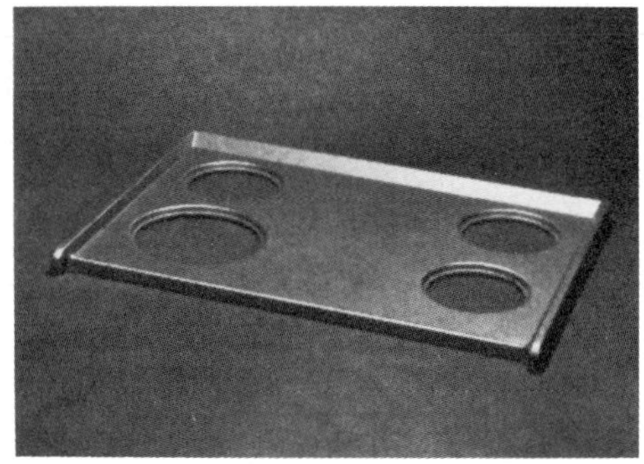

Enameling steels used to produce modern range tops such as the one shown above must possess both good ductility and high strength. Photo at left showing the spill-guard edge configuration for this range top exemplifies why the application requires superior ductility obtained with enameling steels.

duced full hard coils of enameling iron must be annealed to produce a soft ductile product. The annealing treatment consists of feeding sheets or strip through a continuous furnace. The metal is rapidly heated to 1800° F and air cooled. Pickling is then required to remove the scale formed on the surface during the cycle. To produce drawing quality, an additional box anneal at 1100°-1200° F is necessary. The mechanical properties listed in Table I show the effect of the box anneal operation in improved ductility and lower hardness.

Advantages of using enameling iron instead of cold-rolled steel in porcelain enameling are chiefly two-fold. First, enameling iron, due to the low carbon and manganese levels, retains its formed shape better during fusion of the porcelain enamel. This is often described as improved sag resistance. Second, due to the low carbon content, enameling iron is less subject to carbon boiling in the enamel coating. Carbon boiling occurs when carbon in the sheet combines with oxygen in the air during enamel fusion. The result is formation of carbon gases which boil through the enamel coating, leaving pin holes and blisters on the surface.

Decarburized enameling sheets are another specialty product developed for direct application of cover coat porcelain enamels to the base metal. No ground coat is needed. The melted composition of decarburized enameling steel is quite similar to that of cold-rolled steel—0.08% carbon and 0.30% manganese. Processing through the cold reduction step is also similar to that for cold-rolled sheets. Cold reduced coils are decarburized in open coil annealing where the laps of the coils are separated by twisted wire. This allows the decarburizing gases (such as steam, hydrogen and nitrogen) to sweep the entire surface area. The carbon content of the sheet surface is reduced to 0.008% max. Annealing at 1300° F also recrystallizes the grains, and a soft ductile material is the result. This is shown by the mechanical properties in Table I. The extra-low carbon content allows direct applications or porcelain enamel cover coats without carbon boiling. In addition, decarburized enameling sheets are more resistant to enamel fishscale than cold-rolled sheets or enameling iron. Sag resistance is equivalent to that of enameling iron.

Some of the uses are for freezer and refrigerator liners and pans; range tops, panels and ovens; dishwasher tubs, doors and bodies; and clothes dryer tops.

Applications for Stainless

As finishing costs rise with increased labor rates, some stainless steel parts may prove more economical than carbon steel parts that require plating. This is especially applicable to parts made of 0.020 in. and lighter thicknesses, where square foot finishing costs are a high percentage of total part cost.

Stainless steel in various types and finishes also is used for appearance and corrosion resistance in appliance parts. Mechanical properties of five different stainless

CHEMICAL COMPOSITION OF STAINLESS SHEET STEELS

Type	% Carbon	% Chromium	% Nickel	% Other
301	0.10	17.0	7.0	—
304	0.06	18.5	8.5	—
430	0.05/0.08	17.0	—	—
Armco 400	0.03	12.5	—	0.2 Al
409	0.05	11.5	—	0.5 Ti

Table II—Chemical compositions are indicated for five types of stainless sheet steels. Note that carbon and chromium in varying percentages are common to all five steel types, whereas nickel is present only in the 300 Series materials.

steels are listed in Table I. While these steels have higher strengths and hardness than plain carbon steels, some are very ductile and capable of deep draws such as sinks, dishwasher tubs and clothes washer tubs. The product described as Armco 400 does not have an AISI designation, but most producers have proprietary designations all of similar composition. Its advantage over Type 409 is in its consistently better surface quality. Both cost substantially less than the other stainless types listed.

Chemical compositions of the various types are listed in Table II. Note that the 300 series contains nickel and chromium, while the 400 series has only chromium.

Stainless steels are available with a variety of finishes. No. 1 strip finish or 2D sheet finish, which is a dull pickled finish, is supplied for drawn parts such as sinks and dishwasher tubs, or for parts that will be polished by the manufacturer after fabrication. Stainless supplied for trim parts usually has a No. 2 strip or No. 2B sheet finish which is achieved by temper rolling with bright rolls. Sometimes a No. 3 or 4 polish finish is supplied for trim parts. These finishes are made by grinding and polishing the surface.

Types 301, 304, 430 and Armco 400 are used in sinks, dishwasher tubs, clothes washer tubs, and other parts where the corrosion resistance of stainless steel can be sales features. Type 430 is also specified for dishwasher racks, disposal housings, and trim. Type 400 is used for range burner rings and sink trim. Type 409 is ideal for unexposed parts such as brackets and clips. In addition, a new steel, 18SR, shows promise in scaling resistant applications at elevated temperatures.

Pre-Coated Materials

Color-coated steels are among the newest members of the appliance sheet steel family and offer promise of substantial savings to appliance producers. Paints, vinyl laminates, printed patterns and strippable films are applied to steel coils at the mill. As availability of these products expands, significant increases in their use in the appliance industry is anticipated. Paint quality, adherence, uniformity of coating and color contrast are superb as processed on the newly constructed coil coat lines. These units clean the steel, phosphatize the strip, coat, cure and recoil. While there are some limitations in fabricating and joining techniques, coil-coated steel is currently used in refrigerator liners, various external cabinet parts and unexposed primed-only parts on all kinds of appliances.

In summary, sheet steels have contributed to the economy of major household appliances because of their plentiful supply, low cost, high strength, adaptability to high production fabricating, joining and finishing methods. Due to technology available today in steelmaking such as strand or continuous casting and vacuum degassing, improved grades of sheet steel will constantly appear. They will have even more uniform mechanical properties and be capable of deeper draws and more intricate configurations. ——— AE

M. Nakano and O. Kinpara

An Investigation on the Stamping of
High Strength Steel Sheet

Stretching has a correlation with the mechanical properties of the steel sheet: the steel sheet having a higher n-value, a lower yield stress and a higher ductility is more favorable. There is available the experimental formula[1),2)] proposed by H. Kubotera et al. covering mild steel sheets. The stretch formability is higher accordingly as the stretching height, H, becomes larger. The author carried out an experiment on high strength steel under these conditions:

$$H = 0.10\ El(\%) - 0.29YP(kg/mm^2) + 26.1$$

According to the results of the present experiment, this experimental formula is well applicable to the high strength steel sheet at a high accuracy (Fig.1). This suggests the possibility of estimating the stretch formability of high strength steel by controlling the elongation and yield point.

It is widely known that the stretch-flanging property tends to be affected not only by the mechanical property but also by the second-phase particles, since forming is often regulated by the deformation concentrated along the steel sheet edge. For mild steel sheet in which the matrix is rich in ductility and inclusions, carbides and other second-phase particles have little effect, the stretch-flanging property can be predicted with the use of r-value, elongation and other mechanical properties.[1),2)]

When the increase in strength brings about a lower ductility of the steel sheet, the effect of second-phase particles is more apparent. It would be necessary to consider the effect of inclusions as second-phase particles in addition to that of mechanical properties, when studying the stretch-flanging property of high strength steel. By the utilization of elongation and r_{min} as mechanical properties and [S] and [O] as factors expressing inclusions among second-phase particles, the authors successfully formulated the following experimental equation applicable at a high accuracy. A larger value of the hole expansion, λ , based on KWI test, of course corresponds to a more excellent stretch-flanging property (Fig.2):

$$\lambda = 1.80El(\%) + 41r_{min} - 522[S](\%) - 213[O](\%) - 29 \qquad (2)$$

It is understood, from Eq.(2), that, on the assumption of non-improvement of elongation of high strength steel sheet, improvement of stretch-flanging formability cannot be expected unless a steel sheet is adopted,

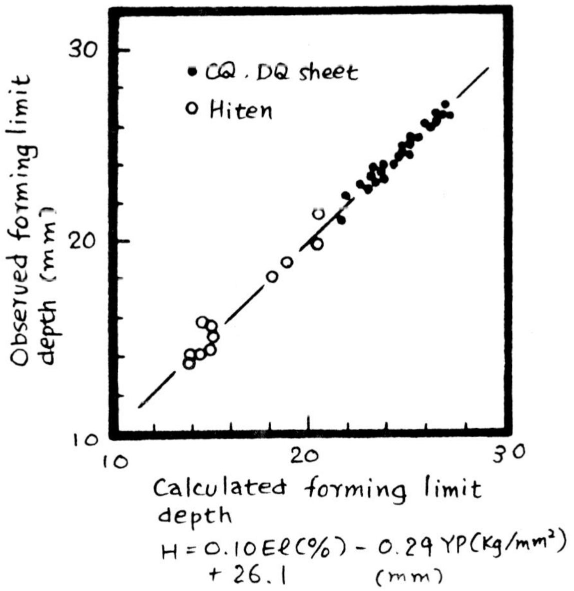

Fig.1 Effect of mechanical properties on stretching limit depth

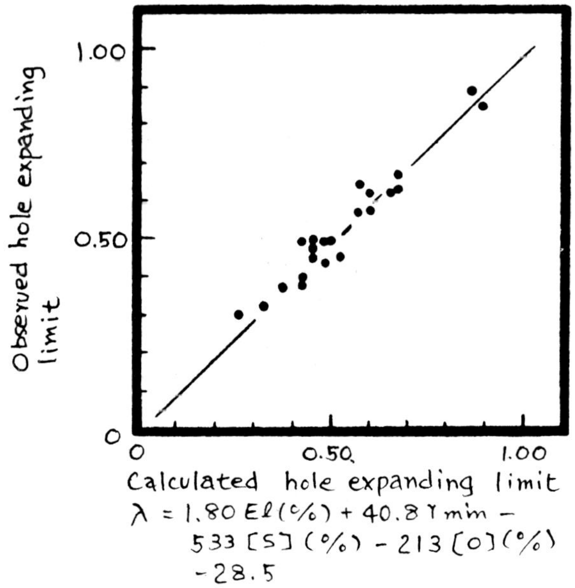

Fig.2 Effect of material characteristic on hole expanding ratio

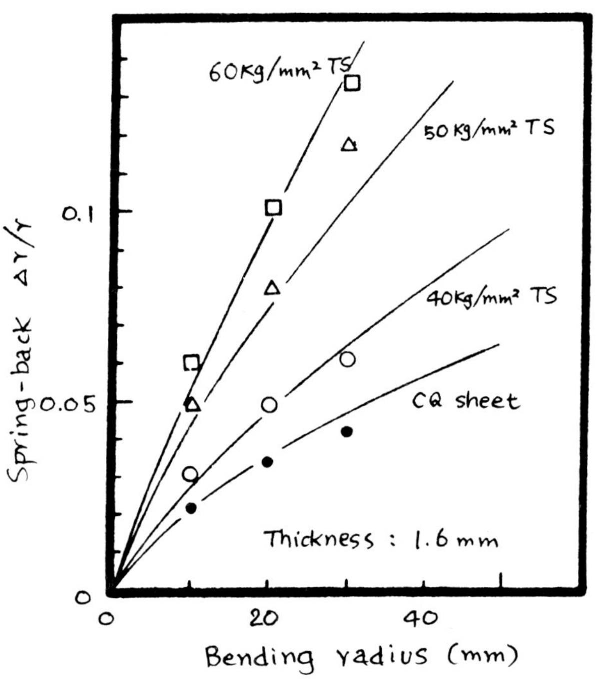

Fig.3 Relation between observed value and theoretical value on spring-back

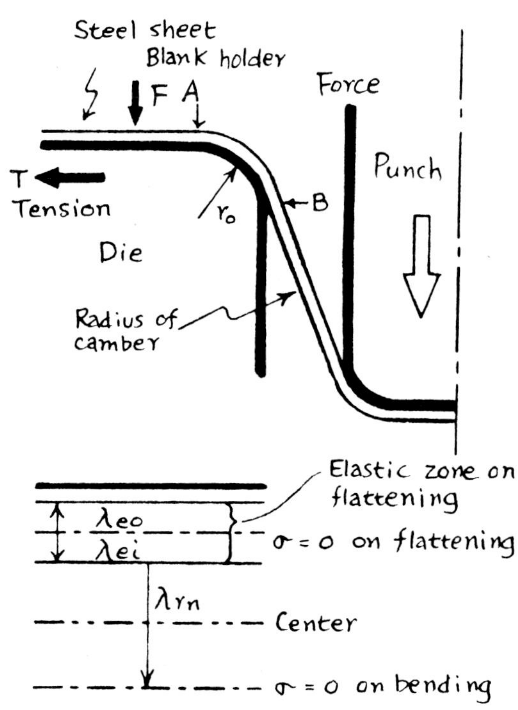

Fig.4 Schematic diagram of camber of wall

in which r-value is high, and inclusions and other second-phase particles are properly controlled.

In inverse proportion to the increase in strength of steel shett, the elongation and n-value are lowered, and there is also the decrease in r-value in general. It is evident that the decrease in these mechanical properties is unfavorable for the stretch formability and the stretch-flanging property, whereas it is possible to improve somewhat the stretch-flanging property by raising elongation, r-value and properly controlling inclusions and other second-phase particles.

3. Shape fixing ability

Spring-back caused in bending has so far been studied by many authors, and results well agreeing with observed values have been obtained.[3)] There are however available only a few experimental data on high strength steel. Under an applied tension stress of null, a three-point bending experiment was carried out on a high strength steel sheet, to see how far the present theory was applicable. According to the results, the following formula gives results satisfactorily agreeing with experimental results and are considered to be convenient in application:

$$\frac{\Delta r}{r} = \frac{\Delta \theta}{\theta} = \frac{3F}{(2+n)E}\left(\frac{2r}{t}\right)^{1-n} \qquad (3)$$

where, bending angle and radius: θ and r;
spring-back angle and radius: $\Delta\theta$ and Δr;
E: Young's modulus; t: thickness

The steel sheet deformation is approximated by $\sigma = F\varepsilon^n$ and the theory of beam is applied.

This formula and observed values are plotted in Fig.3, which suggests a very good agreement.

In the following, camber of pressed parts caused in press-forming is a very serious defect in high strength steel sheet, but only a few examples of analysis are available. On the assumption that this camber was caused by the bending/flattening deformation at the die shoulder, therefore, analysis was carried out and the results were compared with experimental results. The steel sheet applied with tension T by blank holder force F is assumed to be bent to radius r_0 at die shoulder A, and to be flattened to radius r_1 at point B. To this simultaneous tensile bending/flattening deformation, the following equations are applicable (Fig.4):

For the simultaneous bending with tension:[4)]

$$\frac{M_o}{bt^2} = \frac{F}{3(2+n)}\left(\frac{3}{2\sqrt{3}}\right)^{1-n}\left(\frac{t_o}{2r_n}\right)^n\left(\lambda_o^{2+n} \pm |\lambda_i|^{2+n}\right) - \frac{T}{bt}\frac{\lambda n}{2} \tag{4}$$

$$\frac{T}{bt^2} = \frac{2F}{3(1+n)}\left(\frac{3}{2\sqrt{3}}\right)^{1-n}\left(\frac{t}{2r_n}\right)^n\left(\lambda_o^{1+n} - |\lambda_i|^{1+n}\right) \tag{5}$$

$$\frac{t}{2r_n} = \frac{1}{4}\left\{\left(1+\frac{t}{2r_n}\lambda_o\right)^2 - \left(1-\frac{t}{2r_n}\lambda_i\right)^2\right\} \tag{6}$$

For the simultaneous tensile flattening:[3)]

$$\begin{aligned}\frac{M}{bt} &= \frac{1}{2}\int_{-\lambda_i}^{\lambda_o}\sigma d\lambda \\ &= \frac{1}{2}\int_{-\lambda_i}^{\lambda_o}\sigma d\lambda = \frac{1}{2}\int_{-\lambda_i}^{-\lambda r_n}\left\{F\cdot(|\varepsilon| + |\varepsilon_o|)^n\right\}d\lambda \\ &+ \int_{-\lambda r_n}^{\lambda ei} F(|\varepsilon| + |\varepsilon_o|^n d\lambda + \int_{-\lambda ei}^{\lambda eo} E\cdot\varepsilon d\lambda + \int_{\lambda eo}^{\lambda_o}\left\{-F\cdot(|\varepsilon|+|\varepsilon_o|)^n\right\}d\lambda\end{aligned} \tag{7}$$

$$\frac{M_1}{bt^2} = \frac{1}{4}\int_{-\lambda_i}^{\lambda_o}\sigma\lambda d\lambda - \frac{T}{b^2}\cdot\frac{\lambda m}{2} \tag{8}$$

If the sheet has radius r after spring-back upon flattening to radius r_1:

$$\frac{1}{r} = \frac{1}{r_1} - \frac{M_1}{ET} \tag{9}$$

Radius r after spring-back is therefore obtained with the use of Eqs.(4), (5), (6) and (7).

Fig.5 represents the relation between the theoretical solution and the amount of camber obtained by experiment, suggesting a satisfactory correlation. With the use of this equation, the authors calculated the effect of thickness of high strength steel sheet under no-tension condition. Results are shown in Fig.6. The amount of camber increases accordingly as the tensile strength of sheet becomes higher and the thickness becomes smaller. The tension in bending has only a limited effect of reducing camber within the range not exceeding the yield stress of steel sheet. In this respect, the high strength steel sheet is disadvantageous in that a high blank holder force is required. This fact suggests the possibility of estimating the amount of camber of high strength steel sheet

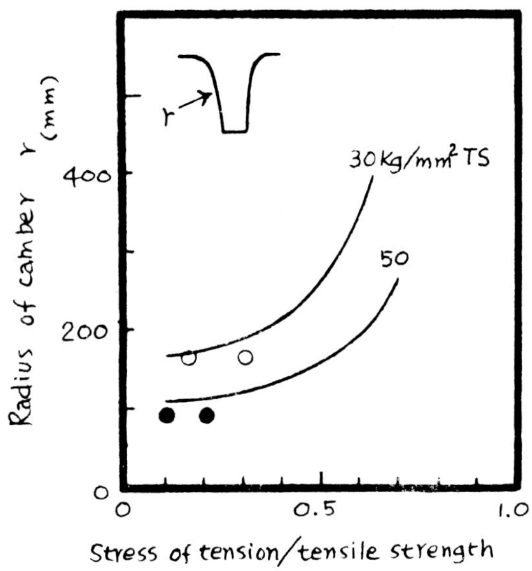

Fig.5 Effect of tension and material strength on camber

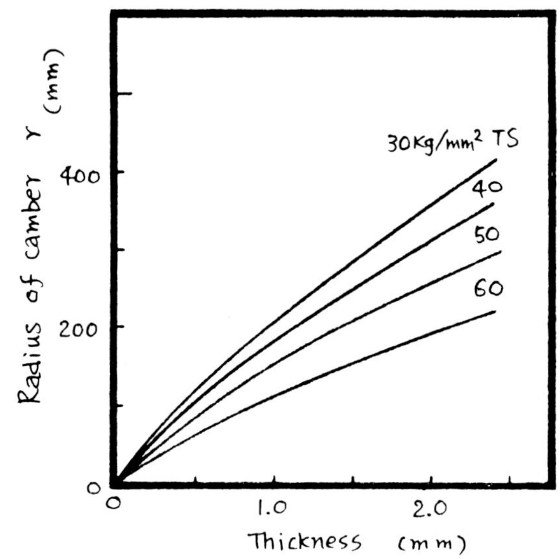

Fig.6 Effect of tensile strength and thickness on camber

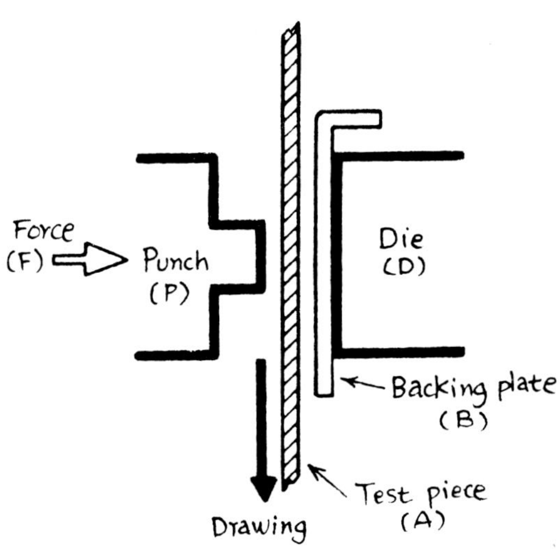

Fig.7 Showing method in galling simulation test

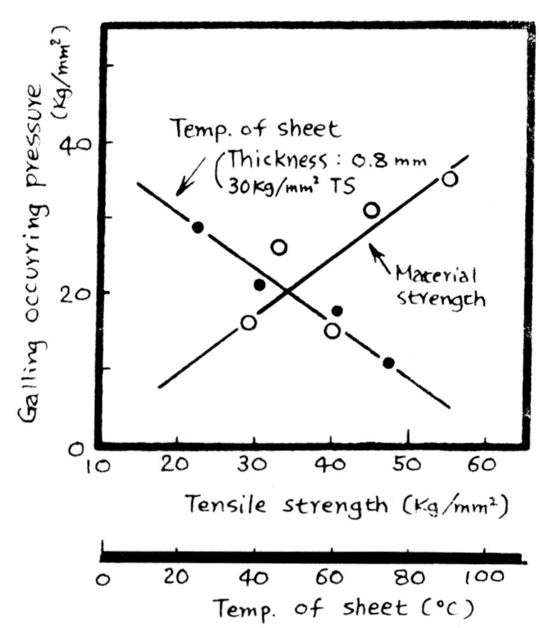

Fig.8 Effect of material strength

by calculation.

4. Galling simulation test

It is a usual practice to raise the blank holder force in pressing high strength steel sheets for obtaining a higher formability as well as a more satisfactory shape property. This causes galling between the press die and the product, thus constituting a serious trouble in press-forming. A galling simulation test apparatus was therefore manufactured to analyze the effect of such factors as steel sheet strength, temperature and lubricant on the occurrence of galling.

The test apparatus is represented in Fig.7. Test piece (A) is clamped by backing plate (B) placed on die (D) and punch (P), and the test piece is pulled by applying an oil hydraulic surface pressure (F) between the punch and the die. Backing plate (B) was defined as die (D) in actual case. Test piece was pulled at a speed of 200mm/min over a distance of 200mm. The face pressure was applied on a portion of punch cross-section of 7mm x 8mm. Galling resistance was evaluated in terms of face pressure (P) at which galling was initiated.

The effect of material strength of test piece was studied by employing a constant material (CQ steel sheet with a hardness of H_R30T 53) for backing plate (B) corresponding to the die in actual pressing. Results are indicated in Fig.8. According to these results, a higher material strength of test piece raises the face pressure (F) initiating galling and gives a satisfactory galling resistance. These results are based on the constant material quality of die. It is necessary to obtain experimental results taking into consideration the hardness and grade of steel of die as well, whereas these conditions were omitted in the present experiment.

Temperature has an important effect on galling. As shown in Fig.8 representing the effect of sheet temperature on galling, the temperature effect in fabricating the material cannot be disregarded.

An experiment was conducted on the lubricants and the surface roughness of steel sheet, to determine the critical pressure initiating galling. Results are indicated in Table 2. According to the results of the present experiment, the lubricant of a higher viscosity gives a more satisfactory galling resistance. The lubricant containing extreme-pressure materials is considered more effective in preventing galling. As shown in the Table, for example, the galling resistance is improved in a mill cleaning sheet

Table 2 Effect of lubricant surface roughness in galling (CQ steel sheet)

Factor		Critical pressure on occurrence of galling
Lubricant (roughness: RMS=35 in)	Rust preventive oil (viscosity: 10.1cst at 100°F)	18.3 kg/mm^2
	Low viscosity lubricant (viscosity: 31.1cst at 122°F)	28.6
	High viscosity lubricant (viscosity: 744cst at 122°F)	Non-occurrence
Surface roughness (lubricant: rust preventive oil)	RMS: 51 μin	17.4
	RMS: 47	16.1
	RMS: 40	13.4
Carbon contination (roughness: RMS=40μin lubricant: rust preventive oil)	Existant (mill cleaning sheet)	22.4
	Non-existant	13.4

slightly stained with carbon. A more extensive surface roughness of steel sheet corresponds to a higher galling resistance, because of the holding of lubricant in a larger quantity between the die and the steel sheet.

5. Buckling strength of panel

Automotive outer panels are required to have a high buckling strength and a property giving a small dent upon collision with a body. The deflection and buckling resistance of a curved sheet are dependent only on the physical properties of the material and the dimensional shape of curved sheet because of the elastic deformation, and have no relationship with the mechanical properties of the material.[5)]

Then, an experiment was conducted with a view to determining the impact strength of panel, as follows.

Press-forming was applied to a square cylinder with a side of 350mm and a depth of 70mm, the bottom surface of which formed a curved surface having a curvature radius of 1600mm. The experiment as shown in Fig.9 was carried out on this bottom surface of sample. By dropping a steel ball of 4kg onto this bottom, the relation between dropping height H and dent depth d was investigated as shown in Fig.9. The sharp increase in dent depth is due to the occurrence of plastic buckling throughout the entire bottom. If the buckling strength of panel is defined in terms of critical dropping height H_o at which plastic deformation takes place in the bottom, a larger values of H_o is evidently more favorable for the panel strength. Dropping of the steel ball at the bottom center causes a plastic dent at the hit point and corners. This dent d increases accordingly as dropping height H is raised, and propagates toward the center. At critical dropping height H_o, the entire bottom buckles and presents a dent diagonally as shown in Fig.9. The dropping energy is mostly consumed as the bending moment in the formation of dent caused diagonally at the center. It is therefore considered that the plastic buckling of the entire bottom is caused when a plastic strain giving a dent not capable of being restored in shape occurs. Bending moment Me at the initiation of plastic region is expressed for elastic, perfectly plastic material by:

$$Me = b\sigma_e t^2/6 \qquad (10)$$

Comparison of values obtained by this equation and observed ones showed a

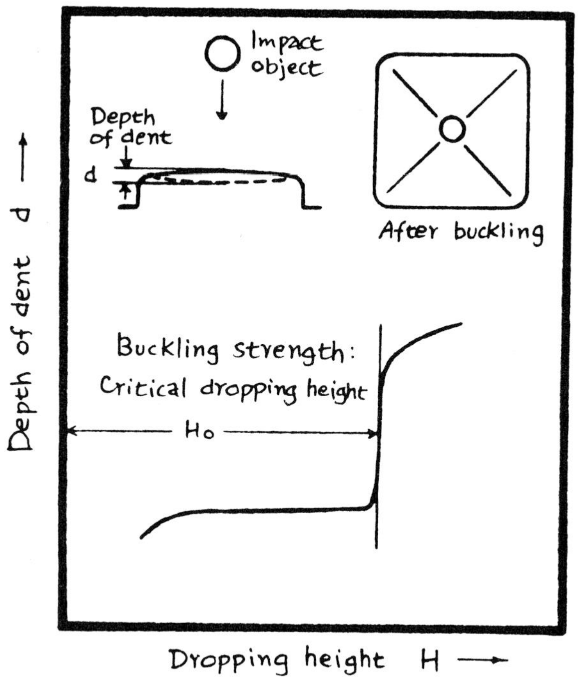

Fig.9 Experimental method and definition of buckling strength of panel

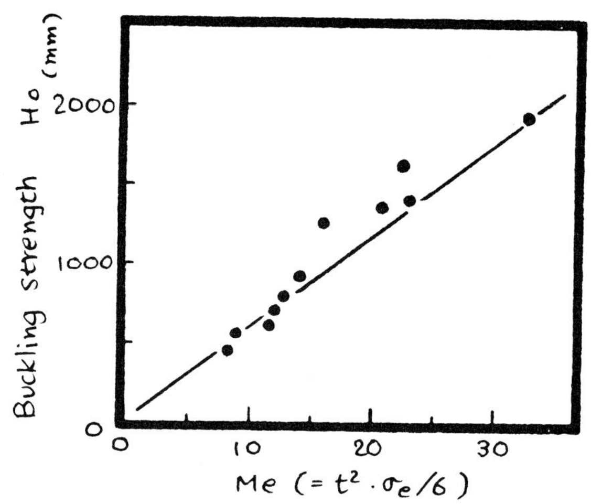

Fig.10 Relation between buckling strength and bending moment at the initiation of plastic deformation (thickness t: 0.6-1.2mm tensile strength $_B$: 29 - 60kg/mm^2)

Table 1 Mechanical property of steel sheet on camber

No.	Yield stress (kg/mm^2)	Tensile stress (kg/mm^2)	n-value	F-value (kg/mm^2)
1	27.0	39.7	0.166	59.9
2	37.6	52.7	0.164	79.7
3	39.9	54.5	0.164	87.3
4	38.7	58.1	0.123	84.4
5	24.6	32.1	0.172	54.0
6	26.0	34.9	0.210	58.3
7	30.1	37.5	0.190	62.0

satisfactory agreement with experimental values as shown in Fig.10. The buckling strength of panel is proportional to the second power of the thickness, increasing proportionally to the yield stress. When an impact is applied to the panel, therefore, the steel sheet having a larger thickness and a higher strength is more favorable as panel.

6. Conclusion

Considerations were made of the formability, shape-property, troubles in pressing and product strength of high strength steel sheet, and the following conclusions were reached:

a. Formability:

The conventional formula of forming limit using the elongation and yield stress is applicable to the stretch formability. As to the stretch-flanging property, however, another highly accurate forming limit formula was worked out, taking into account the effect of [O] and [S] in addition to the material mechanical properties.

b. Shape fixing ability:

Defective shape was studied by dividing it into the spring-back and the camber. Comparison of results of theoretical analysis and observed values showed a very satisfactory agreement. Defective shape is more serious accordingly as the material strength becomes higher, and less serious as the thickness increases.

c. Galling simulation test:

Various experiments were carried out on factors causing galling. According to the results, the use of high-viscosity lubricant is effective in avoiding galling. Furthermore, it is also effective to use mill cleaning sheet.

d. Buckling strength of panel:

Under an impact load, the buckling resistance increases in proportion to the material strength and to the second power of the thickness.

References

1) H. Kubotera, M. Nakano et al., Nippon Kokan Technical Report-Overseas, June, 1973

2) H. Kubotera et al., Proceedings ICSTIS, Section 5, Suppl. Trans. ISIJ, vol.11, 1971

3) M. Masuda, Bending forming of sheet metal, Seibundo-Shinko-sha, Japan

4) Mizuno et al., Journal of the Japan Society for Technology of Plasticity, 6, 50

5) K. Abe, Journal of the JSME, vol.22, No.600

6) K. Kato et al., Nippon Kokan Technical Report-Overseas, May, 1968

W. Yu, Cold-formed steel structure, McGrow-Hill Book

D.G. Adams et al., SAE Automotive Engineering Congress, Detroit, Mich., Feb., 1974

Section III: Metallurgical Properties of Steel Sheet

Campbell Memorial Lecture

Sheet Steel—Micrometallurgy by the Millions

D. J. BLICKWEDE

Two reasons make it particularly appropriate that this forty-third annual Campbell Memorial Lecture be given in Detroit. One has to do with a great industry and its prime material; the other with a great man of science.

The first is that Detroit is the home of the automotive industry. The principal material used by this great industry is low-carbon sheet steel, which is the subject of this lecture.

The second reason, of course, is that the man whom this lecture commemorates, Professor E. D. Campbell, was born in Detroit and attended the nearby University of Michigan. There his inspirational teaching and pioneering research laid the groundwork in this country for the new discipline of physical metallurgy.

Previous lecturers have recounted Professor Campbell's many contributions to metallurgy, most of which came after he became sightless. But perhaps it is less well known that his scientific activities were not limited to chemistry and metallurgy. For instance, in his student years he was a skilled ornithologist and taxidermist and mounted birds and skeletons for exhibition in museums.

All in all, Professor Campbell was a very remarkable man, and I am pleased to pay my respects to him by this lecture.

* * *

The subject of this lecture is low-carbon sheet steel. By this I mean all flat-rolled steel products less than 0.230-in. thick containing less than about 0.15% carbon and less than 1% alloy plus residual elements. This covers a broad range of products in either coils or cut-lengths and includes sheet coated with other metals such as Zn, Al, Pb-Sn, Sn or Cr.

Except for carbon, manganese is the principal alloying element in sheet steel and is normally in the range 0.20 to 0.50%. Small but closely controlled additions of Al, Ti, V, and other elements are often made to obtain certain properties. Usually, the common residual elements, including O, N, P, S, Si, Cu, Cr, Mo, and Ni, are held as low as is consistent with available raw materials and practical steel-making.

Figure 1 is a flow chart of the operations by which sheet steel is made. Detailed descriptions may be found in Ref. 1–3.

Low-carbon sheet steel was not a significant product of the steel industry at the turn of the century when the first horseless carriages were appearing on the scene and Professor Campbell was beginning his research on steel. At that time steel was being used mainly for its strength, rigidity and ductility in applications such as railroads, machinery and weapons.

But the growth of consumer goods industries, with automobiles at the hub, changed all this. These industries found that sheet steel not only provided strength and rigidity but could be formed easily into the intricate shapes desired for consumer products. Today 50% of the weight of an average automobile is sheet steel. It has this preeminence because no other material has yet been able to combine the capabilities of being readily formed at high production rates into intricately shaped parts, of providing an attractive surface after forming, of being easily joined to other parts, of giving strength and rigidity, and of providing safety in the assembled products—all at low cost.

The powerful growth in the consumption of sheet steel in the United States since the turn of the century is seen in Fig. 2. Over the entire period from 1905 to the present, the consumption of sheet steel has steadily grown at the rate of about 5.5% per year.

In 1967 about 45 million tons of the sheet steel products that come within the scope of this lecture were produced with a total selling value of about six billion dollars. For every dollar of sheet steel purchased, about $12 was generated in billings for finished manufactured products. Thus, in 1967 the sales realization from products manufactured from sheet steel was about 70 billion dollars, or 9% of our gross national product in goods and services. Clearly, sheet steel is *the* material of first importance in our economy.

And yet, despite the major importance of sheet steel to all of us, it has received little attention from metallurgists outside the steel industry. As I will show, the reason for this omission is not that the

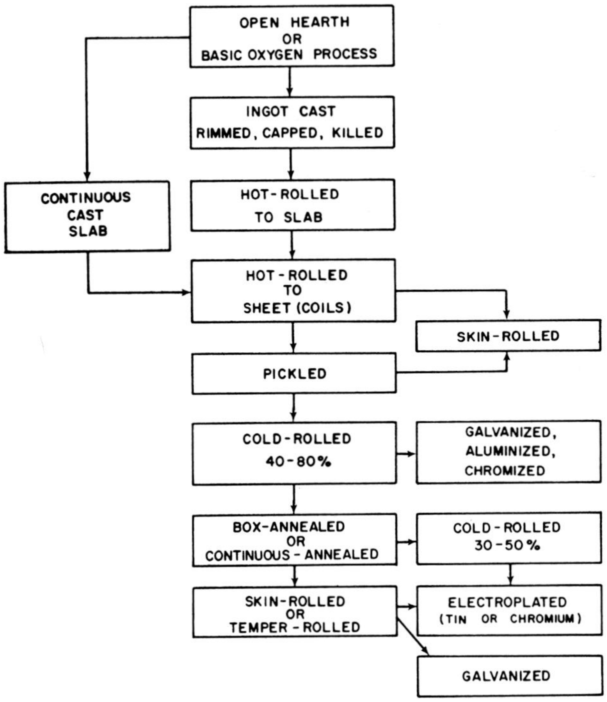

FIG. 1. Simplified flow chart of the production of low-carbon sheet steel.

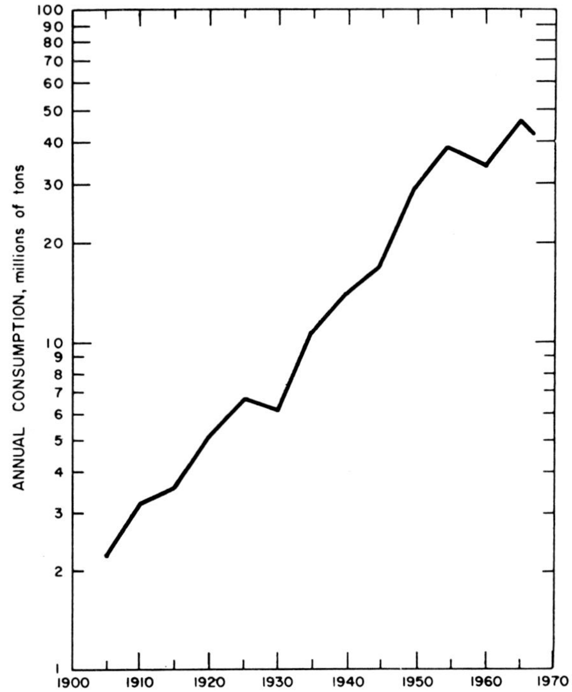

FIG. 2. Annual consumption of low-carbon sheet steel in USA (includes hot-rolled, cold-rolled, coated sheet and tinplate).

metallurgy of sheet steel is somehow lacking in interest or challenge. Far from it. Sheet steel is a very complex material, finely balanced in composition and structure. Small changes toward either greater or lesser purity in composition or in processing strongly affect its properties.

Perhaps the reason sheet steel metallurgy has not received wide attention outside the steel industry is that all the metallurgical characteristics are developed in it at the steel mill; the user has no need to change the composition or structure of sheet steel to be able to utilize it.

The main requirement for most applications of sheet steel is good formability. By formability I mean the capability of being extensively deformed into intricate shapes without fracture or defects in the finished part. The operation by which this is done is variously called press forming, deep drawing or stamping. Accordingly, most of this lecture will deal with the formability of sheet steel and the way in which composition and processing operations influence the basic properties that affect formability.

Another important requirement of sheet steel for some applications is an unblemished surface, capable of accepting a wide variety of finishes after fabrication. At the end of this lecture I will describe some unusual aspects of the composition of the surface of sheet steel.

To put this lecture together I have drawn upon the literature and upon discussions with others in the steel industry; these sources of information will be acknowledged in the body of the lecture. I have also drawn heavily upon the research and developments of my associates at Bethlehem's Homer Research Laboratories. Many have contributed, but special acknowledgment must be given to E. G. Bauer, J. W. Frame, M. M. Goldberg, D. L. Harper, J. B. Horton, W. Jolley, D. A. Karlyn, E. H. Kottcamp, F. H. Laxar, E. H. Mayer, B. S. Mikofsky, E. C. Oren, F. B. Schunk, E. T. Stephenson, R. W. Vieth, R. L. Whiteley and D. A. Witmer. These men have made excellent contributions to the knowledge, understanding and improvements in sheet steel. This lecture includes a description of their most recent work, some of it heretofore unpublished.

THE MECHANICS OF FORMING SHEET MATERIALS

The most common sheet-forming method is press forming, by which a flat blank is formed into a finished shape between a pair of matched dies. There are other methods, but in all of them the mechanics may be resolved into two principal kinds of deformation: drawing and stretching. The properties of the sheet required for good drawability are not the same as those required for good stretchability.

The relative severity of the process in terms of drawing and stretching depends on the shape of the part being formed. It also depends on mechanical factors of the forming operation such as die design and lubrication. As a consequence, the formability of a sheet material cannot be expressed by a single property but is rather a combination of several properties and differs from one part or press operation to the next.

Nevertheless, analysis of the mechanics of idealized

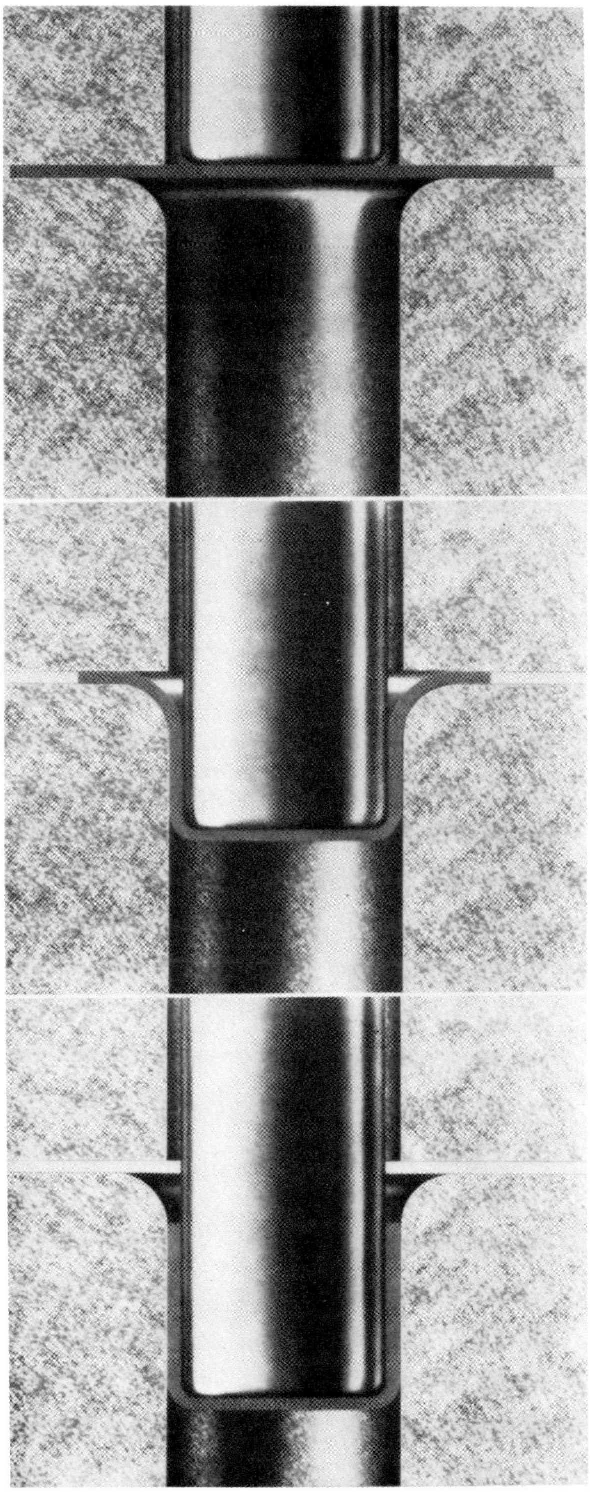

FIG. 3. An idealized drawing operation. The clamping force of the hold-down dies permits the sheet to flow radially inward without wrinkling. All the deformation takes place in the flange and over the lip of the die.

forming operations does serve to point up the properties of the sheet that are of major importance to drawability and stretchability. Consider first the case of drawing:

DRAWING

An idealized forming operation in which drawing is the only component is shown in Fig. 3. Here the clamping force of the hold-down dies is just sufficient to permit the material to flow radially into the die cavity without wrinkling. Deformation of the sheet takes place in the flange and over the lip of the die; no deformation occurs over the nose of the punch.

This process may be seen by imagining a section of the partially drawn cup, shown at the left in Fig. 4, which is then flattened into the two-dimensional piece shown at right of Fig. 4. The flared end of this piece corresponds to the flange of the cup, the straight-sided portion corresponds to the side-wall of the cup, and the pointed portion corresponds to the bottom of the cup. It may be seen that the flange is being compressed circumferentially and pulled radially in the plane of the sheet into the side-wall of the part. This is analogous to wire drawing in that a large cross section is being drawn into a smaller cross section of greater length. That is why this kind of forming process is called drawing to distinguish it from stretching.

The capability of the material to withstand drawing centers on two factors. One is the ability of the material in the flange region to flow easily in the plane of sheet under a condition of pure shear. In other words, it is desirable to have low flow strength in all directions of the plane of the sheet.

The other factor is the ability of the material in the side-wall to resist deformation in the thickness direction. As shown in Fig. 5, the punch prevents the material in the side-wall from changing dimension in the circumferential direction; the only way it can flow is by elongating and becoming thinner. Thus, the ability of the material in the side-wall to withstand the load imposed by drawing down the flange is determined by its resistance to thinning. In other words, high flow strength in the thickness direction of the sheet is desirable.

If we take both factors into account, in drawing operations it is desirable to maximize the ability of the material to flow in the plane of the sheet and also maximize the resistance of the material to flow in a direction perpendicular to the sheet. Low flow strength in the plane of the sheet is of little help if the material also has low flow strength in the thickness direction.

It is difficult to measure the flow strength of sheet in the thickness direction. However, the ratio of strengths in the plane and thickness directions can be obtained by determining the ratio of true strains in the width and thickness directions in a simple tension test. For a given steel strained in a given direction this ratio is a constant called the plastic strain ratio (4) and is expressed as

$$r = \frac{\bar{\epsilon}_{\text{width}}}{\bar{\epsilon}_{\text{thickness}}} \qquad \text{Eq 1}$$

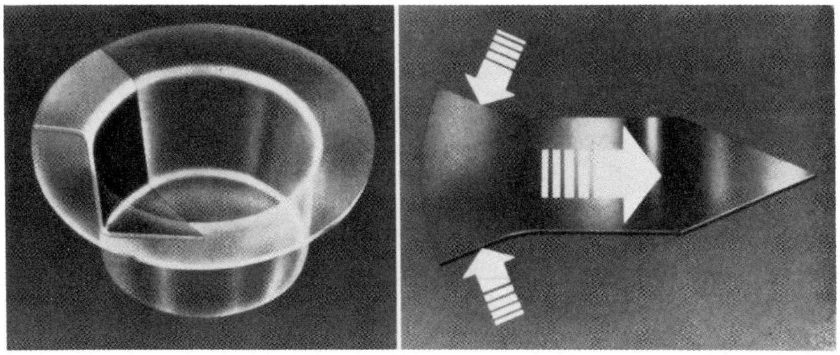

FIG. 4. Section from cup partially formed by drawing, for example, corresponding to Fig. 3 (center). Illustration at right shows the section flattened out. The flange flows under a combination of circumferential compression and radial tension into the side-wall.

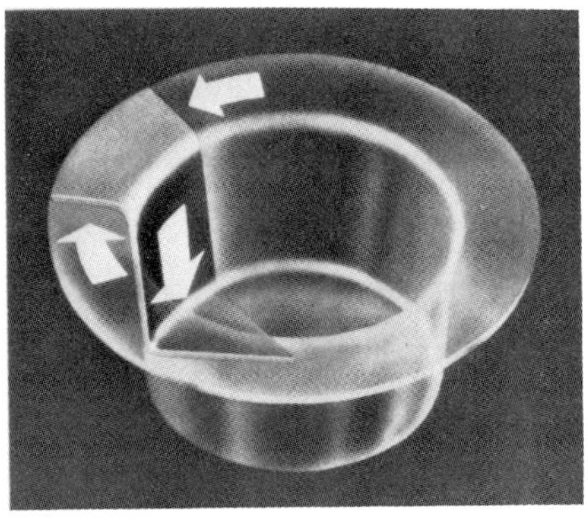

FIG. 5. System of forces in a cup being formed by drawing. Material in the flange region should have low flow strength in the plane of the sheet. In the side-wall region the material is constrained by the punch from flowing in the circumferential direction; it can only flow by becoming thinner. Hence, resistance to thinning is desirable for drawing.

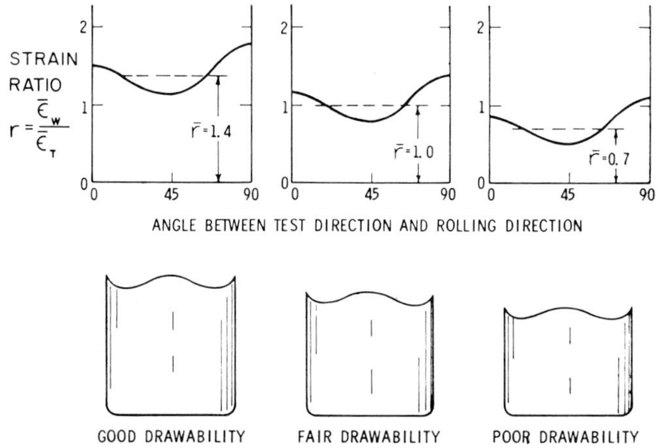

FIG. 6. Upper curves show the typical manner in which the strain ratio r varies with test direction in low-carbon steel. The average strain ratio $\bar{r}$ is a measure of the degree of normal anisotropy. The relationship of this parameter to drawability is indicated by the relative size of cups below each curve. Each cup represents the deepest cup that could be drawn from the material with the average strain ratio indicated.

The properties in the plane of the sheet are usually different in different directions. Therefore, it is necessary to use the average of the strain ratios measured parallel to, transverse to, and 45° to the rolling direction. The average strain ratio is expressed as

$$\bar{r} = \frac{r_L + 2r_{45} + r_T}{4} \qquad \text{Eq 2}$$

An average strain ratio of unity is indicative of equal flow strengths in the plane and thickness directions of the sheet. If the strength in the thickness direction is greater than the average strength in different directions in the plane of the sheet, the average strain ratio is greater than unity. In this case, the material is resistant to uniform thinning.

Experiments by my associate R. L. Whiteley (5, 6) have clearly shown that $\bar{r}$ is directly related to the depth of draw; and the higher the $\bar{r}$ value the deeper the draw. This is illustrated in Fig. 6.

The average strain ratio is a partial measure of the plastic anisotropy of the sheet. Since it gives the ratio of an average flow strength in the plane of the sheet to the flow strength normal to the plane of the sheet, it is called "normal" anisotropy to distinguish it from the variations in the flow strength in the plane of the sheet. On the other hand, the variation of the strain ratio in different directions in the plane of the sheet, Δr, is a measure of the "planar" anisotropy, where Δr is expressed as

$$\Delta r = \frac{r_L + r_T - 2r_{45}}{2} \qquad \text{Eq 3}$$

Thus, a completely isotropic material has a strain ratio of unity in all directions, i.e., $\bar{r} = 1$ and $\Delta r = 0$. The two parameters $\bar{r}$ and Δr are convenient measures of the plastic anisotropy of a sheet material.

STRETCHING

A simple stretch-forming operation is shown in the illustrations at the right of Fig. 7. Here a blank of sheet material is clamped firmly around the periphery, or flange, to prevent the material in the flange from moving into the die cavity as the punch descends. In this case, all the deformation is over the punch.

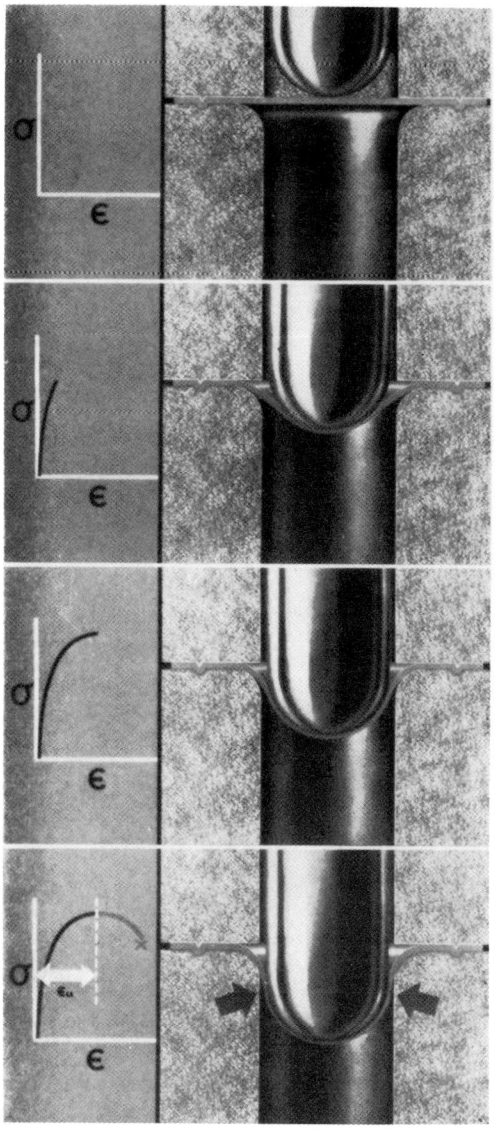

FIG. 7. Idealized stretching operation. Hold-down dies prevent radial flow of the flange, all deformation takes place in the central region where the sheet deforms by elongating and thinning. As shown at the left, if deformation is carried out far enough, deformation will then be localized and fracture occurs in that region. This point is analogous to the uniform elongation in a tension test, ϵ_u.

The stress system in this situation is biaxial, but for illustrative purposes the mode of deformation may be rationalized by a conventional uniaxial stress-strain diagram. The sequence of illustrations at the left side of Fig. 7 shows that the sheet deforms by elongating and uniformly thinning as the punch descends. As in the case of tension testing, if the deformation is beyond the ability of the material to undergo uniform straining, it will be localized and fracture is imminent. As shown, the extension beyond which deformation becomes localized is analogous to the elongation at maximum load in a tension test, a value called uniform elongation, ϵ_u.

For many metals, especially low-carbon sheet steel, the shape of the plastic portion of the tension stress-strain curve, expressed in terms of true stress and true strain, may be closely described by the parabolic equation

$$\bar{\sigma} = K\bar{\epsilon}^n \qquad \text{Eq 4}$$

In this equation it is n, the strain-hardening exponent, that is, the measure of the ability of a metal to resist localized straining and thus withstand complex nonuniform deformation. In fact, if ϵ_u, the uniform elongation, is expressed as true strain it is numerically equal to n.

A metal with a low value of n will undergo localized straining early in the stretching process and failure will occur before much uniform strain has occurred. On the other hand, a metal that has a high n will tend to strain uniformly even under nonuniform stress conditions. Thus, for good stretchability a high strain-hardening exponent, n, is desirable.

The stress system in stretching is biaxial and not uniaxial as depicted in the foregoing simplified example. Keeler and Backofen (7) have shown that under biaxial conditions plastic instability appears as diffuse necking rather than as localized necking. Thus, biaxial conditions increase the over-all capability for nonuniform straining. In spite of these greater complexities, Keeler and Backofen confirm the conclusion drawn from the simple uniaxial tensile case, namely, that a high strain-hardening exponent, n, acts to distribute plastic strain and thereby increases the total stretchability of the material.

Keeler and Backofen also observed that the *local* strain at fracture is nearly a constant value irrespective of the material. Their data are shown in Fig. 8. Keeler (8) and Goodwin (9) have used these observations to develop methods to locate the local regions in a complex stamping that are critical to forming the part without fracture.

COMPLEX FORMING OPERATIONS

In practical press forming operations the stretch-draw situation is usually complex. Critical regions may occur in small areas anywhere over the part. Nevertheless, a number of investigators have shown that the parameters n and $\bar{r}$, in some combination, are important measures of the formability of the sheet. For instance, Nelson and Winlock (10) showed that the performance of sheet steel in operations that were predominantly stretching was related to the strain-hardening exponent, n, with steels having high n values performing the best. Lankford, Snyder, and Bauscher (4) found that steels that had high $\bar{r}$ as well as high n performed best in an unsymmetrical stamping. Later, Whiteley (5, 6) showed that steels with high $\bar{r}$ values performed best in symmetrical as well as unsymmetrical stampings.

Probably the most comprehensive experiments to determine the effect of $\bar{r}$ and n on press performance were conducted by the USA Committee of the International Deep Drawing Research Group, now known as ASTM Committee E-26. This committee

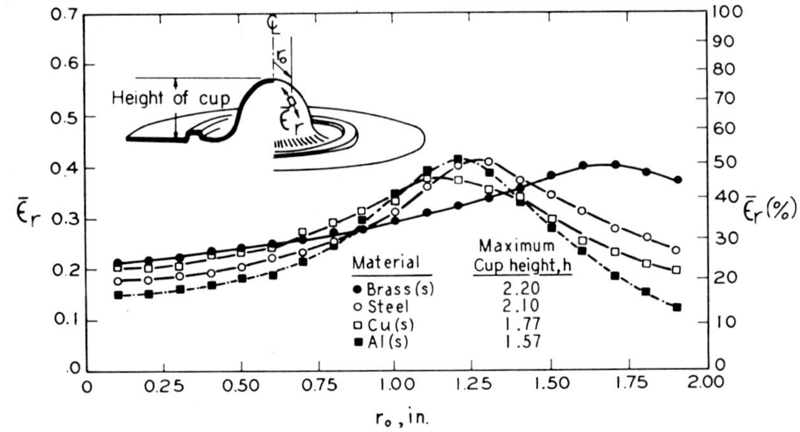

FIG. 8. Distributions of radial and thickness strain at the maximum (fracture) cup height for lubricated, soft materials stretched over a hemispherical punch, showing approximately constant peak strain for all materials. All measurements are plotted at radial positions locating the elements in an unstretched specimen. (After Keeler and Backofen, Ref. 7.)

evaluated 18 different lots of steel in three different stamping operations. The steels were deliberately processed to give: high $\bar{r}$ values with low n values, high n values with low $\bar{r}$ values, and a combination of high $\bar{r}$ and n values.

The stamping operations were selected to represent severe drawings, severe stretching, and a combination of both. Press performance was based on the number of fractures in 1000 stampings. It was found that press performance for the part that was predominantly drawing correlated best with $\bar{r}$. On the other hand, where stretching predominated n correlated best with press performance. Press performance for the part that was a combination of both drawing and stretching correlated with both $\bar{r}$ and n.

Although strain ratio and strain hardening are the principal property parameters that determine the success in stamping a part without fracture, other properties are also important to the acceptability of a successfully formed part. For instance, if the sheet has a large yield elongation, areas of a stamping that are only slightly deformed plastically will show surface markings. These are variously called Lüders lines, stretcher strains, or Piobert lines; they disfigure the surface and may be the cause for rejection of the stamping.

Also, high yield strength in a sheet makes it necessary to increase hold-down pressures on the blank. This can change the draw-stretch situation, lead to buckling, and may even cause fracture.

Much of the remainder of this lecture will be given over to discussion of the effects of composition and structure on the strain-hardening capacity, plastic anisotropy and yielding behavior, and of the way certain steel mill processing operations affect these key properties in sheet steel.

FACTORS THAT INFLUENCE THE STRAIN HARDENING OF SHEET STEEL

We have seen that if sheet steel is to be extensively formed by stretching, it must have a high strain-hardening exponent, n. Unfortunately, a fundamental understanding of strain hardening based on dislocation theory has yet to be perfected for bcc metals. Still, there is enough experimental data to provide a logical working basis for adjustment of the composition and the steel mill processes so as to optimize strain hardening.

EFFECT OF ALLOYING ELEMENTS IN FERRITE

Structure and composition are not always independent, structure often being determined by composition. But because alloying elements partition between excess phases and ferrite, an evaluation of their total effect must consider both their influence on structure and their effect in solution.

Gensamer in his 1945 Campbell Lecture (11) pointed out that alloying additions increase the flow strength and decrease the strain-hardening exponent of carbon- and nitrogen-free ferrite, and that there is a fairly definite relationship between flow strength and strain-hardening exponent. One of my associates, F. B. Schunk, conducted experiments similar to Gensamer's and in general confirmed his conclusions. In addition, Mr. Schunk determined the effect of individual alloying elements in solution in ferrite on the strain-hardening exponent. His data are summarized in Table 1.

TABLE 1. The Effect of Substitutional Alloying Elements on the Strain-Hardening Exponent of Ferrite

Alloying element	Average change in n per wt %*	Approx. solubility in ferrite at R.T., wt %
Cu	−0.06	1
Si	−0.06	15†
Mo	−0.05	3
Mn	−0.04	3
Ni	−0.04	10
Co	−0.04	75
Cr	−0.02	100‡

(n for unalloyed ferrite = 0.30).
* Data corrected for variation in grain size.
† Special case due to slip behavior.
‡ Assumes no sigma phase.

From these data it is clear that substitutional alloying elements in ferrite decrease the strain-hardening exponent of sheet steel. Alloying elements that are least soluble in ferrite, such as Cu and Mo, cause the greatest decrease in n, and elements that are extensively soluble, such as Ni and Cr, have a smaller effect.

Although the effects of interstitial elements on the strain-hardening exponent have not been studied extensively, it was found in two independent investigations (12, 13) that n is not affected by the amount of carbon dissolved in the ferrite of aluminum-killed steel. As far as we know, the literature does not describe the effect of dissolved nitrogen on n. On the basis of the fact that both the dissolved carbon and the dissolved nitrogen have similar effects on tensile properties (14), one would also expect dissolved nitrogen not to affect n since carbon does not affect this parameter. However, the small amount of carbon and nitrogen that remains in solution in box-annealed sheet steels can have a significant effect on strain aging, which, in turn, adversely affects n.

In the usual commercial sheet steels the only alloying element in significant quantity is manganese. Manganese is usually in the range 0.28 to 0.35, which, from Table 1, would reduce n by about 0.01 relative to pure iron. However, some of the manganese is combined as oxides or sulfides and the actual effect on n is probably much less than this.

Thus, if carbon, manganese and residual elements in commercial sheet steels have an effect on strain hardening at all, it is indirect through their influence on strain aging or on the quantity and distribution of the excess phases, i.e., carbides, nitrides, oxides, and sulfides.

THE EFFECT OF EXCESS PHASES

Excess phases decrease the strain hardening exponent (15, 16).

I mentioned previously that Gensamer (11) found a relationship between the strain-hardening exponent and flow strength. He also found that the flow strength was related to the mean free ferrite path. It follows, then, that the strain-hardening exponent is related to the mean free ferrite path. This relationship, as adapted from Gensamer's data, is shown in Fig. 9. Clearly, to increase the strain-hardening exponent it is desirable to have a large mean free ferrite path and hence as low a carbon content as possible.

But sheet steel, which usually contains under 0.10% carbon, does not contain a large volume fraction of carbides or other excess phases. Also, the excess phases that are present are not uniformly dispersed. So, the mean free ferrite path in sheet steels is actually the average grain diameter rather than the spacing between carbide particles. Of course, the specific effects of a grain boundary and a carbide particle are not the same, but their general effects might be considered as similar. If we assume this, it may be deduced from Fig. 9 that increasing grain size would increase the strain hardening exponent.

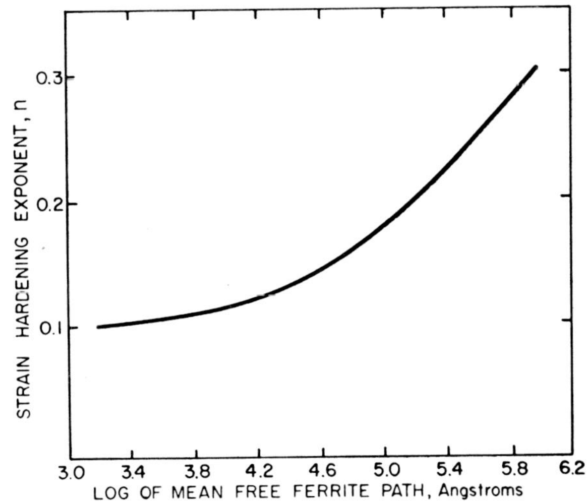

FIG. 9. Relationship between strain-hardening exponent and mean free ferrite path for steels in the range 0.05 to 0.80% carbon. (Adapted from Gensamer, Ref. 11.)

THE EFFECT OF GRAIN SIZE

In fact, increasing grain size does increase the strain-hardening exponent. Morrison (17) has found that the following empirical relationship holds true for mill-produced steels:

$$n = \frac{5}{10 + d^{-1/2}} \qquad \text{Eq 5}$$

where d is the average grain diameter in millimeters. Our own investigations agree with this equation.

The grain size of hot-rolled or annealed sheet steels is usually in the range ASTM 6 to 8 (0.045 to 0.022 mm diam). From Eq 5 this range of grain size would result in a variation of n of about 0.04. Given an n of about 0.22 for most steels, this variation is appreciable and may well spell the difference between success and failure in stretching operations.

Although it is desirable to increase the grain size in order to increase the strain-hardening exponent, grain sizes larger than ASTM 6 are usually detrimental to the surface appearance after forming. There is also evidence that large grain size adversely affects the local strain at fracture under biaxial stress conditions. These practical considerations limit the extent to which large grain size may be used to improve strain hardening.

* * *

The observed effects of composition and structure on n may be summarized by the general statement that in practice the control of the strain-hardening exponent of sheet steels reduces mainly to the control of the grain size. The direct effect of alloying elements and residuals dissolved in ferrite in commercially produced sheet steels is usually not significant.

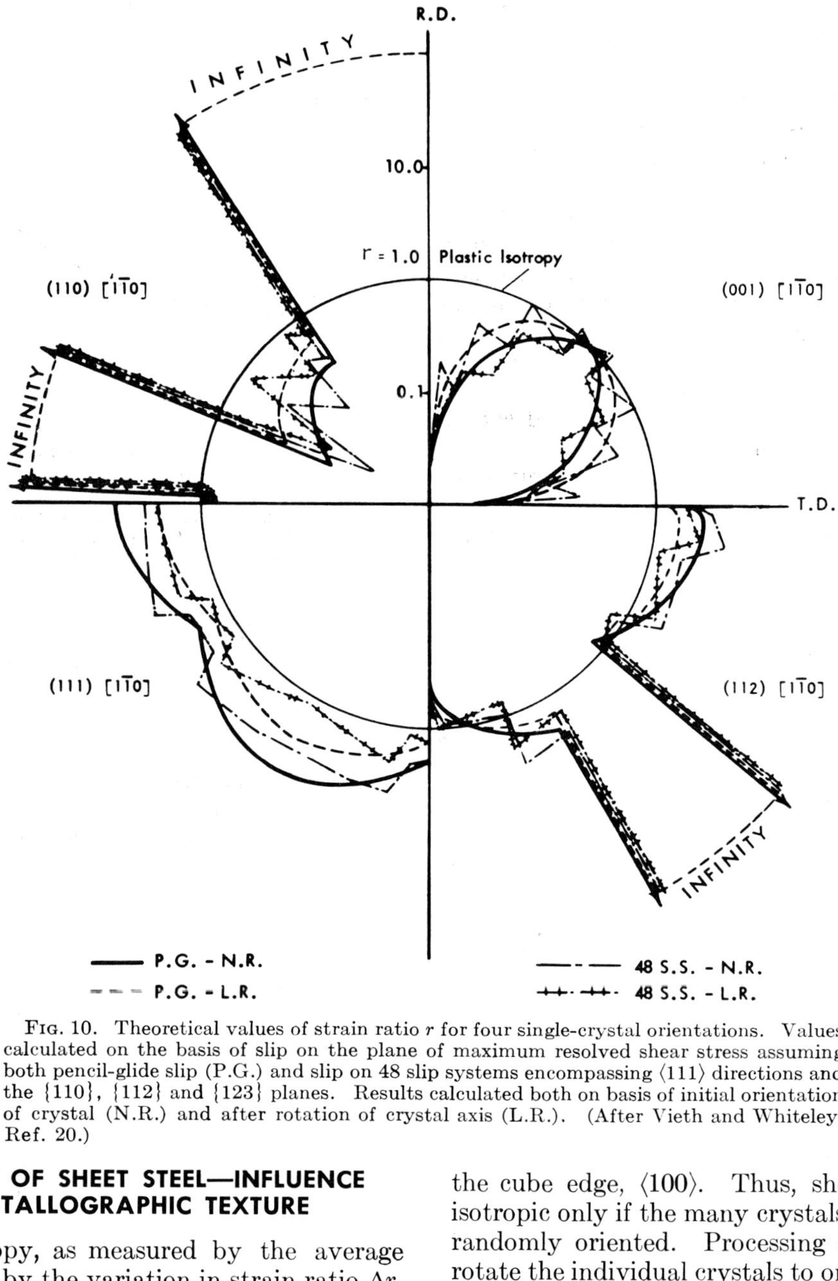

FIG. 10. Theoretical values of strain ratio r for four single-crystal orientations. Values calculated on the basis of slip on the plane of maximum resolved shear stress assuming both pencil-glide slip (P.G.) and slip on 48 slip systems encompassing ⟨111⟩ directions and the {110}, {112} and {123} planes. Results calculated both on basis of initial orientation of crystal (N.R.) and after rotation of crystal axis (L.R.). (After Vieth and Whiteley, Ref. 20.)

ANISOTROPY OF SHEET STEEL—INFLUENCE OF CRYSTALLOGRAPHIC TEXTURE

Plastic anisotropy, as measured by the average strain ratio $\bar{r}$ and by the variation in strain ratio Δr, is important to the success of press forming operations that are predominantly drawing in nature. In fact, the superior formability of aluminum-killed sheet steel is largely the result of the greater anisotropy of this material.

The work of Lankford et al (4) and that of Burns and Heyer (18) clearly demonstrated by two different methods that the strain ratio is related to the crystallographic orientation, or texture, of the sheet. The reason that texture in sheet steel affects the anisotropy of mechanical properties is that the strength of a single crystal of iron varies with the crystallographic direction.

The bcc structure of iron is strongest along the cube diagonal, or ⟨111⟩ direction. It is less strong along the face diagonal, ⟨110⟩, and weakest along the cube edge, ⟨100⟩. Thus, sheet steel would be isotropic only if the many crystals composing it were randomly oriented. Processing operations tend to rotate the individual crystals to one or more common orientations relative to the direction of deformation, thereby affecting the directionality of properties. The type and degree of preferred orientation depend on the mode, magnitude and temperature of deformation, and the subsequent annealing treatment.

STRAIN RATIO

The cube-on-corner texture, i.e., {111}, with the strongest crystal direction oriented normal to the sheet, is favorable to the development of high strain ratios. Conversely, the cube-on-face texture, {100}, with the weakest crystal direction oriented normal to the sheet, is unfavorable to the development of high strain ratios (18, 19).

My associates, R. W. Vieth and R. L. Whiteley,

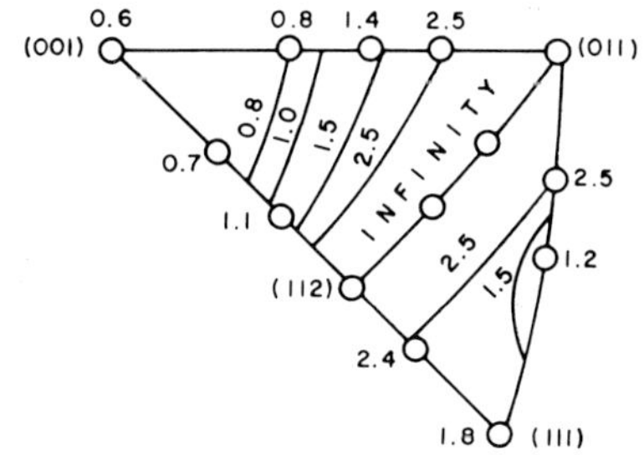

FIG. 11. Average strain ratios, $\bar{r}$ calculated for single bcc crystals of various orientations, assuming pencil-glide slip in $\langle 111 \rangle$ directions.

analyzed theoretically the way plastic deformation affects texture and the average strain ratio, $\bar{r}$, of the sheet (20). Their analysis considered both restricted slip and pencil glide in single crystals. Results are summarized in Fig. 10 which shows the strain ratio, r, as a function of test direction for four different single-crystal orientations. Their experiments with single crystals confirmed these calculations, and just recently Fukuda (21) also confirmed them. The conclusion of Fig. 10 is that the mode of slip has less effect on the average strain ratio than the initial crystallographic orientation of the undeformed material. Also, the results confirm the favorable effect on $\bar{r}$ of a strong $\{111\}$ texture and the unfavorable effect of $\{100\}$ texture. The $\{100\}$ and $\{112\}$ textures also show high r in certain directions. Hence, certain textures may be even more favorable than $\{111\}$ texture. This may be seen in Fig. 11, which shows that $\bar{r}$ values exceeding 2.5 are possible in sheet steel if the right orientation can be developed.

X-ray pole figures also demonstrate the relationship between strong $\{111\}$ texture and higher $\bar{r}$ values. As shown in Fig. 12, a killed steel with an $\bar{r}$ of about 1.9 has a more pronounced $\{111\}$ texture than a rimmed steel with an $\bar{r}$ of 1.3.

A measure of the relative intensities of the cube-on-corner, cube-on-edge, and cube-on-face textures may be obtained by the inverse pole figure method. Such data are useful because they provide a numerical evaluation of texture.

In this connection, consider Fig. 13 in which the intensities of annealed textures are compared with the average strain ratios that result from different

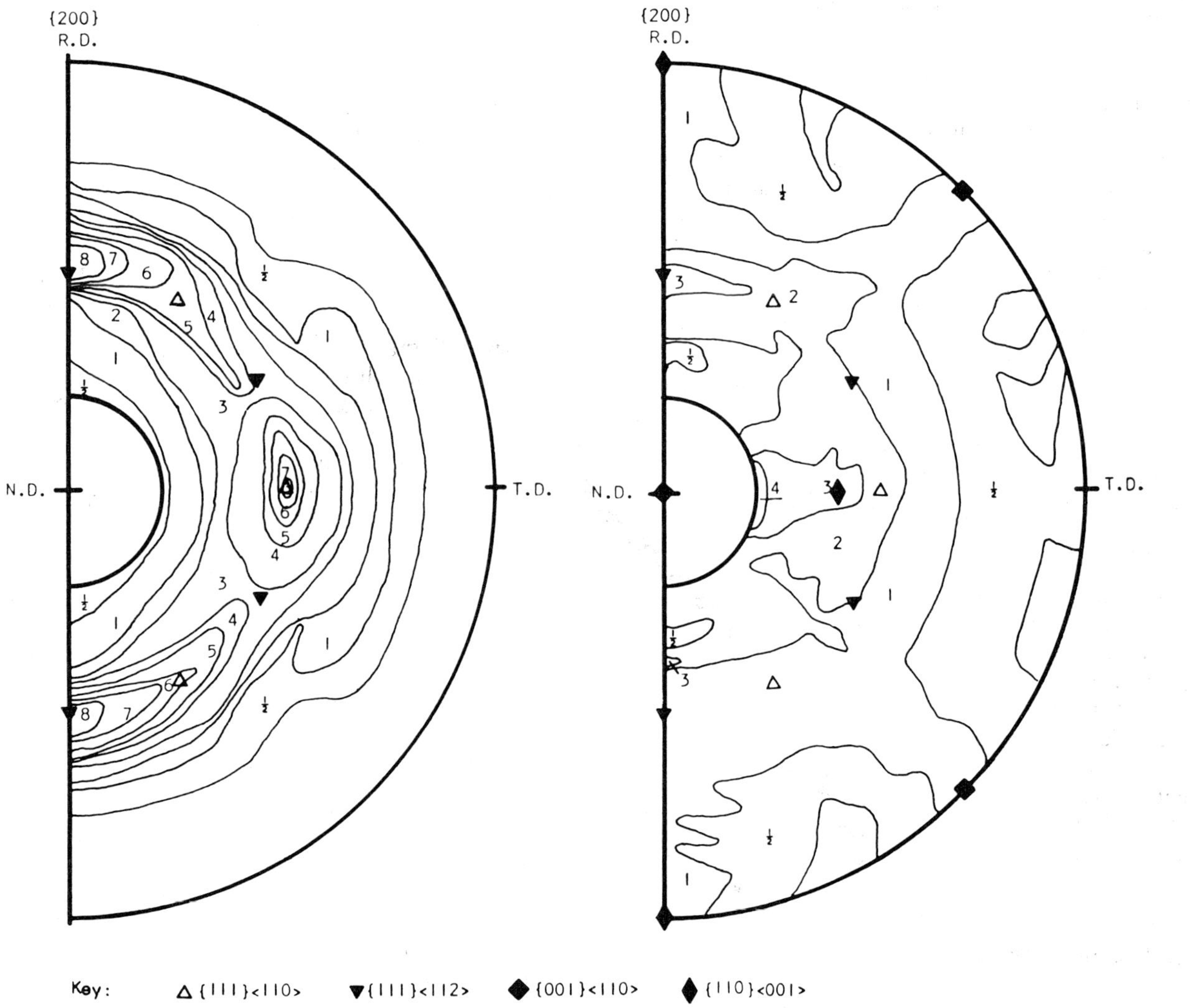

FIG. 12. Effect of steel composition (Al-killed vs rimmed) on texture produced by cold rolling 70% and box annealing 24 hr at 1300 F. (Left) Al-killed steel, $\bar{r}$ = 1.89; (right) rimmed steel, $\bar{r}$ = 1.27.

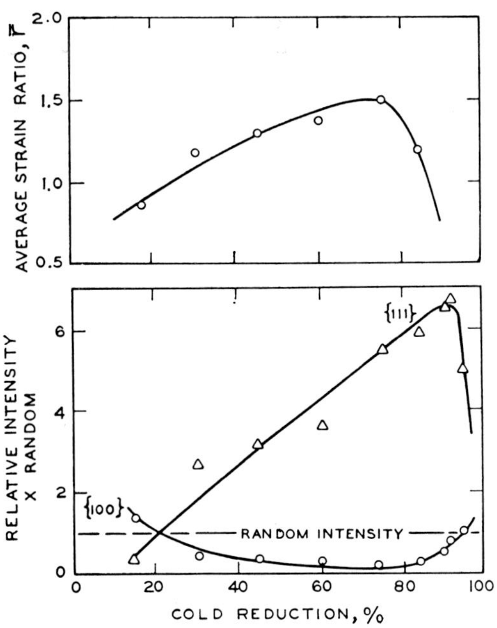

Fig. 13. Relationship between the final annealed texture and the average strain ratio as a function of prior cold reduction. (Adapted from Whiteley and Wise, Ref. 31.)

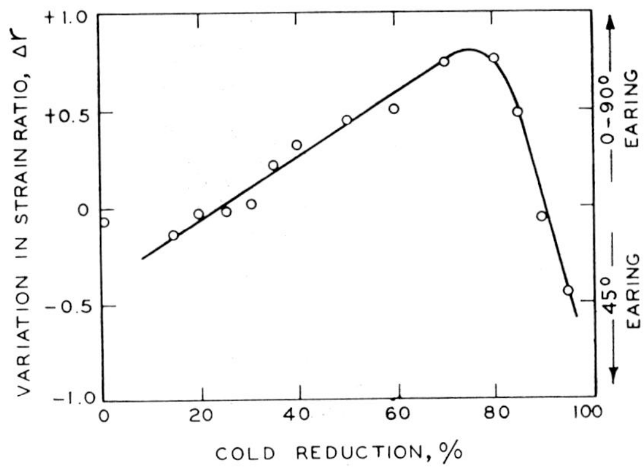

Fig. 14. Effect of cold reduction prior to annealing on the earing behavior of a rimmed steel as indicated by Δr. Two points indicative of non-earing behavior result from reductions of approximately 25 and 90%. Steels given reductions between these limits should exhibit ears at 0 and 90° to the rolling direction. Moreover, steels given light or very heavy cold reductions should develop ears at 45° to the rolling direction; see Fig. 15. (Whiteley and Wise, Ref. 31.)

amounts of cold reduction prior to annealing. It may be seen that the increase in $\bar{r}$ follows an increase in the intensity of the {111}, or cube-on-corner, texture. However, $\bar{r}$ drops off for reductions greater than 75% even though the intensity of the cube-on-corner texture continues to rise with reductions up to about 90%. Perhaps the decrease in $\bar{r}$ with increasing reduction above 75% is a reflection of the increase of the {100}, or cube-on-face, texture. This is because the cube-on-face texture lowers the strain ratio, and the absence of such texture improves the strain ratio in the same way as the presence of a cube-on-corner texture (19, 22). In other words, the change in $\bar{r}$ with increasing cold reduction prior to annealing as shown in Fig. 13 is actually a composite of the effects of a variety of textures.

EARING

Earing, or nonuniform extension in deep drawing, is related to the variation of strain ratio with test direction, Δr. If follows, of course, that Δr is also associated with textural changes. For example, Fig. 14 shows the effect on Δr of the amount of cold reduction prior to annealing. The cups formed from this same series appear in Fig. 15 and illustrate the correspondence between earing and Δr. Crussard et al (23) have shown that the ears occur in the directions where the ⟨100⟩ pole density is the highest.

* * *

To summarize, the ideal deep-drawing steel has both a high average strain ratio, $\bar{r} > 1$, and a Δr of zero. $\bar{r}$ and Δr depend on the crystallographic orientation, or texture, which in turn is affected by the composition and processing operations in making sheet steel. Most sheet steels made specifically for deep drawing are cold reduced from 50 to 70%. This range of cold reduction generally comes close to optimizing $\bar{r}$, somewhat at the expense of low Δr, which may be seen in Fig. 13 and 14.

Excess phases, such as carbides, nitrides, oxides, or sulfides, that are visible under an optical microscope have no direct effect on $\bar{r}$ or Δr. However, as I will point out later, excess phases have an indirect effect on these properties through their influence on the development of favorably oriented grains.

TEXTURE AND STRUCTURE OF COLD-ROLLED AND OF ANNEALED SHEET STEEL

Crystallographic texture and grain size are the principal metallurgical characteristics that affect $\bar{r}$ and n, the properties that determine formability. These characteristics are established by the cold rolling and annealing operations in the production of sheet steel. They are also strongly influenced by small variations in composition and, to a lesser degree, by the prior production step of hot rolling.

In this portion of the lecture the basic characteristics of the cold-rolled and annealed texture and structures are described, to be followed in the next section by a discussion of some of the ways composition affects texture and grain size.

TEXTURE AND STRUCTURE OF THE COLD-WORKED STEELS

Deformation in polycrystalline iron takes place in several slip directions on many slip planes and leads to the formation of dislocation tangles that eventually delineate deformation cell walls. Electron microscope observations have shown that the cells become elongated in the rolling direction and the cell walls become thicker as the deformation continues to increase (24, 25). After 50 to 90% deformation the dislocation density is of the order of 10^{10} to 10^{12}/sq

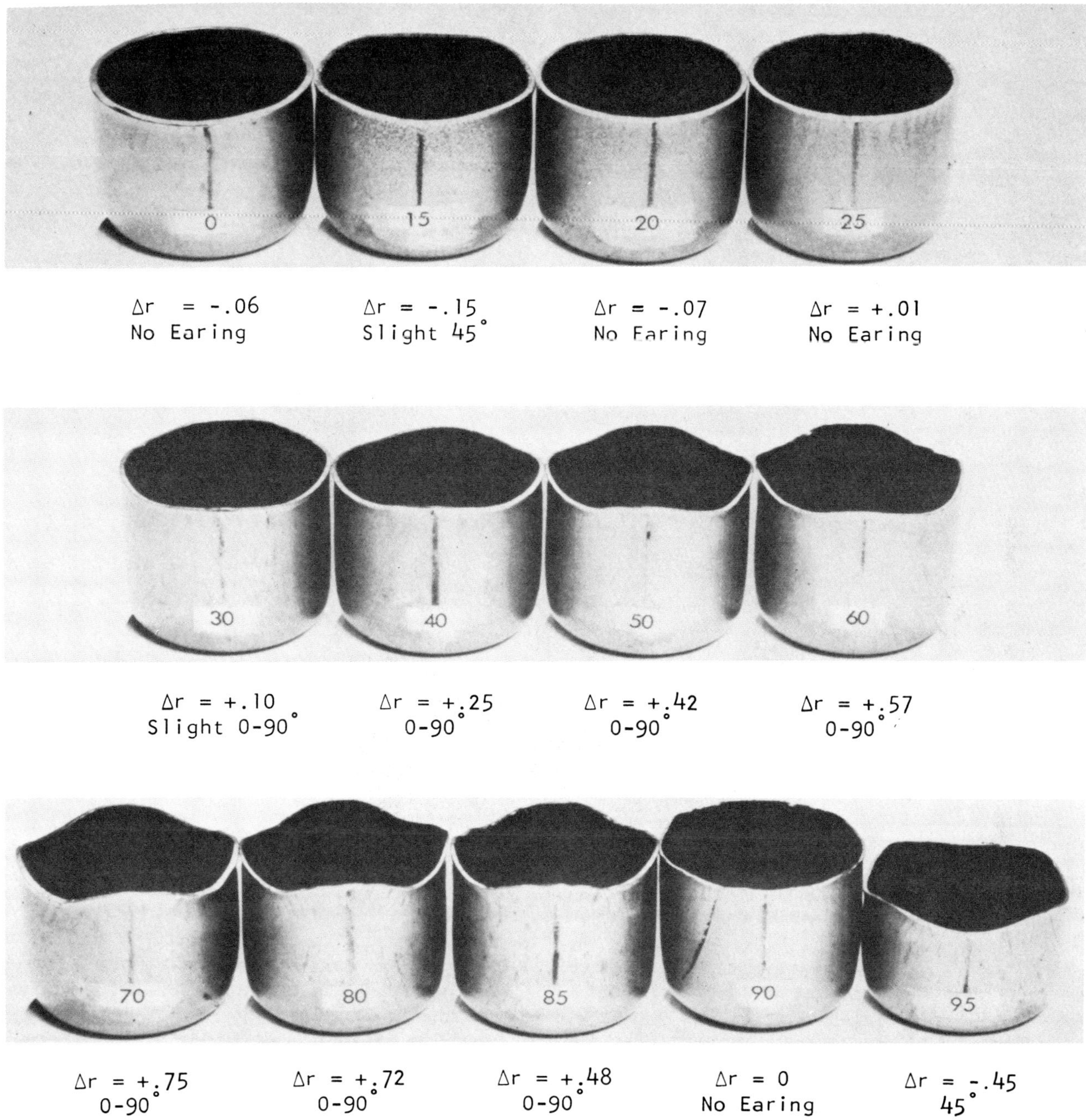

FIG. 15. Demonstration of effect of cold rolling reduction on earing behavior of a steel. Cups shown were drawn from the steel represented by data in Fig. 13. The number on each cup is the percent the steel was cold reduced prior to annealing. The vertical blank line marks the rolling direction in each case. Δr's shown for each cup were obtained from Fig. 14. Because of the very low $\bar{r}$ of the steel reduced 95%, it was necessary to draw a smaller blank than that used in drawing cups from the other steels, hence the smaller cup.

cm and the cells are elongated, their minor axis being one micron or less. Figure 16a is an electron micrograph of the typical longitudinal section of the microstructure of sheet steel after 70% cold reduction.

Texture develops in a way that most easily accommodates slip. As the amount of deformation is increased, an increasingly sharper texture is developed. This is independent of alloy content but depends on the amount of deformation and the temperature at which it is carried out. I will ignore the latter variable because the influence of rolling temperature on deformation texture is not significant at temperatures up to 600 F (3) and commercial cold rolling temperatures are well below this.

Many measurements have been made of the preferred orientation developed during the cold rolling of sheet steels. Transmission pole figure data indicate that the main component of the deformation texture is the {100} ⟨110⟩ with a spread about this orientation that includes the {112} ⟨110⟩, {111} ⟨110⟩ and, in some cases, the {111} ⟨112⟩ (26). On the other hand, inverse pole figure data show

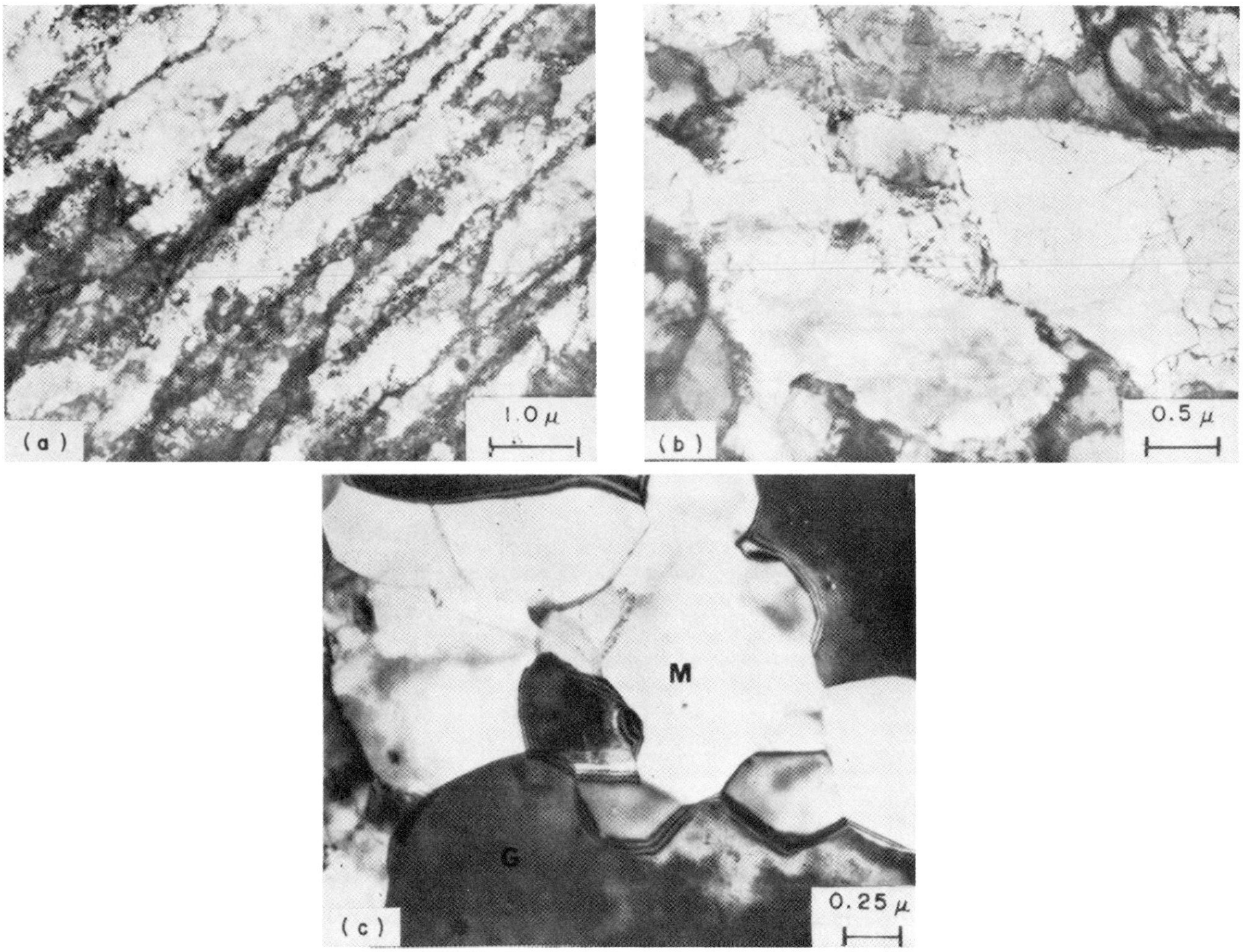

FIG. 16. Transmission electron micrographs of low-carbon sheet steels in the cold rolled, recovered, and partially recrystallized conditions. (a) Cold-rolled 70%. Note elongated deformation cells and high-dislocation density. (b) Cold-rolled 70% and recovered. Note dislocation networks, subgrain structure, and low-angle boundaries. (c) Partially recrystallized. Note dislocation-free recrystallized grain (G) growing into recovered matrix (M) containing subgrains.

that, relative to a random sample, the sum of the {111} textural components is greater than the sum of the {100} textural components. However, as pointed out by Hancock and Roberts (27), this discrepancy between inverse and transmission pole figure data is more apparent than real, and can be accounted for by the fact that the {111} intensity, as measured by inverse pole figure techniques, is made up of the sum of four different {111}-type components whereas the {100} intensity is the result of only one unique {100} component. The four {111} components are separated on a transmission pole figure and appear as four individual intensities, none of which are quite as strong as the one {100} ⟨110⟩ intensity that appears on the same figure.

RECOVERY AND RECRYSTALLIZATION

When cold-rolled sheet steel is annealed, the dislocation density is markedly diminished and the preferred orientation changes from that in the cold rolled condition. This transformation takes place in two stages: recovery and recrystallization. In the recovery stage, the steel reverts to a condition in which the mechanical and physical properties tend toward the undeformed state, but there are no major changes in preferred orientation. In the recrystallization stage, the properties of the undeformed state are attained and their is also a change in preferred orientation.

Recovery itself consists of a progression of changes. In the early stages there is a change in electrical properties but no observable change in microstructure or mechanical properties. As recovery proceeds there is a decrease in dislocation density. At the same time, a dislocation network of subgrains is formed and the strength decreases. Subgrains have low-angle boundaries made up of low-energy dislocation arrays rather than dislocation tangles as in the deformed material (28). After the dislocation rearrangement is complete, the cell size increases by either a process of subgrain boundary migration (25) or coalescence (29). Figure 16b shows typical dislocation networks and subgrains in the microstructure of partially recovered sheet.

Michalak and Paxton (30) have shown that the activation energy for the very early recovery stage is the same as that for vacancy migration, which would account for the change in electrical properties. On the other hand, the later stage has an activation energy like that of self-diffusion of iron, a finding that agrees with the fact that during the early stage of recovery the electrical properties change but the dislocation arrangement does not.

Subgrains in the fully recovered structure are separated by low-energy boundaries. The maximum misorientation between any two subgrains is about 2°, but across any one grain the total misorientation of subgrains may be as high as 10 to 20° (25). During the later stage of recovery, subgrain growth takes place either through isolated subgrains growing at the expense of their neighbors or by coalescence of two or more subgrains. The new subgrain boundaries are higher-angle, higher-energy boundaries than those present before such growth. However, the total boundary area and dislocation density have decreased, and the associated decrease in bulk free energy, minus the increase in surface free energy resulting from the formation of these higher-angle boundaries, is the driving force for subgrain growth. The subgrains slowly increase in size until they have mobile high-angle boundaries, and at this point they can act as the nuclei of recrystallized grains. The period during which this growth or coalescence occurs corresponds to the incubation period for recrystallization.

Recrystallization starts when the high-angle-boundary subgrains begin to grow at the expense of neighbors to form new grains that are relatively free of substructure. Figure 16c shows a recrystallized grain growing into a matrix of subgrains.

CRYSTALLOGRAPHIC TEXTURE AFTER ANNEALING

The texture after recrystallization is, of course, different from that of the cold-reduced steel. In rimmed steel the recrystallized textures have about the same proportion of {111}, {332} and {211} as the corresponding deformation texture but less {100} except at very high reductions (27). In killed steel, however, the change in texture after recrystallization is similar but more pronounced for reasons that will be described in the next section.

Although the amount of cold reduction immediately prior to annealing has a definite effect upon texture, hence $\bar{r}$ (see Fig. 14), the texture prior to cold rolling also affects the final texture. R. L. Whiteley and D. E. Wise found that sheet that is hot rolled at low finishing temperatures often exhibits a fairly strong {110} component (31). The result, in effect, is to lower the $\bar{r}$ vs percent cold reduction curve without affecting its general shape.

Starting textures produced by cold reduction and annealing can give similar effects. For instance, recent experiments by Matsudo et al (32) showed that in two-stage rolling and annealing treatments, the final texture is more a function of the distribution of reduction between the two stages than of the total cold reduction in both stages. Moreover, the amount of cold reduction in the last stage has a greater effect on texture than the reduction in the first stage. Thus, the data in Table 2, taken from Matsudo's results, show that combinations of rolling in two stages with an intermediate anneal do not produce as strong a texture or as high an $\bar{r}$ value after the final anneal as single rolling the same amount. However, double rolling and annealing might provide a way to obtain thin gages without the decrease in $\bar{r}$ that occurs with very large rolling reductions (see Fig. 13).

TABLE 2. Inverse Pole Figure Intensities of Two Components of Recrystallized Texture in Rimmed Steel That Result from Different Combinations of Cold Rolling, Annealing, Re-Cold Rolling and Final Annealing (adapted from Matsudo et al, Ref. 32)

Rolling procedure to 80% reduction		Average stain ratio,	Intensity of diffraction, peak relative to random intensity	
First step	Second step	$\bar{r}$	{111}	{100}
80	0	1.30	3.3	0.8
60	20	1.28	2.0	0.6
40	40	1.18	2.0	0.5
20	60	1.12	1.9	1.1

MECHANISM OF TEXTURE FORMATION

Various theories have been advanced about the mechanism by which the changes in preferred orientation take place. Some hold that the recrystallized texture is established by the formation of nuclei of certain orientations and that the preferentially oriented nuclei then grow without competition from nuclei of appreciably different orientation.

Others believe that nuclei of all orientations are formed and begin to grow, but that some nuclei grow faster than others, depending on their orientation relative to the adjacent recovered matrix. Still others have suggested that the experimental data can be interpreted by a theory involving both oriented nucleation and oriented growth (33, 34). As I shall show, this view seems to best explain the observations.

Dillamore, Smith and Watson at the University of Birmingham and Higgins and Dunn at Liverpool University have described to some of my associates how their recent studies of the relative amounts of stored energy in the components of the rolling texture shed some light on the way the recrystallized texture first forms. They showed by electron diffraction techniques that the relative stored energy in oriented cold-rolled grains of iron decreases in the order {111}, {112}, {100}. They also showed that because the {111}-oriented grains have the highest stored energy, they are the first to recrystallize. Further, the recrystallized orientation is also {111} and is usually nucleated within the grain. This takes place by forming the full {111} orientation around the normal to the plane. At some stage after the {111} grains have started to recrystallize, the remaining orientations recrystallize by the nucleation of new grains mainly at the as-rolled grain boundaries. These boundary-nucleated grains have a wide range of orientations and grow by absorbing the remaining unrecrystallized orientations.

Thus, the intensity of the {111} component remains approximately constant, or only slightly increases during recrystallization, whereas {100} decreases. This concept is supported by the recent work of Kubotera and Nakaoka (35) who noticed that deformed {111} grains tend to form subgrains more readily than deformed {100} grains. As a result, after recrystallization the {100} grains are smaller than the {111} grains and are consumed by the {111} grains during grain growth. Taoka et al (36) have made similar observations in single crystals of 3% silicon iron. It should be noted, however, that this mechanism breaks down with very large cold reductions (see Fig. 13).

Among the many factors in annealing commercial steel that influence recovery-recrystallization behavior are alloy and inpurity content and distribution, second-phase particle size and distribution, recrystallization temperature, heating rate, prior hot and cold rolling history, and composition changes that result from gas-metal interactions during annealing. Some of these effects will not be discussed.

THE EFFECT OF COMPOSITION AND PROCESSING ON THE STRUCTURE OF ANNEALED SHEET STEEL

ALUMINUM-KILLED STEEL

A few hundred parts per million of aluminum in sheet steel results in appreciably higher average strain ratios than the usual rimmed or capped steels. It was once widely held that this was the result of pancake-shaped grains, which in turn were caused by aluminum nitride particles distributed in such a way that grain growth was obstructed in the thickness direction of the sheet. However, as shown in Fig. 17, the average strain ratio is not related to grain shape. Indeed, under certain conditions which will be described later, high average strain ratios may be obtained in rimmed steels with equiaxed grains.

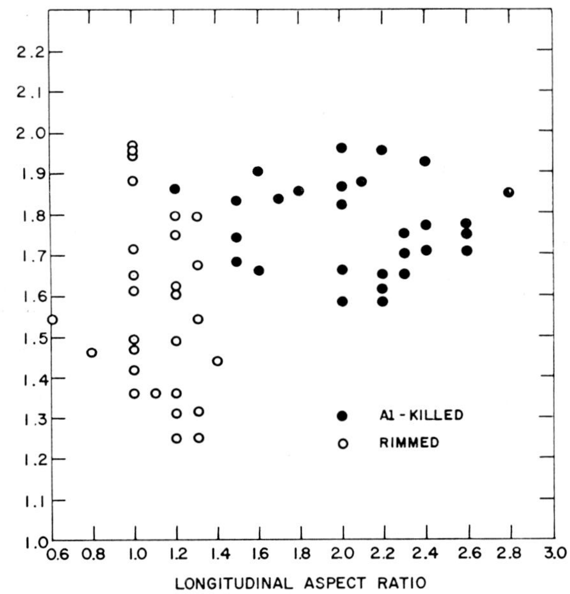

FIG. 17. Average strain ratio, $\bar{r}$, as a function of longitudinal aspect ratio of the grains in cold-rolled and annealed Al-killed and rimmed steel.

The desired texture that leads to a high average strain ratio only appears in annealed aluminum-killed steel that has previously been hot rolled by cooling rapidly from the rolling temperature to the range 1050 to 1150 F before coiling, a practice called "coiling cold." It is generally believed that the rapid cooling avoids significant precipitation of aluminum nitride, with the result that these elements remain in solution to later influence the texture obtained during recrystallization.

In contrast, Hultgren (37) has recently found by internal friction measurements that an appreciable amount of nitrogen is precipitated in aluminum-killed steel that is coiled cold after hot rolling. Presumably, the nitrogen is precipitated as aluminum nitride, although no electron micrographs were shown to support this. In any case, he suggests that the precipitate was formed while the steel was in the austenitic condition during hot rolling and is of a form that is unstable in ferrite. Although it tends to go into solution in ferrite, rapid cooling to below 1150 F prevents this from occurring during the hot rolling and coiling operation. Then, on reheating for annealing, the precipitate does dissolve in ferrite before recrystallization sets in. To continue Hultgren's theory, if the steel is coiled *above* 1150 F after hot rolling, the "austenitic precipitate" dissolves in the ferrite and reprecipitates as "ferrite-stable" aluminum nitride as the coil slowly cools. In this case the precipitate does not go back into solution when the steel is reheated for annealing.

One can agree with Hultgren's observation that an appreciable amount of nitrogen is precipitated in hot-rolled steel which is coiled cold and that this precipitate dissolves in ferrite as the steel is reheated for annealing. However, I believe it is not necessary to assume that the precipitate is formed in austenite and is unstable in ferrite. More likely the precipitate is the usual aluminum nitride that forms in ferrite, except that the particles are very small because they are formed only below 1100 F. If it is assumed that the particle size is smaller than the stable nucleus size at temperatures below that at which recrystallization sets in, i.e., 1200 to 1250 F, then the precipitate would dissolve and reprecipitate as larger particles at higher temperatures or long holding times.

Whether or not either of these precipitation and reversion mechanisms occurs is open to debate and

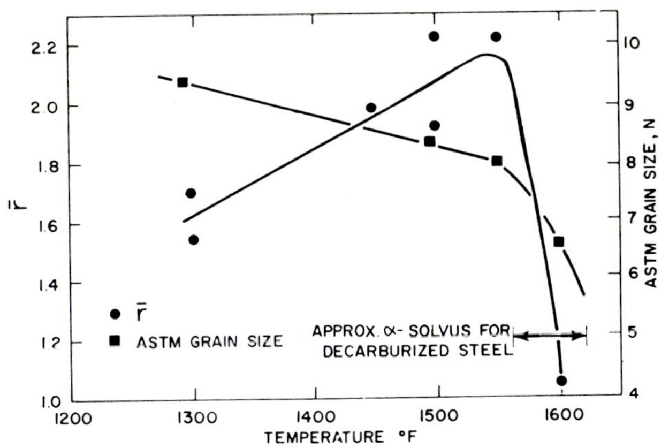

FIG. 18. Effect of annealing temperature on average strain ratio, $\bar{r}$, and grain size of commercial Al-killed steel cold-rolled 70%, decarburized at 1300 F for 24 hr, then annealed at the indicated temperatures for 12 hr.

would require further experimental verification. However, the fact remains that coiling cold is the practical requirement to ensure that aluminum and nitrogen are in solid solution just prior to the recovery and recrystallization of cold-rolled sheet during annealing. The subsequent clustering and precipitation of aluminum nitride particles during the recrystallization process is responsible for the development of good deep-drawing textures in properly processed aluminum-killed steels. Alternatively, if aluminum-killed steel is coiled off the hot rolling mill at temperatures above 1150 F, aluminum nitride particles are formed in the hot-rolled sheet, these particles do not dissolve when the steel is heated to the annealing temperature; hence the recrystallization texture and $\bar{r}$ value are little different from those of rimmed steel.

There have been many investigations into the mechanism by which aluminum influences the crystallographic texture of annealed sheet steel (37–40). It now appears that aluminum in combination with nitrogen affects the final stage of recovery and the subsequent nucleation stage of recrystallization during which subgrains grow and coalesce.

Inverse pole figure data show that on annealing the {111} component of cold-reduced aluminum-killed steels increases and the {100} decreases to almost zero. On the other hand, in rimmed steels the {111} remains approximately constant and the {100} decreases only slightly (41–43). As discussed earlier, in an iron containing no aluminum or nitrogen the {111} component of the recrystallized texture remains at about the same level as in the cold-worked steel, and the {100} approaches random. But in aluminum-killed steel in which aluminum nitride is distributed in the proper form, new grains of random orientation are restricted from nucleating at the remaining as-rolled grain boundaries. Thus, much of the {100} as-rolled component is absorbed by the {111} recrystallized component. This would account for the increase in the {111} and a decrease in the {100} components upon recrystallization.

Investigations of D. E. Wise, D. A. Karlyn and R. W. Vieth at Bethlehem's laboratories have brought out that the $\bar{r}$ value and grain size of aluminum-killed steels are concurrently increased by annealing at high temperatures up to the point of transformation (see Fig. 18). To attain very high annealing temperatures without austenite formation and consequent destruction of the recrystallized texture, they first decarburize-annealed the cold-rolled aluminum-killed steel.

EFFECT OF ALLOYING ELEMENTS OTHER THAN ALUMINUM ON TEXTURE AND STRAIN RATIO

We have seen that a dramatic improvement in strain ratio can be obtained by adding a few hundred parts per million of aluminum to the steel. Certain other alloying elements have been found to similarly affect the crystallographic texture and hence $\bar{r}$ value of annealed sheet steel. For instance, Rickett and Leslie (44) found that copper additions to a rimmed steel produced the desired effect with precipitates of copper apparently causing the development of texture. However, the amount required was large (0.4 to 0.8), and, as shown in Table 1, even a fraction of this amount of copper in solution in ferrite would lower the strain hardening exponent, which is no less important to formability than strain ratio.

E. H. Mayer and D. E. Wise (45) and Yoshida et al (41) claim that 0.05 and 0.08% phosphorus in a sheet steel that is subsequently decarburized and denitrided during the anneal will result in high $\bar{r}$ values. However, such sheet steels often fracture in a brittle manner in heavily deformed regions of a stamping.

Titanium additions in the ratio of 4:1 wt %, or greater, to the carbon content have been found to produce high $\bar{r}$ values (46). The extent of the improvement is also affected by the carbon, nitrogen and oxygen in the steel. For example, sufficient titanium must be present to remove all the carbon and nitrogen from solution in the ferrite. This permits annealing temperatures up to 1670 F without forming austenite, which destroys the desired crystallographic texture. (Incidentally, this ratio of titanium to carbon plus nitrogen also removes the yield point and eliminates strain aging of the steel.) At the same time, the carbon plus nitrogen content should be less than 0.01% to avoid the formation of excessive amounts of titanium carbonitride particles.

Shimizu et al (43, 46) also found oxygen to have an important effect on the efficacy of the titanium additions. As shown in Fig. 19, oxygen contents less than 0.015% result in very strong {111} textures, very weak {100} textures and, consequently, very high $\bar{r}$ values. In view of the sudden change in $\bar{r}$ at 0.015% oxygen, Shimizu reasoned that titanium oxide is present in different forms at low and high oxygen contents and these have different effects on the crystallographic texture and properties devel-

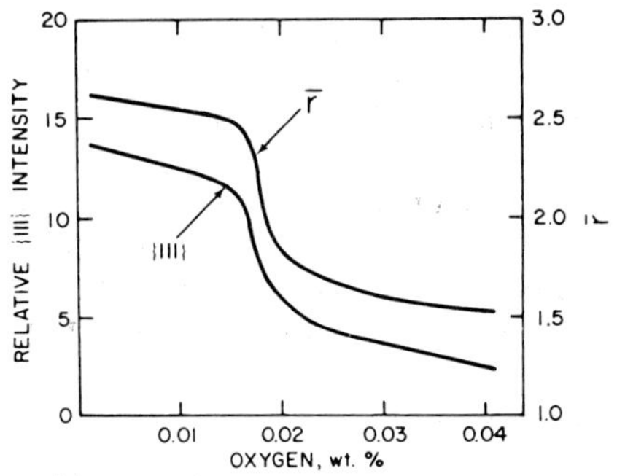

FIG. 19. Effect of oxygen content on $\bar{r}$ value and {111} component of texture in a steel containing approximately 0.007% C, 0.10% Mn, and 0.12% Ti. (Adapted from Shimizu et al, Ref. 46.)

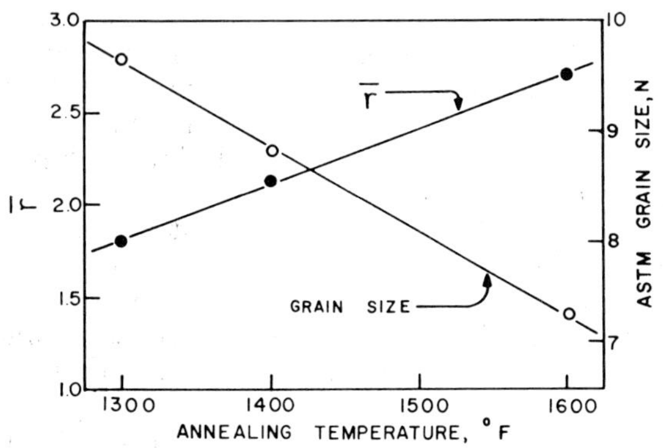

FIG. 20. Effect of annealing temperature on $\bar{r}$ and ASTM grain size for a Ti-bearing steel with Ti:C + N = 12.

oped during recrystallization (46). However, this effect could also be caused by a sudden increase in annealed grain size at oxygen contents lower than 0.015%.

The titanium steels just described tend to recrystallize sluggishly and are fine-grained after annealing at 1200 to 1300 F, the usual range for the rimmed or aluminum-killed steels. I have already pointed out that the strain-hardening capacity is improved with increasing grain size, and in the case of aluminum-killed steels, increasing grain size improved the $\bar{r}$ value as well. Consequently, if the full benefit of the titanium addition is to be attained, the annealing temperature must be increased. Figure 20 illustrates this effect for a very low-carbon steel containing 0.20% titanium (Ti to C + N ratio of 12:1). It is significant that the concurrent change in $\bar{r}$ and grain size is similar to that for the aluminum-killed steel shown in Fig. 18.

EFFECT OF COMPOSITION AND GRAIN SIZE ON STRAIN RATIO AND TEXTURE

The parallel trends of $\bar{r}$ and grain size for two different steels, as shown in Fig. 18 and 20, suggested to my associates D. A. Karlyn, R. W. Vieth and J. L. Forand that grain growth further develops the texture established at the beginning of recrystallization. They investigated the relationship between

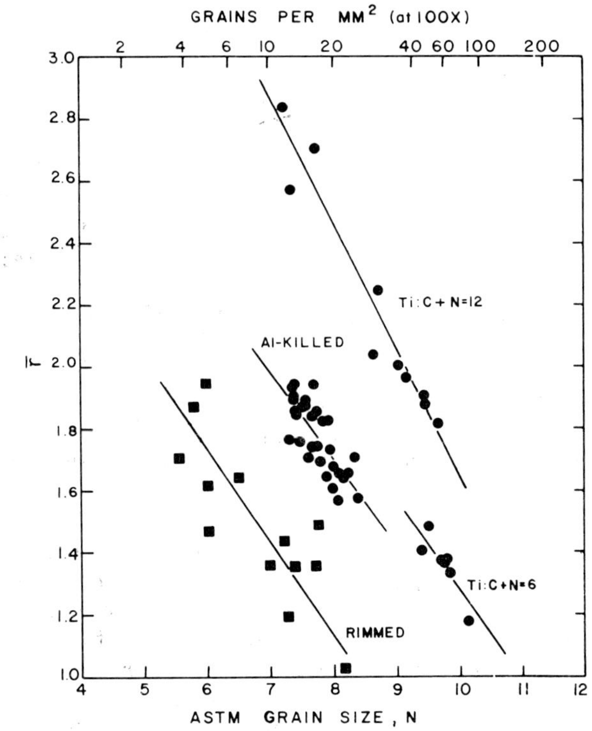

FIG. 21. Variation of $\bar{r}$ with ASTM grain size for four 70% cold-reduced and annealed low-carbon sheet steels.

$\bar{r}$ value and grain size in four different steels: the Al-killed and the 12:1 Ti to C + N steels just described, a 6:1 Ti to C + N steel, and a standard rimmed steel. Details of the treatments used to obtain different grain sizes will appear shortly in Karlyn and Vieth's own publication; it is sufficient for the purposes of this lecture to consider just their results, which are shown in Fig. 21. Here it will be seen that there is a linear relationship between $\bar{r}$ and ASTM grain size (or the logarithm of the grain diameter) for each grade of steel. This may be expressed by the equation

$$\bar{r} = \bar{r}_o - kN \qquad \text{Eq 6}$$

where $\bar{r}_o$ is a constant, different for each type of steel; N is the ASTM grain size number; and k is a constant. Note that the slopes of the lines for different steels are nearly the same; hence k is relatively insensitive to the type of steel.

Equation 6 shows that there are two contributions to the formation of the desired texture for good formability, i.e., high strain ratio. The term $\bar{r}_o$ depends on the composition and the prior processing steps of the steel that affect the orientations formed in the early stage of recrystallization. The second term describes the contribution from grain growth.

It is clear that different alloy additions to sheet steel definitely produce different degrees of the crystallographic texture desired for formability. But in some steels these textures are not fully developed because the grain sizes obtained with the usual subcritical annealing temperatures are too small. Higher annealing temperatures are required to obtain the desired grain size and texture and these can only be attained if the uncombined carbon content of the steel is less than about 0.010, so that austenite is not formed.

The reason for the increase in $\bar{r}$ value with increasing grain size seems clear: recrystallized {111} grains continue to grow at the expense of other orientations, primarily the {100} grains. However, the reason different alloying elements cause the nucleation of different textures in the initial stages of recrystallization is not clear at the present time.

EFFECT OF HEATING RATE ON ANNEALED STRUCTURE

The amount of recovery that occurs prior to recrystallization at a given temperature depends on various factors, heating rate being one of the more important. The more rapid the heating rate the smaller the amount of recovery that occurs prior to recrystallization. Recrystallization under these conditions leads to a fine grain size.

Rapid heating also prevents the full development of the crystallographic texture that provides the high $\bar{r}$ values in aluminum-killed steels (42, 47). This might be attributed to the fine grain size. But grain size cannot be the only cause. There is evidence that at a given grain size the texture that is formed is often not as fully developed as that obtained with slow heating rates. For example, Richards (48) has demonstrated that if aluminum-killed steel is preheated for 1 to 8 hr at 850 to 950 F, the texture and $\bar{r}$ value after subsequent rapid heating to the annealing temperature approach those of the box-annealed steel. This suggests that during rapid heating there is not sufficient time for aluminum and nitrogen to cluster or form pre-precipitates at the as-rolled grain or subgrain boundaries before recrystallization occurs, thereby preventing the establishment of a preferred orientation. The pretreatment at 850 to 950 F before rapid heating to the annealing temperature overcomes this by allowing aluminum and nitrogen to cluster at grain and possibly subgrain boundaries before recrystallization.

Teshima and Shimizu (42) have found that the $\bar{r}$ value attained after annealing aluminum-killed steels at rapid heating rates depends on the uncombined aluminum content of the steel, steels with higher uncombined aluminum having the higher $\bar{r}$ values. They also obtained a fine grain size in steels with high uncombined aluminum, both in the rapidly heated and slowly heated condition. The $\bar{r}$ values and grain sizes of their steels are plotted in Fig. 22 and are in general agreement with Karlyn, Vieth and Forand's data represented by the scatter band taken from the data for aluminum-killed steel in Fig. 21. It would appear that the effect of both the insoluble aluminum content and the heating rate on $\bar{r}$ may be mainly attributed to the way these variables affect grain size. However, the $\bar{r}$ value of the rapidly heated, high uncombined aluminum steel is well below the value obtained with slowly heated steel of the same grain size. Thus, in certain cases heating rate can have an effect on the formation of crystallographic texture beyond the effect on grain size.

In contrast to the effect of rapid heating on $\bar{r}$ value in aluminum-killed steels, Shimizu et al recently reported (46) that very low-carbon titanium-containing steels develop higher $\bar{r}$ values with rapid heating rates and short holding times (4 min) at 1600 F than with slow heating rates and long holding times (16 hr) at 1330 F. However, when their data for both rapidly heated and slowly heated steels are

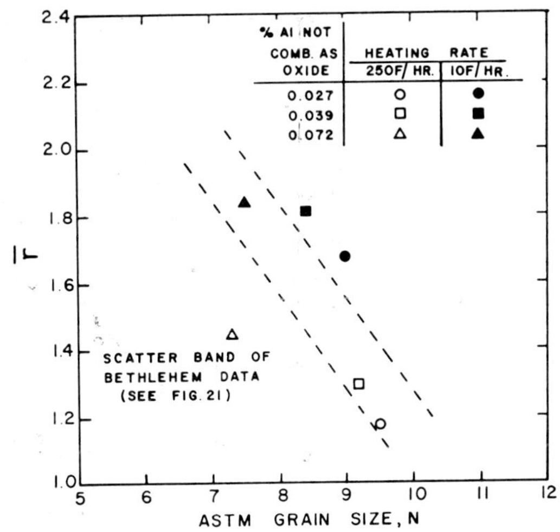

FIG. 22. Comparison of the data of Teshima and Shimizu (42) on the effect of soluble aluminum content and heating rate on grain size and $\bar{r}$ value of Al-killed steel with data of Karlyn, Vieth and Forama on slowly heated Al-killed steels (Fig. 21).

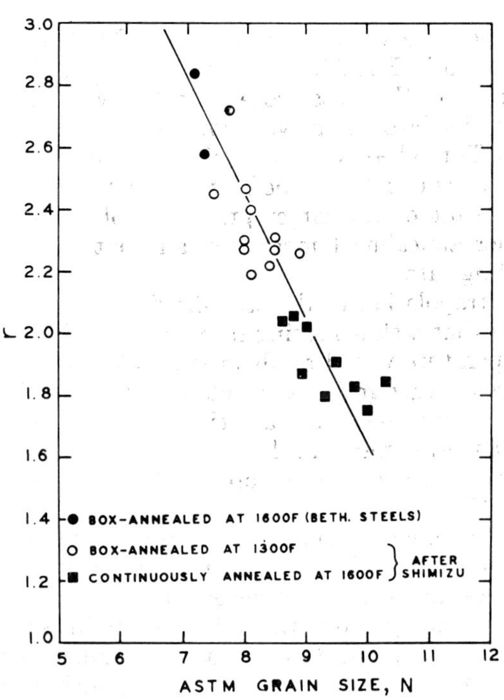

FIG. 23. Average strain ratio, $\bar{r}$, vs grain size of a series of very low-carbon steels containing Ti about in the ratio 15:1 to C + N. Data taken from Shimizu et al (46) fall on the same straight line for a similar steel shown in Fig. 21.

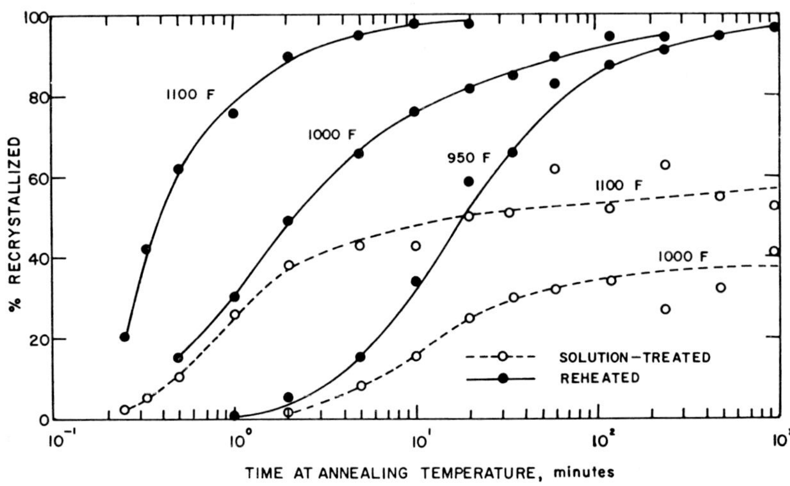

FIG. 24. Isothermal recrystallization curves for steel No. 4 of Table 3. Samples were pretreated by (a) solution treating 2100 F for 30 min and air cooling (ASTM No. 5.1), or (b) solution treating, then reheating 1500 F for 2 hr and air cooling (ASTM No. 5.3).

plotted on a graph of $\bar{r}$ value vs ASTM grain size, as shown in Fig. 23, they form a single straight line. Moreover, the data closely fit Karlyn, Vieth and Forand's data for the very low-carbon steel with the 12:1 Ti to C + N ratio previously shown in Fig. 21. Thus, the effect described by Shimizu et al is simply the result of the larger grain size obtained at the higher annealing temperature and not the result of heating rate.

Extrapolation of the straight line in Fig. 23 suggests that a slow heating rate and long time at high temperature would result in still higher $\bar{r}$ values because a larger grain size would be formed. This was tested by some of my associates and the result obtained was right in line with the prediction—a high $\bar{r}$ value of 2.85 was obtained at a grain size of ASTM 6.

In summary then, heating rate to the annealing temperature definitely affects the crystallographic texture and $\bar{r}$ value of sheet steel. But the effect stems mainly from the different grain sizes produced by different heating rates and annealing temperatures. In some cases, however, such as aluminum-killed steel, the effect may stem from the influence of heating rate on the sequence of the formation of clusters or precipitates prior to recrystallization.

SLUGGISH RECRYSTALLIZATION

Some sheet steels recrystallize very sluggishly when box-annealed at the usual temperatures. Low-metalloid steel, or "enameling iron" as it is called in the trade, is typical. This steel contains lower carbon and lower manganese than regular rimmed or killed steels.

There have been numerous investigations of sluggish recrystallization behavior, and it now appears that interactions between manganese, oxygen and sulfur are responsible (25, 49–51).

According to Leslie et al (50) the manganese content of the steel must be in the 0.02 to 0.15% range for sluggish recrystallization to occur. However, we have found that steels that contain the standard amount of manganese, i.e., about 0.30%, will also recrystallize sluggishly if the oxygen content is high. One of my associates, D. A. Witmer, recently studied this relationship in a series of steels containing different oxygen and sulfur contents (52).

The compositions of four of the steels Witmer studied are listed in Table 3. Of these, only steel 4,

TABLE 3. Compositions of Experimental Steels Used to Study Sluggish Recrystallization After Pretreatment, Cold Rolling, and Annealing. The Last Two Columns Are the Estimated Amounts of Mn and S in Solid Solution After Heating to 2100 F for 30 Min and Air Cooling

Steel	Chemical analyses, %				Estimated % in solid solution	
No.	C	Mn	O	S	Mn	S
1	0.002	0.30	0.008	0.006	0.28	<0.001
2	0.002	0.31	0.011	0.067	0.22	<0.001
3	0.002	0.27	0.071	0.003	0.07	0.003
4	0.003	0.30	0.074	0.073	0.04	0.005

which contained high oxygen and high sulfur, exhibited distinctly sluggish recrystallization. The sluggish recrystallization behavior of this steel could be eliminated by heating in the range 1500 to 1600 F prior to cold reduction and annealing. Heating above or below this range resulted in varying degrees of retarded recrystallization. This is shown in Fig. 24 where the isothermal recrystallization rate for steel 4 preheated at 2100 F prior to cold rolling and annealing is seen to be much slower than when preheated at 1500 F. Figure 25a shows precipitates at ferrite grain boundaries after the 1500 F pretreatment. Witmer identified the precipitates as iron-manganese sulfide by an electron microprobe and electron diffraction investigation of extraction repli-

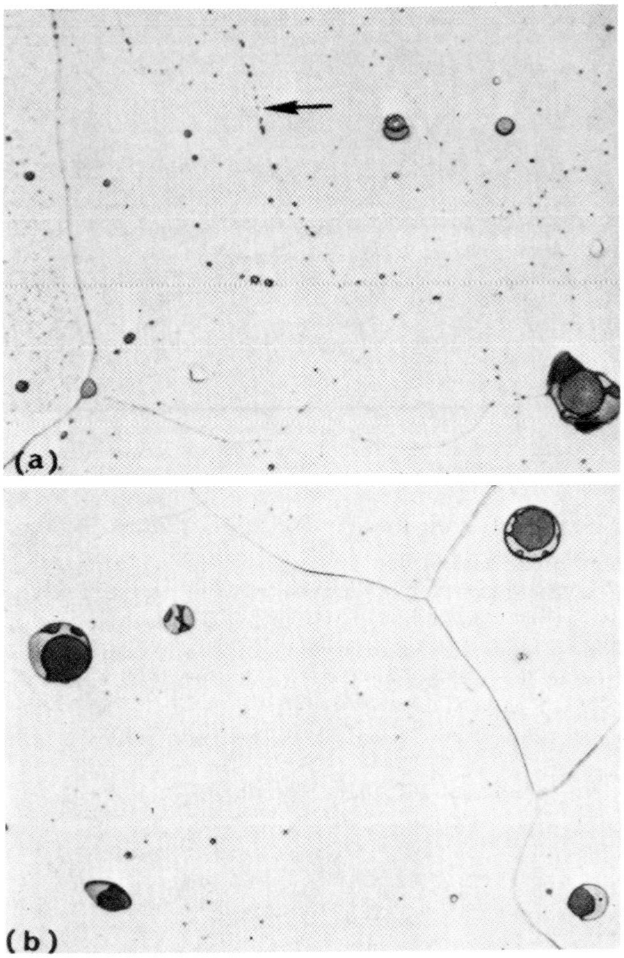

FIG. 25. Effect of thermal history on the microstructure of steel No. 4 (Table 3) before cold reduction. (a) Sample solution-treated, then reheated at 1500 F for 2 hr and air-cooled. Arrow indicates MnS precipitates at former grain boundaries. (b) Sample solution treated at 2100 F for 30 min and air-cooled. Fine precipitates are no longer present. ×1000.

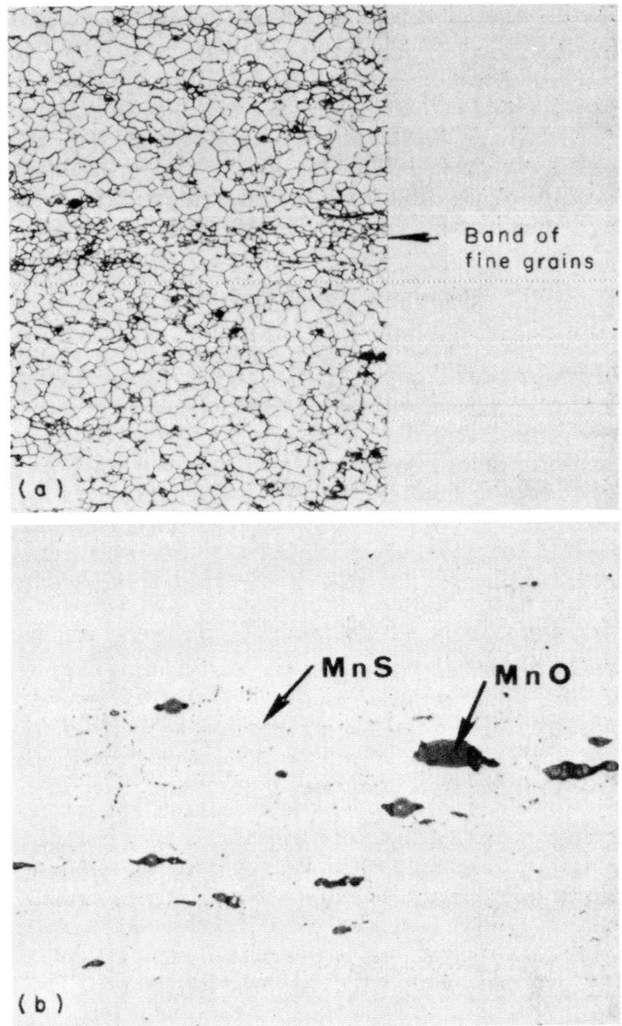

FIG. 26. (a) Cross section of commercial rimmed steel showing a fine-grained striation. Nital etch; ×100. (b) Portion of striated zone of (a) further magnified, showine fine MnS precipitates and large MnO inclusions. Picral etch; ×500.

cas. No precipitates were found when the steel was preheated at 2100 F (Fig. 25b).

From Turkdogan's equilibrium sulfur solubility measurements (53), modified to take oxygen into account, Witmer estimated the amounts of sulfur and manganese in solution in his experimental steels after the 2100 F pretreatment. From these values, given in the last two columns of Table 3, it would appear that low sulfur in solid solution, in this case accompanied by high dissolved manganese, is required for normal recrystallization (steels 1 and 2). On the other hand, high sulfur in solid solution is responsible for sluggish recrystallization (steel 4). Witmer's findings for low-carbon sheet steels containing 0.30% manganese concur with analogous observations by Baird and Arrowsmith (51) who had concluded that dissolved sulfur retards recrystallization in high-purity irons.

Baird and Arrowsmith, working with an annealing temperature of 1390 F, demonstrated further that sulfur retained in solid solution prior to cold working precipitates on the deformation substructure to pin subboundaries and thereby prevent the formation of recrystallization nuclei. However, since Leslie et al (50) and Witmer (52), working with annealing temperatures between 1000 and 1100 F, did not find any well-defined precipitates in their electron microscope examinations, sluggish recrystallization at these lower temperatures would not be attributable to the effect of precipitates. Thus, if sulfide precipitation does not occur at relatively low annealing temperatures, it is reasonable to assume that clusters of sulfur and manganese atoms, as suggested by Leslie (54), or sulfur segregation at grain boundaries are responsible for retarding the movement of subgrain boundaries during sluggish recrystallization.

An illustration of the practical importance of these manganese-sulfur-oxygen interrelationships is found in commercial rimmed steels that, on occasion, have a higher-than-normal oxygen content. Such steels may show bands of fine grains as illustrated in Fig. 26a. These fine-grain bands contain precipitates at the recrystallized grain boundaries, identified by microprobe as manganese sulfides. Microprobe analyses also showed that the ferrite in these bands was depleted of manganese due to the formation of the nearby oxide inclusions (Fig. 26b). Thus, the bands probably contained an excess of dissolved sulfur before cold rolling which latter precipitated during annealing to retard recrystallization.

YIELDING OF SHEET STEEL

A sharp yield point is characteristic of practically all grades of rimmed, capped, or killed commercially produced sheet steels in the hot-rolled or annealed condition. This is because dislocations in these steels are pinned by clusters of nitrogen and carbon in solid solution after cooling from hot rolling or annealing. Thoroughly decarburized and denitrided steels do not exhibit a yield point, nor do steels that contain a strong carbide- and nitride-forming element such as titanium.

It is now believed that the dislocations are so strongly pinned by nitrogen and carbon that they do not move in the initial stages of plastic deformation (55, 56). Instead, a stress is reached at which many new dislocations are formed at sources such as grain boundaries as well as by multiplication. This stress corresponds to the upper yield point. The rapid increase in the number of dislocations at the upper yield point, together with the observation that this dislocation velocity is very strongly stress-dependent, provides a rationale for the yield phenomenon.

During a tension test the specimen elongates at a rate governed by only the applied strain rate. The plastic strain rate of the specimen is given by $\dot{\epsilon} = b\rho v$, where b is the Burgers vector, ρ the mobile dislocation density and the v is the dislocation velocity. Since there is a rapid increase in ρ at the upper yield point where the specimen begins to deform plastically, the same rate of elongation can be maintained by a lower dislocation velocity. A lower dislocation velocity can be achieved by a lower applied stress, because, as stated earlier, dislocation velocity is proportional to the applied stress. This lower stress corresponds to the lower yield point.

In a sheet tensile specimen of a steel that has a yield point, plastic deformation proceeds by the growth of one or more Lüders bands across the width of the specimen at an angle of about 50° to the applied load (57, 58). The Lüders-band fronts then travel along the length of the specimen, the load on the specimen remaining essentially constant. This phase of yielding is called the yield elongation. At the end of this phase strain hardening sets in, the stress to produce plastic deformation increases, and the specimen begins to deform uniformly along the entire length of the test section. A similar sequence of events can occur in portions of a sheet that are subjected to stretching during a stamping operation.

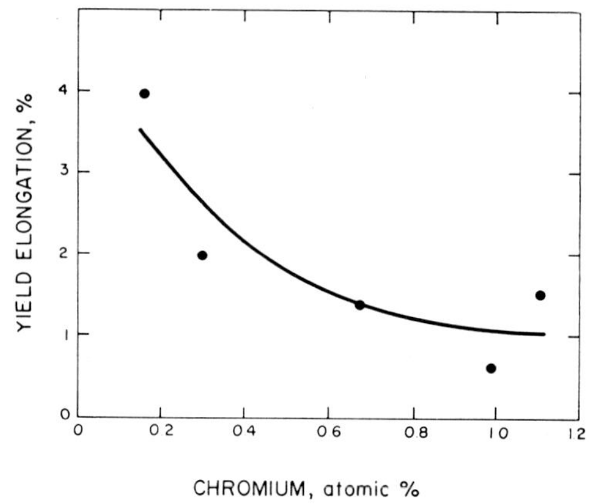

FIG. 27. Effect of chromium on yield elongation of an Fe-0.01% carbon alloy.

The yield elongation is strongly affected by the amount of carbon and nitrogen in solid solution (59). For this reason substitutional elements that combine with these elements will reduce the yield elongation. This is shown in Fig. 27 which shows the yield elongation to be reduced as chromium is added to an iron-carbon alloy. Titanium in amounts about four times the C + N wt % completely removes the yield point.

The yield elongation is also strongly affected by grain size, with finer grain sizes resulting in greater yield elongation (60). Morrison (17) has shown that an ultrafine grain size of 4 to 5 microns, or ASTM 12 to 13, results in a yield elongation as large as 9%. Thus, we again see that in sheet steels it is desirable to avoid fine grain size if formability is to be optimized.

Because of the undesirability of Lüders bands, i.e., stretcher strains, for those applications in which smooth sheet surface after fabrication is important, the yield elongation must be substantially eliminated for such applications. In the steel mill this is usually done by skin rolling, i.e., lightly cold rolling the annealed sheet to a 0.5 to 1.25% elongation. Also, some sheet products are roller-leveled to remove the yield elongation. Generally these mechanical methods of removing the yield elongation are more economical than complete decarburization and denitrification and/or the addition of alloys that are strong carbide and nitride formers.

STRAIN AGING OF SHEET STEEL

As stated, skin rolling effectively removes the sharp yield point present in most commercial annealing steels. But in rimmed or capped steels a new problem appears: if the skin-rolled sheet is stored at normal room temperatures the sharp yield point will reappear. Also, other properties that affect drawability are impaired, particularly the strain hardening exponent and yield strength.

The term given to these changes that take place after straining is strain aging.

TABLE 4. Stages During the Strain Aging of Rimmed Steel As Indicated by Changes in Mechanical Properties

Stage	Change in mechanical property		Mechanistic interpretation: Wilson and Russell (62)	Mechanistic interpretation: Edington et al (63)	Dislocation atmosphere density, atoms per dislocation plane
I	Lower yield point	Increases	Formation of solute atmospheres on dislocation cores	Formation of solute atmospheres on dislocation cores	0 to 1
	Yield elongation	Increases			
	Flow stress	Constant			
	Tensile strength	Constant			
II	Lower yield point	Increases	Formation of solute clusters at favorable sites along dislocations	Continued solute segregation to dislocations and grain boundaries	1 to 10
	Yield elongation	Constant			
	Flow stress	Increases			
	Tensile strength	Constant			
III	Lower yield point	Increases slowly	Formation of extremely fine precipitate on dislocations	Continued solute segregation to matrix dislocations and grain boundaries but decreased segregation in dislocations near boundaries	10 to 60 or more
	Yield elongation	Constant			
	Flow stress	Increases slowly			
	Tensile strength	Increases			
	Strain-hardening exponent	Increases			
	Elongation, total and uniform	Decreases			
IV	Lower yield point	Decreases slowly	Coarsening of precipitates on dislocations	Coarsening of precipitates on dislocations	
	Yield elongation	May increase			
	Flow stress	May decrease			
	Tensile strength	Decreases slowly			
	Strain-hardening exponent	Decreases slowly			
	Elongation, total and uniform	Increases slowly			

Strain aging is the result of the carbon and nitrogen atoms segregating to the dislocations formed in skin rolling (61). As these interstitial atoms segregate to dislocations, forming so-called atmospheres, the mechanical properties of the steel go through a sequence of changes, leading eventually to a sharp yield point. Wilson and Russell (62) suggested this could be divided into the four stages summarized in Table 4. This sequence of stages has been essentially confirmed by others (63–67). However, as Josefsson and Backstrom (67) pointed out, Stages II and III tend to merge into one another. Wilson and Russell found that the strain-hardening exponent was constant during Stages I and II. However, Bergh (66) and Josefsson (67) found that the strain-hardening rate decreased significantly in Stages I and II before increasing in Stage III, and work in our own laboratories confirms the same pattern for n.

The last column in Table 4 shows the approximate atmosphere densities for three of the four stages that were obtained in recent experiments by E. T. Stephenson at Bethlehem's laboratories. He measured changes in the internal friction and mechanical properties of an iron-carbon alloy as a function of time, and also determined dislocation density by transmission electron microscopy. His estimate of the atmosphere density at the end of Stage I (about one atom per dislocation plane) agrees with the estimate obtained by Wilson and Russell (62) by another method.

EFFECT OF CARBON AND NITROGEN

The most important factors affecting the strain aging of sheet steel are the carbon and nitrogen in solution in ferrite and the type of prestrain. As shown in Table 5, all three types of box-annealed

TABLE 5. Amount of Dissolved Carbon and Nitrogen in Various Box-Annealed Steels

Reference	Alloy	Carbon, ppm	Nitrogen, ppm
Laxar, Frame and Blickwede (68)	Killed steel*	2–7	0
Richards and Barratt (69)	Capped steel	3	8
Bergh (66)	Iron	1–4	8–16
Josefsson and Backstrom (67)	Rimmed steel†	1–2	18–30
	Denitrided rimmed	6–9	0
New data (Stephenson)	Killed steel	1–3	0
	Rimmed steel	0–6	6–20

* Cooled at 180 F/hr.
† Soaked at 480 F for one day during cooling cycle.
Note: 10 ppm = 0.001%.

sheet contain very little dissolved carbon, but the rimmed and capped steels do contain significant amounts of nitrogen in solution. Because killed steel has very low dissolved nitrogen and has practically no susceptibility to strain aging, it follows that dissolved nitrogen is the element mainly responsible for strain aging in commercial box-annealed steels. On the other hand, carbon is no less effective than nitrogen in locking dislocations when it is dissolved in ferrite in equal amounts to nitrogen (14).

It is well known that aluminum in sheet steel suppresses strain aging because it combines with nitrogen to form stable aluminum nitrides in the annealed steel. Of course, for aluminum to exercise its beneficial effect on the r value, some small amount of both aluminum and nitrogen must be in solution in the

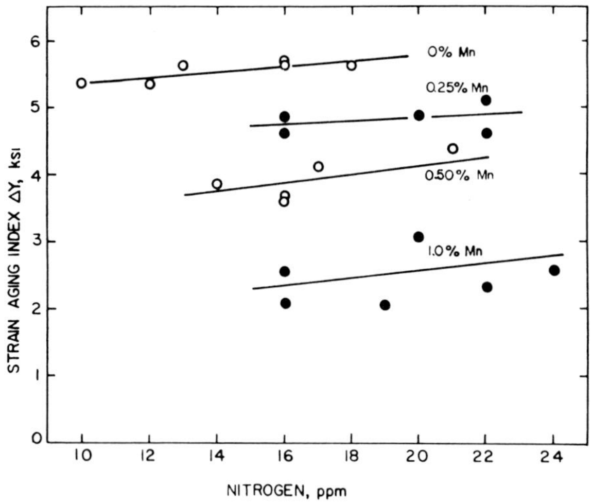

FIG. 28. Effect of manganese and nitrogen on the strain-aging index, ΔY. ΔY is the lower yield point of prestrained and aged specimens minus prestraining stress. Specimens were prestrained 10% and aged at 100 C for 5 min.

cold-worked ferrite before recrystallization sets in. However, no significant amount of nitrogen remains in solution in killed steel after annealing.

V, Cb, Ti, Cr, and B also reduce strain aging by combining with the nitrogen in the steel. Most of these elements affect the rimming action of the steel and several of them suppress it completely. For this reason, and because these alloying elements are costly, they do not normally replace aluminum when the purpose is simply to make the steel non-aging.

E. T. Stephenson at Bethlehem's laboratories has shown that manganese, the major alloying element in commercial rimmed steel, retards strain aging (Fig. 28). However, amounts in excess of 0.50% can adversely affect rimming action and, hence, the surface quality of the sheet.

Thus, we see again that strain aging, another characteristic that strongly affects the properties of sheet steel, is itself affected by microchanges in compositions and in the processing operations used to make sheet steel. Fortunately, the factors that improve strain hardening and strain ratio, such as larger grain size and certain alloying elements, also decrease the consequences of strain aging.

EFFECT OF TYPE OF PRESTRAIN

As stated earlier, skin rolling is the common method of eliminating the yield point. This can also be accomplished by straining in tension, but less strain is required to eliminate the yield point by skin rolling than by tension straining. The difference is believed to be caused by the formation during skin rolling of a pattern of alternating bands of deformed and undeformed metal. These bands are believed to be spaced about 10 to 20 to the inch transverse to the rolling direction (70–73). On a macroscale the bands distribute the small skin-rolling strains throughout the sheet and thus produce uniform macroyielding in a forming operation. On the other hand, the tension prestrain required to produce uniform yielding must be greater than the yield elongation. Thus, more deformation is required to eliminate the yield elongation by tension straining.

The return of the yield point on aging also takes place more slowly in skin-rolled sheet than in tension-strained sheet (74). Moreover, in skin-rolled sheet the return is slower in specimens with the tensile axis parallel to the prestraining direction than in transverse tensile specimens. The reverse is true for the tension-strained sheet (75). These effects may be attributed to the action of heterogeneous patterns of residual microstresses (76).

R. L. Whiteley at Bethlehem's laboratories has developed a mathematical expression that describes the way residual stresses retard the return of the yield point on aging. Whiteley accepts Tardiff and Ball's basic concept that residual stresses do not change the rate of strain aging as measured by the rate of segregation of interstitials to dislocations (75). Instead, residual stresses mask the evidence of this change, which is the return of the yield point. The differential level of residual stresses prevents the whole specimen from yielding at one time.

A definite yield point will be evident only if during yielding the slope of the load-elongation curve becomes negative. On the basis of this criterion, Whiteley calculates that a definite yield point will appear only when

$$(S_U - S_L) > (x_o - x)\left(\frac{-df(x)}{dx}\right) \qquad \text{Eq 7}$$

where S_U and S_L are, respectively, the upper and lower yield strength of the steel in the absence of residual stresses. The right side of the equation represents the masking action of the residual stress pattern described analytically by $f(x)$. The term x is the distance from the point of maximum tensile residual stress to the boundary between the plastically strained and elastically strained regions, which increases to x_o during the loading of the sheet. The term x_o is the half-wavelength of the residual stress pattern. The term $(x_o - x)\ (-df(x)/dx)$ is equal to zero when no residual stress is present and is otherwise always positive.

The difference $(S_U - S_L)$ increases as a function of the rate of segregation of carbon and nitrogen in the steel to dislocations. Where residual stresses are not present the difference $(S_U - S_L)$ is immediately reflected by the reappearance of a definite yield point. However, where residual stresses are present, the reappearance is delayed until the difference $(S_U - S_L)$ becomes greater than the masking action of the residual stresses as given by the function on the right side of the equation.

EFFECT OF STRAIN TEMPERATURE AND STRAIN RATE

Yeo (77) has found that the rate of return of the yield point is also strongly affected by the tempera-

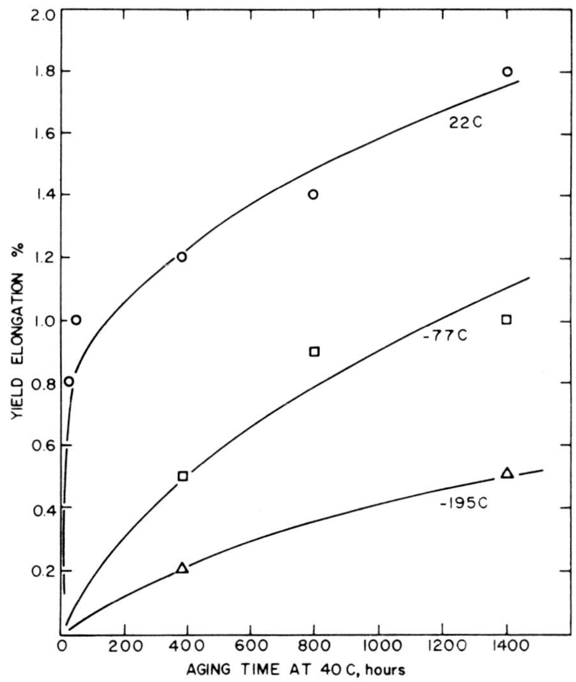

FIG. 29. Effect of 5% reduction by rolling at low temperatures on the return of the yield point in box annealed rimmed steel.

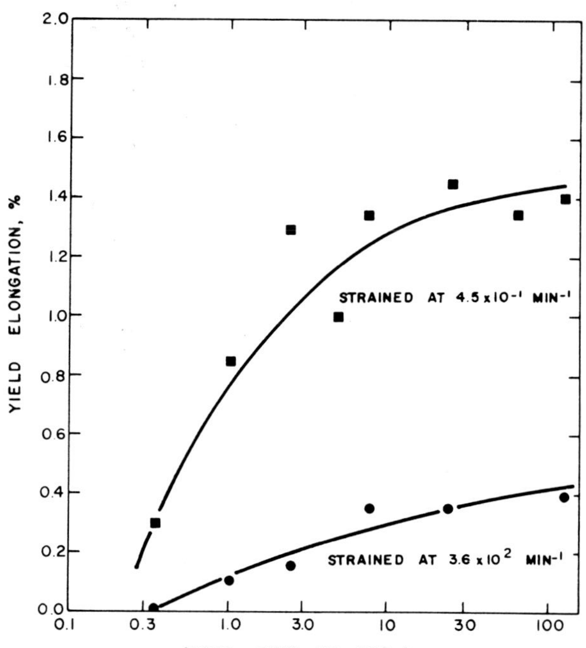

FIG. 30. Effect of prestrain rate at −70 C on the return of the yield point in box-annealed rimmed steel. Specimens were prestrained 4% in tension.

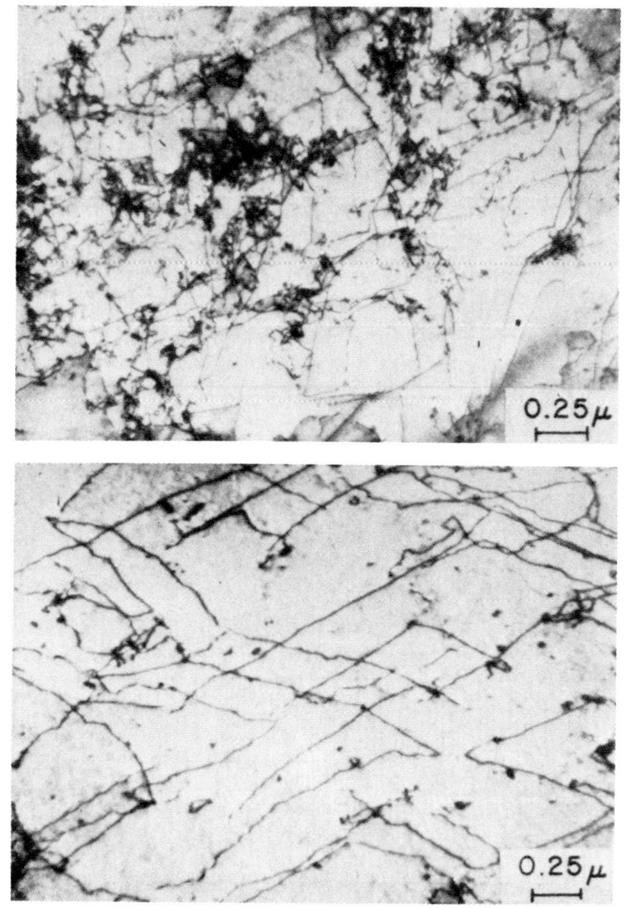

FIG. 31. Transmission electron micrographs of rimmed sheet steel rolled 5% at 22 C (above) and −196 C (below).

ture at which the steel is temper-rolled. My associate, D. A. Karlyn, recently studied this relationship in rimmed steel specimens that, after being reduced 5% in thickness by rolling at room and liquid-nitrogen temperatures, were then aged at 40 C (105 F) and tensile-tested at room temperature. His data appear in Fig. 29: Note that the specimens that were rolled at room temperature developed a yield elongation of 1% after 48 hr, whereas those rolled at the liquid, nitrogen temperature had a yield elongation of only 0.5% even after aging 1400 hr.

It is known that deformation at high strain rates has a similar effect on structure and properties as deformation at low temperatures. With this in mind, another of my associates, E. C. Oren, obtained data (Fig. 30) which show that a high strain rate in tensile deformation also retards the subsequent return of the yield point on aging.

We and others (78, 79) have found less tendency for dislocations in sheet steel to tangle when deformation is carried out at low temperatures and high strain rates. This is clearly illustrated in Fig. 31 which are transmission electron micrographs of the dislocation morphology in a steel rolled 5% at 22 C (70 F) and at −196 C (−320 F). Similar differences were observed in steels deformed at low and high strain rates.

It is tempting to attribute the difference in rate return of the yield point during strain aging to these differences in dislocation arrangement, but it is difficult to explain the orientation dependence of the return of the yield point on this basis. Consequently, strain aging may be less affected by dislocation arrangement than by residual microstresses. In

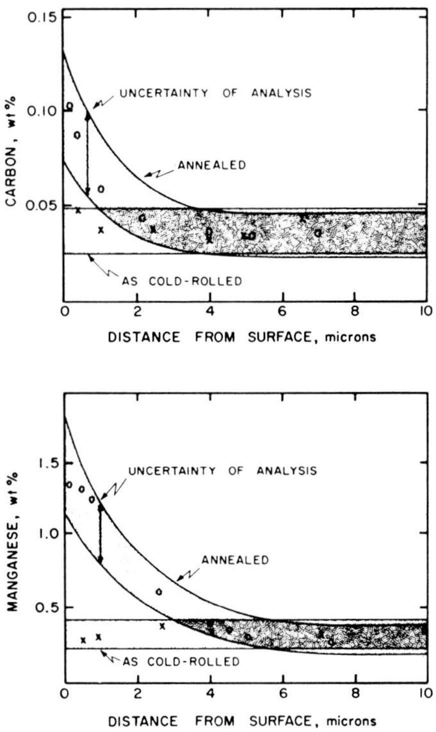

Fig. 32. Carbon and manganese contents in the surface region of rimmed sheet steel before and after box annealing in NH atmosphere.

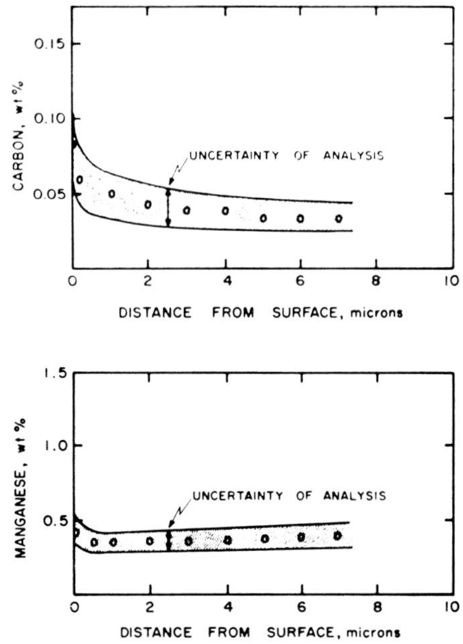

Fig. 33. Carbon and manganese contents at the surface of rimmed steel annealed 1 hr at 1200 F in a vacuum.

support of this interpretation it can be argued that since low temperatures and high strain rates increase yield stress, they also affect the microstresses. But this is conjecture which needs to be developed in more detail and subjected to critical experiments.

THE SURFACE OF SHEET STEEL

I would like to digress now from the structure and properties that affect formability to a brief discussion of the surface of sheet steel. Many characteristics of the surface are established by the operations that follow steelmaking and ingot solidification. Often these are invisible and affect applications such as resistance to rusting, resistance of tinplated sheet to corrosion by certain foods and beverages, and the adherence of inorganic and organic finishes to the surface.

Annealing is the processing step that plays a leading role in establishing these characteristics. For the sake of brevity, I will not discuss the subject of residues that remain or are deposited on the surface during annealing; Refs. 80–83 describe these adequately.

Rather, I want to take this opportunity to describe some new information, obtained by M. M. Goldberg and J. B. Horton at Bethlehem's laboratories, about the changes in composition that occur in the surface region of the sheet itself during annealing. Their findings are not only unusual and unexpected but may have important practical implications.

Goldberg and Horton found that carbon and manganese both strongly concentrate at the surface of sheet steel to a depth of 2 to 10 microns during annealing, the concentration at the very surface being two or three times that of the base steel. For example, Fig. 32 shows the carbon and manganese concentrations in the surface region of a rimmed steel before and after box annealing in an atmosphere of 4% H_2 and 96% N_2. Similar concentrations of carbon and manganese occurred during annealing of aluminum-killed steel. Nitrogen and silicon were also investigated but the amounts of these elements in the usual rimmed and killed steels are so low that determining changes in them is beyond the precision of the method used.

The data presented in Fig. 32 are shown as bands, the width of the band representing the uncertainty in the analysis. The uncertainty is about ±30% of the carbon percentage and ±20% of the manganese percentage, and is a result of a combination of the reproducibility and the calibration factor of the analytical instrument.

The instrument used is called an ion solids analyzer. It determines by mass spectrometry the composition of material sputtered by argon ions from surface layers a fraction of a micron thick. This new instrument was originally developed for geological dating of rocks (84–86). It is a very powerful tool for the determination of many kinds of elements, compounds, and ions on and below the surface of metals. Further details about the information obtained with this instrument will soon appear in Goldberg and Horton's own publication.

At first thought the increase in carbon content shown in Fig. 32 might be attributed to carburization from the gases that exist between the laps of a sheet steel coil during annealing. (Carbonaceous gases are known to be present between laps of a coil even though the protective gas surrounding the coil contains only N_2 and H_2.) However, this cannot be the only source because the same effect occurs in samples degreased and vacuum-annealed in the laboratory, as shown in Fig. 33.

The observation of high concentrations of carbon and manganese at the surface after the annealing seems to be new; at least to the best of my knowledge it has never been reported in the technical literature before. However, a few years ago, my associates J. W. Frame and E. H. Mayer observed by x-ray fluorescence analysis that the manganese content at the surface of annealed sheet steel was appreciably higher than that of the base composition. Furthermore, the manganese concentrated at the surface as a result of the box anneal, as shown in Table 6.

TABLE 6. X-Ray Fluorescence Intensities (Counts in 200 Sec) for Manganese in Rimmed Sheet Steel. (Primary Beam Directed Normal to the Surface of the Sheet.)

	Intensity, counts per 200 sec
As hot-rolled and pickled in H_2SO_4	2480
As cold-reduced	2560
Annealed	3450
Annealed and skin-rolled	3430
Base steel	2000 (approx)

X-ray fluorescence does not permit a determination of the carbon content at the surface. However, indirect evidence that carbon concentrates at the surface may be found in the literature. For instance, Pignocco and Pellissier (87) reported that they were unable to obtain an atomically clean surface for their studies of oxidation of iron unless the iron was carbon-free; the surface of iron containing more than 30 ppm carbon could not be cleaned. Farnsworth and Tuul (88) observed the same behavior in nickel.

Hayfield (89) has observed a surface concentration of iron on copper alloys. He explained this by Gibbs' surface excess effect, which proposes that the surface composition adjusts to minimize the surface energy. However, Aust et al (90) have pointed out that this effect can involve only the first few monolayers of the surface and does not extend several microns deep. Ainslie (91) has suggested that when dislocations migrate to the surface during annealing, impurity atoms may be swept with them. However, it seems unlikely that this would account for the large concentration of carbon and manganese we observed.

Goldberg and Horton's experiments have not yet provided a clear-cut explanation for the mechanism by which carbon and manganese concentrate at the surface during annealing. For instance, diffusion theory requires that there should be a depletion beneath the concentration gradient. Unfortunately, the uncertainty in the analysis does not permit this to be shown with confidence.

Goldberg and Horton have found that a high carbon content is established after fairly short annealing times, e.g., 1 hr at 1200 F, but the manganese concentration seems to follow later (see Fig. 33). This seems to fit the known fact that manganese tends to segregate to high-carbon regions in austenite (92, 93) and it is probably safe to assume that it would do so in ferrite too.

Thus, the key to explaining these observations probably lies in finding an explanation for the concentration of carbon at the surface. Among the possibilities Goldberg and Horton are investigating is that a greater density of dislocations may exist at the surface of the sheet, which would attract and hold carbon atoms. Needless to say, the fact that as the steel recrystallizes dislocations move out of the surface of the cold-worked steel is a strong argument against this possibility.

Regardless of the mechanism, the discovery of such large concentrations of the constituents in sheet steel near the surface raises many questions of fundamental and practical interest. For example, can these surface concentrations be observed by optical or electron microscopy? How can carbon concentrations as great as 0.15% be accommodated in the structure without some readily apparent microstructural manifestation? From the practical point of view, how do these surface concentrations affect the rust resistance, or the adherence of coatings to the sheet? Can the phenomenon be deliberately exploited to attain specific surface properties?

It seems to me that Goldberg and Horton's discovery opens up a new direction for research on sheet steels and, indeed, all metals in which the characteristics of the surface are important to their applications.

* * *

Throughout this lecture I have stressed the theme that small changes in composition and processing operations have large effects on the metallurgical structure and the properties of sheet steel. In fact, the changes that produce important effects are often so small that it is difficult with even the most modern methods to discern the mechanisms involved. Yet, sheet steel is a material that is produced in tremendous quantities to exacting standards.

My hope is that this forty-third Campbell Memorial Lecture about sheet steel has filled a void in the literature, aroused the interest of metallurgists in the most important material of commerce, and demonstrated that this material is complex, challenging and full of opportunities.

REFERENCES

1. Basic Oxygen Steelmaking, 3rd Edition, edited by Gerhard Derge, AIME, New York, 1964; see especially Chapter 9 regarding deoxidation practices.
2. The Making, Shaping, and Treating of Steel, 8th Edition, U. S. Steel (1964).
3. R. L. Rickett and F. C. Kristufek, The Microstructure of Low Carbon Steel, Trans ASM, 41 (1949) 1113.
4. W. T. Lankford, S. C. Snyder and J. A. Bauscher, New Criteria for Predicting the Press Performance of Deep Drawing Sheets, Trans ASM, 42 (1950) 1197.
5. R. L. Whiteley, The Importance of Directionality in Drawing-Quality Sheet Steel, Trans ASM, 52 (1960) 154. (1960 Grossmann Award.)
6. R. L. Whiteley, D. E. Wise and D. J. Blickwede, Anisotropy as an Asset for Good Drawability, Sheet Metal Industries, 38 (1961) 349.
7. S. P. Keeler and W. A. Backofen, Plastic Instability and Fracture in Sheets Stretched Over Rigid Punches, ASM Trans Quart, 56 (1963) 25.
8. S. P. Keeler, Determination of Forming Limit in Automotive Stampings, Sheet Metal Industries (Sept. 1965) 683.
9. G. M. Goodwin, Application of Strain Analysis to Sheet Metal Forming Operations in the Press Shop, SAE Paper No. 680093 (Jan. 1968). (Article will be published in SAE Trans, 77, about Jan. 1969.)
10. P. G. Nelson and Joseph Winlock, Method of Determining the Percentage Elongation at Maximum Load in the Tension Test, Bulletin ASTM (Jan. 1949) 53.
11. M. Gensamer, Strength and Ductility, Trans ASM, 36 (1946) 30 (1945 Campbell Memorial Lecture).
12. W. Heller and M. Nacken, Investigations of the Strain Aging of Mild Carbon Steel Carried Out by Measuring Strength and Internal Friction, Archiv Eisenhüttenw, 31 (1960) 723.
13. E. T. Stephenson, The Effect of Carbon in Solution on the Properties of an Aluminum-Killed Low-Carbon Steel, ASM Trans Quart, 55 (1962) 624.
14. S. Sekino and T. Fujishima, Yield Stress and Other Tensile Properties of Nitrogen or Carbon Charged Pure Iron, J Japan Inst Metals, 30 (1966) 142.
15. D. V. Wilson and Y. A. Konnan, Work Hardening in a Steel Containing a Coarse Dispersion of Cementite Particles, Acta Met, 12 (1964) 617.
16. B. I. Edelson and W. M. Baldwin, Jr., Effect of Second Phases on the Mechanical Properties of Alloys, ASM Trans Quart, 55 (1962) 230.
17. W. B. Morrison, Effect of Grain Size on the Stress-Strain Relationship in Low Carbon Steel, ASM Trans Quart, 59 (1966) 824.
18. R. S. Burns and R. H. Heyer, Orientation and Anisotropy in Low-Carbon Steel Sheets, Sheet Metal Ind, 35 (1958).
19. W. F. Hosford and W. A. Backofen, Strength and Plasticity of Textured Metals, Proc Ninth Sagamore Conference AMRA (1962).
20. R. W. Vieth and R. L. Whiteley, Influence of Crystallographic Orientation on Plastic Anisotropy in Deep Drawing Steel, Anisotropy and Tensile Test Properties and Their Relationship to Sheet Metal Forming, IDDRG, London (1964).
21. M. Fukuda, Mathematical Analysis on the Relation Between Crystallographic Texture and Lankford r Value in Sheet Steels, Trans Iron Steel Inst Japan, 8 (1968) 68.
22. H. Piehler, Plastic Anisotropy of Body Centered Cubic Metals, ScD Thesis, MIT (Aug. 1967).
23. C. Crussard, G. Pomey, D. Lajeunesse and M. Angeli, Effect of the Anisotropy of Sheets on Their Plastic Deformation Under Uniaxial and Biaxial Stresses in Low-Carbon Steels, Mem Sci Rev Met (March 1961) No. 3, 183.
24. A. S. Keh and S. Weissmann, Conference on the Impact of Transmission Electron Microscopy on Theories of the Strength of Crystals, Berkeley, Cal. (1961).
25. W. C. Leslie, J. T. Michalak and F. W. Aul, The Annealing of Cold Worked Iron, Iron and Its Dilute Solid Solutions, John Wiley & Sons, New York (1963) 119.
26. I. L. Dillamore and W. T. Roberts, Preferred Orientation in Wrought and Annealed Metals, Met Rev (1965) 39, No. 10, 271.
27. P. C. Hancock and W. T. Roberts, Influence of Rolling Temperature on Textures in Rimming Steel, J Iron Steel Inst (May 1967) 547.
28. A. S. Keh, Direct Observations of Imperfections in Crystals, Interscience, New York (1962).
29. H. Hu, Direct Observations on the Annealing of a Si-Fe Crystal in the Electron Microscope, Trans AIME, 224 (1962) 75.
30. J. T. Michalak and H. W. Paxton, Some Recovery Characteristics of Zone-Melted Iron, Trans AIME, 221 (1961) 850.
31. R. L. Whiteley and D. E. Wise, Relationship Among Texture, Hot Mill Practice and the Deep Drawability of Sheet, Flat Rolled Products—III, AIME, New York (1962).
32. K. Matsudo, T. Shimomura and Y. Hashimoto, Effect of Two-Stage Cold Rolling-Annealing Process on Deep Drawability of Low-Carbon Rimmed Steel Sheet, Report of Tech Res Inst, Nippon Kokan Ltd. (1966).
33. C. G. Dunn, Cold-Rolled and Primary Recrystallization Textures in Cold-Rolled Single Crystals of Silicon Iron, Acta Met, 2 (1954) 173.
34. J. E. Burke, Origin of Recrystallization Textures, Trans AIME (1952) 263.
35. H. Kubotera and K. Nakaoka, Paper presented at 1967 AIME Mech. Working Conference. (Article will be published in a book entitled Mechanical Working and Steel Processing Conference of AIME held in December, 1967.)
36. T. Taoka, E. Furubayashi and S. Takeuchi, The Formation of Cold-Rolled Texture and Recrystallized Texture in Single Crystals of 3% Silicon Iron, Part I, Trans Iron Steel Inst Japan, 6 (1966) 290.
37. F. A. Hultgren, The Reversion and Re-precipitation of Aluminum Nitride in Aluminum-Killed Drawing Quality Steel, Blast Furnace and Steel Plant 56 (Feb. 1968) 149.
38. R. H. Goodenow, Recrystallization and Grain Structure in Rimmed and Aluminum-Killed Low-Carbon Steel, ASM Trans Quart, 59 (1966) 804.
39. W. Jolley, Effect of Heating Rate on Recrystallization and Anisotropy in Aluminum-Killed and Rimmed Sheet Steels, J Iron Steel Inst, 205 (1967) 321.
40. I. L. Dillamore and S. R. F. Fletcher, Recrystallization, Grain Growth and Textures, ASM, Metals Park, Ohio (1966) 448.
41. H. Yoshida, K. Sasaki and F. Kanzaki, The Effect of Manufacturing Processes on the Preferred Orientation of Deep Drawing Sheet Steel, Inst of Sheet Metal Engineering (June 1964).
42. S. Teshima and M. Shimizu, Recrystallization Behavior of Cold-Rolled Mild Steel, Mechanical Working of Steel II, Gordon and Breach (1964) 279.
43. M. Shimizu, Kawara and Shibata, Method of Manufacture of Cold-Rolled Strip Which Has a Preferred Orientation of (111) Planes in the Rolling Direction, Japanese Patent No. 12348, July 1967.
44. R. L. Rickett and W. C. Leslie, Recrystallization, Structure and Hardness of Low-Carbon Steels Containing Up to 1% Copper, Trans ASM, 51 (1959) 310.
45. E. H. Mayer and D. E. Wise, Deep Drawing Steel and Method of Manufacture, U. S. Patent No. 3,244,565, April 5, 1966.
46. M. Shimizu, K. Matsuda, Y. Sadamura, N. Takahashi and M. Kawaharanda, Process for Manufacturing Cold Rolled Steel Sheets with Excellent Disposition for Press Work, French Patent No. 1,511,529, February 16, 1967.
47. P. N. Richards, The Effects of Heating Rates and Subrecrystallization Heat Treatments on Aluminum Killed Deep Drawing Steel, Recent Developments in Annealing, Iron and Steel Institute Spec. Report No. 79 (1963) 22.
48. P. N. Richards, Low-Temperature Recovery Prior to Recrystallization of Al-Killed Deep Drawing Steels, J Australian Inst Metals, 12 (1967) 2.
49. R. L. Kenyon, Recrystallization and Grain Control in Ferrous Metals, Grain Control in Industrial Metallurgy, ASM, Metals Park, Ohio (1949) 74.
50. W. C. Leslie, J. T. Michalak, A. S. Keh and R. J. Sober, The Effects of a Manganese-Oxygen Interaction on the Recrystallization of Iron, ASM Trans Quart, 58 (1965) 672.
51. J. D. Baird and J. M. Arrowsmith, Recrystallization Behavior of Some High-Purity Irons, J Iron Steel Inst, 204 (1966) 240.
52. D. A. Witmer, Effect of Thermal History on the Recrystallization Behavior of Low-Carbon Irons Containing Manganese,

Oxygen and Sulfur, MS. Thesis, Lehigh University, Bethlehem, Pa. (1968).
53. E. T. Turkdogan and J. Pearson, The Solubility of Sulfur in Iron and Iron-Manganese Alloys, J Iron Steel Inst, 180 (1955) 349.
54. W. C. Leslie, Reply to Discussion of Reference 50 by P. R. Mould, ASM Trans Quart, 59 (1966) 1059.
55. A. H. Cottrell, Discontinuous Yielding, National Physical Lab Symposium No. 15, 2 (1963) 456.
56. D. V. Wilson, Role of Grain Boundaries in the Discontinuous Yielding of Low-Carbon Steels, Metal Science Journal, 1 (1967) 40.
57. W. M. Lomer, The Yield Phenomenon in Polycrystalline Mild Steel, J Mech Phys Solids, 1 (1952) 64.
58. B. Jaoul, Consolidation of Iron Polycrystals and Hypotheses Concerning Origin of Plateau in Traction Curves, J Mech Phys Solids, 9 (1961) 69.
59. J. R. Low and M. Gensamer, Aging and the Yield Point in Steel, Trans AIME, 158 (1944) 207.
60. R. W. Evans, Effect of Prior Cold Reduction on Ductility of Annealed Rimmed Steel, J Iron Steel Inst, 205 (1967) 1150.
61. A. H. Cottrell and B. A. Bilby, Dislocation Theory of Yielding and Strain Aging of Iron, Proc Phys Soc, 62A (1949) 49.
62. D. V. Wilson and B. Russell, The Contribution of Atmosphere Locking to the Strain Aging of Low-Carbon Steels, Acta Met, 8 (1960) 36.
63. J. W. Edington, T. C. Lindley and R. E. Smallman, Strain Aging of Vanadium, Acta Met, 12 (1964) 1025.
64. J. D. Baird, Strain Aging of Steel—A Critical Review, Iron & Steel, 36 (1963) 186, 326, 368, 400.
65. A. S. Keh and W. C. Leslie, Recent Observations on Quench Aging and Strain Aging of Iron and Steel, Materials Science Research, Vol 1, Plenum Press (1963) 208.
66. S. Bergh, Influence of Carbon Content on Strain Aging in Annealed Rimming Deep-Drawing Steels at Different Low Nitrogen Contents, Jern Ann, 150 (1966) 250.
67. A. Josefsson and B. Backstrom, Strain Aging in Carbon-Saturated Ferrite with Various Nitrogen Contents, Iron & Steel, 39 (1966) 43.
68. F. H. Laxar, J. W. Frame and D. J. Blickwede, The Effect of Aluminum on Strain Aging and Internal Friction in Low Carbon Steel, Trans ASM, 53 (1961) 683.
69. P. N. Richards and K. V. Barratt, Solution and Precipitation of Carbon and Nitrogen During Sheet Galvanizing, ASM Trans Quart, 58 (1965) 611.
70. B. B. Hundy, Inhomogeneous Deformation During the Temper Rolling of Annealed Mild Steel, J Iron Steel Inst, 181 (1955) 313.
71. J. F. Butler, Flat Rolled Products—III, Met Soc Conf Vol 16, Interscience, New York (1962) 65.
72. R. D. Butler and D. V. Wilson, The Mechanical Properties of Temper-Rolled Sheets, J Iron Steel Inst, 201 (1963) 16.
73. E. J. Paliwoda and I. J. Bessen, Flat Rolled Products—II, Met Soc Conf, Vol 6, Interscience, New York (1960) 63.
74. B. B. Hundy, Elimination of "Stretcher Strains" in Mild Steel Pressing, J Iron Steel Inst, 178 (1954) 127.
75. H. P. Tardiff and C. S. Ball, The Effect of Temper Rolling on the Strain Aging of Low Carbon Steel, J Iron Steel Inst, 182 (1965) 9.
76. J. D. Baird, R. D. Butler, D. V. Wilson, Discussion to Ref. 64, J Iron Steel Inst, 201 (1963) 764.
77. R. B. G. Yeo, British Patent No. 1,051,307, December 14, 1966.
78. A. L. Keh, Direct Observation of Imperfections in Crystals, Ed. J. B. Newkirk and J. H. Wernick, Interscience Publishers, New York (1962) 213.
79. J. T. Michalak and L. J. Cuddy, Symposium on the Role of Substructure in the Mechanical Behavior of Metals, Technical Documentary Report No. ASD-TDR-63-324, WPAFB (April 1963) 141.
80. R. M. Hudson and G. L. Stragand, Gas-Metal Reactions During Box Annealing of Low Carbon Steel, Trans ASM, 52 (1960) 135.
81. F. V. Schossberger, K. Hattori, and H. Marver, Reaction Mechanism of Black Stain Formation on Annealed Steel Coils, Proc Iron Steel Eng, September 26, 1956, 382.
82. C. R. Lillie and D. W. Levinson, Study of Factors Influencing Surface Staining of Cold Rolled Steel, Iron Steel Eng-September 26, 1956, 377.
83. M. J. Olney and G. L. Smith, Surface Effects Occurring During the Heating and Cooling of Plain Carbon Steels, J Iron Steel Inst, 177 (1959) 107.
84. W. P. Poschenrieder, R. F. Herzog, and A. E. Barrington, The Relative Abundance of Lithium Isotopes in the Holbrook Meteorite, Geochim Cosmochim Acta, 29 (1965) 1193.
85. The Ion Microprobe Mass Spectrometer, Progress in Nuclear Energy Series IX, Anal Chem, Pergamon Press, New York, 7 (1966) 243.
86. The Ion Solids Surface Analyzer is manufactured by the GCA Corp., Bedford, Mass.
87. A. J. Pignocco and G. E. Pellissier, Low-Energy Electron Diffraction Studies of Oxygen Adsorption and Oxide Formation on a (001) Iron Surface, J Electrochem Soc, 112 (1965) 1188.
88. H. E. Farnsworth and J. Tuul, The Influence of Lattice-Defect Density on the Chemisorption and Oxide Formation on a Clean (100) Crystal Face of Nickel as Determined by Low-Energy Electron Diffraction, J Phys Chem Solids, 9, No. 1 (1959) 48.
89. P. C. S. Hayfield, Influence of Residual Impurities in Commercial Purity Copper Upon Low Temperature Oxidation and Corrosion, Proc First International Congress, Imperial on Metallic Corrosion, London, Butterworths (1962) 670.
90. K. T. Aust, R. E. Hanneman, P. Niessen and J. H. Westbrook, Solute Induced Hardening Near Grain Boundaries in Zone Refined Metals, Acta Met 16 (1968) 291.
91. N. G. Ainslie, R. E. Hoffman and A. U. Seybolt, Sulfur Segregation at α-iron Grain Boundaries—I and II, Acta Met, 9 (1960) 523, 528.
92. M. Hillert, T. Wada and H. Wada, The α-γ Equilibrium in Fe-Mn, Fe-Mo, Fe-Ni, Fe-Sb, Fe-Sn and Fe-W Systems, J Iron Steel Inst, 205 (1967) 539.
93. J. S. Kirkaldy, J. von D-Forstmann and R. J. Brigham, Stimulation of Banding in Steels, Canadian Met Quart, 1 (1962) 59.

Process Variables Affecting the *r*-Value of Cold Rolled Steel Sheet

By Kozo YAMADA,* Haruo KUBOTERA, and Kazuhide NAKAOKA****

Synopsis

The effect of conventional alloying elements on the r-value of cold rolled steel sheet was investigated. As for decarburized product of capped steels, manganese, oxygen and sulfur were found to influence the r-value. When contents of these elements are within certain ranges, r-value beyond 2.0 can be obtained. In case of decarburized product of aluminum killed steels, acid soluble aluminum and nitrogen have predominant influence and manganese has little effect. The r-value of capped steel can be improved further, by the application of double rolling and annealing process, in which decarburization is effected at either the first or the second annealing. For a steel of good grain growability, higher r-value can be obtained when sheet is decarburized at the final annealing, while a sheet of poor grain growability shows a higher r-value when it is decarburized at the first annealing.

I. Introduction

Influences of hot rolling, cold rolling and annealing conditions on *r*-value of sheet steel have been investigated by several authors[1)-4)] and it is generally accepted that the sheet product of higher *r*-value is obtained under the following conditions: high finishing temperature and low coiling temperature in hot rolling; cold reduction of 60 to 80 pct; annealing at temperatures high enough, for a period long enough and preferably in a decarburizing atmosphere, to promote grain growth.

As to the effect of alloying elements, additions of Cu, As or P to a steel of conventional chemistry[4),5)] and of Ti subsequently to vacuum degassing of conventional molten steel[6)] were reported to be effective.

Within the limits of the authors' knowledge, however, only a few papers dealt with the influence of conventional alloying elements.[7),8)]

The main purpose of this paper is to investigate the effect of such alloying elements as manganese, sulfur, oxygen, aluminum and nitrogen on the *r*-value of the cold rolled product: the influence of rolling and annealing conditions are also studied in relation to chemistry.

II. Experimentals

Forty heats of capped steel and fifteen heats of Al-killed steel were studied. Chemical composition of the steels is shown in Table 1.

Table 1. Chemistry of steels (wt%)

Capped steels

C	Mn	P	S	O
0.04 ~0.13	0.15 ~0.50	0.011 ~0.016	0.007 ~0.023	0.015 ~0.07

Al-killed steel

C	Mn	P	S	sol. Al	N
0.04 ~0.07	0.15 ~0.36	0.007 ~0.024	0.005 ~0.024	0.027 ~0.062	0.0027 ~0.0073

Hot rolled samples of finishing and coiling temperatures, 800° to 860°C and 550° to 600°C, were cold rolled by 72 pct by means of an experimental mill. Cold rolled samples were heated in a reducing atmosphere to 700°, 720° or 750°C at a rate of 125°C/hr for capped steels and 40°C/hr for Al-killed steels, then soaked for specified time. Large part of the samples was decarburized in the soaking period, in an atmosphere comprizing nitrogen and wet hydrogen, to promote grain growth. Annealed samples were cooled in the furnace and subjected to temper rolling of about 1 pct. Tensile specimens were cut from the temper rolled sheet and *r*-value was measured by the ordinary method by means of the elongation and width strain. The microstructure of several specimens was examined.

III. Result and Discussion

1. *Effect of Conventional Alloying Elements*

1. Capped Steel

In Fig. 1, which was presented at the 10th Mechanical Working and Steel Processing Conference of AIME in 1967,[8)] the abscissa represents the oxygen content and the ordinate the grain diameter after the

* Keihin Works, Nippon Kokan K.K., 2730 Minamiwatarida, Kawasaki, Kanagawa 210, Japan.
** Technical Research Center, Nippon Kokan K.K., 2730 Minamiwatarida, Kawasaki, Kanagawa 210, Japan.

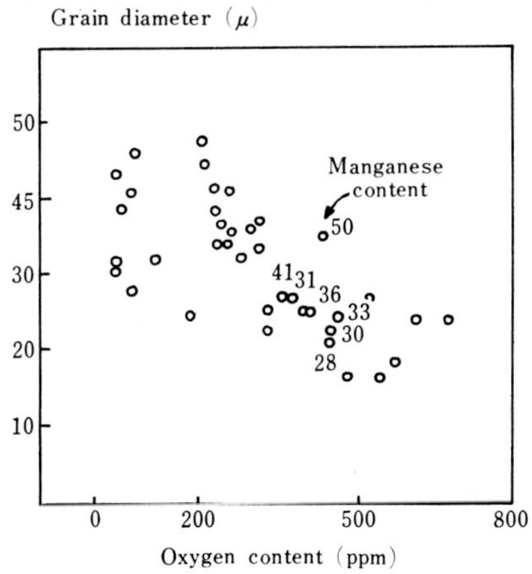

Fig. 1. Effect of oxygen and manganese on grain size of decarburized product

Table 2. Chemistry of steels used for the study of Fig. 2 (wt%)

Steel	C	Mn	S	O
A	0.09	0.36	0.021	0.019
B	0.12	0.37	0.007	0.017

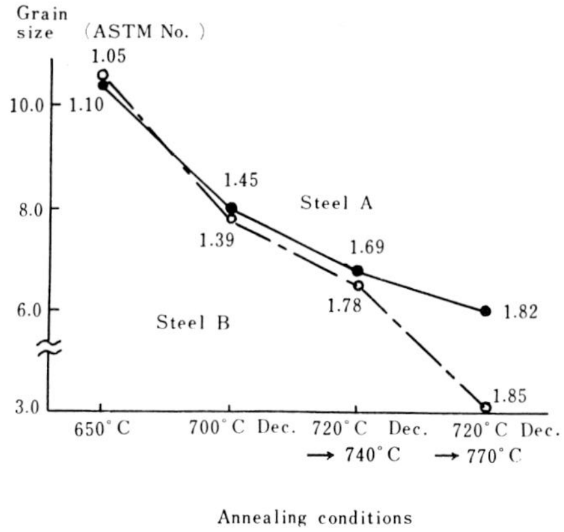

Fig. 2. Effect of sulfur on grain growth
650°C; heated to 650°C and cooled
700°C; decarburized at 700°C
720°C DEC.→740°C; soaked at 740°C after the decarburization at 720°C
720°C DEC.→770°C; soaked at 770°C after the decarburization at 720°C

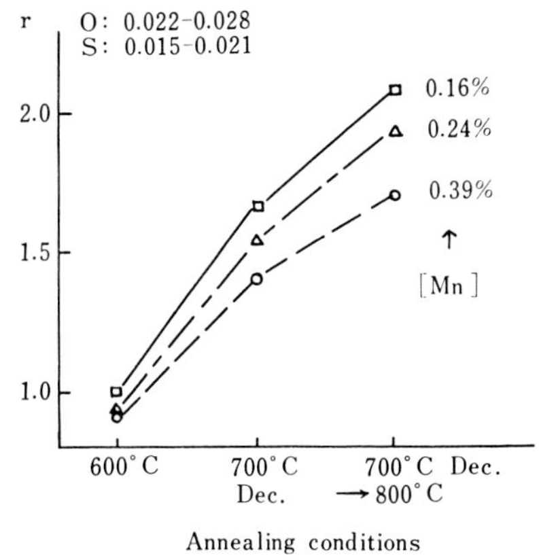

Fig. 3. Effect of manganese on the *r*-value
600°C; heated to 600°C and cooled
700°C DEC.; decarburized at 700°C
700°C DEC.→800°C; soaked at 800°C after the decarburization at 700°C

decarburization at 750°C. It can be seen from the figure that an increase in oxygen content leads to the superior grain growability. This implies that lower oxygen content is favorable to raise *r*-value, with reference to the experimental result of Karlyn *et al.*[9] The influence of oxygen was accounted for by the resistance of oxide inclusions to the grain boundary migration.[8] Since there existed sulfide inclusions, which were also considered to prevent the grain boundary migration, the effect of sulfide content was examined. Two steels were picked up from the stock of samples. Chemistry of them is listed in Table 2; sulfur contents are 0.02 and 0.007 pct for steel *A* and steel *B*, respectively. Manganese and oxygen contents are almost at the same level. The cold rolled steels were subjected to heat treatments, and grain size and *r*-value were measured. The results are shown in Fig. 2. The abscissa represents annealing conditions; 650°C indicates that samples were heated to 650°C, then furnace cooled; 700°C DEC means that samples were decarburized during the soaking period at 700°C, soaking time being 5 hours; 720°C DEC→740°C and 720°C DEC→770°C mean soaking at 740° and 770°C for 3 hours after the decarburization at 720°C, for 4 hours, respectively. It can be seen that a difference between steels *A* and *B* becomes evident when grain growth is promoted at 740° or 770°C after the decarburization annealing. The fact that steel *B* is superior in grain growability to steel *A* indicates that sulfide inclusions act as oxide inclusions, as has been expected.

On the basis of the knowledge about the effect of oxygen and sulfur, the sum of contents of the two elements was taken as a variable relating to the resistance to grain growth.

Another information is being presented in Fig. 1. The numerical figures around several points in Fig. 1 indicate manganese contents corresponding to the data, showing that, for oxygen content around 400 ppm, higher manganese content is favorable for grain growth.

Effect of manganese was studied for lower oxygen content. The result is shown in Fig. 3. The abscissa is annealing conditions and the ordinate $\bar{r}$-value; 600°C means that samples were heated to 600°C, then cooled in the furnace; 700°C DEC and 700°C DEC→800°C indicate that samples were decarburized during the soaking period at 700°C for 6 hours and soaked at 800°C for 3 hours after the decarburization, respectively.

Three kinds of steel were subjected to the study. Chemical compositions are shown in Table 3. Manganese contents are 0.16, 0.24, and 0.39 pct.

In contrast to the previous result shown in Fig. 1,

Table 3. Chemistry of steels used for the study of Fig. 3(wt%)

Steel	C	Mn	P	S	O
A	0.092	0.16	0.011	0.015	0.028
B	0.096	0.24	0.015	0.015	0.022
C	0.088	0.39	0.013	0.021	0.023

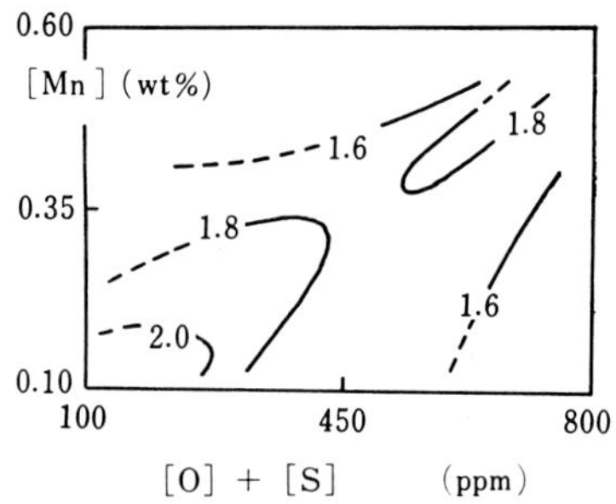

Fig. 4. Dependence of *r*-value of decarburized capped steel on sum of oxygen and sulfur content and manganese content (equi-*r* contours)

it can be seen that lower manganese content is favorable.

The difference in *r*-value corresponding to the manganese level can be distinguishable already at 600°C and it becomes more definite as grain growth is promoted.

It was recognized that, after the promotion of grain growth, the three steels showed almost the same grain size, indicating that there was no difference in grain growability between the steels.

On account of these data, manganese content is taken as another variable.

Figure 4 shows the dependence of the *r*-value of decarburized product, on the two variables defined above.

Samples were decarburized at 700°C for 6 hours and soaked at 750°C for 5 hours.

Equi-*r* countours are drawn in the figure by means of 40 data points.

The *r*-value rises over 2.0 when manganese content is below 0.20 pct and the sum of oxygen and sulfur contents is below 250 ppm.

2. Al-killed Steel

Published data and experience suggest that contents of acid soluble aluminum and nitrogen exert an influence on the *r*-value of aluminium killed steels, since precipitation of aluminum nitride is a main factor affecting the texture of aluminum killed steels.[10)-13)]

In the present study, the effect of manganese was also investigated.

The experimental result is shown in Fig. 5. This figure was obtained by the data for fifteen steels. The abscissa is content of acid soluble aluminum and the ordinate is *r*-value. The data can be divided into two groups; one is for the steels containing nitrogen over 50 ppm (group *A*) and the other is for

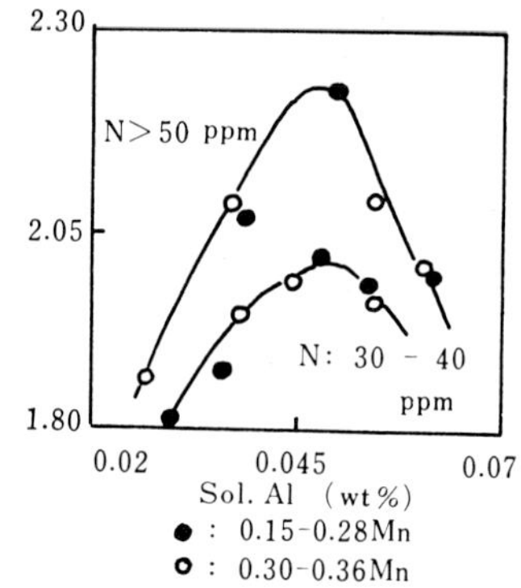

Fig. 5. Dependence of *r*-value of decarburized Al-killed steel on sol. Al, N, and Mn

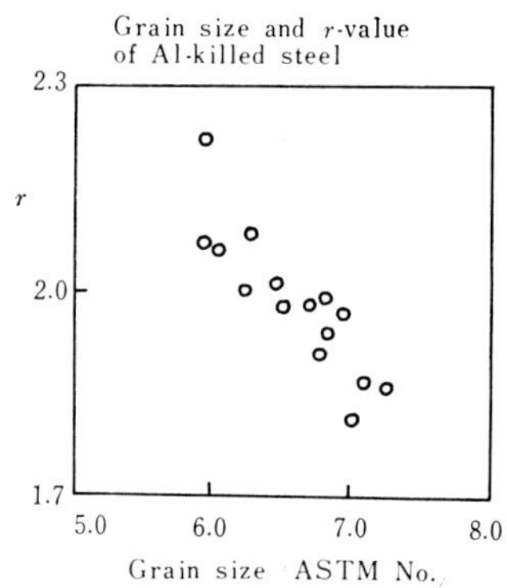

Fig. 6. Grain size and *r*-value of Al-killed steel

the steels nitrogen contents of which are between 30 and 40 ppm (group *B*). Generally, the steels of group *A* show a higher *r*-value for the same level of acid soluble aluminum. Both groups have maximum of *r*-value for the acid soluble aluminum around 0.045 pct.

The open circles indicate manganese content between 0.30 and 0.36 pct and full circles between 0.15 and 0.28 pct.

It can be seen from the figure that the data can be arranged according to the contents of acid soluble aluminum and nitrogen, irrespective of the manganese level.

Relation between grain size and *r*-value of these fifteen aluminum-killed steels is shown in Fig. 6. The *r*-value is strongly correlated to the grain size.

These data indicate that the processes of precipitation and agglomeration of aluminum nitride are main factors affecting the *r*-value of the product.

2. *Effect of Rolling and Annealing Conditions*

As it was expected that *r*-value of capped steel could be improved by the application of double rolling and annealing process,[14)] two kinds of steel were subjected to such treatment.

Chemistry of the steels is shown in Table 4. Steel *A* has a lower oxygen content and consequently better grain growability. Combination of cold reductions was determined with reference to the result of a preliminary investigation as follows:

Table 4. Chemistry of capped steels subjected to the double rolling and annealing process (wt%)

Steel	C	Mn	P	S	O
A	0.084	0.40	0.012	0.020	0.025
B	0.059	0.32	0.011	0.017	0.057

Table 5. Effect of decarburization in double rolling* and annealing process

Steel	*A* (Good grain growth)		*B* (Poor grain growth)	
Annealing	(a)**	(b)***	(a)	(b)
r	2.11	1.61	1.73	2.06

* First red.: 54%, Final red.: 72%
** First ann.: 700°C non-dec., Final ann.: 750°C dec.
*** First ann.: 700°C dec., Final ann.: 750°C non-dec.

first reduction: 54 pct
final reduction: 72 pct

Decarburization was effected at either the first annealing or the final annealing. Result is shown in Table 5. In the decarburization at the final annealing as in (a) the soaking temperature was 750°C and soaking time was 5 hours, while in the decarburization at the first annealing as in (b) the temperature and time were 700°C and 6 hours, respectively.

The table shows that steel *A* gives a higher *r*-value when decarburization is effected at the final annealing, while steel *B* gives a lower *r*-value under this treatment.

Microstructures after the first and the final annealings are as shown in Photo. 1.

Each steel shows almost the same microstructures for the two annealing conditions after the final annealing. After the first annealing, however, steel *A* shows quite different structures between the decarburized and nondecarburized states; decarburized sample has a microstructure with considerably larger grains, indicating the good grain growability, while non-decarburized state shows finer grains.

In case of steel *B*, comparatively small difference is seen between the decarburized and non-decarburized states on account of the poor grain growability.

In the double rolling and annealing process including a decarburization annealing, texture, grain size and carbon content of the steel after the first annealing vary depending on the first annealing condition.

Generally, decarburization improves texture on account of grain growth, makes grain size larger and reduces carbon content.

The texture improvement seems to be inherited to the subsequently cold rolled and annealed product.

On the other hand, according to Hayami *et al.*,[15] finer grains lead to a more frequent nucleation of {111} grains and Fukuda[7] pointed out that lower carbon content led to a larger *r*-value.

It follows that for the improvement of *r*-value after the subsequent rolling and annealing, the grain growth and the reduction of carbon content are the opposing factors. Hence the different behaviours of steels *A* and *B* can be accounted for as follows.

In the case of steel *A*, the grain size after the first annealing is a predominant factor and the first decarburization has little effect in spite of the texture

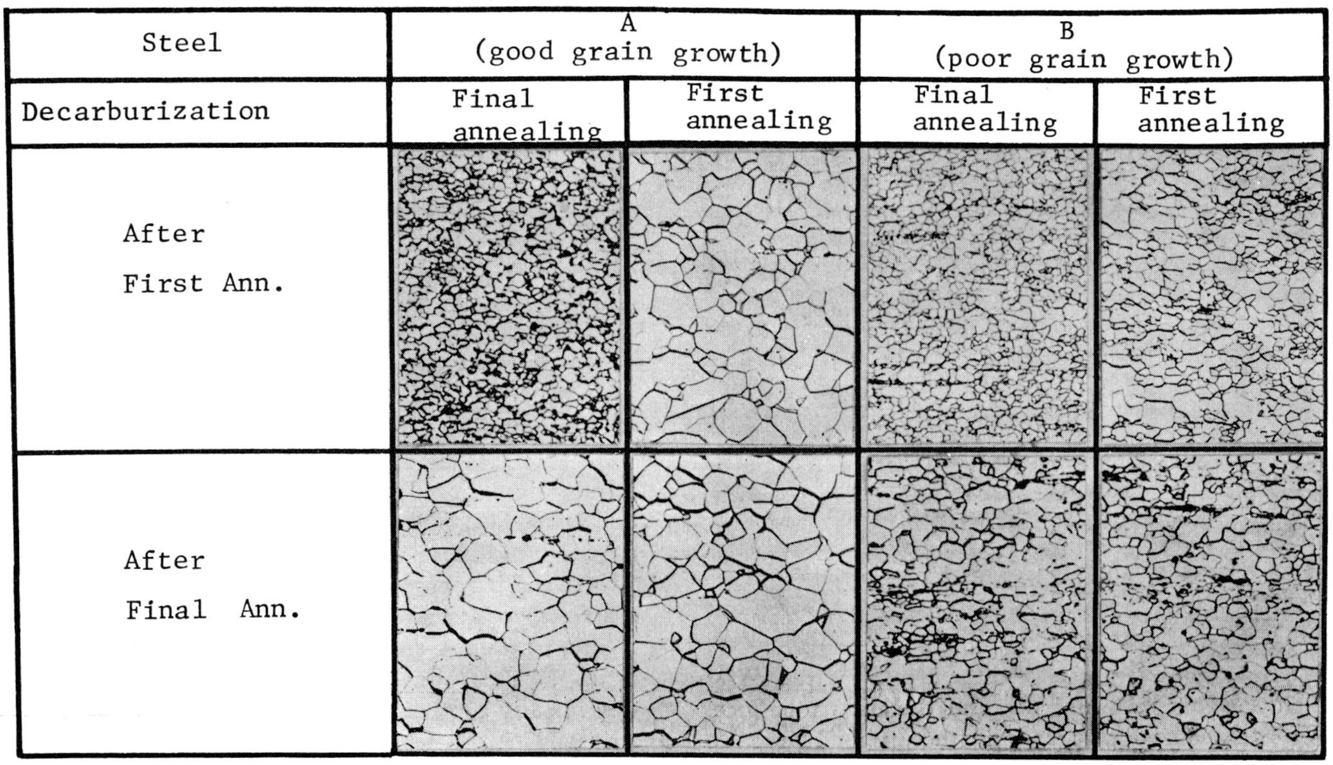

Photo. 1. Microstructure for double rolling and annealing process

improvement and the reduction of carbon.

In the case of steel *B*, the reduction of carbon is a predominant factor and the first decarburization shows the favorable influence.

It has been confirmed that by the application of the double rolling and annealing process to a steel whose manganese content and the sum of oxygen and sulfur contents are in the ranges assuring the *r*-value over 2.0 in Fig. 4, a product with the *r*-value of 2.3 to 2.4 can be obtained.

IV. Conclusion

The *r*-value of decarburized capped steel can be improved by controlling the contents of Mn, O, and S. The *r*-value of Al-killed steel is governed by N and sol. Al and it is independent of Mn. Higher *r*-value can be obtained with the application of double rolling and annealing process, in which decarburization is effected at either the first or the final annealing, depending on the chemistry of steel.

REFFRENCES

1) L. R. Shoenberger: *Trans. Amer. Sos. Metals*, **52** (1960), 164.
2) R. L. Whiteley and D. E. Wise: Flat Rolled Product III (1962), Interscience.
3) R. H. Heyer, D. E. McCabe, and J. A. Elias: Flat Rolled Product III (1962), Interscience.
4) H. Yoshida, K. Sasaki, and F. Kanzaki: IDDRG Colloquium, London, June (1964).
5) P. Waqueze and Van Dale: IDDRG Colloquium, London, June (1964).
6) M. Shimize *et al.*: *Japanese Patent*, **42**–12348; **44**–18066.
7) M. Fukuda: *Tetsu-to-Hagané*, **53** (1967), 559.
8) H. Kubotera and K. Nakaoka: AIME 10th Mechanical Working Conference, Chicago, (1967).
9) D. A. Karlyn, R. W. Vieth, and J. L. Forand: AIME Mechanical Working Conference, Pittsburgh, (1969).
10) W. C. Leslie, J. T. Michalak, and F. W. Aul: Iron and Its Dilute Solid Solutions, (1963), Interscience.
11) R. H. Goodenow: *Trans. Amer. Soc. Metals*, **59** (1966), 804.
12) W. Jolly: *J. Iron Steel Inst.*, **205** (1967), 321.
13) J. T. Michalak and R. D. Shoone: *Trans. AIME*, **242** (1968), 1149.
14) K. Matsudo, T. Shimomura, and Y. Hashimoto: Report of Tech. Res. Inst., Nippon Kokan (1966).
15) T. Hayami *et al.*: Fall Meeting of Japan Inst. of Metals (1964).

Metallurgical Aspects of Sheet Steel for Press Shops

By R.D. BUTLER,* D. NORTH,* and G.M. DAVIES*

Synopsis

Press performance has been discussed from various points of view with particular reference to autobody skin panels and more complex interior panels. The application of limiting strain analysis to the forming of interior panels is considered in relation to the mechanical and metallurgical properties of sheet metals. The influences of metallurgical factors such as grain size on formability, yielding behaviour, tensile strength, work-hardening characteristics, and elongation, are considered. Temper processing is described in relation to temper rolling and roller levelling, and evaluated from material property and practical points of view.

As a further requirement from sheet steel in autobody service, the dent resistance of autobody stampings and a dentability test are referred to. The advantage of rimming over stabilised steel sheet are emphasised, particularly in relation to age hardening after press forming.

Photo. 1. Typical low strain skin panel

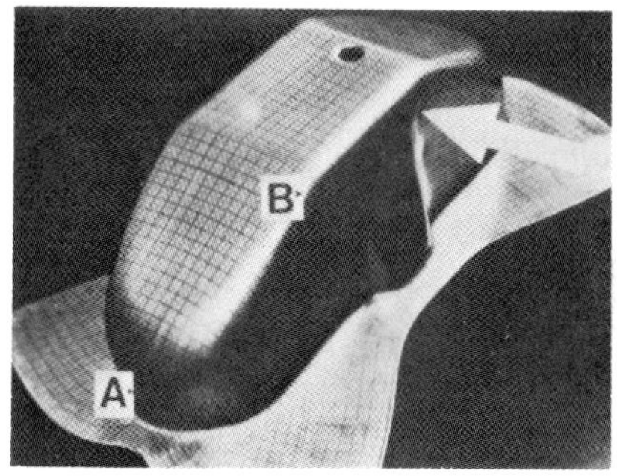

Photo. 2. Complex interior pressing showing high levels of strain
(A) Draw area (B) Stretch area

I. Introduction

Many papers have been published on the subject of general metallurgical requirements relating to sheet metal performance in press shops.[1)-3)] These have emphasized the importance of manufacturing variables, cleanliness, relating to surface imperfections and inclusion content, and overall requirement for consistency of properties. It is not proposed to reconsider those aspects here but to extend the debate to include consideration of engineering and metallurgical requirements related more directly to the performance of autobody pressings. With this in mind the discussion is concentrated on the complexity of levels of strain developed in pressings, the importance of the relatively new techniques of limiting strain analysis in assessing formability performance, and the effect of various metallurgical properties such as yield strength, work hardening, anisotropy and grain size. Yield point suppression is considered and also strain age hardening, particularly in terms of customer performance requirements such as panel denting resistance and energy absorption under heavy impact.

II. Autobody Pressings

These range in character from operations involving simple biaxial stretching and simple deep drawing to the more general situation involving a complex interaction of both stretching and drawing at widely differing levels of overall strain and strain ratio. Autobody stampings can generally be divided into two main categories (a) skin panels (*i.e.* those that are visible on the exterior and interior of the vehicle after all the functional and decorative trim has been added), and (b) interior panels. These are generally widely different in character, the skin panels involving large areas of metal with very low strain (Photo. 1), the strain being largely biaxial stretching. The interior panels are generally much more complicated in form involving higher levels of strain in the finished part, and more complicated strain histories (Photo. 2). The main objective of a press shop fitter in producing the skin panels is to produce the desired shape with the minimum amount of stretch in the pressed panel sufficient to set it so that when it is trimmed to its final shape (typically 40% of the panel is cut off and scrapped) there is a minimum amount of spring back and general panel distortion. To achieve this stretching, the panel is gripped in draw beads and stretched by driving the punch to a predetermined depth. With the large flat panels dictated by modern styling requirements, large areas of the panel are inevitably subjected to small strains, whereas the periphery, *i.e.* those areas strained over the punch and die radii, are heavily strained. When considering interior panels the same criteria applies in some instances. More generally, however, a panel of complex shape is desired with deep form and many breaks in contour. This enables as much of the interior of the body to be achieved in one piece as is reasonable. Strain profiles on these stampings gener-

* Pressed Steel Fisher Division, British Leyland Motor Corporation, Cowley, Oxford, United Kingdom.

ally show only relatively small areas with low strains.

It is not sufficient to define a failed stamping as one which has a split, this term applies equally to stampings exhibiting any severe localised thinning in the metal (necking), and for that matter panels with excessive buckling, wrinkling and general bad shape must also be considered failures.

III. Limiting Strain Analysis

Recently a new approach to the analysis of sheet metal formability in press shops has been introduced by Keeler[4] and Goodwin.[5] An empirical relationship has been determined for autobody steels relating principal strains to a limiting or failure condition for those strain ratios normally encountered. Keeler and Goodwin have emphasized that the forming of panels can be considered in terms of these limiting strains. Keeler's now fairly widely publicised limiting formability diagram shown in Fig. 1 has been clearly shown in many practical instances to give a good guide line to the limiting strain under different strain ratio regimes which can be sustained in actual production stampings.

Photograph 3 shows clearly the results that can be achieved by analysing the performance of stampings in these terms. An interior stamping (Photo. 2) produced a high level of scrap in the press shop when formed from drawing quality sheet with low *r* value. Limiting strains for this condition are shown at point *A* in Fig. 1. This point is above the limiting strain band because the measurement included a significant amount of necked material. By changing metallurgical properties of the sheet, in this case to a higher *r* value, the limiting strain level was reduced to point *B* in Fig. 1, and the press shop scrap was considerably reduced, although many stampings still required rejection because of necking. It was clear that a further improvement in performance was necessary and the only way to achieve this result was to modify the tool geometry slightly such that the strain ratio was changed to a less critical one. Point *C* in Fig. 1 is the result achieved by easing the die radius in a local area of the stamping. Though the same apparently high level of strain was developed in the critical area,

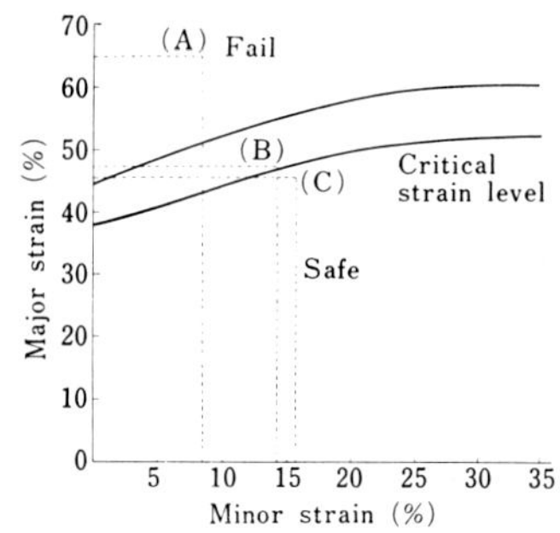

Fig. 1. Positive-positive area of the forming limit diagram (after Keeler[4])

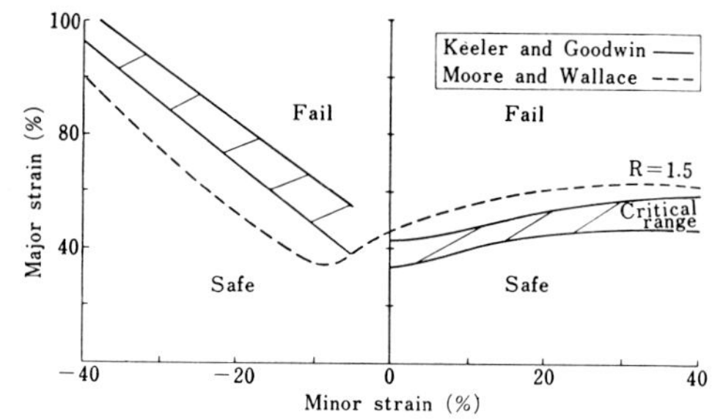

Fig. 2. Goodwin–Keeler empirically determined forming limit diagram with Moore and Wallace theoretical curve (after Brookes[6]) superimposed

this was below the critical strain level at the strain ratio now developed and press shop scrap was effectively eliminated.

Many similar examples of the benefits of this technique of strain analysis can be quoted. It has been particularly valuable since the analysis was extended by Goodwin[5] to include limiting strain analysis for both biaxial extension and for strain conditions involving a combination of tension and compression as shown in Fig. 2.

It must be emphasized that this method of analysis in itself does not provide a complete answer to sheet metal formability questions. It can only be used to examine particular situations and help provide a solution where necessary. It in no way provides explanations for the high strain levels that may be

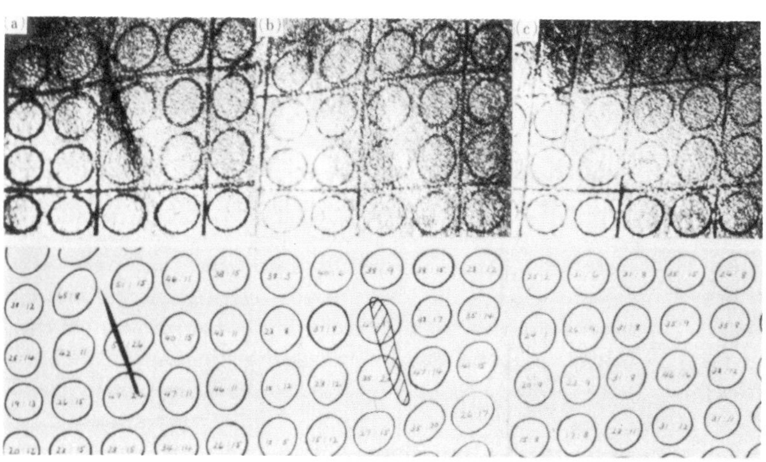

Photo. 3.
Major strain: Minor strain
Strain distribution in critical area of mounting bracket pressing
(a) Using low *r* value sheet
(b) Using high *r* value sheet
(c) Following die modification

present in particular stampings and care has to be taken when considering stampings which may have undergone complex strain histories involving even reversal of strain patterns.

There is considerable debate as to the end points which should be taken in this type of analysis. Autobody stampings can generally be classified as rejects if any form of necking is visible, and it can therefore be argued that instability strains should be the limiting criteria. Keeler's curves are derived from measurement of grid patterns at the nearest whole circles adjacent to a split whilst Goodwin measured circles spanning the fracture. Certainly different conclusions can be drawn depending on the end point selected.

Brookes[6] has observed the match between empirically determined strain curves shown by Keeler and Goodwin and those predicted by the mathematical study by Moore and Wallace[7] using Hill's model[8] and based on instability critieria (Fig. 2). It is clear that there is an extremely good correlation between the two sets of curves emphasizing the success of the analytical model even though some of the basic assumptions invoked by Hill may be in question.[9]

IV. Metallurgical Factors Influencing Formability

In any forming operation both engineering and material variables influence the performance achieved.

It is important to discuss the interaction of metallurgical properties on the strain levels achieved particularly when considering the radically different performance criteria of skin panels and interior panels.

Considering skin panels in general, the mechanical property which correlates best with press shop performance is the yield stress of the material though this may be shown more precisely in terms of yield/ultimate stress ratio. This relationship has been established many times[1),3),10)] and has been generalised in terms of "performance functions" which will not be dealt with in detail here.

Obviously yield stress cannot be considered in isolation from the other properties of the sheet. To simplify the argument, however, it can be stated that it influences the situation in three distinct ways. Firstly yield point plays a particularly important part in determining the acceptability of the surface appearance of a pressed panel. When annealed low carbon steel is pressed, unless strain levels are particularly high in all areas of the stamping, objectionable stretcher strains (Lüders lines) are produced. To overcome this problem, steel to be used in the more visible parts of a vehicle is temper rolled.

The second point is that the press working industry makes active use of the reduction in yield stress achieved as the result of temper rolling. This reduction in yield stress with temper rolling gives considerable advantage in press formability. When forming the relatively flat panels typical of carbody skins it will be recalled that the forming operation involves attempting to deform the sheet virtually at right angles to the direction of punch travel. To achieve this the panel has to be gripped in draw beads and stretched sufficiently to induce plastic deformation throughout the panel the stress being transmitted through the part of the sheet severely deformed over the punch and die radii. The lower the yield stress the easiler it is to achieve this result (Fig. 3). Furthermore, as press tools are designed and manufactured with fixed dimensions, this will result in a more even distribution of strain over all the panel and lower values in those areas most likely to split, *i.e.* the heavily strained areas over the punch radii. Also better panel shape is obtained, "looseness" is more easily avoided and spring back reduced.

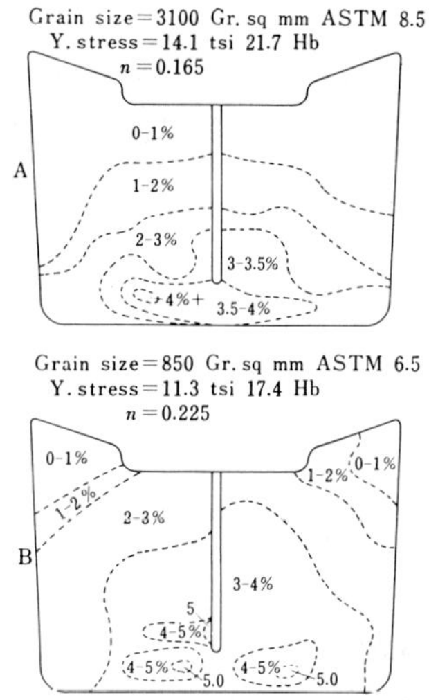

Fig. 3. Strain distribution in bonnet panel, illustrated in Photo. 1, when formed in high and low yield stress sheet

Though the skin panels with their very large relatively flat areas demonstrate the importance of yield stress particularly clearly, this property plays an important part in the majority of stampings.

The third important aspect is attributable to the effect on yield stress of strain age hardening. Strain ageing before press working is particularly disadvantageous because it results in a rapid rise of yield stress, a re-appearance of discontinuous yield phenomena, a slight reduction in elongation, and generally a considerable reduction in formability. Strain ageing can, however, be considered as a positive advantage when it occurs after press working. The effect is to raise the yield stress of the panel and effectively increase its subsequent resistance to damage. This will be discussed more fully later.

There are various metallurgical factors influencing the yield stress of sheet steel. One factor which appears to be receiving less attention than it deserves in recent years is the grain size. The relationships between grain size and yield stress have been clearly shown (*e.g.* the Petch equation).[11),12)] So has the positive relationship with both yield point elongation (Lüders strain) and temper reduction necessary to suppress discontinuous yielding. Also the rate of return of yield point problem, with ageing and the level to which yield stress rises, is grain size dependent.

The optimum grain size for autobody quality steel has been established over the years as around A.S.T.M. 7. This results in a good balance between formability requirements and manufacturing limitations in the steel mill necessary for economic production. In recent years it has been satisfying to note the overall improvement in cleanliness and ductility of much of the steel supplied in the U.K. This is particularly true of CR1 and some grades of CR2 steel where grain size is maintained at about the optimum level. On the other hand it is disturbing to note that the average grain size delivered for CR2, 3 and 4 grades of steel has diminished over the years and is now frequently nearer A.S.T.M. 9. It is felt that a further significant improvement in formability will result if this trend is reversed.

Grain size also affects the work-hardening characteristics of autobody steel which can significantly affect the strain distribution in a pressing at low strains (Fig. 3). Since it affects both yield stress and work hardening exponent n, however, there is more to be gained from a study of the more fundamental grain size effects than n alone, particularly where n is not adequately defined and its application not clearly understood.

Control of grain size can influence tensile strength and to a less significant extent elongation in addition to the yield strength and work-hardening characteristics and it is felt that its control could have a more significant effect on autobody press-shop performance than control over some other properties such as anisotropy. This is not to suggest that the considerable efforts expended in studying anisotropy, particularly planar or "through thickness" anisotropy (r value) has been a waste of time.

The importance of anisotropy, its effect on sheet formability and the uses which can be made of controlled directionality have been very thoroughly reviewed by Wilson.[13] Certainly the relationship between anisotropy and deep drawing is clearly established. Nevertheless, it must be emphasized that anisotropy generally only plays a dominant part in pressings where the overall level of strain is well in excess of the yield strain. A confusion generated in industry and the frequent lack of success when indstury has tried to exploit controlled directionality properties can be directly attributable to the fact that large areas of real production stampings are strained to only very low levels and are therefore influenced very strongly by other properties such as yield stress and work hardening properties at low strains.

V. Yield Point Suppression and Strain Ageing

The mechanism of yield point suppression by temper rolling has been fully described by Butler and Wilson.[14] Light temper rolling deforms annealed sheet inhomogeneously resulting in blocks alternatively of deformed and undeformed material in a "herring bone" pattern through the sheet thickness, as observed by Hundy.[15] The detailed nature of the pattern produced is governed by the grain size and yield point properties of the annealed sheet, sheet and roll surface roughness and by the rolling process variable such as speed, lubrication and roll diameter. Industrially temper rolled sheets subjected to 3/4 to 1% rolling reduction normally contain a substantial proportion (approximately 50%) of undeformed material after rolling.

The low yield stress and freedom from yield point elongation of freshly temper rolled sheet is explained in terms of easy nucleation of yielding at the interfaces between deformed and undeformed blocks in the temper rolling pattern. This gives rise to the simultaneous operation of large numbers of dislocations at the very numerous small Lüders bands and in consequence a low dislocation velocity and slow movement of yield fronts results; thus the low value of yield stress can be attributed primarily to a strain rate effect. This part played by strain rate in lowering yield stress has been illustrated by the strong correlation observed between the number of Lüders bands operating, the relevant yield stress and the yield stress of annealed steel in which a similar number of bands had been artificially induced by introducing yield nucleating spots by centre pop marks.[14]

With strain ageing the number of Lüders bands progressively diminishes as dislocation locking develops until in a strongly aged material the numbers are similar as for annealed material. The yield stress in this case is slightly higher than the original annealed value.

Though macroscopic strain rate effects account in a large part for variations in yield stress as explained above, there are sufficient differences in results to require further explanation. It is now felt that these differences can be largely accounted for in terms of residual stress at the interfaces of yielded and unyielded blocks. This is illustrated by Photo. 4 which shows a section through a temper processed sheet, after coating with brittle lacquer and further plastic deformation applied. It is clear that the new strain has nucleated at the yielded/unyielded block interfaces and initially continued through the originally yielded metal before propogating into the originally unyielded areas. This Bauschinger effect must result from the strain induced by temper processing being markedly different from that induced on restraining in tension. Exactly the same observations are made when temper rolled sheet is similarly examined and therefore the same theory applies.

From a practical point of view the strong dependence of yield point phenomena on grain size is particularly significant. It is clear that a fine grained steel requires a somewhat heavier temper reduction to suppress yield points completely.[14] On the other hand a universally applied reduction sufficient to process satisfactorily fine grained material will result in over rolling of coarse grained material. This may well explain why, on occasion, an EDD temper rolled rimming steel with a large grain size has a higher yield strength than a similar steel with a small grain size.

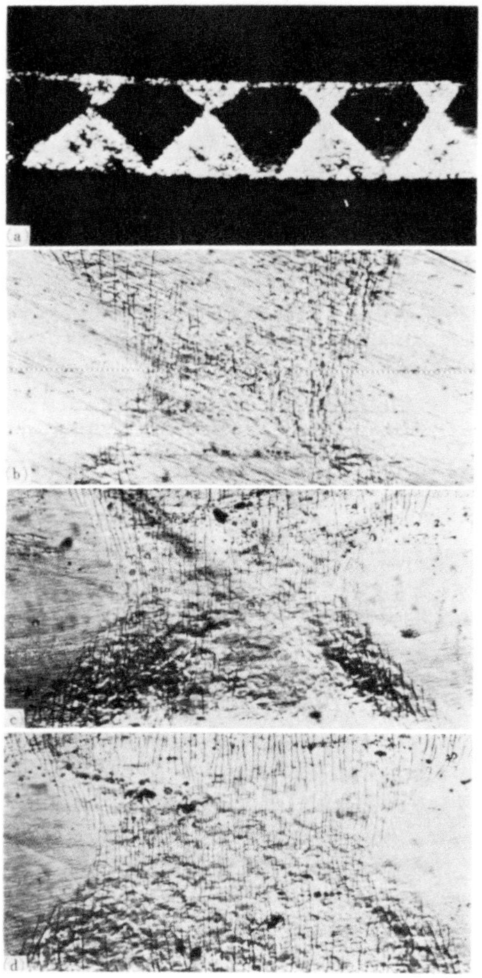

Photo. 4.

(a) Strain pattern observed after lightly temper processing annealed last sheet. Zones of deformed material indicated by roughening of the previously polished edge. ($\times$ 40) ($\times$ 1/3)

(b), (c), and (d) Strain progression observed in processed sheet on restraining in tension, using 'Stresscoat' brittle lacquer thinned approximately 20% ($\times$ 120) ($\times$ 1/3)

Reference has been made above to temper processing as distinct from temper rolling. The temper processing method was devised in an endeavour to achieve one of the ideals of a press shop, that is that sheet for skin panel applications should exhibit freshly temper rolled rimming steel properties just before pressing. The objectives of such a process are to achieve essentially the same pattern of yielded and unyielded blocks as obtained by temper rolling (and therefore the same properties in the sheet) without incurring the problems and expense of a rolling mill located in the press shop. This result can be achieved in several ways and in particular by a controlled bending and unbending process involving rolling the sheet between a hard (steel) and a softer (plastic, aluminium, *etc.*) roll. A mill of this type has many of the characteristics of a roller levelling plant with the addition of a greater control achieved in the bending process by the application of the soft roll. The engineering of such a plant will not be discussed here but the

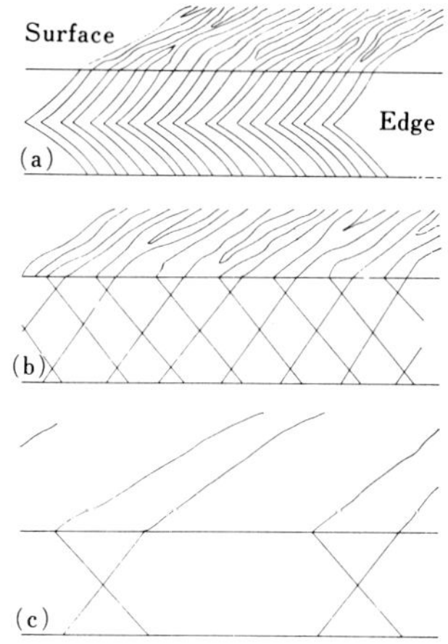

Fig. 4. Schematic representation of:
(a) Temper rolled
(b) Temper processed
(c) Roller levelled sheet structures

metallurgical success can be illustrated by "through thickness" strain patterns obtained. Figure 4 compares the strain patterns obtained from temper rolling, temper processing and roller levelling of annealed sheet. It can be fairly stated that roller levelling can only effectively suppress weak discontinuous yield points typical of those resulting from lightly aged temper rolled steel.

At the other extreme, suitably controlled temper rolling is known to achieve simultaneously an acceptable surface appearance and a very valuable reduction in yield stress (*e.g.* 23 Hbar (15 t.s.i.)$\rightarrow$15.4 Hbar (10 t.s.i.)).

Temper processing achieves the same result as temper rolling with a similar reduction in yield stress. Also in the case of rimming steel the ageing characteristics are very similar. The notable difference is in the shape of the yielded and unyielded zone strain patterns obtained (Fig. 4).

The characteristic "herring bone" strain pattern of temper rolling is replaced by the well known double wedge pattern and neutral axis characteristic of bending operations. It must also be remembered that each wedge represents a yielded area which has been both bent and unbent during processing (this is necessary for sheet to be flat after processing). The part played by the soft roll in this process is to increase the bending moment imposed on the sheet as it passes through the roll gap and in consequence to reduce the pitch between yielded blocks. By suitable control of roll pressure and hardness of the soft roll material a pitch similar to that obtained by temper rolling is achieved.

The practical importance of using temper processing rather than rolling is that the operation can be carried out in a press shop simply and cheaply just before pressing. The sheet thereby retains all the advantages of strain age hardening after press working without incurring the disadvantages due to ageing in storage and transit between the steel mill and the press shop, (sometimes aggravated by the practice of temper rolling while the coil is still warm).

Much work remains to be done from a metallurgical point of view on yield and other properties of the various grades of steel and the effect of subsequent strain ageing. In particular the directional sensitivity of yield stress and return of Lüders strain with strain ageing and the marked sensitivity to mode of pre-strain requires detailed attention. The classic theories for strain ageing in fact successfully explain only the results obtained when steel is pre-strained in tension and then re-strained in the same direction also in simple tension[16] a situation not frequently met in any industrial application. As seen above a fairly successful model also exists for the case of sheet subjected to light temper rolling reductions. However, less is known about the more general situations where prestraining modes and directions differ from those relating to re-straining (*i.e.* most service requirements, *e.g.* a pressed chassis member subjected to severe service loads in a vehicle). Bauschinger effects observed under some of these circumstances and the effect of strain ageing have been discussed by Wilson and Ogram.[17] They consider that the very low yield stresses that they observed can be attributed to stress raisers nucleating yield prematurely. They also account for the protracted ageing times before yield points reappear in terms of a need for long range diffusion of interstitial impurities before Bauschinger effects can be successfully overcome. This is obviously an important area for further study particularly when considering the reliance placed by engineers on the yield stress of a material in designing any structure.

VI. Other Requirements from Sheet Steel in Autobody Service

An acceptable stamping for autobody applications must not only meet the dimensional requirements of the panel so that it fits exactly to its adjacent parts on assembly but it must also meet various engineering and other requirements such as ability to support service loads, weldability, and paintability.

The importance of surface finish and chemistry have been adequately dealt with elsewhere[18),19] and freedom from macroscopic surface blemishes is continually being emphasized.

Weldability, particularly in resistance welding, presents no problems as long as careful control is maintained over carbon levels. To preserve sufficient ductility in the weld it is generally desirable to keep below 0.08%C for normal body gauges. On the other hand a greater emphasis is being placed on freedom from major defects in the sheet such as laminations and large inclusions particularly where components require to be welded and the welds have a direct bearing on the safe performance of the vehicle in service.

The advent of the modern stringent regulations for safe performance of vehicles in service has also resulted in greater consideration being given to the properties of the sheet after pressing. Meeting the new statutary regulations is critically dependent on the amount of energy absorbed by the vehicle on crashing. Obviously the major factors influencing the success in meeting these requirements are engineering considerations such as panel gauge, shape, size, and configuration (i.*e.* whether a beam is bent or straight, open or closed; number and location of welds, *etc.*). On the other hand the mechanical properties, notably the yield and tensile strengths of the pressings must also be considered. A fully aged rimming steel (as long as it is free from embrittlement) obviously offers advantages over stabilised steel particularly when bearing in mind the Bauschinger effects mentioned earlier. As energy absorption in vehicle crash situations involves severe straining of panels advantages of yield stress alone must not be overemphasized and the designer must consider advantages in strength over all the work hardening curve.

The yield stress of a panel is of particular importance when considering resistance to more casual damage such as that resulting from pressing too hard on the panels when slamming boot lids and doors in normal service. This point has been clearly emphasized by a test designed specifically to examine panel dentability (briefly described in the Appendix).

Results from this test indicate that the ageing characteristics of rimming steel offer considerable advantage over non-ageing varieties (Fig. 5) particularly where the level of strain in the panel is low. As low strains are evident in large areas of the majority of skin panels this observation is very important. Surface strains of 0—4% are typical of central areas of such panels and hydraulic bulge samples strained equivalent amounts showed markedly similar strain ageing characteristics to tensile specimens prestrained to the same levels (Figs. 6(a) and (b)).

From the data given in Fig. 5 it can be seen that at equivalent values of low strain, rimming steel has an effective resistance to denting of up to 100% in excess of stabilised steel (or any other non-ageing type of steel).

It can be argued that in a real production situation the difference in dentability will be less than shown in Fig. 5 because a stabilised steel with its lower yield stress will deform more readily and be strained to a

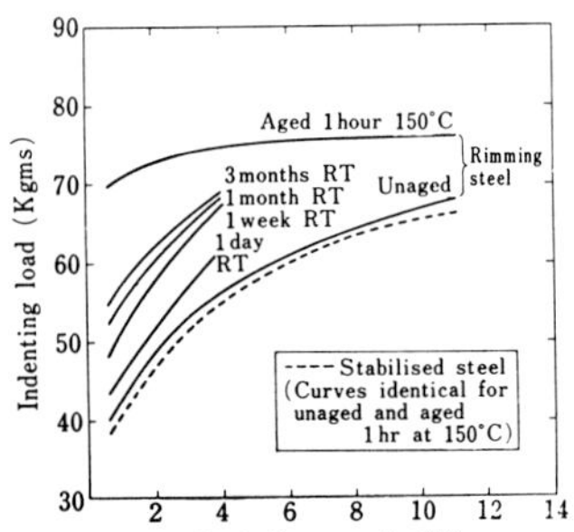

Fig. 5. Effect of ageing on dent resistance of 0.9 mm drawing quality steel (1.3 mm deep dent on 250 mm diameter preformed bulge)

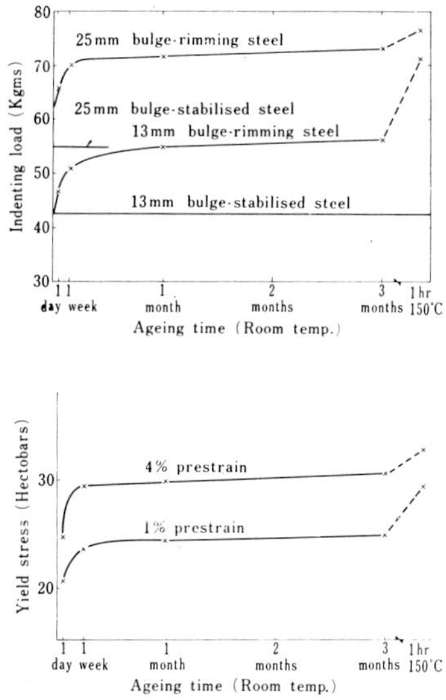

Fig. 6. (a) Effect of ageing on the dent resistance of 0.9 mm (0.035 in.) thick steel panels: 1.3 mm dent. Sheet temper rolled 1% prior to prestraining by hydraulic bulging. Bulge diameter = 250 mm.
(b) Effect of ageing on the yield stress of prestrained tensile test pieces (0.9 mm thick steel-temper rolled 1% prior to tensile prestrain)

higher level over the flat areas of the panel during forming. This would be true if stabilised and rimming steel sheets were pressed one after the other on the same tool at the same tool settings.

However in a production environment the above situation can only be artificially generated for particular experiments. In general a press fitter sets a panel to the same level of strain, *i.e.* just sufficient to set the part to the desired shape, regardless of material grade. He makes use of differences in material properties only by exploiting the ease with which a satisfactory result can be obtained. Thus the comparison of performance of the same strain values is the more realistic.

A more practical illustration of this point is shown in Photo. 5 which highlights the difference between the two materials from a dent resistance point of view. Not only is the dent produced more easily in the stabilised steel panel the damage is considerably more severe. The dent is spread over a much wider area and is therefore more difficult to knock out and reblend to the original curves of the panel in repair.

The difference in dentability results can obviously be attributed to the beneficial effects of strain age hardening. The very remarkable rise in performance (Fig. 5) cannot be explained, however, in terms of the normally observed 25–30% rise in yield stress with strain ageing generally described by the classical theories. In fact, the results more closely follow those of Wilson and Ogram[17] concerning the effects of restraining in directions other than the prestrain direction and clearly Bauschinger effects and the interaction with ageing play a large part.

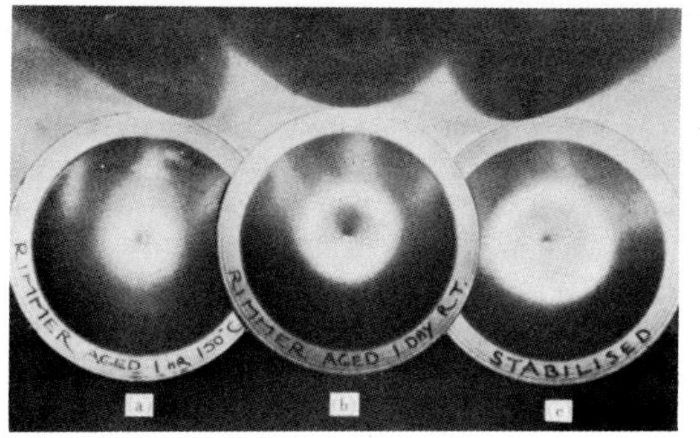

Photo. 5. Rimming and stabilised bulge samples following denting tests
(a) Rimming sheet aged for 60 min at 150°C
(b) Rimming sheet aged for 1 day at R.T.
(c) Stabilised sheet aged for 60 min at 150°C
Bulge diameter = 250 mm.

The rapid rise in dent resistance with room temperature ageing is of special benefit on the vehicle body assembly floor and results in minimising accidental damage and rework costs. In addition the normal painting process for vehicle bodies involves baking at between 150°–200°C, the entire baking cycle in a phosphating and painting systems involving as long as two hours at elevated temperatures. Where rimming steel is used this imparts extra dent resistance and engineering strength to the panel. In this way direct benefit is obtained from a deliberate and careful use of age hardening properties.

VII. Summary

The paper is concerned with the widely different characteristics of autobody skin and interior pressings and the properties required of the sheet both from the pressworking point of view and also taking into account final service requirements.

The major importance of certain properties notably yield point and its effect on the ability of sheet to meet all the above requirements is made abundantly clear. Also the effect of grain size on yield stress is emphasized and it is suggested that there is considerably more benefit to be derived from efforts to achieve a consistent and optimum grain size than from control of other variables including anisotropy.

Apart from the normally desirable features of cleanliness, good surface properties, good work hardening properties throughout the stable plastic working range, high elongation and controlled "through thickness' anisotropy (high $\bar{r}$ and low Δr) the ideal material for press shops would have a low yield stress before pressing and a high value as rapidly as possible thereafter.

In general these requirements can best be achieved with a strain ageing variety of steel by carrying out a process such as temper processing just before pressworking. Temper processing was devised to achieve this result but in the absence of commercially available units (and there are many steel supply problems yet to be solved) these requirements have to be considered within present limitations.

As a general statement skin panels are the easier to form because of their less complex shape though they are particularly sensitive to yield stress both before and after pressing (dentability resistance, *etc.*) For these reasons a rimming steel with its strain age hardening properties is the ideal material as long as it can be pressed before excessive ageing has occurred. On the other hand, if the panel is known to be heavily strained all over during pressing, a properly temper rolled non-ageing (*i.e.* stabilised type) steel may prove acceptable. In this case the panel is probably difficult to form anyway and the more controlled properties of a stabilised type steel may prove to be necessary.

For non skin applications involving severe demands on formability with little call on dentability resistance the use of stabilised type steels can prove advantageous. Examples can also be cited where the higher '*r*' values of these materials add particular benefit.

At first sight there are also many autobody non-skin pressings where formability demands are low and almost any grade of steel appears suitable. However, with the ever increasing pressure for improved vehicle performance and safety, many of these parts cannot be dismissed so lightly and properties such as yield and tensile strength and work hardening properties after pressing, assembly, and painting, must be given serious consideration. Particular attention must be given to Bauschinger effects and any strength advantages offered by strain ageing.

Appendix

Dentability Tests

The effects of strain ageing on dent resistance were studied by means of a denting test applied to panels formed by hyraulic bulging to simulate panel geometry and strain levels normally associated with conditions encountered in skin panel pressings. Strains in the range 0–10% were studied and particular emphasis was placed on panels strained in the the 0–1% range as this range covered the levels of strain observed in the large relatively flat areas of the majority of stampings examined from a wide survey of skin panels for a variety of currently produced vehicles. Hydraulic bulging was used rather than press forming to obtain better control of strain levels in the panel, and better consistency of results by eliminating frictional characteristics from the forming operation.

After forming, the bulges were subjected to a denting test using the apparatus illustrated below (Fig. A). The essential features of this piece of equipment are a round nose Tufnol indentor 50 mm in diameter powered by a pneumatic ram, a load cell working on the proving ring principle and an extensometer measuring deflection *via* a thread attached to the indentor.

This apparatus enabled load/displacement curves to be produced for panels subjected to different prestrain and strain ageing histories. Though it was possible to produce such curves under continuous forming conditions, it was discovered that the contribution of elastic deflection in the panel during testing was playing too large a part in the test results and was an undesirable element of the test. The procedure was therefore modified to enable indenting load to be compared with dent size after the applied load had been reduced to zero. This was achieved in the test by incrementally increasing load, reducing this to zero to allow elastic recovery to take place and then measuring displacement. Figure B shows typical results from this test.

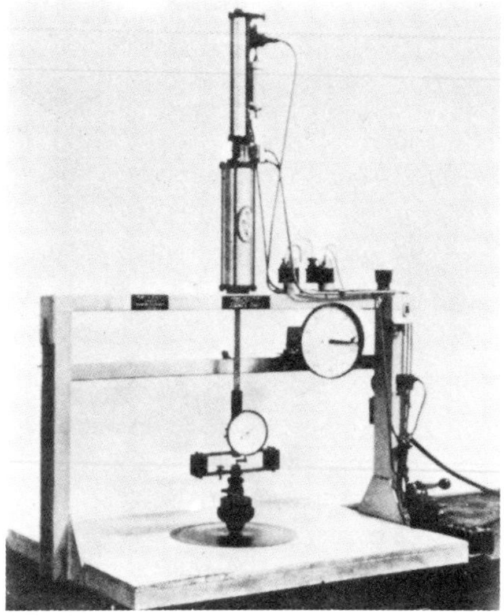

Fig. A. Indentometer used in dentability resistance tests

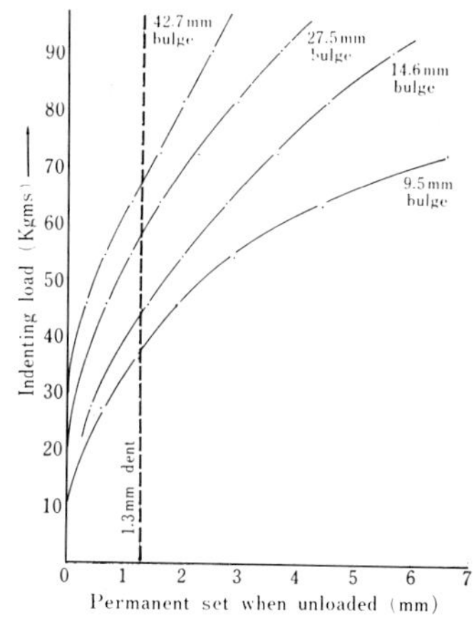

Fig. B. Effect of bulge height on dent resistance of 0.9 mm stabilised sheet

To compare results from panels with different strain histories it was arbitrarily decided to compare denting loads necessary to produce a dent with an unloaded permanent set of 1.3 mm (0.050″). Other values could be selected but this value was sufficient in practice to be reproducably accurate and it was felt adequately reflected the normal minor denting damage that is considered objectionable on the modern motor car.

The curves in Fig. 5 of the paper were determined using the above test procedure.

REFERENCES

1) R. D. Butler and J. F. Wallace: *I.S.I. Special Report, No.*

79, (1963), 131.
2) A. J. F. Honeyman: *Sheet Metal Ind.*, **34** (1957), 51.
3) G. Murray: *Sheet Metal Ind.*, **35** (1958), 657.
4) S. P. Keeler: *S.A.E. Report*, No. 650535 (1965). Also Machinery, (May 1968).
5) G. M. Goodwin: *S.A.E. Report*, No. 680093, (1968).
6) I. C. Brookes: Private communication.
7) G. G. Moore and J. F. Wallace: *J. Inst. Metals*, **93** (1964–65), 33.
8) R. Hill: The Mathematical Theory of Plasticity, Oxford Univ. Press (1950).
9) I. L. Dillamore: *Trans. A.S.M.*, **58** (1965), 150.
10) R. D. Butler: *Sheet Metal Ind.*, Sept. (1964), 705.
11) N. J. Petch: *J.I.S.I.* **174** (1953), 25.
12) J. Winlock and R. Leiter: *Trans. A.S.M.*, **25** (1937), 163.
13) D. V. Wilson: *Metals & Materials, Metallurgical Reviews*, No. 139, **3** (1969), 175.
14) R. D. Butler and D. V. Wilson: *J.I.S.I.*, **201** (1963), 16.
15) B. B. Hundy: *J.I.S.I.*, **181** (1955), 313.
16) D. V. Wilson and B. Russell: *Acta Met.*, **8** (1960), 468.
17) D. V. Wilson and G. R. Ogram: *J.I.S.I.*, Sept. (1968), 911.
18) R. D. Butler: I.S.I. preprint 125, Sept. (1969), 216.
19) D. H. Lloyd: *Sheet Metal Ind.*, April (1966), 307.

TABLE A

LENGTH—INCHES AND MILLIMETERS—EQUIVALENTS OF DECIMAL AND BINARY FRACTIONS OF AN INCH IN MILLIMETERS

From 1/64 to 1 Inch

½'s	¼'s	8ths	16ths	32ds	64ths	Millimeters	Decimals of an inch	Inch	½'s	¼'s	8ths	16ths	32ds	64ths	Millimeters	Decimals of an inch
					1	= 0.397	0.015625							33	=13.097	0.515625
				1	2	= .794	.03125						17	34	=13.494	.53125
					3	= 1.191	.046875							35	=13.891	.546875
			1	2	4	= 1.588	.0625					9	18	36	=14.288	.5625
					5	= 1.984	.078125							37	=14.684	.578125
				3	6	= 2.381	.09375						19	38	=15.081	.59375
					7	= 2.778	.109375							39	=15.478	.609375
		1	2	4	8	= 3.175	.1250				5	10	20	40	=15.875	.62500
					9	= 3.572	.140625							41	=16.272	.640625
				5	10	= 3.969	.15625						21	42	=16.669	.65625
					11	= 4.366	.171875							43	=17.066	.671875
			3	6	12	= 4.762	.1875					11	22	44	=17.462	.6875
					13	= 5.159	.203125							45	=17.859	.703125
				7	14	= 5.556	.21875						23	46	=18.256	.71875
					15	= 5.953	.234375							47	=18.653	.734375
	1	2	4	8	16	= 6.350	.2500			3	6	12	24	48	=19.050	.75
					17	= 6.747	.265625							49	=19.447	.765625
				9	18	= 7.144	.28125						25	50	=19.844	.78125
					19	= 7.541	.296875							51	=20.241	.796875
			5	10	20	= 7.938	.3125					13	26	52	=20.638	.8125
					21	= 8.334	.328125							53	=21.034	.828125
				11	22	= 8.731	.34375						27	54	=21.431	.84375
					23	= 9.128	.359375							55	=21.828	.859375
		3	6	12	24	= 9.525	.3750				7	14	28	56	=22.225	.8750
					25	= 9.922	.390625							57	=22.622	.890625
				13	26	=10.319	.40625						29	58	=23.019	.90625
					27	=10.716	.421875							59	=23.416	.921875
			7	14	28	=11.112	.4375					15	30	60	=23.812	.9375
					29	=11.509	.453125							61	=24.209	.953125
				15	30	=11.906	.46875						31	62	=24.606	.96875
					31	=12.303	.484375							63	=25.003	.984375
1	2	4	8	16	32	=12.700	.5000	1	2	4	8	16	32	64	=25.400	1.0000

Source: ASTM

Effects of Processing Conditions on the Deep Drawability of Cold Rolled Steel Sheet

By Kazuo MATSUDO* and Kazuhide NAKAOKA**

1. Introduction

In 1950, W.T. Lankford et al. found out the possibility of judging the adaptability of cold rolled steel sheet to press forming into front fenders, with the use of the work hardening exponent, n, and plastic strain ratio, r.(1) Later in 1958, R.S. Burns et al. clarified the close relationship between r-value and the crystallographic orientation.(2) In 1960-1962, R.L. Whiteley et al. revealed that the normal anisotropy ($\bar{r}$-value) represented the pure deep-drawability: a high $\bar{r}$-value corresponds to more {111} in parallel with the sheet plane and less {100}, which mean a better deep-drawability; the planar anisotropy (Δr-value) indicates the earing behavior: earing occurs at 0° and 90° to the rolling direction at $\Delta r > 0$, and at 45° to the rolling direction at $\Delta r < 0$; a larger $|\Delta r|$ value leads to the occurrence of larger earing.(3) They investigated for the first time the effects of the hot rolling conditions and cold rolling conditions on the deep-drawability of cold rolled steel sheet.(4) Many researchers and metallurgists have since studied the effect of processing conditions on the $\bar{r}$-value and crystallographic orientation. Although research on this subject seems to have almost reached its goal now, detailed review clarifies some disagreements in results between authors. The present authors will therefore summarize the results of research so far obtained while comparing the respective achievements.

2. Effect of hot rolling conditions

The deep-drawability of cold rolled steel sheet depends on the finishing and coiling temperatures in most cases.

2.1 Effect of finishing temperature

R.L. Whiteley found a lower finishing temperature to lead to less {111} and more {100} in cold rolled sheet, thus resulting in a lower $\bar{r}$-value. He attributes this to the fact that, when finishing at a high temperature (hot rolling completed at a temperature higher than Ar_3 point), the texture of hot rolled sheet becomes almost at random in the $\gamma \rightarrow \alpha$ transformation after the completion of hot rolling, whereas, when finishing at a low temperature (hot rolling completed at a temperature lower than Ar_3 point), the texture consists of more {110} and less {111}.(4) Results are represented in Fig. 1.

In contrast to this, H. Takechi considers that, because the {100} <011> orientation prevailing at the thickness center of hot rolled sheet is

---

* Fukuyama Research Laboratories, Technical Research Center, Nippon Kokan K.K.

** Technical Research Center, Steel Products Sec., Nippon Kokan K.K.

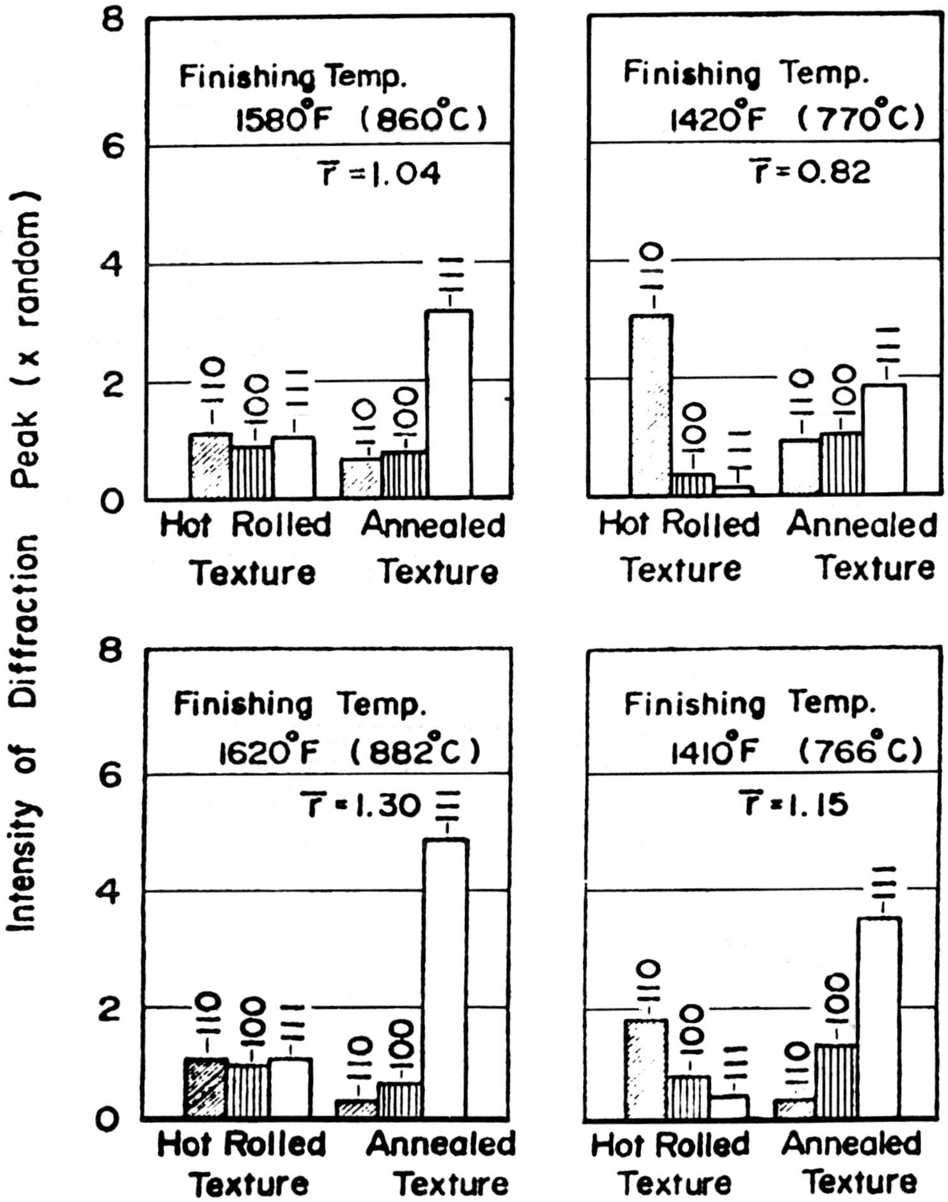

Fig. 1. Effect of finishing temperature on the hot-rolled texture and subsequent cold-rolled and annealed texture. (upper: rimmed steel, lower: aluminum killed steel)

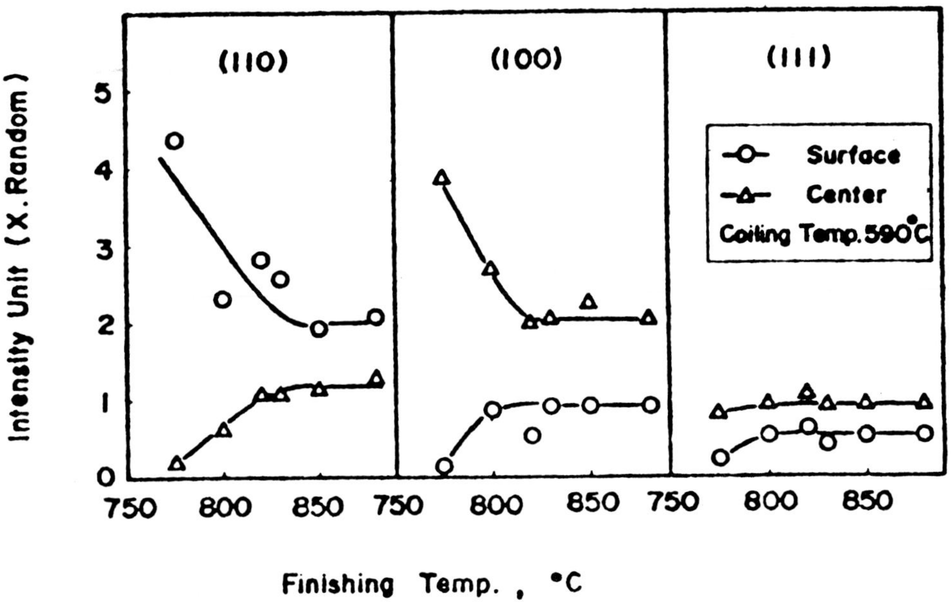

Fig. 2. Effect of finishing temperature on the hot rolled texture of the surface and center. (rimmed steel, C: 0.04, Mn: 0.25, P: 0.011, S: 0.021, N: 0.0020, O: 0.057)

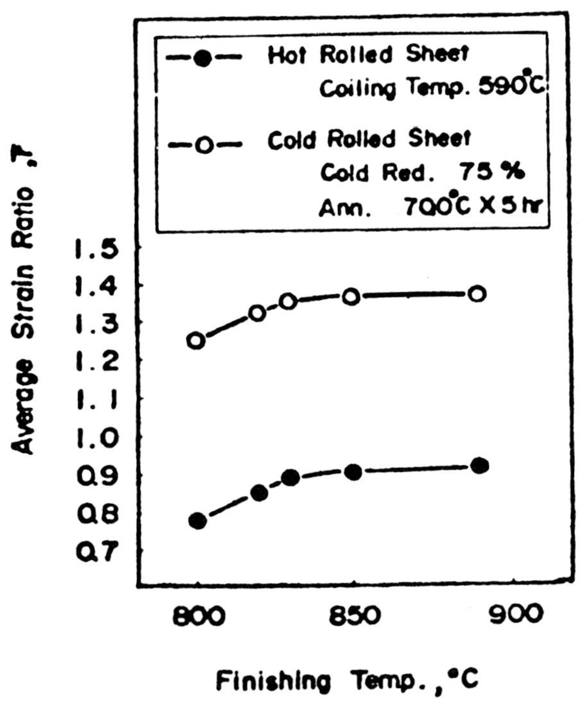

Fig. 3. Effect of finishing temperature on $\bar{r}$-value of the hot-rolled sheet and subsequent cold-rolled and annealed sheet. (the same rimmed steel as that shown in Fig. 2)

hard to break by cold rolling and annealing, cold rolled steel sheet tends to have an inferior deep-drawability.[5] Results of experiment carried out by the authors on low carbon rimmed steel are indicated in Figs. 2 and 3.

According to these results, the texture of hot rolled sheet has more {110} and less {111} in the surface layer, more {100} in the center layer, and shows a lower $\bar{r}$-value accordingly as the finishing temperature decreases from 830 C. At a finishing temperature of higher than 830 C, no change is observed in the texture, and hence no change takes place in the $\bar{r}$-value. The $\bar{r}$-value of cold rolled sheet very well corresponds to that of hot rolled sheet. The ferrite structure of hot rolled sheet is of coarse grain only in the rim layer at a finishing temperature of 830 C, and is of uniform grain at higher temperatures; coarse grains cover also the thickness center layer as this temperature becomes lower, suggesting that the temperature of 830 C substantially agrees with the Ar_3 point.

H. Yoshida et al.[6] and H. Kubotera et al.[7] observe that the $\bar{r}$-value of cold rolled steel sheet reaches its maximum when hot rolling is finished at a temperature immediately above Ar_3 point. Typical results are illustrated in Fig. 4. Effect of the hot rolling finishing temperature on the $\bar{r}$-value and texture of cold rolled steel sheet depends not only upon the $\bar{r}$-value and texture of hot rolled sheet, but also upon the ferrite grain diameter: a smaller ferrite grain diameter has been ascertained to correspond to more {111} and a higher $\bar{r}$-value of cold rolled sheet.[8][9][10] It is therefore well conceivable that results as shown in Fig. 4 should be obtained, depending on the composition, the kind, quantity, and size of impurities.

According to S. Nagashima, the fact that the texture of hot rolled sheet finished at a low temperature contains more {100} in its thickness center layer is considered attributable to the presence of deformed ferrite remaining during coil cooling.[11] According to the present authors, however, difference in texture between surface and center layers is considered to be due to the difference in crystal rotation around the transverse direction caused by the inhomogeneous deformation in hot rolling between these layers. D.S. Hoddinott has reported a similar idea in his paper.[12]

2.2 Effect of rolling temperature

R.L. Whiteley et al. observes that, except for cases of hot rolling at extremely low temperatures, the texture of hot rolled sheet shows almost no change with the coiling temperature. The relation between the coiling temperature and the $\bar{r}$-value of cold rolled sheet, as they observed, is represented in Fig. 5.[4] According to this, the coiling temperature has almost no effect on the $\bar{r}$-value of cold rolled sheet in low carbon rimmed steel. In low carbon Al-killed steel, a lower coiling temperature leads to a higher $\bar{r}$-value of cold rolled sheet. The $\bar{r}$-value is higher in Steel B (0.036% Al) than in Steel A (0.07% Al): this suggests the remarkable effect of the amount of Al. In Al-killed steel, {111} increases and {100} decreases by the AlN precipitation during annealing. The $\bar{r}$-value therefore becomes higher in the state of cold rolled sheet if coiling is conducted at a low temperature ($<$ 600 C) so as to permit presence of Al and N in soluble phase.

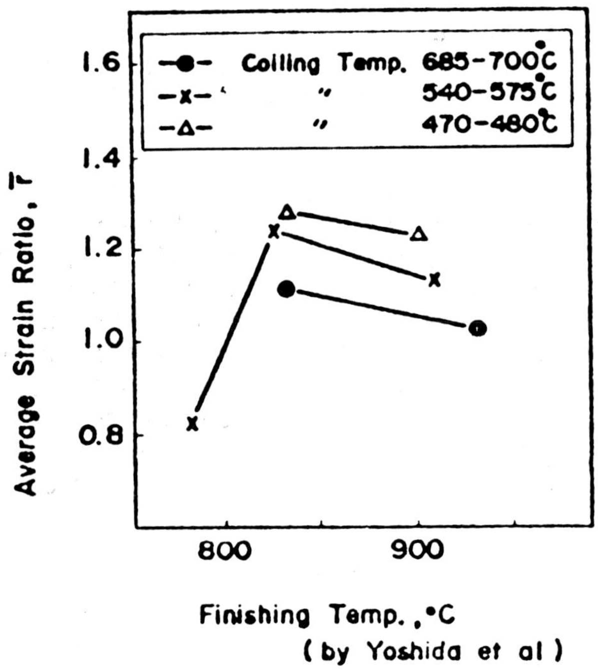

Fig. 4. Effect of finishing temperature on r̄-value of rimmed steel

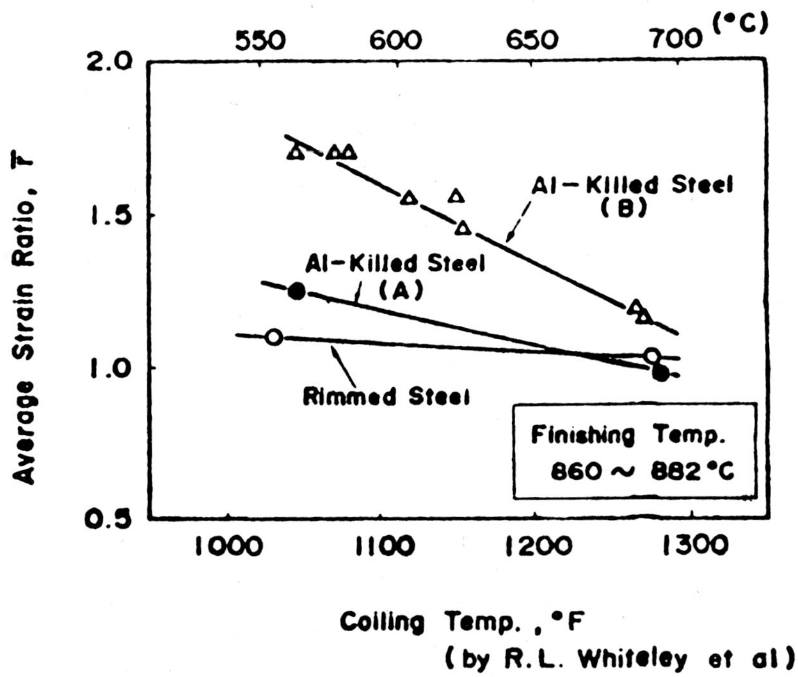

Fig. 5. Effect of coiling temperature on r̄-value of rimmed and aluminum-killed steels. (Aluminum content A steel: 0.07, B steel: 0.36)

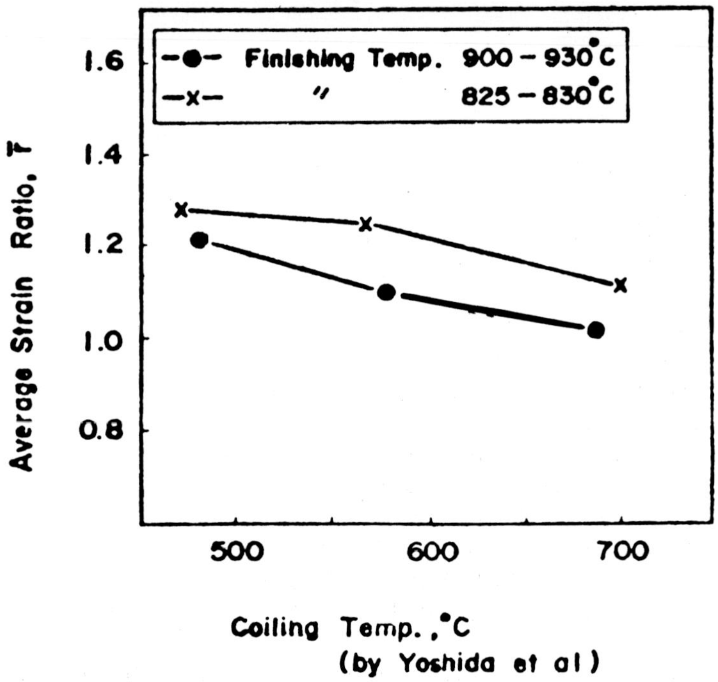

Fig. 6. Effect of coiling temperature on $\bar{r}$-value of rimmed steel.

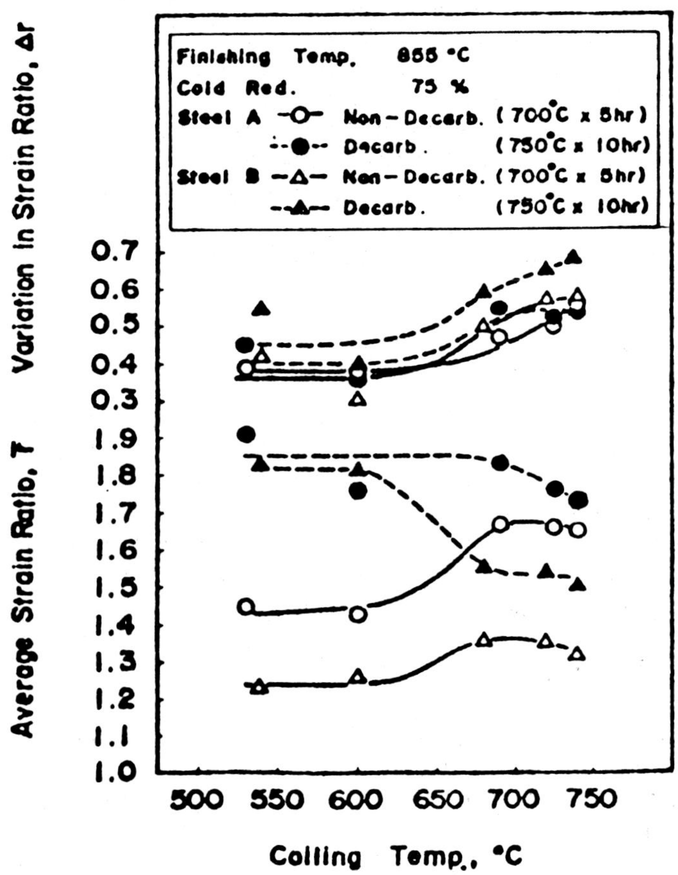

Fig. 7. Effect of coiling temperature on $\bar{r}$-value and Δr-value of rimmed steels (steel A, C: 0.04, Mn: 0.27, P: 0.011, S: 0.023, N: 0.0017, O: 0.051, steel B, C: 0.07, Mn: 0.35, P: 0.012, S: 0.024, N: 0.0016, O: 0.035)

Yoshida et al. have reported that, in the case of low carbon rimmed steel, a lower coiling temperature evidently resulted in a higher $\bar{r}$-value of cold rolled sheet.[6] This is shown in Fig. 6. Also in this case, no change in texture caused by the coiling temperature is observed.

Data by the present authors include those showing quite different tendencies for low carbon rimmed steel, which are indicated in Fig. 7. As is clear from this figure, the $\bar{r}$-value of cold rolled sheet sharply rises at a coiling temperature of 600-680°C in non-decarburized annealing at 700° C, and slightly decreases thereafter. No change is observed at temperatures below 600°C. In decarburized annealing at 750°C, the tendency observed is substantially contrary to that in the non-decarburized annealing at 700°C. Under annealing conditions in all cases, the Δr-value tends to show no change at coiling temperatures below 600°C, rise at 600-680°C and rise further accordingly as the coiling temperature becomes higher.

The rise of $\bar{r}$-value of cold rolled sheet at a coiling temperature of 600-680°C in non-decarburized annealing is attributable chiefly to the aggregating of cementite in hot rolled sheet and the resultant decrease in the amount of soluble carbon in recrystallization. The difference in behavior between decarburized and non-decarburized annealing is caused by the difference in ferrite grain growth during annealing: growth of ferrite grains is far easier in low-temperature coiled materials than in those coiled at a high temperature. The case where the material is finished at a low temperature in hot rolling and coiled at a high temperature would provide the most unfavorable conditions for the deep-drawability of cold rolled sheet. In fact, when a slab from the same heat as in Fig. 3 was hot rolled at a low finishing temperature of 800°C and coiled at 650°C, the resultant cold rolled sheet had an extremely low $\bar{r}$-value of 0.90. This is due to the fact that ferrite grains grow considerably coarser than in the case of finishing at 800°C and coiling at 590 C, thus resulting in a further unfavorable texture.

3. Effect of cold rolling conditions

The deep-drawability of cold rolled sheet depends considerably upon the cold reduction and the rolling lubricant ability. This is illustrated in Fig. 8. The $\bar{r}$-value, rising with the increase in the cold reduction, reaches its maximum and then decreases. At a cold reduction kept constant, the $\bar{r}$-value is higher at a better rolling lubricant ability. The $\bar{r}$-value shows its maximum at a higher cold reduction when the lubricating ability in rolling is better: it is about 50% for non-lubrication rolling, about 60% for water-lubrication rolling, and about 80% for water-soluble oil lubrication rolling.

In oil lubrication rolling, the crystallographic orientation throughout the thickness of the sheet is substantially homogeneous, and the change in the texture with the cold reduction very well agree with changes in $\bar{r}$-value as shown in Fig. 9.[4] Non-lubrication rolling results in an inhomogeneous deformation: even after annealing, the surface layer contains more $\{110\}$ and less $\{111\}$ and $\{100\}$ than in the center layer. The center layer is identical with that in the oil lubrication rolling. This may be due to the difference in crystal rotation between surface and center layers as a result of too low lubricating ability in rolling. Similar results have been reported by J.F. Held.[13]

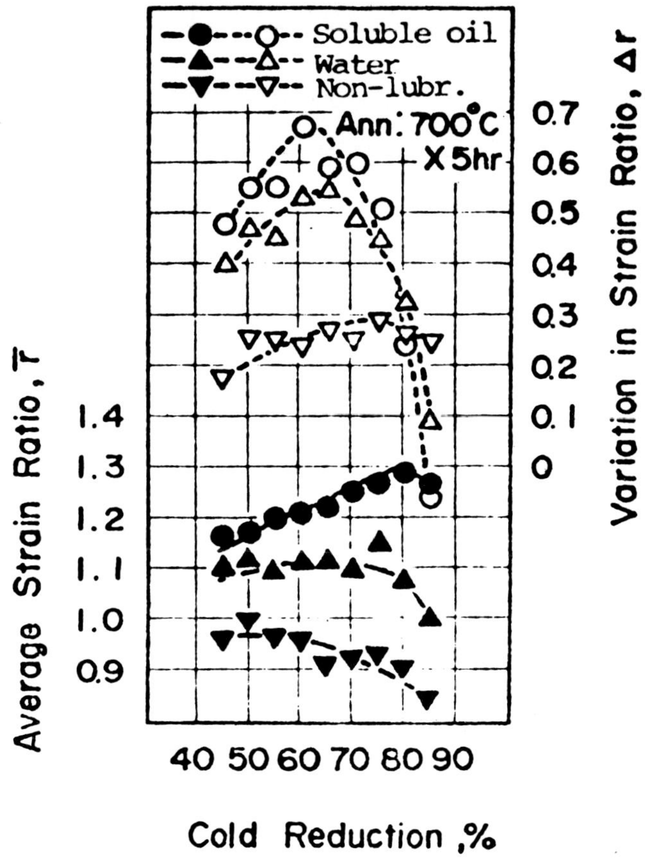

Fig. 8. Effect of lubricating conditions and cold reduction on $\bar{r}$-value and Δr-value of rimmed steel. (C: 0.05, Mn: 0.32, P: 0.012, S: 0.023, N: 0.0017, O: 0.054)

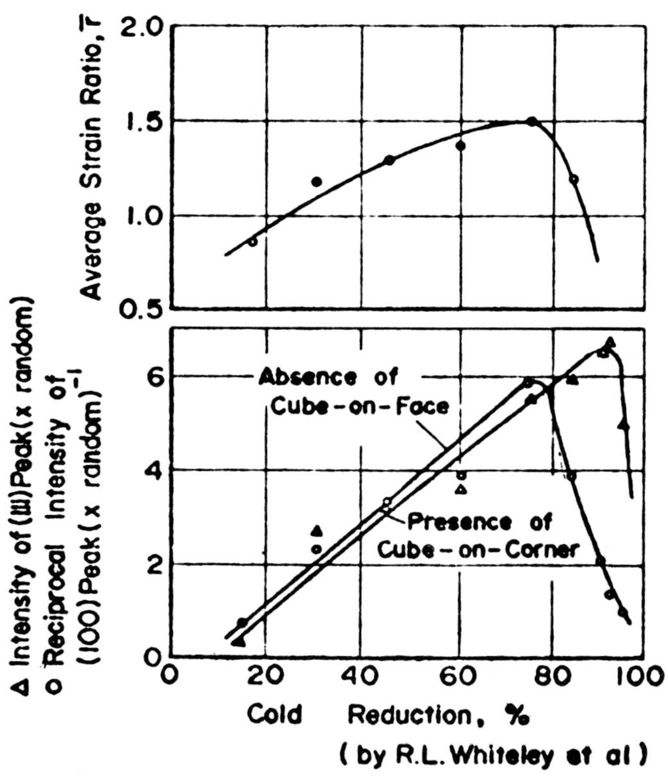

Fig. 9. Relationship between the changes in $\bar{r}$-value with cold reduction and the changes in the annealed texture as a function of cold reduction of aluminum-killed steel.

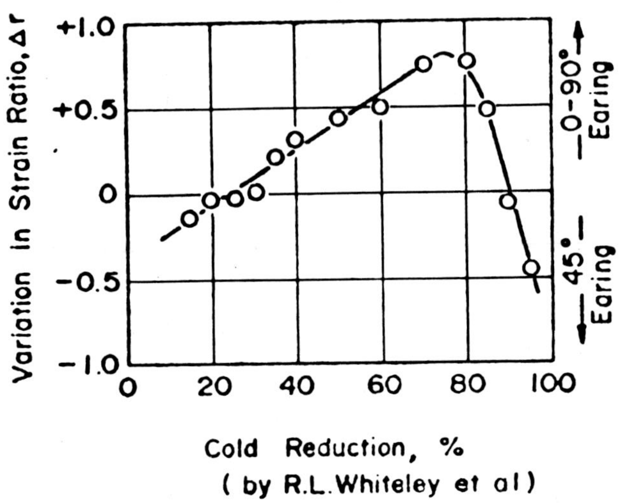

Fig.10. Effect of cold reduction on Δr-value of rimmed steel.

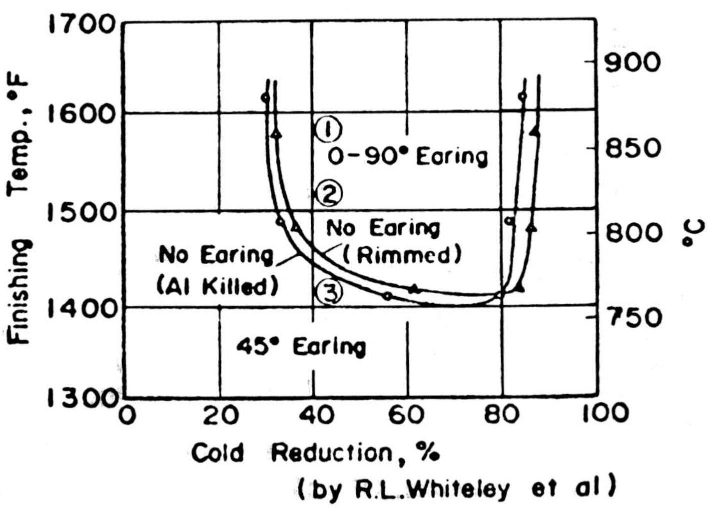

Fig.11. Range of practice to produce non-earing properties for rimmed and aluminum-killed steels.

The cold reduction corresponding to the maximum $\bar{r}$-value depends also on the carbon content: a lower carbon content causes this cold reduction to move on the higher reduction side.[14] It varies also between rimmed and Al-killed steels. According to R.L. Whiteley et al., it is about 60% for low carbon rimmed steel, and is about 75% for low carbon Al-killed steel.[4] In view of its considerable variations with the amount of Al and N and annealing conditions, as is revealed in the authors' experience, differences in results between authors would be conceivable. However, the present authors have constantly observed a cold reduction of about 80% for low carbon rimmed steel, unlike that of about 60% reported by R.L. Whiteley et al. Results of many other authors lie in the range between about 60 and 80%. Causes of this difference are not as yet known at present.

Fig. 10 shows the relation between the cold reduction and Δr-value.[4] The Δr-value is negative at low cold reductions, and accordingly as the cold reduction rises, it is raised through $\Delta r = 0$ up to its positive maximum at 70-80%, lowering again, at further higher cold reductions, through $\Delta r = 0$ until it becomes negative again. This means that there are two cold reductions at high and low levels at which no earing appears in deep-drawn articles. This cold reduction depends also on the finishing and coiling temperatures. Effect of the finishing temperature is indicated in Fig. 11.[4] The cold reduction corresponding to the positive maximum of Δr-value and the vertical position of Δr-value curve vary with various conditions. As an example, changes caused by rolling lubrication conditions are represented in Fig. 8. As is clear from Fig. 8, the cold reduction corresponding to the maximum $\bar{r}$-value does not agree with that corresponding to $\Delta r = 0$. At a cold reduction of about 85%, $\Delta r = 0$, with a considerably high $\bar{r}$-value. Under the effect of the rolling mill capacity and other restricting conditions of the equipment, however, hot rolled sheets of 2.3-3.2 mm thick are actually often employed in general for the manufacture of 0.8 mm (cold reduction: 65-75%) thick cold rolled sheets, for example.

4. Effect of annealing conditions

Heating rate, annealing temperature, soaking time and annealing atmosphere also exert influence on the deep-drawability of cold rolled steel sheet.

4.1 Effect of heating rate

Relation between the heating rate and $\bar{r}$-value is shown in Fig. 12.[15] According to this figure, the $\bar{r}$-value almost linearly decreases with the increase in the heating rate in low carbon rimmed steel. In low carbon Al-killed steel, the tendency varies with the amount of Al: at a low Al content, the $\bar{r}$-value decreases monotonously as in the case of low carbon rimmed steel (but at a higher rate), but accordingly as the Al content becomes higher, a peak $\bar{r}$-value appears, and the heating rate corresponding to the peak moves toward the higher heating rate side with the increase in Al content. This implies that, when precipitation of AlN and nucleation of recrystallization take place in good timing in Al-killed steel, much $\{111\}$ are produced and $\bar{r}$-value becomes higher.

A recently released paper has reported that the heating rate corresponding to the maximum $\bar{r}$-value varied not only with the Al content but also with the

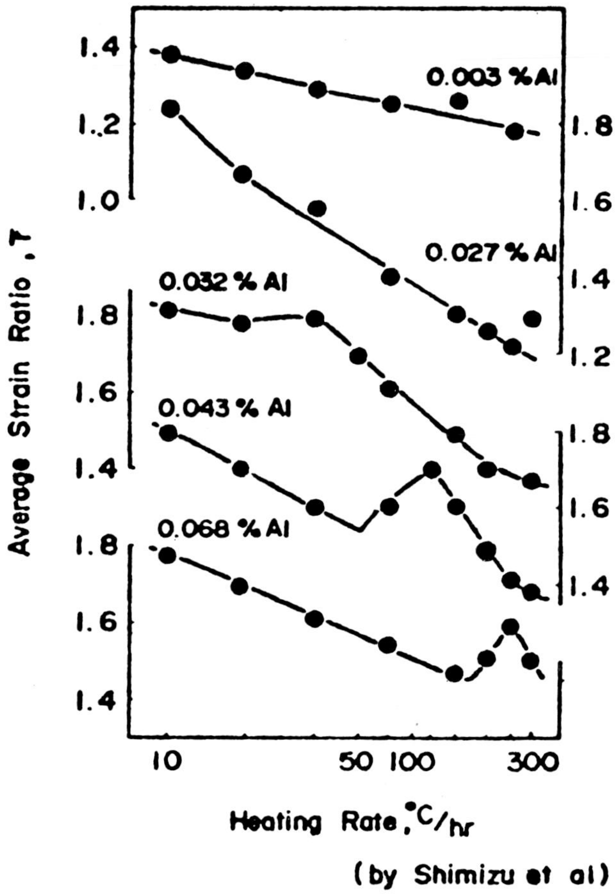

Fig.12. Effect of heating rate and aluminum content on $\bar{r}$-value.

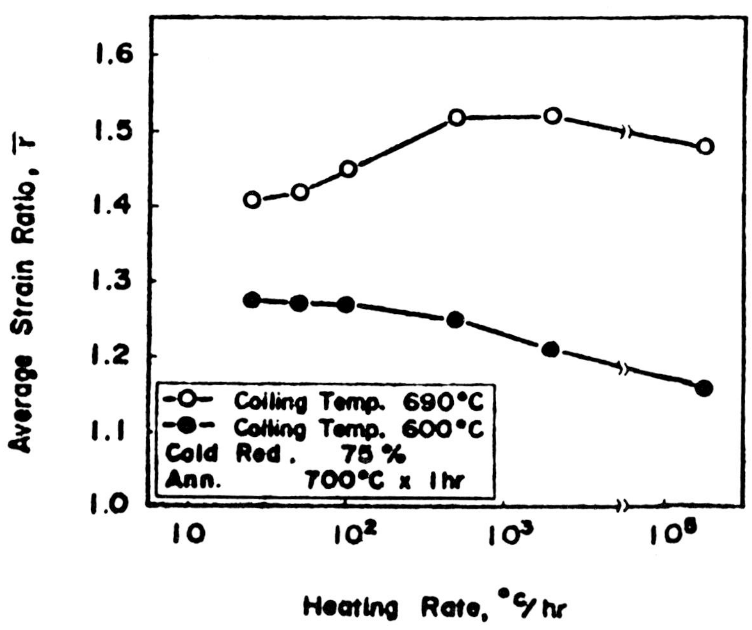

Fig.13. Effect of heating rate and coiling temperature on $\bar{r}$-value of rimmed steel.

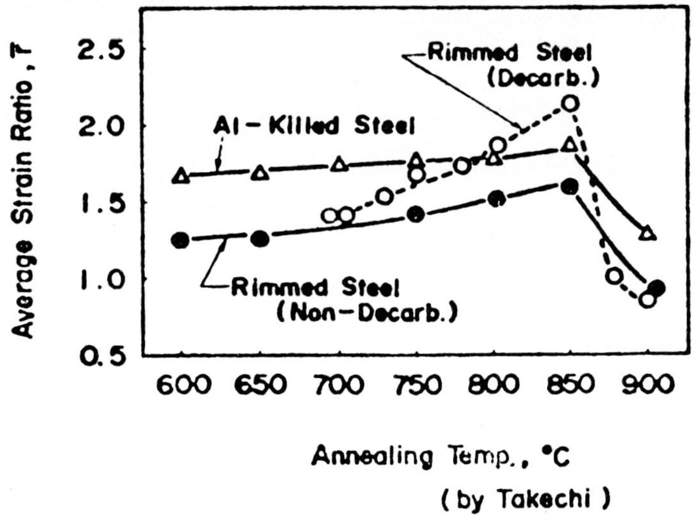

Fig.14. Effect of annealing temperature on $\bar{r}$-value of rimmed and aluminum-killed steels.

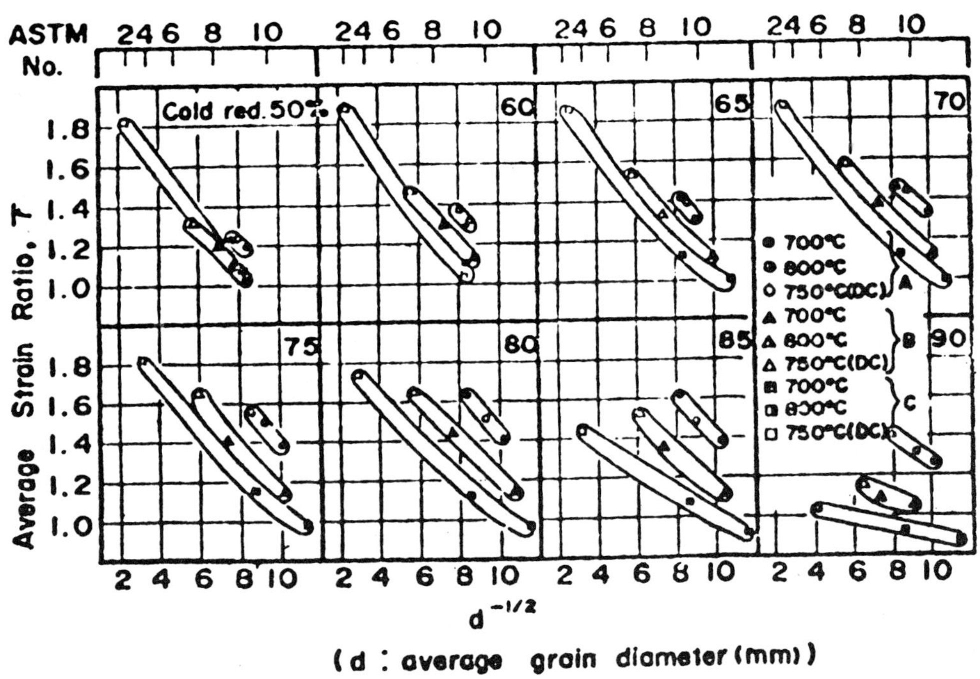

Fig.15. Increase in $\bar{r}$-value by ferrite grain growth after recrystallization of rimmed steels of various cold reductions. (steel A, C: 0.05, Mn: 0.28, P: 0.016, S: 0.022, N: 0.0016, O: 0.068; steel B, C: 0.09, Mn: 0.45, P: 0.012, S: 0.026, N: 0.0022, O: 0.034; steel C, C: 0.17, Mn: 0.53, P: 0.015, S: 0.020, N: 0.0022, O: 0.021)

N content, Mn content and cold reduction.[16]

For cases of coiling at a high temperature, however, quite a different tendency is observed: $\bar{r}$-value increases with the increase in the heating rate, and decreases after once reaching its maximum. Results for low carbon rimmed steel are represented in Fig. 13.

When raising the heating rate from 25°C/hr to 500-2,000°C/hr, the decrease in the amount of soluble carbon in recrystallization caused by the delay in dissolution in heating of coarse cementite (positive effect) takes place at the same time as the increase in the nucleation frequency (negative effect: only this effect is present in case of low-temperature coiling), and the $\bar{r}$-value is raised under the effect of the former being exceeding the latter. At a higher heating rate, at which the latter effect should be more apparent, the saturation of the former effect is considered to bring about decrease in $\bar{r}$-value. Manufacture of drawing quality cold rolled sheet is actually practiced in continuous annealing furnace by the use of this phenomenon.[17][18]

4.2. Effect of annealing temperature, soaking time and annealing atmosphere

Fig. 14 shows changes in $\bar{r}$-value with the annealing temperature.[5] Rise of the annealing temperature leads to the increase in $\bar{r}$-value, which however decreases suddenly at temperatures above 880°C. This is attributable to the random texture caused by transformation $\gamma \rightarrow \alpha$. Increase in $\bar{r}$-value is the result of increase of $\{111\}$ and decrease of $\{100\}$, since the growth of $\{111\}$ grains is achieved by the expense of $\{100\}$ grains.

Fig. 15 indicates results of investigations of the relation between the ferrite grain growth and the rise of r-value for rimmed steel containing 0.05-0.17% carbon at various cold reductions. An almost linear correlation is observed between the two factors, which can be expressed as follows:[14]

$$\bar{r} = \bar{r}_0 + k\ (d^{-\frac{1}{2}} - d_0^{-\frac{1}{2}})$$

where, k: cold reduction kept almost constant at 50-80% ($k < 0$)
$\bar{r}_0$: value at the completion of recrystallization
d, d_0: average ferrite grain diameter (mm) at the completion of annealing and recrystallization, respectively.

This suggests that the $\bar{r}$-value depends upon the $\bar{r}_0$-value at the completion of recrystallization and the subsequent growth of ferrite grains. This relation has been found to be applicable not only to the rimmed steel, but also to the low carbon Al-killed steel as well as to extra-low carbon Ti-added steel.[19]

A longer soaking time results in a more remarkable growth of ferrite grains, thus leading to a higher $\bar{r}$-value. The effect of annealing in the decarburizing atmosphere lies in the acceleration of ferrite grain growth.[6] This is evident also from Fig. 15. The maximum ferrite grain size so as not to cause orange peel in stamping is at about ASTM No. 6. This forms therefore the limit for the increase in r-value caused by annealing.

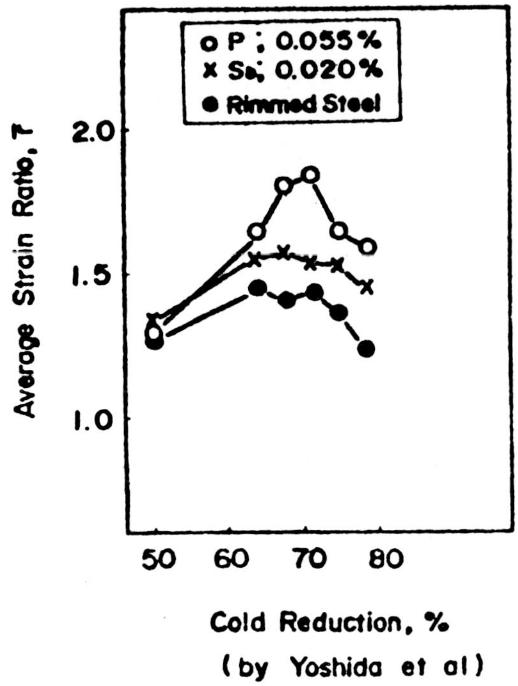

Fig.16. Effect of additional elements and cold reduction on r̄-value of rimmed steel. (Decarburized)

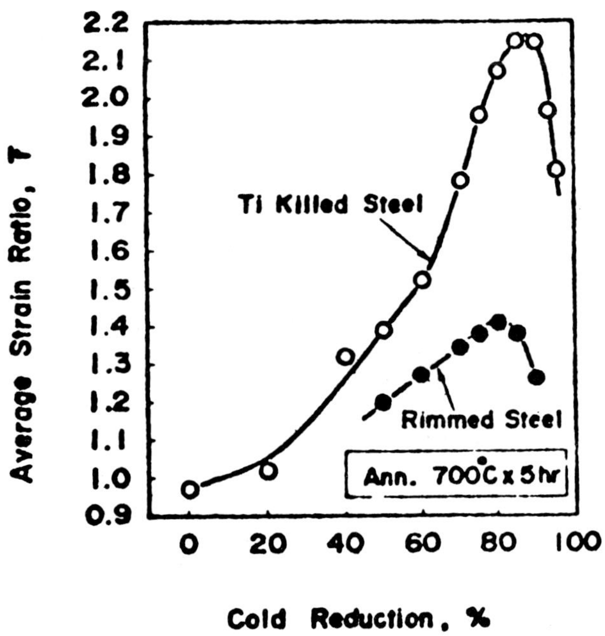

Fig.17. Effect of cold reduction on r̄-value of rimmed and titanium-killed steels. (rimmed steel, C: 0.046, Mn: 0.29, P: 0.008, S: 0.019, N: 0.0015, O: 0.049; titanium-killed steel, C: 0.006, Mn: 0.14, P: 0.017, S: 0.020, N: 0.0032, O: 0.004, Al: 0.046, Ti: 0.15)

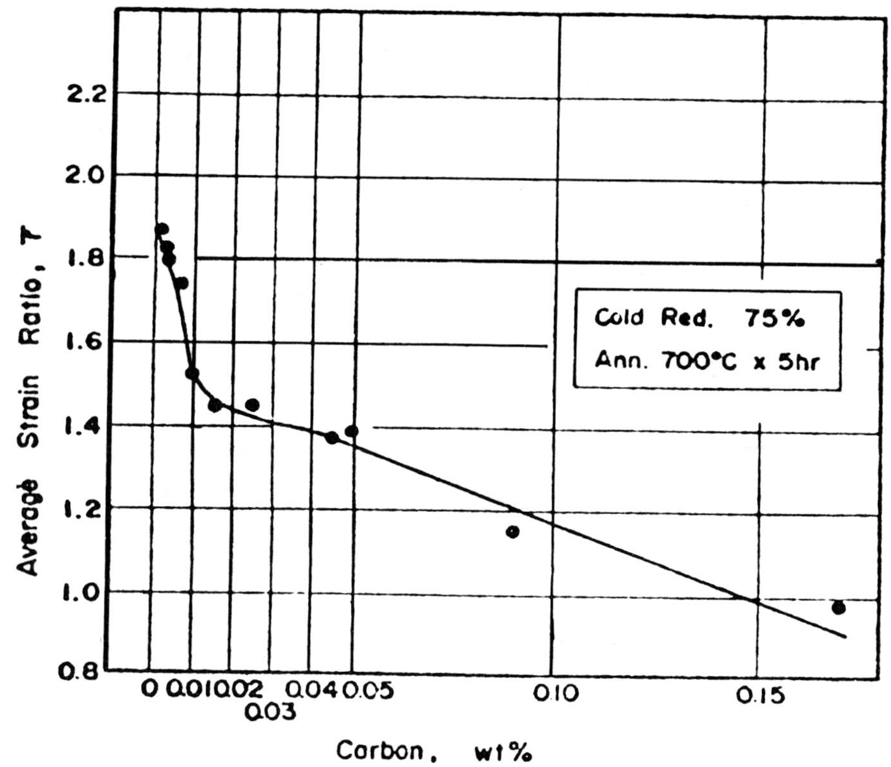

Fig.18. Effect of carbon content on r̄-value of rimmed steel.

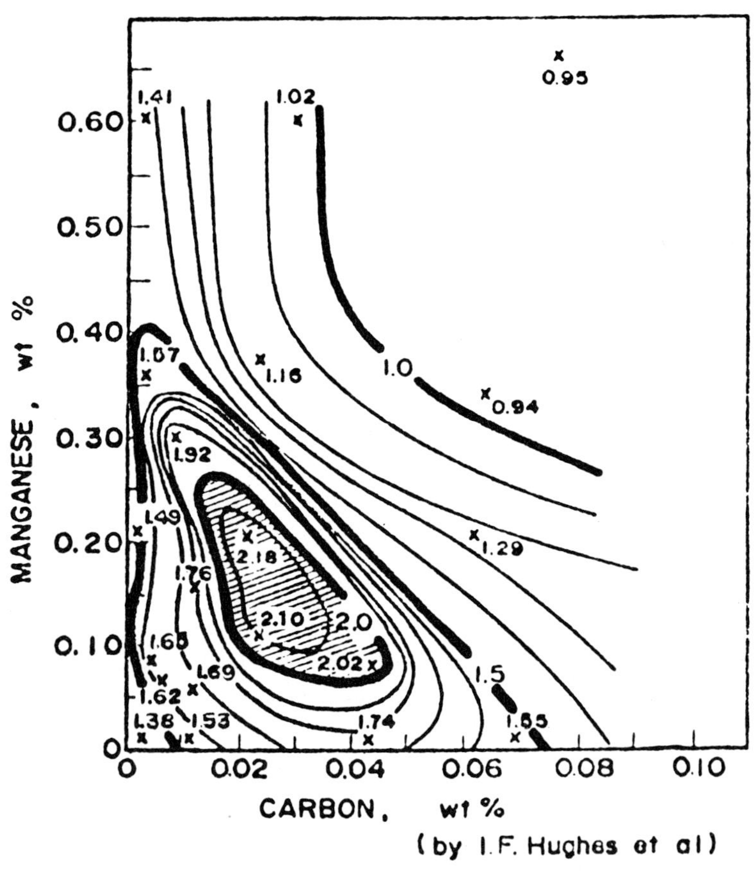

Fig.19. Iso-r̄-value curves showing the effect of manganese and carbon contents on r̄-value. (cold reduction: .85%; annealing condition: 704°C x 24 hr.)

5. Improvement of deep-drawability

With a view to producing cold rolled sheets with an improved deep-drawability, many research efforts have been reported on the effect of additional elements and impurity elements other than the various manufacturing conditions mentioned above. The two-stage cold rolling-annealing process developed by the present authors is a special process in this connection.

5.1 Effect of minor elements

Improvement of the deep-drawability is attempted by adding or eliminating appropriate elements in refining. In addition to Al, effective additive elements reported include Cu, Sb, Bi, Pb, P, Mo, As, Ti, Nb, Zr, W and Ta. Examples of addition of Sb and P are represented in Fig. 16.[6], and that of Ti, in Fig. 17.

The effect of P is reportedly considered to be its contribution through the change in precipitation of cementite.[20] With regard to the effect of Ti, some authors advocate that TiC having finely precipitated throughout the matrix exerts a favorable effect in recovery and recrystallization,[20] some others, that the precipitation of TiC causes decrease in soluble carbon in the matrix,[21] bringing about changes in the sub-structure in the cold rolled state,[22] and further some others, that solute Ti favorably acts in recovery and recrystallization.[23] Details of this effect is not however known to the general satisfaction.

In the addition of Sb and P, the decarburization annealing is applied, and in the addition of Ti, the carbon is rendered to the extra-low level through vacuum degassing. Thus, in many cases, the $\bar{r}$-value improving effect is further raised not only by adding effective elements, but also by decarburizing in advance or during annealing.

As to the beneficial effect of lowering C and Mn contents, several investigations have been conducted. Fig. 18 represents the relation between the C content and $\bar{r}$-value.[24] For the low-carbon region, the hot rolled sheet of 0.05% C was decarburized prior to cold rolling so as not to change factors other than C content. The $\bar{r}$-value increases with the decrease in C, the rate of this increase differs between carbon levels as divided at about 0.01%. The slow increase in $\bar{r}$-value caused by the decrease in C at C contents of more than 0.01% is considered to be due to the decrease in the amount of cementite, whereas the sharp increase at C contents below 0.01% may be attributable to the decrease in the content of solute carbon.

Fig. 19 shows the relation between the content of C and Mn and the $\bar{r}$-value.[25] This figure, representing the results of experiment on vacuum melt materials, reveals that the $\bar{r}$-value reaches its maximum at about C: 0.02-0.03% and Mn: 0.10-0.20%. In spite of many explanations of the cause so far released, there is as yet no generally acceptable one.

Also in rimmed steel refined under the atmospheric pressure, a high $\bar{r}$-value is obtained, not by simply lowering C and Mn during refining, but at appropriate C and Mn contents. The reason is as follows. Too low C and Mn contents bring about a higher O content, causing conversion of most of Mn into MnO and leading to the depletion of Mn to form MnS.

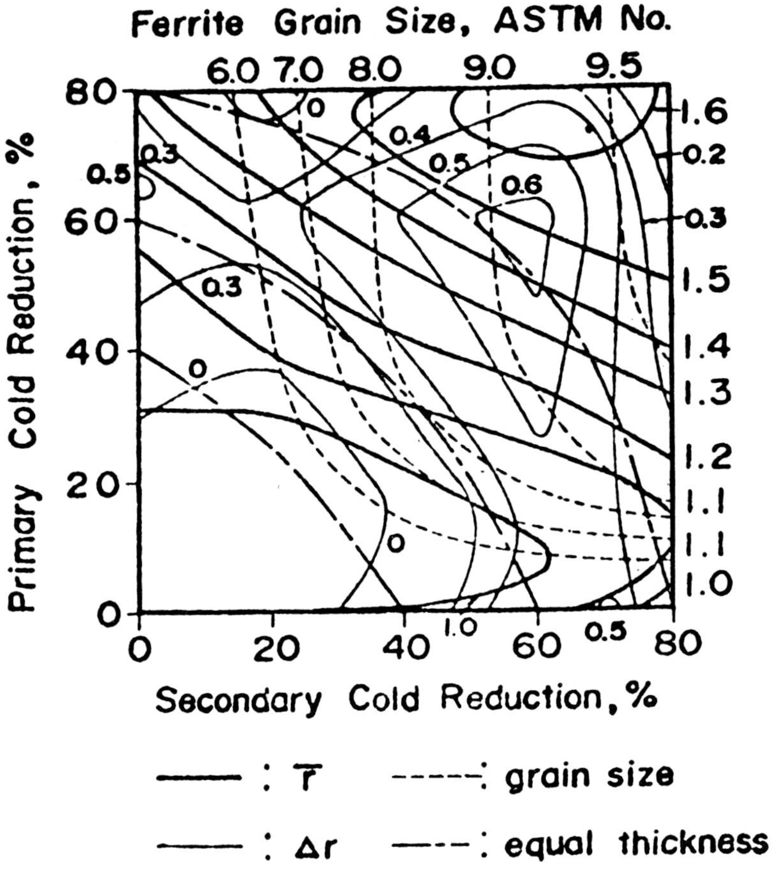

Fig. 20. Iso-$\bar{r}$-value, Δr-value and ferrite grain size curves showing the effect of the primary and secondary cold reduction of two stage cold rolling-annealing process of rimmed steel. (C: 0.05, Mn: 0.28, P: 0.015, S: 0.026, N: 0.0015, O: 0.054)

Thus fine (Fe, Mn)S precipitates during annealing. These changes in composition are considered to largely inhibit the ferrite grain growth in decarburizing annealing. For the acceleration of ferrite grain growth, it is desirable to keep the S content as low as possible.

5.2 Two-stage cold rolling-annealing process

Fig. 20 indicates changes in $\bar{r}$-value, varying with the combination of primary and secondary cold reductions in two-stage cold rolling-annealing process as applied for low carbon rimmed steel. This suggests the existence of a range of primary and secondary cold reductions giving high $\bar{r}$-values impossible in an ordinary process. This represents a case where non-decarburizing annealing was applied at 700°C both in primary and secondary anneals.[26] Manufacture of cold rolled steel sheets for extra-deep drawing is possible by further selecting appropriate conditions for composition adjustment, and primary and secondary anneals. The remarkable effect of the two-stage cold rolling and annealing is attributable to the fact that the $\bar{r}$-value after the primary annealing, in the stage corresponding to the hot rolled sheet in an ordinary process, is higher than that of the ordinary hot rolled sheet. It is however necessary to select conditions so as not to cause coarsening of ferrite grains after the primary annealing.

Table I compares deep-drawability of cold rolled steel sheets between various manufacturing processes. Apart from the production cost considerations, this comparison indicates the capability of giving the highest deep-drawability in the application of the two-stage cold rolling-annealing to extra-low carbon Ti-added steel. For low-carbon rimmed steel finished and coiled at a high temperature in hot rolling, a more satisfactory deep-drawability is obtained by treating in a continuous annealing furnace, as compared with low-carbon rimmed steel batch annealed by an ordinary process. Annealing time can also be considerably reduced in this case.

6. Conclusion

Various manufacturing conditions having effects on the deep-drawability of cold rolled steel sheet have been outlined above. In manufacturing cold rolled steel sheets for press-forming in Japan, in most cases, low-carbon rimmed steel and low-carbon Al-killed steel are finished at a high temperature and coiled at a low-temperature in hot rolling, cold rolled to about 70%, and non-decarburization-annealed at a temperature immediately below A_1 point, or decarburizing-annealed at a temperature between A_1 and A_3 points. At some specific producers, cold rolled steel sheets for extra-deep drawing are manufactured by the use of extra-low carbon Ti-added steel and the two-stage cold rolling-annealing process to serve as parts to be subjected to especially severe press-forming. Manufacture of sheets for drawing in a continuous annealing furnace, now applied only by a limited number of producers, is considered to form one of the future methods to improve steel texture orientation because of its excellent advantages including the labor saving, low cost and homogeneous product quality.

Table 1. Mechanical properties of cold rolled steels produced by various processes.

Mark	Type of Steel	Y.S. (kg/mm^2)	T.S. (kg/mm^2)	El. (%)	$\bar{r}$	Er. (mm)	C.C.V. (mm)
A	Rimmed Steel	22.1	32.0	46.5	1.31	10.8	38.0
B	〃	22.0	32.9	45.0	1.51	10.1	37.8
C	〃	15.8	28.8	53.0	1.78	12.0	36.5
D	〃	15.2	28.7	54.0	2.00	12.0	35.0
E	〃	14.8	27.3	56.6	2.30	12.1	draw through
F	Al Killed Steel	17.2	31.1	46.0	1.71	11.5	36.9
G	〃	16.0	28.4	50.1	2.10	12.1	36.2
H	Ti Killed Steel	14.6	30.1	52.0	2.20	12.2	draw through
I	〃	12.3	28.9	54.8	3.00	12.1	draw through

A: conventional
B: continuous annealing
C: decarburizing annealing
D: process of mark C with the optimum composition
E: steel of mark D by two stage cold rolling-annealing
F: conventional
G: steel of mark F by decarburizing annealing
H: conventional
I: steel of mark H by two stage cold-rolling annealing

REFERENCES

1. W.T. Lankford, S.C. Snyder and J.A. Bauscher: Trans. ASM, 42 (1950), 1197
2. R.S. Burns and R.H. Heyer: Sheet Metal Ind., 35 (1958), 261
3. R.L. Whiteley: Blast Furnace and Steel Plant, 48 (1960), 1245
4. R.L. Whiteley and D.E. Wise: Flat Rolled Products III, 16 (1962), 29
5. H. Takechi: JJSTP, 12 (1971), 85
6. H. Yoshida, K. Sasaki and F. Kansaki: I.D.D.R.G. Colloquium, London, (June 1964)
7. H. Kubotera: 9th Nishiyama Memorial Technical Lecture, (August 1969)
8. C. Dasarathy and R.C. Hudd: Z. Metallkde, 60 (1969), 853
9. K. Matsudo and T. Shimomura: Tetsu-to-Hagane, 55 (1969), S 536
10. M. Matsuol, S. Hayami and S. Nagashima: proceedings ICSTIS, Suppl. Trans. ISCJ, Vol. II (1971)
11. S. Nagashima: Texture, (1966)
12. D.S. Hoddinott and G.J. Davies: JISI, (1972), 624
13. J.F. Held: AIME, 239 (1967), 573
14. K. Matsudo and T. Shimomura: Trans. of ISIJ, 10, (1970), 448
15. M. Shimizu, K. Matsukura, N. Takahashi and Y. Shinagawa: Tetsu-to-Hagane, 50 (1964), 2094
16. M. Takahashi and A. Okamoto: Tetsu-to-Hagane, 59 (1973), 196
17. T. Kurihara: Tekkokai, March, (1973), 22
18. K. Toda, S. Ando and K. Sakurai: Tekkokai, July, (1974), 50.
19. D.A. Karlyn, R.W. Vieth and J.L. Forand: AIME Mechanical Working Conf. Pittsburgh, Jan. (1969).
20. N. Ohashi, M. Konishi and Y. Arima: Kawasaki Steel Corp. Technical Report, 5 (1973), 6
21. N. Fukuda and M. Shimizu: JJSTP, 13 (1982), 841
22. N. Takahashi, H. Takechi and M. Shimizu: JJI, 69th Conf., (1971), 104
23. R.H. Goodenow and J.F. Held: Met. Trans., 1 (1970), 2507
24. K. Matsudo, T. Shimomura and P. Kobayashi: JJI, 69th Conf., (1971), 105
25. I.F. Hughes and E.W. Page: Met. Trans., 2 (1971), 2-67
26. K. Matsudo, T. Shimomura and Y. Hashimoto: JJSTP, 8 (1967), 381

Section IV: Formability Tests for Steel Sheet

NEW CRITERIA FOR PREDICTING THE PRESS PERFORMANCE OF DEEP DRAWING SHEETS

By W. T. Lankford, S. C. Snyder and J. A. Bauscher

Abstract

In the past it has been generally believed that an isotropic material is more suitable for all types of sheet metal forming operations than an anisotropic material. Evidence has been developed in the present investigation, however, which refutes this general belief. Although it is still believed that an isotropic material is best suited for symmetrical formations, it has been demonstrated that, for certain unsymmetrical formations, material having a considerable degree of plastic anisotropy of a favorable nature results in the best press performance. By considering favorable plastic anisotropy together with favorable strain hardening characteristics as requisites for good press performance, it is possible to discriminate very successfully between materials having quite similar ordinary mechanical properties but which exhibited wide differences in press performance in unsymmetrical fender draws. A good correlation has been found between plastic anisotropy and magnetic anisotropy, indicating that the plastic anisotropy arises from preferred crystallographic orientations. These new findings suggest that for optimum press performance the plastic anisotropy characteristics of the material must be suited to the symmetry of the forming operation.

A CRITICAL examination of all available information on the formability of sheet metals clearly indicates that the material characteristics responsible for good drawing performance are not well understood. Furthermore, it can be inferred from occasional observations of unexplainably high breakage in severe sheet metal forming operations that conventional testing methods are failing to reflect important aspects of the material behavior which are operative under actual forming conditions. This situation is of particular concern in connection with very severe forming operations, such as automobile fender draws, in which such great demands are placed on the

A paper presented before the Thirty-first Annual Convention of the Society, held in Cleveland, October 15 to 21, 1949. The authors, W. T. Lankford, S. C. Snyder and J. A. Bauscher, are associated with the Research and Development Division, Carnegie-Illinois Steel Corp., Pittsburgh. Manuscript received May 7, 1949.

material that very subtle differences can result in wide variations in press performance.

Because of the great need for a better understanding of the basic properties which are essential for optimum press performance in deep drawing sheets, a comprehensive, long-range program of research in this field was inaugurated by the Carnegie-Illinois Steel Corporation about four years ago. In this program it was planned to conduct studies of all aspects of the plastic flow and ductility characteristics believed to be of importance in sheet metal formability. It was expected that the information developed in such a research program would not only indicate the most fruitful directions for improvements in deep drawing sheets, but would also provide a basis for the development of more discriminating practical tests for the evaluation of deep drawing quality.

One phase of this comprehensive investigation of sheet metal formability has involved studies of several lots of deep drawing sheets which exhibited wide differences in press performance and yet were quite similar as judged by the customary tests. The results of these studies have been very revealing and are believed to provide a new point of view for considering the material requirements for good drawing performance. In particular, it appears that prevalent beliefs that directionality in deep drawing sheets is generally undesirable must be abandoned in the light of the new findings. Material possessing pronounced plastic anisotropy of a favorable nature was found to exhibit press performance in unsymmetrical fender draws greatly superior to that of material having a minimum of anisotropy. Since plastic anisotropy is not measured in any of the customary tests applied to sheet metals, it is not surprising that materials considered to be satisfactory by the usual criteria are sometimes observed to exhibit poor performance in the press.

No attempt will be made here to review or criticize the large number of tests which have been proposed for predicting formability, since several excellent surveys (1, 2, 3)[1] of this nature are already available. It can be said, however, that the correlation of laboratory tests with press performance, in general, has been disappointing. Some of these tests may be capable of distinguishing very poor material which will almost certainly exhibit poor press performance; the fact that a material is judged favorable by such tests, on the other hand, does not necessarily insure that the performance will be good.

[1]The figures appearing in parentheses pertain to the references appended to this paper.

Most of the customary tests applied to sheet metals are empirical in nature and, while very useful as rapid checks of probable uniformity of product, do not measure directly the properties which may be controlling the performance in a particular draw.

The most widely used mill control tests for deep drawing sheets include hardness tests, cupping tests for ductility, and an examination of microstructure. Where more complete information on mechanical properties is desired, these tests may be supplemented by tension tests involving determinations of yield point, tensile strength, and percentage elongation. Although all of these tests are extremely useful for control purposes, it is believed that they not only fail to reflect certain important properties, but they also may not be sufficiently sensitive to detect small differences in the measured properties which would lead to significant differences in press performance.

In the early phases of the Carnegie-Illinois formability program, attention was focused on the tension test, since it was believed that more basic information could be derived from this test than the customarily determined properties such as yield point, tensile strength, and elongation. Tensile strength, for example, is a fictitious stress value, since it is based on the initial cross sectional area of the specimen rather than on the actual area at the instant the maximum load is measured. The actual area at maximum load will, of course, depend on the amount of strain or contraction the specimen undergoes before necking down. The percentage elongation in 2 inches is likewise an ambiguous quantity, because it includes not only the uniform elongation which the specimen undergoes up to the maximum load, but also the localized deformation associated with necking down. Of the three properties ordinarily determined in the tension test, it can be seen, therefore, that only the yield point is of fundamental significance.

What information can be obtained from the tension test, then, which will have a more basic significance and which can possibly be related more directly to behavior in forming operations? The point of view adopted in the present investigation is that the tension test should be considered first of all as a means of studying the work hardening characteristics of a metal. It has long been recognized that, in forming operations involving stretching, the ability of the material in the critical areas of the blank to work harden and pass along deformation to adjoining material is conducive to improved performance. In other words, a high capacity for work hardening retards localization of deformation and reduces breakage by permitting a larger area of material to contribute to the over-all stretching required to make the part.

One criterion of work hardening which is sometimes used is the separation between the yield point and the tensile strength, or the yield-tensile ratio. The larger the difference between the yield point and tensile strength, or the lower the yield-tensile ratio, the higher the work hardening capacity is said to be, and the more favorable the material is considered to be for drawing. The yield-tensile ratio is not a precise measure of work hardening, however, largely because the tensile strength, as usually defined, is a fictitious stress. Furthermore, even in materials having similar stress-strain curves beyond the yield point, the observed value of the yield point may be subject to considerable variation arising from the effects of small differences in the amount of prior cold work, for example as experienced in temper rolling. It is much more informative, therefore, to consider work hardening in terms of "true" stresses and strains, wherein both stress and strain are defined in terms of instantaneous specimen dimensions during the course of the test, rather than in terms of initial specimen dimensions. The shape of the true stress-strain curve accurately reflects the work hardening characteristics of a material. The actual rate of work hardening is a function of the amount of strain and can be determined directly from the slope of the true stress-strain curve for any particular strain, high slopes representing high rates of work hardening.

Fortunately, it is possible to quantitatively describe the shapes of the true stress-strain curves for most steels in terms of an empirical analytical relationship between true stress and true strain. Several investigators (4-8) have demonstrated that true stress-strain curves for most steels conform quite closely to the following relationship:

$$\text{True stress} = k\ (\text{True strain})^n,$$

where "k" and "n" are constants characteristic of the material and are referred to as the strength coefficient and the strain hardening exponent, respectively. These two constants completely describe the shape of the true stress-strain curve. The "n" value is believed to be a characteristic of major importance in stretching operations, since it provides a measure of the ability of the material to distribute deformation uniformly and can be shown to be numerically equal to the uniform elongation in the simple tension test, i.e., the elongation up to maximum load. The "k" value, on the other hand, provides some indication of the level of strength of the material and, hence, of the magnitude of the forces required in forming.

In the course of developing experimental techniques and obtaining test data for the determination of "k" and "n" values for a number of sheet steels, it was observed by Jackson, Smith and Lankford (9) that, in some of the materials being studied, the percentage contraction in the width dimension as the test specimen was plastically strained in tension was not equal to the contraction in the thickness dimension. The ratios of width strain to thickness strain were found to be different, not only among different materials, but also in some instances between longitudinal and transverse tests of the same material. This behavior was recognized as a manifestation of plastic anisotropy, probably arising from preferred crystallographic orientations, since a specimen of an isotropic or randomly oriented material plastically strained in simple tension should exhibit equal strains in the width and thickness directions. In some of the sheet steels which have been tested, the ratios of width strain to thickness strain have shown a very pronounced deviation from the value of unity characteristic of an isotropic material, with values as high as 2.8 and as low as 0.7 being observed. The phenomenon of unequal strains in the width and thickness directions in simple tension tests has been observed and reported by other investigators in certain nonferrous metals, e.g., Baldwin (10) in soft copper and Lankford, Low and Gensamer (11) in aluminum alloys.

Directionality in deep drawing sheets has long been of major concern and has been the subject of many experimental investigations. It is recognized that directionality in sheets can be of two types. The first type is that associated with the development of a "fibrous" structure which results from the tendency for nonmetallic inclusions or segregates to become elongated and strung out during rolling of the sheet. Directionality resulting from these effects is most likely to be manifested as a difference in ductility in the rolling and cross rolling directions of the sheet. The second type of directionality is that associated with the preferred crystallographic orientations characteristic of cold-reduced, box-annealed sheets. Preferred orientations in sheets are known to result in directional variations in plastic properties, the formation of ears in cup drawing being one of the most familiar manifestations of this effect. The latter type of directionality and its relationship to the plastic behavior of sheets in forming operations are of major concern in the present investigation.

Since the materials in this investigation exhibited considerable departure from isotropic behavior as measured by their ratios of width strain to thickness strain in the simple tension test, it seemed logical to expect that the distribution of deformation in biaxial stretching would likewise be greatly influenced by this anisotropy. If this were the case, the performance in actual drawing operations might likewise be expected to be strongly dependent upon the nature of the plastic anisotropy of the material being formed. It has been recognized in the past that directionality in deep drawing sheets is an important factor in press performance, but it has been commonly believed that pronounced directionality always has a deleterious effect on performance (1, 12). The results of the present investigation refute this general belief and indicate that pronounced plastic anisotropy of a favorable nature actually leads to better performance in unsymmetrical forming operations. It is believed that these findings form the basis of a new concept of material requirements for deep drawing operations, i.e., for optimum press performance the plastic anisotropy characteristics of the material must be suited to the symmetry of the forming operation.

Information has also been developed which emphasizes the importance of work hardening as a factor in controlling deep drawing performance. By considering good work hardening characteristics, as measured by the strain hardening exponent ("n" value), together with favorable plastic anisotropy as a combined criterion of good drawing quality, an excellent correlation with press performance was established.

Materials

The materials studied in this investigation include samples from forty-six lots of fully-aluminum-killed, low carbon, deep drawing sheets which exhibited wide variations in drawing performance in the plants of several producers of deep drawn parts. All of the material was in the cold-reduced, box-annealed, and temper-rolled condition. Twenty-nine of the lots exhibited high die scrap, whereas seventeen lots were representative of good performance. Materials of good and poor performance in four different fender draws are represented, each of these formations involving severe biaxial stretching of an unsymmetrical nature. The distribution of sample lots among these formations, which have been designated as Formations A, B, C and D, is illustrated in Table I. Samples of material which performed satisfactorily were selected as soon as possible after trials of poor material, in order that die conditions would be quite similar; i.e., on each occasion of obtaining a poor lot or group of poor lots, a sample of good material was also selected at the same time whenever possible for comparison. For this investigation a special effort was made to obtain materials having poor performance which exhibited quite normal mechanical properties.

Table I
Distribution of Sample Lots Among Various Fender Formations

	Number of Lots	
Formation	Good Performance	Poor Performance
A	10	12
B	4	13
C	1	1
D	1	2
..	1*	1*

*Two lots for which no actual performance data were available were included in the investigation. These lots were experimental materials purposely processed in such a way as to produce a difference in microstructure and a difference in the plastic anisotropy characteristics. The performances indicated for these two lots refer to unsymmetrical types of forming operations and have been assumed on the basis of extensive previous experience with the same types of material.

Two of the poor lots were experimental materials purposely processed in such a way as to produce a more isotropic material than that known to give best drawing performance in the formations studied. These lots were produced in order that additional information could be obtained on the effects of plastic anisotropy on the strain distributions in unsymmetrical draws.

Experimental Methods

True stress-strain curves were determined by a precision technique similar to that described by Low and Garofalo (7), the only difference being that a considerably greater accuracy in load measurement was attained in the present investigation. Loads were measured by means of an elastic weighing ring accurate to about 0.20% over the range used. All tests were conducted with a rate of crosshead movement of approximately 0.03 inch per minute.

The true stress-strain curves were analyzed in terms of the relationship:

$$\sigma = k\delta^n,$$

where σ is the true stress and δ is the true strain, each being defined in the usual manner. As already mentioned, the constants "k" and "n" will be referred to as the strength coefficient and strain hardening exponent respectively. If this relationship is valid, a log-log plot of stress-strain data should result in a straight line having a slope of "n". A plot of this kind, which is typical of the data obtained in the present investigation, is shown in Fig. 1; it can be seen that the conformity to a straight line is very good. In a series of reproducibility tests conducted on fifty adjacent longitudinal specimens of 20-gage, fully-aluminum-killed, deep drawing steel sheet, it was found

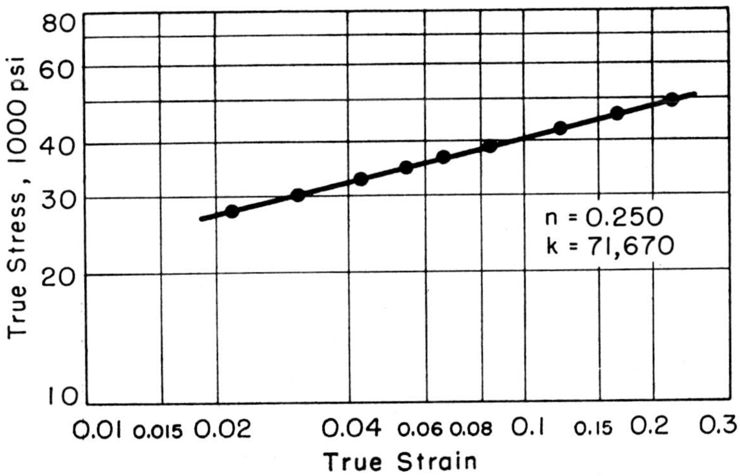

Fig. 1—Typical Logarithmic Stress–Strain Curve for Temper Rolled, Fully–Aluminum–Killed, Deep Drawing Steel Sheet.

that the "n" values determined by the above experimental technique only varied between 0.230 and 0.238. The "n" values reported in subsequent sections of the paper represent the averages of triplicate tests made on both longitudinal and transverse specimens.

In the early work on the ratio of width strain to thickness strain, the ratio was determined from a plot of true width strain against true thickness strain, these strains being measured at several levels of elongation. It was found that such a plot results in a straight line; the determination of the slope of this line provides an average measure of the ratio of width strain to thickness strain, hereinafter referred to as the "R" value. In a large number of determinations by this method, it was found that the points corresponding to the width and thickness strains at or just beyond maximum load always fell on the line drawn through the points determined at lower strains, indicating that the ratio of width strain to thickness strain does not change with increasing strain, at least up to the maximum load. On the basis of this observation it was decided to discontinue the practice of measuring strains at several levels and to simply make one set of measurements at maximum load. This procedure has the added advantage that the same specimen can be used for both the determination of the true stress-strain curve and the ratio of width strain to thickness strain, since the extensometer employed for elongation measurements is removed at maximum load. Each tension test, therefore, was interrupted at maximum load, and width and thickness measurements were made on the specimen at ten locations within the central 2-inch gage length. The ratios of the width strain to thickness strain at these ten locations were then averaged. It was also found that a fairly satisfactory determination of the "R" value could be made on a fractured specimen from measurements of the width and thickness dimensions about 1 inch from the fracture. The principal disadvantage of this procedure is the difficulty of obtaining reliable measurements when the specimen "cups" or warps across the width direction. It is interesting to note that gross differences in "R" values can be readily detected simply by visual comparison of the minimum widths of fractured tensile specimens. Even though the total elongations of two specimens having different "R" values are equal, the final width of the specimen having the higher "R" value will obviously be less than that of the specimen having the lower "R" value.

As previously mentioned, it seemed likely that the observed differences in "R" values among different materials resulted from differences in preferred crystallographic orientations. Since magnetic anisotropy as indicated in the magnetic torque test is known to reflect preferred orientations (13), magnetic torque curves were obtained for all the materials in this investigation. These curves were determined on a new recording type of torque magnetometer developed at the U. S. Steel Corporation Research Laboratory (14). In this investigation, rather than attempting to analyze the magnetic torque curves in terms of actual preferred orientations, the curves for the various materials studied have been classified on the basis of their characteristic shapes.

The ordinary mechanical properties were determined by conventional techniques. "Yield points" were determined either by the dividers method or from the proportional limit on a recorded stress-strain curve using a strain magnification of four. It will be understood that, since the steels tested did not exhibit a sharp yield point, the "yield point" values reported actually represent an offset yield strength, the offset corresponding very closely to 0.2%.

Experimental Results

A—Investigation of First Group of Materials Exhibiting a Wide Range in Press Performance Associated with Differences in Plastic Anisotropy

The first group of lots of deep drawing sheets of known press performance obtained for testing included twenty lots which had exhibited extreme variations in press performance in the difficult fender formation designated as Formation A. Since this group of samples provided the first evidence of the importance of plastic anisotropy as a factor in controlling press performance, this portion of the investigation will be discussed before considering the results from the entire group of forty-six lots. The poor lots in the first group of materials resulted in 100% breakage in every attempt to form them, although no significant differences could be found by the customary criteria between these lots and those exhibiting good performance. The grain sizes of the poor lots tended to be somewhat larger than those of the good lots, although this difference was not consistent. In Table II the similarity in the ordinary mechanical properties of typical lots of good and poor material can be readily seen. Precision determinations of "k" and "n" values from true stress-strain curves likewise failed to discriminate between good and poor material, as can be seen from Table III. The "n" values of both the good and the poor materials are so similar that no inference can be made regarding the influence of the "n" value in controlling press performance in this particular formation. From data which will

Table II

Comparison of Ordinary Mechanical Properties of Lots Exhibiting Satisfactory Press Performance with Those for Lots Exhibiting 100% Breakage in Formation A

Lot	Rockwell "B" Hardness	Olsen Value	Test Direction	Yield Point	Tensile Strength	% Elongation in 2 Inches
			Satisfactory Performance			
A	36	444	L	22,700	44,600	41.5
			T	24,900	44,200	40.5
D	30	473	L	17,500	40,400	39.0
			T	18,700	41,200	42.0
E	40	460	L	20,800	43,300	43.5
			T	21,600	43,500	43.5
			Poor Performance			
B	28	445	L	21,700	41,500	39.0
			T	21,600	42,400	37.5
C	27	448	L	19,700	40,400	40.5
			T	21,800	41,100	38.5
F	34	450	L	20,300	41,500	45.0
			T	20,500	42,500	42.5

Table III

Comparison of "k" and "n" Values for Lots Exhibiting Satisfactory Press Performance with Those for Lots Exhibiting 100% Breakage in Formation A

Lot	"n" Value	"k" Value
	Satisfactory Performance	
A	0.251	78,430
D	0.274	75,460
E	0.266	77,770
	Poor Performance	
B	0.242	73,350
C	0.250	73,860
F	0.269	76,710

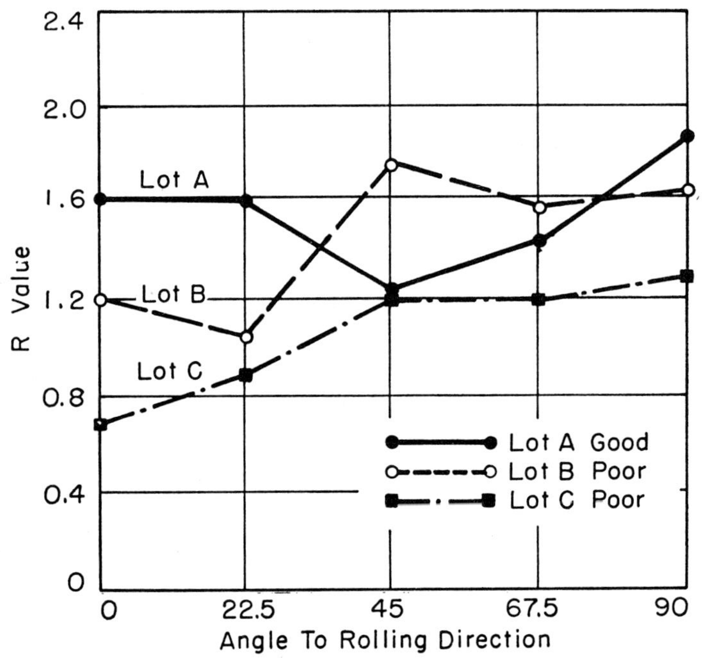

Fig. 2—Variation of "R" Value with Direction of Testing for Lots Exhibiting Good and Poor Press Performance in Formation A.

subsequently be presented, however, it will become apparent that the "n" values for both the good and the poor lots are at a favorable level.

Since it was believed that variations in plastic anisotropy might possibly lead to marked differences in the plastic behavior under the unsymmetrical stretching conditions involved in this fender formation, "R" value determinations were made on the lots under consideration. Very large differences in the "R" values were found between material having good press performance and that having very poor press performance. Typical results of the "R" value determinations on samples from fender blanks are summarized in Fig. 2, from which it can readily be seen that large differences exist among the good and poor materials. Tests were conducted at 0, 22.5, 45, 67.5 and 90 degrees to the rolling direction, and it can be seen that the patterns of variation of "R" value with testing direction are quite different. The greatest differences in the level of the "R" values occur in longitudinal tests where the "R" value varies from 0.67 to 1.62.

If the longitudinal "R" values (R_L) for all twenty lots in this group are considered, it is possible to successfully discriminate seven

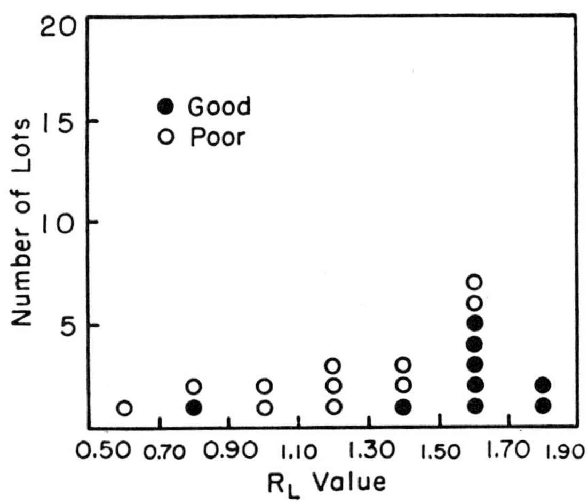

Fig. 3—Distribution of "R_L" Values for Original Group of Twenty Lots of Known Press Performance in Formation A.

of the nine good lots from nine of the eleven poor lots. In other words, the performance of sixteen of the twenty lots can be successfully predicted on the basis of "R_L" values alone, based upon the premise that a minimum "R_L" value of 1.50 is required for good performance. Fig. 3 shows the distribution of "R_L" values for the various lots tested.

As already pointed out, it seemed likely that the observed differences in plastic anisotropy were associated with differences in preferred crystallographic orientations. Magnetic torque curves were therefore obtained for the entire group of samples in order to ascertain whether such differences actually existed. Not only were differences in preferred orientations found to exist, but, in addition, a very high degree of correlation between drawing performance and type of magnetic torque curves was found. The types of magnetic torque curves which were observed in this series of tests are illustrated in

Fig. 4. All of the good lots except one exhibited torque curves of the "A" type, whereas all of the poor lots with but two exceptions exhibited curves of types other than "A". Fig. 5 shows the distribution of various curve types among the lots tested. If the various types

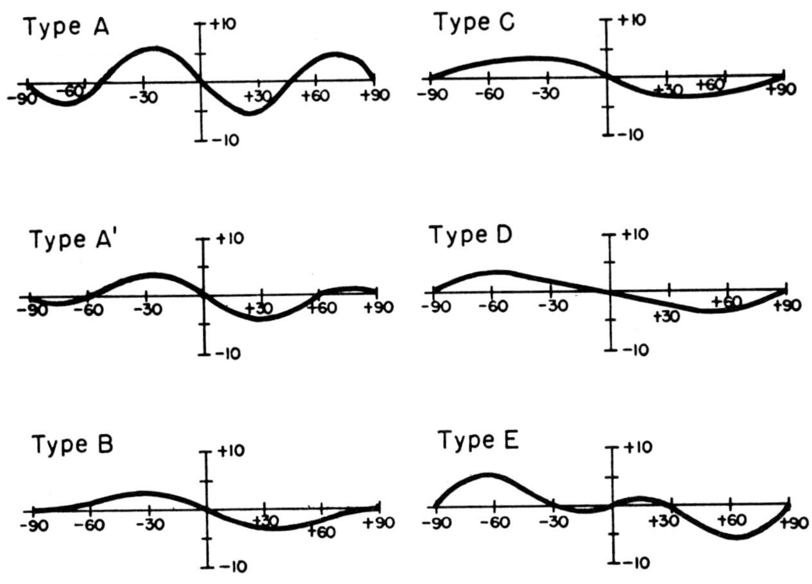

Fig. 4—Various Types of Magnetic Torque Curves Observed in This Investigation.

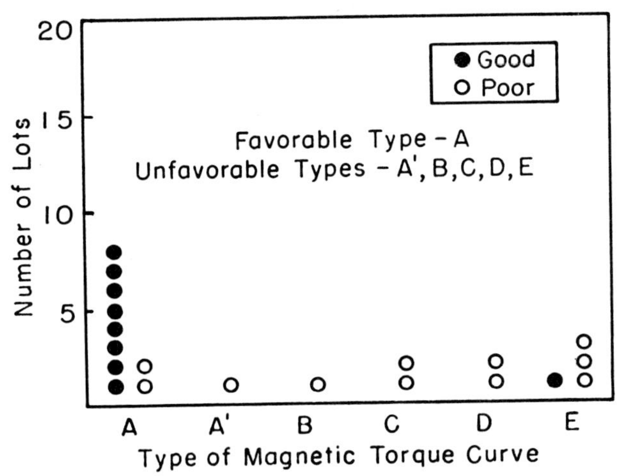

Fig. 5—Distribution of Types of Magnetic Torque Curves for Original Group of Twenty Lots of Known Press Performance in Formation A.

of torque curves are compared with the corresponding "R_L" values, it is found that a high degree of correlation exists. As can be seen in Fig. 6, lots exhibiting favorable magnetic torque characteristics also exhibited high "R_L" values, whereas those lots having unfavorable magnetic torque characteristics exhibited low "R_L" values, with but a single exception.

The results of this phase of the investigation conclusively demonstrated the importance of favorable plastic anisotropy as a factor in establishing good press performance in a severe unsymmetrical draw. Since all other properties were very similar, the degree of plastic anisotropy can be said to have been the critical or controlling factor

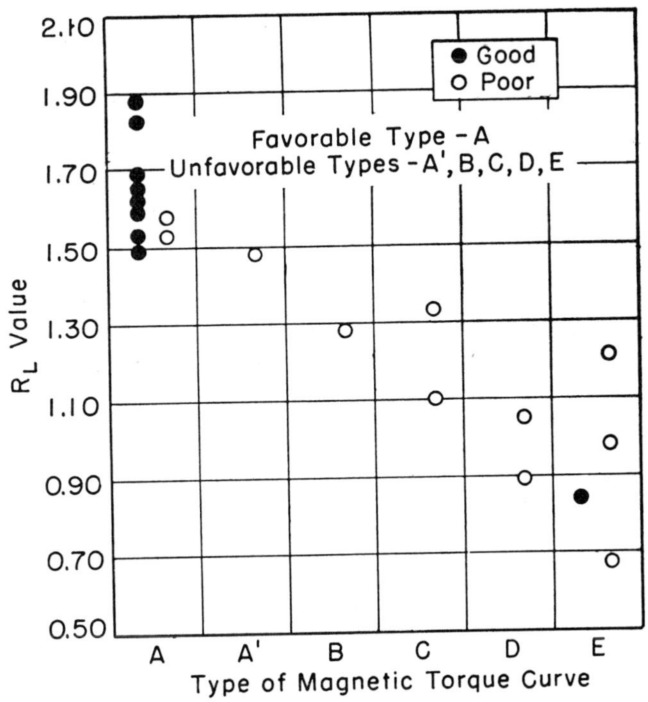

Fig. 6—Qualitative Correlation Between "RL" Value and Type of Magnetic Torque Curve for Original Group of Twenty Lots of Known Press Performance.

in the performance of the lots tested. Since this characteristic is not reflected in any of the tests customarily applied to deep drawing sheets, it is not surprising that the reason for the high breakage of the poor material was not immediately apparent.

B—Summary of Results for All of the Materials Investigated

As already mentioned, samples from a total of forty-six lots of known drawing performance have been tested, including the twenty lots discussed in the previous section. In this section no attempt will be made to consider each group of lots separately; rather, all of the forty-six lots will be considered as a single group for purposes of analysis. The various properties of all of the lots investigated and the correlation of these properties with press performance have been analyzed and interpreted in terms of frequency distribution plots in which the data are presented in such a way that the distribution of each property for good and poor lots can be distinguished. Such plots have been prepared not only for "R_L" values, "n" values, and types of magnetic torque curves, but also for the customarily determined properties such as Rockwell "B" hardness, yield point, yield-tensile ratio, per cent elongation, and Olsen test values.

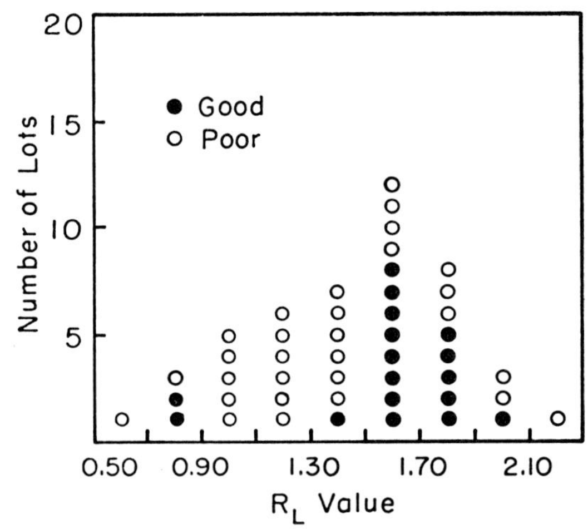

Fig. 7—Distribution of "R_L" Values for All Lots Tested.

First of all, an examination of Fig. 7, in which the distribution of "R_L" values is plotted, reveals that the poor performance of all the bad lots cannot be accounted for on the basis of unfavorable "R_L" values. It should be noted, however, that below an "R_L" value of 1.50 only three good lots appear out of a total of twenty-two lots. In other words, nineteen out of twenty-two lots having an "R_L" value lower than 1.50 exhibited poor drawing performance, confirming the previous observation of the importance of favorable plastic anisotropy. The occurrence of several poor lots with "R_L" values higher than 1.50 suggest, however, that some other property must be taken into consideration along with the "R_L" value. From the point of view of the distribution plots, this means that a distribution should be sought wherein preponderantly poor lots are found beyond a certain limit of the property being considered. By comparing such a distribution with the distribution of "R_L" values, it could then be ascertained whether unfavorable values of the property being considered would account for the performance of those poor lots having favorable "R_L" values.

The next distribution to be examined was that of the "n" value, or strain hardening exponent, since it has already been pointed out that this property is of great importance in formations involving stretching. Furthermore, since the "n" value and "R_L" value are not directly related, it seems logical that both should be favorable

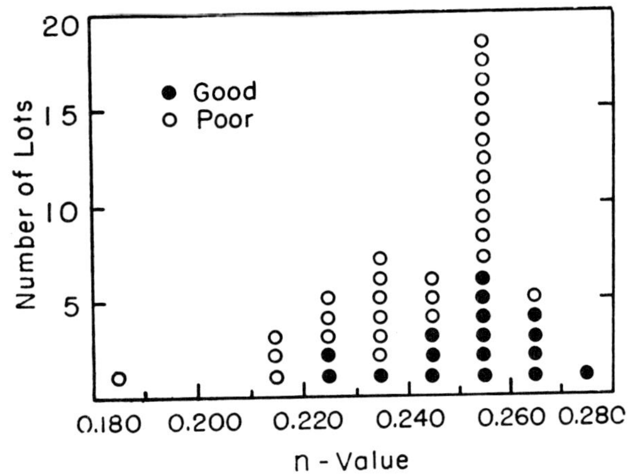

Fig. 8—Distribution of "n" Values for All Lots Tested.

for good drawing performance. Upon examination of the "n" value distribution shown in Fig. 8, it is found that below an "n" value of 0.240, thirteen out of sixteen lots exhibited poor drawing performance. If it is then arbitrarily assumed that a minimum "R_L" value of 1.50 and a minimum "n" value of 0.240 are required for satisfactory press performance, it is found that twenty-five of the twenty-nine lots indicated as having poor press performance correspondingly failed to come up to these assumed requirements for good performance. Similar treatment of the test results of the good materials indicates that fourteen of the seventeen lots fulfill the assumed requirements for satisfactory press performance. This rather startling separation of good and poor materials is shown in Fig. 9 in which "n" values for all of the lots have been plotted against the corresponding "R_L" values. Obviously, the region of this plot lying above the horizontal line drawn at $n = 0.240$ and to the right of the vertical line drawn at $R_L = 1.50$ corresponds to good performance in accordance with the criteria assumed above. The curved line shown in Fig. 9 represents the separation of good and poor materials which can be made on a somewhat different basis as will be explained in the following paragraphs.

It is also possible to set up a single quality index which takes into account both the "R_L" and the "n" values by arbitrarily assuming that these two characteristics contribute equally to the press performance. It is first assumed that the "R_L" and "n" values are normally distributed and the standard deviations are computed. The

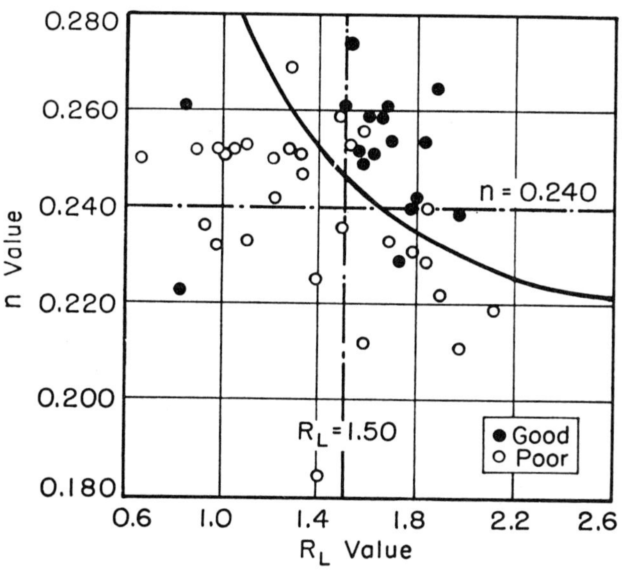

Fig. 9—"n" Value Versus "R_L" Value for All Lots Tested.

knowledge of the standard deviations makes possible a calculation of the probable maximum and probable minimum values, using regular statistical methods. The following designations may then be employed:

R_{max} = probable maximum value
R_{min} = probable minimum value
$R_{range} = R_{max} - R_{min}$
R_i = the measured value of R_L

and corresponding designations for the values of "n". The quality index can then be expressed as follows:

$$(R \times n)\ \text{index} = \frac{R_i - R_{min}}{R_{range}} \times \frac{n_i - n_{min}}{n_{range}} \times 1000$$

It is clear that this quality index can range from 0 to 1000, because, if either R_i or n_i is at the minimum, then one of the fractions is zero; and if both R_i and n_i are at the maximum, then both fractions are equal to 1. The (R x n) quality index was calculated for all of the lots in this investigation, and the distribution of values is shown in Fig. 10. It can be seen that the poor and good lots fall almost completely into two separate distributions. Out of twenty-seven lots having indices lower than 300, twenty-four lots exhibited poor press performance, whereas fourteen of the nineteen lots having indices above 300 exhibited good press performance.

The separation between good and poor lots which can be effected

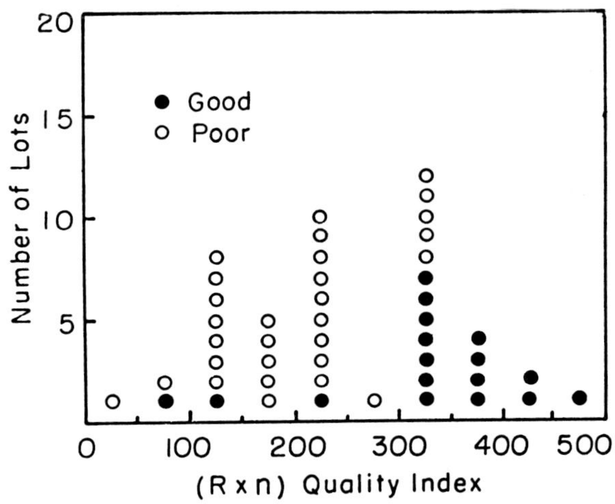

Fig. 10—Distribution of Values of (R x n) Quality Index for All Lots Tested.

by the use of the (R x n) quality index is illustrated in another manner in Fig. 9 wherein, as already mentioned, "n" values are plotted against corresponding "R_L" values for all the lots tested. The heavy curved line was calculated on the basis of an (R x n) index of 275 and represents the dividing line between good and poor quality. According to this criterion, the region of the plot lying above the calculated curve represents good quality. It can be seen that the points for twenty-four of twenty-nine poor lots lie below the curve and those for fourteen of seventeen good lots lie above the curve. It should also be noted that the calculated curve conforms closely to that which would be drawn by inspection of the plotted points in such a way as to effect the best separation between good and poor materials.

The question can now be raised as to whether an equally good criterion for predicting press performance could be obtained by considering some property other than the "n" value in combination with the "R_L" value. Obviously, the "R_L" value must be retained in the criterion, since it measures a characteristic which is not reflected in the customary tests. Proceeding on this basis, the yield-tensile ratio was next considered. Since, as previously indicated, this property reflects to a certain extent the strain hardening characteristics of a material, it might be expected to provide a correlation with performance similar to that obtained with the "n" value. The relationship between yield-tensile ratio and "n" value can be seen from the data plotted in Fig. 11 which indicate that, although there is considerable scatter, the yield-tensile ratio tends to decrease with increasing

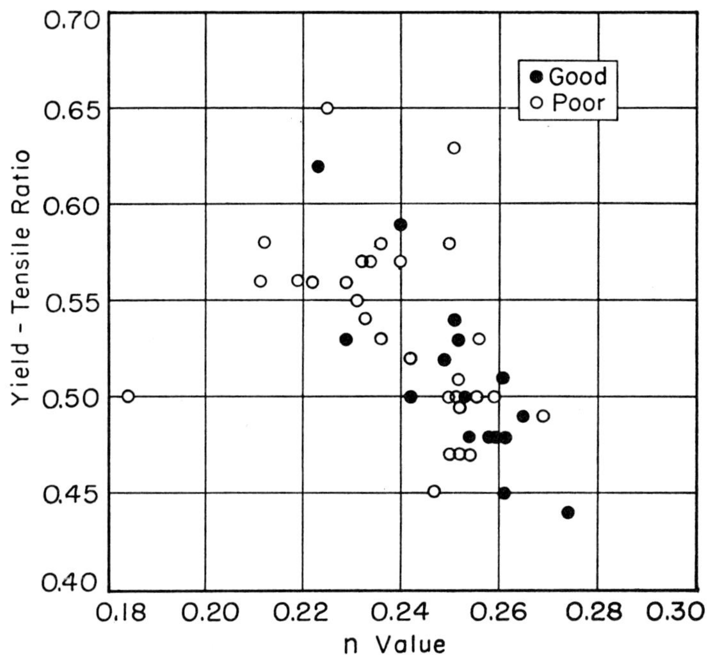

Fig. 11—Relationship between Yield–Tensile Ratio and "n" Value.

"n" value. It is evident, therefore, that lower yield-tensile ratios are indicative of better work hardening characteristics. In Fig. 12 the distribution of yield-tensile ratio values is shown. It can be seen that above a ratio of 0.55, thirteen out of fifteen lots exhibited poor drawing performance. Since this separation is very similar to that observed with the "n" value, a combined criterion of favorable "R_L" value and favorable yield-tensile ratio was tested. The results can be seen in Fig. 13 in which yield-tensile ratio is plotted against "R_L" value. If it is assumed that an "R_L" value of 1.50 or greater, together with a yield-tensile ratio of 0.55 or less, is required for good press performance, it is found that twenty-three out of twenty-nine lots failed to reach these requirements. Similar treatment of the

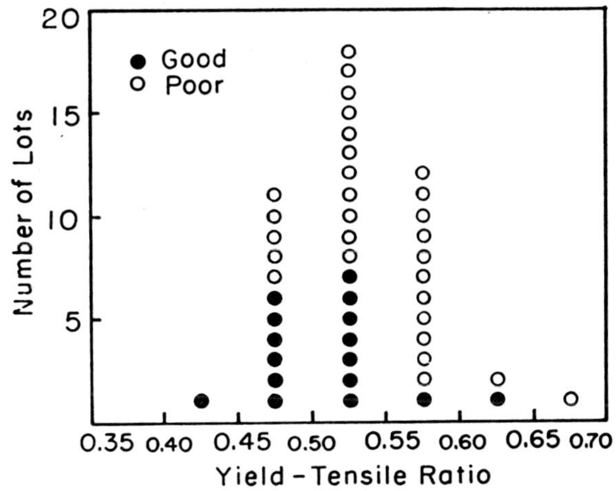

Fig. 12—Distribution of Yield–Tensile Ratios for All Lots Tested.

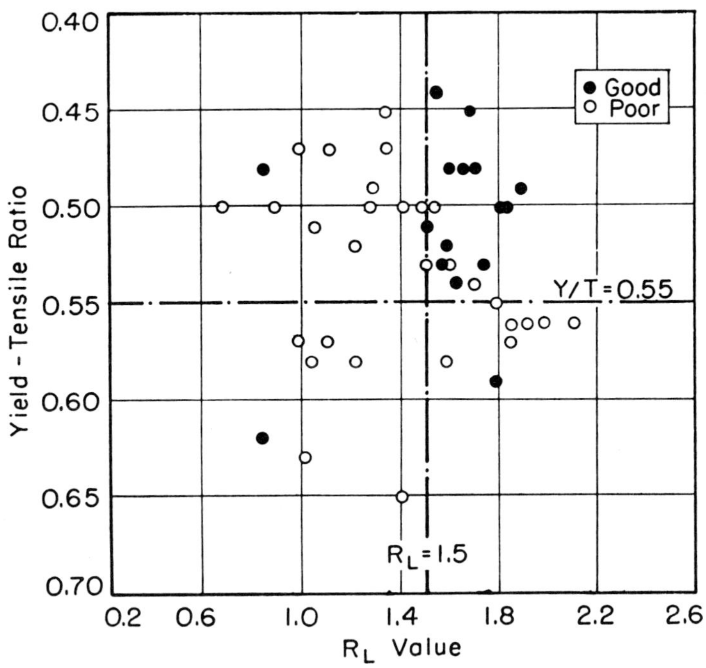

Fig. 13—Yield-Tensile Ratio Versus "R_L" Value for All Lots Tested.

good lots indicated that thirteen out of sixteen lots fulfilled the assumed requirements for good press performance. It is evident, therefore, that this criterion results in a separation of good and poor lots almost as good as that obtained when the "R_L" and "n" values were considered in combination. It is evident from the distribution of the data in Fig. 13, however, that a single quality index combining

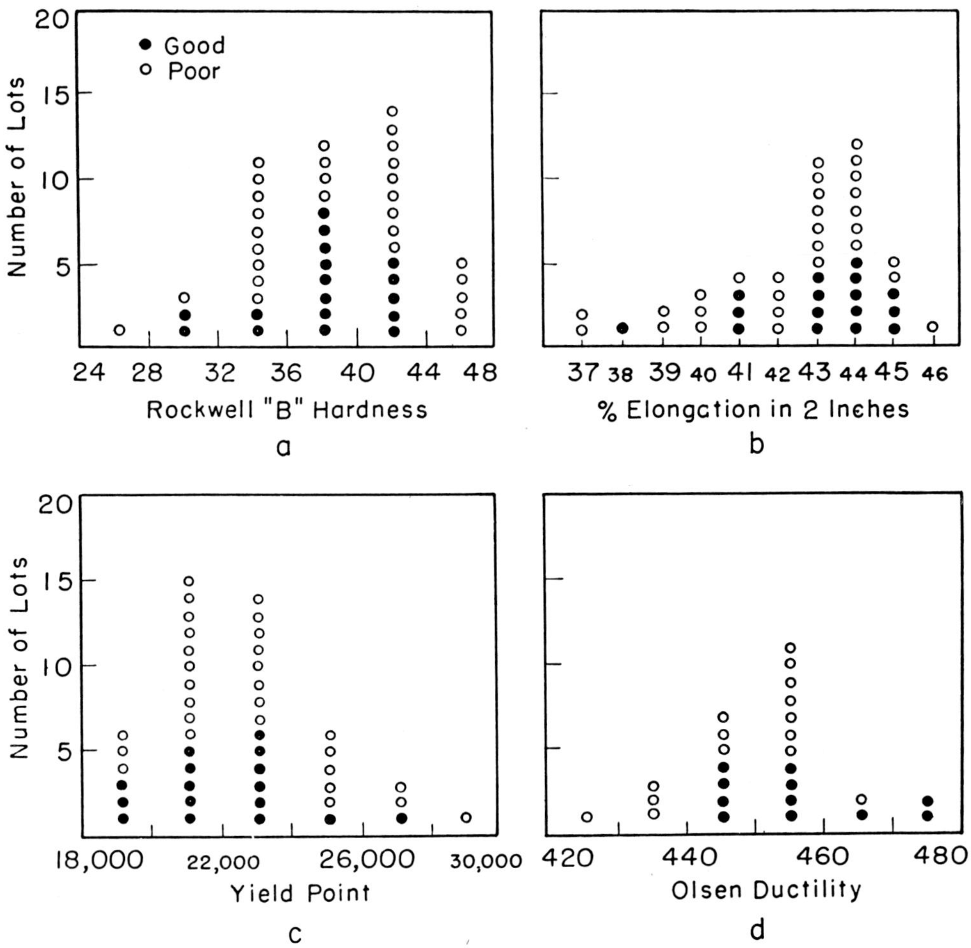

Fig. 14—Distribution of (a) Rockwell "B" Hardness, (b) Per Cent Elongation in 2 Inches, and (c) Yield Point Values for All Lots; (d) Distribution of Olsen Ductility Values for Twenty-Six Lots.

the "R_L" value and yield-tensile ratio could not effect as complete a separation as was obtained by the use of the (R x n) quality index described previously.

From the distribution of hardness values shown in Fig. 14a, it can be seen that fourteen out of nineteen lots having hardness values above Rockwell B-40 exhibited poor press performance. By arbitrarily considering a maximum hardness value of 41 and a minimum "R_L" value of 1.50 as requisites for good press performance, it is found that twenty-five of twenty-nine poor lots failed to come up to these assumed requirements. When the seventeen good lots are considered, however, it is found that only twelve actually fulfill the above assumed requirements for good press performance. It is evident that this separation is considerably poorer than that obtained by considering the "R_L" value in combination with either the "n" value or the yield-tensile ratio.

The question might also be raised as to whether the total elongation in the tension test could be considered in combination with the "R_L" value as a criterion for predicting performance. The distribution of elongation values is shown in Fig. 14b, from which it can be seen that twelve out of sixteen lots having elongation values under 43% exhibited poor press performance. When elongation is considered in conjunction with "R_L" value, however, the separation which is possible between good and poor material is very unsatisfactory, with six good lots and six poor lots being incorrectly rated. In order to obtain even this degree of separation, it is necessary to use a rather high elongation value as a minimum requirement for good press performance.

Distributions of yield point values and Olsen cupping test values are shown in Figs. 14c and 14d. Olsen values for only twenty-six of the lots were available, but it is evident that this test does not discriminate satisfactorily between good and poor material. Inspection of the distribution of yield point values also reveals that these values are distributed in such a way that no effective separation of good and poor lots can be made.

It is important to point out that in only a very few instances in the present investigation did material having poor press performance exhibit a definitely unfavorable microstructure. It is generally considered that the elongated grain structure characteristic of cold-reduced and box-annealed aluminum-killed steel which has been rapidly cooled following finish rolling of the hot strip (15) is a favorable characteristic for best performance in severe fender draws. The equiaxed grain structure, on the other hand, which can result from slow cooling of the hot strip after finish rolling, is generally associated with poor press performance in the types of "forming" operations considered in this investigation. All of the good lots which were examined possessed an elongated grain structure; twenty-three of the poor lots also had an elongated grain structure, with the remaining six poor lots exhibiting an equiaxed grain structure. Among the twenty-three poor lots with an elongated grain structure, no excessively small nor excessively large grain sizes were found, indicating that neither grain shape nor grain size could account for the poor performance of more than six out of a total of twenty-nine poor lots. It should be pointed out that all of the lots having an equiaxed grain structure exhibited low "R_L" values. It should not be inferred from this observation, however, that elongated grain material will always possess high "R_L" values. As a matter of fact, not only did quite a number of lots having elongated grain structure exhibit low "R_L" values, but also an elongated grain lot had the lowest "R_L" value of any observed in this investigation.

Magnetic torque curves were obtained for all of the forty-six lots of known drawing performance. An examination of Fig. 15, in which the distribution of the various types of torque curves is plotted, reveals that fifteen out of seventeen good lots are characterized by type "A" curves. However, it is also apparent from Fig. 15 that the performance of all the poor lots cannot be accounted for on the basis

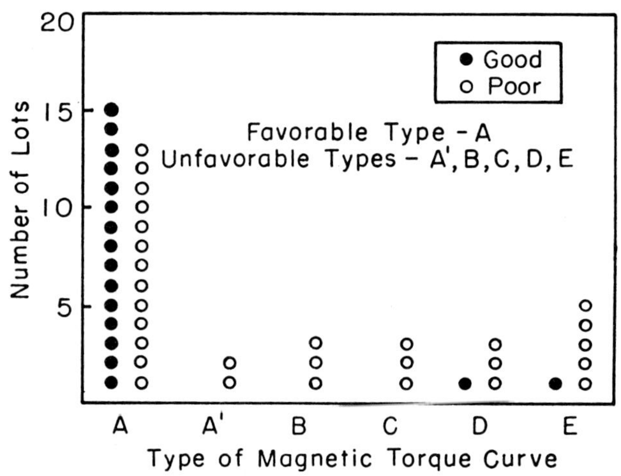

Fig. 15—Distribution of Types of Magnetic Torque Curves for All Lots Tested.

of unfavorable types of torque curves alone. This observation is not unexpected, since it has previously been shown that the "R_L" values, which are related to the magnetic torque characteristics, likewise could not account for the performance of all of the poor lots. It will be recalled that the best separation of good and poor lots was made by considering the "n" value in combination with the "R_L" value. If it is now assumed that a type "A" torque curve, together with a minimum "n" value of 0.240, is required for good press performance, it is found that fifteen of the seventeen good lots and twenty-five of the twenty-nine poor lots can be correctly rated. The relationship between "R_L" value and the type of magnetic torque curve for all forty-six lots investigated is shown in Fig. 16. It can be seen that the correlation is about the same as that previously shown in Fig. 6 for the first group of twenty lots.

C—Performance Trial of Materials Processed so as to Have Different Plastic Flow Characteristics

As already pointed out, all of the formations considered in the above correlations of various properties with drawing performance

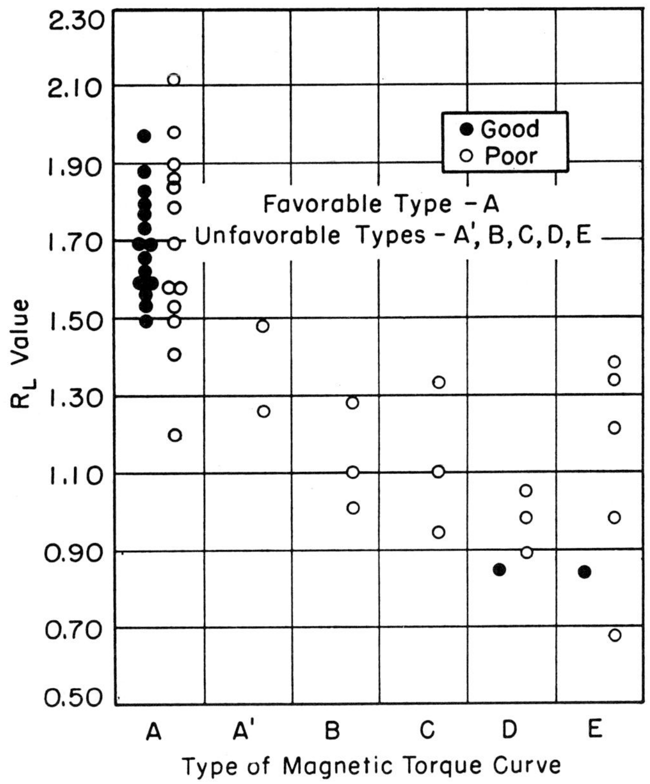

Fig. 16—Qualitative Correlation between "R_L" Value and Type of Magnetic Torque Curve for All Lots Tested.

involve severe biaxial stretching of an unsymmetrical nature. In view of the fact that most of the poor lots studied in this investigation exhibited considerably lower "R_L" values than those characteristic of good material, there can be little doubt that plastic anisotropy is a major factor in controlling press performance in the forming operations considered. It seems likely that anisotropy of a favorable character can lead to a better general distribution of strain in an unsymmetrical draw by minimizing localization of deformation.

In order to obtain more specific information of the effects of plastic anisotropy on press performance in an unsymmetrical type of

Table IV
Comparison of "R_L" and "n" Values for Trial Lots "L" and "X"

Lot	"R_L" Value	"n" Value
L	1.56	0.252
X	0.93	0.236

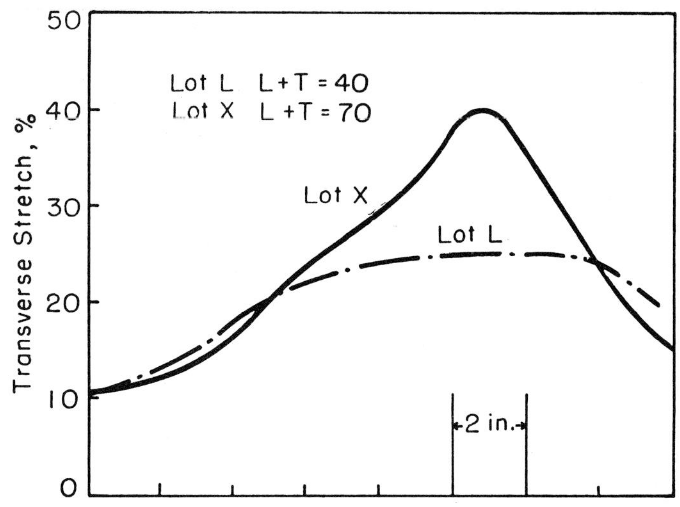

Fig. 17—Distribution of Strains in Fenders (Formation C) Formed from Trial Lots L and X.

forming operation, two special lots were produced for trial in Formation C. The first of these lots was processed in such a way that its properties were considered optimum for best performance in this formation, whereas the second lot was processed in such a way as to produce a more isotropic material having ordinary mechanical properties closely similar to those of the first lot. Press runs involving approximately seventy sheets from each of these lots were then made.

The "R_L" and "n" values for these two lots, which were designated as "Lot L" and "Lot X", respectively, are shown in Table IV. In the performance trials, the first lot ran with no breakage, whereas the second lot ran with 18% breakage. Study of the strain distributions in the noses of fenders formed from gridded blanks indicated that much more severe localization of deformation and much higher maximum strains occurred in material from Lot X than in material from Lot L. Since the maximum strain in fenders formed from Lot X was much greater than that in fenders formed from Lot L, the higher breakage in Lot X obviously cannot be attributed to a lower ultimate ductility, but, rather, must be accounted for on the basis of a less favorable distribution of strain, in which a local area of material makes a disproportionately large contribution to the overall stretch required to form the fender. It is believed that the favorable plastic anisotropy of Lot L is largely responsible for the more favorable strain distribution obtained in this unsymmetrical draw. In Fig. 17 a comparison can be made of the strain distributions obtained in fenders formed from the two lots. The curves shown represent the transverse strains measured along a transverse section passing through the point of maximum strain in the nose of the fender. It can be seen that the maximum transverse strain in the fender formed from Lot L is 25%, whereas the corresponding maximum strain in the fender formed from Lot X is 40%. The sums of the longitudinal and transverse stretches at these points are also indicated in Fig. 17 and amount to 40 and 70% for Lots L and X, respectively. Whether this difference in strain distribution is a consequence of difference in plastic anisotropy alone cannot be stated at the present time, since the "n" value for Lot X is somewhat lower than that for Lot L. This difference in "n" value could well have contributed to the observed difference in strain distribution, although it is believed that the major effect is that due to the difference in plastic anisotropy.

General Discussion

As indicated previously, it has been generally believed that material substantially free from directionality in plastic properties would result in the best press performance in any type of sheet metal forming operation. Apparently this belief has its origin in difficulties experienced in the formation of symmetrical shapes, such as cylindrical cups, where it is well known that pronounced plastic anisotropy produces earing and frequently leads to breakage. Since it has come to be recognized so widely that these difficulties in symmetrical types of forming operations can be eliminated through the use of material processed so as to have a minimum of anisotropy, the assumption has apparently been made rather generally that such materials would likewise be best suited for other types of forming operations, and, indeed, numerous statements to this effect have been made in the literature.

As the present investigation proceeded, it became evident that the prevalent ideas regarding the desirability of using isotropic material in all types of deep drawing operations should be modified. Although it is still believed that an isotropic material is best suited for symmetrical forming operations, it has been demonstrated that material having a considerable degree of plastic anisotropy of a favorable nature results in the best press performance in the particular unsymmetrical formations studied in this investigation. On the other hand, it has been shown that material having a minimum of plastic anisotropy results in very poor press performance in these same unsymmetrical formations. The superior performance of the anisotropic material in these formations is believed to be a consequence of a more favorable distribution of strain than that obtained with isotropic material in the same formations. The evidence obtained in the strain distribution studies on fenders formed from Lots L and X, described previously, supports this hypothesis. In this experiment, it was shown that fenders formed from the nearly isotropic material represented by Lot X exhibited a very severe localization of strain, whereas fenders formed from the markedly anisotropic material represented by Lot L exhibited a more widespread distribution of strain and consequently much lower maximum total strains. This difference in maximum strain was, of course, reflected in the higher breakage observed for Lot X. It should be emphasized that Lot X was purposely processed in such a way as to be ideally suited for a symmetrical draw, i.e., so as to have a minimum of anisotropy. This lot is thus representative of the type of material which is widely used to obtain best results in symmetrical forming operations.

On the basis of this evidence, it appears that the material requirements for best drawing performance for all types of formations must be re-evaluated. In particular, information developed in this investigation strongly indicates that the performance in many deep drawing operations can be improved by suiting the plastic anisotropy of the material to the symmetry of the formation.

Summary and Conclusions

1. Sheet steels having excellent press performance in difficult unsymmetrical fender draws exhibit marked plastic anisotropy as measured by the ratio of width strain to thickness strain ("R" value) in the simple tension test.

2. This plastic anisotropy has been found to be associated with preferred crystallographic orientations by correlating "R" values with magnetic torque characteristics, the latter being known to reflect preferred orientations.

3. A large number of examples of apparently anomalous drawing performance in severe fender draws can be explained on the basis of unfavorable plastic anisotropy, thus demonstrating the importance of favorable anisotropy as a factor in good press performance in this type of draw.

4. A quality index based on a combined requirement of favorable plastic anisotropy ("R_L" value) and favorable strain hardening exponent ("n" value) gives a very high degree of correlation with drawing performance in tests of forty-six lots having wide differences in press performance but exhibiting very similar ordinary mechanical properties.

5. The yield-tensile ratio can be used as an alternate to the "n" value as a measure of strain hardening, although somewhat poorer correlation with drawing performance is obtained when the yield-tensile ratio and the "R_L" value are considered in combination than when the "n" value and the "R_L" value are considered in combination.

6. Material processed so as to give good performance in a symmetrical draw, i.e., in such a way as to minimize anisotropy, is shown to result in high die scrap in an unsymmetrical fender draw.

7. It appears that the prevalent belief that a randomly oriented, isotropic sheet is best for any type of forming operation, other things being equal, should be abandoned in favor of the concept that the plastic anisotropy characteristics of the material must be suited to the symmetry of the forming operation.

Acknowledgments

Many of the data used in this study were obtained in a cooperative research project supported by the Carnegie-Illinois Steel Corporation at the Pennsylvania State College. The valuable contributions of Dr. J. R. Low in the planning and direction of this project are greatly appreciated. Important contributions were also made by members of the staff of the U. S. Steel Corporation Research Laboratory, as well as by numerous individuals in the Operating and Research and Technology Departments of the Carnegie-Illinois Steel Corporation. The authors wish to acknowledge especially the encouragement and assistance of Messrs. M. W. Lightner and M. A. Grossmann in the analysis and interpretation of the data and in the preparation of the paper.

References

1. J. Dudley Jevons, "The Metallurgy of Deep Drawing and Pressing," 1940, p. 437. John Wiley and Sons, Inc., 440 Fourth Avenue, New York City.
2. J. R. Low, Jr., "Mechanical Tests for Deep Drawing Sheets," The Modern Industrial Press, November 1940.
3. M. Asimow and J. N. Crombie, "Problems in the Drawability of Deep Drawing Sheets," TRANSACTIONS, American Society for Metals, Vol. 30, 1942, p. 639.
4. J. R. Low, Jr., and T. A. Prater, "Plastic Flow of Aluminum Aircraft Sheet under Combined Loads," Office of Scientific Research and Development, Report No. 4052, Aug. 22, 1944.
5. M. Gensamer, "Strength and Ductility," TRANSACTIONS, American Society for Metals, Vol. 36, 1946, p. 30.
6. J. H. Hollomon, "Tensile Deformation," *Transactions,* American Institute of Mining and Metallurgical Engineers, Vol. 162, 1945, p. 268.
7. J. R. Low, Jr., and Frank Garofalo, "Precision Determination of Stress-Strain Curves in the Plastic Range," *Proceedings,* Society for Experimental Stress Analysis, Vol. IV, No. 2, 1947, p. 16.
8. P. G. Nelson and Joseph Winlock, "Method of Determining the Percentage Elongation at Maximum Load in the Tension Test," Bulletin, American Society for Testing Materials, January 1949, p. 53.
9. L. R. Jackson, K. F. Smith and W. T. Lankford, "Plastic Flow in Anisotropic Sheet Steel," *Metals Technology,* American Institute of Mining and Metallurgical Engineers, August 1948, T.P. 2440.
10. W. M. Baldwin, Jr., T. S. Howald and A. W. Ross, "Relative Triaxial Deformation Rates," *Transactions,* American Institute of Mining and Metallurgical Engineers, Vol. 166, 1946, p. 86.
11. W. T. Lankford, J. R. Low, Jr., and M. Gensamer, "The Plastic Flow of Aluminum Alloy Sheet under Combined Loads," *Transactions,* American Institute of Mining and Metallurgical Engineers, Vol. 171, 1947, p. 574.
12. J. K. Stanley, "Orientation in Low Carbon Deep Drawing Steel," *Transactions,* American Institute of Mining and Metallurgical Engineers, Vol. 158, 1944, p. 354.
13. C. S. Barrett, "Structure of Metals," 1943, p. 450. McGraw-Hill Book Co., Inc., New York City.
14. "Recording Torque Magnetometer," *Mechanical Engineering,* Vol. 71, 1949, p. 510.
15. R. L. Rickett, S. H. Kalin and J. T. MacKenzie, Jr., "Recrystallization and Microstructure of Aluminum-Killed Deep Drawing Steel," *Journal of Metals,* American Institute of Mining and Metallurgical Engineers, Vol. 1, 1949, p. 242.

DISCUSSION

Written Discussion: By James W. Halley, chief research engineer, Inland Steel Co., Chicago.

The authors are to be congratulated on demonstrating a new and important variable in the drawing quality of steel.

The amount of effort and perseverance necessary to collect the 46 lots of variable performance can be appreciated only by those who have tried to get similar samples from production press shops.

The importance of the "R" value in drawing fenders is thoroughly demonstrated, but the authors have little confidence in its importance in symmetrical drawing operations. Though no data are given for symmetrical drawing operations, consideration of the significance of the "R" value would indicate that it should apply to all types of drawing operations.

The "R" value represents anisotropy of strain between the thickness and plane of the sheet. This is not necessarily related to anisotropy in the plane of the sheet. Fig. 2 of the paper shows this. Lot A has an "R" value greater than 1.2, when tested in various directions, and in no direction shows an "R" value as low as Lot C.

A high "R" value represents a marked resistance of the sheet to thinning and a strong tendency to pull metal from the sides. As shown in Fig. 2, this tendency is not confined to tensile stresses in any one direction in the plane of the sheet. A high "R" value should, therefore, be as beneficial in a symmetrical drawing operation as in an unsymmetrical drawing operation.

If the "R" value remains constant during uniform elongation, it should be possible to obtain a single index from the tensile test, which will represent the ability of the sheet to distribute plastic strain. If the thickness strain is expressed as a natural strain, and plotted against the actual stress, on log-log paper, a straight line should result. The slope of this line will give an "n" value that represents the resistance of the sheet to becoming thinner when strained in the plane of the sheet. This index, which could be called "n_t", would be an index of the drawing quality of the sheet.

Written Discussion: By R. H. Heyer and R. L. Solter, Research Laboratories, Armco Steel Corp., Middletown, Ohio.

The test methods developed by the authors are of great interest to a large group of producers and fabricators of steel sheets and strip, and it is hoped that the tests will prove to be indicative of drawing or stretching quality when put to more general use.

The objective of the work reported in this discussion is a comparison of the methods of the authors with other procedures reported in the literature and with certain variations of these methods which seemed promising.

Two lots of deep drawing steel were used, one of which was satisfactory and the other unsatisfactory for a difficult symmetrical drawing operation. The mechanical properties of these materials are:

	A (Satisfactory)	B (Unsatisfactory)
Thickness, inches	0.0425	0.0415
Yield Strength, psi	22,700	29,000
Tensile Strength, psi	42,500	43,800
Elastic Ratio	0.53	0.66
Elongation in 2 Inches, %	46	43
Rockwell B	34	38

Four-inch parallel gage section specimens similar to those used by the authors were machined. Ten lines were lightly scribed at 0.2-inch spacing in the center two inches of the specimen. Width and thickness were measured by hand micrometer at each line. The specimen was stretched in a hydraulic testing machine (2000-pound range) using Templin grips. Huggenberger tensometers were placed on the edges of the specimen for alignment purposes, using loads well under the yield point. The Huggenbergers were removed and a ×10 KBY autographic extensometer applied to the center section of the gage length. The test was continued to maximum load, which occurred at 25 to 30% extension. The specimen was removed from the testing machine and the width and thickness remeasured. The test was then taken to fracture. The rate of strain beyond the yield was 0.1 inch per inch of gage length per minute. After fracture, the width and thickness were measured at those lines which were 1 inch or more from the fracture.

The "n" and "R" values obtained by the following methods are summarized in Table IV.

Table IV
Summary of Experimental Values of n and R for Two Materials: A (Satisfactory Draw), B (Unsatisfactory Draw)

		A_1	A_2	B_1	B_2
n_1	Slope of true stress-strain curve	0.237	0.244	0.239	0.236
n_2	Nelson and Winlock[2] at 0.05 strain	0.285	0.285	0.277	0.277
n_3	Nelson and Winlock[2] at 0.10 strain	0.265	0.270	0.270	0.268
n_4	Ln (A_o/A) at maximum load	0.242	0.241	0.245	0.245
n_5	Ln (A_o/A) after fracture	0.311	0.318	0.252	0.288
n_6	Krupkowski and Kawinski[3]—ln $(1 + e')$ after fracture	0.311	0.321	0.254	0.288
n_7	Krupkowski and Kawinski[3] at maximum load	0.243	0.247	0.244	0.245
R_4	Authors' method	1.40	1.54	1.19	1.23
R_5	Alternate method proposed by authors	1.37	1.33	1.21	1.23
R_6	Krupkowski and Kawinski[3]—e'_h/e'_b after fracture	1.45	1.69	1.23	1.27
R_7	Krupkowski and Kawinski[3]—e'_h/e'_b at maximum load	1.48	1.57	1.21	1.26

n_1—From the autographic load extension curve, true stress was calculated at strain intervals of 0.025 between a strain of 0.05 and the strain at maximum load. In this discussion, strain intervals indicated are expressed in percentage extension; i.e., 0.025 equals 2.5% elongation.

$$\text{true stress} = \frac{\text{load}}{A_o}(1 + e)$$

True strain was calculated from the extension values

$$\text{true strain} = \ln(1 + e)$$

where A_o is the original area

e is the measured extension (nominal)

n_1 was determined from the slope of the true stress – true strain curve as in Fig. 1. Elastic strain is neglected in this method.

n_2—The determination of n_2 is based on the method of Nelson and Winlock[2] using the load reading at 0.05 strain and the maximum load reading from the autographic curve.

n_3—Same as n_2 except based on 0.10 strain.

n_4—It has been shown that n is numerically equal to the uniform elongation; i.e., the true strain at maximum load.[2]

$$n_4 = \text{true strain at maximum load} = \ln\frac{A_o}{A}$$

[2]Paul G. Nelson and Joseph Winlock, "A Method of Determining the Percentage Elongation at Maximum Load in the Tension Test," *Bulletin,* American Society for Testing Materials, No. 156, January 1949, p. 53.

where A is the average cross sectional area at maximum load from width and thickness readings at ten locations.

n_5—Same as n_4 except A is the cross sectional area from width and thickness measurements made 1 inch or more from the fracture.

n_6—Krupkowski and Kawinski[3] have recently published a method for determining plastic anisotropy of metals based on measurements made after the fracture. Their coefficient K_a is somewhat similar to the authors' R.

$$K_a = \frac{e'_h}{e'_b}$$

where e'_h and e'_b are coefficients of partial uniform elongation defined as:

$$e'_b = 0.5\left[\left(\frac{b_o}{b_n}\right)^2 - 1\right]$$

$$e'_h = 0.5\left[\left(\frac{h_o}{h_n}\right)^2 - 1\right]$$

where b_o and b_n are initial and final thickness
h_o and h_n are initial and final width

$$e'_b + e'_h \cong e'$$

where e' is the coefficient of total uniform elongation (equivalent to e as used in determining n).

$$n_6 = \text{true strain} \cong \ln(1 + e')$$

n_7—Same as for n_6 except that the width and thickness measurements at maximum load were used.

R_4—The principal method used by the authors is based on average width and thickness at or near the maximum load. It is not clear to us whether the authors used nominal or true width and thickness strains. In this discussion

$$R_4 = \frac{\ln \frac{w_o}{w_m}}{\ln \frac{t_o}{t_m}}$$

R_5—Same as R_4 except measurements of width were made after fracture.

R_6—$R_6 = K_a = \frac{e'_h}{e'_b}$ according to Krupkowski and Kawinski.[3]

R_7—Same as R_6 except measurements were made after fracture.

The values n_1 and R_4 are most directly comparable to the n and R_L values reported by the authors.

The n_2 and n_3 values by the simplified method proposed by Nelson and Winlock have been found to be higher than n_1, both in the tests here reported and others conducted in our laboratory. However, the n_2 and n_3 values are quite reproducible.

It is interesting to note that the reported numerical equality between n and the total true strain to maximum load is quite well confirmed by the experimental values of n_4.

When this method was modified by using "after fracture" instead of "maximum load" measurements, the resulting values of n_5 were found to be high compared with n_1, and not closely reproducible.

---

[3]A. Krupkowski and S. Kawinski, "The Phenomenon of Anisotropy of Annealed Polycrystalline Metals," *Journal*, Institute of Metals, July 1949, p. 869.

Using the same "after fracture" measurements in an extension of the Krupkowski and Kawinski relationships, the values of n_6 check n_5 closely.

A further adaptation using measurements at maximum load gave n_7 values which are in good agreement with n_4 and reasonably close to n_1.

Reviewing the various n values, there appears to be no significant difference between the two materials. However, the R values determined by various methods all indicate superiority of the A (satisfactory) material.

The methods yielding n_4 and n_7 are the most promising of the rapid methods for determining the work hardening rate. However, these are both more difficult to calculate than the Nelson and Winlock method yielding n_2 and n_3.

Experimentally, n_5 and n_6 are the most rapid. Although the results are not in agreement with n_1, with more experience they may be found to be significant.

The R_6 value can be obtained with the simplest test procedure and calculations, hence should be given consideration as an alternate to the method for R_4.

Our data are not extensive enough to justify any of the modified procedures reported here. We would be interested, however, in the authors' evaluation of the several methods.

Written Discussion: By T. S. Howald, chief chemist, Chase Brass & Copper Co., Inc., Midwestern Division, Euclid, Ohio.

It is encouraging to know that others are thinking along the line that directionality in deep drawing metal is not to be regarded as completely detrimental. For some time, we have mused upon this very thought but, because of the great efforts that are made in mill practice to eliminate directionality and, likewise, because of the aversion which the press shop operator has for directional metal, we have not pursued this idea to any great extent. In our own files, we have a very good example of where it was possible to decrease the blank size and make a difficult draw in 70/30 brass by virtue of the fact that the material had anisotropic properties.

The concept of evaluating some simple quantity from a simple tension test that would correlate with the deep drawing property of metal is not entirely new. In the second half of our paper,[4] we demonstrated a high degree of correlation between slope σ plots (which are reciprocal R values described in this paper) and the earing or drawing properties of copper having varying degrees of directionality. Since this time, this work has been further extended and the same correlation has been found in the case of gilding metal,[5] cartridge brass,[6, 7] aluminum, nickel, aluminum-killed and rimming steel.[7] The present paper is an excellent confirmation

---

[4]Wm. M. Baldwin, Jr., T. S. Howald and A. W. Ross, "Relative Triaxial Deformation Rates," *Transactions*, American Institute of Mining and Metallurgical Engineers, Vol. 166, 1946, p. 86.

[5]T. S. Howald, "Directionality in Gilding Metal," Master's Thesis, Case Institute of Technology, 1947.

[6]M. L. Fried and R. E. Spear, "A Correlation of Triaxial Deformation Rates with Earing Effects in 70/30 Brass," Undergraduate Thesis, Case Institute of Technology, 1946.

[7]Unpublished work, Research Department, Midwestern Division, Chase Brass & Copper Co., Inc., Euclid, Ohio.

of this work and demonstrates that the test can be very easily applied on a commercial basis to determine drawing properties.

There is one point to re-emphasize and that is that, as yet, there is no exacting fundamental explanation for the correlation that exists between the slope σ curves and the position of ears on cups drawn from circular blanks of anisotropic metal.

Written Discussion: By L. R. Jackson, research supervisor, Battelle Memorial Institute, Columbus, Ohio.

The authors are to be congratulated, not only for their excellent analysis of the data, but also for being so fortunate as to have so many heats in which press performance data and detailed mechanical property data were both available.

Assuming that some of the material is still available, I wonder if the authors have considered the idea of extending the comparison between "R" values and magnetic to torque curves with respect to the following considerations.

The "R" values certainly represent plastic anisotropy. While the plastic anisotropy may be determined by preferred orientation, it might also be affected by other variables. That is, one might imagine a material that would have random crystallographic orientation of the grains and would be elastically isotropic but which might still be plastically anisotropic for other reasons. The magnetic torque measurements are probably governed mainly by preferred crystallographic orientation, but these also might be affected by the other variables that control plastic anisotropy.

In view of the fact that the magnetic torque test is potentially a more rapid test than the determination of "R" values, it would seem desirable to have a clear-cut understanding of the relation between magnetic torque results and "R" values. If a satisfactory relationship could be established, the practical usefulness of the torque test could be fully realized.

Authors' Reply

The authors wish to express their appreciation for the very constructive discussion presented to this paper.

Mr. Halley has raised a very interesting question regarding the mechanism by which plastic anisotropy influences behavior in the press. He suggests that a material such as Lot A, having "R" values in all directions in the plane of the sheet greater than one, would give better performance in a symmetrical drawing operation than an isotropic material. It is difficult to refute this suggestion directly because most of the symmetrical draws with which we have had experience involve an added complication, namely, the hazard of earing. Material such as Lot A is known to be subject to earing, and in some instances may ear to such an extent as to prohibit successful formation of the part. Under such conditions, an isotropic material is obviously superior. If, for the sake of argument, earing is omitted from the discussion, we still do not believe that the problem is as simple as visualized by Mr. Halley. It is true that a high "R" value represents a marked resistance of the sheet to thinning and a strong tendency to pull metal from the sides in a simple tension test. When the material is subjected to biaxial stretching, however, the picture is not so simple. If one stretches a sheet uniformly in all directions in the plane of the sheet, it would appear that the amount of thinning would hardly depend on the "R" value.

Mr. Halley has also suggested a single quality index, "n_t", apparently with the thought of combining the work hardening and directionality attributes into a single parameter. If we understand the suggested procedure correctly, it is not evident that any quantity other than the "n" value as described in the paper would be determined. If, as Mr. Halley assumes, and as was stated to be the case in the paper, the "R" value remains constant during uniform elongation, then the thickness strains and the width strains are fixed fractions of the axial strains (elongations). The only effect of plotting thickness strain rather than axial strain as was done in the paper would be to cause a displacement of the straight-line log-log stress-strain curve parallel to itself, and the "n" values determined by the two methods would be identical. In the paper, the power law relationship was written

$$\sigma = k\delta^n$$

with δ referring to the true axial strain, which shall now be designated as δ_L. The "R" value is defined as

$$R = \frac{\delta_W}{\delta_T}$$

where δ_W and δ_T are the true width and thickness strains. Therefore,

$$\delta_W = R\delta_T$$

Making use of the constancy of volume condition

$$\delta_W + \delta_T + \delta_L = 0$$

it can readily be shown that

$$(R + 1)\,\delta_T = -\delta_L$$

Considering absolute values of strain, δ_L can be replaced in the power law relationship with the following result:

$$\sigma = k\,(R + 1)^n\,\delta_T{}^n$$

The factor $(R + 1)^n$ is constant for any given test and only acts to displace the log-log plot of stress versus strain parallel to itself. It can be seen, therefore, that although the displacement of the stress versus thickness strain plot would be a function of the "R" value, the slope of the log-log plot would remain unchanged and two parameters would still be required to describe both the strain hardening characteristics and the plastic anisotropy.

The discussion presented by Messrs. Heyer and Solter is appreciated not only because data confirming the importance of the "R" value are presented, but also because of the valuable comparisons of various modified procedures for determining the "n" and "R" values. At the present time, we are not in a position to add to the evaluation of the several methods discussed, although certain of them are currently being investigated. In the investigation described in the paper, it was our belief that the utmost attention should be paid to experimental methods and that the most precise techniques available should be employed. In this way, the greatest significance could be attached to observed differences in the properties of various materials. Once these significant differences are isolated, modifications of a practical nature in testing procedure can be considered more rationally.

Mr. Howald's interesting discussion includes some additional confirmation of the concept of favorable directionality based on observations of 70-30 brass. With regard to Mr. Howald's statement, "The concept of evaluating some simple quantity from a simple tension test that would correlate with the deep drawing property of metal is not entirely new," it was not our intention to imply that the evaluation of directionality by measurement of the "R" value is new. Reference was made to the excellent paper by Baldwin, Howald and Ross (Ref. 10), in which measurements of this type are reported. The correlation with drawing performance in that paper, however, was concerned specifically with the earing tendency only, and not with the problem of breakage and the more general aspects of the influence of plastic anisotropy in both symmetrical and unsymmetrical draws.

Mr. Jackson has raised an important question regarding the relationship between plastic and magnetic anisotropy. Theoretically, this relationship could best be rationalized in terms of a knowledge of the actual preferred crystallographic orientations existent in the material. Practically, however, this is an exceedingly difficult problem for two reasons: (a) the degrees of preferred crystallographic orientation in the materials discussed in the paper are relativey low; (b) the crystallographic system of slip in alpha iron is so complex that even if the exact description of the preferred orientations were available, a direct relationship with the plastic behavior would be very difficult to establish. For the time being, therefore, it seems likely that the "R" values and magnetic torque curves must be correlated on an empirical basis, in which case a limited generality must be placed on any observed correlation.

It should be emphasized, however, that in the tests reported in the paper, clear-cut differences in magnetic torque curves were always found when large differences in "R" values existed. As stated in the paper, this observation is interpreted as indicating that the type of plastic anisotropy discussed in the paper is associated with preferred crystallographic orientations.

THE IMPORTANCE OF DIRECTIONALITY IN DRAWING QUALITY SHEET STEEL

By R. L. Whiteley

While directionality which produces earing during drawing is generally objectionable, another component of directionality representing differences in properties perpendicular and parallel to the plane of the sheet can be an asset to good drawability. This component of directionality can be measured in a simple tensile test by the ratio of transverse strains. The relationship between drawability and directionality as measured by this strain ratio is demonstrated in a simple cup drawing operation. The results of the experimental work are explained by a theoretical analysis of cup drawing which allows for the anisotropy or directionality of the metal. (ASM-SLA Classification: Q23q, G4; ST, 4–53)

Directionality in drawing quality steels has long been of major interest to fabricators and the subject of several experimental investigations (1,2,3).[1] Usually it is recognized only in a two dimensional sense as indicated by differences in properties measured in the plane of the sheet. This "planar" directionality is, of course, readily observed in the press shop as "ears" in cup drawing operations, the height of the ears being indicative of the severity of the differences in properties. As such, directionality is usually considered objectionable in drawing quality steels.

Absence of earing during drawing does not preclude other components of directionality in the sheet, however. Directionality is a three dimensional effect and its absence in the plane of the sheet does not assure that properties measured in a direction perpendicular or normal to the sheet are the same as the properties measured in the plane of the sheet. The "normal" directionality of a material is often more pronounced than its planar directionality, and while not visually observed in the press shop, is nevertheless important to the press performance of the material. Moreover, unlike the planar directionality, normal directionality can be beneficial to the press performance of the material.

The importance of normal directionality has probably not been recognized, because it is not visually observed in the press shop and the obvious difficulty of measuring properties perpendicular to the plane of the sheet. However, tensile specimens of sheet material on plastic

[1] The figures appearing in parentheses pertain to the references appended to this paper.

The author, R. L. Whiteley, is Supervisor of Mechanical Metallurgy, Physical Metallurgy Division, Research Department, Bethlehem Steel Company, Bethlehem, Pennsylvania. Manuscript received April 13, 1959.

deformation often exhibit different amounts of strain in their width and thickness dimensions. This behavior is a manifestation of normal directionality and can be demonstrated (4) to be a consequence of a difference in the strength of the material perpendicular and parallel to the plane of the sheet. The ratio of the width and thickness strains measured in a tensile specimen at any strain level is almost a constant value. This constant strain ratio, R, has previously been suggested as a measure of directionality (2,3).

If R is measured in several directions in the sheet, an indication of both the planar and normal components of directionality can be obtained. Planar directionality is indicated by a difference in R in the different directions, while the normal directionality is indicated by the average value of R. A completely isotropic material would have a value of R equal to one in all directions. A material with an average value of R greater than one would be generally more resistant to thinning and puckering during forming than a perfectly isotropic material. Such a material should be desirable for most drawing operations.

The strain ratio was first related to drawability by Lankford, Snyder and Bauscher (2) on the basis of the press performance of forty-six lots of aluminum-killed steels applied to unsymmetrical fender draws. They found that materials with high values of R showed better press performance than materials with low R values. However, they felt the correlation established could only be explained by the unsymmetrical nature of the fender draws, and would not persist in a symmetrical drawing operation. Discussions of their paper by Halley, and by Heyer and Solter suggested that directionality indicated by high values of R should be as desirable for symmetrical drawing operations as for unsymmetrical drawing operations. The present study offers an experimental and theoretical demonstration that a relationship between drawability and directionality for a symmetrical drawing operation does exist.

Experimental Method

The experimental work was designed to evaluate the drawability of several materials of widely varying properties and directionality with the purpose of determining the relative effect of directionality and other properties upon drawability.

Twenty-two drawing materials were tested; sixteen low carbon steels (eight rimmed and eight aluminum-killed), two stainless steels, two aluminum alloys, commercially pure copper and 65/35 brass. These materials ranged in yield strength from 5000 psi. to 45,000 psi in total elongation from 25 to 50%. The average strain ratio, R, ranged from 0.6 to 1.6 indicating a considerable degree of normal directionality in certain materials. The stainless steels and nonferrous metals were all commercially processed, but some of the low carbon steels were specially treated to attain a greater range of properties than ordinarily en-

Table I
Description and Mechanical Properties of Materials Tested

Code No.	Material	Condition	Yield Strength psi.	Tensile Strength psi.	% Elongation in 2 inches
10	Rimmed steel	Annealed[1]	23,200	38,500	43.6
11	Rimmed steel	Annealed	31,600	43,600	41.5
18	Rimmed steel	Annealed	31,300	42,900	41.0
22-1	Rimmed steel	As skin rolled	25,700	45,500	39.3
22-2	Rimmed steel	Aged 43 days	31,300	45,200	36.7
23	Rimmed steel	As skin rolled	26,700	44,400	36.4
24	Rimmed steel	As skin rolled	24,700	43,000	40.1
25	Rimmed steel	As skin rolled	26,900	44,600	39.7
1	Killed steel	As skin rolled	23,200	45,100	36.3
2	Killed steel	Annealed[1]	14,400	41,400	42.0
14	Killed steel	As skin rolled	24,600	43,900	40.6
19	Killed steel	As skin rolled	24,600	45,100	40.1
20	Killed steel	As skin rolled[2]	27,300	45,600	37.8
21	Killed steel	Annealed	31,300	44,700	40.6
34	Killed steel	As skin rolled	19,800	42,500	42.2
38	Killed steel	As skin rolled	23,700	44,700	40.5
3	430—Stainless	Annealed	45,400	78,200	25.2
6	301—Stainless	Annealed	43,200	93,900	50.7
4	2SO—Aluminum	Annealed	5,200	13,300	34.6
5	52SO—Aluminum	Annealed	12,800	28,200	25.8
7	Copper[3]	Annealed	12,300	31,300	43.7
9	65/35 Brass	Quarter hard	38,600	55,500	37.0

Notes:
[1] Annealed in wet H_2
[2] Skin rolled—2%
[3] Commercially pure copper

countered commercially. A brief description of the materials and their properties is given in Table I.

The drawability of each material was evaluated by determining the largest blank of the material which could be successfully drawn into a 2-inch flat-bottomed cup like that shown in Fig. 1. This drawing operation was carried out on a small drawing press (5) under conditions proposed by the Swift-Cup-Forming Sub-Committee of BISRA (6). The procedure followed was to draw successive blanks of increasing diameters until the blanks failed to draw. The drawability of the material was then expressed as the ratio of D, the diameter of the largest blank successfully drawn, to d, the diameter of the drawn cup, (in this case two inches). The ratio D/d is called the drawing ratio.

The directionality of each material was determined by measuring the strain ratio, R, at different angles to the rolling direction. This was done by measuring the longitudinal and width strains of a ¾-inch wide parallel sided specimen elongated approximately 20%. The longitudinal strain was measured over a 2-inch gage length while the width strain was the average of five values measured within the 2-inch gage length. On the assumption that the volume of the specimen is constant during plastic deformation, the strain ratio, R, was then calculated from these values of width and length strains (3). Strain ratios determined by this procedure were found to be more accurate and reproducible than values obtained by measuring the thickness strains directly. Other mechanical properties were determined by standard tests.

Fig. 1—Initial Test Blank and Final Cups Formed in a Successful and an Unsuccessful Cup Drawing Test. Earing of the successfully formed cup is a result of the planar directionality of the material.

Experimental Results

Considering the extreme range of properties in the materials tested, the range of drawabilities as measured by the drawing ratio was not great. The best material, an aluminum-killed steel, had a drawing ratio

Table II
Drawability and Directionality of Materials Tested

	Drawability		Directionality	
Code No.	Maximum Drawing Ratio D/d	Direction of earing, Angle to R. D.	Average Strain Ratio $R_{avg.}$	Direction of Maximum R, Angle to R. D.
10	2.175	0°, 90°	1.01	0°, 90°
11	2.15	0°, 90°	1.12	0°, 90°
18	2.20	0°, 90°	1.11	0°, 90°
22-1	2.175	0°, 90°	1.13	0°, 90°
22-2	2.175	0°, 90°	1.10	0°, 90°
23	2.175	0°, 90°	1.30	0°, 90°
24	2.20	0°, 90°	1.28	0°, 90°
25	2.175	0°, 90°	1.10	0°, 90°
1	2.25	0°, 90°	1.48	0°, 90°
2	2.30	0°, 90°	1.62	0°, 90°
14	2.225	0°, 90°	1.64	0°, 90°
19	2.25	0°, 90°	1.35	0°, 90°
20	2.25	0°, 90°	1.34	0°, 90°
21	2.225	0°, 90°	1.40	0°, 90°
34	2.225	0°, 90°	1.54	0°, 90°
38	2.20	0°, 90°	1.34	0°, 90°
3	2.15	0°, 90°	1.13	22½°, 90°
6	2.175	No ears	0.93	No max.
4	2.10	0°, 90°	0.65	0°, 90°
5	2.125	0°, 90°	0.58	0°, 90°
7	2.225	No ears	0.95	No max.
9	2.15	No ears	0.74	No max.

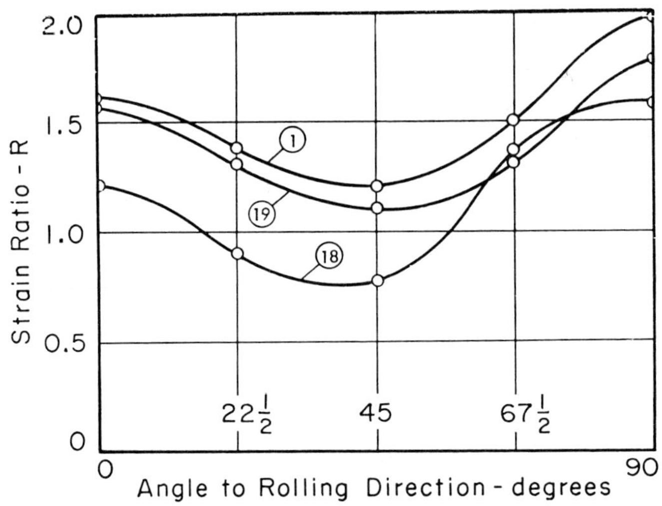

Fig. 2—Typical Directional Variation of the Strain Ratio, R Observed in All the Low Carbon Steels Tested. Steels 1 and 19 are drawing quality aluminum-killed steels while steel 18 is a drawing quality rimmed steel.

of 2.30, while the poorest material, 2S0 aluminum, had a drawing ratio of 2.10. While this is only a 10% difference in drawing ratio, it represents nearly a 30% difference in the depth of cup formed, a perhaps more significant parameter. In general, the aluminum-killed steels showed the best drawabilities, while the stainless steels and nonferrous metals had the poorest (See Table II). The drawing ratios for the aluminum-killed steels and rimmed steels were similar to those reported by Kemmis (6). While most of the materials produced ears when drawn, the occurrence of earing did not appear to affect the relative drawability of the materials.

On the basis of the measured R values none of the materials were perfectly isotropic (See Table II). Copper and the austenitic stainless steel which showed the least directionality had strain ratios slightly less than one in all directions. The 65/35 brass showed little planar directionality but had an average strain ratio of 0.74 indicating a significant degree of normal directionality. All other materials possessed considerable planar directionality as indicated by the different values of strain ratio measured in different directions. Fig. 2 shows the directional variation in R for three low carbon steels, these variations being typical of that found in almost any box-annealed low carbon steel. The correspondence between earing and the directional variations in R was quite good; the height of the ears produced generally varied according to the degree of fluctuation in R and in almost all cases the ears formed in the directions in which R was a maximum. All the low carbon steels tested exhibited ears at 0 ° and 90 ° to the rolling direction, and the only materials which did not ear (brass, copper, and austenitic stainless steel) were those which showed only little or no directional variation in R.

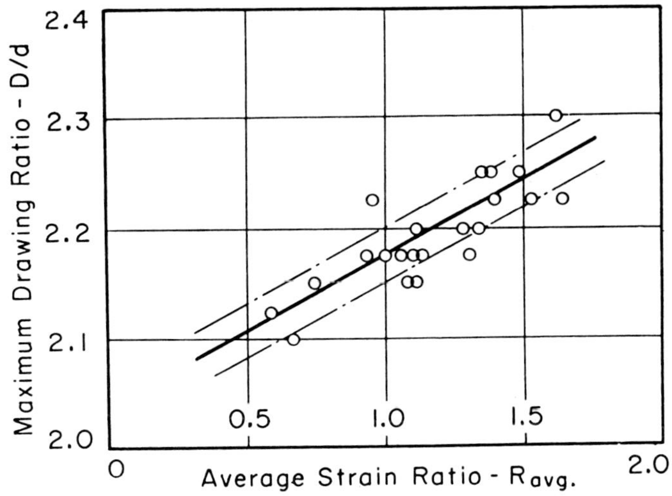

Fig. 3—Effect of the Average Strain Ratio, $R_{avg.}$, On the Maximum Drawing Ratio, D/d. The dashed lines represent the limits of experimental error.

Table III
Correlation of Drawing Ratio with Mechanical Properties

Mechanical Property	Coefficient of Correlation—r[1]	Significance of Correlation
Yield Strength	−0.12	Not significant
Tensile Strength	+0.07	Not significant
Yield-Tensile Ratio	+0.16	Not significant
Elongation in 2"	+0.42	Not significant
Olsen Ductility	+0.44	Not significant
Strain Hard. Exp.-n	+0.04	Not significant
Strain Ratio—$R_{avg.}$	+0.83	Highly significant[2]
Factor—R × n	+0.73	Highly significant[2]

[1] For perfect correlation $r = \pm 1$, a negative coefficient indicating an inverse correlation.
[2] Probability that correlation is due to chance is less than 0.1%.

The normal directionality of each material was determined by averaging the strain ratios measured in all directions. Fig. 3 shows the relationship between the drawing ratios and the average strain ratios of the materials. While the data show considerable scatter, a definite relationship exists between drawability and normal directionality. If we assume the experimental accuracy of the drawing ratio is ±.025, a value suggested by Kemmis, a direct straight-line relationship is indicated between the drawing ratio, D/d, and the average strain ratio, R_{avg}.

The relative dependence of drawability on the strain ratio and other mechanical properties was determined by treating the data statistically to determine coefficients of correlation between the drawing ratio and each of the properties measured. The results of this statistical analysis are summarized in Table III. The only single property to show a significant relationship to drawability was the average strain ratio. The best indication of press performance found by Lankford, et al., was the factor R × n, where n was the strain-hardening exponent of the steel. While a similar correlation can also be established for the present data, the correlation is not as good as that between the drawing ratio and the value of R alone. The present data indicates that drawability is determined largely, if not entirely, by the normal directionality of the material.

Discussion

The singular dependence of the drawing ratio upon R found by the above experiments is predicted by an analysis of the forces acting during drawing. The analysis, described in the Appendix, shows that the strength of the wall of the forming cup increases with the value of the strain ratio of the material. Because a material with a high strain ratio can support a greater punch load than an isotropic material of the same tensile strength, a larger diameter blank can be drawn into the die without failure in the wall of the cup. Other properties which act to strengthen the wall of the cup (for instance, tensile strength) also act proportionately to increase the punch force required to draw a blank of a given diameter. As a result these properties do not affect the overall drawability of the material.

The relation between press performance and the factor $R \times n$ found by Lankford, et al., is probably due to the character of the forming operations they studied. The drawing of a flat-bottomed cup involves almost no stretching, but the forming of a fender usually involves a considerable degree of stretching as well as drawing. In a pure stretching operation the forming limit is determined by the useful ductility (uniform elongation) of the material. This can usually be related to the strain-hardening exponent, n, of the material. Thus in a forming operation which involves both stretching and drawing, such as a fender draw, the press performance would more than likely depend on both R and n.

The superior press performance of aluminum-killed steels over rimmed steels recognized by fabricators for some time was also observed in the present experiment. Past efforts to correlate this superior performance with a simple material property have heretofore been unsuccessful. However, in the present experiment the superior drawing performance of the aluminum-killed steels can be attributed entirely to their higher average strain ratios, i.e., their greater directionality.

Conclusions

Directionality is an important property of drawing quality steels not always detrimental to press performance. While planar directionality is usually objectionable, normal directionality indicated by a high average strain ratio is desirable for good drawability even in symmetrical drawing operations. The general superiority of aluminum-killed steels over rimmed steels for deep drawing applications can be attributed to the greater normal directionality of the aluminum-killed steels.

Appendix

In a simple cup drawing operation, the portion of the blank beyond the die opening is drawn inward toward the punch. As each circular element reaches the die opening it is bent down over the die profile and pulled through the die opening to form a portion of the cup wall. The operation is completed when the periphery of the blank is reduced to the circumference of the cup wall.

The punch load required for this operation is determined by the forces required to plastically deform that portion of the blank outside the die opening. This force acts from the perifery of the die opening in the form of radial tensile stresses. These stresses are comparable to the internal stresses required to plastically deform a hollow cylinder whose outer diameter is equal to the initial blank diameter, D, and whose inner diameter is equal to that of the final cup diameter, d. The solution of this problem for a directional or anisotropic (but nonstrain hardening) material has been carried out by Hu (7). The radial stress is:

$$\sigma_r = \sqrt{\alpha_{33}/G}\,\kappa \ln(D/d) \qquad \text{Equation 1}$$

where α_{33} and G are anisotropic parameters
κ = the effective strength
σ_r = the radial stress acting at r = d/2.

This stress multiplied by πdt (t = thickness of the blank) is the portion of the punch load required to plastically deform the material itself. Willis, (8) summarizing the work of Swift, states that this load must be increased by a factor of $(1 + \eta)$ to account for frictional forces. The parameter η is constant for given conditions of geometry and friction, and usually has a value between 0.2 and 0.3. The total punch load required to draw a blank of diameter D, is thus:

$$P_r = (1 + \eta)\pi dt\sqrt{\alpha_{33}/G}\,\kappa \ln(D/d) \qquad \text{Equation 2}$$

However, the maximum diameter blank that can be drawn is limited by the axial (punch) load which can be supported by the material forming the wall of the cup. Using Hu's (7) equations for plane stress, we can also determine this load in terms of the same anisotropic parameters. Hu expresses the effective strength of the material in plane stress by the equation:

$$\alpha_{11}\sigma_{11}^2 - 2\alpha_{12}\sigma_{11}\sigma_{22} + \alpha_{22}\sigma_{22}^2 = \kappa^2 \qquad \text{Equation 3}$$

where $\alpha_{11}, \alpha_{12}, \alpha_{22}$ are anisotropic parameters
σ_{11} = the principal stress in the 11 direction
σ_{22} = the principal stress in the 22 direction

Associating 11 and 22 with the axial and circumferential directions of the cup wall we must impose the restricting condition that the circumferential strain, $de_{22} = 0$. Applying this restriction to Hu's stress-strain increment equations and solving Equation 3 for σ_{11}, the axial strength of the cup wall is found to be:

$$\sigma_a = \sqrt{\alpha_{22}/G}\ \kappa \qquad \text{Equation 4}$$

This strength multiplied over the cross-sectional area of the cup wall πdt gives the maximum punch load which can be supported by the cup wall:

$$P_{max} = \pi dt \sqrt{\alpha_{22}/G}\ \kappa \qquad \text{Equation 5}$$

The maximum diameter blank which can be drawn is that which requires a punch load just equal to the maximum load which can be supported by the wall of the cup. Thus equating 2 and 5, the maximum drawing ratio D/d is found to be determined only by the anisotropic parameters of the material for given conditions of geometry and lubrication, i.e., for $\eta =$ constant.

$$\ln (D/d) = 1/(1 + \eta) \sqrt{\alpha_{22}/\alpha_{33}} \qquad \text{Equation 6}$$

Note that κ, the effective strength of the material, does not effect the maximum drawing ratio.

If we assume a material with only normal directionality (rotational isotropy) the above anisotropic parameters can be expressed in terms of the average strain ratio, R. Such a material requires that:

$$de_{33}/de_{22} = R_{11} = R_{22} = de_{33}/de_{11}$$

Applying this restriction and that of symmetry in the 11 and 22 directions, and arbitrarily setting $\alpha_{11} = 1$, the following relationships are found:

$$\begin{aligned} \alpha_{11} &= \alpha_{22} = 1 \\ \alpha_{33} &= 2/(1 + R \\ \alpha_{12} &= R/(1 + R) \\ \alpha_{13} &= \alpha_{23} = 1/(1 + R) \\ G &= (1 + 2R)/(1 + R)^2 \end{aligned}$$

Using these relationships, Equations 2, 5, and 6 can be rewritten in terms of the strain ratio, R.

$$P_r = (1 + \eta)\ \pi dt \sqrt{(2 + 2R)/(1 + 2R)}\ \kappa \ \text{In}\ (D/d) \qquad \text{Equation 2-R}$$

$$P_{max} = \pi dt \sqrt{(1 + R)^2/(1 + 2R)}\ \kappa \qquad \text{Equation 5-R}$$

$$\ln (D/d) = \sqrt{(2 + R)\ /\ (2 + 2\eta)} \qquad \text{Equation 6-R}$$

Equation 6-R indicates a nearly linear relationship between D/d and R over the range of values usually experienced in practice. The effect of R on the maximum drawing ratio is primarily due to the effect of R on the strength of the cup wall as defined by 5-R, the effect R on Pr only being slight.

A similar analysis of the forces obtained in cup drawing was carried out by the author in which the strain-hardening properties of the material were also considered. The relationship between the drawing ratio and the strain ratio was found to be approximately the same as that expressed by Equation 6-R.

Acknowledgments

The author wishes to acknowledge the suggestions and encouragement given this project by Dr. D. J. Blickwede and Mr. J. W. Frame, and to acknowledge the assistance of Mr. J. W. Covino in performing the various tests.

References

1. J. K. Stanley, "Orientation in Low Carbon Deep Drawing Steel," *Transactions*, American Institute of Mining and Metallurgical Engineers, Vol. 158, 1944, p. 354.
2. W. T. Lankford, S. C. Snyder, and J. A. Bauscher, "New Criteria for Predicting the Press Performance of Deep Drawing Sheets," TRANSACTIONS, American Society for Metals, Vol. 42, 1950, p. 1197 and Discussion p. 1225.
3. R. S. Burns and R. H. Heyer, "Orientation and Anisotropy in Low Carbon Steel Sheets," *The Annealing of Low Carbon Steels*, 1958, p. 29. Lee Wilson Engineering Company, Inc., Cleveland.
4. R. Hill, "Mathematical Theory of Plasticity," *Oxford Engineering Science Series*, 1950, p. 315. Oxford University Press, New York.
5. E. Coupland, "A Power Operated Press for the Swift Cup-Forming Test," *Sheet Metal Industries*, Vol. 34, 1957, p. 345.
6. O. H. Kemmis, "The Assessment of the Drawing and Forming Qualities of Sheet Metal by the Swift Cup-Forming Test," *Sheet Metal Industries*, Vol. 34, 1957, p. 203, 251.
7. L. W. Hu, "Studies of Plastic Flow of Anisotropic Metals," *Journal of Applied Mechanics*, Vol. 23, 1956, p. 444.
8. J. Willis, "Deep Drawing," *Butterworths Scientific Publications*, 1954, p. 28. Butterworth Publications Ltd., London.

THE CUP-DRAWING TEST METHOD FOR THE DETERMINATION OF THE DEEP-DRAWING PROPERTIES OF SHEETS AND STRIPS[1]

by Dr. Ing. Helmut Beisswänger
(translated from the German)

Introduction

For the testing and acceptance of sheets, numerous methods have been developed or taken over from other fields of materials testing. Besides the determination of hardness, strength, surface conditions and the like, these methods frequently serve to allow the workability of the sheet metals to be judged. For the determination of the drawing or deep-drawing quality, in Germany the Erichsen method, known since 1913, is mainly employed. Abroad, similar methods are in use[2] (i.e., Olsen cup test).

From a paper published by E. Siebel[3], it may be inferred that the workability, in the case of a chipless forming method, is not a property exclusively dependent on the material. It depends largely on the forming method; i.e., on the nature and amount of stresses prevailing in the deformed piece of work.

The state of stress present in the Erichsen test is not identical or similar to all drawing and deep-drawing procedures. There is, therefore, only a limited possibility of using the characteristic values of the Erichsen test for predicting drawing or deep-drawing quality of a sheet metal. For this reason it is not surprising that criticism of the Erichsen method started early[4], and that attempts are being made, in various publications up to the present, to limit the area of application of this method, or to work out its differences from production methods.[5] Since the Erichsen method became known, further sheet-metal testing methods have been recommended continually. Only recently, Petrasch[6] has reported on a new method, the so-called impact-deep drawing test method with subsequent expansion test. It may be assumed, therefore, that the methods known up to now are not fully satisfactory.

Part of the sheet metal testing methods has been incorporated in the Materials Standards for Steel, Iron and Non-Ferrous Metals, including the Erichsen method and the tensile test. Prompted by the requirements of a fairly large metal fabricating enterprise with a versatile finishing program, there has been an investigation some time ago regarding which testing methods are suitable especially for the selection of deep-drawing sheet metals. The result was that the AEG method, or cup-drawing test, supplies a suitable workability index for all the many deep-drawing procedures which it resembles in the mechanics of deformation. Furthermore, this test permits a judgment with respect to the tendency of the sheet metals to produce flow figures, scallops and ruptures. Further, the surface appearance of the drawn cup, as well as the appearance of the fracture of the ruptured cup, give an indication of the nature of the material. A disadvantage of the AEG method in its present form is its requirement for a large apparatus. For this reason, the method has been frequently called uneconomical, compared to the very simple Erichsen test.

As a result, there has been work in progress by various sheet-metal producing and fabricating firms, the Speciality Standards Committee for Materials Testing, and the Research Association for Sheet Metal Fabrication, directed towards the development, i.e. the simplification, of the cup test. During the execution of these experiments it became known that work on the introduction or the improvement of this method has been

[1]Footnotes are given at the end of this article.

in progress also in the East Zone of Germany and abroad; consequently there appears to be a general demand for such a method for the determination of deep-drawing quality.

The attempted simplification now appears to have been successful. On the basis of the experimental results, there was first submitted to the Specialty Standards Committee for the preparation of an industry standard, a working proposal for a cup-drawing test method for sheet thicknesses from 0.5 to 2.0mm. The standards proposal is to be extended later to thinner and to thicker sheets. As soon as a reasonably large quantity of numerical results has been produced by a substantial trial in industry, further standards proposals - if advisable, in connection with DIN 1623, the draft for DIN 1624, DIN 1774 and DIN 1788 - are to be established for the minimum blank diameters for the various materials and quality groups, which are to be workable without fracture.

In the following, the mechanics of deformation in deep drawing, and the old AEG procedure, will first be dealt with. It will be shown that this procedure may be extensively simplified. Subsequently, the possibility of applying the depth of cup obtained with a small tool and low punch speed to larger fabricating tools will be treated. Finally, the mechanics of deformation of various sheet metal testing methods will be compared with those of deep drawing; the test results of this method will be compared with those of the cup method.

I. The Mechanics of Deformation in Deep Drawing

It will suffice to mention the mechanics of deformation briefly, since it is to be treated at length

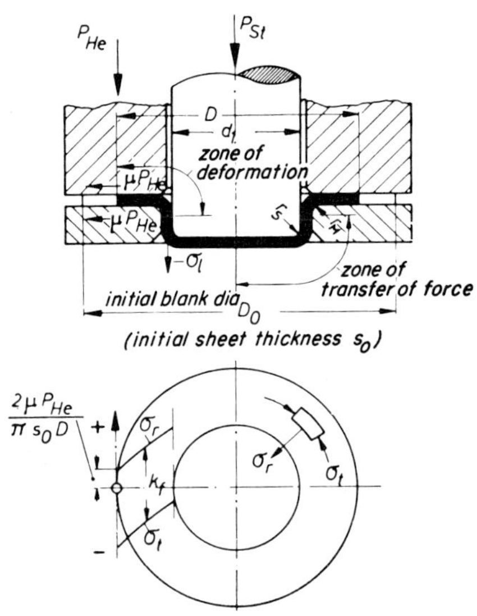

Fig. 1 (left)
Forces and Stresses in the Drawn Part, in Drawing Round, Cylindrical Parts

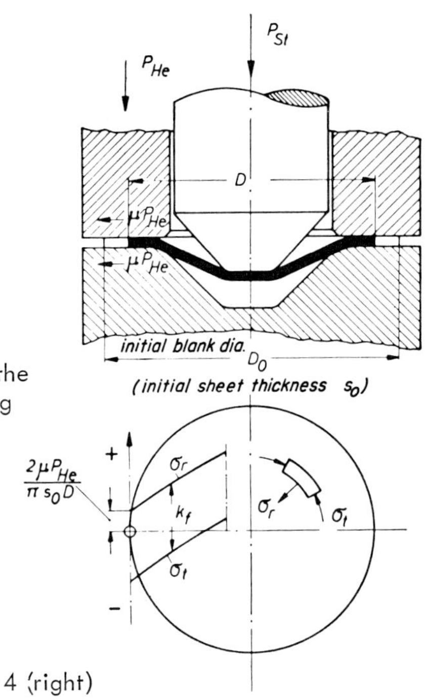

Fig. 2 (right)
Forces and Stresses in the Drawn Part, in Drawing Conical Parts

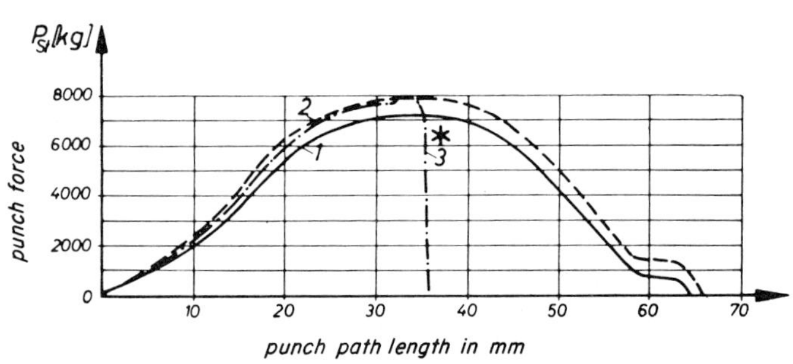

Fig. 3. Plot of Punch Force vs. Punch Part Length, for the Initial Draw of Round, Cylindrical Parts Ms 63, d_1 = 50 mm.

Fig. 4 (right)
Bottom Rupture in the Initial Draw of Round, Cylindrical Parts

I. The Mechanics of Deformation in Deep Drawing

It will suffice to mention the mechanics of deformation briefly, since it is to be treated at length in another publication, where the terms and symbols of chipless forming methods, used in what follows, will be dealt with in more detail also.

By "deep-drawing" procedures will be understood in the following those procedures where the deformation - similar to the initial or redrawing of round objects - occurs essentially by a narrowing down of the outer regions of the blanks or drawn parts without a thinning out of the wall of the work piece. This applies also to the usual initial drawing or redrawing of elliptical, rectangular and similar parts, as well as to the drawing of concial, parabolical and similar shapes. Stretch-drawing methods, where the forming consists essentially in an extension of the sheet, will not be counted as deep-drawing methods in this article.

In the initial drawing of cylindrical and prismatic parts, the deformation occurs in the plane between die and blank holder, and along the radius of the die, under radical tension and tengential compression (Fig. 1), where due to the compression, σ_t there arises an increase in sheet thickness largest at the rim of the drawn part. On its bottom and along the radius of the punch, however, the sheet becomes weaker.

On redrawing, the deformation continues, under the same load conditions, in front of and along the die radius.

On the initial draw of conical or similar parts, furthermore, the diameter in the region between die radius and punch radius is more or less reduced. With relatively thin sheets, there may result in this region special folds because of the slight compres sion, σt , present there (Fig. 2). To avoid this formation of folds, this compressive force σt may be reduced, removed, or even changed into a tangential tensile force, as the case may be, by means of an increased blank holding pressure or by increase in the initial blank diameter. In that case, the radial tension, σr , in this region is increased, while the tangential compressive force is increased or changed into a tensile force.

The punch force is transferred over the bottom and the side of the drawn part into the zone of deformation. It varies in the course of the drawing process, where at a certain stage a punch force maximum is reached. Fig. 3 shows graphs of punch force vs. punch path for an initial draw of round cylindrical parts. Somewhere in the middle of the punch path there occurs the punch force maximum, for the case of pure deep drawing without thinning out. As shown by Fig. 3, under otherwise equal conditions the magnitude of this maximum punch force depends on the initial blank diameter. Starting with a definite maximum blank diameter, there occur bottom ruptures (Fig. 4) when the maximum tension $_{1\ max}$, reaches the bottom tensile strength of the drawn part.

Fig. 5 shows again the dependence of the maximum punch force and the bottom rupture force on the blank diameter. The largest attainable blank diameter reached thus depends on the one hand on the relationships in the zone of deformation, and on the other

Fig. 5
Maximum Drawing Force and Bottom Rupture Force, as a Function of Blank Diameter, in Drawing of Round, Cylindrical Parts, in the Initial Draw.

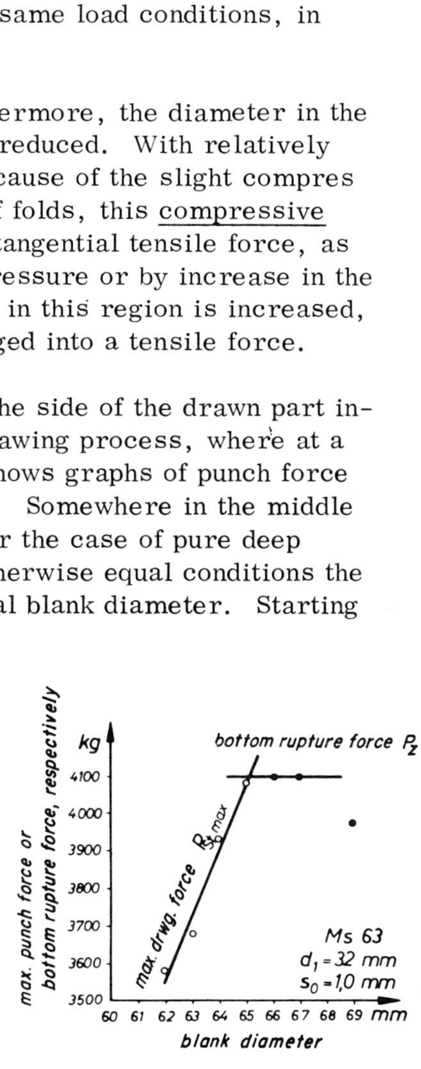

hand on the relationships in the zone where the force is transferred. Of influence in the zone of deformation are, besides the characteristic values of the drawing sheet metal, the holder pressure, the lubrication, the surface quality of sheet and tool, and the radius of the die; in the zone of transfer of force, besides the characteristic values of the sheet metal, the radius of the punch. When the clearance between die and punch is less than 1.4 times the initial sheet thickness, the maximum punch force is increased as a consequence of the thinning out of the wall; the bottom rupture force however appears to increase, because a part of the punch force is immediately directed into the zone of deformation through the effects of friction.

The greatest possible blank diameter is mostly referred to the punch diameter, and the value thus obtained is termed the maximum drawing ratio. Section III B shows how this drawing ratio can be related to other punch diameters.

E. Siebel,[3] in the study mentioned above, has demonstrated that this maximum drawing ratio depends on the ratio of the average strength of the zone of deformation, existing at the stage of maximum punch force, to the strength with respect to bottom rupture of the drawn part at the punch radius. The larger this maximum drawing ratio is, the smaller is the ratio of strengths; i.e., the less the cold work hardening resulting in the material from the deep drawing. The same result is reached by H. Kostron.[8] According to E. Siebel,[3] corresponding results apply to the drawing of round parts in redraws. With all deep-drawing procedures the limits of workability determined by bottom ruptures depend in the same way on the conditions in the zone of deformation and on those in the zone of transfer of force.

The limit of workability in deep drawing of round parts, in the initial draw, thus is determined by the interaction between two different regions. The same is true of the other deep-drawing procedures mentioned in the beginning. Hence it seems reasonable to choose quite generally the limit of workability of round parts in the initial draw (the so-called maximum drawing ratio) as the characteristic value for deep-drawing quality, as is done in the cup drawing test methods. As will be shown later, the workability in other procedures can also be judged by means of this drawing ratio.

The depth of cup of the Erichsen test (essentially a ductility value) as well as the extension at break of the tension test are measures of the workability of materials under certain boundary conditions. There exists no connection between these extension values and the maximum drawing ratio, because the mechanics of the initial draw prescribe completely different boundary conditions. The same is true of the "maximum drawing ratios" of the other deep-drawing procedures.

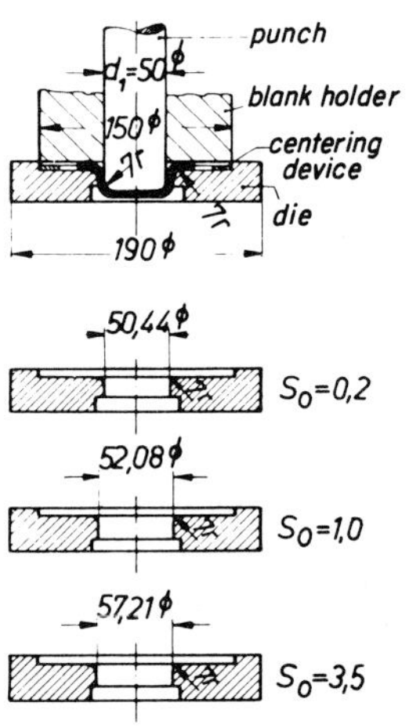

20 dies for intial sheet thicknesses S_0 from 0.2 to 3.5mm

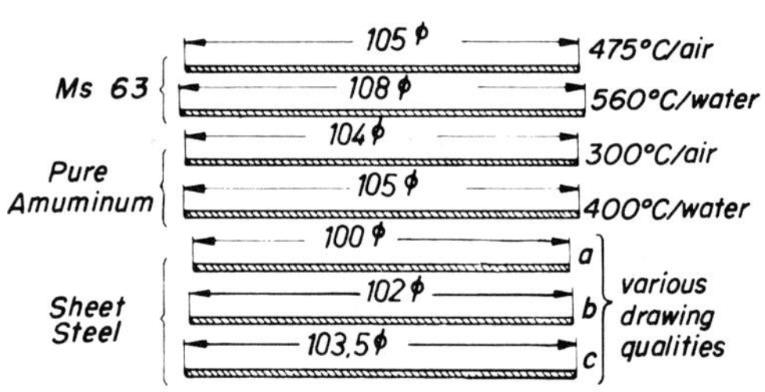

Fig. 6 (left)

AEG Method for Testing of Minimum Blank Diameters

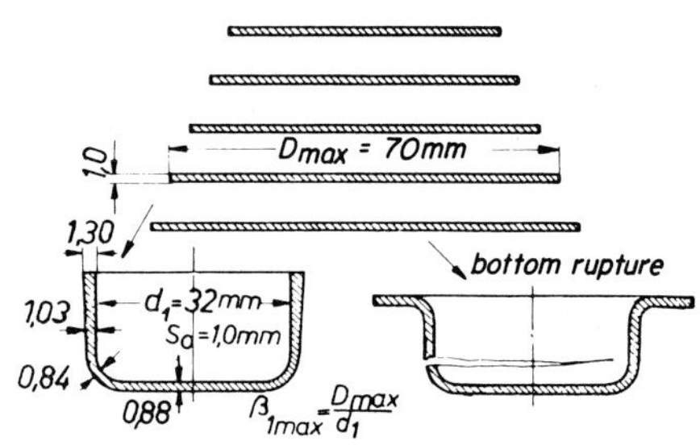

Fig. 7 (right)

Cup Drawing Test Method for the Determination of Maximum Blank Diameter, or Drawing Ratio

II. The Cup Drawing Test Method

It shall now be shown how the drawing ratio was determined in the older cup drawing test methods, and how it may be determined according to newer investigations.

A) The AEG Method

Fig. 6 represents the original tool set-up of the AEG method.[4] One initial drawing tool was employed. The punch diameter was 50 mm, punch and die radii each were 7 mm. Punch and holder are the same for all sheet thicknesses; the inside diameter of the die, however, depends on the sheet thickness. For sheet thicknesses from 0.2 to 3.5 mm, 20 different dies were provided. The testing tool was inserted into industrial drawing presses.

The AEG, according to G. R. Fischer,[9] established test blank diameters which had to be workable without failure for various materials and pre-treatment methods. (Fig. 6).

As shown in Fig. 7, the cup method can also be carried out by determining the maximum blank diameter by trial and error. In the absence of prior knowledge of the test results to be approximately expected, one requires about six different diameter blanks. In order to simplify the procedure and to reduce the consumption of material, M. Schmidt (Kapfenberg) therefore suggested a modified cup method.

B) Method According to M. Schmidt (Kapfenberg)[10]

As previously illustrated in Fig. 5, in the region of maximum blank diameter the maximum punch force increases approximately linearly. Schmidt has shown that the maximum drawing force increases linearly over a major region of blank diameters when it is plotted against the logarithm of the blank diameter (Fig. 8). He proposed that the maximum blank diameter be determined from the point of intersection of the bottom rupturing force (found from a bottom rupture) and the straight line representing the punch force (determined by two points representing draws with smaller blank diameter).

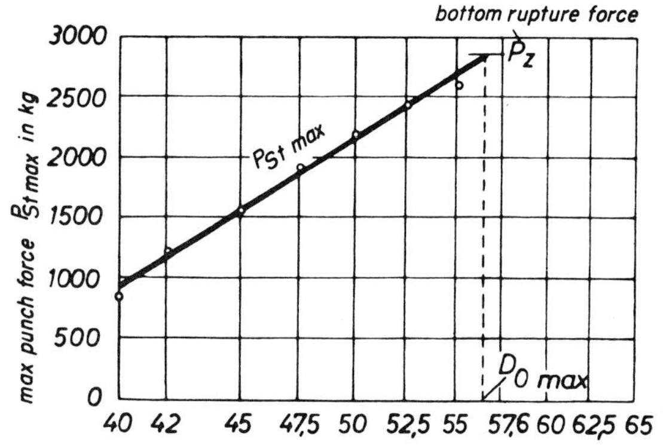

Fig. 8. Maximum Punch Forces as a Function of the Logarithm of the Blank Diameter, D_o (according to Schmidt-Kapfenberg)
Punch diameter, d_1 = 30 mm, deep drawing sheet steel, S_o = 0.75 mm.

On the basis of numerous tests by the author, the method of Schmidt-Kapfenberg does not appear to be sufficiently accurate. With the usual variations in quality of sheets, and with a test punch diameter of about 32 mm, the differences in maximum blank diameter amount to only a few millimeters. The maximum blank diameter therefore needs to be determined with a precision of $\pm$ 0.5 mm, to allow a reliable statement regarding the deep-drawing quality.

With respect to this method it should be mentioned also that drawing test instruments with adequate provision for force measurement are considerably more expensive than instruments without such provision. Although it is feasible to attach accessories to universal material testing machines, the larger testing machines suitable for this purpose, with a range of force in excess of 10 metric tons, are available to only a few organizations in the sheet-metal fabricating industry.

C) The Trial-and-Error-Method

The cup drawing method may also be performed by determining the maximum blank diameter by trial. This was the procedure in many of the author's experiments, in spite of the availability of force-measuring instruments. With some practice, and an approximate knowledge of the expected maximum diameter, blanks of only a few diameters are required. These may be produced in a simple manner by cutting on a lathe (Fig. 9). The cutting arrangement consists in essence of a hardwood or plywood disc held in the three-jaw chuck and turned square; the samples, in the form of squares, are pressed against the surface of the disc.

By means of the cup method, the sheet metal producers can use the analysis carried out each time to estimate the expected maximum diameter, and then to determine it with a few diameter blanks. This maximum blank diameter is taken into consideration when ranking the sheets in quality groups.

It has been mentioned above that, following the introduction of the cup method and its trial, efforts have been made to establish for the various methods and quality groups minimum blank diameters that are workable without fracture. The criteria for use named up to now in the materials standards are not adequate for a selection of sheet metals. It is therefore proposed now to name in place of the present metal uses (e.g., "simple drawing parts," "normal drawing parts," "for deep-drawing loads") possible degrees of deformability. One such degree of deformability would be the possible drawing ratio.

Testing the mentioned minimum blank diameter with the cup method requires approximately the same number of tests as does the Erichsen method. The preparation of test samples with the cup method is only little more complicated than with the Erichsen method. In the following it will be shown that the cup test tool may be extensively simplified, so that the tool requirements also become comparable with those of the Erichsen method.

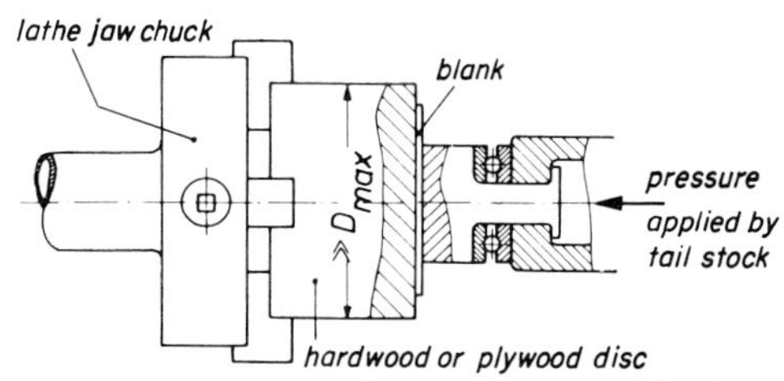

Fig. 9. Setup for Cutting out Blanks.

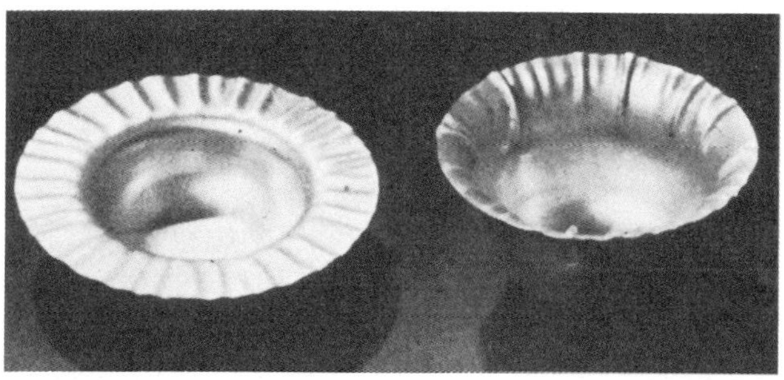
Fig. 10a. Formation of Folds in Deep Drawing (Initial Draw) Formation of Folds between Die and Holder.

Fig. 10b. Formation of Folds along Die Radius. Formation of Folds between Punch and Die Radius. Formation of Folds (unsymmetrical) near Rim of Drawn Part.

D) The New Cup Drawing Test Method

a) Construction of Tool and Conditions for Draw

The region of sheet thickness for the first testing tool was chosen so that the sheets used the most, of about 1 mm thickness, could be tested.

The first tool was designed for sheet thicknesses from 0. 5 to 2. 0 mm. It is planned to develop further tools for sheets of less than 0. 5 mm and more than 2. 0 mm thickness.

The AEG method employed a punch diameter of 50 mm. This means that for a maximum drawing ratio of 2. 3 blanks of up to 115 mm diameters are needed. The consumption of material would then be considerably greater than with the Erichsen method. This expenditure needs to be considered, since costly, highly alloyed sheet metals will have to be tested also.

In a later section it will be shown that the maximum drawing ratio, may in a wide region, be independent of the punch diameter. For sheet thicknesses from 0. 5 to 2. 0mm, this applies to punch diameters from 10 to 70 mm. For this reason a relatively small punch diameter of 32 mm was chosen in imitation of an instrument already commercially available; the consumption of material in testing the mentioned minimum blank diameter then corresponds roughly to that of the Erichsen method.

With the AEG method[4], the clearance between die and punch in each case equalled the sum of sheet thickness and the maximum tolerance for sheet thickness. Oehler[2] however proposes for the test method a clearance depending on both material and sheet thickness; this would require a larger number of dies.

H. Beisswanger and S. Schwandt[11] have found that the maximum drawing ratio increases if the clearance is decreased from 1. 3 times to 1. 1 times the sheet thickness. This increase occurs because with a smaller clearance a portion of the force is transferred by the punch directly through the wall of the drawn part into the zone of deformation, relieving the force on the drawn part at the die radius. The clearance chosen therefore must be large enough so that the maximum drawing ration cannot be influenced by variations in sheet thickness.

Occasionally there have been published claims that excessive clearance would give rise to folds in the drawn part. Experiments by the author have shown that with the proposed sheet thickness range from 0. 5 to 2. 0 mm, it is possible to work with one single die, where the inside diameter of this die was set at 38 mm, according to the largest sheet thickness, and taking into account the sheet thickness tolerance, and the thickening of the rim of the drawn part of about 40%. The clearance between die and punch thus amounts to 3. 0 mm; despite this, there occurs no formation of folds which would influence the accuracy of the procedure, even with metal 0. 5 mm thick. Die and punch radii also were chosen so that no disturbing folds would occur. For a better understanding of this discussion, Fig. 10 illustrates the various kinds of folds that occur in deep drawing. When the holder pressure is too small, or the fixed clearance between die and holder is too large, folds will arise on the flange of the drawn part; if the die radius is too large, folds arise along this radius during the last stage of drawing. When the clearance between die and punch is unusually large, and with large die and punch radii, in the case of relatively thin sheets there arise folds also between the die and punch radii, similar to the familiar folds in the case of drawing hemispheres, cones and similar shapes.

The folds thus far enumerated, which occur in rotational symmetry, are determined by the procedure and by the sheet thickness. Beyond this, dissymmetries may cause dissymetrical fold formation, particularly near the edge of the drawn part. (Fig. 10b, right). With the dimensions chosen for the testing tool, one has to count on such fold formation toward the end of the drawing process with sheet thicknesses of about 0. 5 mm, but the test result is not affected by this.

The essential dimensions of the testing tool are recognizable from Fig. 11. Regarding the choice of die and punch radii, it should be remarked here that the AEG method employed a single die or punch radius of 7 mm. According to Oehler[2], the radius should be chosen from the well-known graph of Kaczmarek. But for the proposed range of sheet thicknesses this would again result in a larger number of testing dies, since according to this graph the die radius depends on the sheet thickness. Experiments of Herrmann and Sachs[12] and experiments of the author indicate that the radii reported by Kaczmarek are not to be considered the most favorable ones; radii of 10 times the sheet thickness allow greater drawing ratios than do radii from the Kaczmarek diagram.

In choosing the die radius, one could take into account that maximum drawing ratios are reached with a punch radius approximately 1/3 of the die diameter[12]. However, it is not necessary that one work with the tool of optimum dimensions and drawing conditions, because in fabricating one cannot always work under optimum conditions, for a variety of reasons. In the following, it is shown that the drawing ratios obtained with the testing tool differ only little from those obtained with optimum tool dimensions and drawing conditions. Fig. 12 shows optimum dimensions and conditions, such as were employed in the comparison tests (with exception of the punch diameter). The holder pressure applied with a pneumatically actuated holder in this case exceeds slightly the required minimum holder pressure.

With the testing tool, however for the sake of simplicity of operation, the pneumatically applied holder pressure was established at a fixed value for all thicknesses of a given material, subject to the material and the minimum sheet thickness of 0. 5 mm. The cup drawing method can therefore be performed in the same simple manner as the Erichsen method, except for the somewhat more complicated preparation of the test samples.

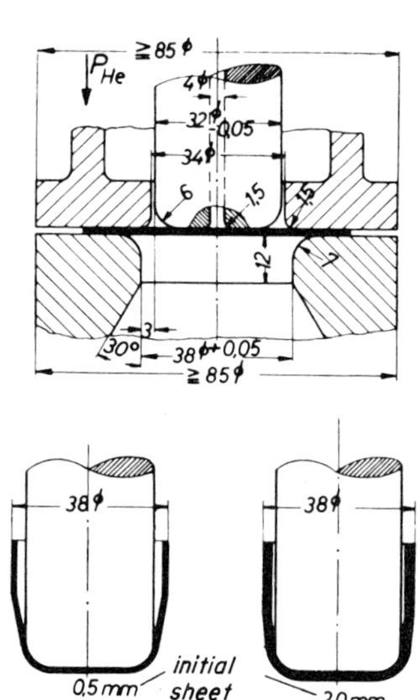

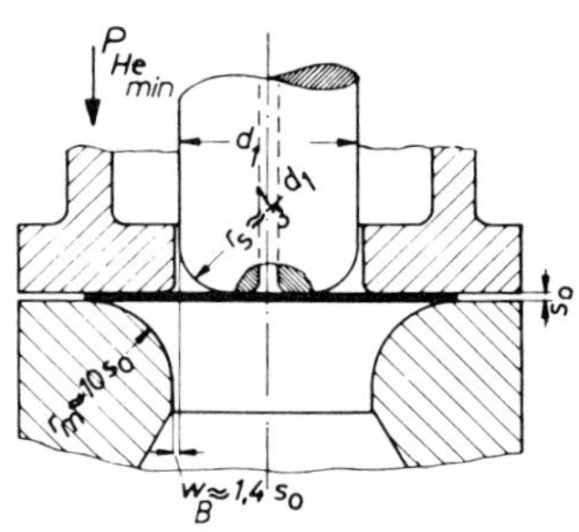

Fig. 12
Tool with Optimum Dimensions and Drawing Conditions

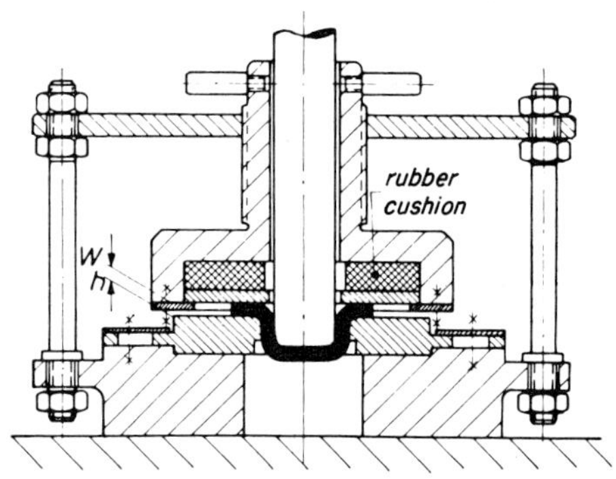

Fig. 11 (left)
Testing Tool (for Pneumatic Holder) for Sheet Thicknesses from 0.5 to 2.0 mm

Fig. 13
Test Apparatus with Rigid Holder

Upon the suggestion of Prof. Dr.-Ing. Siebel, the question was also investigated whether the test could be performed also with a rigid holder and with fixed clearance between die and holder, so that the Erichsen machine available in many plants could be used also for the cup drawing test method. It should presumably be possible to work also with an arrangement according to Fig. 13, with die and punch diameters according to Fig. 11, where the clearance, w_h, between die and holder is to be adjusted according to the sheet thickness, and possibly also the material. By a suitable choice of die and punch radii one may presumably assure that the testing tool with rigid holder attains the same maximum blank diameter values as that with pneumatic holder.

b) Determination of the maximum drawing ratio of sheets

The experiments described in what follows were carried out on a universal testing machine in which was inserted an accessory instrument as in Fig. 14. The punch velocity was approximately 0.1 m/min. The influence of punch velocity on the maximum drawing ratio will be dealt with in a later section.

In all experiments the blanks were uniformly lubricated with machine oil. It is common knowledge that the maximum drawing ratio depends on the lubricant. The influence of the lubricants however is difficult to estimate by quantitative measurement, since the various methods allow a determination of the surface quality of the blanks and the drawing tools only at individual points, but not as a whole. Drawing tests of the author and of others, with different lubricants, have in some cases revealed no influence, and in other cases a considerable influence, of the lubricant on the maximum drawing ratio. It is assumed that this behavior is to be attributed to the different surface quality of the drawing sheet metals tested respectively. But this would also mean that the maximum drawing ratio depends materially on the surface quality of the drawing sheet metal. In standardizing the cup drawing test method, the lubricant will have to be specified, as in the Erichsen method. Here, however, in contrast to the Erichsen Standard DIN 50 101, the lubricant ought to be defined more strictly, and test methods for its control ought to be stated.

DIN 50 101 prescribes that the working surfaces of the tools are to be ground and given a high polish. Unfortunately, the known surface testing methods are not adequate to establish exactly the surface quality of the tool in the cup method either. According to the mechanics of the initial draw, which will be treated in detail elsewhere, the effect of friction or surface condition of the die and holder plane is small with relatively thick sheets; but the friction effect of the die radius is relatively large. It would therefore be necessary first of all to measure and to specify the surface quality of this radius; but this is hardly feasible with present measuring methods. Hence one must attempt to minimize the friction effect by means of high quality lubricants. For this, mixtures containing chalk appear suitable, among others.

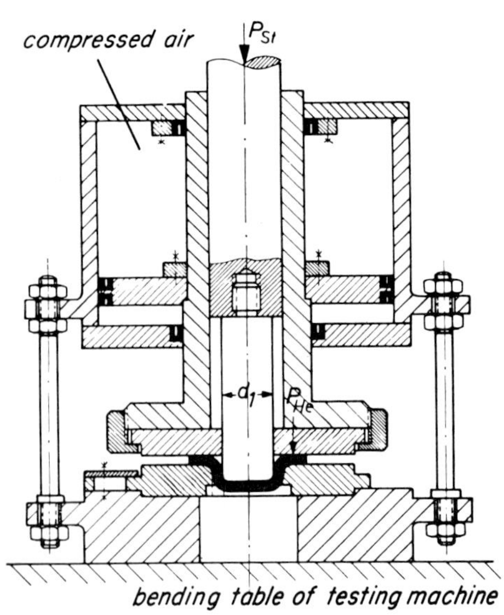

Fig. 14. Test Apparatus with Pneumatic Holder

Fig. 15 represents the experimental results obtained with different materials and with the testing tool with pneumatic holder of Fig. 11. For a comparison, draws were also made with the test punch and dies with optimum die radii, and clearances between die and punch of about 1.4 times the initial sheet thickness. The holder pressure was slightly in excess of the minimum holder pressure and was applied pneumatically.

Employed were Ms 63, a stainless steel (18% Cr and 8% Ni), St 4.24, and Al 99.5; the various sheet thicknesses of the same material were taken from the same lot and were treated by the rolling mill in such a way that the properties of the material each time were largely the same. The values of the tension tests with Ms63, St 4.24 and the stainless steel agreed well for the different sheet thicknesses. With Al 99.5, it was not possible to perform tension tests. With a sheet thickness of 0.5 mm, this attains a substantially smaller blank diameter, with the same test apparatus, and even with optimum die dimension and drawing conditions. It is suspected that in this case the properties of the material differ more considerably from those of the thicker sheets.

As the experiment shows, and as may be explained from the mechanics of the initial draw, the results with the new testing tool, at constant holder pressure, depend negligibly on the sheet thickness and differ little from the values attainable with optimum tool dimensions. By means of the new method, therefore, it is possible also to compare conveniently sheets of different thickness.

It may be interpolated here that there have been earlier attempts to determine the maximum drawing ratio with the cup drawing test method independently of the sheet

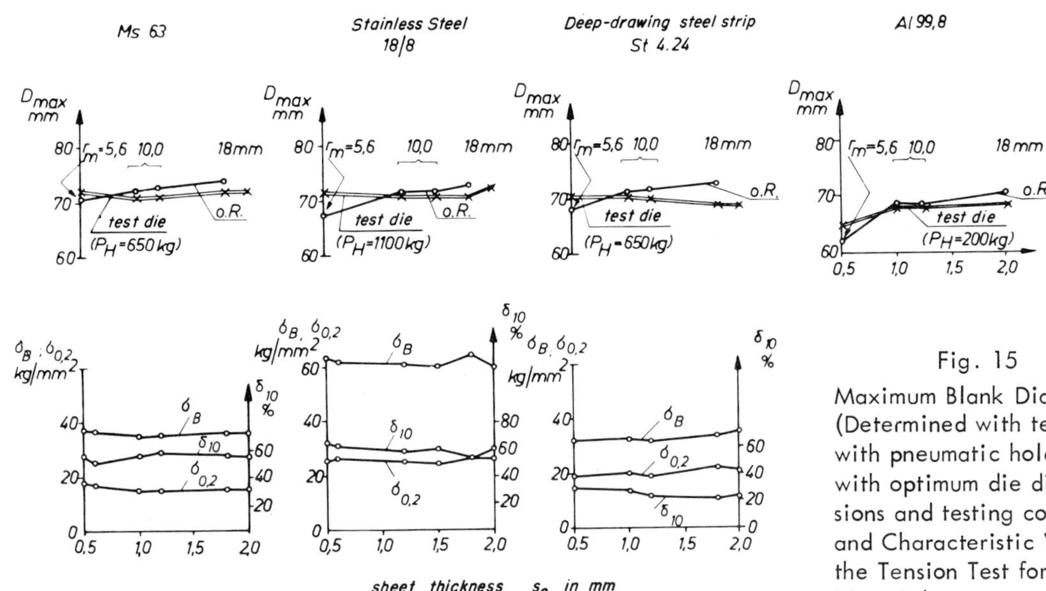

Fig. 15
Maximum Blank Diameters (Determined with test tool with pneumatic holder or with optimum die dimensions and testing conditions) and Characteristic Values of the Tension Test for Various Materials

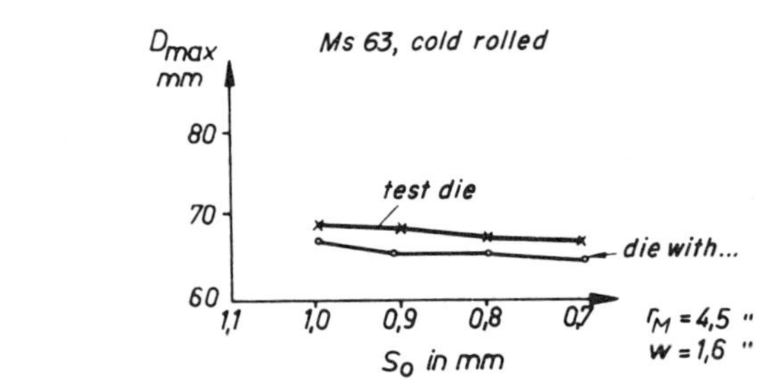

Fig. 16. Comparison of Drawing Results between the Testing Tool and a Tool of Ordinary Dimensions

Table I

Comparison of Maximum Blank Diameters for Test Die with those of Ordinary Die

Sheet Steel According to DIN 1623	D_o max (r_M=4.5; w_B=1.4s_o) mm	D_o max with test die of mm	
St II 23	62 mm	66 mm	66/62 = 1.07
St VIIIc 23	65 mm	69.5 mm	69.5/62 = 1.07

$s_o \approx 0.9$ mm

thickness. For this purpose, H. Esser and H. Arend[13] suggested a punch radius equal in every case to three times the sheet thickness.

The new method here described does not by any means result in a levelling down of the results, as Fig. 16 and Table I show. With cold rolled Ms 63 this testing tool resulted in approximately the same change in the maximum blank diameter as did the test punch and a die of conventional dimensions, with a holder pressure corresponding - approximately to the minimum holder pressure. The same results were found with sheets of different quality grade according to DIN 1623.

The experiments described here were carried out in cooperation with several firms, with good agreement of results. Thus there appears to have been found a simple cup drawing test method, which may be set alongside the Erichsen method. On the basis of the preceding material, a new working proposal was prepared, which has already been under consideration by a subcommittee of the Specialties Standards Committee for Materials Testing.

With sheet thicknesses less than 0.5 and more than 2.0 mm, one uses smaller or larger diameter test punches, respectively. By a choice of die and punch radii one can predetermine that the maximum drawing ratio will be independent of the test punch diameter.

E) Judgment of Tendency to Formation of Flow Figures, Scallops and Tears, as well as Sensitivity to Aging of Sheets, by means of the Cup Test Method

It has been mentioned already initially that the Cup Method allows an estimation of the tendency of sheets to form flow figures, scallops and tears.

a) Judgment of tendency to form flow figures

H. Beisswanger and H. Lammle[14] have reported on the different kinds of flow figures arising with deep-drawing sheet steels. R. Ergang and S. Welz[15] have treated the appearance of flow figures in brass and nickel silver. Beyond this it is known that flow figures can arise also with light metal sheets.

With deep-drawing sheet steels two different types of flow figures have been noted. With sheets having a well-defined yield point in the tension test, small amounts of deformation lead to flow figures which disappear again as the deformation proceeds. Fig. 17 shows such flow figures in the form of spirals on the flange of the drawn part. The same kind of flow figure has been found by Ergang and Welz also with brass and nickel silver sheets. In the tension test, these sheets also had a well-defined yield point. The flow figures run in the direction of greatest shear stress; i.e., in just the direction in which the material actually flows in chipless forming. This type of flow figure formation can be eliminated by a slight amount of cold rolling, or by means of the so-called fullering, or by a stretching of the sheets under so-called stretching hammers. As is well known, the tendency to form flow figures is suppressed only for a limited time by these measures.

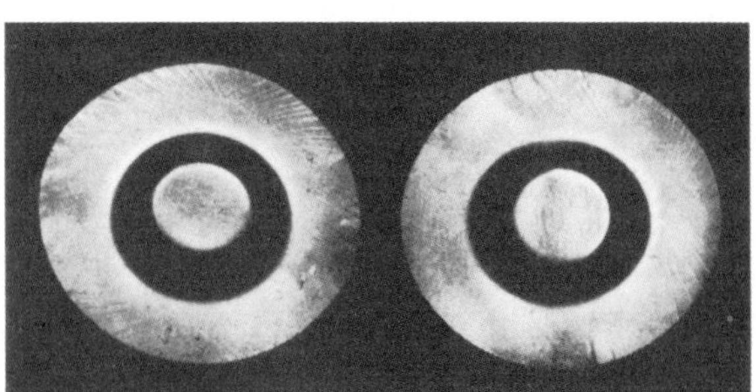

Fig. 17. Flow Figures at Rim and Bottom of Drawn Part.

A second type of flow figures across the direction of rolling of the sheet has been found on the bottom of the drawn part in deep-drawing sheet steel (Fig. 17). These flow figures are caused by buckling of the sheet following the last annealing, and perhaps also by uneven cold rolling traceable to strongly convex rolls.

The tendency to both types of flow figures can be demonstrated with the cup drawing test method, either together with a test for the minimum blank diameters (or the determination of maximum blank diameter), or in a special series of experiments. In the first case, the drawing process must be interrupted several times, which is possible without great effect on the drawing result. In the other case, blanks of the same diameter are predrawn to different depths; to determine the first type of flow figures one may then work with very large blank diameters with which the bottom breaks through in an early stage of drawing.

b) Evaluation of tendency to form scallops

The tendency of sheet metals to form scallops can be described numerically by testing the minimum blank diameter or by determining the maximum blank diameter. Fig. 18 shows such scallops; the tendency to scallop formation can be expressed by the number of scallops and by the ratio of the average scallop height to the average depth of cup.

c) Evaluation of susceptibility to tearing

The susceptibility of drawing metals to tearing can also be evaluated by means of the cup drawing test method. Apart from the bottom ruptures caused by the extent of the

Fig. 18. Cups with and without Scallop Formation

Fig. 19. Cups with Longitudinal and Transverse Tears

Fig. 20

Internal Stress Tears in a Cup of Deep-Drawing Sheet Steel (appeared after first redraw)

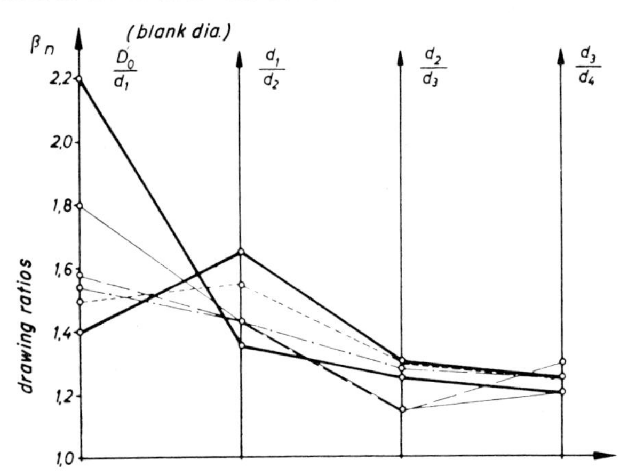

Fig. 21. Limiting Drawing Ratios for Ms 63 and Redraws without Intermediate Annealing Ms 63, original sheet thickness: S_o = 1.10 mm Punch diameter for first four draws: d_1 = 45.7 mm, d_2 = 34.3 to 27.5 mm, d_3 = 27.5 to 21.4 mm, d_4 = 22.7 to 17.4 mm, for d_2 to d_4, 2a = 90°.

deformation, there occur other tears during and after the draw. Inclusions of slag or special structural formations can give rise to tears during the draw. In the stainless Cr - Mn steel cups represented in Fig. 19, the tears running crosswise to the circumferential lines are caused by slag inclusions. These tears occurred already during the drawing.

The circumferential tears developed after the drawing. They are caused by internal strains developed during the drawing, and by aging phenomena in the drawn part. The time elapsed until the occurrence of these tears depended on the drawing ratio in each case; this time increased with decreasing ratio of draw. Such internal strain tears develop also in other materials, frequently in connection with corrosion effects (Fig. 20).

d) Evaluation of susceptibility to aging

According to F. Eisenkolb, the susceptibility of sheet metals to aging can also be tested by the cup method. It is known, and has already been mentioned here, that sheet steels undergo property changes upon long-term storage in a cold-worked state. In particular, one may then note a more or less great embrittlement. As a consequence of this embrittlement, when predrawn parts are redrawn after excessive storage, tears occasionally occur in the direction of the circumferential lines.

As is known, the aging occurs appreciably more rapidly at higher temperatures. Eisenkolb therefore proposes for the testing of deep-drawing sheet steels that predrawn cups be heated to 250° for two hours and then redrawn with a redrawing tool. Cups of steels susceptible to aging are then supposed to rupture. For the redraw, a drawing ratio of 1.25 is recommended.

With the test for aging susceptibility proposed by Eisenkolb it should be observed that his testing method must be related to the actual aging time and the subsequent strains. Again at 200°C with one-half hour storage corresponds, according to K. P. Harten[17], to an ordinary aging at room temperature and approximately one-half year storage. The same aging effect occurs with one-day storage at 100°C.

It seems necessary to suit the aging test to the actual conditions in each case. Beisswanger and Schwandt[11] have noted that there was no appreciable aging effect on the drawing ratio attainable in the redraw, after the usual storage times, and fabrication and storage of steel sheets at room temperature.

F) The Workability of Sheets in Redrawing of Round Parts and in Production of Elliptical, Conical and Rectangular Parts

a) Redrawing of round parts

Investigations of Beisswanger and Schwandt[11] have shown that the maximum redrawing ratio depends, on the one hand, on the shape of the redrawing tool and, on the other hand, on the drawing ratio in the initial draw. Fig. 21 shows this for Ms 63 and a total sequence of four draws, with the same shape tool in each redraw. The greatest overall drawing ratio is attained by beginning the initial draw with the greatest possible blank diameter. The attainable drawing ratio then decreases sharply in the first redraw, but only slightly in the further redraws. The limit of workability is then set by bottom ruptures also for the redraws.

On the basis of the experiments mentioned, it may be assumed that the maximum drawing ratio attainable in the redraw varies only slightly within the same materials and quality grades, so that its determination may be dispensed with for the acceptance testing of sheets. Besides, the testing of the drawing ratios in redraws would require an entire set of finely graded tools. It may therefore be considered sufficient if the dependence of the drawing ratios for redraws on the number of draws and on the tool shape has been determined generally for the various materials and quality grades at maximum initial drawing ratios. For the grading of the sequence of draws, it is however still essential to know the maximum initial drawing ratio.

b) Drawing of elliptical parts

The maximum drawing ratio of round cylindrical parts in the initial draw may serve to judge the drawing of elliptical parts in the initial draw. H. Oettinger[18] has investigated the deep drawing of elliptical parts in initial and redraw. In the initial draw the axis ratio b/a of the elliptical blank was changed from 0.4 to 0.6 (Fig. 22); ordinary round parts of axis ratio 1.0 were included in the investigation. The length of axis, 2a, in all cases was 50 mm. On the basis of the procedure developed by H. Oettinger[19], the blanks were so determined that there always resulted drawn parts with plane rims.

Fig. 22 shows that the graph of punch force vs. punch path length in the initial draw of elliptical parts corresponds to that of round parts. In a similar way, the maximum punch force also increases upon increase of the blank or the depth of the drawn part. The maximum depth of draw varies continuously with the axis ratio. Hence the maximum depth of draw of elliptical parts may be determined from this axis ratio and the maximum drawing ratio of round parts.

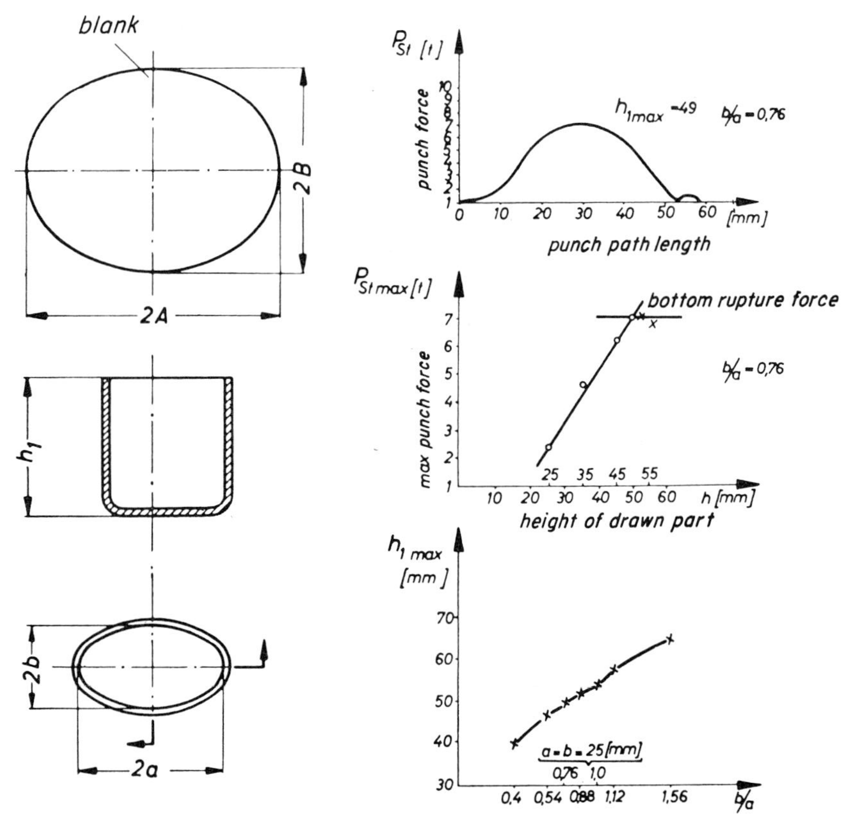

Fig. 22 (right)
Investigation of the Drawing of Elliptical Parts in Initial Draw
Ms 63, s_o = 1.25 (mm);
Punch Dimensions:
$2 \cdot a$ = 50 (mm); r_S = 10 (mm)
r_M = 10 (mm)

c) Drawing of conical and rectangular parts

In a similar manner it may be shown that the drawing of conical parts in the initial draw upon decreasing the cone angle changes continuously into the drawing of cylindrical parts; hence here also the maximum drawing ratio of the cup test method permits inferences with respect to the drawing of conical parts. For rectanuglar parts, it is also expected that the workability into steps may be inferred from the maximum drawing ratio. G. Oehler and F. Kaiser[20] have already shown a connection between this maximum drawing ratio and the drawing of rectangular parts, though in connection with the Erichsen cupping values. This connection appears doubtful in view of the introductory discussion above.

III. The Effect of Test Velocity and Test Punch Diameter on the Test Result (The possibility of transferring test results to industrial tools)

The cup method, as does the Erichsen method, employs relatively small tool dimensions; in the Erichsen method, the test velocities are small, especially in manually powered machines. With the cup drawing test method, the test velocity depends on the testing installation chosen. With the cup drawing test machines containing force-measuring instruments, the drawing velocity amounts to approximately 0.2 m/min; however, the test may also be performed at greater velocity by means of industrial presses, if suitable accessory equipment is available.

A) Effect of Drawing Velocity

The effect of drawing velocity has already been investigated in connection with the drawing of round cylindrical parts[21]. Then, and in more recent investigations, it was found that the drawing velocity has no effect on the maximum drawing ratio in the case of steel and brass deep-drawing sheets and aluminum sheets, in the region from 0.1 to about 30 m/min; with velocities of about 240 m/min, the drawing ratio decreases slightly. In the case of stainless steel (18% Cr, 8% Ni) it was found, under not yet perfected lubrication conditions, that the maximum drawing ratio decreased from 2.12 at 0.1 m/min to 2.03 at 15 m/min and 1.94 at 240 m/min.

It has been variously pointed out that at higher drawing velocities sheets are worked more poorly. It seems necessary to investigate if, and in how far, difficult and irregularly shaped drawn parts with very uneven stress distributions are more poorly worked at higher velocity. Possibly there is an appreciably stronger development of peak stresses in the zone of force transfer, at raised velocity - particularly, raised initially velocity - than at low velocities. In testing steel, brass and aluminum sheets, the test velocity accordingly would be expected to be without effect. With materials sensitive to velocity, the test result has to be stated together with the testing velocity.

B) Effect of Test Punch Diameter

W. Engelhardt[22] has recently attempted to show that the maximum drawing ratio in each case depends in a simple way on the punch diameter. He has referred here to older experiments by J. Blume[23]. Since experiments by the author were in contradiction to those of Engelhardt and Blume, the effect of punch diameter on maximum drawing ratio was investigated theoretically and experimentally.

Theoretically, the punch force for the initial draw was determined following the studies of G. Sachs[24] and E. Siebel[25]. The derivation of the punch force equation would lead too far in this connection; hence the essentials only are here excerpted.

The punch force is determined by the deformation of the blank between die and holder, as well as along the die radius. (By deformation is meant here only the reduction of the blank diameter to the punch diameter.) The punch force is increased by friction of the blank between ring and holder, and by friction along the die radius, as well as by the bending of the sheet around this radius. It was explained already in the beginning that the maximum drawing ratio depends on the ratio of the average strength against deformation at the stage of the punch force maximum, to the strength against bottom rupture. For this reason, we start in the following from the punch force maximum and investigate in what way the maximum drawing stress, $\sigma'_{1\,max}$, (see Fig. 1) depends on deformation, friction and bending. Then:

$$\sigma'_{I\,max} = e^{\mu\frac{\pi}{2}} \Big(\underbrace{1{,}1\, k_{f_{mI}} \cdot \ln \frac{D_P}{d_i}}_{\substack{\text{Deformation} \\ \text{(reduction in diameter)} \\ \text{of blank}}} + \overbrace{\frac{2\mu P_{He}}{\pi D_P \cdot s_0}}^{\substack{\text{Friction} \\ \text{between die} \\ \text{and holder}}} \Big) + \underbrace{k_{f_{mII}} \cdot \frac{s_0}{2 r_M}}_{\substack{\text{Bending} \\ \text{around} \\ \text{die radius}}}$$

(Friction along die diameter: the factor $e^{\mu\frac{\pi}{2}}$.)

Here $d_i = d1 + s_o$ is the average diameter of the completed drawn part; D_p is the outside diameter of the blank at the stage of the punch force maximum; $k_{f_{mI}}$ and $k_{f_{mII}}$ are the mean strengths against deformation in the region between D_p and d_i, and along the die radius, respectively; μ is the coefficient of friction; P_{He} is the holder force; e is the basis of the system of natural logarithms.

The results of theory and of measurement are included in Table II, along with the dimensions of the tools used. The die radius of the tools is $r_m \approx \frac{1}{7}\,d1$; the punch radius, $r_s \approx \frac{1}{3}\,d1$. These two ratios of radii were chosen on the basis of some ratios of the same kind present originally. The clearance, W_B, between die and punch is 1.4 to 1.5 mm.

Table II

Maximum Possible Drawing Ratio as Function of Punch Diameter. Effects of Deformation, Friction and Bending. Effect of Tool Dimensions on Maximum Tensile Stress and Bottom Rupture Strength

Tool Dimensions and Drawing Conditions							Calculated Stress Components							Results of Measurements		
Ratio of Punch Diameters ϕ	d_1 mm	r_S mm	r_M mm	w_B mm	$P_{He\,min}$ kg	$p_{e\,min.}$ kg/cm²	β_1	u	$\sigma_{I\,V}$ kg/mm²	$\sigma_{I\,H}$ kg/mm²	$\sigma_{I\,K}$ kg/mm²	$\sigma'_{I\,max}$ kg/mm²	$\sigma_{I\,max}$ kg/mm²	$\sigma_{I\,max}$ ka/mm²	σ_z kg/mm²	$\beta_{max.}$
X1.0	11.6	4.0	1.25	1.4	0	0								32	32	2.2
X1.75	20.4	6.3	2.5	1.4	35	2.7	2.21	1.2	22	0.1	9	35.0	35.5	35.5	36	2.21
X2.75	32	10.0	4.5	1.5												2.22
X5.3	61.2	20.4	7.5	1.5	2180	20	2.19	1.2	25	1.8	3.5	35.7	36.5	40	40.6	2.24
X6.3	73.2	23.0	10.3	1.4										40.5	40.8	2.17
X10.3	(120)		(15)			(28)				7	1.5					(~2.2)

Ms 63, s_o = constant = 1.0 mm

For Dies of Ge
Blanks of Ms 63
and Machine Oil as Lubricant

The abbreviations contained in Table II for the computation of the stress components of are:

$$u = e^{\mu \frac{\pi}{2}}$$ (component due to friction along die radius),

$$\sigma_{1v} = 1,1\ k_{f_{mI}} \cdot \ln \frac{Dp}{d_i}$$ (component due to deformation)

$$\sigma_{1H} = \frac{2 \mu p_{He}}{\pi\, Dp.\, s_o}$$ (component due to friction between die and holder),

$$\sigma_{1k} = k_{f_{mII}} \cdot \frac{s_o}{2r_M}$$ (component due to bending along die radius).

Experiment and theoretical calculations refer to Ms 63 of 1.0 mm thickness; but the calculations were carried out only for punch diameters, d1, of 20.4 and d1 = 61.2 mm, where the drawing ratio, $\beta 1 = Do/d1$, was 2.21 and 2.19, respectively. (Do is the initial blank diameter, $\beta 1$ the drawing ratio for the initial draw, $\beta 2$ that for the first redraw, etc.) For d1 = 20.4, the drawing ratio used as a basis for the calculations was the maximum one.

At the stage of the drawing force maximum, the procedure in each case was discontinued, the flange diameter Dp was measured, and the original diameter of the inside margin of the deformation zone (at the transition of the die radius into the cylindrical portion of the die) was determined by means of concentric circles scratched into the blank prior to the draw. From this it was possible to calculate the deformations of the diameters Dp and d1; and from these values and the tensile curve[7] for Ms 63, the applicable strength moduli, $k_{f_{mI}}$ and $k_{f_{mII}}$ could be found.

By calculation of $\sigma'_{1\ max}$ and measurement of $\sigma_{1\ max}$, the conditions in the zone of deformation were described. Beyond this, by measurement of the bottom rupture tension, σ_z, conditions in the zone of transfer of force were described also.

The measured drawing stress, $\sigma_{1\ max}$, agrees well with the calculated stress, $\sigma_{1\ max}$. With both the punch diameters used as a basis in the calculations, the contribution of the deformation predominates, with 22 and 25 kg/mm^2, respectively. Each time, the minimum holder pressure force was employed. With d1 = 20.4 mm, the holder pressure, referred to the area between initial blank and punch diameter (the so called specific minimum holder force, $p_{e\ min}$) amounted to about 2.7 kg/cm^2; with d1 = 61.2 mm, $p_{e\ min} \approx 20$ kg/cm. The stress contribution of the friction thus caused between ring and holder thus became 0.1 and 1.8kg/mm^2, respectively.

On the basis of these measurements it may be stated that this minimum holder pressure would amount to about 28kg/cm^2 for a punch diameter of d1 = 120 mm and Ms 63 of 0.1 mm thickness; with a drawing ratio of about 2.2, the contribution of friction between die and holder then would amount to 7 kg/mm^2. This contribution of friction thus approaches the contribution of deformation (at constant sheet thickness and increasing punch diameter); at the same time, the contribution of bending decreases, since with $r_m \approx \frac{1}{7} d1$ this radius increases also. With larger punch diameters, the increase in the contribution of holder pressure friction, for relatively thin sheets, is larger than the decrease in the contribution of bending. On account of the thus caused greater increase in the maximum drawing stress, $\sigma_{1\ max}$, the maximum drawing ratio for relatively thin sheets must therefore decrease with increasing punch diameter. Experiment has shown that, with the chosen tool dimensions and drawing conditions, the maximum drawing ratio remains nearly constant over a wide area of punch size, since the changes in the contributions of the various stresses in the deformation zone nearly compensate for each other, and in addition to the maximum drawing stress $\sigma_{1\,max}$, the bottom rupture strength σ_z also increases.

E. Hauff[26] has used the experimental results just stated as well as some of his own to discuss the publication of W. Engelhardt. According to his studies, with relatively large, constant die and punch radii the maximum blank diameter decreases only slightly with increasing punch diameter.

Blume[23], unfortunately, has not stated details of the materials, sheet thicknesses, and punch and die radii used in his experiments. On the basis of a numerical example given in his work, one might assume that he has worked with relatively thin sheets. As previously demonstrated, with these the contribution of friction between die and holder increases rapidly with increasing punch diameter. Hauff has juxtaposed the three experimental results (Fig. 23a). It is to be noted here that these presumably are results from partially very different experimental conditions, from sheet thickness to punch diameter. On the basis of experiments of the author's and the calculation performed in this connection, the relationship shown schematically in Fig. 23b is to be expected. The plot shown there corresponds in its three segments approximately to the three curves of Fig. 23a; this means that no contradiction is notable between the three experimental results.

It is necessary now to show how the formability with industrial tools can be inferred in a simple manner from the characteristic value of the cup method, with due consideration of the tool dimensions and the sheet thickness. For this, special investigations are planned which are to be reported on at a later time.

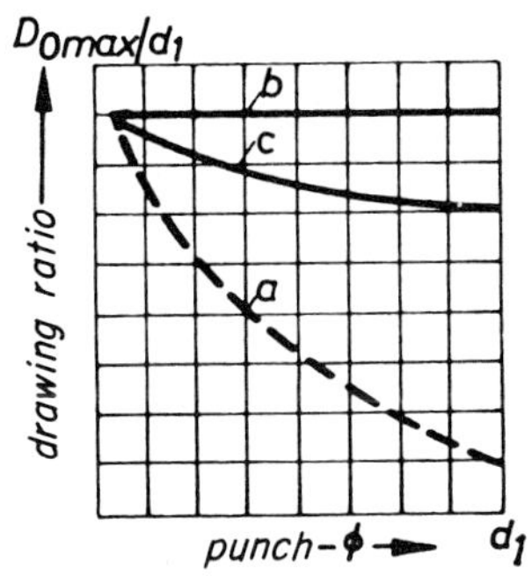

Fig. 23a
Maximum Drawing Ratio as Function of Punch Diameter, according to Experiments and Theory of Blume (a), according to Experiments of Beisswanger (b), and according to Experiments of Hauff (Cited by E. Hauff[26])

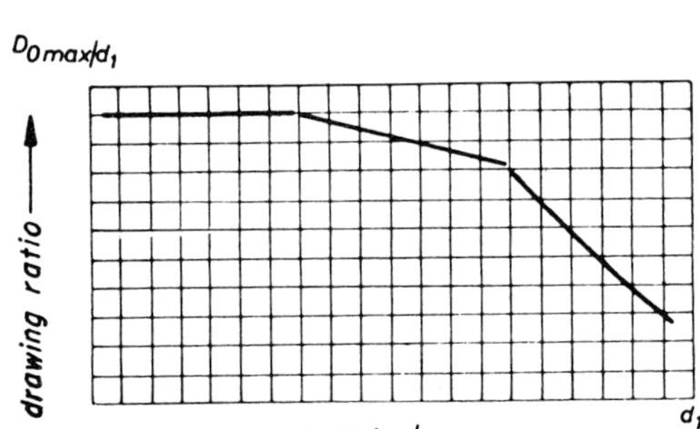

Fig. 23b
Presumable Dependence of Maximum Drawing Ratio, β_{max}, on Punch Diameter, at Constant Sheet Thickness (Schematic Representation)

The maximum drawing ratio, according to Table II, was independent of the punch diameter, at a sheet thickness of 1.0 mm and punch diameter from 11.6 to 73.2 mm. The ratio of sheet thickness to punch diameter at the same time decreased from 0.86 to 0.0137. The chosen test punch diameter of 32 mm lies within the region investigated; the same is true of the ratio of sheet thickness to punch diameter for the sheet thickness region from 0.5 to 2.0 mm and the punch diameter of 32 mm. The choice of the relatively small test punch diameter thus appears justified.

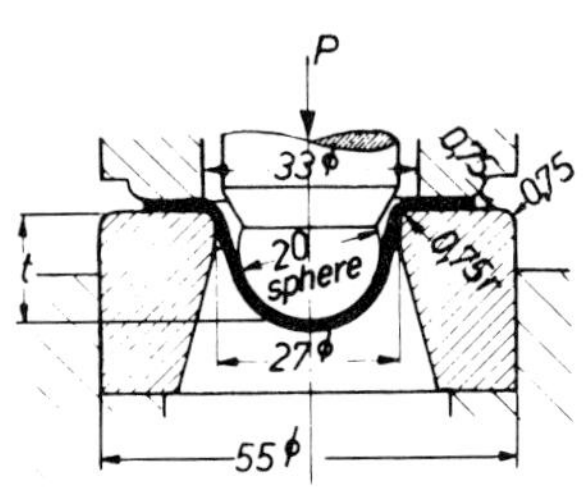

Fig. 24
Tool for Erichsen Test for Sheet Thicknesses up to 2.0 mm

Fig. 25 (right)
Erichsen Impression Values for Steel Sheets under DIN 1623

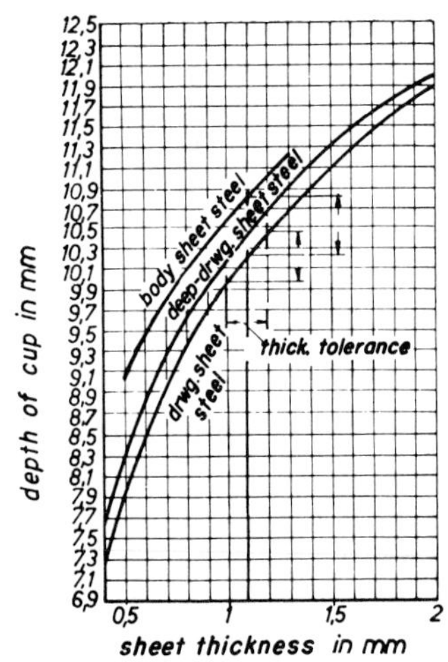

IV. Additional Sheet Testing Methods
(Comparison with the cup test method)

For the testing of drawing and punching sheets, there are in use at the present, according to the standards, the following methods:

A) the Cupping Test according to DIN 50 101,
B) the Tension Test, with reference to DIN 50 114
C) the Folding Test, with reference to DIN 1605,
D) the Double Folding Test[16] (according to DIN 1623, for steel sheets under 3mm),
E) the Sphere Impression Test according to DIN 50 351,
F) chemical analysis[16] (with brass and aluminum sheets according to DIN 1774 and 1788).

For the evaluation of formability, in addition to the previously treated cup drawing method there are used or recommended, among others, the following methods:

G) The Deep Drawing-Expansion Test according to Siebel and Pomp,
H) The Wedge-Drawing Test according to Sachs,
I) The Impact-Deep Drawing Test Method with subsequent Expansion Test, according to Petrasch.

The methods named will be dealt with in regard to their suitability for evaluating the deep-drawing quality of sheets; if necessary, the stress conditions and the deformations present in the various methods will be touched on briefly.

A) The Cupping Test according to DIN 50 101 (Erichsen Test)

The Erichsen Test is performed for sheets up to 2mm thick with the tool shown in Fig. 24. It is used to test strips of at least 70 mm width; the center of the impression must be at least 35 mm distant from the edge, and 70 mm from the center of the next impression.

One measures the depth of impression, t, reached before a tear appears. As Fig. 25 shows, the depth depends strongly on the sheet thickness. At 1.0 mm thickness, the difference in depth of impression caused by the thickness tolerance is of about the same magnitude as the difference in depth of impression between body sheet and drawing sheet. According to DIN 50 101, the depths of impression are larger if, instead of strips, square blanks of 70 mm edge, or round blanks of 70 mm diameter are employed. The depth of cup also depends on the lubrication of the test sample.

In the Erichsen test, in contrast to deep drawing, there flows hardly any material around the die radius (Fig. 26).

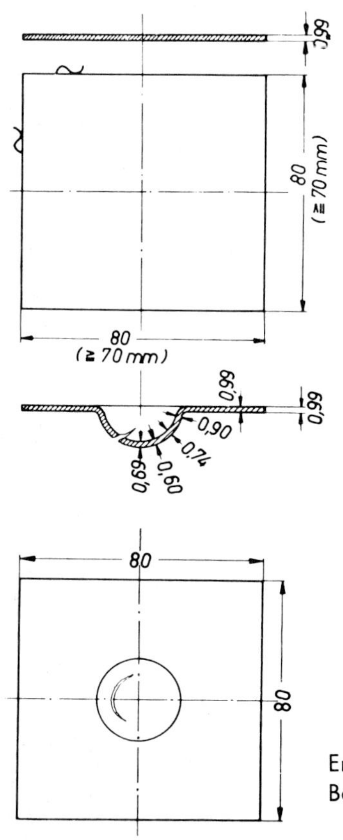

Sample before Test

Sample after Test

Fig. 26
Erichsen Test Sample, Before and After Test

The impression arises by stretching of the sheet radially and tangentially, with simultaneous decrease in thickness, as found by Guth[5]. Thus the cupping occurs under radial and tangential tensile stress in the region between die and punch radii. This stress condition and flow of material is present only in exceptional cases in sheet metal drawing, as has been discussed above in connection with the mechanics of drawing of conical parts. E. Siebel and A. Pomp[27] have shown that the Erichsen Method is a type of tensile test with short length of test sample. In this test, a special elongation at break is determined as the characteristic value.

A connection between Erichsen Impression and maximum drawing ratio in the initial draw has not been possible to be proven and would, in fact, contradict the general laws of deformation. It is therefore understandable that the Erichsen method frequently is not sufficient for the selection of sheets for deep-drawing purposes. Like all other sheet metal testing methods, the Erichsen Method is suitable for the selection of sheets only when the conditions of stress and the course of deformation or the working process correspond to those of the testing method.

At this point it may be added that, with the Erichsen test, the surface appearance of the test sample and the shape of the fracture following the impression give an indication of the grain structure.

B) The Tension Test, with Reference to DIN 50 114

The Tension Test is performed with standardized test sample rods which have to be carefully and exactly finished in the region of elongation. With thin sheets, the performance of the Tension Test is difficult. As a rule, tensile strength, yield point and elongation at break are used to judge the deep-drawing quality. Occasionally also it is recommended that, in place of the elongation at break, the so-called "uniformly distributed elongation" be used; "uniformly distributed elongation" is defined as the (uniform) elongation over the entire measured length of the sample rod, corresponding to the highest load. As soon as the maximum load is reached, as is well known, there occurs a local constriction; at this point the fracture occurs eventually. "Uniformly distributed elongation" and constricting elongation together make up the elongation at break.

According to the Schuler pocket book[28], the best drawing results are obtained with sheets of high elongation and tensile strength, and low yield point. This would lead to the obvious conclusion that high tensile strength means the possibility of transferring a relatively high punch force around the punch radius, while there occur only relatively small resistances to deformation in the zone of deformation, because of a low yield point and high extensibility; both together lead to a greater initial drawing ratio. This is however only approximately true for, as previously alluded to, the deformation processes and also the demands on strength in drawing occur under completely different conditions of stress than in the tension test or the Erichsen test. Greater elongation or cupping, in the comparison of different sheets, do not therefore permit the conclusion that a higher drawing ratio has been reached.

C. Arbel[29] has investigated in what way the tension test should be evaluated to determine the drawing qualities. To this end he uses several tension tests to determine the "flow curve"[7] of the material under test and then determines - similar to Kostron[8] - from the "flow curve" a characteristic which he terms the "drawing coefficient." This, according to Kostron, characterized the embrittlement tendency of materials. According to Kostron, the ratio of "strength with respect to deformation to initial tensile strength," which decides the maximum drawing ratio, is solely dependent on this drawing coefficient.

This means that this drawing ratio can be found by way of the drawing coefficient. Arbel found experimentally the relationship illustrated in Fig. 27. Very gradually different materials have approximately the same maximum drawing ratios.

This drawing coefficient, furthermore, is supposed to equal the maximum value of deformation at break. C. Arbel, therefore, attributes to it considerable physical significance. According to E. Siebel[3] however, such deformation values are dependent on the procedure. In conclusion, one may remark regarding the procedure of Arbel that the maximum drawing ratio is determined from the tension test and a special evaluation of the latter; but it can, as previously shown, be found directly in a simple manner by way of the cup drawing test method.

With respect to the actual drawing test one should add that - as in the Erichsen test - the surface appearance of the sample, and the broken area, give an indication of grain properties. Further, the tendency of sheets to form flow figures can be estimated[14]. This flow figure formation has been discussed in some detail in connection with the cup drawing method.

C) to F) Folding and Double Folding Test-Sphere Impression Test. Chemical Analysis

These four methods can be dealt with together briefly, since none of them are applicable for determining deep drawing quality. The Folding and the Double Folding Test are employed when the suitability for bending and curling jobs is to be tested.

In place of the Sphere Impression test, nowadays the Pyramid Impression test according to Vickers is frequently undertaken. From the hardness numbers of either test, one can determine approximately the tensile strenth. From this, one can generally infer the formability; this however allows a determination of the maximum drawing ratio just as little as does the tension test.

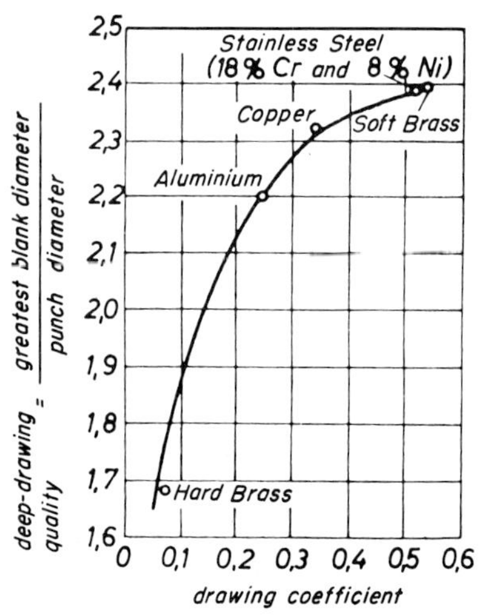

Fig. 27 Relationship between Deep-Drawing Quality and Drawing Coefficient (according to: Claude and Arbel, Stahl and Eisen 70 (1950), No. 24, pp. 118-19)

The contents of alloying and accompanying elements are fixed within definite limits for the various drawing materials and qualities. The chemical analysis is regularly performed by the producer of the sheet metal. On the basis of the chemical analysis only a general statement is possible with respect to formability, because only scattered investigations into the

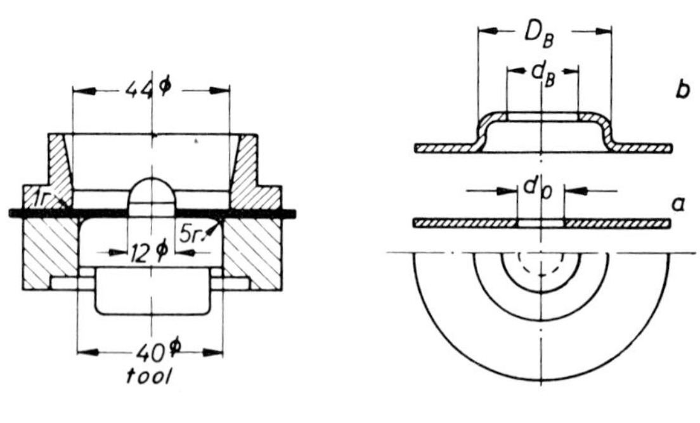

Fig. 28. Deep Drawing-Expansion Test, according to E. Siebel and A. Pomp
a = Sample before expansion
b = Sample after expansion

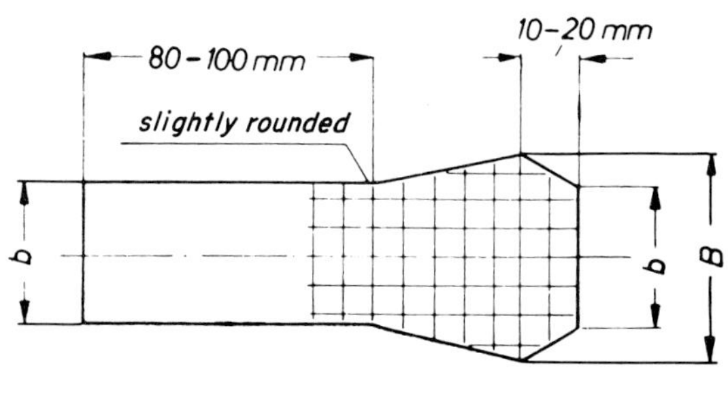

Fig. 29. Form and Dimensions of Wedge-Drawing Samples.

connection between the contents of alloying and accompanying elements and the maximum drawing ratio are available. Of at least the same importance as the chemical is the metallurgical report regarding grain size, type of structure, etc. Still, it is largely impossible to make any definite statement regarding the effect of grain size on drawing ratio.

G) The Deep Drawing Expansion Test according to Siebel and Pomp[30]

In this test a tool as in Fig. 28 and a round test sample with a hole are employed. The bore here must be cleanly machined. In contrast to deep drawing, the sample here is firmly clamped between holder and die. Upon descent of the punch, the wall of the resulting cup is formed by the material flowing in from the bottom, around the punch radius. The test is ended when the hole shows tears. In this cupping and stretching, there arise radial and tangential tensile stresses on the bottom of the samples; the state of stress in this zone of deformation thus does not correspond to that in the zone of deformation in the usual deep drawing procedure.

In sheets with a texture, i.e. with an oriented structure in the direction of rolling, tensile strength, elongation and yield point depend on the position of the test rods with respect to the direction of rolling and vary in size depending in the position of the test rod with respect to the rolling direction. This dependence of properties on the rolling direction in sheets with a texture has as a consequence that the extension of the hole does not occur evenly. The presence of texture, which in deep drawing leads to the formation of scallops already discussed, can be recognized in this way. With strong texture, however, it becomes more difficult to measure an expansion that deviates from roundness.

In the Deep Drawing-Expansion Test, the tensile strength can be determined by measurement of the maximum punch force. However, from the expansion of the hole, (that is, a special elongation), and the tensile strength, it is not possible to determine the maximum drawing ratio, as has already been discussed in describing the tensile test. The procedure is suited mainly for selecting sheets for the deep drawing-expansion processes which resemble this testing method, as well as a gradation of these processes.

H) The Wedge Drawing Test according to Sachs[31]

In the Wedge-Drawing Test, wedge-shaped samples are drawn through an orifice, as may be recognized from Fig. 29 (in schematic representation). Here the width, B, is increased, at constant wedge angle, until the sample, in being drawn, tears. The value Bmax thus found is referred to the final width b. This ratio, Bmax/b, independently of the wedge angle corresponds to the maximum drawing ratio of the initial draw. The shape of the sample is however of decisive influence on this possibility of comparison.

As in deep drawing, the deformation in wedge drawing occurs under "radial" tension and "tangential" compression. It must be observed, though, that in contrast to deep drawing there is a great amount of pressure between surfaces at the wedge faces of the orifice, and hence a great friction force. This surface pressure is, on the average, equal to the initial tensile strength of the sheet metal. According to this, the test result should be more strongly dependent on the lubrication of the sample.

In the wedge drawing method, a characteristic value corresponding to the maximum drawing ratio is thus determined from several samples of different size. The surface appearance and the appearance of the fracture here also give an indication of the nature of the grain. Besides, as in the tension test, the tendency of the sheets to form flow figures can be estimated.

In the wedge drawing method different amounts of deformation are attained in the sample. The formation of large grains which occasionally occurs during a recrystalizing annealing is known to depend on the amount of the preceding cold working. With the help of this method it is possible in a simple way to determine the critical amount of deformation decisive for this large grain size.

J) The Impact-Deep Drawing Test Method with subsequent Expansion Test, according to W. Petrasch[8]

In the Impact-Deep Drawing Test method a tool for initial drawing of round parts is employed. Tool and blank diameters remain unchanged; the punch diameter is approximately 37 mm; the blank diameter, 75 mm. The drawing ratio thus is approximately 2.0. The tool is inserted in a drop hammer, with which the work done in drawing can be measured in a simple way. The work is referred to the volume of the blank (mkg/cm^3) and is termed "specific deep-drawing work." Here a smallest possible deep-drawing work is considered favorable.

If one knows the specific deep-drawing work, one may no doubt make some general statement about the formability, just as is possible from a knowledge of the Erichsen impression or the elongation at break of the tension test. Although the state of stress and the effect of material in the Impact-Deep Drawing test resemble the usual initial draw, the maximum drawing ratio cannot be obtained from the test. The amount of deep-drawing work is determined only by the conditions in the zone of deformation, the maximum drawing ratio on the other hand by the conditions in the zone of deformation and in the zone of transfer of force.

According to Petrasch, the cups drawn in the first test may, following this impact-deep drawing test, be expanded. To eliminate possibly present scallops, the cups are cut off to the same height prior to expansion. For the expansion, a tapered mandrel is impressed by means of a spindle; at the appearance of tears, the expansion is discontinued. The amount of increase in diameter, referred to the initial diameter of the cup, serves as the characteristic value.

E. Siebel[3] has pointed out the publication of F. Korber, A. Eichinger and H. Moller[32] on the behavior of upset metal in tensile stress. There it is shown that the formability of metals in pure tension is low if they have been previously upset by an appreciable amount. Initial draws with drawing ratios of 1.8 and more result in an upset in the outer portions of the blank (i.e. the cup rim) which permits only small amounts of expansion. If the tools are of correct form, such an expansion can be avoided in the redraws[11]. It occurs only with unfavorable tool shapes, in the last drawing stage; the maximum tension is increased and the formability thus made poorer. The expansion therefore must be avoided in redrawing. According to what has just been discussed, it is therefore not possible to judge the formability of sheets in redrawing by means of the expansion test proposed by Petrasch. By means of this test it may be possible, according to Petrasch, to judge the sensitivity of sheets to aging after preceding cold working.

K) Combinations of Sheet Metal Test Methods

Besides the Impact-Deep Drawing Test with subsequent Expansion Test dealt with above, other combinations of sheet metal test methods have been proposed. According to F. Eisenkolb[30], a strip is to be taken out of the shell of deep drawn cylindrical parts and cupped according to Erichsen. From other quarters[34] comes the recommendation to first deform a test strip in the wedge drawing apparatus and subsequently to cup it according to Erichsen. In either case, the Erichsen test cannot be related to the redrawing, with respect to stress conditions and progress of deformation.

The Erichsen test occurs under a two-dimensional tensile stress; according to Korber, Eichinger and Moller[35], the extensibility of a material after a 50% cold-work deformation is only slight. In deep drawing, however, the material exists in a compression-tension field that is known to be favorable for formability. These combined methods therefore can at most allow some estimate of the possibilities of redrawing.

L) Comparison of the Test Values of the Tension and the Erichsen Test with those of the Cup Drawing Test Method

The test values from the Tension and the Erichsen tests were compared with those of the Cup Drawing Test Method, for sheets of various materials, quality groups and pretreatments.

In the tension test, sample rods of 50 mm length were used in accordance with DIN 50114 for the determination of elongation of break. The Erichsen tests were performed with square blanks of 80 mm edges, and according to DIN 50101. For the Cup Drawing test, the accessory tool of Fig. 14 was used with a universal testing machine. The holder was pressed down pneumatically. The holder force, P_H, was barely above the minimum holder force. The tool dimensions are separately stated for each case. They do not yet correspond to those of the standard tool, since the test results reproduced in the following were obtained previous to the development of this tool.

Table III juxtaposes the test values of the Tension, Erichsen and Cup Drawing tests, for St VIII 23, Ms63, A1 99.8 and stainless steel (18% Cr, 8% Ni). Sheet thickness for all materials was about 1.2 mm. Among these four materials, tensile strength, yield point, elongation at break, and Erichsen impression varied greatly, although the maximum drawing ratios differed little from one another.

Table IV contains depths of Cup and maximum drawing ratios for sheet steels according to DIN 1623. It appears that satisfactory drawing conditions can be reached already with St. II 23, whereas St III 23 partly exceeds St. VIII 23 as regards drawing ratio. The variation in Erichsen depth of cup within the comparison groups does not always occur parallel to the maximum drawing ratio; in certain instances this has been emphasized by the inserted arrow.

Table III

Characteristic Values from the Erichsen and Tension Tests, and from the Cup Method

Material	s_o mm	$\beta_{1\,max}$	t mm	$\sigma_{0.2}$ kg/mm^2	B kg/mm^2	δ %	H_v k/gmm^2
St VIII 23	1.25	2.23	10.5	22.9	36.9	34.3	110
304 (USS 18-8)	1.2	2.16	13.3	21.4	60.2	67	165
Al 99.8	1.25	2.10	11.6	3.5	6.8	34-40	19
Ms 63	1.25	2.23	12.9	15.9	36.5	55-59	81.1

$\beta_{1\,max}$ = Maximum Drawing Ratio d_1 = 45.7 mm r_M = 10 mm

t = Depth of Cup r_s = 15 " w_B = 1.4 s_o

Table IV

Depth of Cup and Maximum Drawing Ratio for Sheets of Various Quality Groups According to DIN 1623

d_1 = 32 mm r_M = 7.1 mm r_s = 6.3 mm w_B = 3.0 mm

Quality Grade	Type of Annealing	Type of Smelting*	C %	Si %	Mn %	P %	S %	s_o mm	t mm	β_1 max
II	kasten	Thomas	-	-	-	-	-	0.88	9.0	1.94
VIII c	normal.	SM	0.08	0.05	0.33	0.021	0.034	0.94	10.1	2.05
VIII c	kasten	SM	0.08	0.05	0.30	0.018	0.032	0.88	10.5	2.03
II	kasten	Thomas	-	-	-	-	-	1.03	9.0	1.94
III	-	-	-	-	-	-	-	1.04	9.6 ↓	2.06 II
V	normal.	SM	0.07	Sp.	0.37	0.026	0.025	1.01	10.6 ↓	2.07
VII a	normal.	SM	-	-	-	-	-	1.08	10.3	2.01
VIII c	normal.	SM	0.09	0.05	0.35	0.030	0.032	1.03	10.2	2.06
VIII c	kasten	SM	0.09	0.05	0.27	0.026	0.028	1.09	10.8	2.06
III	normal.	Thomas	-	-	-	-	-	1.95	12.0 ↓	2.13 ↑
VII a	normal.	SM	-	-	-	-	-	2.07	13.0 ↓	2.06 ↑
VII b	normal.	SM	0.08	Sp.	0.32	0.027	0.034	2.05	10.8 ↑	2.05 II
VIII c	normal.	SM	0.10	0.07	0.31	0.017	0.027	2.20	11.9 ↓	2.08 ↓
VIII c	kasten	SM	0.10	0.07	0.31	0.017	0.027	2.12	12.7 ↓	2.12 ↓

*Thomas refers to the Thomas-Gilchrist process; SM refers to the Siemens-Martin process.

Table V

Depth of Cup and Maximum Drawing Ratio for Ms 63, after Cold Rolling Loss of Thickness up to 30% Ms 63 (Initial Sheet Thickness 1.0 mm)

Sheet Thickness mm	t mm	β_1 max	
1.0	11.7	2.11	
0.91	9.3	2.07	Rolled Down
0.80	7.8	2.07	Rolled Down
0.80	7.8	2.07	Rolled Down

d_1 = 23 mm; w_B = 1.6 mm; r_s = 6.3 mm, r_M = 4.5 mm

Table V lists the depths of cup and the maximum drawing ratios of an Ms 63 sheet which had been cold rolled down to 70% of its initial thickness. The Erichsen depth of cup changes greatly, whereas the drawing ratio decreases only slightly. It is known that such a rolling down increases strongly the yield point and the tensile strength, but decreases the elongation at break.

The test results represented by Tables III to V confirm that the Erichsen and the tension test furnish no basis for a statement as to maximum drawing ratio. The tables demonstrate further that combined methods (e.g., a combination of the Sachs Wedge Draw with an Erichsen impression, or a cold rolling with an Erichsen impression) do not enable one either to make any statements as to initial drawing.

In various places, attempts have been made to determine the suitability of the sheet testing methods experimentally. In this, the procedure has often been to test sheets of different materials and quality groups by different test methods. On the basis of the various test methods, there resulted each time a certain quality ranking of the sheets. Furthermore, the sheets were used to fabricate deep drawn and formed parts of various forms in plants, under the there prevailing conditions. On the basis of the formability of the sheets with the various procedure one then again obtains a quality ranking of the sheets. The groupings found in the plant are then compared with those found by the test method, and that test method is then considered as the one best suited whose quality ranking agrees best with that of the plant results.

In such comparison test, the test conditions must agree largely with those of the fabricating conditions. In the present publication, this has been pointed out repeatedly. Defects noted in testing must be examined separately. They have to be considered in the selection and the fabrication of the sheets, on their individual basis. However, in the comparison tests just mentioned it was found that the sheets are ranked according to various defects; i.e., different limits of deformation were compared to one another so that one ought to expect that these limits are determined not only by the deforming properties and deficiencies of the sheet metal, but also by deficiencies in the industrial installations. It is therefore necessary to use for most comparisons only those test and fabricating methods which agree in the mechanics of their deformation. In evaluating the results it is further necessary to separate the limits of formability due to the procedures (i.e. the material characteristics determined by them) from defects of materials and fabricating methods.

Summary

By means of the mechanics of deformation in deep drawing it has been shown that the limiting drawing ratios (deformation limits) in the various deep drawing methods caused by bottom ruptures do not depend on the absolute values of tensile strength, yield point and elongation at break. It is thus not possible to determine these limiting drawing ratios by means of the test values of the tension and Erichsen test which have been accepted by the Materials Standards; this has been corroborated by the author's experiments.

By contrast, it was possible to estimate the formability by means of the maximum drawing ratio determined with the AEG or cup drawing test method in various deep drawing procedures, while with the Erichsen or Tension test, sheets could be selected, for instance, for stretch drawing procedures. The Cup method furthermore permits determination of the tendency of sheets to form flow figures, scallops and tears.

It was shown experimentally that one can employ the cup method for a sheet thickness range from 0.5 to 2.0 mm in a simple manner, with a punch of 32 mm outside diameter and a die of 38 mm inside diameter. The holder pressure, with pneumatically actuated holder, depends only on the material. It is probably feasible to test also with a rigid holder, in which case the clearance between holder and die will depend on the sheet thickness, and possibly on the material.

The maximum blank diameter and the maximum drawing ratio determined with this testing tool depend only slightly on the sheet thickness, for identical materials and material pretreatments. Sheets of different thickness thus are easily compared with respect to their deep drawing properties. With this testing tool differences in maximum blank diameter, in testing sheets of different kinds, are of about the same magnitude as those in ordinary drawing tools.

These experiments demonstrate further that the maximum drawing ratio is independent of velocity, within the range of velocities common in practice. Only with stainless steels, there occurs a small deterioration for larger velocities. Under certain conditions, the maximum drawing ratio is independent, over a wide range, of the punch diameter. However, it remains to be shown in what way the maximum drawing ratio depends on the size of the punch diameter, at constant sheet thickness.

First, it is planned to standardize the test method. A corresponding work sheet is presently under consideration by the Specialty Standards Committee for Materials Testing. The method is at present being employed by a number of plants as a part of their current materials testing. This has already shown that satisfactory drawing ratios may sometimes be obtained even with sheets of lower quality grades. It is planned to establish later, on the basis of this current testing, minimum blank diameters for the various materials and quality grades in additional Standards. These minimum blank diameters must be formable without fracture. Beyond this, the maximum blank diameters for the selection of sheets for difficult drawn parts may be determined in a simple way by a small variation of the blank diameter. The graphic or arithmetic approximation method making use of maximum drawing force for each blank diameter is necessary any longer.

With sheets below 0.5 mm or above 2.0 mm, one can draw test punch diameters below and above 32 mm, respectively. By a proper choice of die and punch radii one can assume that the maximum drawing ratio for identical sheets is independent of the test punch diameter.

This article was originally published in the September 15, 1952 edition of "Communications for the Members of the Research Association for Sheet Metal Fabrication," Dusseldorf, West Germany, and has been translated and printed with the kind permission of the author and publisher.

FOOTNOTES

1. Expanded version of a lecture given at the meeting of the Research Association for Sheet Metal Fabrication at Düsseldorf, on December 6, 1951.

2. G. Oehler, Testing of Deep-Drawing Workability of Sheets. VDI-Zeitschrift [Periodical of the Society of German Engineers] (1950), pp. 371-74.

3. E. Siebel, Limits of Workability. Mitteilungen Forschungsgesellschaft Blechverarbeitung (1952) No. 16, pp. 177-84.

4. Special Testing Setups, AEG-Mitteilungen [Communications] (1927), pp. 419-20.

5. H. Güth, Local Deformation in Deep Drawing and Impression Drawing, Measured in Three Dimensions, Die Technik (1951), pp. 397-405.

6. W. Petrasch, The Impact Deep-Drawing Test Method with Subsequent Expansion Test, Mitteilungen Forschungsgesellschaft Blechverarbeitung (1951) No. 17, pp. 209-11.

7. See also: E. Siebel, Pressure Experiments with Thin Sheets for the Determination of the Flow Curve. (To be published shortly.)

8. H. Kostron, Contribution to the Technological Mechanics of Deep Drawing, Archiv für das Eisenhüttenwesen [Archive for Iron Affairs] (1951), pp. 205-13.

9. G. R. Fischer, AEG Deep-Drawing Test Method, AEG-Mitteilungen [Communications] (1929), pp. 483-86.

10. M. Schmidt (Kapfenberg), Testing Deep Drawing Sheet, Archiv für das Eisenhüttenwesen (1929-30), pp. 213-22.

11. H. Beisswänger and S. Schwandt, Investigations of the Effect of Tool Shape on the Maximum Drawing Force and the Maximum Drawing Ratio Attained on Redraw of Cylindrical Shells. (To be published shortly in these communications.)

12. L. Herrmann and G. Sachs, Investigations of Deep Drawing, Metallwirtschaft [Metal Industry] (1934), pp. 687-92 and 705-10.

13. H. Essen and H. Arend, The Deep Drawing Testing of Sheets. Archiv für das Eisenhüttenwesen (1940-41), pp. 223-31.

14. H. Beisswänger and H. Lämmle, Investigations of the Appearance of Flow Figures in Deep Drawing Sheet Steels. Mitt. Forschungsgesellschaft Blechverarbeitung (1952), pp. 90-102.

15. R. Ergang and S. Welz, Origin and Prevention of Flow Figures in Nickel Silver and Brass, Zeitschrift für Metallkunde [Journal of Metallurgy] (1952), pp. 45-49.

16. F. Eisenkolb, Testing of Thin Metal Sheets. Stanzereischriften Folge 5 [Communications on Punching, Series 5], München [Munich], 1949.

17. K. P. Harten, Problems in Metal Fabricating. Maschinenbau/Der Betrieb [Machine Building/The Plant] (1937), pp. 73-77.

18. H. Oettinger, Deep Drawing of Elliptical Parts. (To be published shortly in Mitt. Forschungsgesellschaft Blechverarbeitung.)

19. H. Oettinger, Determination of Blanks for Elliptical Drawing Parts. Mitt. Forschungsgesellschaft Blechverarbeitung (1952), No. 15, pp. 165-74.

20. G. Oehler and F. Kaiser, Cutting, Punching and Drawing Tools. Berlin, 1949.

21. H. Beisswänger, The Effect of Drawing Speed on the Maximum Drawing Ratio in the Initial Draw and on the Change in Sheet Thickness. Mitt. Forschungsgesellschaft Blechverarbeitung (1951), No. 1, pp. 10-12.

22. W. Engelhardt, The Drawing Value. Werkstatt und Betrieb [Shop and Plant] (1951), pp. 232-33.

23. J. Blume, Determination of Drawing Quality of Punching Sheets. Metallbörse [Metal Exchange] (1922), pp. 1945-46.

24. G. Sachs, Chipless Forming of Metals. Separate Print XVI of Mitt. d. Deutschen Materialprüfungsanstalten [Communications of the German Materials Testing Institutes]. Berlin, 1930.

25. E. Siebel, Forming in the Plastic State. Düsseldorf, 1932.

26. E. Hauff, Drawing Ratio and Absolute Cup Diameter in the Initial Draw. Werkstoff und Betrieb [Shop and Plant] (1951), pp. 527-28.

27. E. Siebel and A. Pomp, About the Progress of Force in Deep Drawing and the Cupping Test. Mitt. K.W.I. für Eisenforschung [Kaiser Wilhelm Institute for Iron Research] (1929), No. 126.

28. Pocket Book for Industrial Sheet Metal Fabrication, 3rd ed., 1937, ed. by L. Schuler, A.G., Göppingen/Wörtt.

29. C. Arbel, A Relationship between the Deep Drawing Properties of Metals and the Results of the Tension Test. Rev. Métall., Mém., 47 (1950), pp. 388-94. According to E. Marke, Stahl und Eisen [Steel and Iron] (1950), pp. 1118-19.

30. E. Siebel and A. Pomp, Testing Thin Metal Sheets by the Deep Drawing Expansion Test. Mitt. K.W.I. Eisenforschung 12 (1930), pp. 115-25.

31. G. Sachs, A New Testing Apparatus for Deep Drawing Sheet Metals. Metallwirtschaft [Metal Industry] (1930)

32. F. Korber, A. Eichinger and H. Moller, Behavior of Upset (Compressed?) Metals under Tensile Stress. Mitt. K.W.I. - Institut f. Eisenforschung (1941), pp. 123-33.

33. F. Eisenkolb, Investigations of the Testing of Deep-Drawing Quality of Siemens-Martin Thin Metal Sheets. Stahl und Eisen (1932), pp. 357-64.

34. H. Kayseler, H. Lossek, W. Püngel and E. H. Schulz, Effect of Production Conditions on Properties, Especially Deep-Drawing Quality, of Steel Strip. Stahl und Eisen, 54 (1934), No. 39, pp. 993-98.

35. F. Körber, H. Eichinger and H. Möller, Behavior of Punched Metals under Tensile Stress. Mitt. K.W. - Institut Eisenforschung, 23 (1941), pp. 123-33.

Olsen Ductility Cup Test

Deep Drawing Cup Test

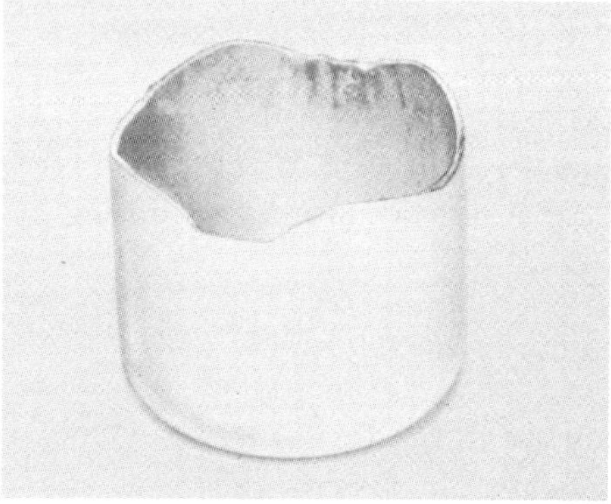

Hole Expansion Test

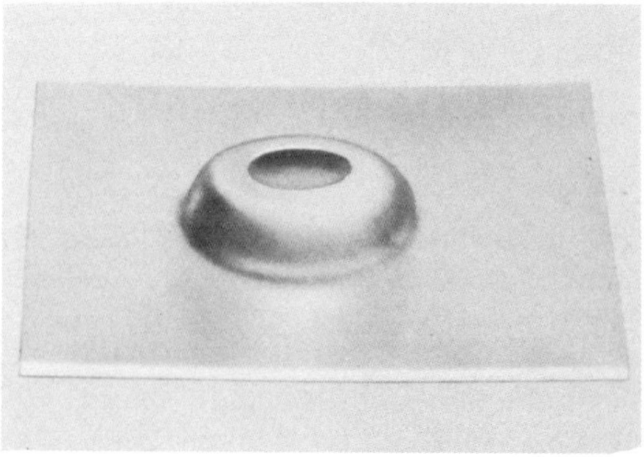

Most of the tests described in this article can be conducted with a Tinius Olsen DucTomatic Sheet Metal Tester. This machine has interchangeable forming tools and dies for making Olsen Ductility Cup, Deep Drawing Cup, Hole Expansion (K.W.I.) or Coating Adhesion Tests. Load is applied by the upward motion of the forming tool, and is indicated on a large dial indicator. Cup height is shown on a separate indicator. An electronic cut-off control (optional) can be set to stop the test automatically as soon as a hairline crack appears in the specimen. An independent hydraulic system applies and holds constant any desired clamping pressure. Details and specifications on the Tinius Olsen line of DucTomatic Sheet Metal Testers are given in Bulletin 68.

Strain Analysis of Formed Sheet Metal Parts

JOHN R. NEWBY

The electrochemically etched grid pattern on sheet metals is useful for tailoring the various grades of steel which are commercially available to the requirements of fabricators. Examples of production parts which have been studied by strain analysis are given. In addition, the relationship between the type of deformation and the mechanical properties of the sheet metal is established. It is concluded that strain analysis can be a powerful diagnostic approach in the solution of sheet metal forming problems.

STRAIN analysis is a simple, effective technique which has been valuable in solving problems of metal forming. When sheet metal is formed into a three dimensional part, as in deep drawing, some areas of the metal are subject to high strains, and may break or wrinkle. By etching a grid onto the sheet surface before forming, the amount of local strain can be measured quantitatively later on the formed part.

Strain analysis has been successful in solving field problems in fabricators' plants. It graphically illustrates how the metal is moving in critical areas, and whether the dies are controlling the metal flow. The technique for applying these grids and the method of measuring the strains after the part has been fabricated will be explained. Then the interpretation of the results will be considered to permit meaningful conclusions to be drawn. Conclusions should be based not on strain analysis alone, but in combination with other contributing factors to metal flow, such as material properties and the effects of lubricants.

In evaluating material requirements, the use of strain analysis is not new. Until now, however, it has not been precise enough to be used in material selection. There was an attempt several years ago to correlate the increase in area of a drawn part with the need for drawing quality or commercial quality steel. The difference in choice of steels was based on whether the increase in surface area was less or greater than 25 pct. This base still applies for a majority of drawn parts, but it is now known that other factors must also be considered.

The scribed square method has been improved by employing electrochemical etching to gain precision in placing gage marks on blanks. This etching technique permits the rapid application of patterns of a desired configuration from prepared stencils. Spacing between the measuring points is accurately fixed. The pattern shown in Fig. 1 is currently used at Armco.[2]

STRAIN ANALYSIS

Strain analysis consists of etching the grid on the surface of the blank in what are suspected to be the critical areas, or over the entire surface if necessary. Although it is best to clean the surface before etching, the grid may be successfuly used on mill-oiled sheets as it will penetrate the film. Grids have even been etched through dry soap films with no need to recoat before forming. The blank is then drawn in the press as any other part would be in normal production. Standard lubricants can be applied after the grid has been etched on the sheet surface.

The parts may be examined after the first draw (Fig. 2), or the sequence may be continued through any desired number of production steps. The amount of metal movement on the surface of the sheet is then measured using the deformed grid of marks, originally equally spaced.

A master stencil of any gage length or design pattern is prepared. The size of the stencil can be controlled by photographic reproduction to provide grids with precise spacings of less than 0.1 in. (0.254 cm) up to 1 in.

JOHN R. NEWBY is Research Associate, Armco Steel Corp., Middletown, Ohio.

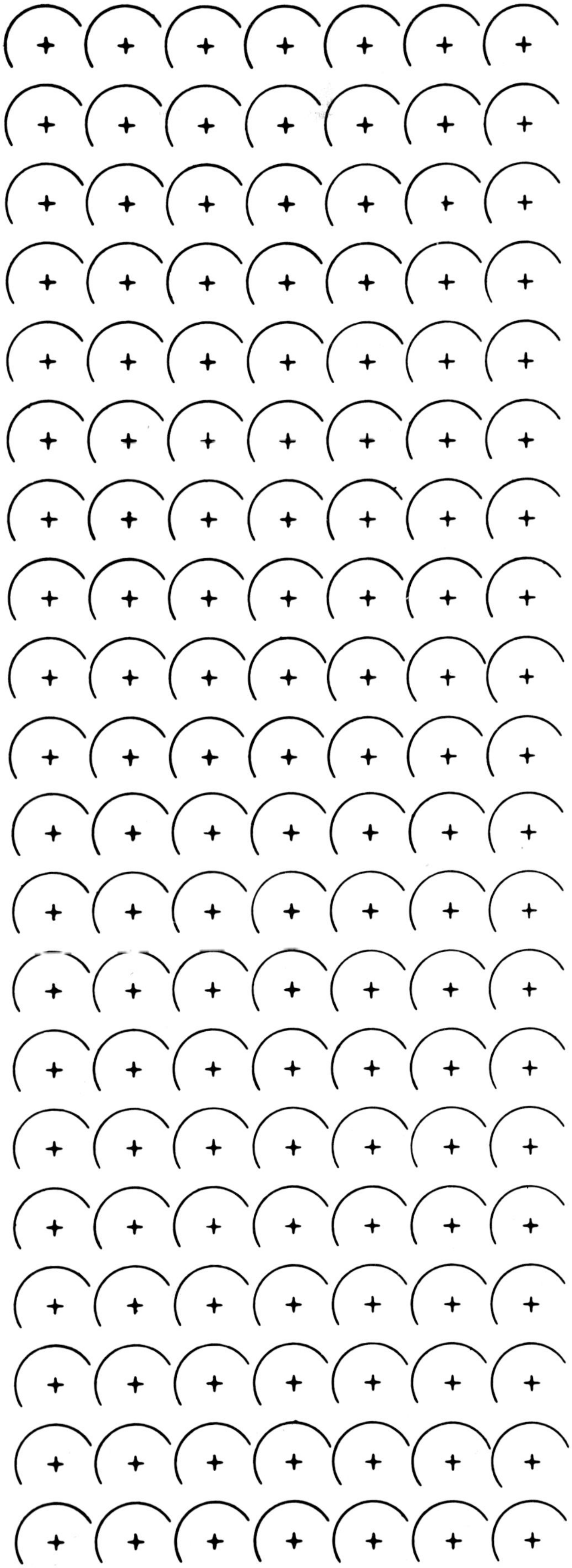

Fig. 1—Pattern of gage marks for determining e_1 and e_2 strains.

(2.54 cm) or larger. A spacing of 0.2 in. (5 mm) is most convenient for field investigations where strains are to be evaluated in the vicinity of the press without carrying the formed part to some distant laboratory. Larger spacings may cloud the maximum strain, and smaller spacings are difficult to measure without the assistance of a magnifier.

ELECTROCHEMICAL ETCHING

Development of strain analysis at Armco initially followed the conventional approach of scribing parallel straight lines on the surface of blanks. Aside from the method being slow, spacing between lines was not always 1 in. (2.54 cm), and spacing of the initial gage had to be measured as well as the distance between gage marks after the drawing operation. In addition, there were many questions this technique did not answer because the spacings were too gross and parts did not always deform in a balanced biaxial manner. Next, a steel-pointed compass was used to scribe circles on sheets to avoid distortions of squares into trapezoids or other odd shapes in extreme drawing conditions, such as the corners of deep-drawn pans. The next development was to use a photogrid to print patterns of squares and circles on the surface of sheets. This method is limited by application equipment size and adherence problems during fabrication. It is still useful for applying extremely small grid patterns on small diam cup tests (Fig. 3).

The pattern of squares and circles was used to make silk screen applied lacquer or enameled grids on sheet metals (Fig. 4). The problems of painting, cleaning up, inflammable solvents, adherence, and the coining effect of the paint film on the sheet surface were being endured. Then, in 1956, we learned of electrochemical etching from the Electromark Corporation. It was initially used to make code marks for identification of steel parts, but we thought it would be a good replacement for the silk screen, and had some stencils of our square and circle pattern made. After checking the reproducibility, accuracy of gage point spacing, effect on the steel, and adherence during forming operations, it was apparent the process was ideally suited for strain analysis work.

Electrochemical etching involves passing an electric current, either AC or DC of low voltage and amperage, through a salt solution (a mixture of sodium chloride and citric acid or other harmless salts) between the steel sheet being etched and a felt pad with a stainless steel backing, which is pressed or rocked over the stencil. Current flows only where perforations of the stencil allow the electrolyte to contact the steel, and it takes only a minute or so to complete grid application.

This simple process has been described in several articles.[3,4,5] Portable kits to do the etching sell for around $250, excluding the stencils, which cost about $4 each and last for hundreds of applications. Commercial etching solutions are available[6] which do not clog the stencils, rust the sheet metal, and are specially formulated to give dark or white lines on steel, brass, zinc, aluminum, and other metals. An article by Roger Pearce[7] gives a detailed analysis of the etch depth, which is minimal around 0.001 in. (0.0254 mm), depending on applied voltage and time, line width, and gage spacing accuracy. The conclusion is that the marks are sufficiently accurate for strain work, and the marks do not influence

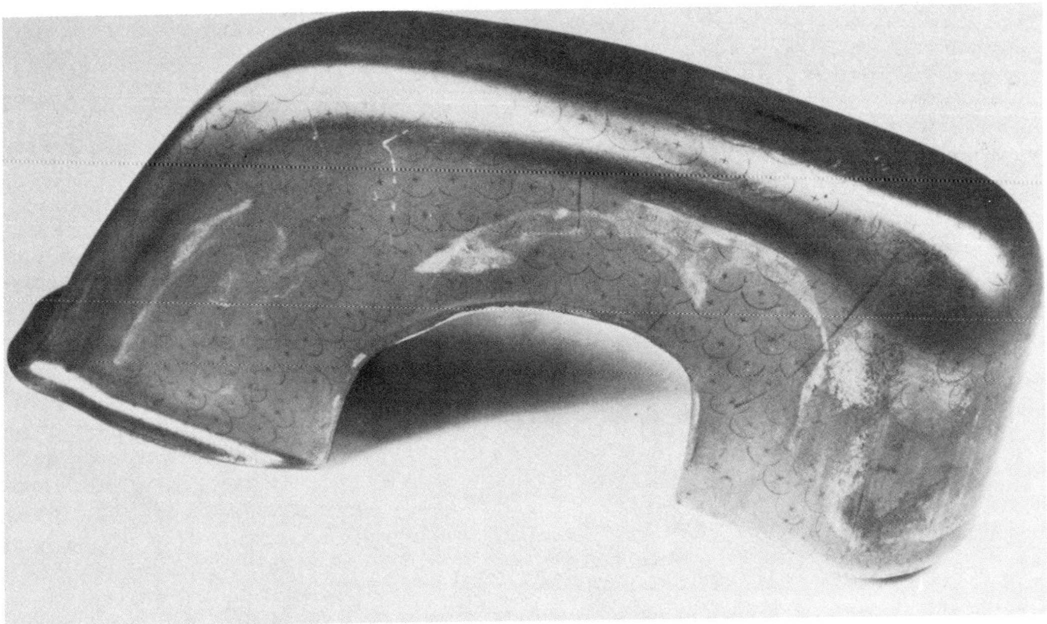

Fig. 2—Formed part showing deformation of electrochemically applied grid pattern.

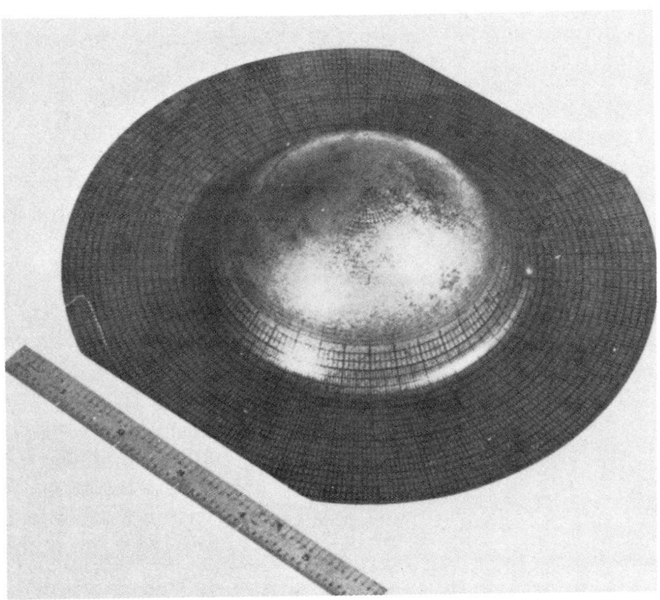

Fig. 3—Stretch cup 4-in. (10.16 cm) diam illustrating the photogrid for strain analysis.

Fig. 4—Drawn part with silk screen painted grid pattern for strain analysis.

the metal's normal behavior. After the grid has been etched on the surface, the sheet metal blank can be immediately drawn, and the strains developed in the forming operation analyzed.

MEASUREMENT OF STRAIN

Though strain analysis itself is not a panacea for forming problems, it does provide a good way of locating the problem and deducing the answer. The conveniently placed gage marks that were spaced 0.2 in. (5 mm) before the part was drawn can now be measured using dividers, a scale, or a clear plastic tapered wedge. This wedge has a pct increase scale printed on, which gives a direct reading of the pct strain (Fig. 6). It is placed in the maximum strain area, and the tapered edges are both lined up with the plus marks, or a plus mark and the arc. The reading is then recorded as pct maximum strain, which is referred to as (e_1). The tapered scale is graduated from −50 to +200 pct. It must be made for use with one gage length, such as 0.2 in. (5 mm).

The wedge is then turned 90 deg to the maximum strain

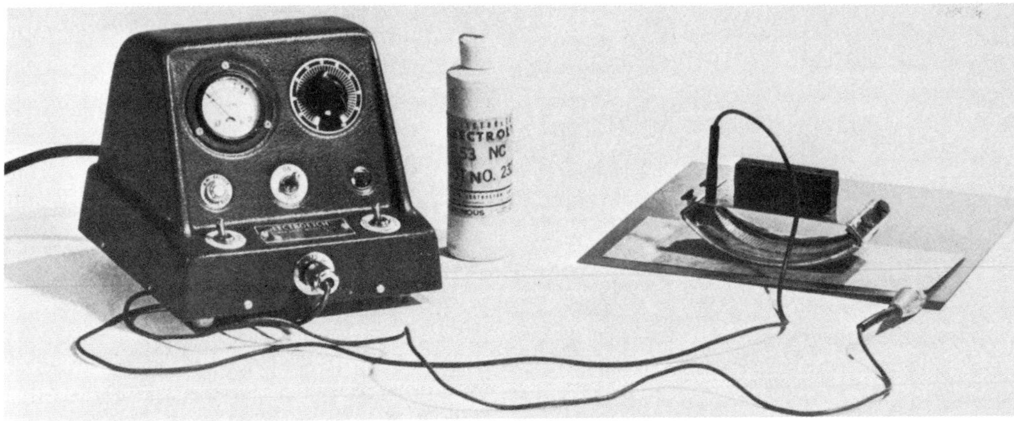

Fig. 5—Equipment for applying grids by electrochemical etching.

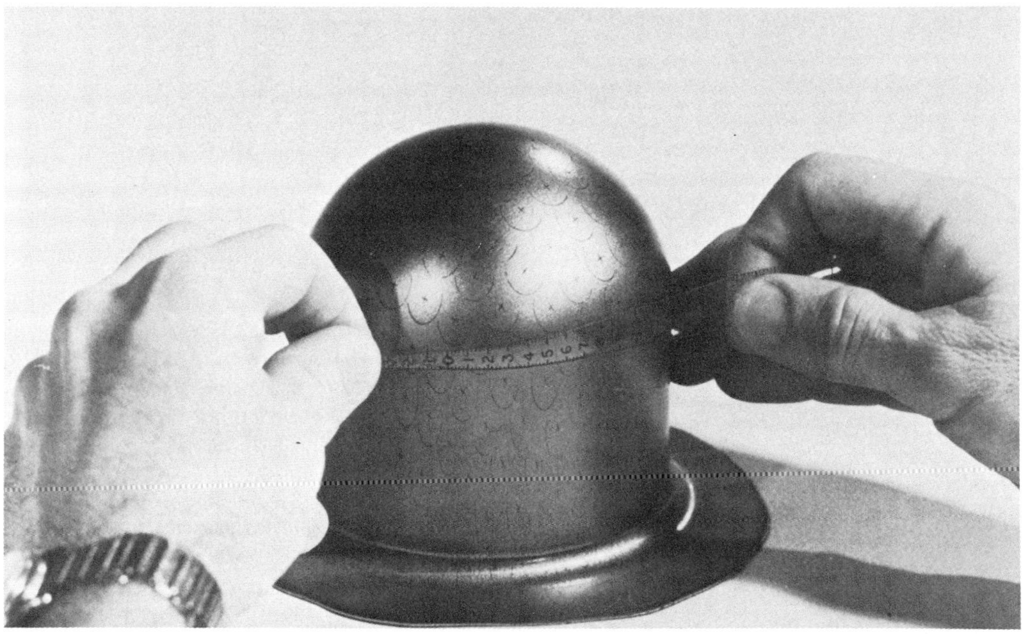

Fig. 6—Measuring e_1 strain on a formed part using a tapered scale. Grid of 0.2-in. (5 mm) half circles.

direction, and the gage spacing is measured to give the minor strain (e_2). This reading may be on either side of the gage marks used to measure the maximum strain. Both values (e_1 and e_2) may be plotted, and compared with similar values for other parts. Data can also be used in other ways—for example, assuming constancy of volume, the area change and the thickness change can be calculated. Another evaluation is a contour plot. Using the maximum strain direction in the individual gage lengths, it is easy to plot contours of equal strain over the surface of the formed part.

First, let us consider the e_1-e_2 strain relationship—that is, the distortion in the longest direction (major strain) vs the distortion at 90 deg to this (minor strain). Fig. 8 shows a plotting of these strains and the associated area increase in "iso-area" curves. The zero axis of the e_2 minor strain is a vertical line in the middle of the figure, and is an important feature in understanding metal failure. Strains to the left of zero are negative, meaning the metal has been compressed or squeezed-in in this direction on the sheet surface. Strains to the right of zero e_2 are positive, meaning the metal has been stretched. It should be recalled that this is the minor strain at 90 deg to the direction of major strain (e_1). All e_1 strains are positive because the force to form the parts is a punch-pushing metal into a die cavity, and inherently this is a uniaxial metal stretching operation. Strains to the left of zero e_2 are, therefore, tension-compression, representing deep drawing. To the right, strains are biaxial tension, representing bulging.

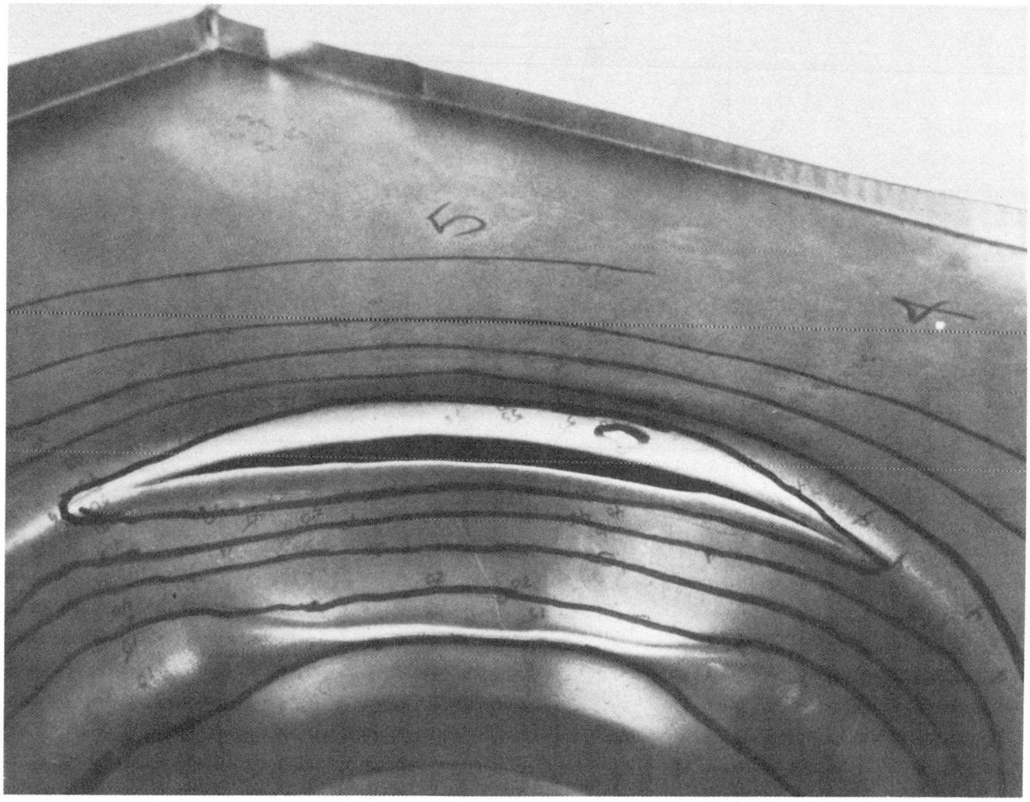

Fig. 7—Location of maximum e_1 and e_2 strains near fracture of a formed part.

SEVERITY CURVE

Some years ago, the Society of Automotive Engineers established a 25 pct increase in area criteria to differentiate between commercial quality and drawing quality steels.[8] Although this was based on a 1-in. scribed square, it also applies to Fig. 8, which is based on gage lengths of 0.1 in. (0.254 cm) up to 1 in. (2.54 cm). Within limits, such as major strains (e_1) below 50 pct, the choice between commercial quality and drawing quality steel sheets could be based on the 25 pct area increase concept.

Forming limits of Fig. 9 were established after analyzing a great number of different parts. It is interesting to note that most deep-drawing operations result in a local area increase around 25 pct. In deep drawing, if the area increases beyond 50 pct and the minor strain (e_2) is negative, the part can be expected to break during the forming operation. There is seldom any local area change less than 0 pct, which means the sheet metal only rarely thickens during the forming operation. Any extreme tendency to force this in the dies is terminated by wrinkling, or by ironing if clearances are tight.

In general, for any deep-drawn part, if the metal strain is below the classic 25 pct increase in area, less trouble will be encountered in the fabrication. Since the minor strain (e_2) component is determined by the dies, the only way the material may be varied to stay below 25 pct area increase would be to lower the major strain component (e_1).

In the vicinity of zero e_2 strain, we find a minimum in the allowable e_1 strain! In this condition, the sheet metal is subject to a high shear force, and metals are weakest in shear. If the metal is stretched in the e_1 direction and restrained in the e_2 direction so that it is neither stretched nor compressed, the maximum allowable e_1 strain before failure is reduced.

There are instances where metal has failed well below the expected maximum due to a highly localized shear condition not detectable by the usual strain analysis.

As shown in Fig. 9, most data for production parts fall within the severity band. Outside of this band, the strain often leads to failure (top curve near 50 pct area increase), or data do not fall (left of 0 area increase curve), or are meaningless (lower right).

MECHANICAL PROPERTIES

To the left of zero e_2 strain, assuming the maximum strain lies within the severity band, an increase in major strain (e_1) requires steel with better formability. This would be a steel with a low yield strength, high elongation, and either a high plastic strain ratio (r_m) or a high strain hardening capacity (n). For reference, an (r_m) of 1.0 is low and 2.0 is high, while an (n) of 0.15 is low and 0.25 is high.

We once felt that a high (r_m) applied left of zero e_2 and a high (n) was needed right of zero e_2. Recent evidence indicates this is not necessarily the situation in some severe forming operations.

If a metal has a high (r_m), it will have a basic structure which resists change through the sheet thickness,[9] which is helpful under tension-compression conditions under moderate strain. Side walls of deep-drawn pans, for instance, are less subject to breakage if the (r_m) values are high. If the major strain (e_1) is excessively high, under either tension-compression or biaxial tension conditions, this resistance to thickness change can work against the deformation process, causing earlier

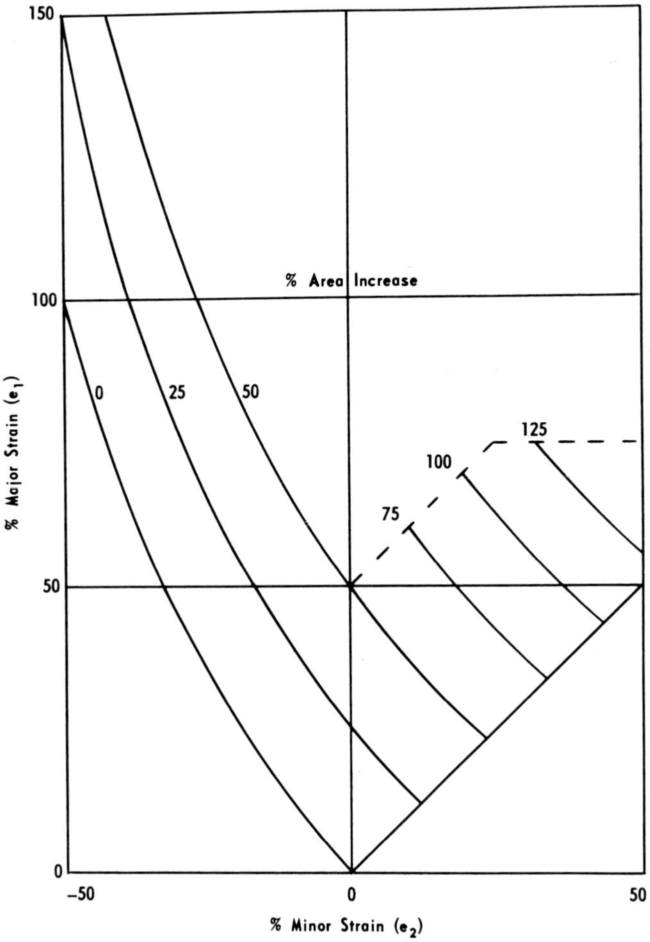

Fig. 8—Relationship of e_1 strain, e_2 strain, and area increase.

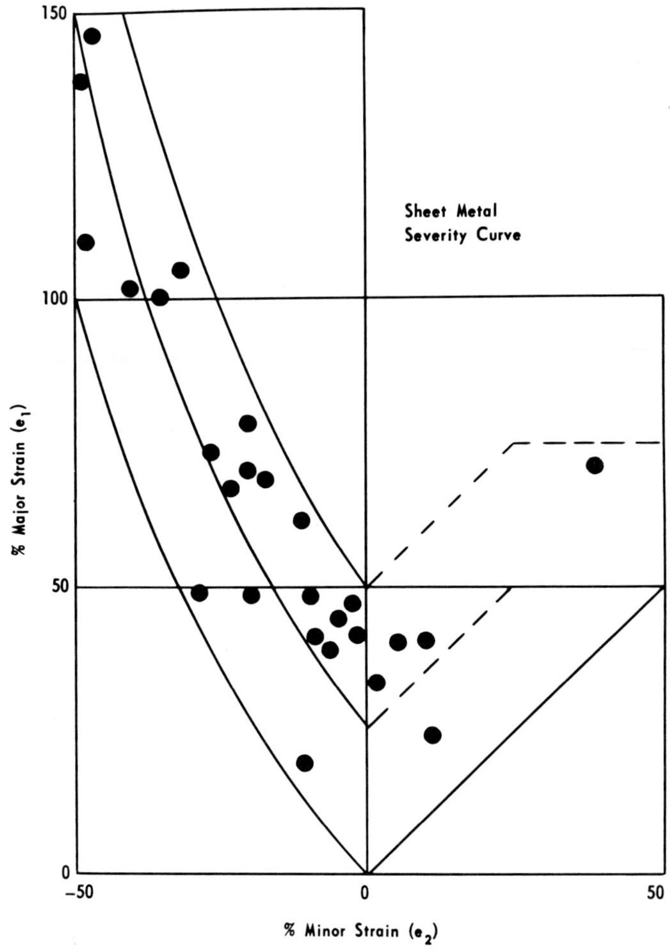

Fig. 9—Severity curve including data from production parts. Note all data lie within 0 to 50 pct area increase when e_2 is 0 or negative.

breakage or an increased major strain (e_1) level in the critical area of the part.

Strain hardening capacity, on the other hand, assists throughout the deformation process. Austenitic stainless is an example of a steel which has a high strain hardening capacity, and these materials can be formed into extremely deep-drawn items. The area of the blank which is deformed first is strengthened by the cold work, and subsequent strains are transferred to adjacent areas. This process continues throughout the forming operation. Two problems which result are difficulty in holding the metal back in the flange areas and wrinkling in flat sidewall areas.

Shear strength is usually 60 pct of the tensile strength. In a tension test, the metal necks inward as the specimen is elongated, following the 25 pct area increase curve. Although the elongation measured over the usual 2-in. (5.08 cm) gage length seldom exceeds 50 pct in low carbon steels, the e_1 strain in a 0.2-in. (5 mm) segment of the specimen can show strains of over 100 pct.[10] If the sides of the tension specimen were restricted from drawing in so that the minor strain (e_2) was maintained at zero pct, it would be in a highly restricted shear condition, and would fail at a lower total elongation.

Increasing the minor strain (e_2) to provide a better balance between the two, as in a biaxial stretch test, results in improved metal flow. This is shown by the increase in the limiting strain to the right of the zero e_2 line. Sheet metals don't mind being stretched in one direction as long as they are simultaneously either squeezed in or stretched out in a direction at 90 deg to the major stretch direction.

DEEP DRAWING

A refrigerator vegetable pan formed to a depth in excess of 4 in. (10.16 cm) is shown in Fig. 10. This is a moderately severe deep drawing with a major strain of 110 pct e_1. In a similar pan drawn to a depth of 6 in. (15.24 cm), a maximum strain approaching 300 pct developed near the flange. Breakage occurred in this area unless an excellent drawing quality sheet was used. Two decarburized grades of steel were supplied for trial in forming this part. One had a plastic strain ratio of 1.2 and an n value of 0.25, while the other had a plastic strain ratio of 2.0 and an n value of 0.20. They both made the part most of the time when the press and lubricant were in good adjustment. It would seem that a high r_m material would be best since the pan is definitely a deep-drawn item.

When the strains were measured, it was found that the high r_m material had a maximum e_1 strain of 270 pct and an e_2 strain of -68 pct. The high n material developed a lower maximum strain of 220 pct e_1 with the same e_2. When pans did break, the high r_m material developed a chevron split in the maximum e_1 strain area near the flange, while the high n material failed catastrophically in the side wall area.

It is interesting to note that the e_2 strains were the

same in this deep-drawn part. The minor strain (e_2) component in deep drawing is determined by the dies, and is independent of material. If trouble is encountered in forming a part, we can reduce the incidence of breakage by modifying the dies, changing the material, or using a different lubricant. We can modify the dies to increase the minor strain (e_2) relative to the major strain (e_1), the material to lower the major strain (e_1), or the lubricant to control the metal flow into the forming areas.

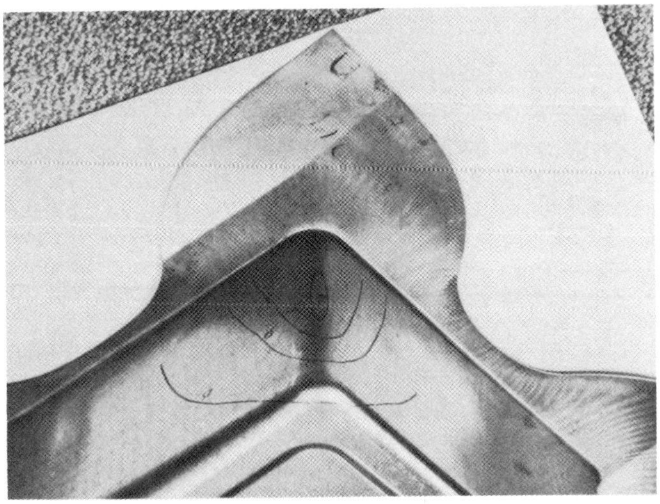

Fig. 10—Refrigerator vegetable pan showing contours of equal e_1 strain in a deep-drawn corner. Maximum e_1, 110 pct; e_2, −45 pct; area increase, 15 pct. Grid of 0.5-in. (1.27 mm) diam circles.

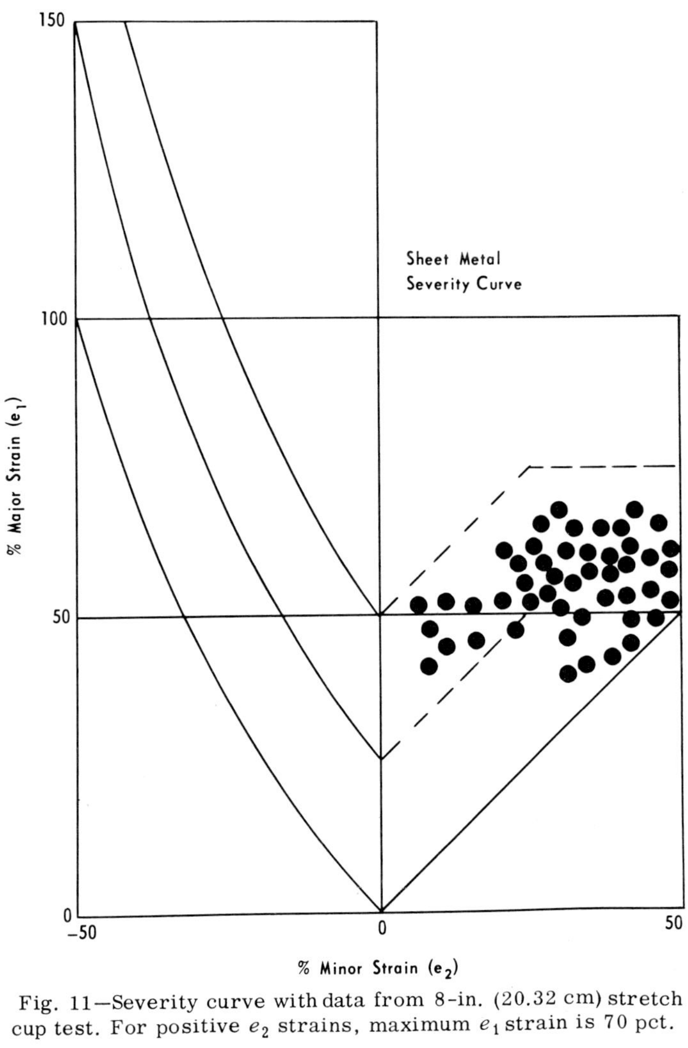

Fig. 11—Severity curve with data from 8-in. (20.32 cm) stretch cup test. For positive e_2 strains, maximum e_1 strain is 70 pct.

BIAXIAL STRETCH

As far as area increase is concerned, the best condition exists when the minor strain (e_2) is positive and high relative to the major strain (e_1). Of course, if the e_2 strain exceeds the measured e_1 strain, e_2 becomes e_1 and their positions reverse. Thus, values to the right of the lower 45 deg line for positive e_2 strains do not exist, as shown in Fig. 11. With the strains in good balance, it is possible (in certain instances with special materials) to obtain area increases as high as 150 pct without failure!

It is difficult to find production parts which fail in biaxial tension. This area of the strain chart is of interest, however, in developing material properties and evaluations of materials. The 8-in. (20.32 cm) hemispherical stretch cup test has proved useful in obtaining e_1 and e_2 data in the biaxial tension region of the strain curve. The metal is stretched until it fractures, and radial and circumferential strains are measured adjacent to the fracture, but not including the locally necked-down metal.

The e_1 and e_2 strains are plotted for 8-in. (20.32 cm) stretch cups of a variety of materials in Fig. 11. Those with high strain-hardening (n) values are located in the upper right, indicating better biaxial stretchability. When better lubrication is used, the data also locate further to the right. As materials with high (r_m) values do not stretch as well under biaxial tension, data for these are in the lower e_1 region. Materials with generally poor mechanical properties fail at low e_1 and e_2 values under biaxial tension.

DIE DESIGN

Most dies are made to match drawings designed to make the desired shape in a minimum of operations. Most designers like to have corners as close to square as possible. Aside from past experience, not much is done in considering how the metal must move to form the part without wrinkling, tearing, or warping. Sheet metal has a texture, is made up of a crystalline structure, and (depending on composition and processing) may be strong or weak, and easy or difficult to form in certain directions relative to the rolling direction of the coil in which it is processed.

Considering steel alone, there is much we can do to tailor the product to the requirements of the fabricator. We also know how to make steel of fair quality for a relatively low cost. If the designer of the dies understands the capability of the sheet metals available, it is quite possible that he can use this knowledge to build dies that do not need the more expensive tailored steels.

The best place to use strain analysis is during die tryout because tooling can be more easily modified at this time. The die designer should be present, and should use the knowledge gained from these trials in his next die design.

STRAIN CONTOURS

Strain contours developed in the formed part can be related to the original flat blank. This locates critical areas and illustrated strain gradients, which can then be reduced by die design if too severe. It is possible to evaluate the need for modification of the dies by observing how rapidly the strains increase over the surface. Such contours also permit locating the critical zone exactly.

Strain contours on a deep-drawn part are shown in Fig. 10. The automobile oil pan of Fig. 12 is a difficult part to fabricate due to a bulge in the bottom, which appears to be under biaxial tension. The e_1 strain adjacent to the fracture, however, was 48 pct and the e_2 strain was -6 pct. These are not biaxial tension nor are they excessively high, so the question is—why did the part break? The reason for the failure is the high shear condition between the biaxially stretched corner and the excess of metal in the side wall just below this corner. It was possible to reduce the incidence of fracture by substituting a material with a high r_m value. Sheet metals are weakest under shear conditions, and simply measuring the maximum e_1 and e_2 strains does not show this condition.

Strain analysis involves more than applying a grid pattern and making some measurements. The answers to press forming problems are not always immediately exposed by this procedure. Assuming the die cannot be changed and the pressman is encountering excessive breakage, strain analysis can be used to indicate the material which should do the best job on the part. To do this, other information is also needed—what is the ultimate use of the part; is it to be painted, enameled, exposed, unexposed, made from coated grades or stainless; does it have to withstand fatigue, heat, impact, or weight?

Resistance to thinning in low carbon steel grades can be varied by adjusting the crystallographic orientation, measured by the plastic strain ratio (r_m) value. Strain-hardening capacity measured as the (n) value can also be varied. These two properties are less well-known than the hardness, yield and tensile strength, elongation, or even the Olsen cup value which some fabricators try to use in material selection. How do they relate strain analysis? In general, if one property is found to be more important, others must be accepted at lower values in the acceptability listing. A material with a high plastic strain ratio (r_m) will have a lower strain hardening capacity (n). High strength is associated with lower ductility; an unflawed surface requires closer inspection and higher cost.

Fig. 12—Automobile oil pan with contours of equal e_1 strain in a biaxial stretch area. Maximum biaxial strain, e_1, 16 pct; e_2, 8 pct; area increase, 25 pct. Maximum strain on part was e_1, 48 pct; e_2, -6 pct. Grid of 0.25 in. (6.35 mm) radius circles.

SUMMARY

It is evident that strain analysis is a useful tool in material selection for a formed part. However, it requires more than measuring the stretch in the e_1 and e_2 directions, and plotting these on a strain analysis graph. We need to know how the material has been processed, as well as many properties which can be measured only in a well-equipped testing laboratory. Strain analysis helps the metallurgist make a recommendation of grade within the limits set by the die design and ultimate use requirements.

Extensive use has been made of the technique in the laboratory, and in sheet metal fabrication plants to assist in assessing breakage problems. It is the simplest and best method we have found for field evaluations. It is objective, certainly being more factual than observation followed by opinion. Application of strain grids and their measurement to give accurate values of e_1 or principal strain, and e_2, the strain at 90 deg to e_1 on the surface of the formed part has been described. Examples of e_1-e_2 strain relations for different materials in forming a deep-drawn part have been shown.

CONCLUSION

In conclusion, we feel that strain analysis can be a powerful diagnostic approach in the solution of sheet metal forming problems. It ties together information from the pressroom with that from laboratories in which tests for composition and mechanical properties are made.

ACKNOWLEDGMENT

Data for this paper were obtained by the Armco Steel Corporation Research and Technology division. The author wishes also to acknowledge the counsel of Messrs. G. E. Selby, S. T. Belsheim, R. E. Fender, and W. E. Newby.

1. *The Selection of Low Carbon Steel Sheet for Formability*, p. 319-35, ASM Metals Handbook, 8th Ed., vol. 1, 1961.
2. S. P. Keeler: *Circular Grid System—A Valuable Aid for Evaluating Sheet Metal Formability*, SAE Paper 680092, Automotive Engineering Congress, 1968.
3. R. Loewen: *Make an Electric Stencil to Etch Metal*, pp. 154-55, Popular Science, Nov. 1968.
4. D. R. Palmer: *Electro-Chemical Marking of Grids in Autobody Press Working*, vol. 45, pp. 198-204, Sheet Metal Industries, March, 1968.
5. B. W. Mayer: *Handbook of Electrochemical Marking.*
6. (a). The Electromark Corp., Willoughby, Ohio. (b) The Lectroetch Co., E. Cleveland, Ohio. (c). Monode, Inc., Cleveland, Ohio.
7. Roger Pearce: *Some Aspects of Electrochemical Marking*, vol. 45, pp. 751-55, Sheet Metal Industries, October, 1968.
8. SAE Recommended Practice—J86a—*Methods of Determining Plastic Deformation in Sheet Steel Stampings*, Iron and Steel Handbook Supplement HS 30, p. 80, 1969.
9. J. R. Newby: *Sheet Metal Testing for Formability*, SME Technical Paper MF 69-515, 1969.
10. J. R. Newby: *Analysis of Stretch Forming Variables*, SME Technical Paper MF 70-503, 1970.
11. R. H. Heyer and J. R. Newby: *Effects of Mechanical Properties on Biaxial Stretchability of Low Carbon Steels*, SAE Paper 680094, Automotive Engineering Congress, 1968.

Section V: Blanking

How to improve metal stamping die performance

By HARDING R. HUGO, product manager, Punchrite Components, Danly Machine Corporation.

The word 'performance' implies some form of productivity comparison with other stamping dies as well as successive runs of the same die assembly. Two methods of measurement are in general use, viz., number of parts produced for press run and number of parts per sharpening grind. But for accurate performance comparisons the measurement system must be based on burr heights related to specific part quality levels.

Le mot 'performance' suppose une certaine sorte de comparaison de la productivité de plusieurs matrices d'emboutissage, et des cycles successifs de la même presse. Deux méthodes sont utilisées généralement: on compte le nombre de pièces produites par cycle de marche, et le nombre de pièces produites entre affûtages. Mais pour comparer les performances avec exactitude, le système d'évaluation doit tenir compte de la hauteur des barbes, en fonction des niveaux de qualité exigés pour des pièces spécifiques.

Der Ausdruck 'Leistungsfähigkeit' legt einen irgendwie gearteten Produktivitätsvergleich mit anderen Stanzwerkzeugen wie auch von aufeinanderfolgenden Fertigungsreihen mit ein- und derselben Werkzeuganordung nahe. Die beiden allgemein benutzten Messverfahren vergleichen die Anzahl der je Pressen-Fertigungsreihe bzw. je Nachschleifperiode hergestellten Teile. Für genaue Leistungsvergleiche muss das Messsystem auf speziellen Güteniveaus der Teile, z.B. Grathöhen, basiert sein.

La palabra performance' (resultado) denota alguna forma de comparación de productividad con otros troqueles estampadores, así como con otras series sucesivas del mismo conjunto de troquel. Se emplean generalmente dos métodos de comparación: según el número de piezas producidas por serie y según el número de piezas por rectificación de afilado. Aunque para comparaciones precisas de 'performance' debe adoptarse un sistema basado en la altura de rebabas en relación con los niveles de calidad de una pieza concreta y determinada.

THE WORD 'Performance' implies some form of productivity comparison with other stamping dies as well as successive runs of the same die assembly. Obviously, then, a basic system of productivity measurement must be used for accurate comparisons.

Two methods of measurement are in general use and both have built-in errors. One is based on the number of parts produced per press run; the other on the number of parts produced per sharpening grind.

The 'Duration of Run' method generally fails to recognize the difference between a terminated run and an interrupted run. Termination occurs when the required number of stampings has been produced. Interruption occurs when the quality level of the parts is objectionable or when components fail.

The 'Per Sharpening Grind' approach eliminates the need to isolate differences in quantity requirements. However, it does not recognize significant differences in the required quality levels of various parts. For example, the number of parts produced per sharpening grind can be very substantial when the permissible burr height is 0.006in maximum; the die is said to be a high-performance tool. Conversely, the same die will exhibit poor performance characteristics if the permissible burr height is reduced, necessitating more frequent sharpenings.

For accurate performance comparisons, the measurement system must be based on burr heights related to specific part quality levels. Five classes of cut-edge characteristics are recommended for defining part quality levels and these are illustrated in the chart 'Punch-to-Die Clearances' (*see* page 132). By limiting productivity measurements in terms of burr height for each of the five classes, the individual effect of the many contributing factors can be properly evaluated. This method also provides a common denominator for accurate comparisons.

The stamped part

The desired dimensions on the periphery of the blank of hole are called the 'Feature' sizes. There is an established relationship between the feature sizes of the blank or hole and the punching components. The hole feature size establishes the punch size, and punch-to-die clearance is added into the die cavity. Conversely, the blank feature size established the die-cavity size, and clearance is subtracted to determine the punch size. Although these are generally accepted rules of application, they are subject to variations.

The punch size is not necessarily the same as the desired hole size; it may produce a smaller hole or a larger hole depending upon the clearance between punch and die, the type of stripper used, the condition of the cutting edge, and the type of material being punched. Conventional per side clearances of five per cent of the stock thickness for ferrous materials and 3-4 per cent of the stock thickness for non-ferrous materials usually result in producing a part hole approximately 0.0005in smaller than the punch. Doubling these conventional values results in holes that are equal to or about 0.0005in larger than the punch which produced them.

A spring-loaded stripper restricts the movement of the top surface of the material and will generally result in producing a part hole size which is different from that produced with a channel-type stripper.

Dull cutting edges require deeper penetration than the sharp edges to effect fracture and, therefore, have a minor effect on the resultant hole size. The material being punched is harder worked and this gives rise to a greater work-hardening effect which influences hole size.

Fig. 1 Punched holes and blanks; nomenclature

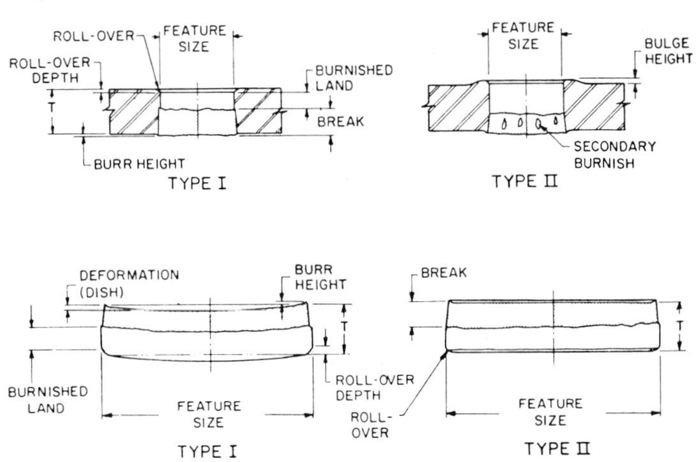

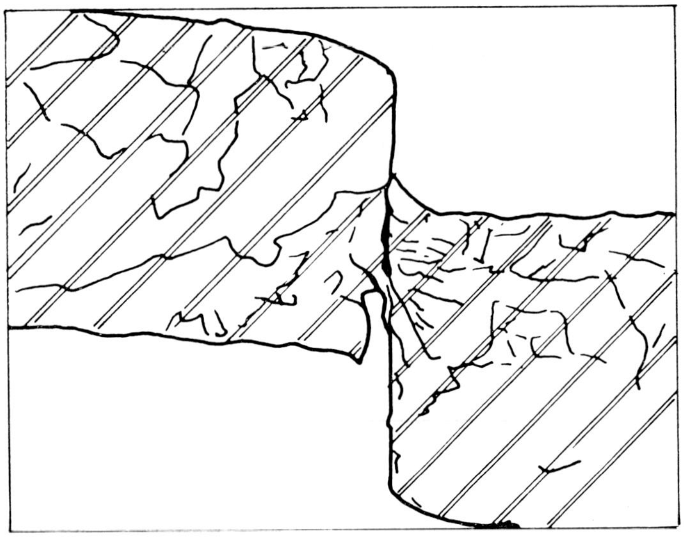

Fig. 2 (a) Burr formation, lamination steel, 0.025in thick.

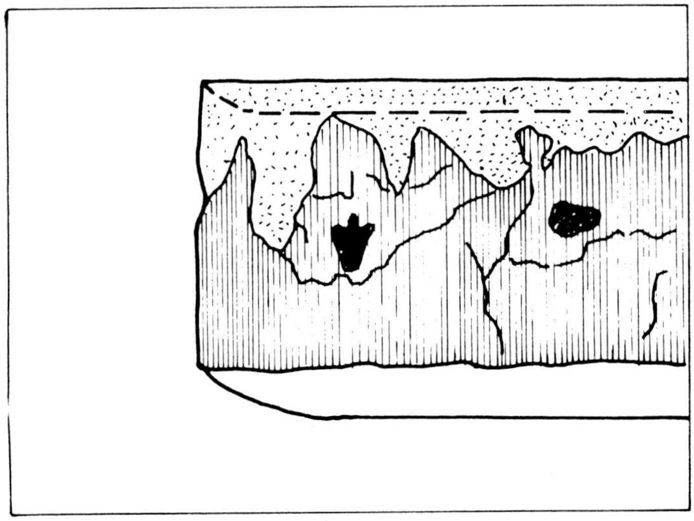

Fig. 2 (b) Characteristics of cut edge.

The proportion of blank size to stock thickness may be large enough to introduce an oil-can effect. Unless the part is supported in the cavity, the blank may be larger than the cavity due to deformation.

CUT-EDGE CHARACTERISTICS AND NOMENCLATURE

The American Metal Stamping Association has prepared a series of Design Guide Sheets. These guides are intended to provide a basic understanding of stamped parts and to define practical limits for general purpose applications. One of the first sheets to be developed covered common terminology of the cut edges.

The standardized cut edge characteristics and terminology may be applied to both the hole and the blank or slug. The sketch 'Punched Holes . . . Nomenclature' (Fig. 1) illustrates the elements that comprise the cut edge.

Roll-over

Roll-over is the rounded edge caused by the plastic flow of the material under a compression load. Since some degree of clearance between the punch and die is required to effect fracture, the resulting torque causes the material to roll as the punch initially penetrates the stock and the underside of the stock is displaced into the die cavity. Bulging or lateral displacement is generally combined with the roll-over when the punch-to-die clearance is insufficient and the punched material is soft or ductile. This edge characteristic is illustrated in Type II of punched holes (Fig. 1).

Roll-over depth is the distance from the surface where the rolling begins to the intersection with the burnished land.

Burnished land

Burnished land is the relatively straight portion of the edge produced by the plastic deformation of the material being punched. The length of the land is the distance from the intersection with the roll-over to the point where fracture or separation occurred. Burnished land may also be referred to as the useful length of the feature size or as the sheared length. For general purpose holes, the burnished land length is said to be one-third of the stock thickness.

Angular break

The angular break is the edge portion that results from sudden separation due to fracture. Under the punching load, the material confined between the punch flank and the die land gradually stretches and becomes thinner. When its ability to withstand the imposed load is exceeded, the fibres separate suddenly and fracture occurs. The texture of the break area is dull and appears to be pock-marked. When insufficient punch-to-die clearance is utilized, small burnished areas will become apparent as illustrated in the Type II punched hole.

It is important to emphasize that the magnitude of the angle is very small. For example, a total clearance value of 0.003in for 0.030in thick stock in a general-purpose hole results in an angle of only 0.0015in in 0.020in length. Unfortunately the angle is so small that it can only be indicated by exaggerating it. The resulting illustration is retained in one's mind and tends to supplant the actual magnitude.

Burr height

Burr height is the distance from the top of the burr to the surface of the material from which it projects. Because the material failed in tension, the fibrous stringers that result will extend beyond the surface where fracture occurred. The burr on the hole edge results from the fracture generated by the cutting edge of the die cavity. The burr on the slug or blank is caused by the fracture at the punch cutting edge. This phenomenon is illustrated in the sketch 'Burr Formation' (Fig. 2(a)).

Influencing factors

Each of these characteristics is affected by the mechanical properties of the material being punched, the punch-to-die clearance, the cutting edge conditions, the type of stripper, and the ratio of feature size to stock thickness. The sketch 'Characteristics of the Cut Edge' (Fig. 2(b)) illustrates the effect of worn cutting edges on the punch. It is evident that a considerable amount of smearing took place because of the plastic metal that accumulated below the corner radius on the punch.

Recommended specifications for stamped parts

To reduce the cost of tooling and stamping production, several basic recommendations are presented:

1. Feature size should be measured only within the burnished land.
2. Shape deviations within the feature size limits should be permissible.
3. The length of the burnished land should not be specified unless the part function will be impaired. If it is necessary to specify this dimension, only the minimum value shall be given.
4. Burr height is to be specified only as a maximum, and the largest value consistent with part function shall be specified.

FACTORS THAT INFLUENCE PERFORMANCE

The stamping operation may be subdivided into two major groups to simplify the isolation of influencing factors. These groups are based on the operational sequence and are known as the Shearing Factors and the Stripping Factors.

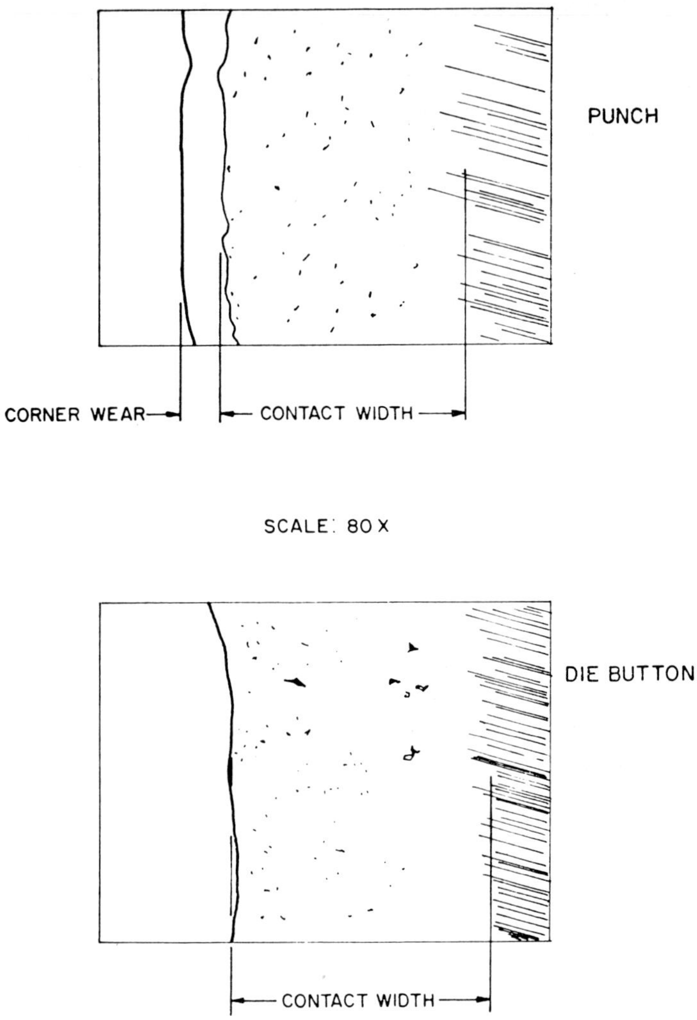

Fig. 3 Typical face wear patterns.

Shearing factors

Productivity is based on the ability of the punch to withstand the stresses to which it is subjected. The factors involved are shock, edge pressures, tensile strength, and deflection.

The punch is subjected to at least two impact forces. Initial impact occurs when the punch first penetrates the stock. Only a slight entry is required to build up sufficient resistance to overcome the weight of the ram. As soon as this level of resistance is developed, the punch stops its downward travel while the clearances in the bearings of the various moving parts are being taken up. When the clearances in the main bearings, the crank bearings, and the Pitman connection are taken up, the punch slams into the stock once again. As soon as the fracture occurs, the punch, which had been foreshortened by the opposing loads, attempts to regain its original length. These alternating cycles of compression and tension are violent and rapid. Obviously, the punch material must be capable of withstanding these violent alternations.

Edge pressure calculations are usually made on the basis of full face contact. As a consequence, edge chipping or flaking is blamed on the punch material rather than on edge loading. In reality, only a small ring or band of contact exists and this must be considered as the area over which the punching load is distributed. Any failure to observe this fact will eliminate any chance of improving performance.

Deflection is the most harmful factor of all because it almost always results in punch breakage. Deflection may be caused by misalignment of the punch to the die, by unbalanced loads relative to the centre of pressure, and by misfeeds. Careful consideration of each of these factors is absolutely essential to their control and will result in a substantial productivity increase.

Stripping factors

Component wear can only be caused by frictional forces. Starting with the formula for friction, one can isolate the contributing factors.

$F = \mu N$ Where: F = Force of Friction
μ = Coefficient of Friction
N = Normal Force

It is important to note that the punch wears at two or three times the rate of the die cavity. The basic explanation is that the punch goes through the stock twice while the part of slug goes through the cavity only once. Although this is a statement of fact, it must be extended in terms of the friction generated during the shearing process relative to that of stripping. Based on the results achieved in industry when controls were instituted that primarily affected stripping friction, it is apparent that stripping friction generates approximately twice the wear rate of the shearing friction. For example, the burnished land is produced by plastic deformation of the material being punched. When the stripping cycle begins, the burnished land has become work-hardened and results in an increase of the coefficient of friction.

Until very recently, there was not enough knowledge relative to the factors that influence the force of friction. As a consequence, performance improvement was limited to a change in the punch material. With a fixed value of friction, the only course of action was to select a new tool steel with greater abrasion resistance. Unfortunately, the shock resistance of tool steel is reduced as the abrasion resistance is increased. As a consequence, it was not unusual to exchange the component failure from premature wear to that of breakage. It is quite obvious that this approach offers very little opportunity for performance improvement.

FACTORS THAT INFLUENCE THE COEFFICIENT OF FRICTION

The best approach to problem-solving relative to premature wear lies in reducing the frictional forces generated during the stripping cycle. Any reduction in the coefficient of friction will result in a reduction of friction itself. Among the contributing elements are:

Hardness Disparity
Lubrication
Finish
Alignment
Deflection
Cutting Edge Condition.

Careful control of these elements will afford a substantial increase in stamping die performance.

FACTORS THAT INFLUENCE THE NORMAL FORCE

The normal force is influenced by four factors:

Ratio of Hole Size to Stock Thickness
Proximity of Adjacent Holes
Type of Stripper
Punch-to-die Clearance.

Punch-to-die clearance offers the greatest potential for increasing stamping die performance. Since larger than conventional clearance values generally result in slug-pulling problems, there must be a change in the slug size. An expansion of this idea indicates that there may be a hole size change as well. Investigation and research revealed that larger clearances produced holes equal in size to the punch or slightly larger than the punch. In effect, the normal force could equal zero which practically eliminates the stripping friction! At the very least, it provides the most potent solution to the premature wear problem.

The problem of slug-pulling can be resolved in several ways. Ejector pins installed in the perforated punch will strip the slug. Other methods include an air blast through the punch, a rubber insert in the punch, or a vacuum bushing placed below the cavity when the punch point is too small to incorporate an ejector pin.

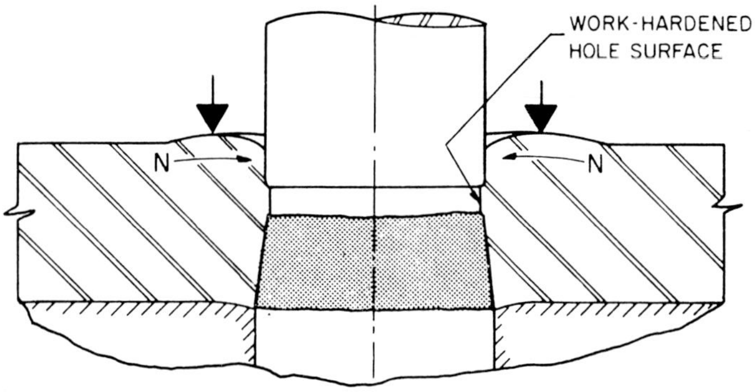

Fig. 4 Cutting-edge breakdown.

Punch-to-die clearances

The subject of punch-to-die clearances is very complex. Although most stampers are inclined to think of most punching or blanking operations in terms of the punch being smaller than the die cavity, this is not always the case. Essentially, there are three conditions of punch-to-die clearance:

1. Negative Clearance
2. Neutral Clearance
3. Positive Clearance.

The discussion of these various conditions must be predicated on the basis of burr height considerations. Most objections to the use of increased or optimum clearances are the result of the mistaken notion that the burr height would become quite objectionable. In order to substantiate that increased burr height will result only from tight clearances rather than from increased values, one must examine the only situation where both kinds of clearance can exist simultaneously—a misaligned punch and die cavity.

A study of the resultant hole will reveal that the length of the burnished land is not uniform. Further, the burr height is not uniform around the periphery of the punched hole. Correlating these two factors, one will discover that the highest burr is in the same relative position as the longest burnished land. By relating the hole to the punch and die, it will reveal that the tightest side produced both the longest burnished land and the highest burr. It is indeed unfortunate that it has taken so long a time to look at the other side where the clearance is actually considerably larger than it was intended to be!

Since burr height is such a burning issue, this discussion will be oriented to this factor. By way of discussion and definition, the burr will be any projection of material below the bottom surface of the stock being punched.

Negative clearance

As Fig. 5 indicates, this is the condition where the punch is larger than the die cavity. The term applied to the hole produced with this clearance is a half-hole or 'squirt'. The punch only penetrates a portion of the stock thickness and fracture will not occur.

As the load is imposed on the material, the elastic limit is exceeded and the material becomes plastic or easily deformed. In this plastic state, volume of material displaced by the punch penetration is forced to flow into the die cavity. The material that projects below the bottom surface of the stock may then be called a compressive burr because compression was the origin of its displacement. It is well to remember that the punching load for this type of operation is generally a maximum.

Neutral clearance

Many die builders pride themselves on their ability to produce die assemblies where there is supposedly no clearance between the mating components. Line-to-line or metal-to-metal fits are possible although they are not very probable—especially if punch penetration into the die cavity is desired. This area is entirely too argumentative for this section and will not contribute to the accomplishment of the stated objectives. Suffice to

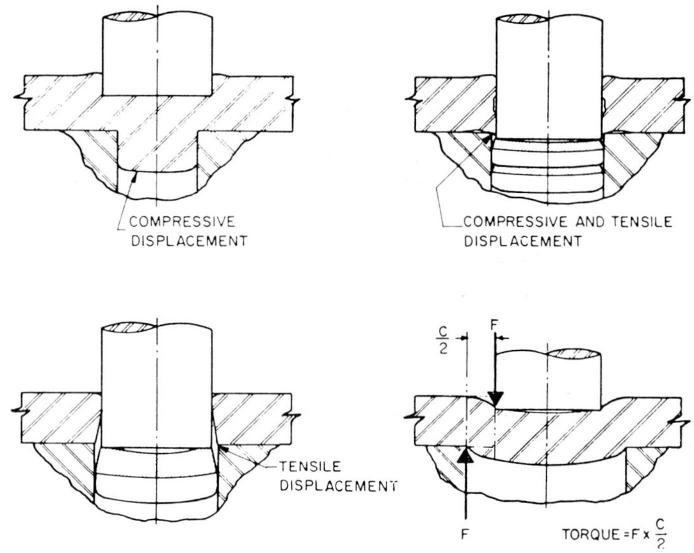

Fig. 5 (Top row) Negative and neutral clearance. **(Bottom row)** Positive clearance and shear diagram.

state that, for purposes of discussion, the metal-to-metal concept will be assumed to be entirely practical.

The punching load will of necessity be quite large to cause the plastic deformation of the material being punched. The burr will be relatively high because it consists of a compressive element and a tensile element. Just as the burr is caused by plastic deformation in the case of negative clearance, the burr in the neutral clearance condition will exhibit some of the same characteristics. The material in this case has failed in tension and the residual string-ends will also project beyond the bottom of the stock. Thus the burr is somewhat hybridized in nature and is quite large.

The stripping load will be substantial because of the long burnished land and the heavy lateral displacement of material around the periphery of the punch.

Positive clearance

This clearance condition results when the punch is made smaller than the die cavity. Depending upon the clearance or the space between the punch and the die cavity, the compressive squirt may or may not be present. If compressive displacement is present, however, it will be less than that of the first two clearance conditions.

Conventional clearances are positive clearances in that the punch is usually smaller than the die. The fact remains, however, that the resulting burr does not contain an element of plastic displacement. As the clearance is increased, the amount of plastic squirt will be physically reduced until, with sufficient clearance, the plastic squirt will be entirely eliminated leaving a burr composed only of the string-ends. This tensile burr is the smallest burr that can be obtained when punching any given material.

Burr height is not only the result of the clearances used; it is influenced by the cutting edge condition and the properties of the material being punched. Dull punches or die cavities tend to smear the metal when it is in the plastic state and cause longer burnished lands and increased burr heights. The correlation between these two factors will be demonstrated in another section. Ductile materials tend to be plastically deformed for a considerable distance before fracture occurs and, as a result, increase the height of the burr. Brittle materials, on the other hand, cannot be plastically deformed to any major degree and fracture rather readily. As a result, the burr tends to be very small.

Shear diagram

The shear diagram illustrates the effect of the punch-to-die clearance that is used. By definition, a couple is the system of forces that can only result in producing torque. It consists of two equal and opposite forces that are applied parallel to each other and separated by some distance. The resulting torque value can be calculated by applying the following listed formula:

$$\text{Torque} = F \times \frac{C}{2} \quad \text{where: } F = \text{One of the forces}, \quad \frac{C}{2} = \text{Perpendicular distance between the forces.}$$

In the case of punched holes, the perpendicular distance between the forces is the per-side clearance. Obviously, the greater the clearance, the greater the torque that will be generated. Torque is a rolling action and is as destructive to the material being punched as the compressive load. Torque also stresses the material and helps the compressive load to strain the material beyond its ultimate strength. Because the torque is a product of the compressive load, it doesn't require additional energy to create it; the space between the punch and die permits it to be developed. Since torque does not require the addition of energy and contributes to the failure of the material, it would seem that larger clearances should effectively reduce the punching loads required.

The punching load curve

Fig. 6 graphically shows the reduction in press energy that is required to punch a hole when the clearance is increased. The solid lines indicate the burnished land length and minimal roll-over that results from conventional clearance values. The phantom lines indicate the corresponding characteristics when larger clearances are utilized. By using the hole edge characteristics to indicate penetration, the load curve can be drawn.

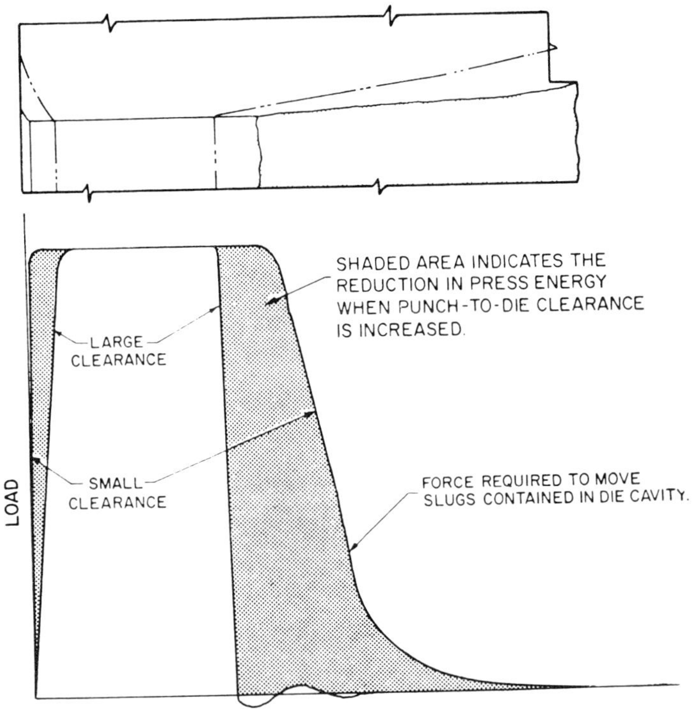

Fig. 6 Punching load curve.

The maximum load must be applied at the point where the roll-over meets the burnished land. By drawing a line from the 0-0 reference to this point, the rate of load application will be shown. The load maximum must continue to be applied until fracture occurs; this point is located at the end of the burnished length. A horizontal line drawn the length of the burnished land represents the duration of the application of the maximum load. At fracture, the slug is separated from the stock and the only load necessary is that required to push the slugs out of the die cavity.

In the case of the conventional clearances, the back line is somewhat sloped with a large sweeping curve at the end. This is intended to indicate the force necessary to move the slugs and that the required force is not a consistent one. Overcoming the inertia of the slugs contained in the die cavity requires a larger force than that necessary to keep them in motion, hence the straight line and the sweeping curve.

The larger clearances, however, do not tend to permit the slugs to remain in the cavity; instead, each slug is 'shot' out of the cavity by the slug ejector action. This is the reason for the almost vertical nature of the back slope. It will be noted that the depth of the roll-over is greater with increased clearance resulting in slower rate of load buildup. The burnished land is also shorter and this reduces the time that the maximum load is applied. Remember, the torque effect of larger clearances is also very destructive!

By comparing the load curve 'envelopes', one can readily see that the conventional clearances have generated a load curve that encompasses a very substantial area. The shaded areas illustrate the greater energy requirements for tight clearances. Obviously, the larger the clearance, the less energy will be required.

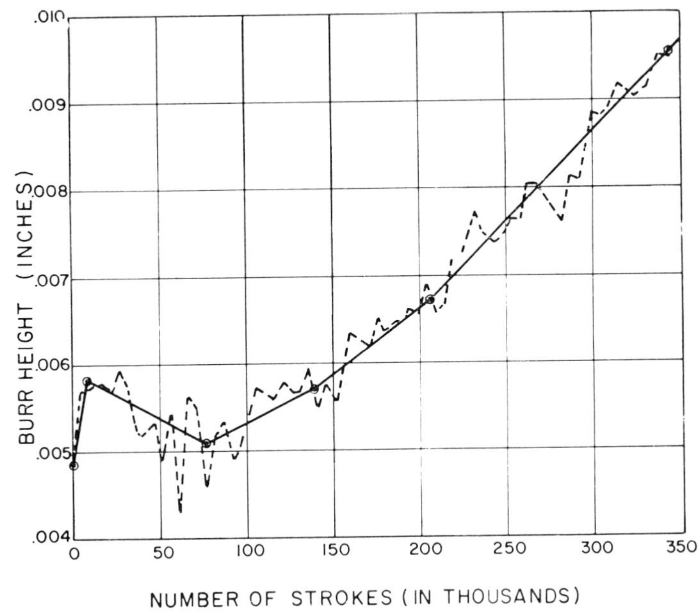

Fig. 7 Graph of burr height *versus* number of strokes. Material—Lamination steel 0.025in thick × 2.25in wide.

Optimum clearances

Optimum clearance values may be defined as follows: 'The largest clearance that can be used without impairing the part function.' The accompanying chart shows five edge conditions and the recommended clearance percentages to achieve them in various materials. By selecting the desired edge conditions and with a knowledge of the material to be punched, one can easily establish the proper clearance to use. All of the recommended values are based on cutting edges that have a 0.002 to 0.006in radius to stimulate wear so that the illustrated edge characteristics will be the average condition over the production run.

As a precautionary note, the edge conditions that will require positive slug ejection devices are denoted in the chart headings.

Calculation of punching loads

The formula to use for such calculations is shown in 'Case Report No. 1' along with actual examples of its application. To simplify the calculation, the chart of Shear Strengths for various materials has been included with the clearance recommendations. It is especially interesting to note that the punching load formula does not have any provisions for the effect of various punch-to-die clearances. This indicates that the formula is designed to incorporate large safety factors so that the proper press can be selected without too much worry that it will be big enough.

However, it is wise to recall that the point on the stroke where impact will occur cannot be overlooked!

ASTME DIE CLINIC REPORT

Subject: Burr Height vs. Number of Strokes

For greater clarity, each chart was redrawn to the same scale. In addition, the 'Number of Strokes' scale was changed from counter readings to direct readout in thousands of strokes.

The dotted lines indicate the actual plot of the curves and the solid lines represent the average of the plotted values.

Objectives of Tests

Test procedures were based on the hypothesis that a relationship exists between burr height and punch wear. In effect, as the punch wears, the burr height increases. Other factors could influence burr height, but the identification and effects of these factors were not clearly defined.

Various values of punch-to-die clearance were used plus two types of die hole configuration. At intervals of 5000 strokes, slugs were measured relative to burr height and the average reading was recorded. In an effort to restrict the length of runs, all tests were made without lubricants.

Runs were to be terminated when the average burr height was between 0.005in and 0.006in.

Test Results

Fig. 7

Total clearance 0.001in or 4 per cent of the stock thickness (T)

Die Button: No straight land—taper relieved 0.013in/in, included.

This test was made to record the pattern of burr height increase relative to the number of strokes. No effort was made to restrict the burr height to the pre-determined maximum.

The rise and fall in the curve over the first 75 000 strokes has not been satisfactorily explained. One theory has been advanced that is based on logic. The punch face had not been ground at assembly; and this omission is contrary to general industry practice. As a consequence, the punch face retained the somewhat radial peaks and valleys produced by an abrasive cut-off wheel. It is believed that the grooves were chipped away by the punching loads and that new but still irregular edges were exposed. After about 75 000 strokes, the contact area of the punch face was apparently cleared of surface irregularities and normal wear patterns were generated.

Note: Initial burr height: 0.0048in approx.

With the exception of the first few thousand strokes, the burr always exceeded 0.005in in height.

Fig. 8

Total clearance 0.005in or 20 per cent of T

Die Button: No straight land—taper relieved 0.013in/in, included.

The curve fluctuations for the first 75 000 strokes are again apparent. The commentary relative to Fig. 16 may be applied here.

Note: Initial burr height: 0.0029in approx

The burr height remained less than 0.006in for approx 190 000 strokes.

Fig. 9

Total clearance 0.003in or 12 per cent of T

Die Button: No straight land—taper relieved 0.013in/in, included.

The initial rise and fall pattern of the curve is not apparent. Since the pattern is quite distinct for both larger and smaller clearance values using the same die hole configuration, it is reasonable to suppose that the run was prematurely terminated.

Note: Initial burr height: 0.0041in approx

Duration of run: 80 000 strokes approx

Fig. 8 Graph of burr height *versus* number of strokes.

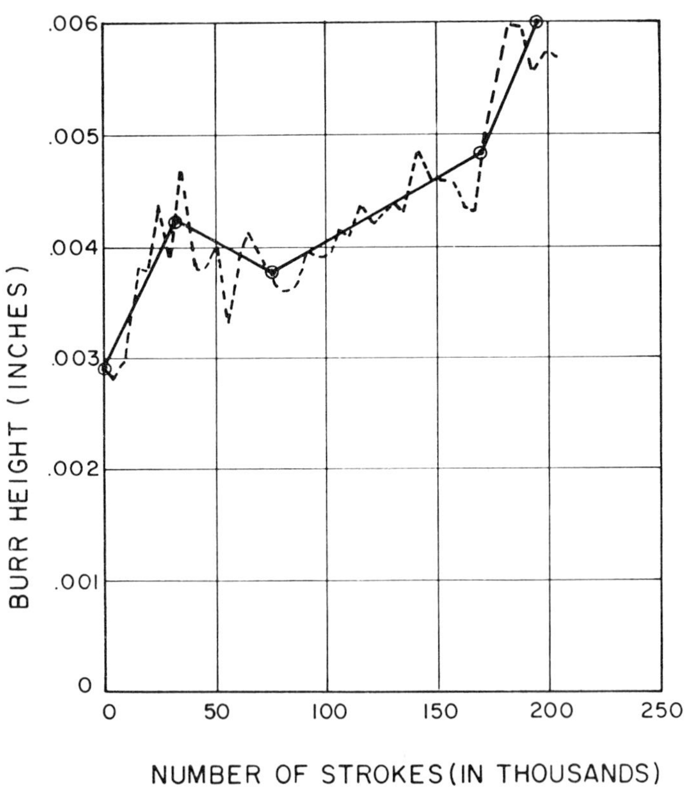

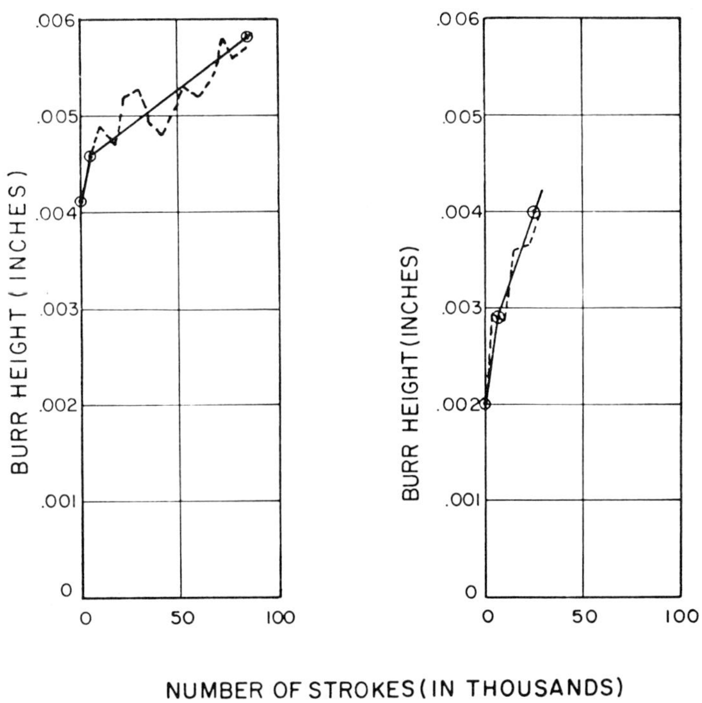

Figs. 9 and 10 Graphs of burr height *versus* number of strokes. **(Fig. 10)** Punch chipped after taking first set of measurements.

Fig. 10

Total clearance 0.003in or 12 per cent of T

Die Button: ⅛in straight land—taper relieved 0.013in/in, included.

The test was terminated after approx 25 000 strokes. The punch had chipped after the first set of measurements and was apparently not reground. A subsequent test was recorded in Fig. 23.

Note: Initial burr height: 0.0020in approx

Duration of run: 25 000 strokes

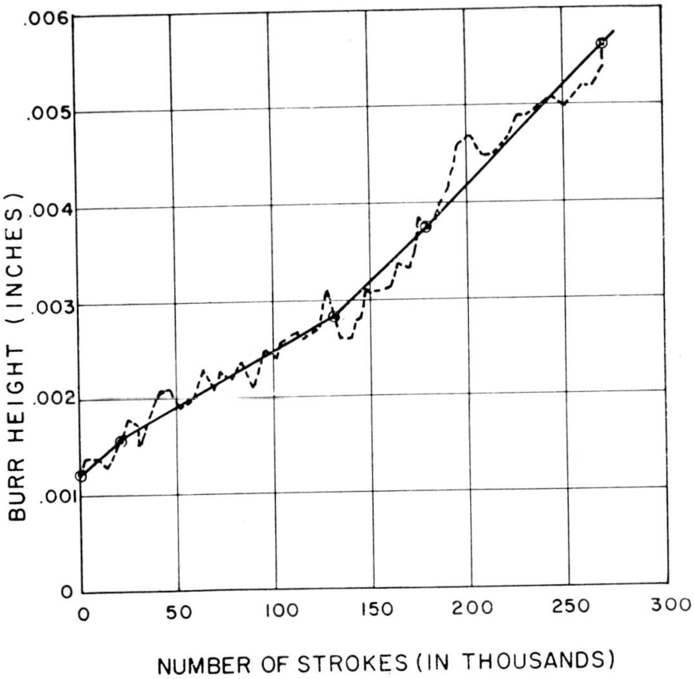

Fig. 11 Burr height *versus* number of strokes.

Fig. 11

Total clearance 0.005in or 20 per cent of T

Die Button: $\frac{1}{8}$in straight land—taper relieved 0.013in/in, included.

This is the best combination of clearance and die button internal structure that was tested.

Note: Initial burr height: 0.0012in approx

Duration of run: 272 000 strokes approx

Fig. 12 *and Fig.* 13

Total clearance 0.001in or four per cent of T

Die Button: $\frac{1}{8}$in straight land—taper relieved 0.013in/in, included.

Although the internal structure of the die buttons and the total clearance values are the same, each test was made with a different punch and die combination. It is significant to note that punch chipping was observed in each test. The punches and die buttons were reground after each chipping episode. The effects of the improved finish on the punch face is readily apparent. The initial burr height is substantially reduced as punch face finish is improved.

In Fig. 12, the run was terminated at 50 000 strokes despite the fact that the burr height criteria had not been attained. No explanation for this action was provided.

The test recorded in Fig. 13 was also terminated while the burr height was less than 0.005in, again without explanation.

The curves are very similar for both tests. The burr height increased from approximately 0.001in initially to 0.0035in in 50 000 strokes in each case.

Figs. 12 and 13 Burr height *versus* number of strokes.

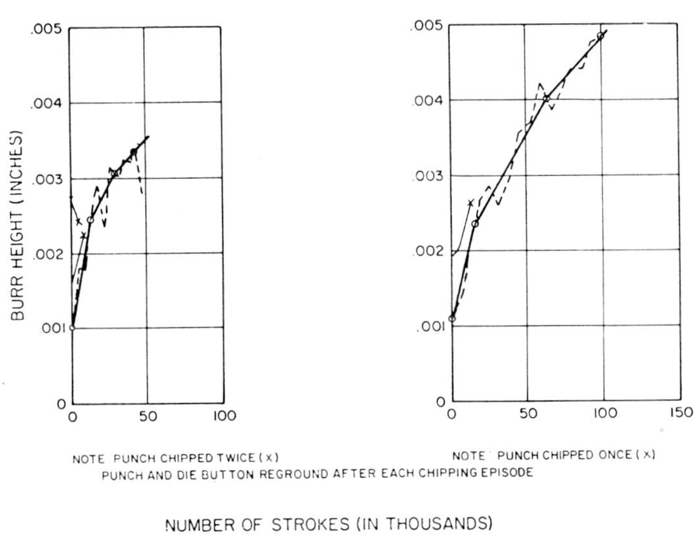

Fig. 14

Total clearance 0.003in or 12 per cent of T

Die Button: $\frac{1}{8}$in straight land—taper relieved 0.013in/in, included.

The punch and die button faces had been reground to less than 30 micro-in (RMS).

Initial burr height: 0.0014in approx

Duration of run: 175 000 strokes approx

Composite Chart (*Fig.* 15): *Burr Height vs. No. of Strokes*

This is a composite of all the individual test curves to facilitate comparisons.

Conclusions

1. The initial burr height is considerably influenced by the internal construction of the die button. It is minimal whenever a straight land is used and is not affected by the clearance utilized.

   When a tapered relief is used instead of a straight land, the initial burr height is inversely proportional to the clearance. As the clearance is reduced, the initial burr height is increased.

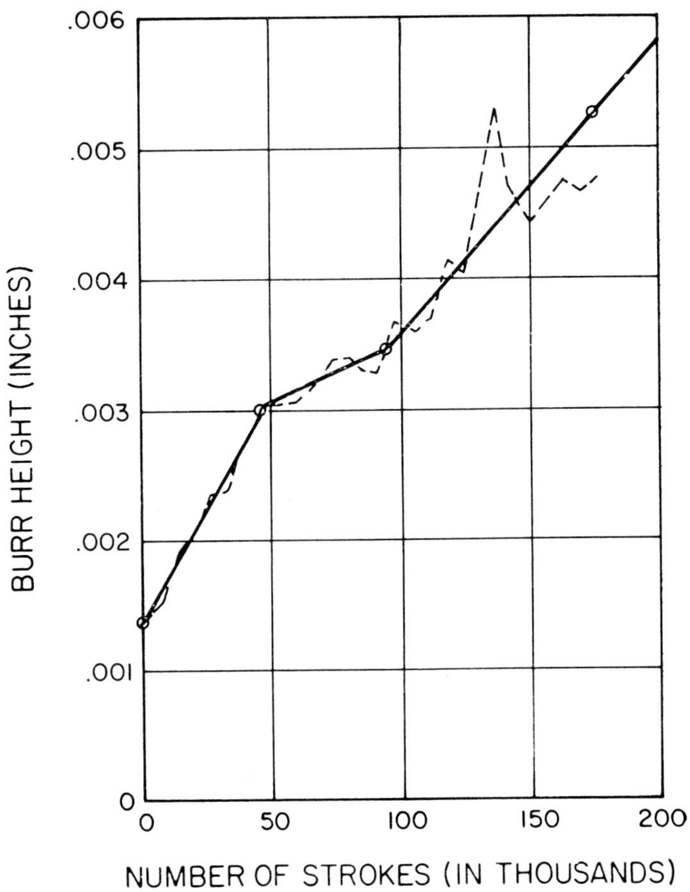

Fig. 14 Burr height *versus* number of strokes.

Punch face finish influences the initial burr height. Poor finish tends to increase the initial burr. It is also apparent that this effect is limited. For example, when the punch face was reground to less than 30 micro-in, the resulting initial burr height was not much different than that produced with a normal surface ground face finish.

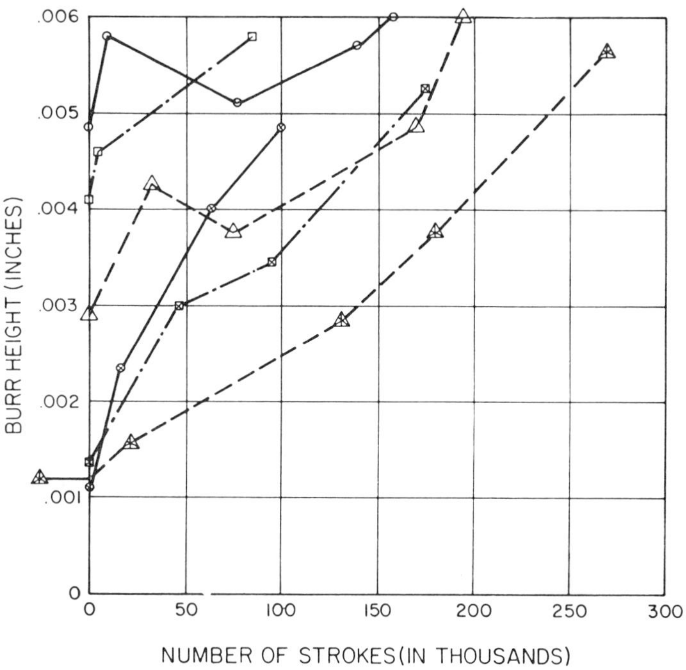

Fig. 15 Composite chart of burr height *versus* number of strokes.

2. The rate of burr height growth is inversely proportional to the punch and die clearance. As the clearance increases, the rate of burr height growth decreases.
3. For this particular material, the optimum clearance is probably greater than 20 per cent of T.

Comments:

1. The burr height produced at the bottom of the part hole is likely to be considerably less than that on the slug.
   (a) The stock is rigidly supported by the die button surface. The slug receives no external support and is subject to bending or deformation.
   (b) The die cutting edge remains sharper than the punch. It will, therefore, generate much greater stress causing rapid fracture. The burr on the hole is caused by fracture at the die button cutting edge as opposed to the burr on the slug resulting from fracture at the cutting edge of the punch.
2. Addition of the proper lubrication will reduce the rate of cutting edge breakdown. The increase in performance will not be an additive—it will be a multiplier.

CASE REPORT NO. 1

Material

Type 302 Stainless Steel . . . Hardness: Rc 39-40
Ultimate tensile strength: From 189 000 to 198 000lb/in^2
Thickness: 0.060 for structural type angles
0.090 for roll-formed corrugated sheets.

Operation

Punch 12 to 34 holes (0.215 dia) in line on one rolled section of the corrugated sheet. The parts range in length from 12in to 72in. The original intent was to run both parts across the same die assembly.

Punch press

125-ton OBI press with back-gear and air cushion.
Stroke: 6in Speed: 60spm
Parts are fed manually.

Lubrication

A composition of sulphur and lard oil is brushed on to the top and bottom surfaces of the stock.

Design of the die assembly

The punches were made of D2 steel (HCHC); the punch points were self-supporting with vee-shaped notches ground into each face. The grooves provided shear to reduce the punching load. Punch-to-die clearance was 0.0045 per side (5 per cent T). The channel-type stripper was fastened to the lower die assembly.

Initial performance: At the first press stroke, seven (7) punches were broken.

Analysis of the problem

The total tonnage required to punch 34 holes was calculated to be approximately 160 tons. The vee-shaped grooves reduced the initial tonnage requirement to 1.7 tons per punch or about 58 tons total which was safely within the capacity of the press.

The initial result—broken punches—could only be the result of punch overloading or point deflection. The area of one of the two chordal sections was calculated to be 0.00872in^2 resulting in a stress load of approximately 390 000lb/in^2—well in excess of its load-carrying abilities.

Not all of the punches were broken, however, which indicated that punch point deflection may have also occurred. This is a distinct possibility because of the curvature due to the part corrugations.

Recommendations

Install guide or support bushings in the stripper with a nominal clearance of 0.0005in/side between the bushing and the punch point to reduce any tendency for lateral deflection.

Increase the punch-to-die clearance to 0.0067in/side (7 per cent T) so that the punching load would be further reduced and the impact softened.

Eliminate the vee-groove in each punch face and stagger the lengths of the punches. One half of the punches were alternately shortened by 0.045in (half of the stock thickness).

Install ejector punches made of A2 steel; the ejectors to provide positive assurance against slug-pulling and the A2 steel to provide greater toughness.

All of the recommendations were adopted and the die assembly was reworked accordingly. These results were reported:

After 26 hits, the punch points had bulged 0.0005in dia at the cutting edge. One (1) punch point broke after 108 hits while the other punches each produced 2100 holes without further interruptions. At that time, galling was observed and the punch points began to break off during the stripping cycle.

Second Analysis

Bulging of the punch points revealed that the elastic limit of A2 had been exceeded. The galling also indicated that the coefficient of friction had gradually increased until seizure had developed. It was recommended that the ejector punches be made of T15 because of the higher resistance to abrasion. Due to the lower toughness value of T15, the punch-to-die clearance should be increased to 0.012in/side (13 per cent t) in order to further reduce the impact effect.

Again the die assembly was reworked to incorporate the recommendations with the following reported results:

Each punch had produced over 4000 holes and was still in operation. Periodic inspections revealed no signs of bulging, galling, or cutting edge breakdown.

CASE REPORT NO. 1 CALCULATIONS

Punching load (Per punch)

$P = LTS$ where: P = Punching load (lb)
L = Length of cut (in)
T = Stock thickness (in)
S = Ultimate shear strength (lb/in^2)

Because of the high hardness values, let S = 80 per cent of the average tensile strength:

$$S = 0.8 \times 0.5\,(189\,000 + 198\,000)$$
$$= 0.8 \times 0.5 \times 387\,000 = 0.8 \times 193\,500$$
$$= 154\,800\text{lb/in}^2$$

PUNCH-TO-DIE CLEARANCES

CUTTING EDGES .002 TO .006 CORNER BREAKDOWN BURR AND BURNISHED LENGTH INCREASES AS CUTTING EDGES BREAK DOWN DURATION OF RUN LONGER RUNS AS CLEARANCE INCREASES	A BURR: TENSILE AND COMPRESSIVE MIN. ROLL-OVER	B BURR: TENSILE AND COMPRESSIVE MIN. ROLL-OVER	C BURR: TENSILE MODERATE ROLL-OVER	D BURR: TENSILE LARGE ROLL-OVER	E BURR: TENSILE AND EXTRUSION VERY LARGE ROLL-OVER
	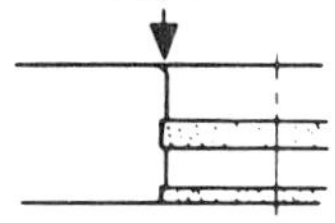	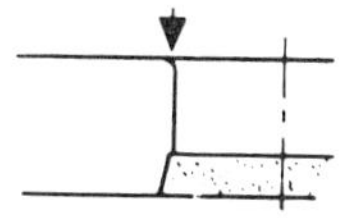	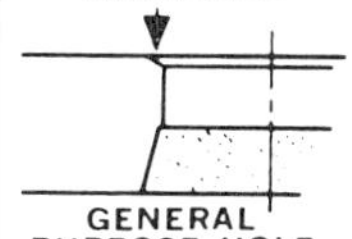 GENERAL PURPOSE HOLE	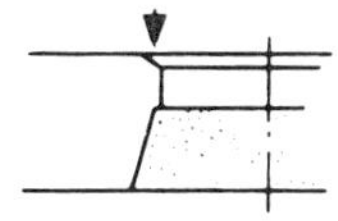	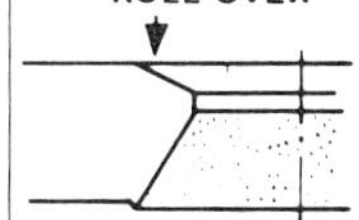
	DOUBLE SHEAR	SECONDARY SHEAR	EJECTOR PUNCHES REQUIRED		
MATERIAL		TOTAL CLEARANCE IN PERCENT OF MATERIAL THICKNESS			
ALUMINUM—HARD —SOFT	1— 2% 1— 2%	10—12% 4— 8%	18—20% 12—16%	25—28% 16—20%	40% Max. 34% Max.
BRASS—½ HARD —ANNEALED	1— 2% 1— 3%	6—10% 4— 6%	12—16% 12—16%	18—22% 16—20%	48% Max. 42% Max.
BRONZE, PHOSPHOR	3— 5%	7—10%	20—24%	25—27%	50% Max.
COPPER—½ HARD ANNEALED	2— 4% 1— 2%	6—10% 4— 8%	12—16% 10—14%	18—22% 16—18%	50% Max. 50% Max.
LEAD	3— 5%	8—12%	13—15%	16—20%	44% Max.
MAGNESIUM	1— 2%	3— 5%	7— 9%	10—14%	32% Max.
STAINLESS STEEL	2— 4%	6—10%	18—22%	25—27%	46% Max.
STEEL, CRS (1020) HIGH CARBON	2— 4% 5—10%	10—14% 22—26%	16—20% 28—32%	23—25% 34—38%	42% Max. 50% Max.

SHEAR STRENGTH CHART

MATERIAL DESCRIPTION	HARDNESS	ULTIMATE TENSILE P.S.I.	ULTIMATE SHEAR P.S.I.
STEELS			
Low Carbon, H. R. Sheet ASTM A-415	Rb 70	60,000	50,000
Low Carbon, C. R. Sheet Special Killed Drawing Quality	Rb 50	50,000	40,000
Low Carbon, C. R. Sheet			
(Soft)	Rb 45-60	53,000	42,000
(¼ Hard)	Rb 60-75	60,000	45,000
(½ Hard)	Rb 70-85	72,000	50,000
(Hard)	Rb 80-95	92,000	61,000
.40—.50% Carbon Steel H. R. Sheet	BHN 200	100,000	80,000
SAE 1074 C. R. Annealed Spring Steel	Rb 90	95,000	75,000
SAE 1095 C. R. Annealed Spring Steel	Rb 95	100,000	80,000
SAE 1074 or 1095 Spring Steel Hardened to Spring Temper	Rc 45-50	260,000	200,000
Abrasion—Resisting H. R. Sheet Steel	BHN 200/245	120,000	100,000
Cor-Ten Steel	BHN 140	70,000	55,000
Tri-Ten Steel	BHN 120	60,000	50,000
T-1 Steel (Types A & B) 100,000 P.S.I. Y.S.	BHN 260	130,000	105,000
STAINLESS STEELS			
202—Annealed	Rb 95	110,000	90,000
302, 303, 304—Annealed	Rb 85	95,000	75,000
310—Annealed	Rb 90	105,000	90,000
316, 321, 430—Annealed	Rb 90	95,000	75,000
410—Annealed	Rb 85	85,000	75,000
ALUMINUM BASE*			
ALLOY—TEMPER			
1100—0	BHN 23	13,000	9,000
—H14	BHN 32	18,000	11,000
2024—0	BHN 47	27,000	18,000
—T3	BHN 120	70,000	41,000
3003—0	BHN 28	16,000	11,000
—H14	BHN 40	22,000	14,000
—H16	BHN 47	26,000	15,000
3105—H25	BHN 47	26,000	16,000
5005—H34	BHN 41	23,000	14,000
5052—0	BHN 47	28,000	18,000
5052—H32	BHN 60	33,000	20,000
6061—0	BHN 30	18,000	12,000
—T6	BHN 95	45,000	30,000
7075—0	BHN 60	33,000	22,000
—T6	BHN 150	83,000	48,000
COPPER BASE			
**ALLOT—TEMPER			
110 —Electrolytic Tough—Pitch Copper			
—.050mm G.S.	Rf 40	32,000	22,000
—½ Hard	Rb 40	42,000	26,000
—Hard	Rb 50	50,000	28,000
220 Comm. Bronze, 90% —½ Hard	Rb 58	52,000	35,000
230 Red Brass, 85% —¼ Hard	Rb 55	50,000	35,000
260 Cartridge Brass, 70% —.035mm G.S.	Rf 68	49,000	34,000
—½ Hard	Rb 70	62,000	40,000
—Spring	Rb 91	94,000	48,000
280 Muntz Metal —⅛ Hard	Rb 55	60,000	42,000
342-A High Leaded Brass —½ Hard	Rb 70	61,000	40,000
675 Manganese Bronze, A —Soft Anneal	Rb 65	65,000	42,000
MISC. METALLIC MATERIALS			
Gold—14 Carat, Soft		58,000	42,000
Lead—		5,000	3,500
Silver—Sterling—Soft		37,000	27,000
—½ Hard		65,000	45,000
Tin—Block		5,000	4,000
Zinc—Drawing		20,000	16,000
Titanium—Unalloyed	Rc 23-29	60-110,000	50-60,000
Magnesium—AZ31B-F (Extruded)	BHN 46-49	36-38,000	19,000
Platinum (Annealed)	DPH 40	28-30,000	19-21,000

*500 Kg Load 10mm Ball

**Copper Development Association No.

P = 3.1416 × 0.215 × 0.090 × 154 800
= 9410lb/punch

log 3.1416 = 0.49715
log 0.215 = 9.33244–10
log 0.090 = 8.95424–10
log 154 800 = 5.18977

log P = 23.97360–20 (P = 9410lb)

P (tons) = 9410lb divided by 2000lb per ton
= 4.705 tons per punch
P (tons) for 12 punches = 4.705 × 12 = 56.46 tons
P (tons) for 34 punches = 4.705 × 34 = 159.97 tons
Note: Shear was not considered in the foregoing calculations

Area calculations

Punch face with Vee-groove

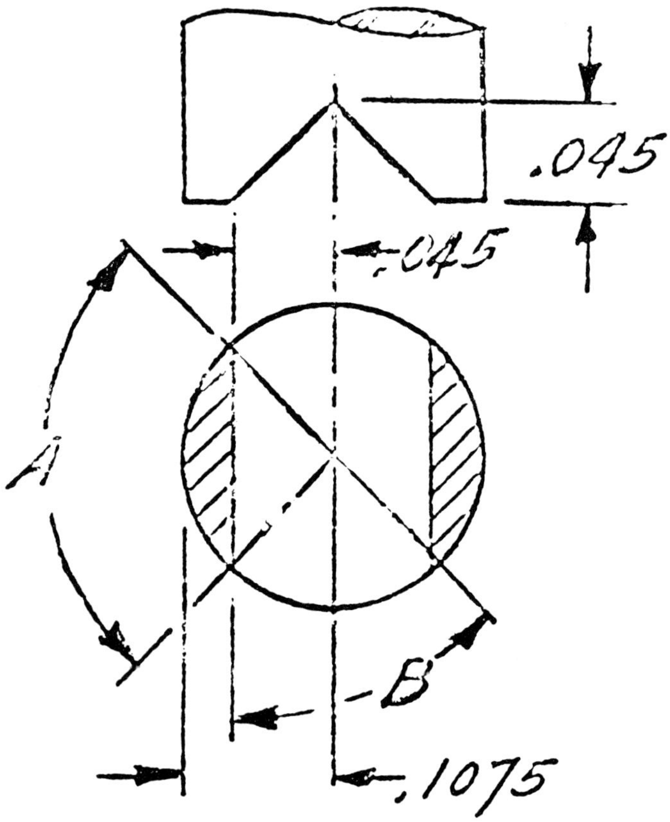

Segment:
L = 3.1416 × 0.215 × A divided by 360 deg
= 3.1416 × 0.215 × 130 deg divided by 360 deg
= 0.2439in
A = 180 deg – (2 × B) = 180 deg – 49.5 deg
= 130.5 deg or 130 deg
B = $\sin^{-1}$ 0.045 divided by 0.1075
= 24 deg 45 min

Segment:
log 0.045 = 18.65321–20
log 0.1075 = 9.03141–10

log sin B = 9.62180–10

B = 24 deg 44 min 47 sec or 24 deg 45 min
log 3.1416 = 0.49715
log 0.215 = 9.33244–10
log 130 = 2.11394

11.94353–10
log 360 = 2.55630

log L = 9.38723–10
L = 0.2439

Area of Segment
A = 0.7854 × 0.215 × 0.215 × 130 divided by 360
= 0.01311in^2
log 0.7854 = 9.89509–10
log 0.215 = 9.33244–10 × 2 = 18.66488–20
log 130 = 2.11394

30.67391–30
log 360 = 2.55630

log A = 28.11761–30
A = 0.01311

Area of Chordal Section = Area of Segment – Area of Triangle
= 0.01311 – 0.00439
= 0.00872in^2
Area of Triangle = 0.5 × 0.045 × Base
= 0.5 × 0.045 × 0.1953
= 0.00439in^2
Base = $2(0.1075^2 - 0.0452)^{\frac{1}{2}}$
= $2(0.011556 - 0.002025)^{\frac{1}{2}}$
= $2 \times 0.009531^{\frac{1}{2}}$
= 0.1953
log 0.1075 = 9.03141–10
log 0.1075 = 9.03141–10

log 0.1075^2 = 18.06282–20
0.1075^2 = 0.01155635
= 0.011556
0.045^2 = 0.002025

Sum = 0.009531
log 0.009531 = 7.97914–10
divide by 2 = 3.98957–5
log 2 = 0.30103

log Base = 4.29060–5
Base = 0.195255in
log 0.1953 = 9.29060–10
log 0.045 = 8.65321–10
log 0.5 = 9.69897–10

log Area = 27.64278–30
Area = 0.0043932in^2

Load Distribution over Chordal Section

Stress = L × T × S divided by the Area
= 0.2439 × 0.090 × 154 800 divided by 0.00872
= 389 691lb/in^2

log 0.2439 = 9.38723–10
log 0.090 = 8.95424–10
log 154,800 = 5.18977

log Load = 23.53124–20
Load = 3398.2lb

log Load = 13.53124–10
log 0.00872 = 7.94052–10

log Stress = 5.59072
Stress = 389 691lb/in^2

CASE REPORT NO. 2

Material

Cold Rolled Steel—Commercial Quality
Ultimate tensile strength: 53 000lb/in^2
Ultimate shear strength: 42 000lb/in^2
Thickness: 0.042in

Operation

Part contains more than 3300 holes with a minimum web thickness of 0.047 between adjacent holes. Over 800 holes

(0.290in dia) are produced at each press stroke in a 4-position shuttle die.

Punch Press

Only the rated capacity of 200 tons is known.

Lubrication

No information was provided.

Design of the die assembly

The punches were Light Duty Ball Lock type and were made of M2 steel. The stripper was cam-actuated and functioned much like the spring-loaded types. Punch points were self-supporting and punch-to-die clearance was 0.0034/side (4 per cent T). The clearance was specifically reduced from the conventional five per cent t value in order to minimize the risk of slug-pulling and the part damage that could be generated.

The punch lengths were staggered and the three lengths were distributed uniformly throughout the die assembly. The length increments were based on 0.5 T multiples.

Initial Performance: Each punch produced an average of 88 000 holes (22 000 parts) between sharpening grinds. The parts had a tendency to warp across the large flat surface that contained many of the holes, and the burr height was considered to be excessive. Severe press frame deflection was noted during the punching cycle which indicated a substantial overload—especially at the point of impact.

Analysis of the problem

The punching load for 825 punches was calculated to be 663 tons. Because of the three staggered lengths, the tonnage requirement is approximately 221 tons per length—apparently a 10 per cent overload! The unusually close clearance may be the cause of the excessive burr extension due to the plastic 'squirting' under the heavy compressive loading. It may also be generating a lateral thrust around the periphery of the hole that results in deformation of the top surface.

Recommendations

Test punch the material utilizing a variety of punch-to-die clearance values and evaluate the results. These tests were performed and the best results were achieved with clearance values of 12.5 per cent T and 17.5 per cent T/side. The smaller value was selected on a play safe basis.

The die assembly was reworked to incorporate ejector punches and the following results were reported:

The productivity average was increased to 228 000 holes/punch (57 000 parts) between each sharpening grind. The burr height was reduced to approximately 0.0005 over the average production run, and the surface warp was noticeably reduced.

It was estimated that the tonnage required was reduced by about one-third because press frame deflection could not be observed.

It is also interesting to note that the ejector punches were made of A2 steel because this material was provided in all ejector punches furnished by the punch manufacturer.

CASE REPORT NO. 3

Material

Type 5086 Aluminium—Full Hard
Tensile and Shear strength values not available
Thickness: 0.018 +0.0005, −0.0000in

Operation

Perforate, Rib, Blank, and Assemble. The resulting assembly is the tear tab and lid for an aluminium beverage can. Parts were required in large quantities and each must be as free of burrs as possible. Because the stamped parts are assembled in a multi-stage progressive die assembly, the individual parts cannot be deburred in a separate operation.

Punch press

Straight side press operating continuously at 320 spm.
No other details were made available.

Lubrication

Alcoa's M-73 lubricant was the best performer tested. All other normally used lubricants did not reduce the friction to the point where seizure would not be likely to occur. The M-73 lubricant demonstrated some unusual characteristics. It is applied as a liquid and does an excellent lubrication job. After about 24 hours, the liquid changes to a powder-like substance and evaporates completely—leaving no residue.

Design of the die assembly

The progressive die is assembled on a series of ball bearing die sets. Several stages are combined on a single die set; in all, there are two ball bearing die sets mounted on a single master-type ball bearing die set.

The punch-to-die clearance was experimentally determined. Starting at the generally accepted clearance value of four per cent T/side, the actual clearance was increased in increments of 0.001in until the optimum value was reached. The decision was based on the part requirement of a burnished land between 0.003in and 0.005in long. It was further conditioned by the need to grind into the taper relieved section of the die cavity. The particular punch-to-die clearance selected for this die was 11 per cent T to 15 per cent T/side.

The performance records for several die assemblies was reported as follows:

Part production averaged 20 000 000 pieces between each sharpening grind.

Conclusions

The phenomenal results achieved by several of these die assemblies cannot be attributed only to the larger clearances that were utilized. Several other influencing factors were isolated, tested experimentally, and ultimately controlled.

Lubrication, for example, posed a rather knotty problem. Until M-73 was tested and proved, many other more conventional lubricants were tried. Invariably the results were essentially the same: After approximately 150 000 strokes, the abrasion had degenerated to the point of galling and seizure. On one occasion, the blanking punch actually froze in the die cavity with rather serious consequences. The force of the seizure was so great that four socket head cap screws ($\frac{5}{16}$in dia) were completely stripped out of the punch holder!

Rigidity and strength were the basic considerations for press evaluation. The precise nature of the fitted moving parts and the frame rigidity are such that it is possible to stand a coin on edge, anywhere on the press or feed while it is operating at 400 strokes/min, and the coin will not roll, wobble, or fall!

Component finish was not overlooked despite the excellent results provided by the special lubricant. All of the punches, pilots, and die cavities were diamond-lapped to a 2-microinch finish so that the coefficient of friction would be reduced to its lowest value.

Alignment of the mating punches and die cavities was maintained within 0.0001 to assure uniform clearance at every point on their periphery.

The perforating punches were equipped with slug-ejectors, but the blanking punch was left solid because it pushed the blank into the correct assembly position and staked it in place. Any slug-ejector action at the assembly point would disturb the alignment of the blank with its mating part.

The punches and pilots were made of A2 steel and the blanking punch was made from D3 tool steel. The change in material was based on the relatively larger mass of the blanking punch and its position relative to the edges of the stock strip.

CASE REPORT NO. 4

Material:

Type 5086 Aluminium—Full Hard
Ultimate tensile and sheer strength values not available
Thickness: 0.018in +0.0005, −0.0000in

Operation

Perforate, Rib, Slit, Carry, and Assemble.

Punch press

Straight side press operating continuously at 280 spm
No other details available.

Lubrication

Alcoa's M-73 lubricant applied in liquid form.

Design of the Die Assembly

The progressive die is mounted similarly to that in Case Report No. 3; the only exception is the fact that the blanking punch does not enter the die cavity. Component materials, finish, size control, and alignment are the same as that of the die assembly discussed in report No. 3; even the clearance has been maintained.

The performance records, however, are surprisingly superior. This die assembly has been in operation 20 hr/day, 6 days/week for more than eight months. Productivity has been conservatively estimated at over 70 000 000 parts and the components have never been sharpened.

The difference in production volume is easily explained. In case No. 3, the blanking punch entered the die cavity approximately 0.50in to effect the final assembly. None of the punches in case No. 4 even enter the die because the blank is pushed back into the stock strip.

CASE REPORT NO. 5

Material

Cold Rolled Steel—No. 4 Temper
Ultimate tensile strength: 72 000lb/in^2
Ultimate shear strength: 50 000lb/in^2
Stock thickness: 0.036in
Used in coil form; stock strip was 4.835in wide.

Operation

Punch 29 holes (0.125 dia) on 0.1875 centres which left a nominal web thickness between adjacent holes of 0.0625in.

Punch press

The press was continuously operated at 120 spm; a hitch feed provided the necessary stock progression.

No other press information was made available.

Lubrication

The coil strip passed between felt rollers which applied a light oil to the top and bottom surfaces.

Design of the die assembly

The punches and die buttons were made of A2 steel; the punches were not supported at the points. Punch-to-die clearance was established at 0.001in/side (2.8 per cent T).

Initial Performance: The average productivity/punch before replacement became necessary was 1 346 172 parts or holes. The number of sharpening grinds/punch were recorded and the average was 20 sharpenings/punch. By calculation, the average productivity between sharpening grinds was 67 000 holes/punch.

Analysis of the problem

Despite the unusually high performance attained by using catalogue components and conventional punch-to-die clearances, the output was insufficient to supply subsequent operations. The options available for increasing productivity, short of building another die assembly, lay wholly within the province of abrasion resistance. Rather than change the tool steel, it was suggested that the die clearance be increased to 0.0045/side (12.5 per cent T). This rather drastic increase in clearance would be very likely to generate slug-pulling problems and ejector punches were recommended.

It was also suggested to use A2 ejector punches so that the effects of clearance could be isolated. The recommendations were accepted and the following results were reported:

Average productivity/punch was increased to 3 260 840 parts with an estimated life of 250 000 hits remaining. The average productivity between sharpening grinds was recorded at 203 000 hits for each of 16 sharpenings.

The documentation of the results were so thorough that even the time spent in sharpening was recorded. For the initial close clearance, the sharpening time per million strokes was 41 hours. After increasing the punch-to-die clearance, the sharpening time/million strokes was reduced to 12.5 hr. Not only was the maintenance time reduced by approximately two-thirds, but the die remained in production 28.5hr/m hits more than it had with the closer clearance. This extra production time enabled the stamper to produce about 200 000 more parts in the time he had previously lost to die maintenance.

Blanking of Low-Carbon Steel

By the ASM Committee on Press Forming of Steel

A BLANK is a shape cut from flat or preformed stock. Ordinarily, a blank serves as a starting workpiece for a formed part; less often, it is a desired end product. This article deals with the production of blanks from low-carbon steel (such as 1008 and 1010) sheet and strip, in dies in a mechanical or hydraulic press. The blanking and piercing of magnetically soft materials and of high-carbon steel are discussed in separate articles in this volume.

Improving the quality of blanked edges by shaving is discussed later in this article. Fine-edge blanking is treated in a separate article beginning on page 56 in this volume. Shearing, a method of making blanks that does not use a die, is dealt with in the articles "Shearing of Plate and Flat Sheet", which begins on page 265, and "Slitting and Shearing of Coiled Sheet and Strip", which begins on page 271.

Methods of Blanking in Presses

Cutting operations that are done by dies in presses to produce blanks include cutoff, parting, blanking, notching and lancing. The first three of these operations can produce a complete blank in a single press stroke. In progressive dies, two or more of these five operations are done in sequence to develop the complete outline of the blank and to separate it from the sheet, strip or coil stock.

Trimming, or cutting off excess material from the periphery of a workpiece, usually is done in dies, and is similar to blanking. Often it is the final operation on a formed or drawn part.

Applications of these methods are described in examples throughout this article. For other examples of the use of these methods of producing blanks, see the articles "Piercing of Low-Carbon Steel" (page 44), "Blanking and Piercing of Magnetically Soft Materials" (page 60), "Press Forming of Low-Carbon Steel" (page 112), "Press Forming of High-Carbon Steel" (page 132), and "Deep Drawing" (page 162).

Cutoff is cutting along a line to produce blanks, without generating any scrap in the cutting operation, most of the part outline having been developed by notching or lancing in preceding stations. The cutoff line may have almost any shape — straight, broken or curved. After being cut off, the blanks fall onto a conveyor or into a chute or container.

A cutoff die may be used to cut the entire outline of blanks whose shape permits nesting in a layout that uses all of the material (except possibly at the ends of the strip), as shown in Fig. 1. Alternating positions can sometimes be used in nesting, as shown in the middle strip in Fig. 1, to avoid the production of scrap except at strip ends.

Cutoff also is used to cut blanks from strip that has already been notched to separate the blanks along part of their periphery, as described in the example that follows.

Example 6. Use of Cutoff to Separate Blanks Partly Outlined by Notching (Fig. 2)

Figure 2 shows the layout for cutoff of blanks for body-bolt brackets of an automobile. The brackets were completed by piercing and forming (see Examples 43 and 152).

The coiled strip, of galvanized hot rolled 1006 steel, was 0.075 in. thick by 19.5 in. wide. This was wide enough for two blanks, as shown in Fig. 2. The blanks were alternated in the layout for greater ease in trimming and piercing.

The strip was fed into a progressive die, in which the first stations notched and semi-notched it (making a small amount of scrap). In the stations that followed, straight cutoff punches made four blanks at each stroke, without producing any additional scrap.

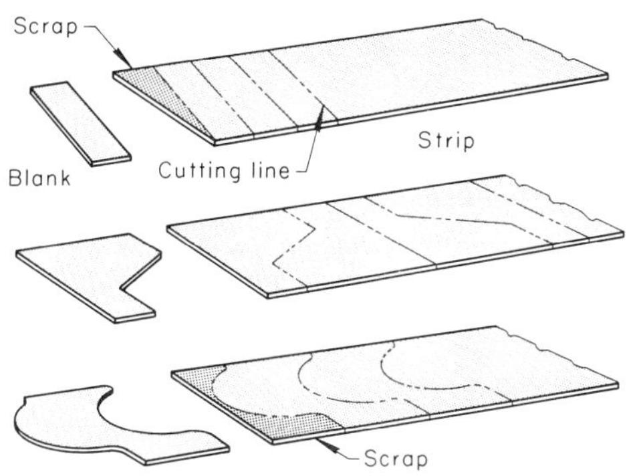

Fig. 1. Nested layouts for making blanks by cutoff

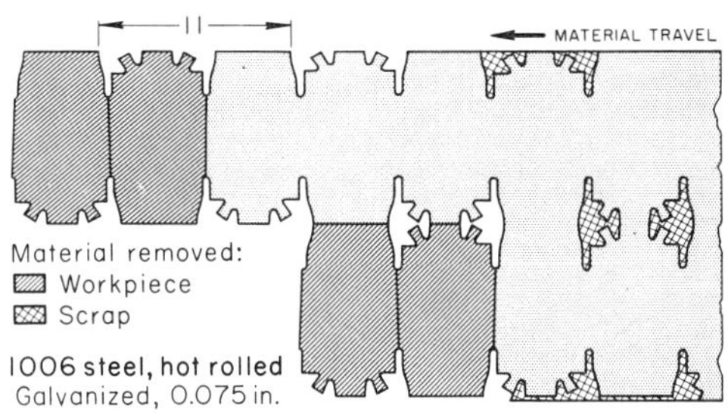

Fig. 2. Layout for cutoff of four blanks at each press stroke from notched and semi-notched strip (Example 6)

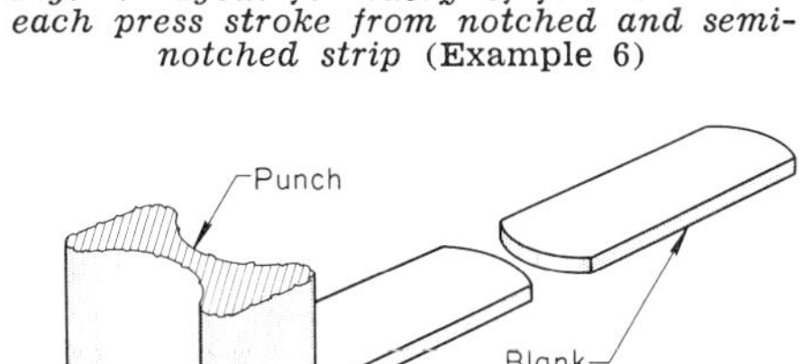

Fig. 3. Use of a parting punch to make blanks not having mating adjacent surfaces

The work was done in a 400-ton single-action mechanical press with an air cushion. The press was equipped with a double roll feed and made 50 strokes (200 blanks) per minute.

Advantages of cutoff in making blanks include:

1 The die has few components and is relatively inexpensive.
2 Waste of material in blanking is minimized or eliminated.
3 The die can be resharpened easily, and maintenance costs are low.

Disadvantages of cutoff include:

1 It can be used only to make blanks that nest in the layout without waste.
2 Cutting on one edge causes one-way deflection and stress.
3 Accuracy may be affected adversely by the method of feeding.

Parting (Fig. 3) is the separation of blanks by cutting away a strip of material between them. Like cutoff, it can be done after most of the part outline has been developed by notching or lancing. It is used to make blanks that do not have mating adjacent surfaces for cutoff (Fig. 3); or blanks that must be spaced for ease of handling, to avoid distortion, or to allow room for sturdy tools. Some scrap is produced in making blanks by parting, which is therefore less efficient than cutoff in the use of material.

Blanking (also called punching) is the cutting of the complete outline of a workpiece in a single press stroke. Usually a scrap skeleton is produced, so that blanking causes some waste of material. However, blanking is usually the fastest and most economical way to make flat parts, particularly in large quantities.

The skeleton left by blanking sometimes has only scrap-metal value, but many shops have organized programs to maximize the use of cutouts and sizable scrap skeletons in making other production parts (see Example 12). Waste of material is avoided completely in the use of the scrap skeleton that remains from certain blanking operations to provide perforated stock for items such as air filters for forced-air furnaces.

Piercing (with a flat-end punch), often also called punching or perforating, is similar to blanking, except that the punched-out (blanked) slug is the waste, and the surrounding metal the workpiece. Piercing is discussed in the article that begins on page 44 in this volume.

Notching is an operation in which the individual punch removes a piece of metal from the edge of the blank or strip (Fig. 4). Reasons for notching include the following:

1 To free some metal for drawing (Fig. 4a) and for forming (Fig. 4b) while the workpiece remains attached to the strip
2 To remove excess metal before forming (Fig. 4c)
3 To cut part of the outline of a blank that would be difficult to cut otherwise (Fig. 2 and 34).

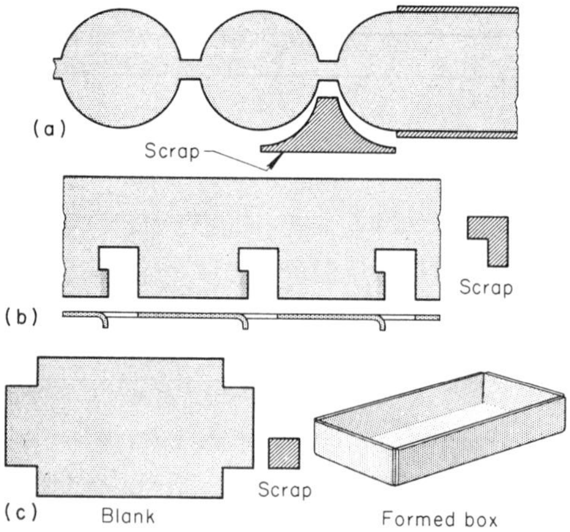

Fig. 4. Notched work illustrating use of notching for freeing of metal before drawing (a) and before forming (b), and for removing excess metal before forming (c)

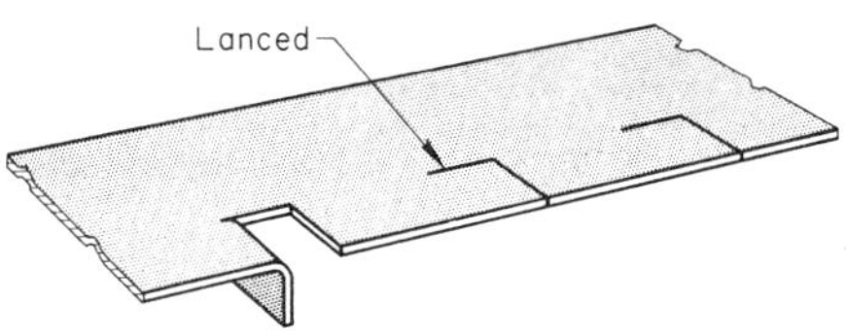

Fig. 5. Strip lanced to free metal for forming

Piercing of holes of any shape in a strip to free metal for subsequent forming, or to produce surfaces that later coincide with the outline of a blanked part, is sometimes called seminotching. The pierced area may outline a portion of one part, or of two or more adjacent parts in a strip. Progressive die layouts incorporating seminotching are illustrated in Fig. 2, 18 and 31.

Lancing is a press operation in which a single-line cut or slit is made part way across the strip stock, without removing any metal. Generally, lancing is done to free metal for forming, as shown in Fig. 5, or as in forming louvers. The cut does not have a closed contour and does not release a blank or a piece of scrap. Besides its use in freeing metal for subsequent forming, lancing also is used to cut partial contours for blanked parts, particularly in progressive dies.

Trimming is an operation for removing excess metal (such as deformed and uneven metal on drawn or formed parts) and metal that has been needed in a previous operation (such as a blankholding flange for a draw operation). Trimming is done in several ways, depending on the shape of the workpiece, on the accuracy required, and on production quantity.

Figure 6 illustrates the tooling for trimming a horizontal flange on a drawn shell in a separate operation. The drawn shell is set on a locating plug for trimming. After scrap from a sufficient number of trimmed shells has accumulated, the piece of scrap at the bottom is severed at each stroke of the press by the scrap cutters shown in Fig. 6, and falls clear. Except that the die must be constructed to accept and locate the drawn shell, the operation is identical with the blanking of a flat workpiece and produces square edges of the same accuracy and quality.

A drawn shell or formed part may be trimmed in a press without leaving a flange on the completed part, by one of three methods: pinch trim, shimmy trim, or trim and wipe-down.

Pinch trimming, shown as a separate operation in a push-through die in Fig. 7, is done only on a part that has at least a narrow flange as-formed. The shell must be free from wrinkles at or near the trimming line. The trimmed edge is not square with the sidewall, but has the general shape shown in the lower right corner of Fig. 7. Accuracy of height resulting from pinch trimming is affected by variation in wall thickness and flange radius. To be sure of an even pinch-off and to avoid sharp or rough edges, clearance between punch and die must be held to a minimum, and the punch must be kept sharp.

Pinch trimming is also done without a blankholder or hold-down, using a die otherwise similar to that shown in Fig. 7. The scrap rings may be blown off the die at each stroke. In another method, the scrap rings climb the punch until they are severed by being compressed against a scrap cutter, after which they are spread apart and allowed to run out along a track for disposal. Pinch trimming without a blankholder is particularly well suited for use in high-volume production of eyelets and other small parts.

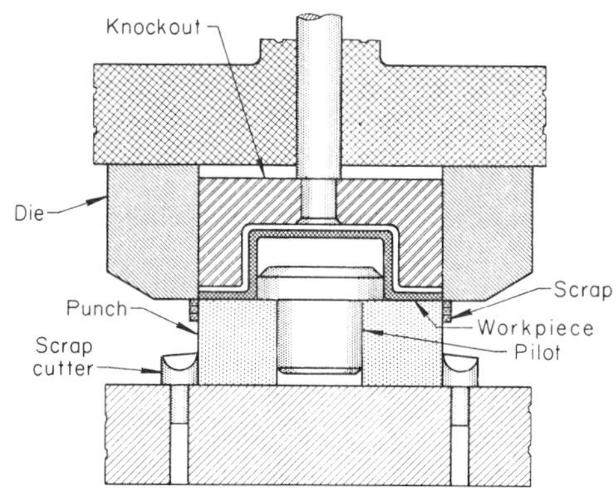

Fig. 6. Single-operation die for trimming a horizontal flange on a drawn shell

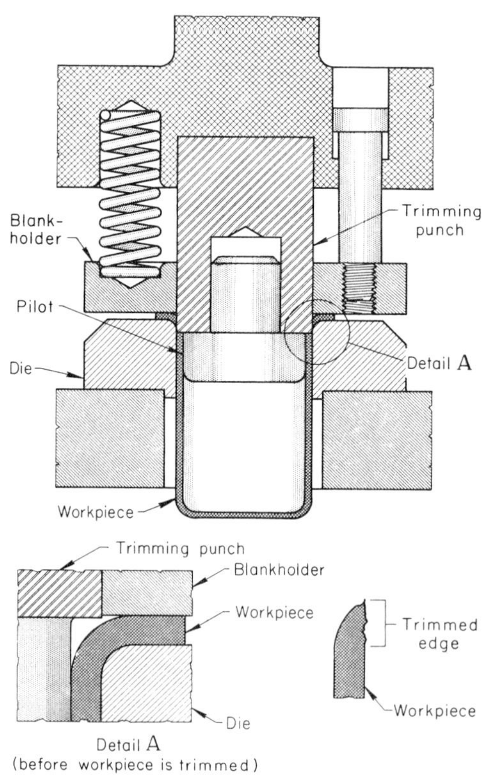

Fig. 7. Pinch trimming a drawn shell in a push-through die

Pinch trimming is primarily a mass production method. The production rate is high, because only one stroke of the press is required to complete the trim. The method often is combined with drawing in a compound draw-and-trim die, to reduce production cost still further.

Disadvantages are excessive burrs, sharp cut edge, and high die maintenance.

Shimmy Trimming. In trimming with a shimmy die (known also as Brehm or model trimming), the drawn shell is held in a close-fitting die of the exact shell height, and is trimmed in segments by successive horizontal oscillations of an internal cam-actuated punch toward the outside of the shell. The resulting trimmed edge is square and closely resembles the conventional blanked edge on a flat part. Shell height is more accurate than with pinch trimming.

Besides its application to shells that must have square, accurate edges, shimmy trimming is used on shells that have a wrinkled or otherwise nonuniform top edge as-drawn (cut-off is done below the defects), and on shells that cannot be produced economically with even the narrow flange needed for pinch trimming.

Tooling cost for shimmy trimming is much higher than for pinch trimming. Also, shimmy trimming is slower, because it requires four or more oscillations of the punch in one press stroke and cannot be combined with other operations in a compound die.

Shimmy dies are inexpensive to maintain, because they remain in alignment and thus are not likely to wear by shearing or chipping.

Trim and Wipe-Down. In this type of trimming, a flange is cut to width with a die such as that shown in Fig. 6, and then wiped or straightened into line with the sidewall of the shell or formed part. Because of narrow flange width, trimming and wiping down may be two operations.

The edge is square with the sidewall, but the shell height may be slightly irregular because of the forming characteristics of the metal. Also, a ring may be visible at the original location of the flange radius.

Trimming, other than shimmy trimming, is frequently combined with one or more other operations in a compound die. Trim stock often is left on a drawn or formed workpiece so that it can be trimmed to size in a second operation. This is done to get the most accurate relation of some other feature, such as a pierced hole, to the trimmed outline of the workpiece.

Characteristics of Blanked Edges

The sheared edges of a blank produced in a conventional die are not smooth and vertical for the entire thickness of the part, but instead show the characteristics represented on an exaggerated scale in Fig. 8. The blank is shown in the position in which it would be cut from the work metal by the downward motion of the punch. A portion of the stock remaining after removal of the blank is shown at the top of the illustration.

Rollover on the lower edges of the blank develops by plastic deformation of the work metal as it is forced into the die by the punch. Compression of the metal above the rollover zone against the walls of the die opening burnishes a portion of the edge of the blank, as shown in Fig. 8. As the punch completes its stroke, the remaining portion of the blank edge is broken away or fractured (resulting in "die break"), and a tensile burr is formed along the top of the blank edge.

The angle of the fractured portion of the edge is identified in Fig. 8 as the breakout angle. The breakout dimension of the blank and the burnish dimension of the hole in the scrap skeleton are approximately equal to the corresponding punch dimension, and the burnish dimension of the blank is very close to the corresponding die dimension. Thus, the punch determines the hole size and the die governs the blank size.

Penetration depth, or the amount of penetration of the punch into the work metal before fracture occurs, is shown on the edge of the remaining stock or scrap skeleton in Fig. 8. This depth is approximately equal to the sum of rollover depth and burnish depth on the blank, except when low die clearance produces secondary burnish. It is usually expressed as a percentage of the work-metal thickness.

The percentage of penetration (before fracture) depends on the properties of the work metal, as shown in Table 1, which gives approximate values for various steels and nonferrous metals under typical blanking conditions. Percentage penetration affects energy consumption and cutting force in blanking, as described under "Calculation of Force Requirements", on this page.

Die Clearance

The terms clearance, die clearance, and punch-to-die clearance are used synonymously to refer to the space between punch and die. Clearance is important for reliable operation of the blanking equipment, quality and type of cut edges, and life of punch and die. In general, the effects of clearance on these factors in blanking are the same as in piercing, and are discussed on page 44 in the article "Piercing of Low-Carbon Steel".

The edge characteristics of slugs produced in piercing holes are described on pages 44 and 45 in the article on Piercing, and are illustrated in Fig. 2 of that article. The data in that illustration can serve as a guide for selecting clearances for blanking.

All clearance values given in this article are per side, except where indicated otherwise.

Optimum blanking clearance may sometimes be less than optimum piercing clearance. This is partly because the blanked edge generally is close to the stock edge, and material expansion therefore is less restricted. A piercing tool must move a great deal of material away from its cutting edge and, for longest life, the clearance should be selected to eliminate as much compressive loading on the work metal as possible.

A part blanked using clearance much greater than normal may show double shear, which ordinarily is evident only with extremely small clearance (see edge types 4 and 5 in Fig. 2 on page 45). Also, a part blanked using large clearance will be smaller than the die opening (except for a deeply dished blank), and it is difficult to correct the tooling to compensate for this. In some applications, retaining the blank becomes almost as great a problem as expelling the slugs into a die cavity after piercing, because of the increased clearance.

Relief in a blanking die (Fig. 9) is the taper that is provided so that the severed blank can fall free. The relief angle may range from ½° to 2° from

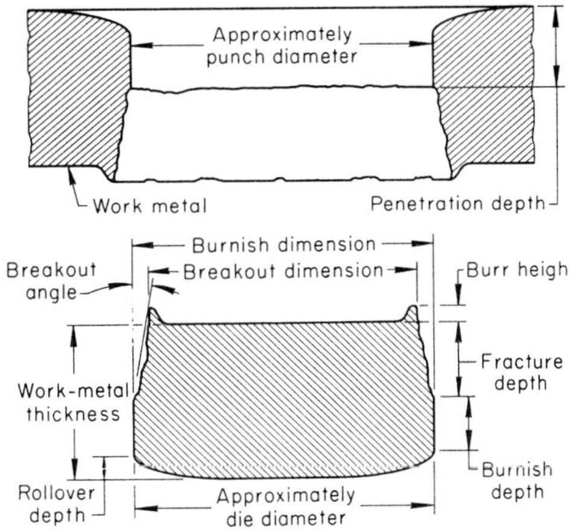

Fig. 8. Characteristics of the sheared edges of a blank. Curvature and angles are exaggerated for emphasis.

Table 1. Approximate Penetration of Sheet Thickness Before Fracture in Blanking

Work metal	Penetration, %	Work metal	Penetration, %
Carbon steels(a):		**Nonferrous Metals**	
0.10% C, Ann	50	Aluminum alloys	60
0.10% C, CR	38	Brass	50
0.20% C, Ann	40	Bronze	25
0.20% C, CR	28	Copper	55
0.30% C, Ann	33	Nickel alloys	55
0.30% C, CR	22	Zinc alloys	50
Silicon steels	30		

(a) Ann = annealed; CR = cold rolled

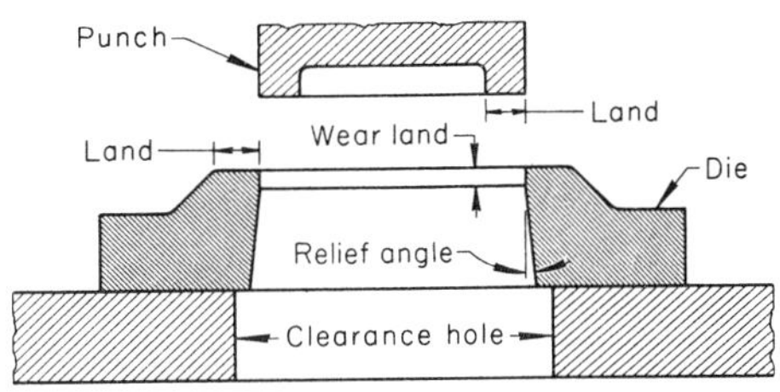

Fig. 9. Relief in a blanking die

the vertical wall of the die opening. Sometimes, relief in a die is called draft or angular clearance. In some dies, the relief may start at the top of the die surface and have a taper of only 0.002 in. per inch per side (see page 64 in the article "Blanking and Piercing of Magnetically Soft Materials", in this volume). In other dies, there is a straight, vertical land between the top of the die and the relief.

Calculation of Force Requirements

Calculation of the forces and the work involved in blanking gives average figures that are applicable only when: (*a*) the correct shear strength for the material is used; and (*b*) the die is sharp and the punch is in good condition, has correct clearance, and is functioning properly.

The total load on the press, or the press capacity required to do a particular job, is the sum of the cutting force and other forces acting at the same time, such as the blankholding force exerted by a die cushion.

Cutting Force: Square-End Punches and Dies. When punch and die surfaces are flat and at right angles to the motion of the punch, cutting force can be found by multiplying the area of the cut section by the shear strength of the work material:

$$L = S_s t l$$

where L is load on the press, lb (cutting force); S_s is shear strength of the stock, psi; t is stock thickness, in.; and l is length or perimeter of cut, in. Shear strengths of various steels and nonferrous metals are given in Table 2.

Cutting Force: Dies With Shear. For cutting large blanks, shear may be applied to the face of the die by grinding it at an angle to the motion of the punch (Fig. 10), but shear is not used in cutting small blanks. Shear reduces shock on the press, blanking noise, and blanking force, but the same amount of work is done as with a flat die surface.

The most common type of shear used on the die is convex (Fig. 10a). The apex of the die face is rounded slightly to avoid initiating a crack in the work metal. Concave shear (Fig. 10b) is somewhat more difficult to grind on the die, but holds the work metal more securely during blanking. A radius approximately equal to that of the grinding wheel is produced where the ground surfaces meet.

A third type of shear, sometimes used on a die for a large blank, consists of a wavy or scalloped surface around the die opening. This technique uses several convex and concave shear surfaces around the die opening. The punch load is distributed over the entire die surface, thus minimizing punch shift.

The amount of shear in a die can be less than or greater than stock thickness. Shear that is equal in depth to the stock thickness (or is greater than the stock thickness) is called full shear.

Table 2. Shear Strengths of Various Steels and Nonferrous Metals at Room Temperature

Metal	Shear strength, psi
Carbon steels:	
0.10% C	35,000 to 43,000
0.20% C	44,000 to 55,000
0.30% C	52,000 to 67,000
High-strength low-alloy steels	45,000 to 63,700
Silicon steels	60,000 to 70,000
Stainless steels	57,000 to 129,000
Nonferrous Metals	
Aluminum alloys	7,000 to 46,000
Copper and bronze	22,000 to 70,000
Lead alloys	1,825 to 5,870
Magnesium alloys	17,000 to 29,000
Nickel alloys	35,000 to 116,000
Tin alloys	2,900 to 11,100
Titanium alloys	60,000 to 70,000
Zinc alloys	14,000 to 38,000

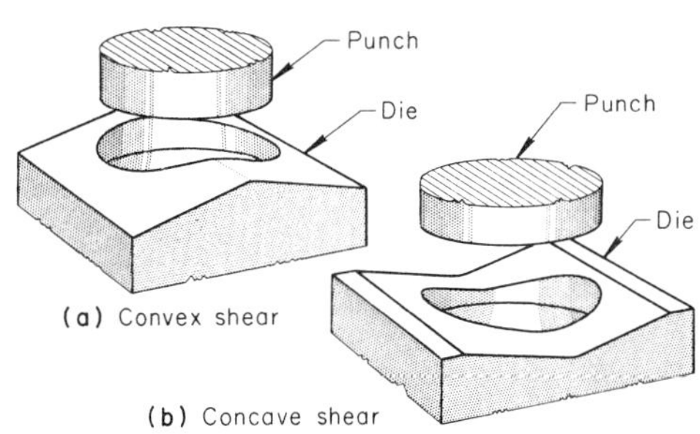

Angle and depth of shear are exaggerated for emphasis. Normally, depth of shear does not greatly exceed stock thickness.

Fig. 10. Convex shear and concave shear on blanking dies

Table 3. Scrap Allowance for Blanking

Work metal	Scrap allowance when length of skeleton segment between blanks or along edge is: 2t or less — Thickness of stock (t), in.	2t or less — Edge of stock to blank, in.	2t or less — Between blanks in row, in.	Greater than 2t — Thickness of stock (t), in.	Greater than 2t — Edge of stock to blank, in.	Greater than 2t — Between blanks in row, in.
Metals in general:						
Standard strip stock	Up to 0.021 ..	0.050	0.050	Up to 0.044 ..	0.050	0.050
	0.022-0.055 ..	0.040	0.040	Over 0.044 ..	0.9t	0.9t
	Over 0.055 ..	0.7t	0.7t			
Extra-wide stock and weak scrap skeleton	Up to 0.042 ..	0.060	0.050	Up to 0.033 ..	0.060	0.050
	Over 0.042 ..	1.4t	1.2t	Over 0.033 ..	1.8t	1.6t
Stock run thru twice	Up to 0.042 ..	0.060	0.050(a)	Up to 0.033 ..	0.060	0.050(a)
	0.043-0.055 ..	1.4t	0.040	0.034-0.044 ..	1.8t	0.040
	Over 0.055 ..	1.4t	0.7t	Over 0.044 ..	1.8t	0.9t
Stock run thru twice; blanks in rows 1 & 2 interlock	Up to 0.042 ..	0.060	0.050(b)	Up to 0.033 ..	0.060	0.050(b)
	Over 0.042 ..	1.4t	1.4t	Over 0.033 ..	1.8t	1.8t
Stainless, silicon and spring steels	Up to 0.042 ..	0.060 min	0.060 min	Up to 0.033 ..	0.060 min	0.060 min
	Over 0.042 ..	1.4t	1.4t	Over 0.033 ..	1.8t	1.8t
Ni-base magnetically soft alloys	All	0.060	0.060	All	1t(c)	1t(c)

(a) Allowance between blanks in the same row and also between blanks of the first and second rows. (b) Allow 0.060 in. between blanks at first and second rows. (c) When the blank edge is parallel to the edge of the stock, or when the length of the skeleton segment between blanks is more than 4t, scrap allowance is 1.8t.

Cutting force for a die with shear can be calculated after first finding the work done (energy used) in blanking. Work done in blanking equals the force required in blanking (load on the press) multiplied by the distance that the force acted:

$$W = Ls$$

where W is work done in blanking, in.-lb; L is load, lb; and s is distance the load acts (thickness multiplied by percentage of penetration before fracture), in. To obtain accurate work values, percentage penetration must be known accurately.

Cutting or blanking force is reduced by the use of angular shear in the die, the amount of reduction in force depending on the depth of the angular shear. The reduced average cutting force on the press is:

$$L_{sh} = \frac{W}{s + s_1}$$

where L_{sh} is average cutting force, lb, with angular shear; W is work done in blanking, in.-lb; s is the distance, in., that the load acts (thickness multiplied by percentage penetration before fracture); and s_1 is the depth of angular shear, in.

In simplified practice, some plants ignore partial shear in calculating cutting force for blanking. When full shear is used, force is calculated as without shear and then reduced by 30%.

Stripping force is the force that is needed (when drop-through is not used) to free the blank from the die or the strip from the punch when they stick or jam because of springback. Stripping force can be calculated from the following formula:

$$L_{st} = kA$$

where L_{st} is stripping force, lb; k is a stripping constant, psi; and A is area of the cut surface, sq in. (stock thickness, t, in., multiplied by length or perimeter of cut, l, in.). Approximate values for the constant, k (as determined by experiment for low-carbon steel), are: 1500 for sheet metal thinner than 0.062 in., when the cut is near an edge or near a preceding cut; 2100 for other cuts in sheet thinner than 0.062 in.; and 3000 for sheet more than 0.062 in. thick.

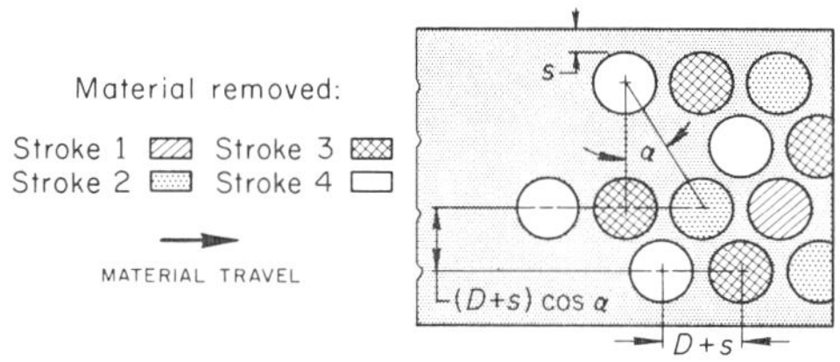

Fig. 11. Strip layout for blanking four circles per stroke with minimum waste of material

Factors That Affect Processing

Factors that affect the processing of blanks include the following:

1 Size and shape of the blank
2 Material for blanking
3 Form in which the material is supplied
4 Thickness of the blank
5 Production quantity and schedule
6 Quality specifications
7 Availability of equipment and tools
8 Number and type of subsequent operations required for completing the work.

Size and shape of the blank affect the form and handling of the material blanked, the blanking method, and the handling of the completed blank.

Thickness of the blank affects the press load required (see the preceding section, "Calculation of Force Requirements"), and the selection of equipment and the choice of blanking and handling methods (see the section "Effect of Work-Metal Thickness", on page 42 in this article).

Production quantity and schedule determine the choice of equipment. Usually, a total production of less than 10,000 pieces is considered a short run, 10,000 to 100,000 pieces a medium run, and more than 100,000 pieces is considered a long run.

Quality specifications and tolerances for thickness, camber, width, length, flatness and finish affect the handling of the material.

The availability of single, double or triple-action presses (rated at various tonnages, sizes, speeds, lengths of stroke, strokes per minute, and shut heights) affects the selection of the processing method. The availability and capacity of auxiliary press equipment can have an effect on the selection of a tooling system, and on whether a part can be made in-plant.

Operations that follow blanking also affect the choice of equipment, the processing method and the handling procedures. Such subsequent operations may include piercing, bending, forming, deep drawing, machining, grinding or finishing. Only rarely is the blank a final product.

Selection of Work-Metal Form

Work metal for blanking in presses ordinarily is in the form of flat sheets, strip, or coil stock. Less frequently, steel plate is blanked in presses (see the section "Effect of Work-Metal Thickness", in this article). In some applications, the metal is preformed before blanking.

Special preparation of the work metal ordinarily is not required for the blanking operation itself. However, annealing, leveling or cleaning often is needed because of subsequent forming operations on the blank, as discussed in the article "Press Forming of Low-Carbon Steel", in this volume.

Sheet or Strip. Flat sheet usually is the work metal for large blanks, such as automobile roofs. Square-sheared sheet can be used as a blank, or it can be blanked in a die. Small quantities of blanks, regardless of size, usually are made from straight lengths of sheet or strip.

Coil stock is used for mass production, whenever possible. In continuous production, the use of coil stock can save as much as one-third of the time needed for producing an equal quantity from flat stock. Besides, there are fewer scrap ends when coil stock is used.

Sheet metal is least expensive when it is supplied in large coils from the mill. For most applications, the coil must be slit to the proper width for blanking, and some edge material must be trimmed off. Parts sometimes can be made most economically in a progressive die by using coil stock that is the width of the developed blank, as in Example 28 in the article on Piercing.

Blank Layout

In medium and high production of medium-size blanks, the cost of material is 50 to 75% of the total cost of the blank; for large blanks, it may be more than 95% of the total cost of the blank. Substantial savings in net material cost often can be achieved by coordinating blank layout with the selection of stock form and width to minimize the amount of scrap produced.

Several trial layouts may be needed to find the width of stock and the layout that use the material most efficiently, while taking into account possible effects of orientation of parts on subsequent operations. The layout must include the minimum workable scrap allowance between blanks, providing just enough material to support or hold down the strip during blanking. Scrap allowances, based on the use of well-maintained equipment and good shop practice, are shown in Table 3.

The percentage of scrap in a strip layout can be calculated as:

$$100\ (1 - A_B/A_S)$$

where A_B is area of blanks produced in one press stroke, and A_S is area of strip consumed by one press stroke, or strip width times feed length.

Round blanks can be staggered in rows, at the same spacing as for hexagons (Fig. 11), for the most efficient blanking from a long strip. With such a layout, 20 to 40% more blanks can be made from a given amount of material than by blanking each circle from a separate square. With the layout shown in Fig. 11, each press stroke (after the third stroke) produces four blanks—spaced so as to provide enough room for mounting the punches and dies.

The percentage of scrap loss for the layout of Fig. 11 can be calculated using the general formula given above and the following:

$$A_B = n\,\pi D^2/4$$
$$A_S = wl$$
$$l = D + s$$
$$w = (D + 2s) + (n - 1)\,(D + s)\cos\alpha$$

where n is number of rows of blanks across strip width; D is blank diameter, in.; l is feed length (one blank made in each row per press stroke), in.; w is strip width, in.; s is scrap allowance from edge of strip to blank and between blanks, in.; and α is angular displacement between blanks, degrees. The area of the holes pierced in a blank is not considered in these calculations, because it does not affect the efficiency of the layout.

Rectangular blanks generally can be laid out more easily than other shapes. The following example compares the cost of producing notched rectangular blanks from two different layouts, by three methods.

Example 7. Reducing Cost by Improving Layout and Blanking Method (Table 4)

Table 4 shows the costs for producing a notched rectangular blank in quantities of 10,000, 100,000 and 500,000 by three different methods and from two different layouts. The blanks were made from cold rolled, commercial quality 1010 steel, 0.036 in. thick (20 gage). The blank and layouts are illustrated in Table 4.

In method A, the blanks were made from sheets purchased to 36½-in. width by 78½-in. length. The sheet was sheared to 36 by 78 in., and then notched and pierced in a punch press. As shown in Table 4, nine blanks were produced from each sheet.

For method B, the blank was turned 90° in relation to the layout for method A, and was produced from a sheet sheared to 60¼-in. length from 26¼-in.-wide coil stock. Five blanks were made from each sheet by notching, trimming ends, piercing and cutoff, in a press. Method A cost less than method B for 10,000 blanks because of lower tool cost; for larger quantities, method B cost less (see Table 4).

The layout for method C was like that of method B, but the blanks were produced continuously from coil stock roll-fed to a blanking-press line, instead of from sheared lengths. The lower cost of labor was the chief reason that method C cost slightly less than method B.

Odd-shape blanks generally are more difficult to lay out for the greatest economy. The following example shows three different layouts, the most economical of which would require a change in the shape of the blank to save material.

Table 4. Comparative Costs for Producing the Same Blanks by Three Different Methods (Example 7)

Cost item	Cost for producing blanks in quantities of: 10,000	100,000	500,000
Method A (Nine Blanks From Sheared Sheet)			
Material	$2,512	$25,120	$125,600
Labor(a)	320	3,205	16,025
Setup(b)	100	100	100
Tool(c)	1,800	1,800	1,800
Total cost	$4,732	$30,225	$143,525
Cost per blank	$0.4732	$0.30225	$0.2870
Method B (Five Blanks From Narrower Sheet)			
Material	$2,437	$24,370	$121,850
Labor(a)	109	1,090	5,450
Setup(d)	75	75	75
Tool(c)	2,500	2,500	2,500
Total cost	$5,121	$28,035	$129,875
Cost per blank	$0.5121	$0.28035	$0.2597
Method C (Blanks From Coil Stock)			
Material	$2,427	$24,270	$121,350
Labor(a)	55	550	2,750
Setup(b)	100	100	100
Tool(c)	2,500	2,500	2,500
Total cost	$5,082	$27,420	$126,700
Cost per blank	$0.5082	$0.2742	$0.2534

1010 steel, cold rolled, commercial quality, 0.036 in.

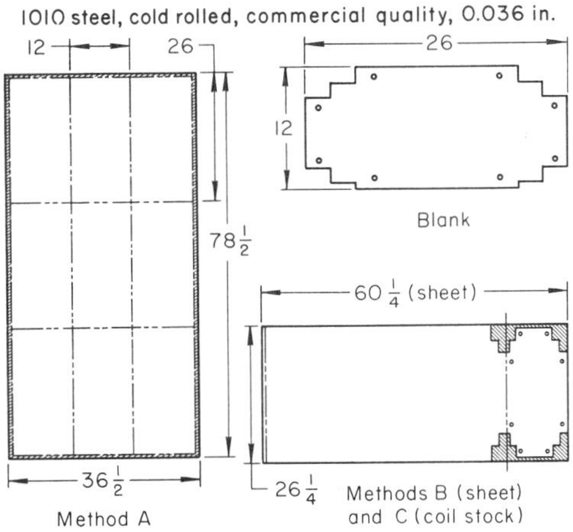

(a) At $5 per hour. (b) Ten setups of 2 hr each, at $5 per hour. (c) Amortized over 1 year. (d) Ten setups of 1.5 hr each, at $5 per hour.

Example 8. Effect of Blank Layout on Cost (Fig. 12)

In determining the most economical procedure for manufacturing a simple odd-shape blank, three different blank layouts were considered (Fig. 12). The strip used was hot rolled, commercial quality 1010 steel, 0.12 in. thick.

Layout A in Fig. 12 was simple but wasteful; a blank could be produced with one stroke of the press.

Layout B would use less material, but the die would have to blank two pieces per stroke, which would increase tooling cost, or the material would have to be passed through the press a second time, which would add to labor cost. For flat lengths of strip, this would add 10 to 15% to labor cost; coil stock would have to be re-coiled for a second pass, causing still higher labor cost.

Layout C would have been the simplest and least expensive. It would make two blanks with one press stroke. Because layout C called for a change in the design of the blank, it could be used only if the shape were not functional.

The table accompanying Fig. 12 summarizes costs for materials for the three layouts.

Nesting (interlocking of blanks in the layout to save material) should be done wherever the shape of the blank permits; this is possible with many irregular blanks.

Figure 13 shows a layout in which irregular blanks are nested so that an appreciable amount of material is saved. With such a layout, a double die may be used, blanking two pieces per stroke. A single die can be used for short runs; the strip is turned around after the first pass and fed through the die again.

Another way of nesting blanks is a layout such as the one illustrated in Fig. 14. With this layout three punches cut four pieces per stroke (after the third press stroke), in a shearing action that produces no scrap except at the ends of the strip.

Other strip layouts in which the blanks have been nested are shown in the articles "Press Bending of Low-Carbon Steel", which begins on page 89, and "Press Forming of Low-Carbon Steel", which begins on page 112.

Use of Full Stock Width. Blanks that have two parallel sides sometimes can be made most economically in a layout that uses the full width of the stock. The remaining outline of the blank is produced by shearing, lancing, notching or parting (or a combination of these operations). (See layout C in Fig. 12, and Example 28, page 51, in the article on Piercing.)

Effect of Rolling Direction. For blanks that must later be bent or formed, consideration must be given in layout to the orientation of the blanks with respect to the direction of rolling (grain direction).

Ideally, blanks should be laid out so that severe bends are made with the bend axis at right angles to the direction of rolling, or, if this is not practical, with the bend axis at an

1010 steel, hot rolled, commercial quality, 0.12 in.

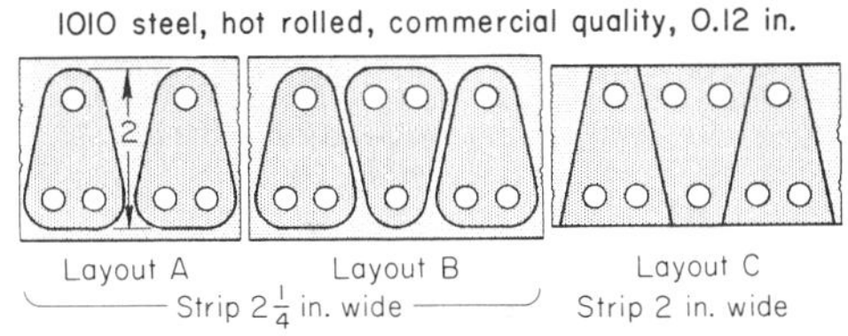

Item	Layout A	B	C
Metal per blank, sq in.	3.105	2.54	2.00
Cost of metal per 1000 blanks (at 10¢/lb)	$10.82	$8.85	$6.97

Fig. 12. Three blank layouts and comparison of metal cost for each (Example 8)

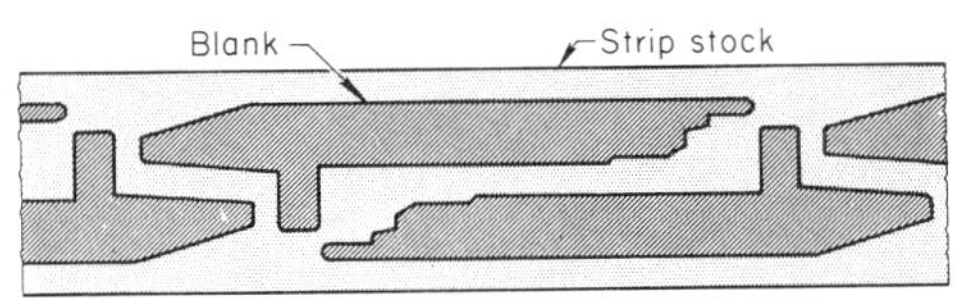

Fig. 13. Nesting of irregular blanks in layout to save material

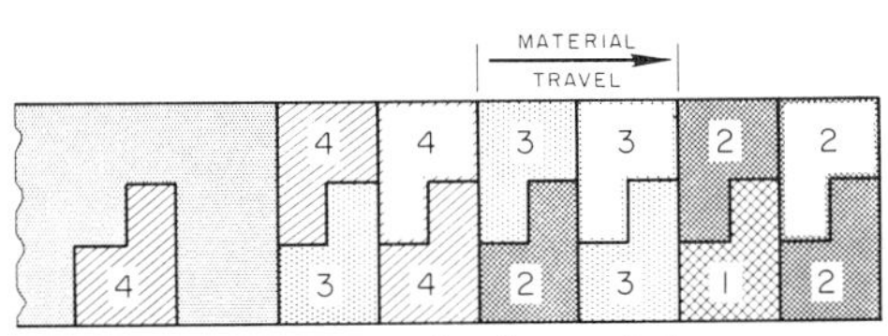

Numbers show order of press strokes.

The three shaded shapes for each press stroke denote the blanks produced directly by the three punches. The unshaded shape for each press stroke denotes a blank produced by the action of adjacent punches.

Fig. 14. Layout for scrapless blanking of four blanks per press stroke, using three punches

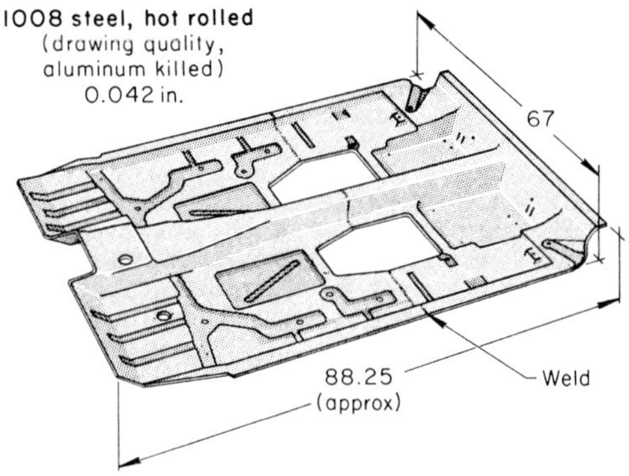

Fig. 15. Floor panel formed from a lap-seam welded two-piece blank at 7% less cost than from a one-piece blank (Example 9)

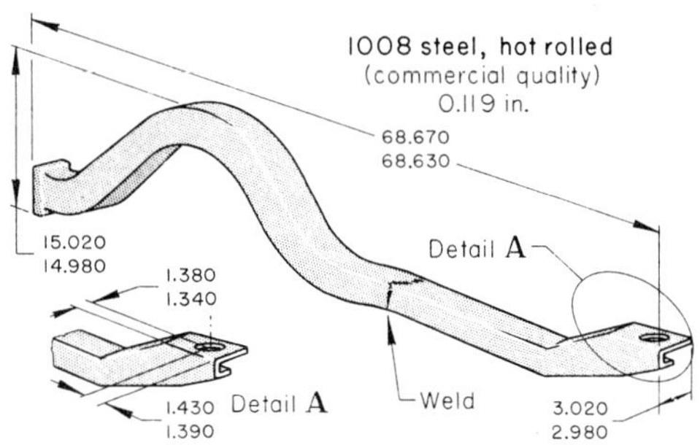

Fig. 16. Automobile-frame side rail that was formed from two blanks joined by MIG *welding* (Example 10)

angle to the direction of rolling. Stretching should be in the direction of rolling, whenever possible. (See the articles "Press Bending of Low-Carbon Steel" and "Press Forming of Low-Carbon Steel" for examples and illustrations of blank layouts made to take advantage of the direction of rolling.)

Welded Blanks

Welded blanks are used when they have advantages over one-piece blanks, as in the following situations:

1 The welded blank may cost less than an equivalent one-piece blank if scrap or other low-cost metal can be used to make the welded blank, or if tooling and production for the one-piece blank cost more than for the welded one.
2 Stock for a welded blank may be more readily available than stock for an equivalent one-piece blank.
3 The blank may have a shape that would waste more material if it were made in one piece instead of being welded. Sometimes material can be saved by welding projecting portions, such as tabs and ears, to simpler shapes.
4 The welded blank, when used in subsequent forming operations, may reduce the cost of tooling. Flat or simple shapes are welded in a layout designed to avoid the presence of seams in certain portions of the blank, and to permit automatic welding, if possible.

Large blanks that would cost extra because of width or for other reasons if they were made in one piece can sometimes be made at less cost by welding, as shown in the following example.

Example 9. Use of a Large Lap-Seam Welded Blank (Fig. 15)

Originally, the blank for a large press-formed floor panel (Fig. 15) was made from a single sheet of hot rolled, drawing quality, aluminum-killed 1008 steel, 0.042 by 75 by 99 in. Because of the large size, the sheets had to be obtained from a distant supplier at extra freight charges.

The added costs for shipping and extra width were saved by making the blank from two pieces of 0.042-in. sheet stock, 56 by 75 in. and 44 by 75 in., joined by lap-seam welding. The weld seam was in a portion of the formed panel that was not subject to severe deformation.

The one-piece blank cost $8.19. Cost of the two-piece blank, including $0.45 for welding, was $7.63 — a saving of $0.56 per blank.

Difficult shapes that would waste a considerable amount of material if they were made in one piece can sometimes be made by welding two or more simple blanks together. In the application described in the following example, two developed blanks were welded and then formed into a bent channel.

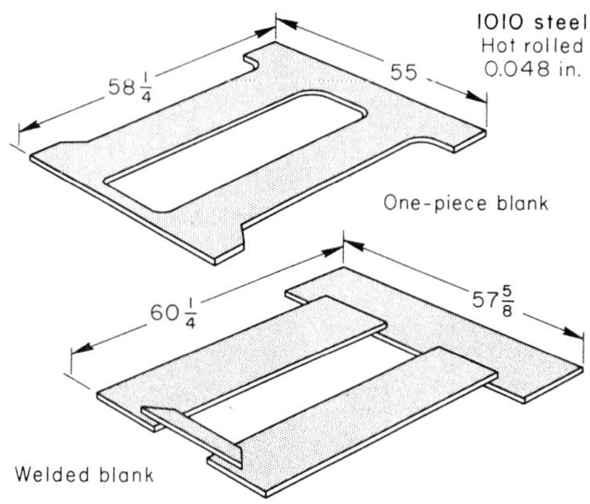

Cost item	Cost per blank One-piece(a)	Welded(b)
Material	$3.6980	$2.6354
Labor and overhead	0.6758	1.0879
Total	$4.3738	$3.7233
Saving		$0.6505

(a) Entire method consisted of shearing sheet, blanking sheared piece, trimming and piercing, and flanging. (b) Entire method consisted of shearing four pieces from sheet, spot welding, blanking weldment, trimming and piercing, and flanging.

Fig. 17. One-piece blank, and more economical spot welded blank, for a truck floor pan (Example 11)

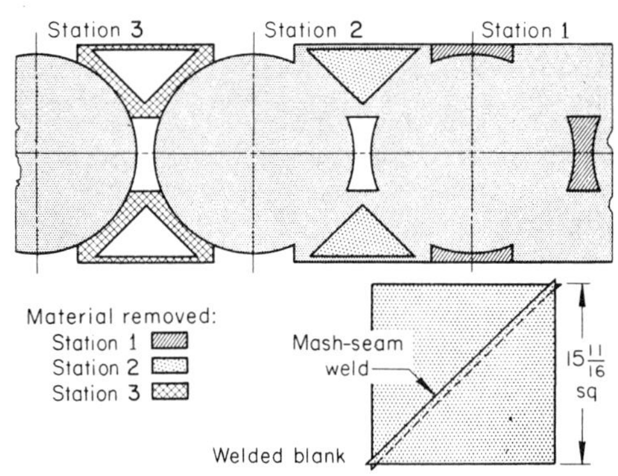

Cost item	Cost per blank One-piece	Welded
Material	$0.2248	$0.0292(a)
Labor	0.0030(b)	0.0337(c)
Total	$0.2278	$0.0629
Saving		$0.1649

(a) Scrap value of metal was 1¢/lb. (b) For slitting and shearing. (c) For welding ($0.0303) and dipping in rust preventive ($0.0034).

Fig. 18. Strip layout that yielded triangular "scrap" pieces from which a square blank was made by mash-seam welding at 28% of the cost of a one-piece blank (Example 12)

Example 10. Use of Two MIG-Welded Blanks to Form an Automobile-Frame Rail (Fig. 16)

Figure 16 shows a side rail for an automobile frame that was formed from 0.119-in.-thick hot rolled, commercial quality 1008 steel. The blank for this rail had a shape that could be made in one piece only with excessive waste of metal in scrap. The two pieces used were blanked from nested layouts with little scrap waste.

Blanking was done in a die that made both pieces in a single press stroke. The two portions were butted and joined by high-speed automatic MIG welding and then formed. Tolerance on all dimensions shown in Fig. 16 was ±0.020 in.

The saving in metal was more than the cost of welding, but if production needs had been fewer than 10,000 pieces, the saving in metal would not have paid for the welding and the blanking dies.

The production quantity was 300,000 rails made in lot sizes of 20,000 pieces.

Open Shapes. A blank with a large cutout can sometimes be made at less cost than a one-piece blank by welding simpler pieces together, as demonstrated in the next example.

Example 11. Use of a Spot Welded Four-Piece Blank to Reduce Cost (Fig. 17)

The blank for a large floor pan used in a truck originally was made in one piece by blanking in the usual manner and then trimming a large inner portion from the middle of the blank. Because the trimmed material was scrap, the method was changed to make the blank at lower cost by spot welding four pieces together, as shown in Fig. 17. Both blanks were completed in the same way — trimming and piercing, and then flanging. Costs for floor pans produced by the two methods are compared in the table accompanying Fig. 17.

Waste metal sometimes can be joined by welding to make a blank that costs less than a one-piece blank, as shown in the following example.

Example 12. Blank Produced by Mash-Seam Welding of Pieces Salvaged From Another Blanking Operation (Fig. 18)

A large circle was blanked from coiled strip in three stations of a progressive die, as shown in Fig. 18. Some of the scrap metal was saved for later use by blanking two triangles in station 2 at each stroke of the press. The triangles then were mash-seam welded to make a 15 11/16-in.-square blank (Fig. 18). This square was drawn and trimmed to make a blower cover. A comparison of costs for producing the square blank from one piece of metal and from two scrap triangles by mash-seam welding is shown in the table with Fig. 18. The metal was cold rolled low-carbon steel or enameling iron.

Welding methods used in making blanks include resistance welding methods (lap-seam, spot, foil butt-seam, mash-seam, flash, and high-frequency butt) and fusion welding.

Lap-seam welding (wheel electrodes) often is used (see Example 9). Tooling is simple, if the components are joined first by tack welding. The disadvantages of lap-seam welding are loose edges, and joints that are double the thickness of the work metal.

Spot welding, which was used in Example 11, is fast, and needs only simple, inexpensive tooling. The disadvantages of spot welding are loose edges, and joints that are not tight and are not as strong as those made by other methods.

Foil butt-seam welding is a fast method that makes smooth, tight joints with no loose edges. Its disadvantages include the cost of adding foil to one or both sides of the seam, and the need for using starting tabs to make strong joints.

Mash-seam welding makes smooth joints that often need no grinding, and tight joints that have no loose edges. It was used in Example 12. Disadvantages include the short life of electrodes, and the high cost of tooling that results from difficulty in maintaining the small overlap.

Flash welding uses simple tooling; the joints are tight and free from loose edges. However, the length of joint that can be produced by flash welding is limited, and the joints are rough and therefore must be ground before the weldment can be worked in a die.

High-frequency butt welding is fast and can be used to join two dissimilar metals. Electrode life is good. It requires costly equipment, however, and is generally suitable only for mass production.

Fusion welding methods include metal-arc inert-gas (MIG) and tungsten-arc inert-gas (TIG). MIG welding was used in Example 10.

Presses

Most blanking is done in single-action mechanical presses.

Some dies can be used only with a particular type of press; usually the die is made to suit a specific press. The capacity or tonnage rating must be adequate for the work, and must be well above the calculated cutting force (see page 33). Press capacities are given in tons at a certain distance above the bottom of the stroke. This distance must suit the die and the operation. Most blanking is done near the bottom of the stroke where the available force is largest. In compound dies, blanking may be done near midstroke, where the available force is much lower than at the bottom.

Size of bed, shut height, stroke length, and speed must all be suitable for the die and the work. Some types of dies can be run at high speeds and some need moderate or slow speeds, as discussed in the section "Construction and Use of Conventional Dies", which starts on the next page.

Construction and Use of Short-Run Dies

Small and medium quantities of blanks often are produced in punch presses by the use of inexpensive short-run dies. These include steel-rule dies, template dies (sometimes called plate dies or continental dies), and subpress dies. Although most applications of such dies are for production quantities of a few hundred to 10,000 pieces, suitably constructed dies of these types have been used for quantities of 100,000 pieces or more.

Short-run dies are used to a limited extent for blanking initial quantities of parts that are to be mass-produced. Because they can be made and put into operation more quickly than conventional dies, short-run dies make it possible to expedite delivery of completed parts.

In addition, short-run dies are used to produce trial lots of parts that may be subject to extensive changes in design. If the trial lots show that die design changes are needed, the changes can be made at less cost before the conventional die is completed. After the conventional die has been set up, either the entire production can be transferred to it, or both dies can be used.

For small quantities (fewer than 100 pieces), even the most inexpensive short-run blanking die may not be justified. Such small quantities of blanks generally are cut at less cost with standard tools, such as a nibbler, a squaring shear, or a rotary shear. Also, small quantities of blanks may be made by contour band sawing, routing, gas cutting, filing or machining.

Steel-rule dies are simple, inexpensive dies that are made by setting thin, bevel-edge strips of high-carbon tool steel on edge to outline the blank. The rule is set tightly into a slot in plain or impregnated plywood, and the plywood is backed by a steel subplate, as shown in Fig. 19. The die plate or template is attached directly to a steel subplate, and both upper and lower subplates are fastened to master die shoes, which are mounted in a conventional press.

Steel-rule dies are used for blanking, piercing, notching, and shallow forming. For work on flat blanks or on flat, sheared stock of low-carbon steel up to about ½ in. thick, a steel-rule die usually can be made more rapidly and at less cost than any other kind of die. Blanks as large as 4 by 7 ft have been made in steel-rule dies.

Die. The steel rule that is used as the die is made of high-carbon steel or of a tool steel such as W1 or W2, in spring temper and other hardnesses. It is available in stock lengths in several thicknesses, from 0.014 to 0.170 in., and in widths of 1.25 in. and narrower. Actual printers' rules of thickness from 1 point (0.014 in.) to 12 points (0.170 in.) are sometimes used. The finished rule usually has a square back edge, and a cutting edge that is ground to a 45° bevel or to a V-edge.

The back edge is fitted tightly into sawed slots in hard plywood as shown in Fig. 19, so that it will cut the outline of the blank. The steel subplate is used to back up the steel rule and support the plywood.

Punch. For blanking low-carbon steel, a die plate (high-carbon or tool steel template), of the same shape as the required blank, is used as the punch, opposing the steel rule. Other punch elements and die parts, as needed, are added to the die for piercing holes and slots at the same time that the blanking is done. Solid steel blocks, instead of a steel rule, may be used in the die to cut sharp corners and notches in the blank.

For some work, including the cutting of paper and other soft materials, the punch may be a block of hard wood, or a thick sheet of rubber or other soft material, with a working surface that extends beyond the area enclosed by the steel rule.

Stripper. For cutting paper and leather, the steel-rule die is stripped by elastic material such as sponge rubber, which is added to the die. In blanking low-carbon steel, blocks of tougher solid rubber may be used as strippers. Positive spring-loaded steel stripper plates are also used.

Accuracy. The accuracy of the blanks produced in steel-rule dies depends mainly on the skill of the diemaker and the care used in their construction. For noncritical parts blanked in steel-rule dies, the tolerance may be as large as ±1/32 in.; for more critical parts, the work can be located accurately to maintain a tolerance of ±0.005 in. Closer tolerances on blanks can be obtained at increased cost by using rotary-head millers or jig boring machines in constructing the dies.

Because holes and slots made by steel-rule blanking are pierced with conventional punch and die elements that are added to the steel-rule die, they can be produced to the same tolerances as in conventional blanking.

Steel-rule dies commonly blank laminations with burrs only 0.002 in. high.

Cost. A steel-rule die, made to blank low to moderate quantities of low-carbon steel, generally costs about 20% as much as a conventional die that is made for mass production of similar work.

The following example describes the use of a steel-rule die for stopgap and trial production of an automobile part. A die change was made cheaply; then, after being proved successful in the steel-rule die, the same change was included in the design of the conventional die for production use.

Example 13. Use of a Steel-Rule Die for Temporary Production (Fig. 20)

A steel-rule die was used to blank a part for an automobile frame, to begin production without waiting until the conventional die could be delivered. It was expected that the steel-rule die would have to produce 325 blanks before production could be changed to the conventional die.

The die, shown in Fig. 20, was used in a 250-ton straight-side press to blank annealed cold rolled low-carbon steel, 0.156 in. thick by 9 by 19 in., with three 9/16-in.-diam holes and three round-end slots 5/8 by 1½ in. Tolerances were ±1/32 in. on the blank outline, and ±0.005 in. on the pierced holes and slots. No burr limits were specified.

The steel rule was made of 12-point rule stock (0.170 in. thick by 1¼ in. wide), set full depth into hard plywood 5/8 in. thick. A die plate of steel, which fitted inside the rule, was used as the punch. The die was made by measuring a developed formed blank.

The ease and low cost of making a change in a steel-rule die proved important in this

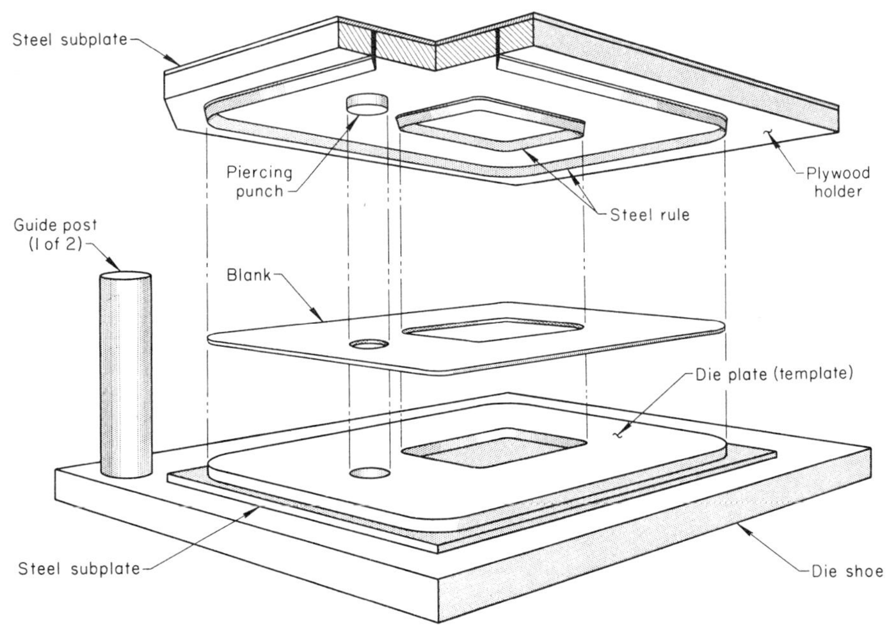

Fig. 19. Exploded view of a steel-rule die. See text for discussion.

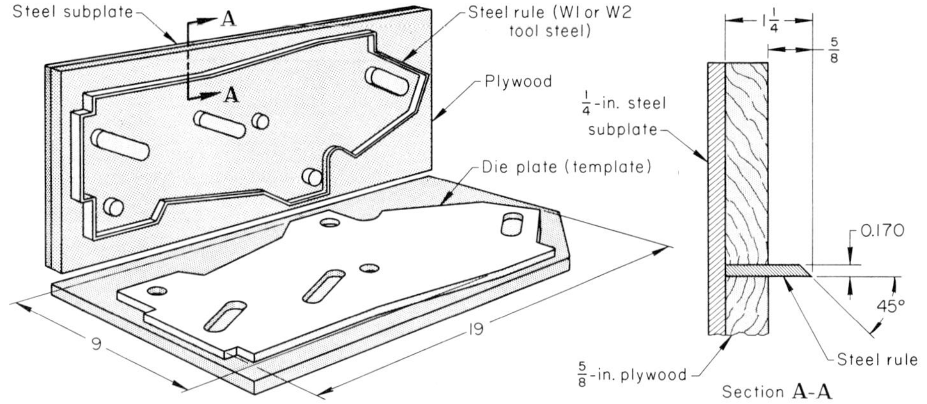

Fig. 20. Steel-rule die that was used as temporary tooling for blanking of low-carbon steel sheet 0.156 in. thick (Example 13)

application, because it was decided that one of the holes would not be needed. The punch that had been added for that hole was simply removed from the steel-rule die. After tryout, that hole was also eliminated in the conventional die before it was completed. The change, if made after the conventional die had been completed, would have cost much more.

The steel-rule die cost less than \$800, which was 20% of the cost of the conventional die. The steel-rule die was fed and unloaded by hand, so that production was only a few pieces per minute. More than 1000 blanks were made in this steel-rule die.

Template dies (also called plate dies) are competitive with steel-rule dies, both in cost and in the quantity that they can produce.

An exploded view of the elements of a template die is shown in Fig. 21. Punches and die elements can be added, as in steel-rule dies, to combine piercing with blanking.

Punch. The punch or template (Fig. 21) is made to fit the outline of the blank to be produced. Ordinarily the punch is made of medium-carbon steel plate (1040, 1050 or 4140) or of ground flat stock such as O2 tool steel, on which the edges may be flame hardened. In more difficult applications or where longer life is needed, the punch may be made of D2 or equivalent tool steel and hardened.

Die. The die usually is assembled of doweled hardened blocks of steel, ground to fit the punch with proper clearance. The same materials are used as for the punch.

Construction. Typical clearance for template dies is 0.003 in. The construction shown in Fig. 21 is satisfactory for blanking low-carbon steel in the same types of applications as steel-rule dies. For severe blanking, a template die can be made stronger by using one-piece construction or adding pins to prevent the die blocks from spreading, and by nesting the die or die segments into a recess in the die backing plate (subplate).

Operation. A continuous strip of stock can be fed into a template die, feeding the stock against a stop for each press stroke, using a side guide for the stock.

Usually blocks of tough rubber are used as strippers (Fig. 21), pushing the blank back into the strip, so that the blank is removed from the die by the feed motion of the strip.

Typical applications of template dies are described in the two examples that follow.

Example 14. Blanking 0.060-In.-Thick Laminations in a Template Die

Motor laminations of 0.060-in.-thick pickled hot rolled low-carbon steel were blanked in a template die. The blank was 12.5 by 5.5 in. The punch was made of a one-piece template, except for inserts that were provided where changes might be needed. The die was made with 0.003 in. clearance all around the punch (template).

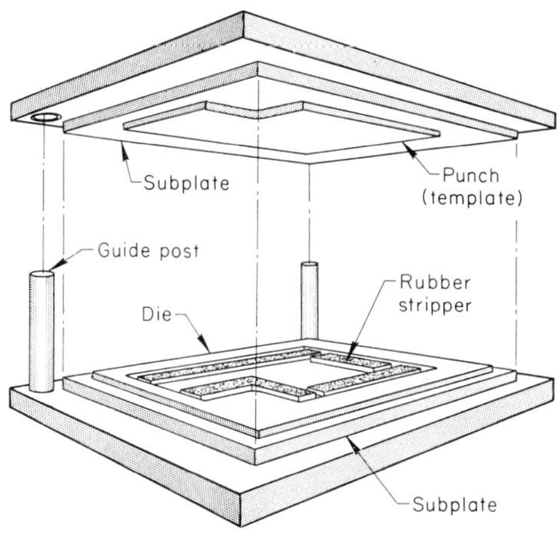

Fig. 21. Exploded view of a template die

The die was made of D2 tool steel. Rubber blocks were used for strippers, which pushed the blank back into the skeleton, in the vertical press, so that the blank was unloaded by the feeding of the stock. Feed by hand produced fewer than ten pieces per minute.

The burr was about 0.002 in., well below the specified limit of 0.004 in. Estimated die life was 100,000 pieces.

Example 15. Blanking Stock 0.130 In. Thick in a Template Die

Small quantities of a blank for a tractor part were made from 0.130-in.-thick hot rolled low-carbon steel in a template die. The blank measured 3½ by 10¾ in.; burr limit was 0.005 in. The die blocks for this heavy-duty blanking were nested into the die backing plate (subplate). The work metal was hand fed to the die. Production rate was a few pieces per minute; annual production was less than a thousand pieces.

Subpress dies are short-run die sets that are attached to the press bed, but in which the punch shoe is not attached to the ram. They require less setup time, but have a lower production rate, than conventional dies. The length of stroke of a subpress die is limited, because springs are used to raise the punch.

Subpress dies sometimes are used to blank the precision parts used in instruments and timepieces.

Self-contained notching tools can be purchased ready to install as subpress tools. Notching units, consisting of both punch and die mounted in individual C-frame units, are available in a variety of standard-corner, V, and square-edge sizes for notching low-carbon steel in thicknesses up to ¼ in. Special shapes of irregular outline can be incorporated in standard notching units. The notching units can be used singly or in groups, and can be used in combination with piercing tools of similar construction.

Each unit is self-contained and self-stripping by means of springs. The punches are held in close alignment and are not attached to the ram of the press. Each unit is located, pinned, and bolted to a die plate, template, or T-slot plate, and mounted on the bed of any type of press or press brake of adequate shut height.

This type of subpress tool can be used to make small blanks, but it is more commonly used to notch and pierce precut blanks. The units can be reused to produce parts of different shapes by relocating the tools.

Construction and Use of Conventional Dies

Conventional blanking dies consist basically of one or more mating pairs of rigid punches and dies, and are the standard tooling for production blanking of sheet metal in a press. Mating pairs of metal punches and dies are combined in various ways; and additional components are added to make up compound, progressive, transfer and multiple dies.

Conventional dies are costly, specialized tools, used generally for only one product, but they are so efficient, accurate, and productive that ordinarily they are the best method of mass production at the lowest cost per piece. Occasionally they are used for short-run production when tolerances are exceptionally stringent, or when other reasons make the use of short-run dies impractical.

Conventional dies are more accurate than most short-run tooling, and they retain their accuracy for a greater number of pieces. Also, they generally can be resharpened after wear has affected their action or the quality of the work. Before dies are worn out, they ordinarily have been resharpened many times. It is common for conventional dies to produce several million blanks before they must be replaced.

Tool materials commonly used for blanking low-carbon steel sheet in conventional dies include, in order of increasing lot size for which they are recommended: 1020 steel; W1, O1, A2 and D2 tool steels; and, for extremely long runs, carbide. For long runs on steel thicker than about ¼ in., M2 tool steel is used instead of carbide, because of the limited shock resistance of carbide. Type D2 tool steel is probably the most commonly used and most widely available tool material for mass-production blanking of steel and other metals.

Cold rolled sheet and hot rolled pickled-and-oiled sheet are far less damaging to tools than grit-blasted or hot rolled unpickled surfaces. Tool materials that have a high resistance to abrasion, such as A2 or D2 tool steel, are recommended for use in tools for the blanking of sizable production lots of hot rolled unpickled steel. (For detailed information on the selection of tool material and on tool life, the reader is referred to the article on Selection of Materials for Blanking and Piercing Dies, which begins on page 69 in this volume.)

Single-Operation Dies. The simplest conventional blanking dies are single-operation dies. They are used as separate units to produce blanks or as parts of more complex dies that do several operations on a workpiece. The separate stations of a progressive die are similar to single-operation dies (although integrally constructed), and transfer presses use many single-operation dies.

One of the commonest types is a drop-through die (Fig. 22), in which the severed blank is forced through the die opening by the downward motion of the punch, and drops through into a chute or container. This type of blanking die has a minimum number of parts and is relatively inexpensive. Another major advantage of drop-through blanking dies is their simple and reliable blank-ejection system, which is usually compatible with the use of this type of construction in progressive or transfer dies. In other types of dies, the ejection system may be more complicated than the die itself.

Disadvantages of drop-through dies for blanking include the following:

1. Unless parallels are placed between die and bolster, blanks must be small enough to go through the hole in the bed.
2. Blanks may distort by dishing.
3. Some shapes make drop-through difficult.
4. The die must be on the lower shoe and the punch on the upper shoe of the die set.

Two other types of single-operation blanking dies can be used when, because of size or susceptibility to damage, blanks or workpieces cannot be unloaded by dropping through the die, but can be removed between the die and punch faces. These are inverted and return dies.

Inverted dies (Fig. 23) have the punch on the lower shoe and the die on the upper shoe. A knockout pin releases the blank from the die, and the blank is removed mechanically or manually from the top of the punch. A scrap cutter is usually included, so that the scrap can be blown or knocked away. Sometimes the scrap is allowed to stack up in successive collars around the punch and is stripped off manually after a number of pieces have been blanked.

Return dies (Fig. 24) are made in the usual way, with the punch on the upper shoe and the die on the lower shoe. The punch shears the blank and presses it into the die cavity, as with other types of blanking dies. A spring-loaded pressure plate or die cushion acts as an ejector for the die, returning the blank to the surface of the die, where it can be picked off manually or mechanically. A spring-loaded plate on the upper shoe acts as a blankholder on the downstroke and as a punch stripper on the upstroke.

Like many inverted dies, a variation of the return die in Fig. 24 includes a scrap cutter (not shown in Fig. 24) that parts the ring of scrap so that it can drop away, be blown away, or be removed mechanically.

Inverted and return dies have the advantage of not needing a clearance hole to let the workpiece or blank drop through. The main disadvantages of inverted and return dies are (*a*) they are more expensive than drop-through dies, because they have more parts; and (*b*) they may require careful adjustment and synchronization of external ejectors and air blasts, which adds to setup costs. Of the two types, inverted dies

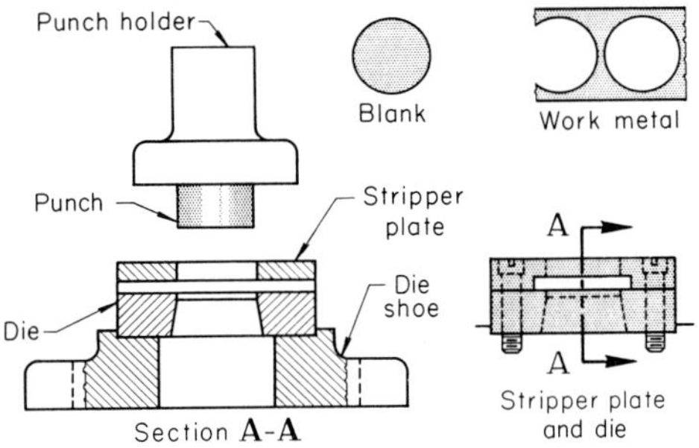

Fig. 22. Elements of a conventional drop-through blanking die

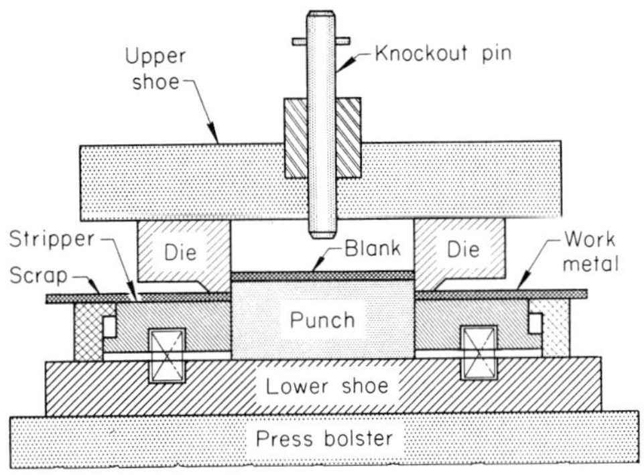

Fig. 23. Inverted blanking die, with the punch on the lower shoe and the die on the upper shoe

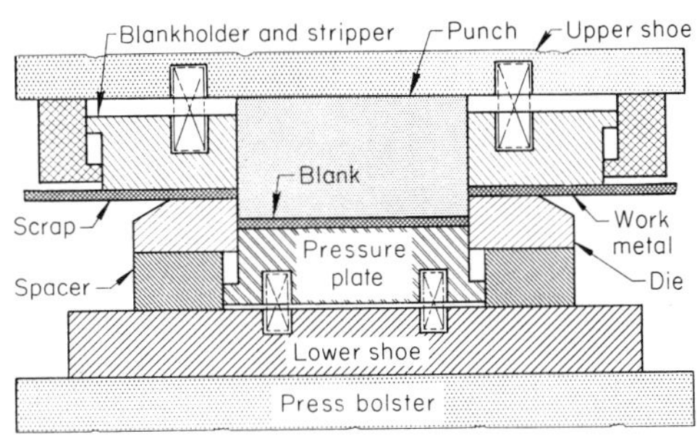

Fig. 24. Return blanking die, with spring-loaded pressure plate that acts both as a die cushion and as an ejector for the die

generally are simpler to construct and have less complicated knockout mechanisms than the pressure-plate or die-cushion ejectors in return dies.

Return dies are better suited to continuous strip operation, because the strip remains in line and is not pressed down by the die over the punch. If the workpiece must be clamped before blanking, the blankholder or the pressure plate (or both) in a return die holds the workpiece or blank through the entire working stroke.

Instead of being removed by the methods described in the preceding paragraphs, the severed blank sometimes is pushed back (completely or partly) into the strip, to be removed later, as sometimes is done in a progressive die. Pushback also may be used for other purposes, such as to provide knockouts for fitting in electrical panels or junction boxes and in other sheet-metal products. Sometimes a flattening operation is added to help in pushback.

Compound dies do several operations on the same workpiece in the same stroke of the press, such as blank and pierce, or blank, pierce, and form. The elements of a compound die for simultaneous blanking and piercing are shown in Fig. 25. In this die, the blanking punch is in the bottom; a hole in the punch is used as the piercing die. The piercing punch and the blanking die are in the top.

Compound dies generally are more economical in mass production than a series of one-operation dies, and they usually are more accurate. For instance, a compound die that blanks and pierces a workpiece can hold the spacing between pierced holes or the relation of the pierced hole to the edge of the blank more accurately than would be possible if the individual operations were done in separate dies. This is because of possible variation in locating the blank for piercing, or in locating a prepierced strip for blanking.

Because the complexity of operation causes greater difficulties in unloading the workpiece, compound dies usually run slower than single-operation dies; the maximum speed of a compound die is about 250 strokes per minute. Other disadvantages of compound dies in comparison to single-operation dies are that they are more specialized, so that a change in the product is more likely to make the die obsolete; and both initial and maintenance costs are higher. An advantage of compound dies is that, because of their slower operation, they generally produce more pieces per sharpening than do single-operation dies.

Sometimes a complex compound die can be more economical than two simpler compound dies in making the same part. This is illustrated by the application described in the next example.

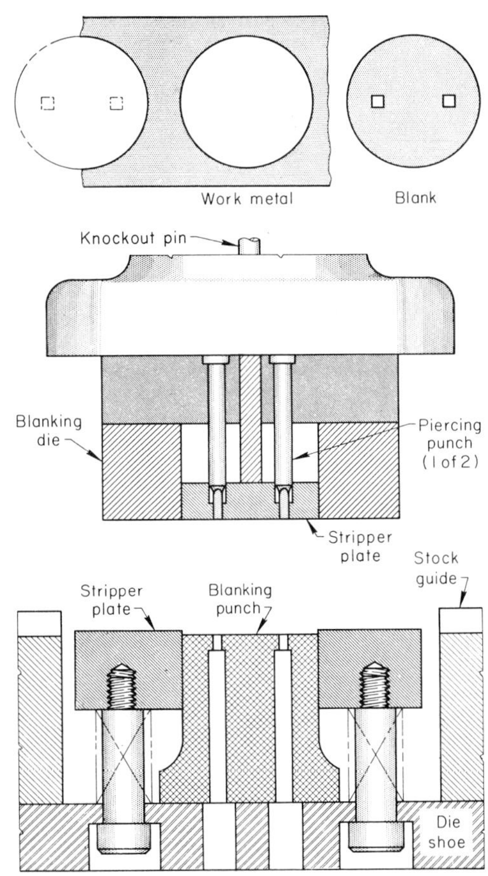

Fig. 25. Elements of a compound die for simultaneous blanking and piercing

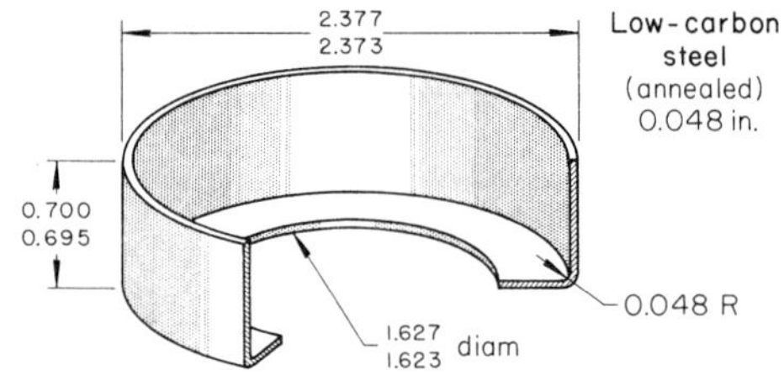

This part was previously made in two compound dies: blank and draw, and trim and pierce.

Fig. 26. Cup that was blanked, drawn, pierced and trimmed in a compound die (Example 16)

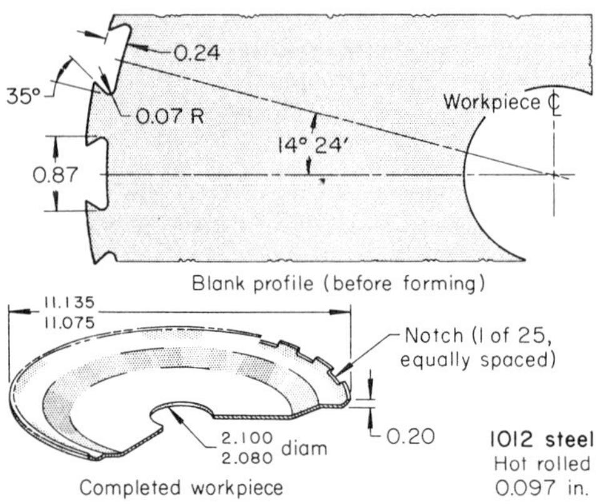

Fig. 27. Brake-drum back that was produced in two press operations, in a blank-and-notch die and in a restrike-and-form die (Example 18)

Example 16. Replacing Two Two-Operation Dies With a Four-Operation Die (Fig. 26)

Originally, the cup shown in Fig. 26 was produced from annealed cold rolled low-carbon steel in two compound dies, in separate press operations. The first die was a blank-and-draw die; the second was a pierce-and-pinch-trim die. Two separate dies had been used on the assumption that the large pierced hole in the cup would not leave enough tool thickness to sustain all four operations. A 45-ton open-back inclinable press was used for each of the two operations, at a production rate of 500 pieces per hour.

A restudy of the procedure led to the design of a die to blank, draw, pierce and trim — all in one press stroke. The same press was used at the same speed, thus cutting the labor cost per piece in half. Tooling cost also was reduced, because the cost of the new die (less than $1000) was less than the combined cost of the original two dies.

The two examples that follow describe production applications in which compound dies were used for blanking and other operations because of their inherent accuracy.

Example 17. Maintaining Extremely Accurate Hole Position in Blanking and Piercing

A part for a business machine was a rectangle of 1008 or 1010 cold rolled steel, 0.093 by 4 by 5 in., with two notched corners. The part was blanked, and six round holes and twenty rectangular holes were pierced, in a blank-and-pierce compound die.

The positions of all the holes were specified within a few thousandths of an inch, both hole-to-hole and hole-to-edge. Most of the holes had to be in accurate position within ±0.002 in. Extreme accuracy of hole position was assured by guiding the piercing punches in the stripper and by trimming the blank in the compound die.

Anticipated demand was 2000 to 3000 pieces per month for an indefinite number of years. The part was case hardened, and ground on both faces to remove burrs. The die was made of D2 tool steel and was designed for a life of 500,000 pieces.

Example 18. Blanking, Forming and Piercing a Brake-Drum Back in Compound Dies (Fig. 27)

An automobile brake-drum back, shown in Fig. 27, was formed from hot rolled 1012 steel in two press operations. A compound die was used to blank and notch the part in a 200-ton press, at the rate of 17½ strokes per minute, making one part per stroke. The tolerances shown in Fig. 27 were maintained in production with ordinary shop practice, as was the tolerance on the 14° 24′ angle (±0° 30′). The die, made of vanadium tool steel, required reworking after 50,000 pieces. Chlorinated oils were used as lubricant.

In the second press operation, a compound die in a 400-ton press formed the workpiece, pierced the center hole, and flattened the workpiece at the notches. This die was made of tungsten oil-hardening tool steel. Production rates, lubrication and die life were the same as in the blank-and-notch operation.

Progressive Dies. In a progressive die, the workpiece, while attached to the strip (or to the scrap skeleton), is fed from station to station at each stroke. At each stroke, the die performs work at some or all of the stations. The workpiece is cut off and unloaded at the last station. Each station may be simple or compound. The principal parts of a two-station blank-and-pierce progressive die are shown in Fig. 28.

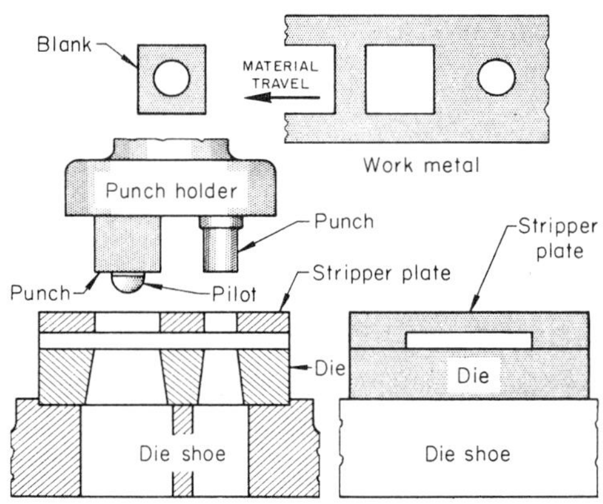

Pierced blank and work metal (steel strip) are shown at the top. The use of a pilot in the pierced hole assures accuracy to within a few thousandths of an inch.

Fig. 28. Principal components of a two-station blank-and-pierce progressive die

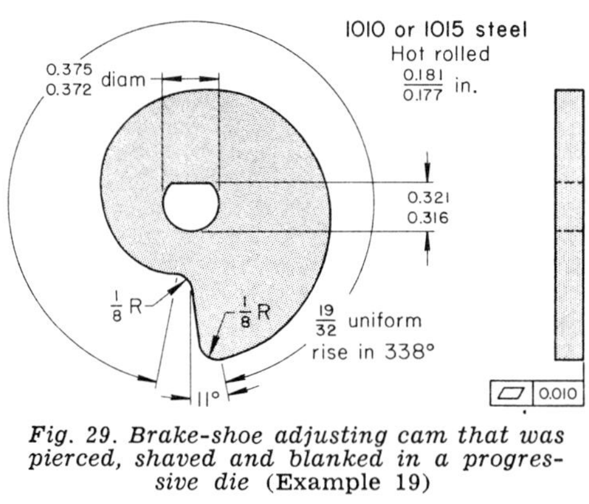

Fig. 29. Brake-shoe adjusting cam that was pierced, shaved and blanked in a progressive die (Example 19)

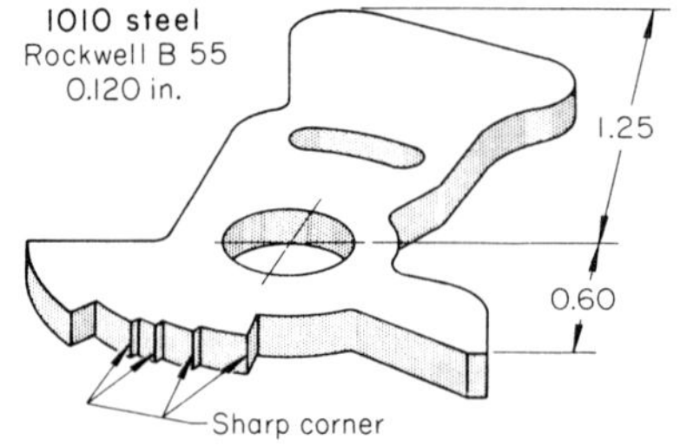

Fig. 30. Fast-idle cam on which sharp corners were cut in a progressive die (Example 20)

In producing the simple blank shown in Fig. 28 from coiled strip in a progressive die, the round hole is pierced at the first station. The strip is then fed left, to the next station. There the pilot enters the hole as the blanking punch moves down to complete the blank. Accurate relation of the hole to the outline of the blank depends on accurate fit of the pilot in the hole. The completed part is not separated from the strip until the last operation, regardless of the number of operations. After the first piece, one piece is completed at each stroke.

The cost of a progressive die is high, and usually it is set up in an automatic press with scrap cutter, feeder, straightener and uncoiler, so that the total cost of auxiliary equipment is also high. Other disadvantages of progressive dies are: (*a*) the part cannot be turned over between operations; and (*b*) material may be wasted, because the workpiece may not nest well in the strip layout.

Coil stock (or, less often, flat strip stock) is used. Many operations on small and medium-size parts can be done in conjunction with blanking in a progressive die, but the planning may be complicated. Soft or thin stock may be troublesome, because the pilot may distort the locating holes. Setup and maintenance may be difficult.

The example that follows describes a high-production application of a progressive die in which piercing and shaving were combined with blanking, to make an adjusting cam for a brake shoe.

Example 19. Piercing, Shaving and Blanking a Cam in a Progressive Die (Fig. 29)

The brake-shoe adjusting cam shown in Fig. 29 was pierced, shaved in two operations, and blanked — all in a progressive die. The work metal was hot rolled 1010 or 1015 steel, pickled and oiled, 0.179 in. thick.

Hole diameter was held within 0.003 in. by shaving. A satisfactory cam surface without shaving was maintained by resharpening the die after blanking 15,000 to 20,000 pieces. The maximum punch-to-die clearance was 10% of stock thickness for piercing, 1% for shaving and 5% for blanking.

The punch and die were made of D2 tool steel and hardened to Rockwell C 59 to 60. The die was run in a 90-ton press at 55 strokes per minute.

The example that follows describes an application in which a progressive die was preferred to a compound die because it was too difficult to cut sharp corners in the compound die.

Example 20. Change From a Compound to a Progressive Die to Cut Sharp Corners (Fig. 30)

Figure 30 shows a fast-idle cam on which the corners had to be sharp. Because it was difficult to make and maintain sharp corners in a compound die, a progressive die was used. The part was blanked from coiled 1010 steel strip, 0.120 by 2⅜ in., with maximum hardness of Rockwell B 55. A seven-station progressive die cut the steps in the cam.

The punches were made of M2 tool steel and hardened to Rockwell C 60 to 63, and the die and the punch holders were made of O1 tool steel and hardened to Rockwell C 56 to 58. About 10,000 to 15,000 cams were made between sharpenings of the progressive die, using a punch-to-die clearance of 0.004 in. This die was mounted in a 60-ton press that was run at 76 strokes per minute.

In another operation, the steps in the cam were shaved to remove die break, so that surfaces were flat across the thickness. Tools were of the same materials as in the progressive die, and had a life of 5000 to 6000 pieces between sharpenings. The shaving was done at 16 strokes per minute in a 40-ton press.

Finally, the cam was liquid carburized 0.003 to 0.006 in. deep and oil quenched to file hardness. Zinc plating (0.0002-in. minimum thickness) and a chromate postplating treatment followed.

In the next example, three blanks were cut from coil stock at each press stroke. The layout resulted in minimum scrap loss and maximum utilization of the press capacity. Less than ½ in. of the coil width (based on minimum coil width), and less than ¼ in. at the end of each blank, was scrap.

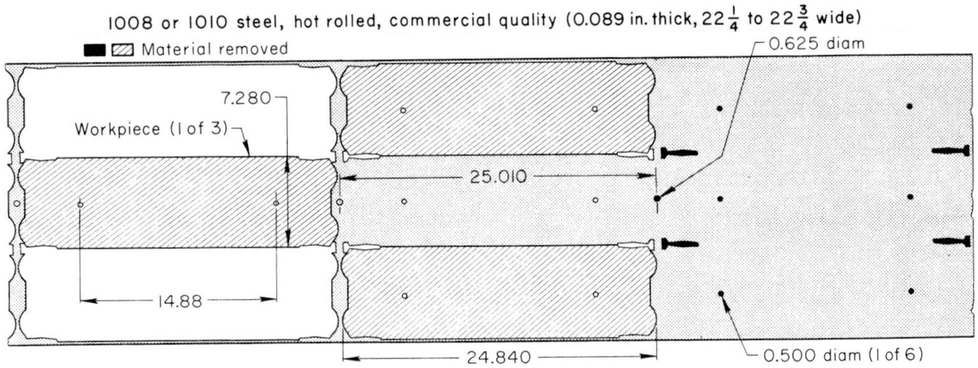

Fig. 31. Strip development in the production of three blanks at a time in a three-stage progressive die (Example 21)

Example 21. Making Three Blanks at Each Press Stroke in a Progressive Die (Fig. 31)

The strip development for making three blanks at a time is shown in Fig. 31. These blanks were formed into parts in a transfer die (see Example 138, in "Press Forming of Low-Carbon Steel"). The blanks were made in a three-station progressive die, from 0.089-in.-thick hot rolled, commercial quality 1008 or 1010 steel coiled strip, pickled and oiled. The coil had a slit edge and was 22¼ to 22¾ in. wide.

In the first station, six 0.500-in.-diam holes, one 0.625-in.-diam pilot hole and four cutouts were pierced. Next, the two side blanks were cut from the strip to fall through the die. The remaining middle strip was the width of the third blank, which was cut off in the third station. The total scrap width was 0.41 to 0.91 in.

The die was made of W2 tool steel and hardened to Rockwell C 60 to 62. The die shoe was 36 by 84 in. A 600-ton mechanical press with a 66-by-120-in. bed, operating at 30 strokes per minute, was used. Clearance per side between the punches and dies was 0.012 in., or 13.5% of stock thickness. Die life was 200,000 strokes for each resharpening.

A hydraulic-actuated roller stock pusher kept the work metal against roller guides in the first station. Hardened tool steel guides were used in stations 2 and 3.

As the blanks were cut, 750 to 900 of them were loaded automatically into containers. The containers were moved to the transfer press, where the blanks were loaded automatically into the transfer die.

Transfer dies, in which separate workpieces are fed from station to station by transfer fingers, are used for blanking only when coil stock is used. Blanking is done at the first station, and is followed by other operations.

Transfer dies ordinarily are used for additional operations on precut blanks made in a separate press (blanks that permitted close nesting for best utilization of the stock). When equally high utilization of stock can be obtained, coil stock may be used in a transfer die, with blanking being done at the first station.

Like progressive dies, transfer dies and their related equipment (presses, special attachments, and feeding devices) are expensive and are best suited for mass production. Production rates are high.

Multiple dies (also called multiple-part dies) make two or more workpieces at each stroke of the press. The workpieces may be pairs of right-hand and left-hand parts, duplicate parts, or unrelated parts. Punch height may be staggered to reduce shock and blanking noise. Such dies are used in mass production.

Multiple dies may be multiples of single-operation dies or multiples of compound dies. They generally cost only a little more than similar dies that make only one part per stroke. (A die that makes two parts per stroke may cost only 5% more than a die that makes only one.)

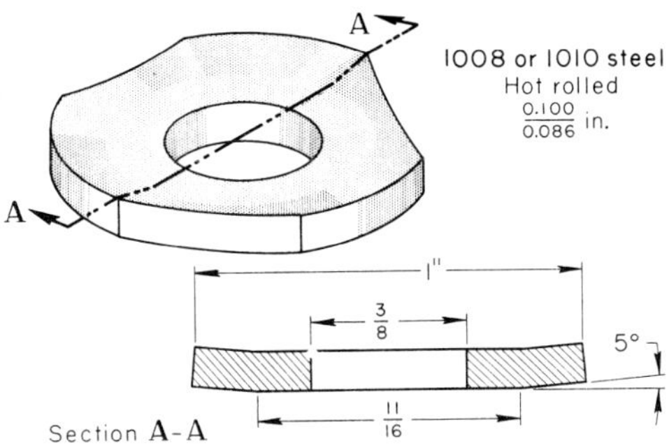

Fig. 32. Dished washer that was blanked, pierced and formed, three pieces per stroke, in a multiple compound die (Example 22)

Such dies are used mostly for blank-and-form sequences. Draw operations are more difficult to combine with blanking or other operations, because of blankholder needs and slower draw operation. Unloading of the work is sometimes difficult.

The use of multiple dies depends on size and shape of the workpiece, size of the press, production quantity, possible savings in material and labor, and costs for setup and maintenance.

Advantages of multiple dies may include savings in material by better blank layout, reduced cost of labor, and increased production. Disadvantages of multiple dies may include an increase in the costs of setup and maintenance.

Often it is better to increase the production of a single-part die by some simple change, such as putting the die in a faster press, rather than to replace the single-part die with a multiple die. Also, a multiple die may have to be run slower than the simpler die, if a press of greater tonnage must be used to provide ample force.

The use of a multiple compound die to blank, pierce and form three pieces per stroke is illustrated in the following example.

Example 22. Blanking, Piercing and Forming Three Pieces per Stroke in a Multiple Compound Die (Fig. 32)

A multiple compound die was used to make a thick dished washer with three flats equally spaced on the edge circle, as shown in Fig. 32. The part was made of hot rolled 1008 or 1010 steel, 0.093 ± 0.007 in. thick.

Operations in the compound die were blank, pierce and form. The blanks were nested so closely in the stock layout that the pilot holes (half-circles) had to be notched in the edge of the stock. Three parts were made with each stroke of the die, and the parts were pushed partly back into the scrap skeleton so that they were carried out of the press for unloading.

Production was 500,000 pieces per month. The die, made of D2 tool steel, needed to be resharpened after 150,000 strokes (450,000 pieces), and required reconditioning (replacement of some parts) after 3.5 million pieces.

Operating Conditions

In order to achieve high productivity and low unit cost, most blanking is done in high-speed mechanical presses. Speeds as high as 1200 strokes per minute are used. The equipment for high-speed blanking ordinarily includes a short-stroke press, automatic feed devices, and dies designed for bottom ejection.

In most blanking operations, press speed is limited by the length of feed, which is governed by blank size, or by the relation between press capacity (tonnage) and load. The combination of blanking with forming or drawing in compound dies also restricts press speed. Blanking speed may be as low as ten strokes per minute in producing blanks that are extremely large or that present handling problems for other reasons.

Regardless of the number of strokes per minute, the velocity of the punch always approaches zero near the bottom of the stroke. Within the usual range for production work in conventional blanking dies, the speed of the press has little practical effect on the speed of the punch during the blanking portion of the stroke. This effect, however, is critical for fine-edge blanking, in which punch speed during the interval while the punch is cutting through the work metal is usually about 0.3 to 0.6 in. per second (see "Fine-Edge Blanking and Piercing", which begins on page 56).

Lubrication requirements generally are less critical for blanking than for forming or deep drawing; stock to be blanked is often fed into the press with no lubrication other than the residue remaining on the stock from the lubrication at the mill. Sometimes the stock is coated with a light mineral oil or a light chlorinated oil. However, lubrication is important in dies that have close clearance between punches and stripper. At speeds of 40 strokes and more per minute, such dies must be lubricated constantly with a spray of light mineral oil to prevent galling of the punches in the stripper.

For a more complete discussion of lubrication requirements in blanking, the reader is referred to the article on Selection and Use of Lubricants in Forming of Sheet Metal, which begins on page 23 in this volume.

Effect of Work-Metal Thickness

Stock thickness affects the selection of material for dies and related components, as well as the selection of die type and design. The amount of shear and relief (angular clearance or draft) built into a blanking die, and the amount of clearance between punch and die, all depend on blank thickness.

Work-metal thickness also is a factor in the selection of blanking method, handling procedure and handling equipment. Blanking in a punch press usually is the fastest and most economical way of producing blanks less than about ¼ in. thick, in medium or large quantities.

Plate stock, in thicknesses of ¼ to 1 in., is less frequently blanked in presses than is sheet or strip. Blanks of such thick material are often made by gas cutting, sawing, nibbling or routing, instead of by shearing or by press operations, with the selection of method depending primarily on plate thickness and production quantity.

Nearly all blanks thinner than ⅛ in., except for intricate shapes chemically blanked from foil, are produced with conventional dies in mechanical or hydraulic presses. In only four of the examples of commercial practice presented in this article was work metal thicker than ⅛ in. (Example 13, 0.156 in.; Example 15, 0.130 in.; Example 19, 0.179 in.; and Example 23, 0.187 in.)

Because of its strength and rigidity, material thicker than ⅛ in. seldom is blanked from coil stock or in a progressive die. On the other hand, because of its lack of strength and extreme flexibility, material thinner than 0.020 in. generally requires special handling techniques.

The articles in this volume on Piercing of Low-Carbon Steel (page 44), Blanking and Piercing of Magnetically Soft Materials (page 60), and Press Forming of High-Carbon Steel (page 132) contain additional information on the effect of work-metal thickness on processing. For instance, the blanking and piercing of a perforated comb for an electric shaver from 0.0045-in.-thick steel is described in Example 195 on page 134. The examples that appear in the articles on bending, forming, and deep drawing of low-carbon steel, and on press-brake forming, describe a number of parts that were blanked from thick material.

Example 33 in the article on Piercing describes the blanking of steel shims from thin (0.010-in.) coil stock in a compound die.

In blanking complex shapes from thin low-carbon steel sheet by repeated strokes of a notching die, distortion is often a problem. Distortion of such parts can be minimized by the use of hardened stock and by doing the entire blanking operation in one stroke in a single die.

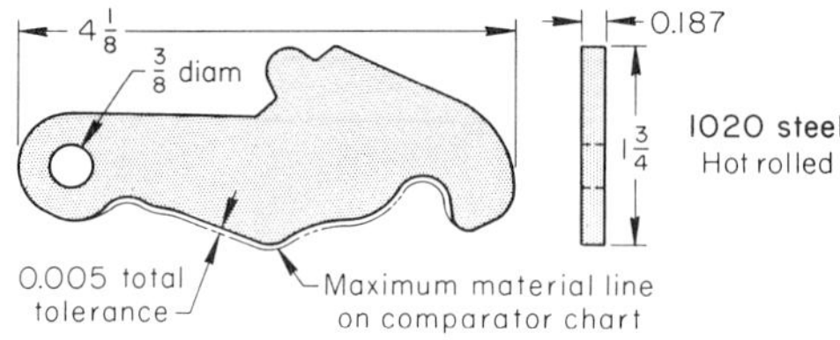

Fig. 33. Cam that was blanked and pierced in a compound die within an envelope tolerance of 0.005 in. TIR (Example 23)

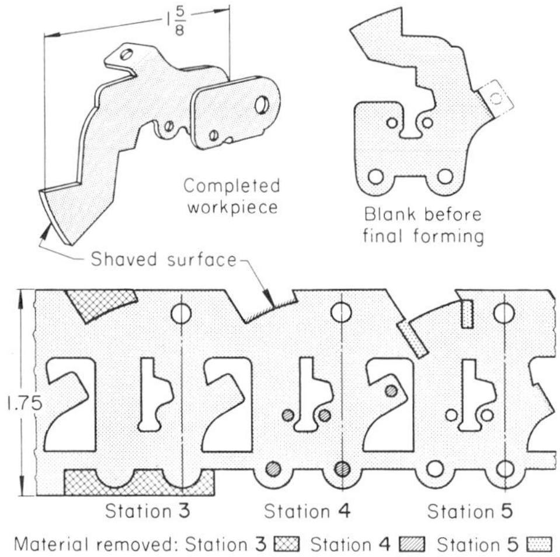

Fig. 34. Formed part on which a blanked edge was made smoother and more accurate by shaving (station 4) after notching (station 3) in a progressive die (Example 24)

Accuracy

Blanking in conventional dies readily produces parts within a total tolerance of 0.002 to 0.010 in., depending on the accuracy of the dies and the condition of the press.

The tolerances given in Table 4 in the article "Piercing of Low-Carbon Steel" (page 51) generally apply also to blanks. The total tolerances listed under the column head "Location" apply to the relation of a point on the periphery of the blank to a hole or other reference feature on the blank; the values listed under "Size" apply to a diameter for round blanks, or to a similar control dimension for other blank shapes.

Production of blanks to these tolerances is illustrated by the examples in this article and the article "Piercing of Low-Carbon Steel". The example that follows describes the use of a compound die to maintain a total (envelope) tolerance of 0.005 in. on the relation of a cam surface to a hole.

Example 23. Blanking and Piercing a Cam to 0.005-In. Total Tolerance on Cam Surface in Relation to Hole Position (Fig. 33)

A compound die was used to blank and pierce the cam shown in Fig. 33, so that the hole would be in accurate relation to the cam surface. The die was made of A2 or equivalent tool steel and hardened to Rockwell C 62, and had a clearance per side equal to 10% of the stock thickness.

The cam, used in the hinge mechanism of an automobile door, was blanked from hot rolled 1020 steel, 0.187 in. thick. Samples of the cam were inspected by means of an optical comparator, which compared the relation of hole and cam surface to an outline that showed the full 0.005-in. tolerance (as illustrated in Fig. 33).

In another operation, the cam surface was machined to remove die break, so that the edge would be square. Then the cam was case hardened.

The part was produced in a 160-ton open-back inclinable press at 60 strokes per minute. A sulfur-base lubricant was used. Die life was 40,000 pieces per sharpening.

Examples 17 and 18 in this article describe other applications of compound dies in blanking and piercing to conventional tolerances.

For parts made in a progressive die, the relation of the blank outline to features of the part produced in other stations of the die depends on the accurate fit of the pilot in the pilot hole. Transfer dies ordinarily give better accuracy than progressive dies, because the positive location of separate parts can be more precise than roll feed of coil stock with pilots.

Holding close tolerances on parts made in a progressive die is particularly difficult on soft or thin material, because distortion of the locating holes by the pilot is more likely. Handling problems that might contribute to excessive variation in location or dimensions are not usually encountered in transfer dies, except with blanks of extremely thin material.

Shaving (see the section that follows) is used to improve the accuracy of blank outlines to meet close tolerances or to improve edge quality.

Short-run dies generally are less accurate than conventional dies. By using more accurate methods of constructing short-run dies, closer tolerances on blanked work can be obtained, but at some increase in die cost (see the section on short-run dies in this article).

Generally, making blanks by methods other than the use of dies in presses, except for machining or grinding, results in a lower level of accuracy.

Shaving

Shaving is an operation that may be used after blanking, to give a smooth, square edge and greater accuracy than can be achieved in ordinary blanking.

Shaving removes only the blanked edge — cutting away the deformed, broken and burred edge that was left in blanking. The elimination of these irregularities and the removal of locally work-hardened metal minimizes breakage of the work metal during subsequent flanging, particularly flanging of holes. The scrap produced in shaving is so thin that it resembles the chips produced in finish machining, rather than the usual scrap that is produced in a press.

When shaving is planned, a small amount of extra stock is left on the workpiece, to be removed in shaving. Shaving can be done in a separate operation, or it can be included in one station of a progressive die.

The shaving operation produces a straight, square edge, generally to about 75% of the metal thickness. Two shaves make a better, straighter edge (to about 90% of the metal thickness) than does a single shaving operation. To eliminate rollover from blank edges, which requires the removal of a greater amount of stock, it may be better to consider machining the workpiece rather than shaving.

Punch-to-die clearances range from ero to 1½% of stock thickness per ide. Sturdy guideposts in a heavy die et are necessary to maintain the close lignment needed to prevent damage to ne punch and die.

Shaving causes more wear on a die han ordinary blanking, so that the die roduces fewer parts per grind and eeds more frequent maintenance. Slivrs of shaving scrap (chips) can jam eeding mechanisms, become embedded 1 workpieces or mar the punch and die urfaces if not removed after each ress stroke.

Because of the problems described in he preceding paragraph, special attenion must be given to die design when having is included among the operaions done in a progressive die, as lustrated in the example that follows.

Example 24. Shaving in a Progressive Die (Fig. 34)

In evaluating methods for high-volume production of the shaved low-carbon steel part nown in Fig. 34, the use of a single proressive die for all cutting and forming operations was projected as the most economical nethod, principally because this method ould entail fewer operations and less handing than the other methods considered. owever, two major problems were anticipated: a) the life of the shaving tools was expected be much shorter than that of the other tools the progressive die (which would have esulted in costly interruption of production sharpen or replace the shaving tools), and b) it appeared likely that misfeed could ocur from jamming of the feeding mechanism y slivers of shaving scrap.

By designing the die and feeding mechaism to eliminate difficulties from these two ources, efficient and economical production as obtained. The tools were of A2 air-hardning tool steel, hardened to Rockwell C 54 to 8 for the forming sections and to Rockwell 60 to 62 for the cutting sections. The shavg section was made with a replaceable inert to minimize downtime when sharpening as needed. Damage to the progressive die rom an accumulation of shaving scrap was revented by including in the die a stock top (station 3 in Fig. 34) and misfeed and ouble-thickness protectors.

The shaving was done in station 4, as nown in Fig. 34. Production rate was 80 ieces per minute in a 60-ton mechanical ress having a 2-in. stroke. Tolerance on most ections of the part was ±0.005 in.

The use of shaving to hold the diamter of a hole in a brake-shoe adjusting am within a total tolerance of 0.003 in. s described in Example 19 in this artile. For additional examples of the use f shaving to maintain close tolerances nd to improve the quality of hole valls, see the article "Piercing of Low-arbon Steel" in this volume.

Shaving allowance, or the amount of tock to be removed from the workiece, depends on the hardness and hickness of the blank. Generally, the mallest amount of stock that will produce the desired result is left for the having operation. Shaving allowances ecommended by one manufacturer are hown in Table 5.

When shaving only one edge of a lank, shifting of the blank can be reuced by shaving the opposite edge as vell, even if not required for function.

Setups for Shaving. Shaving requires hat the blank be accurately located ver the die or the punch, because only few thousandths of an inch of metal s removed by the operation (see Table). Piloting pins, projecting from the unch, can engage holes in the blank to assure proper location. If the holes are not included in the original design, it may be permissible to add them for locating purposes only.

If adding holes is not permitted, a locating device such as that shown in Fig. 35 can be used. The clamping arms engage the blank at suitable nesting points. When the punch descends, the shaved blank falls through the die. The position of the clamping arms is fixed by the two stop pins.

Operation of this die can be improved by the use of a spring-loaded ejector and pressure pad within the die opening. As the punch ascends, the ejector lifts the shaved workpiece above the die face, thereby eliminating the fall through the die block, which may result in dents or other surface defects.

Table 5. Shaving Allowances Recommended by One Manufacturer

Blank thickness, in.		Allowance per side, in., for steel with Rockwell B hardness of: 50 to 66	75 to 90	90 to 105
First Shave (or a Single Shave)				
3/64 (0.047)	...	0.0025	0.003	0.004
1/16 (0.062)	...	0.003	0.004	0.005
5/64 (0.078)	...	0.0035	0.005	0.006-0.007
3/32 (0.094)	...	0.004	0.006	0.007-0.008
7/64 (0.109)	...	0.005	0.007	0.009-0.011
1/8 (0.125)	...	0.007	0.009	0.012-0.014
Second Shave (Add to First Shave)				
3/64 (0.047)	...	0.00125	0.0015	0.002
1/16 (0.062)	...	0.0015	0.002	0.0025
5/64 (0.078)	...	0.00175	0.0025	0.0030-0.0035
3/32 (0.094)	...	0.002	0.003	0.0035-0.0040
7/64 (0.109)	...	0.0025	0.0035	0.0045-0.0055
1/8 (0.125)	...	0.0035	0.0045	0.006-0.007

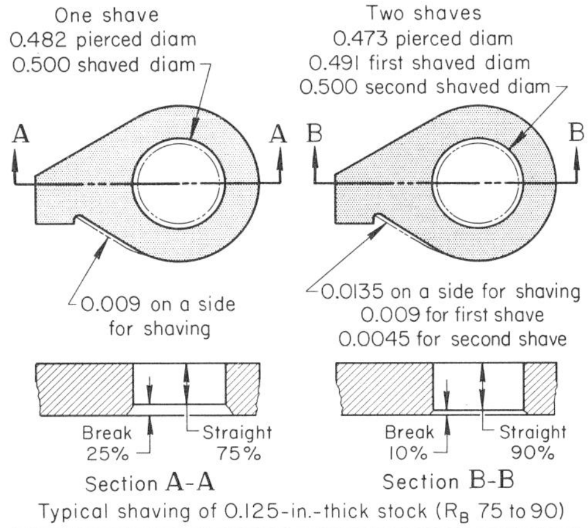

Burr Removal

The shape, height and roughness of burrs must be controlled to some degree in nearly all blanking operations. Complete elimination of burrs is not possible, but their formation can be minimized by the use of proper clearance between punch and die (page 33), and by good maintenance. (See the section on burrs in the article on Blanking and Piercing of Magnetically Soft Materials, page 66.)

Exposed burrs on the finished part can be unsafe and unsightly. Burrs on some blanked work can cause difficulties in forming and can increase the rate of workpiece breakage and die wear. Burrs can be removed by grinding, which generally removes the burr together with a portion of the work-hardened edge. Tumbling in a barrel is a common method of deburring small parts. Other deburring methods include chemical and electrolytic deburring, belt grinding, polishing and ultrasonic methods, as described in Volume 3 of this Handbook. Hand scrapers can be used to remove burrs from irregular shapes or soft metal parts.

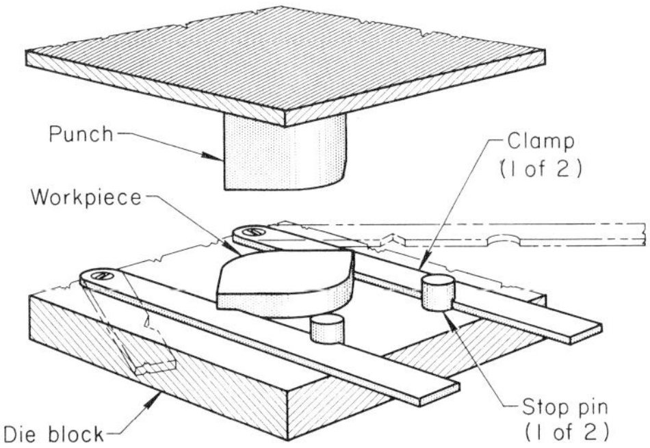

Fig. 35. Shaving die with a device for locating a blank with no holes for piloting

Blanking in Presses vs Alternative Methods

Alternative methods of making blanks include fine-edge blanking, milling, chemical blanking, contour band sawing, and gas cutting.

Fine-edge blanking (see the article that begins on page 56) is used primarily where die break is unacceptable and would require removal by subsequent shaving if conventional blanking were used. In fine-edge blanking there is no die break, and the entire wall surface of the cut is burnished.

Milling is applicable chiefly for cutting stacked parts, for short runs, and for making parts that are subject to frequent design change; it substitutes an inexpensive template for a conventional punch and die.

Chemical blanking may be competitive with blanking in presses for intricate parts that are only a few thousandths of an inch thick. On page 243 in Volume 3 of this Handbook, Table 3 shows a comparison of costs for making a small steel part from 0.008-in.-thick steel by these two methods. Combs for electric shavers, such as illustrated in Example 195 in the present volume, are more frequently made by chemical blanking of stainless steel than by the mechanical blanking method shown.

Contour band sawing and gas cutting may be competitive with blanking for stacked parts and thick material.

Safety

In all blanking operations, as in all press operations, there are hazards to operators, repairmen, and personnel in the vicinity. No press, die, or auxiliary equipment should be considered operable until these hazards are eliminated by installing necessary guards and other safety devices. The operator and all persons working around the blanking operation should be instructed in all precautions for safe operation before work is started. For safety information, see the article that begins on page 1 of this volume and the references listed at the end of that article.

DECIMALS OF A FOOT
For each 32nd of an inch

Inch	0	1	2	3	4	5	6	7	8	9	10	11
0	0	.0833	.1667	.2500	.3333	.4167	.5000	.5833	.6667	.7500	.8333	.9167
1/32	.0026	.0859	.1693	.2526	.3359	.4193	.5026	.5859	.6693	.7526	.8359	.9193
1/16	.0052	.0885	.1719	.2552	.3385	.4219	.5052	.5885	.6719	.7552	.8385	.9219
3/32	.0078	.0911	.1745	.2578	.3411	.4245	.5078	.5911	.6745	.7578	.8411	.9245
1/8	.0104	.0938	.1771	.2604	.3438	.4271	.5104	.5938	.6771	.7604	.8438	.9271
5/32	.0130	.0964	.1797	.2630	.3464	.4297	.5130	.5964	.6797	.7630	.8464	.9297
3/16	.0156	.0990	.1823	.2656	.3490	.4323	.5156	.5990	.6823	.7656	.8490	.9323
7/32	.0182	.1016	.1849	.2682	.3516	.4349	.5182	.6016	.6849	.7682	.8516	.9349
1/4	.0208	.1042	.1875	.2708	.3542	.4375	.5208	.6042	.6875	.7708	.8542	.9375
9/32	.0234	.1068	.1901	.2734	.3568	.4401	.5234	.6068	.6901	.7734	.8568	.9401
5/16	.0260	.1094	.1927	.2760	.3594	.4427	.5260	.6094	.6927	.7760	.8594	.9427
11/32	.0286	.1120	.1953	.2786	.3620	.4453	.5286	.6120	.6953	.7786	.8620	.9453
3/8	.0313	.1146	.1979	.2812	.3646	.4479	.5313	.6146	.6979	.7813	.8646	.9479
13/32	.0339	.1172	.2005	.2839	.3672	.4505	.5339	.6172	.7005	.7839	.8672	.9505
7/16	.0365	.1198	.2031	.2865	.3698	.4531	.5365	.6198	.7031	.7865	.8698	.9531
15/32	.0391	.1224	.2057	.2891	.3724	.4557	.5391	.6224	.7057	.7891	.8724	.9557
1/2	.0417	.1250	.2083	.2917	.3750	.4583	.5417	.6250	.7083	.7917	.8750	.9583
17/32	.0443	.1276	.2109	.2943	.3776	.4609	.5443	.6276	.7109	.7943	.8776	.9609
9/16	.0469	.1302	.2135	.2969	.3802	.4635	.5469	.6302	.7135	.7969	.8802	.9635
19/32	.0495	.1328	.2161	.2995	.3828	.4661	.5495	.6328	.7161	.7995	.8828	.9661
5/8	.0521	.1354	.2188	.3021	.3854	.4688	.5521	.6354	.7188	.8021	.8854	.9688
21/32	.0547	.1380	.2214	.3047	.3880	.4714	.5547	.6380	.7214	.8047	.8880	.9714
11/16	.0573	.1406	.2240	.3073	.3906	.4740	.5573	.6406	.7240	.8073	.8906	.9740
23/32	.0599	.1432	.2266	.3099	.3932	.4766	.5599	.6432	.7266	.8099	.8932	.9766
3/4	.0625	.1458	.2292	.3125	.3958	.4792	.5625	.6458	.7292	.8125	.8958	.9792
25/32	.0651	.1484	.2318	.3151	.3984	.4818	.5651	.6484	.7318	.8151	.8984	.9818
13/16	.0677	.1510	.2344	.3177	.4010	.4844	.5677	.6510	.7344	.8177	.9010	.9844
27/32	.0703	.1536	.2370	.3203	.4036	.4870	.5703	.6536	.7370	.8203	.9036	.9870
7/8	.0729	.1563	.2396	.3229	.4063	.4896	.5729	.6563	.7396	.8229	.9063	.9896
29/32	.0755	.1589	.2422	.3255	.4089	.4922	.5755	.6589	.7422	.8255	.9089	.9922
15/16	.0781	.1615	.2448	.3281	.4115	.4948	.5781	.6615	.7448	.8281	.9115	.9948
31/32	.0807	.1641	.2474	.3307	.4141	.4974	.5807	.6641	.7474	.8307	.9141	.9974

AMERICAN INSTITUTE OF STEEL CONSTRUCTION

The Selection of Material for Blanking and Piercing Dies

By the ASM Committee on Blanking Dies

THIS ARTICLE deals with the selection of material for punches and dies to blank, pierce and shave metallic and nonmetallic sheet and plate in a stamping press, at the lowest over-all production cost. Parts and examples are chosen to cover the ordinary ranges of size, shape, tolerance, work material, thickness, tooling and operations. The examples deal with categories of shapes and sizes that may determine or limit the selection. Recommendations are based on actual performance of tool materials in similar tooling circumstances.

All tool steels discussed here are assumed to be conventionally hardened and tempered to their highest usable hardness. No emphasis is given in this article to special techniques such as marquenching or the oil quenching of water-hardening types of tool steel.

Selection Tables

Recommendations for tool materials are summarized in a series of selection tables herewith. Normal good quality tool making, heat treating, grinding practice, tool design, tool maintenance, press equipment and manufacturing procedures are assumed in all the recommendations. For instance, commonly used clearances are as-

Table 1. Recommended Punch and Die Material for Blanking Part 1 and Similar ¾-In. Parts from 0.050-In. Sheet Materials

Material to be cut	Total quantity of parts to be made: 1,000	10,000	100,000	1,000,000	10,000,000
Aluminum, copper and magnesium alloys	Zn(a), CS(b), W1	W1, O1	O1, A2	D2	Carbide
Carbon and alloy steels, up to 0.70% C, and ferritic stainless steel	CS(b), W1	W1, O1	O1, A2	D2	Carbide
Stainless steel, austenitic, all tempers	W1	W1, A2	A2, D2	D4	Carbide
Spring steel, hardened, Rockwell C 52 max	A2	A2, D2	D2	D4	Carbide
Electrical sheet, 0.025-in., transformer grade	W1, A2	W1, A2	A2, D2	D4	Carbide
Paper, gaskets, and similar soft materials	W1(c)	W1(d)	W1(d), A2(e)	W1(d), A2(e)	D2
Plastic sheet, not reinforced	CS(b), W1	W1, O1	O1, A2	D2	Carbide
Plastic sheet, reinforced	O1(f), A2(g)	A2(g)	A2(g)	D2(g)	Carbide

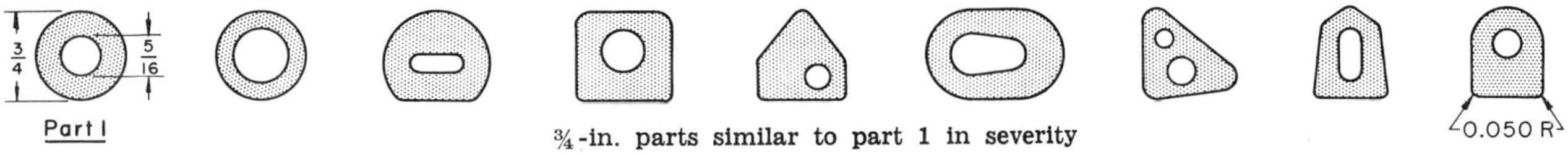

¾-in. parts similar to part 1 in severity

NOTE: Although carbide is recommended in this table only for 10 million pieces, it should usually be considered also for runs of 1 to 10 million pieces.

(a) Zn refers to zinc alloy plate for the die, sheared in, using a punch made of hardened tool steel. (b) CS refers to mild steel, such as 1018 or 1020, carburized to a depth of 0.010 to 0.020 in. or cyanided to a depth of 0.004 to 0.008 in. (c) For punching 1000 parts, the W1 punch and die would be left soft and the punch peened to compensate for wear if necessary. (d) The W1 punch may be left soft, or a hardened W1 punch and die may be selected. (e) A2 tool steel is preferred if compound tooling is to be used for quantities of 100,000 to 1,000,000. (f) This O1 punch may have to be cyanided to make even 1000 pieces. (g) The punch and die should be gas nitrided for 12 hr at 1000 to 1050 F.

Table 2. Recommended Punch and Die Material for Blanking Part 2 and Similar 3-In. Parts from 0.050-In. Sheet Materials

Material to be cut	Total quantity of parts to be made: 1,000	10,000	100,000	1,000,000	10,000,000
Aluminum, copper and magnesium alloys	Zn(a), O1, A2	O1, A2	O1, A2	D2	Carbide
Carbon and alloy steel, up to 0.70% C, and ferritic stainless steel	O1, A2	O1, A2	O1, A2	D2	Carbide
Stainless steel, austenitic, all tempers	O1, A2	O1, A2	A2, D2	D4	Carbide
Spring steel, hardened, Rockwell C 52 max	A2	A2, D2	D2	D4	Carbide
Electrical sheet, transformer grade, 0.025-in.	A2	A2, D2	A2, D2	D4	Carbide
Paper, gaskets, and similar soft materials	W1(b)	W1(b)	W1(c), A2(d)	W1(d), A2(d)	D2
Plastic sheet, not reinforced	O1	O1	O1, A2	D2	Carbide
Plastic sheet, reinforced	O1(e), A2	A2(f)	A2(f)	D2(f)	Carbide

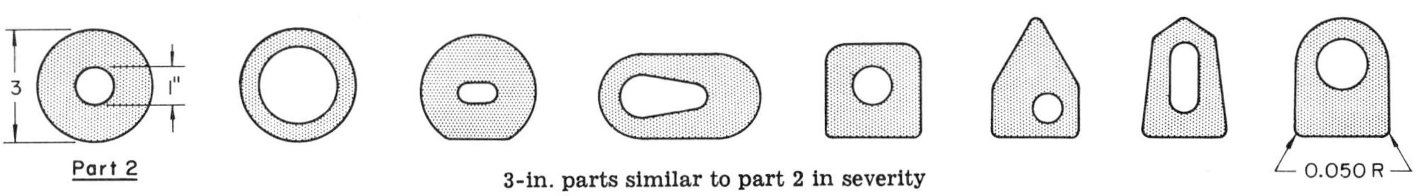

3-in. parts similar to part 2 in severity

NOTE: Although carbide is recommended in this table only for 10 million pieces, it should usually be considered also for runs of 1 to 10 million pieces.

(a) Zn refers to a die made of zinc alloy plate and a punch of hardened tool steel. (b) For punching up to 10,000 parts, the W1 punch and die would be left soft and the punch peened to compensate for wear if necessary. (c) For punching from 10,000 to 1,000,000 pieces, the W1 punch can be soft so that it can be peened to compensate for wear, or it can be hardened and ground to size. (d) Of the two alternatives listed, A2 tool steel is the preferable selection if compound tooling is to be used for quantities of 10,000 to 1,000,000. (e) This O1 punch may have to be cyanided 0.004 to 0.008 in. deep to make even 1000 pieces. (f) For the application indicated, the punch and die should be gas nitrided 12 hr at 1000 to 1050 F.

Table 3. Recommended Punch and Die Material for Blanking Part 3 and Similar 12-In. Parts from 0.050-In. Sheet Materials

Material to be cut	Total quantity of parts to be made: 1,000	10,000	100,000	1,000,000	10,000,000
Aluminum, copper and magnesium alloys	Zn(a), 4140(b)	4140(c), A2	A2	A2, D2	Carbide
Carbon and alloy steel, up to 0.70% C, and stainless steels up to quarter hard	4140(c), A2	4140(c), A2	A2	A2, D2	Carbide
Stainless steel, austenitic, over quarter hard	A2	A2, D2	D2	D2, D4	Carbide
Spring steel, hardened, Rockwell C 52 max	A2	A2, D2	D2	D2, D4	Carbide
Electrical sheet, transformer grade, 0.025 in.	A2	A2, D2	A2, D2	D2, D4	Carbide
Paper, gaskets, and similar soft materials	4140(d)	4140(d)	A2	A2	D2
Plastic sheet, not reinforced	4140(d)	4140(c), A2	A2	D2	Carbide
Plastic sheet, reinforced	A2(e)	A2(e)	D2(e)	D2(e)	Carbide

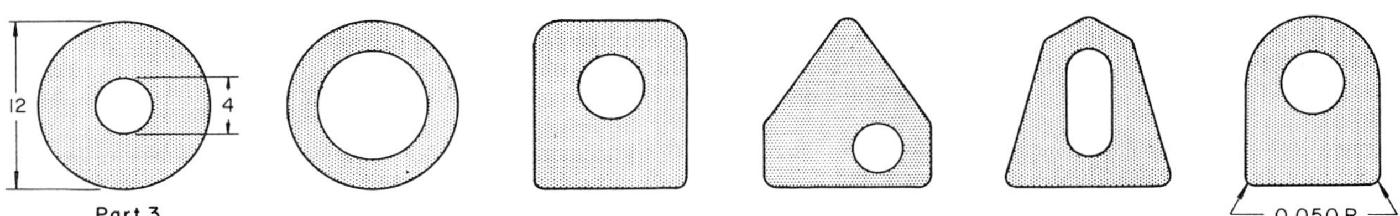

12-in. parts similar to part 3 in severity

NOTE: Although carbide is recommended in this table only for 10 million pieces, it should usually be considered also for runs of 1 to 10 million pieces.

(a) Zn refers to zinc alloy plate for the die, sheared in, using a hardened tool steel punch. (b) Soft. (c) Working edges of 4140 are flame hardened in this application. (d) 4140 may be soft or flame hardened. (e) For necessary wear resistance in the application indicated, the punch and die are gas nitrided 12 hr at 1000 to 1050 F.

sumed; small clearances used to produce minimum burr would decrease the tool life. Inferior practices such as off-center punches, erroneous heat treating, or poor lubrication of sheet stock will result in die life shorter than indicated by the tables.

It is understood that some types of work require more precise tool making than others, and this has been allowed for; however, unusually close tolerances have not been. In general, a tolerance of ±0.005 in. (not ±0.005 in. per in.) is assumed for all part dimensions. The tool materials shown in all tables are intended to provide tools of minimum cost per piece for the listed quantities of production. Where more than one tool material is shown for a given application, the one listed first will provide the least expensive dies.

This criterion of least cost has the effect of eliminating high speed steels from the selection tables. However, high speed steel could be selected instead of D2 in many places where D2 is indicated in the tables. The comparability of high speed steel and D2 is evident in Fig. 1, which shows data on punch wear and progressive increase in burr height for two high speed steels and three other punch materials. These data are for carefully controlled piercing of ¼-in. holes with punches ground after heat treatment. The work material was electrical grade silicon steel (0.052% C, 1.38% Si) from one heat, with a bright pickled surface.

Carbide has been shown at quantities of ten million but not one million in the selection tables. This is an arbitrary simplification and, as is pointed out in a later section of this article, carbide should usually be considered for runs of one million pieces.

Tables 1 to 3 deal with the selection of tool material for three sizes of simple parts, most of them disk-shaped. Size of die, kind of material and total quantity of production are the principal variables affecting the selection.

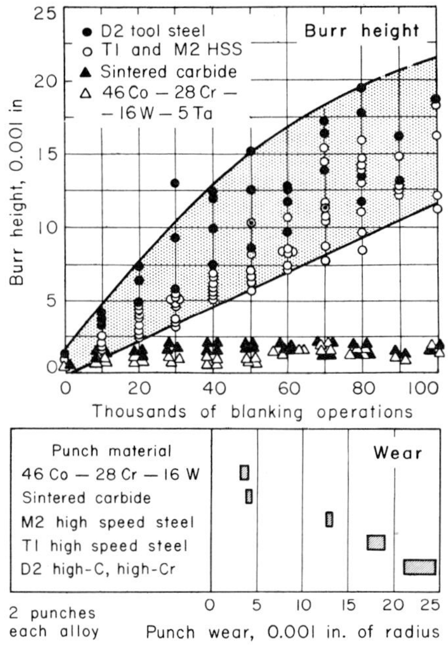

Fig. 1. Burr height and punch wear developed during piercing of ¼-in. holes in silicon steel strip (0.052% C, 1.38% Si) with bright pickled surface (A. T. Hamill, American Machinist, November 27, 1950, with additional data from the same author)

Table 1 recommends tool materials to blank and pierce the ¾-in. part 1 and other small parts illustrated. The selections apply to dies not ground after hardening except on the top and bottom faces. In this small size, the cost of tool material is such a small fraction of the total die cost that selection of a more expensive material will have little effect on cost. Speed of completing the tooling, total die life, life between grinds, and other considerations will dominate the selection.

The recommendations are intended for both punch and die. However, for making up to 100,000 pieces from the first two listed materials, a W1 stock perforator punch can be used, as discussed later.

W1 tool steel is shown in 14 out of 40 places in this table because the die is so small that size change from hardening is not a serious consideration and there is little danger of cracking.

The case depth of a carburized ⁵⁄₁₆-in. diam punch should not exceed 0.020 in.; however, it would probably be as cheap and quicker to make the small punch from W1 drill rod and cyanide it, using an oil quench for hardening.

Table 2 recommends material for punches and dies to make the simple 3-in. part 2 and similar parts four times as large in diameter as part 1, but the same in sheet thickness. Because of the larger size of the dies required for part 2, the change in dimensions during hardening is more important than for part 1; accordingly, water-hardening steels are not recommended, except for the blanking of paper gaskets, for which service the tools are not hardened. For shapes as simple as part 2, cracking of the dies is not a problem, so that W1 could be used as indicated for part 1, where tolerances are large enough — for instance, a tolerance of about twice the value of ± 0.005 in. assumed in this article.

Table 3 deals with tool material selection for punches and dies to make the 12-in. part 3 and similar shapes four times as large as part 2. Except where noted, the selections are for both punch and die. For this size tooling the selection of material is predominantly a matter of economics, because the cost of the tool

Table 4. Recommended Punch and Die Material for Blanking Part 4 and Similar 1½-In. Parts from 0.050-In. Sheet Materials

Material to be cut	Total quantity of parts to be made: 1,000	10,000	100,000	1,000,000	10,000,000
Aluminum, copper and magnesium alloys	Zn(a), A2(b), O1	A2(b), O1	O1, A2	D2	Carbide
Carbon and alloy steel, up to 0.30% C, and ferritic stainless steel	A2(b), O1	A2(b), O1	A2, O1	D2	Carbide
Carbon and alloy steel, 0.30 to 0.70% C	A2	A2	A2, D2	D2	Carbide
Stainless steel, austenitic, up to quarter hard	A2(b), O1	A2	A2, D2	D2	Carbide
Stainless steel, austenitic, over quarter hard	A2	A2, D2	D2	D4	Carbide
Spring steel, hardened, Rockwell C 52 max	A2	A2, D2	D2, D4	D4	Carbide
Electrical sheet, transformer grade, 0.025 in.	A2	A2, D2	A2, D2	D4	Carbide
Paper, gaskets, and similar soft materials	W1(c)	W1(c)	W1(d), A2	W1(d), A2	D2
Plastic sheet, not reinforced	O1, A2	O1, A2	A2	A2, D2	Carbide
Plastic sheet, reinforced	A2(e), D4	A2(e)	D2(e)	D4	Carbide

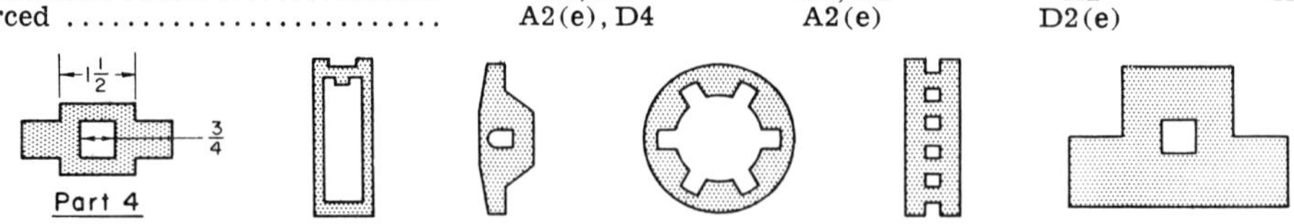

1½-in. parts similar to part 4 in severity

NOTE: Although carbide is recommended in this table only for 10 million pieces, it should usually be considered also for runs of 1 to 10 million pieces.

(a) Zn refers to zinc plate for the die, sheared in, using a hardened tool steel punch. (b) A2 is shown as preferred over O1 because some of the O1 dies probably would be cracked in conventional hardening. However, if the O1 dies were martempered, the cracking hazard would be eliminated and, neglecting the extra cost of martempering, O1 would be preferred. (c) For punching up to 10,000 parts, the W1 punch and die would be left soft and the punch peened to compensate for wear, if necessary. (d) For punching from 10,000 to 1,000,000 pieces, the W1 punch would be left soft so that it can be peened to compensate for wear, and the die would be either a solid die of A2 or a sectional die of W1. (e) For necessary resistance in the application indicated, the punch and die should be gas nitrided for a period of 12 hr at 1000 to 1050 F.

material may vary from 10% of the total tool cost, for 4140 and zinc tools, to more than 50%, for tool steels such as D2 and M2. The tool steels recommended in this table are all through-hardening forged steel rings, used as fitted inserts in a die holder or punch holder.

Table 4 shows recommended tool material for dies to blank and pierce part 4 and other moderately difficult parts, as illustrated. The sharp corners inside and outside on part 4 require tool steels that will not crack in quenching or chip in service. The greatest wear is localized at outside corners of the female die.

Alloy steel 4140 is not used for tools to make this part, because of the difficulty of flame hardening such a shape uniformly. If part tolerances were liberal, through-hardened 4140 could be used for small quantities.

Table 5 lists recommended tool materials to blank parts for which the punch or die may crack in heat treating. Part 5 and the others shown in the table are typical of such designs. In blanking plastic sheet, parts of the most intricate shapes illustrated would probably crack or split during the blanking operation, even if the sheet material and the tools were heated. The selections are suitable for parts up to about 3 in. maximum dimension.

Tool steels for dies to make part 5 must have high dimensional accuracy and safety in hardening. For this reason and because of its wear resistance, A2 is recommended in 19 of the 40 applications, although it is a little more expensive than the O1 oil-hardening steel shown, and is more difficult to machine. It is also satisfactory for integral one-piece dies because of its accuracy during air quenching or because of its ample wear resistance for notching-dies to run the same number of parts, even though four strokes per part would be required.

Carburized steel is not recommended on account of distortion in heat treatment and chipping in operation. Flame hardened dies are not recommended because of the problem of hardening intricate shapes uniformly. A zinc alloy die sheared in with a hardened A2 tool steel punch might be used for production of 1000 soft metal parts. If segmented dies are made, any tool steel with the wear properties to run the required quantity can be used.

Table 6 shows recommended blanking die materials for making part 6 and other 12-in. parts of the shapes illustrated. All tool steels recommended are air-hardening, to

Table 5. Recommended Punch and Die Material for Blanking Part 5 and Similar 3-In. Parts from 0.050-In. Sheet Materials

Material to be cut	Total quantity of parts to be made: 1,000(a)	10,000	100,000	1,000,000	10,000,000
Aluminum, copper and magnesium alloys	O1, A2	O1, A2	O1, A2	A2, D2	Carbide
Carbon and alloy steel, up to 0.70% C, and ferritic stainless steel	O1, A2	O1, A2	O1, A2	A2, D2	Carbide
Stainless steel, austenitic, all tempers	A2	A2, D2	A2, D2	D2, D4	Carbide
Spring steel, hardened, Rockwell C 52 max	A2	A2, D2	D2, D4	D2, D4	Carbide
Electrical sheet, transformer grade, 0.025 in.	A2	A2, D2	D2, D4	D2, D4	Carbide
Paper, gaskets, and other soft materials	W1(b)	W1(b)	W1(c), A2	W1(c), A2	D2
Plastic sheet, not reinforced	O1	O1	A2	A2, D2	Carbide
Plastic sheet, reinforced(d)	O1(e)	A2(f)	A2(f)	D2(f)	Carbide

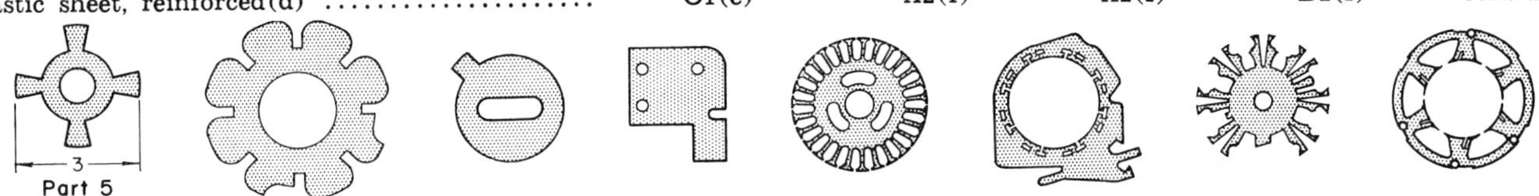

3-in. parts similar to part 5 in severity

NOTE: Although carbide is recommended in this table only for 10 million pieces, it should usually be considered also for runs of 1 to 10 million pieces.

(a) Tool steels given in this column are recommended for both notching dies and compound dies. (b) For punching up to 10,000 parts, the W1 punch and die would be left soft and the punch peened to compensate for wear if necessary. (c) For punching from 10,000 to 1,000,000 pieces, the punch would be W1, left soft so that it can be peened to compensate for wear, and the die would be O1, hardened. (d) Some of the parts among those shown above are susceptible to cracking during blanking. (e) Cyaniding of the punch is advisable, even for 1000 pieces. (f) For necessary resistance in the application indicated, both punch and die should be gas nitrided for a period of 12 hr at 1000 to 1050 F.

Table 6. Recommended Punch and Die Materials for Blanking Parts 6, 7 and Similar 12-In. Parts from 0.050-In. Sheet Materials

Material to be cut	Total quantity of parts to be made: 1,000(a)	10,000	100,000	1,000,000	10,000,000
Aluminum, copper and magnesium alloys	A2	A2	A2, D2	A2, D2	Carbide
Carbon and alloy steel, up to 0.70% C, and ferritic stainless steel	A2	A2	A2, D2	A2, D2	Carbide
Stainless steel, austenitic, up to quarter hard	A2	A2	A2, D2	D2, D4	Carbide
Stainless steel, austenitic, over quarter hard	A2	D2	D2	D2, D4	Carbide
Spring steel, hardened, Rockwell C 52 max	A2	A2, D2	D2	D2, D4	Carbide
Electrical sheet, transformer grade, 0.025 in.	A2	A2, D2	D2	D2, D4	Carbide
Paper, gaskets, and other soft materials	W1(b)	W1(b)	W1(c)	W1, A2	D2
Plastic sheet, not reinforced	A2	A2	A2	A2, D2	Carbide
Plastic sheet, reinforced(d)	A2(e)	A2(e)	D2(e)	D2(e)	Carbide

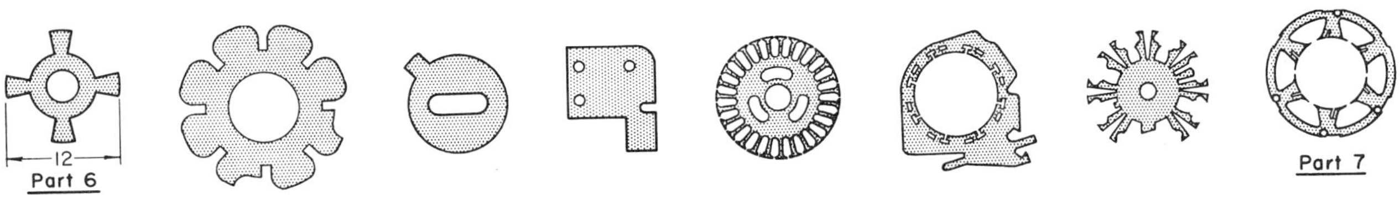

12-in. parts similar to part 6 in severity

NOTE: Although carbide is recommended in this table only for 10 million pieces, it should usually be considered also for runs of 1 to 10 million pieces.

(a) The recommendations in this column are for either notching dies or compound dies. (b) For punching up to 10,000 parts, the W1 punch and die would be left soft and the punch peened to compensate for wear if necessary. (c) For punching from 10,000 to 1,000,000 pieces, the W1 die would be hardened and the W1 punch would be soft, so that it can be peened to compensate for wear. (d) Some of the reinforced plastic parts shown in the illustrations might crack during the blanking operation. (e) The punch and die should be gas nitrided 12 hr at 1000 to 1050 F for punching any quantity of reinforced plastic.

minimize distortion or the danger of cracking, except in the few applications where the material is not hardened, or where it is used in simple shapes in a sectional die. A zinc die and A2 punch would be economical for blanking and piercing 1000 pieces from aluminum.

The recommendation of A2 for 1000 pieces may seem extravagant. However, the nondeforming qualities of A2 are necessary in a compound die and good wear characteristics are desirable in a notching die because of the additional number of hits necessary.

Part Tolerances

Table 7 shows how recommended selections are affected by dimensional part tolerances. The selections are based on considerations of expected distortion of the dies during heat treating, assuming that only the top and bottom faces would be ground after hardening.

Dimensional tolerance sometimes is the crucial factor in deciding between two possible steels or their exact heat treatment. The bar graph of Fig. 2 indicates relative size change for 18 combinations of steel and quenching temperature. Additional data on distortion in tool steels are given on pages 645 and 654 in this Handbook.

Part Shape

The recommendations in all the selection tables include consideration of quench cracking during heat treatment. For instance, in part 8 of Table 7, steel O1* is sufficiently resistant to quench cracking and also will meet most of the listed tolerances after heat treating. However, if the (3/20)A dimension (0.075 in. for the ½-in. part) were to become (1/10)A instead, the safety of O1 in hardening would become questionable, and A2 would be the preferred steel. Radii at corners are important; a sharp corner will usually demand an air-hardening die and, regardless of the steel selected, die life will be only about half that for the same part with a 1/16-in. radius.

Sheet Thickness

Table 8 shows the change in recommended punch steels as the thickness of the steel being pierced is increased. These changes depend on compromises between chipping and wear resistance of the punch, in addition to the increase in wear with increased thickness. To pierce quantities as low as 10,000 pieces from thick plate (0.250 to 1-in. thick), the shock-resisting steels such as S1 and S5 have adequate resistance to wear and adequate strength to resist breakage. It is improbable that 100,000 piercings

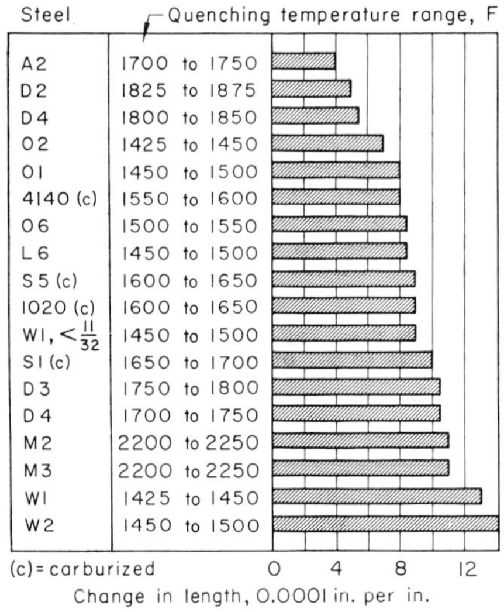

Fig. 2. Length change in the quenching of steels used for blanking dies. These values may be used to estimate differences among steels but not to predict the magnitude of size change of a particular shape of a given steel.

could be accomplished in ½-in. plate with one punch. M2 punches would average about 1500 pieces per grind. Punches made of M3 or M4 high speed steel, liquid nitrided, would probably do better.

Punch life in piercing thick plates is unpredictable because punch breakage, rather than wear, is the most common form of failure. An M2 punch, conventionally hardened, would probably fail to survive 100,000 piercings of 1-in. plate. However, liquid nitrided M2 or M3 would have a good chance of achieving this quantity.

Sheet Material

Table 9 lists sheet metals that cause die wear at about equal rates for the same part. Thus, die material recommended in Tables 1 to 6 for blanking the materials indicated therein should also be suitable for blanking similar parts from the other sheet materials listed as similar in Table 9.

The least wear on the dies is experienced in blanking leaded alpha brasses (1% Pb or more). Unleaded brasses or those with an alpha-beta structure impose more wear on the tools.

The nonmetallic sheet materials, omitted from Table 9, are covered in Tables 10 and 11, which recommend punch and die steels for blanking part 1 from various natural and synthetic materials of the thickness indicated. In selection of tool materials it should be remembered that the life of a given die material decreases with increasing sheet thickness. The materials and shapes shown in Tables 10 and 11 are frequently cut on a printing press using the knife-edge type of steel rule dies. A newer proprietary process employs a bevel-edged steel rule made from a modified O2 tool steel at about Rockwell C 55 for blanking steel up to 0.250 in. thick.

Tool Wear in Piercing

In a recent investigation of tool wear in piercing, it was determined that wear is significantly influenced by the amount of slip between the sheet material and the faces of the punch and die.* It was found that the amount of slip varied with (*a*) the roughness and tensile strength of the sheet, (*b*) the roughness of the tool surfaces, and (*c*) lubrication. Based on measurements of slip distance, it was demonstrated that a rough-faced tool lowered slip dis-

*O. and W. Kienzle, "Tool Wear in the Cutting of Thin-Gage Steel Sheets", Stahl und Eisen, June 12, 1958

Table 7. Recommended Die Steel for Blanking Steel Parts to Different Tolerances
(For 100,000 to 1 million pieces)

Tolerance on dimension A, ±, in.	Recommended die material for — Part 8	Part 9	Part 10
Dimension A, ½ In.; Sheet Thickness, 0.020 In.			
0.008	W1	O1	A2, D2
0.003	W1	O1	A2, D2
0.001	O1	O1	A2, D2
Dimension A, 2 In.; Sheet Thickness, 0.040 In.			
0.008	W1	W1	A2, D2
0.003	O1	O1	A2, D2
0.001	A2	A2	A2, D2
Dimension A, 8 In.; Sheet Thickness, 0.060 In.			
0.015	O1	O1	A2, D2
0.008	A2	O1	A2, D2
0.003	A2	A2	A2, D2
0.001	A2	A2	A2, D2

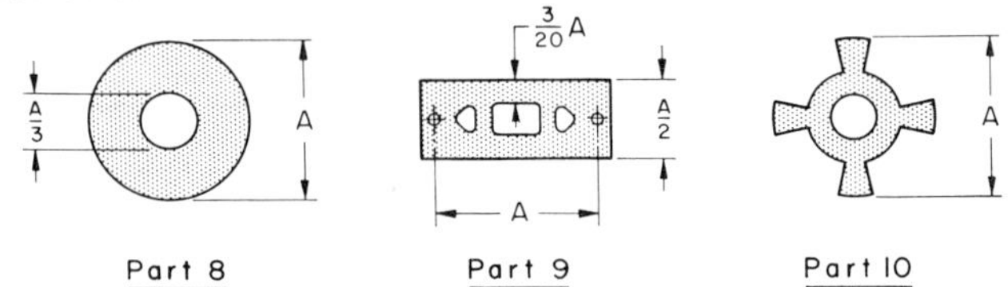

Where O1 is shown, the recommendation is based on its better machinability and grindability than A2, rather than on the cost advantage of $0.04 per lb.

*Where O1 or O2 is mentioned in this article, O6 may be substituted, with some increase in wear resistance and slightly more distortion during heat treatment.

tance and reduced the slip zone from that obtained with a smooth, polished tool.

To account for this reduction in slip distance as a function of tool surface roughness, the investigators reviewed each step involved in the shearing process. They found that the initial stage in piercing entails pure deformation alone — that is, deformation without separation or cracking. This is followed by gradual separation and further deformation in the cutting zone. Cracking occurs at the under side of the sheet (the side not in contact with the punch) because it is subjected to the greatest tensile stress from the stretching of the outside fibers. On the under side, only material from the slug is available for stretching in a radial direction. On the top side of the sheet, material surrounding the slug is free to flow inward, thus reducing localized stresses.

Burr Formation. Punch and die wear alter this normal cutting process. As cutting edges become rounded, the outside fibers of the

Table 8. Recommended Punch Steel for Piercing 3-In. Holes in Steel Sheet and Plate (Rockwell B 75 to 85) of Seven Different Thicknesses

Thickness of steel, in.	Total quantity of parts to be pierced: 1,000	10,000	100,000	1,000,000
0.010	W1	W1	W1	W1, A2
0.031	W1	W1	W1	D2
0.062	W1	W1	W1	D2
0.125	W2	W2	W2, A2	D2
0.250	S5(a)	S5(a)	A2, S1(a)	D2, M2
0.500	S5(a), S1(a)	S5(a), M2	M2	D2(b), M2(b)
1.000	S1(a)	S1(b), M2	M2(b)	....

(a) Shock-resisting steels carburized to obtain 0.70 to 0.75% C to a depth of 0.010 to 0.020 in. (b) It is doubtful whether this quantity can be made with one punch, even with successive regrindings.

Table 9. Classification of Sheet Metals That Require Similar Die Materials

Sheet metals for which blanking die materials are shown in Tables 1 to 6	Other sheet metals for which the same blanking die materials are recommended
Aluminum, copper and magnesium; ferritic steels to 0.30% C; ferritic stainless steel	Zinc, tin, lead and noble metal alloys; SAE 5050 alloy steel; beryllium copper, annealed
Carbon and alloy steel, up to 0.070% C	2024-T4; 10 and 18% nickel silver, alloys A and B, hard and spring tempers; phosphor bronze, grades A and C, spring and extra spring tempers; type 420 stainless steel, annealed; Monel, hard tempers; K-Monel, annealed
Stainless steel, austenitic, up to quarter hard	Mumetal, annealed; 4750 alloy (50% Ni); 45% Permalloy; beryllium copper, condition A; nickel-palladium bimetal; 17-7 PH stainless steel, annealed
Stainless steel, austenitic, over quarter hard	Beryllium copper B194, all tempers (acid cleaned); type 410 stainless steel, heat treated; 4-79 Moly Permalloy
Spring steel, Rockwell C 52 max	Beryllium copper, half hard and harder; vanadium Permendur; 17-7 PH stainless steel, hardened
Electrical sheet, transformer grades(a)	M-22 dynamo grades, HR and CR; AISI transformer grades as follows: M-6, grade 66, CR; M-7, grade 73, CR; M-8, grade 80, CR; M-17, grade 65, HR, PQ, and grades 58 and 65, HR; and M-19, grade 72, HR and CR, and grade 58 HR, PQ

(a) Other AISI types give up to ten times the die life, depending approximately on silicon content and hardness. These include field grade, CR, 0.25% Si; M-27 motor grade, HR and CR, 2.80% Si; M-36 electrical grade, HR, 1.30% Si, and CR, 1.60% Si; and M-43 armature grade, HR, 0.70% Si. CR means cold rolled; HR, hot rolled; and PQ, punching quality.

Table 10. Recommended Punch and Die Materials for Blanking Part 1 from Various Natural and Synthetic Nonmetallic Materials(a)

Material to be blanked	Total quantity of parts to be blanked: 1,000	10,000	100,000	1,000,000
Varnished cotton cloth, cambric and nylon fiber cloth; untreated cotton fabric, nylon, silk, dacron, rayon, orlon and linen fabric; canvas, felt and wool, up to 0.020 in. thick	1018(b), W1(c)	1018(b), W1(c)	W1(d)	W1(d)
Fish paper, rag paper and wood-pulp paper, up to 0.031 in. thick	1018(b), W1(c)	1018(b), W1(c)	W1, A2	W1, A2
Nonrigid plastic film and rigid plastic sheets of all types, up to 0.031 in. thick	1018(b), W1	W1, A2	W1, A2	A2, D2
Pressboard, hard fiber, vulcanized fiber, cork and rubber, up to 0.031 in. thick	W1	W1	W1, A2	W1, A2
Untreated asbestos fabric, plain or reinforced, up to 0.020 in. thick	W1(d)	W1, A2	W1, A2	A2, D2
Glass cloth, varnished, and silicone, up to 0.031 in. thick		A2, D2	A2, D2	D2, D4

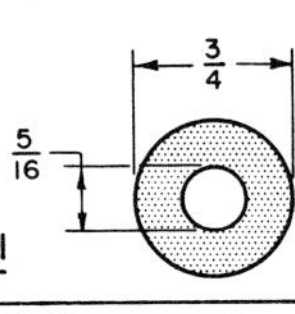

Part 1

(a) Parts of intricate shape to be made from flexible or rigid nonabrasive materials are frequently cut with a knife-edge steel-rule die made of water-hardening tool steel. (b) Carburized mild steel such as 1018 or 1020. (c) For punching up to 10,000 parts,, the W1 punch and die would be left soft and the punch peened to compensate for wear as necessary. (d) The W1 punch may be left soft and peened to compensate for wear, or it may be hardened and used in the conventional manner.

Table 11. Recommended Punch and Die Materials for Blanking Part 1 from Heated, Laminated or Reinforced Plastic Sheet 0.050 In. Thick

Material to be blanked	Total quantity of parts to be made: 1,000	10,000	100,000	1,000,000
Phenolic, paper-base and fabric-base	O1, A2	A2, D2	A2, D2	D2, D4
Phenolic, nylon-base	O1, A2	A2, D2	A2(a), D4	D2(a), D4
Melamine, glass-base; silicone, glass-base; polyester, glass mat and pasted mica, flexible or rigid	A2(a), D2(a)	A2(a), D2(a)	D2(a), D4	D2(a), D4
Silicone or organic varnish, glass-mica-glass, mica-glass or paper-mica-glass	A2(a), D2(a)	D2(a), D4	D2(a), D4	D2(a), D4(b)

(a) The punch and die should be gas nitrided 12 hr. (b) Sintered carbides may be used for this quantity of the most abrasive materials such as the glass-filled sheet.

Part 1 same as in Table 10 above

material being cut develop a lower tensile stress, and the pressure counter to the grain direction is increased. As a result, the amount of cold working and work hardening increases, and separation or cracking is delayed. When a break does occur, it nucleates from both sides of the sheet simultaneously, and a burr develops on both the slug and the surrounding area of sheet from which it was cut.

Burr formation is not the direct result of the rounding of cutting edges, since even worn edges remain fairly sharp. The formation of a burr is much more dependent on the work hardening that results from cold working. Since the ability of the material to deform is greatest at the location of highest pressure, tearing does not occur at the punch edge but at a certain distance from it. This accounts for the formation of a burr, the size of which increases with increasing tool wear.

Tool Roughness. Edge rounding is not the only wear to which piercing tools are subjected. There is also slip wear on the faces and on the side surface of the punch. For this reason, tool wear can be reduced by increasing the roughness of the tool face and thereby lessening the amount of slip. The lapping or polishing of punches is both superfluous and detrimental to the life of the tool. The roughness of a conventionally ground surface (80 to 120 micro-in. finish) provides satisfactory tool performance.

In the piercing tests performed to evaluate tool wear, O2 (2% Mn) and D3 tool steels were used; both were hardened and tempered to Rockwell C 61 and were run without lubricant. Clearance angle on punches, die clearance, and surface roughness of the tools were held constant for all tests. Punch press speed (60 strokes per minute), stroke setting, striking velocity, punch and die design, and automatic feed apparatus were also standardized for all tests.

Tool wear for punches and dies made from the two tool steels is compared in Fig. 3, and is related to the three types of sheet pierced. With the low-alloy tool steel, O2 (2% Mn), the wear on the dies was from 9 to 46% greater than the wear on the punches. With the D3 tool steel, however, the wear on dies and punches was almost equal in most applications, and punch wear sometimes slightly exceeded die wear.

Tool Durability. The cutting durability of the D3 tool steel was about twice as great with electrical sheet, and about 7 to 10 times as great with stainless steel sheet, as that of the low-alloy tool steel. The ratios show a greater variation with carbon steel sheet. Carbon steel sheet produces the least face wear, stainless sheet produces the most, and electrical sheet falls somewhere between the two extremes. The presence of aluminum in the electrical sheets that were aluminum killed

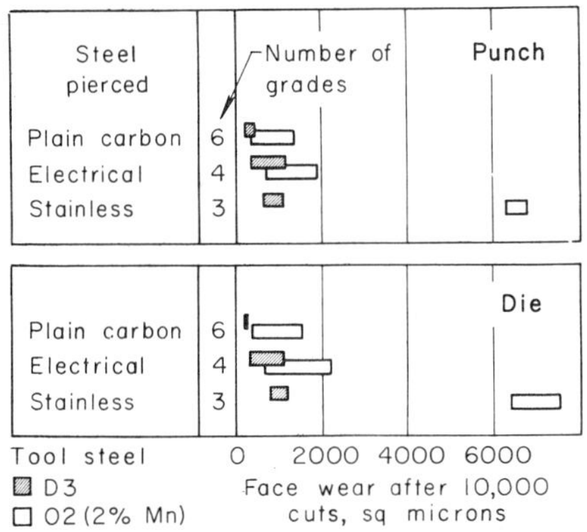

Fig. 3. Face wear on punches and dies made of O2 and D3 tool steels after 10,000 cuts. Holes were approximately ¼-in. diam; sheet thickness was 0.039 in. for plain carbon and stainless steels. Thickness of electrical sheet ranged from 0.014 to 0.020 in. (O. and W. Kienzle, Stahl u Eisen, Je 12, 1958)

showed no significant effect on tool wear.

A total of 13 grades of steel were included in the three types. The six grades of carbon steel sheet tested contained from 0.06 to 0.11% C and from 0.31 to 0.47% Mn. Carbon content in the electrical sheets ranged from 0.04 to 0.13%; the range in silicon content was 0.59 to 4.38%. The stainless grades comprised a standard type 302, a lower-carbon (0.06%) type 302, and type 321 containing 0.06% C.

In another series of tests connected with the same program, the influence of tool hardness on face wear was evaluated for the same low-alloy tool steel at hardness levels of Rockwell C 59 to 61. It was shown that face wear depends markedly on tool hardness and that an increase in hardness can result in a 25% decrease in face wear. An increase in impact speed from 0.26 to 0.65 ft per sec *decreased* tool wear by about 8%. This is a significant indication that the wear of tool steel blanking tools does not increase with higher impact speeds.

Burr Measurements. Depending on quality requirements, burr height is normally limited to a predetermined maximum. In these tests, maximum burr height was measured on five slugs and five holes after every 5000 cuts. Figure 4 summarizes the results for both the slugs and the perforated sheet. The burr height increased proportionately with the number of cuts, up to 50,000. From 50,000 to 75,000 cuts, burr growth proceeded at a slower rate. From 75,000 to 85,000 cuts, the rate of burr growth increased, attaining the initial rate of growth. The rate continued to increase beyond this point. The height of the burr on the slug was always a few microns greater than that on the perforated sheet.

Based on these test results, the investigators proposed a procedure for evaluating the life of various tool materials and the punchability of various grades of sheet material. Because burr size is more dependent on die clearance than on tool wear, they concluded that measurement of tool wear was a more accurate indicator of both tool life and punchability. Although face, side, and longitudinal edge wear are all available for measurement, the authors determined that measurements of face wear provided the most dependable reference because the cross section of the volume worn away is directly related to the number of cuts. In addition, measurements of worn cross section are independent of the length of cutting line and the diameter of the punch. A test procedure based on face-wear measurements was proposed.

Die Life Examples

In general, a life of about 50,000 parts per grind would be expected from W1 or O1 dies in running 0.062-in. steel parts and about 100,000 per grind for brass parts where burr height is held between 0.003 and 0.006 in. On the same basis (steel parts), A2 dies will run 100,000, D2 dies double that, and sintered carbides from 5 to 10 times as much as D2.

Die life between grinds where 0.015-in. cold rolled steel was blanked with a 0.002-in. permissible burr, in an actual case, was about 200,000 pieces with D2 dies, and 700,000 with carbide dies. In another case, 100,000 steel parts 0.125-in. thick are blanked between grinds on a D2 die when the burr height is over 0.005 in.

In blanking smoothly contoured parts from 0.110 to 0.120-in. cold rolled steel with an A2 die that gives a burr height of 0.002 in. at the start, production between grinds has averaged:

Burr height	Press strokes
0.003 in.	5,000 to 10,000
0.005	20,000 to 25,000

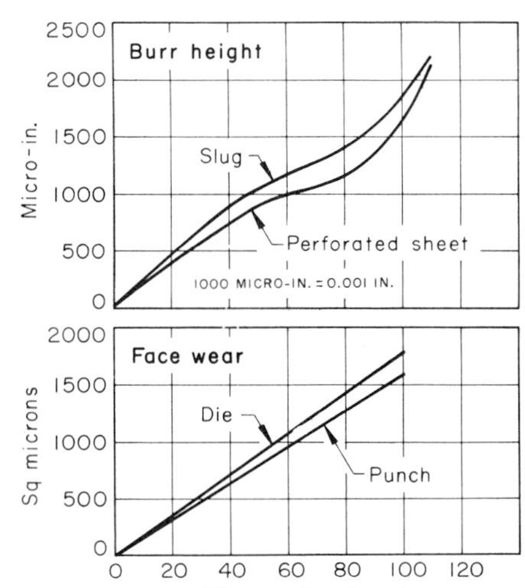

Fig. 4. Adjusted curves correlating burr height and face wear in piercing carbon steel sheet with D3 tools. Burr height was measured on five slugs and five holes after every 5000 cuts. Holes were approximately ¼-in. diam; sheet thickness was 0.039 in. (Source: same as that for Fig. 3)

Cold rolled spring steel parts 0.010 in. thick which produce a 0.001-in. burr at the beginning of production on D4 dies have developed average burr heights as shown below:

Burr height	Press strokes
0.002 in.	10,000 to 15,000
0.003	65,000 to 85,000

Brass parts approximately 0.080 in. thick were blanked and pierced on a D2 die and have produced the following results:

Burr height	Press strokes
0.001 in.	40,000 to 60,000
0.002	80,000 to 120,000

Type 302, quarter-hard stainless steel parts 0.060 in. thick that start off production on high speed steel dies with a 0.001-in. burr have averaged the following production between grinds:

Burr height	Press strokes
0.002 in.	1,000 to 3,000
0.003	4,000 to 6,000
0.005	8,000 to 12,000

Parts with a burr limit of 0.002 in. made from annealed 301 stainless steel 0.080 in. thick in an actual application ran an average of 20,000 pieces between grinds on dies made of D2 and 110,000 on sintered carbide dies.

The production of 1,000,000 pieces from electrical sheet blanked with burr heights from 0.003 to 0.005 in. would require a thickness allowance of about 3⁄16 in. for regrinding dies made of D2 or D3.

Life between grinds in blanking and piercing 0.025-in. electrical laminations from steel having a silicon content between 2 and 4½% may be estimated as follows:

Burr height	Press strokes: D2 steel	Press strokes: Carbide
0.001 in.	25,000	200,000
0.002	60,000	500,000

In another set of data, the total life of eight D2 dies punching motor grade electrical sheet (2.80% Si) averaged 5,500,000 and of twenty D3 dies punching armature grade (0.70% Si) averaged 6,990,000, both based on an allowance of 1⅜ in. to be ground off the dies during their lifetime. In a similar application, and confirmed in other unrelated tests, T1 dies averaged about 5,800,000 total production under the same conditions.

In controlled tests punching electrical grade lamination stock, dies made of D4 and M2 had about equal life, and at the same burr height exceeded dies of D3 by about 23% and dies of D2 by about 28%. In turn, dies made of carburized M2 had about 35% longer life at the same burr height than both D4 and M2 with standard treatments.

Quantity

In small dies, particularly those smaller than 6 in., the cost of even D2 tool steel is only about 10% of the die cost, and the risk of wearing out and the necessity for rebuilding the tools are more important than savings in the cost of tool steel.

In dies larger than about 12 in., the cost of tool steel is often more than 50% of the total die cost. Material selection is important, whether the quantity is large or small. If the quantity is small, flame hardened 4140 or a die of zinc alloy and a punch of W1, O1 or A2 may be selected, as indicated in the tables. Frequently, a die can be built up from a combination of specially made sections and standard purchased punches and bushings, in order to economize on the cost of material or fabrication, or both.

Kind of Tooling

Table 12 shows how recommendations of tool steels may depend on the kind of tooling employed.

In general, short-run or single-operation dies are used for small-volume production where the cost of tooling must be kept low because it is the major cost factor, and the cheapest tool materials are used. Compound dies are used for medium to large-volume production where intricacy is not the dominant problem and where accuracy often is. Progressive dies are used for medium-volume, and particularly for large-volume, production where

Table 12. Recommended Tool Steels for Making Parts 1 and 6 of 0.050-In. Sheet Steel on Four Kinds of Tooling

Part	Total quantity of parts to be made: 1,000	10,000	100,000	1,000,000
Short Run				
Part 1 (2 operations, 2 dies) ...	W1(a)	....	....	....
Part 6 (6 operations, 3 dies) ...	W1(b)	....	....	....
Single Operation				
Part 1 (2 operations, 2 dies) ...	W1	W1	O1	....
Part 6 (3 operations, 3 dies) ...	O1	O1	O1, A2	....
Compound Tooling				
Part 1 (1 operation 1 die)	A2(c)	A2	A2	D2
Part 6 (1 operation, 1 die)	A2(c)	A2	A2	D2
Progressive Tooling				
Part 1 (2 stations, 1 die)	W1(d)(e)	W1(e)	W1(e), O1	A2(f), D2
Part 6 (2 stations, 1 die)	O1(d)	O1	O1, A2	A2, D2

(a) W1 would be used for making 1000 pieces, especially where this die is not intended for use on future jobs or where stock punches and dies are not available, and for holes of nonstandard size. Stock punches and dies intended for repeated use on successive jobs should be made of D2.
(b) The ID and OD are made on different dies, and the piece is then notched in four operations. All the tools except the notching die are simple enough in shape to be made of W1. The notching die should be made of O1 to avoid the hazard of cracking in quenching.
(c) It is assumed that compound tooling would be used to make 1000 parts only to attain special accuracy. Therefore, the material should be A2, because it distorts little in heat treating and its additional cost, compared with O1, is a small fraction of the total tooling cost.
(d) Progressive tooling may sometimes be used for small quantities where cork or rubber-pad strippers are employed, because stripping requirements are less than for compound tooling.
(e) The use of W1 may not meet the requirements of accurate registry between stations unless used in sections or as inserts. In such applications, O1 or A2 should be used.
(f) D2 is the preferred steel when maximum number of pieces between grinds is desired.

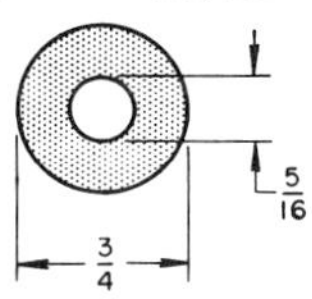

Part 1

Part 6

Table 13. Recommended Material for Perforator Punches

Material to be punched	Total quantity of parts to be punched: 10,000	100,000	1,000,000
Punch Diameters up to ¼-In.			
Aluminum, brass, carbon steel, paper and plastics	M2	M2	M2
Spring steel, stainless steel, electrical sheet and reinforced plastics	M2	M2	M2
Punch Diameters over ¼-In.			
Aluminum, brass, carbon steel, paper and plastics	W1	W1	D2
Spring steel, stainless steel, electrical sheet and reinforced plastics	M2	M2	M2

Schematic illustration of punch

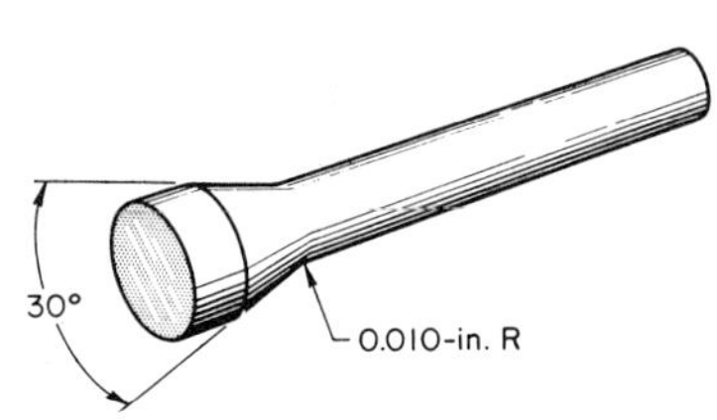

the die is preferably not made as intricately as the part.

In adjacent and simultaneous blanking operations where the die sections are thin or intricate, type A2 tool steel is preferred for runs up to 100,000 parts of most materials. M2, D2, D3 or carbide is preferred for longer runs.

Die Components

Perforator punches of the type illustrated in Table 13 may be made of the materials listed in that table.

The usual limiting slenderness ratio of punch diameter to sheet thickness for aluminum, brass and steel is 2.5-to-1 for unguided punches and 1-to-1 for guided punches. The limiting slenderness ratio of punch diameter to sheet thickness for piercing spring steel and stainless steel is a range from 3-to-1 to 1.5-to-1 for unguided punches, and a range from 1-to-1 to 0.5-to-1 for accurately guided punches.

Where these usual limits are exceeded and breakage cannot be eliminated by stepping the punches, tool steels such as O1, A2 and M2 are used. W1 is used if the diameter is greater than 3/8 in. but less than 3/4 or 1 in. The recommended hardnesses for W1, D2 and M2 perforator punches are given in Fig. 5.

Perforator Bushings. Table 14 shows recommended materials for perforator bushings of all three types (punch holder, guide or stripper, and perforator or die), particularly bushings of the precision type — for instance, where the outside diameter is ground to minus zero, plus 0.0003 in., concentric with the inside diameter within 0.0002-in. total indicator reading. The hardness of the W1 bushing should be Rockwell C 62 to 64, and the D2 bushings should be Rockwell C 61 to 63.

Die plates and die parts that hold inserts are made of Class 50 gray iron, alloy steel or (for heavy work) tool steel, and of cast iron or low-carbon wrought steel for blanking and piercing soft and thin materials.

For stamping thick sheets or hard materials, either gray iron of 40,000 to 60,000 psi tensile strength or 4140 treated to Rockwell C 30 to 40 should be used. Particularly on heavy-gage or hard material and on long runs for which inserts are pressed in, steels like 4340 or H11 are used; when inserts are screwed into the die plate, 4340 is nearly always used.

Die plates for stamping thin or soft sheets may be made of gray iron of 20,000 to 40,000 psi tensile strength, or mild steel.

Punch holders and die shoes for carbide dies are of high-strength gray iron or mild steel plate. Yokes retaining carbide sections are usually made of O1, hardened to Rockwell C 55 to 60.

Backup plates for carbides are preferably made of O1, hardened to Rockwell C 48 to 52.

Stripper plates can ordinarily be made of some low-carbon or medium-carbon steel like 1020 or 1035. Where a hardened plate is used for medium-production work, the preferable steels are 4140 flame hardened, W1 conventionally hardened or, for intricate shapes, W1 cyanided and oil quenched. For carbide dies and high-production D2 or D4 dies, hardened strippers are of O1 or A2, hardened to Rockwell C 50 to 54.

Guides and locator pins can be made from W1 or W2 for most dies, or from alloy steels such as 4140 for short-run low-cost dies. Many commercial guide pins are made from SAE 1117, carburized, hardened, and finished to a surface roughness of 15 micro-in.

Combined operations like blank-and-draw or blank-and-extrude give rise to selection problems best solved by determining which of the operations is the more severe, and selecting for that operation. The applicable article on material selection in this section of the Metals Handbook may be referred to in selecting steel of the required wear resistance in each instance. Selection of material for punch-and-extrude sections of dies, not dealt with elsewhere in this Handbook, should follow the recommendations of this article.

Wear of extrusion or embossing dies can be offset by nitriding A2 and D2 materials. However, nitriding may shorten the life of blanking tools because of chipping of the edges, unless the treatment is confined to dies used for thin and soft sheet.

Table 15 lists the recommended punch and die materials for shaving one thickness (0.050 in.) of various materials, including nonferrous alloys, carbon and alloy steels, and stainless steels. For thicker sheets, the next best material should be selected or shorter life accepted.

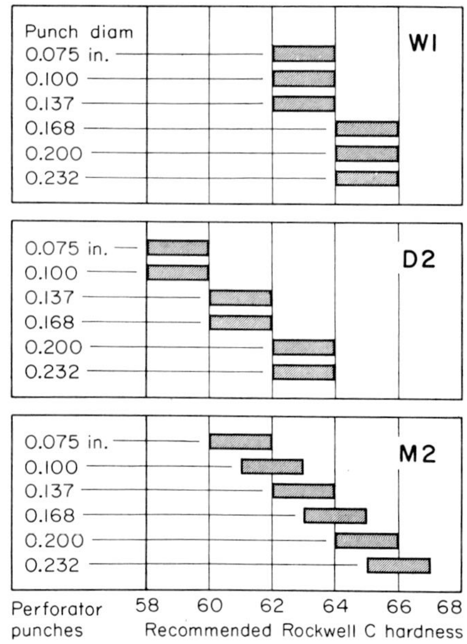

Fig. 5. Recommended hardness for tool steel perforator punches. Punches should be tempered to Rockwell C 56 to 60 when they are subjected to heavy shock or used to pierce thick material.

Cost

Figure 6 compares the relative cost of tools to make parts 2 and 3 when using the type of tooling shown in Fig. 7. Factors that are not affected by the selection of tool material are not included in these cost figures. However, the costs may be used to estimate relative unit cost per part for the various die materials, provided other costs such as die resharpening are also included. All punches shown in Fig. 7 are included in the punch costs of Fig. 6.

Figure 6 shows that cost of tool material in small dies is a minor consideration. Labor cost to fabricate A2 is about 10% more than O1,

Table 14. Recommended Material for Perforator Bushings

Material to be punched	Total quantity of parts to be made: 10,000	100,000	1,000,000
Aluminum, brass, carbon steel, paper and plastics	W1(a)	W1(a)	D2
Spring steel, stainless steel, electrical sheet and reinforced plastics	D2	D2	D2 or Carbide

(a) When bushings are of a shape that cannot be ground after hardening, an oil-hardening or air-hardening steel is recommended.

Table 15. Recommended Punch and Die Materials for Shaving 0.050-In. Sheet

Material to be shaved	Total quantity of parts to be made: 1,000	10,000	100,000	1,000,000
Aluminum, copper and magnesium alloys	O1(a)	A2	A2	D4(b)
Carbon and alloy steel, up to 0.30% C, and ferritic stainless steel	A2	A2	D2	D4(b)
Carbon and alloy steel, 0.30 to 0.70% C	A2	D2	D2	D4(b)
Stainless steel, austenitic, up to quarter hard	A2	D2	D4(b)	D4(b)
Stainless steel, austenitic, over quarter hard, and spring steel hardened to Rockwell C 52 max	A2	D2	D4(b)	M2(b)

(a) It is preferable to use O2 for dies that must be made by broaching. (b) On frail or intricate sections, D2 should be used in preference to D4 or M2. Carbide shaving punches may also be practical for this quantity.

which in turn is 5 to 10% more costly than W1. D2 costs about 35% more to fabricate than O1. For blanking dies up to 6 in. on the maximum part dimension, die steel averages about 6% of the die cost. The cost of O2 and A2 may be somewhat less than 6%, while D2 may average 15%, and D4 slightly more. The D4 dies for part 2 would cost about the same as D2 dies.

Tool Materials

Table 16 shows nominal compositions of the tool steels recommended in the selection tables. All of these steels serve best when used at maximum tempered hardness, particularly in blanking thin material and when shock will be absent. For conditions of shock, the hardness is lowered to produce a tolerable level of breakage. Whether longer die life can be achieved by tempering to a lower hardness or using a tougher steel at full hardness cannot be readily predicted.

W1 and W2 are readily available, readily machinable, wear resistant and highly versatile water-hardening grades, furnished with various carbon contents in 0.10% ranges. W1 and W2 are interchangeable in performance, but W2 is of little advantage except to help prevent development of coarse grain in the steel as a result of overheating.

The depth of hardness of the water-hardening grades is shallow (from about 0.125 to 0.150-in. in a 1-in. diam bar), and for this reason such steels should not be used where grinding of the hard core will be needed to correct for distortion due to heat treating, except for short-run dies. W1 may make a brittle, easily broken punch if less than ⅜-in. diam, but a tough one if the diameter is about ¾ in. Hardness should be the highest obtainable with a tempering temperature of 325 to 375 F — usually Rockwell C 62 to 66.

Shock-resisting tool steels S1 and S5 are used for punches only where the probability of breakage is high. With normal heat treatment they have unacceptable levels of wear resistance, and are economical only if they are carburized to obtain 0.010 to 0.020-in. case containing 0.70 to 0.75% C. S1 should be used at Rockwell C 57 to 60, and S5 at Rockwell C 59 to 62.

Oil-hardening steel O1 is safer to harden and distorts less than W1 steel. O2 is preferable to O1 for dies that are to be made by broaching. It distorts less in hardening. Steel O6 is easier to weld, has consistently better life in blanking and piercing dies than O1, and has reduced regrinding and maintenance by about one half in stamping 1040 and other steels up to ⅜ in. thick. Although less widely available than O1, the usage of O6 steel has increased greatly during recent years. Advantages derived from the use of O6 in die applications relate to its greater resistance to sliding wear and its better machinability, as compared

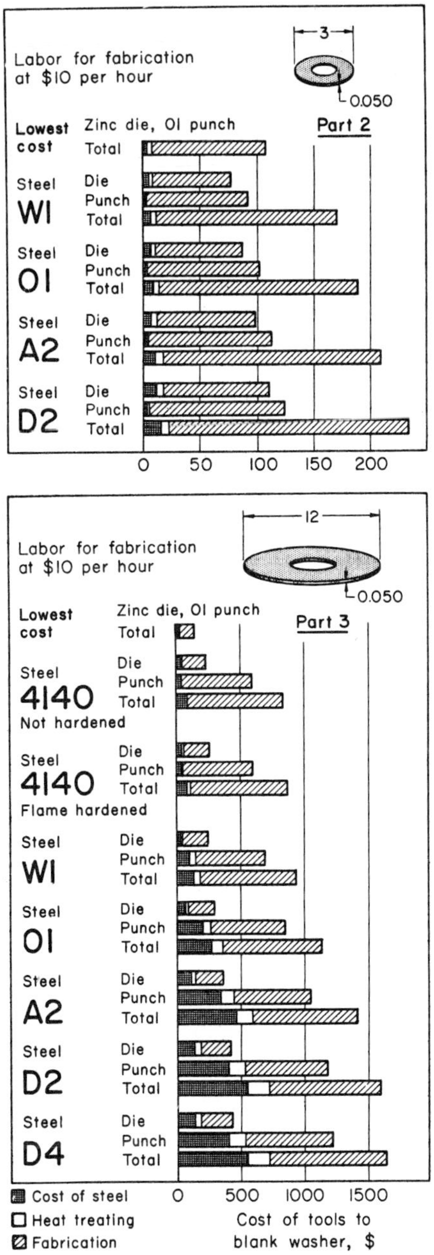

Fig. 6. Relative cost of tooling to make the two parts illustrated. Schematic representations of the tool setups are given in Fig. 7.

with other O grades; however, it may distort more during heat treatment.

A2 air-hardening medium-alloy (5% Cr) tool steel has wear resistance about halfway between oil-hardening steels and D2. Its base price is only $0.04 per pound more than O1 oil-hardening steel and about $0.42 less than D2. It has the least size change and cracking hazard in heat treatment of all the tool steels, followed closely by D2, air-hardened D4, and then by oil-hardened O and oil-hardened S types. Like D2, the A2 steel can be nitrided for dies for thin, soft materials or for punching reinforced plastics, to extend wear, to resist firing of the punch, and for heat resistance.

D2 high-carbon high-chromium air-hardening tool steel is probably the most commonly used and may be the most satisfactory and most widely available tool steel for large-volume production of stampings. It is about the second best steel for high accuracy and safety in heat treatment, and through-hardens in 3-in. sections. Its highest usable hardness of Rockwell C 62 to 63 is recommended for punches and dies where breakage is not a problem, as in dies blanking steel less than 0.062 in. thick and softer than Rockwell B 90. Maximum resistance to breakage may be developed by tempering back to Rockwell C 58 and 60, but only at a sacrifice in wear resistance. For lamination dies, the hardness should be not less than Rockwell C 61 or 62.

D4 high-carbon high-chromium air-hardening tool steel is somewhat more wear resistant than D2 and D3, particularly in punching electrical sheet, where, at Rockwell C 63 to 65, it often wears about 20% less than D2 and D3 and about the same as M2. These high-carbon high-chromium steels should be nitrided to extend die life only for blanking reinforced plastics or for soft and thin materials.

D5 high-carbon high-chromium tool steel has replaced D2, D3, M2 and M3 recently in some plants for the piercing, trimming and blanking of austenitic stainless steel. Metal pickup and scoring have been minimized in such applications by the use of D5, with an increase of 100 to 200% in die life for some stainless steel parts.

M2 high speed steel is the least costly, most used, and most readily available high speed steel for blanking dies and punches. It is equal to, or better than, D4 in wear resistance. For punching electrical sheet, the conventionally hardened M2 is surpassed only by carbide, cast alloys, and carburized M2. Carburized, M2 is about 30% more resistant to punch wear in punching laminations than it is with the standard heat treatment. It is equalled only by D4 and three less widely used high speed steels: M4 carburized, T1 carburized, and T15 carburized. It is recommended in Table 8 with standard heat treatment because it is less likely to break than other steels of equal wear resistance and has much better wear resistance than the shock-resisting steels S1 or S5. In blanking dies, it should be used at the hardness given in Fig. 5.

M3 high speed steel, with its 1% carbon and high vanadium content, is more wear resistant than M2 and the D grades. Its wear resistance can be improved by liquid nitriding. Selection of M3 depends on whether the dies made from it can be ground economically; to reduce the amount of grinding, M3 is generally used only for inserts. M3 is more difficult to grind than M2; caution must be used to avoid "burning" and the formation of surface cracks.

Hot rolled mild steel plate with carbon content from 0.10 to 0.20% may be used for short runs of small parts after it has been surface hardened, either by carburizing to a depth of 0.010 to 0.020 in. or by cyaniding to 0.004 to 0.008 in. Because of distortion in heat treatment, its use is limited to small and symmetrical shapes.

4140 alloy steel is generally available in various sizes of plate of aircraft quality. It is flame hardened to about Rockwell C 50 for blanking long runs on soft materials. However, tools with inside or outside corners will flame harden with soft spots and perform poorly. Flame hardening of the working edge of a large die has the advantage over complete hardening of very little warping or change of size.

Rolled zinc alloy or tooling plate is available in the form of ¼-in. plate from the principal suppliers of zinc-base die-casting alloys and lead-free zinc at about the same cost as W1. Dies of this material are sheared in with a flame hardened O1 punch and invariably use strippers of ⅜-in. sheet cork. Zinc dies cost about ⅕ as much as tool steel dies and are in major use for blanking fewer than 2000 parts from aluminum sheet as thick as 0.064 in.

Sintered carbide tooling is usually considered where production is four or more times the life of a D4 tool steel die, especially where close tolerances and minimum burr are required and a heavier press is available. Partial or complete inserts of carbide in tool steel may be considered for lower quantities or where the tool life between grinds needs to be extended. However, brazed sections are hazardous, and dovetailed or mechanically held sections will approach the cost of a complete carbide die.

Composition and hardness of carbides frequently used in blanking and piercing dies are as follows:

Nominal composition: Tungsten	Carbon	Cobalt	Rockwell A hardness
75.1	4.9	20.0	86
78.9	5.1	16.0	86
81.7	5.3	13.0	88
88.3	5.7	6.0	91

The first material should be used where shock is appreciable. The second of the above combines toughness and wear resistance, and is preferred for heavy-duty service, such as piercing silicon steel. Where close tolerances must be held in piercing silicon steel laminations, the third material is useful. The last of the above materials will be best for guides and guide rolls, or for applications involving very light shock.

Comparison of Tool Steel and Carbide*

Although the cost of carbide has been reduced considerably in recent years, it is estimated that the average carbide die will cost from 2¼ to 3 times as much as its tool steel counterpart. Therefore, justification for the use of a carbide die is normally based on its greater productive capacity. It can be expected to produce from 8 to 12 times as many pieces as a high-carbon high-chromium tool steel die.

*Based on an article "How Do Steel Dies Compare With Carbide Dies" by Joseph P. Schmidt (*Carbide Engineering*, Dec. 1958). This discussion has been considerably extended to include previously unpublished data provided by the author and others.

Table 16. Nominal Composition of Tool Steels Recommended in the Selection Tables

Steel	C	Mn	Cr	Mo	W or V
W1 ...	1.00(a)	...	...	...	...
W2 ...	1.00(a)	...	...	...	0.25 V
S1 ...	0.50	...	1.50	...	2.50 W
S5 ...	0.55	0.80	...	0.40	2.00 Si
O1 ...	0.90	1.00	0.50	...	0.50 W
A2 ...	1.00	...	5.00	1.00	...
D2 ...	1.50	...	12.00	1.00	...
D4 ...	2.25	...	12.00	1.00	...
M2 ..	0.80	...	4.00	5.00	6.00 W(b)

(a) Available carbon range is 0.60 to 1.40%, in 0.10% ranges. (b) Also 2% V.

Initial Costs. The higher initial cost of carbide dies reflects several factors. The cost of raw material in rough carbide sections is approximately 15 times that of tool steel on a pound-for-pound basis. Grinding time is increased by 25% on carbide sections. (Wear on diamond wheels will average about $1 per hour of use.) The cost of support sections is higher, because these sections must be stronger to accommodate carbide dies. Many carbide die designs require more elaborate sectioning to avoid losing an entire complicated contour in case of accident. As a result, more die pieces must be fitted together. These costs, along with the cost of a diamond grinding wheel, are the major extras involved in converting to carbide operation.

Whether or not the initial cost of a carbide die is a determining factor in die material selection will depend primarily on the production quantities required. At some quantity level, there is a break-even point at which either carbide or tool steel dies could be used with equal justification. As quantity of parts is increased above this break-even point, the probability of 8 to 12 times greater production per grind can justify the cost of carbide dies.

Chipping and Breakage. Occasional chipping of carbide dies in normal usage is not uncommon. However, the metallurgical nature of carbide lends itself to shallow flaking, rather than deep fracturing. Proper and careful handling will assure a minimum of chipping. Even in smash-ups caused by carelessness, a multistation progressive die with only a part of the carbide sections in each station is not likely to be damaged to any great extent.

Condition of Press. During the early years of carbide die building, it was assumed that such dies would operate properly only on new presses or on those with extremely close press tolerances. This is an extreme and unwarranted point of view, although it is obviously true that carbide dies cannot be operated in any available press, regardless of its condition. Normally, press requirements for carbide dies include: (*a*) sufficient tonnage, (*b*) average close-fitting slides, (*c*) good ram-to-bed parallelism, (*d*) an accurate feeding device, and (*e*) a satisfactory clutch capable of fast stops. The die should have a misfeed detector, to actuate the fast stop.

In general, a press that functions accurately with tool steel dies without chipping them, and with no excess deflection or abnormal wear of pins and bushings, should be suitable for carbide dies. Exceptions to this rule might apply to dies running thin material (under 0.025 in. thick) with minimum punch and die clearance.

Suggested minimum press tolerances for carbide operation are as follows: (*a*) ram-to-bed parallelism should be within 0.001 in. per foot; (*b*) roll feed accuracy should be within 0.005 in. and even closer for materials less than 0.025 in. thick; (*c*) slide fit should be held within 0.0015 in.

In blanking, a complete break should occur when the punch penetrates the die opening for a distance equivalent to the stock thickness. Failure of the stock to break with this amount of penetration indicates excessive play or press deflection.

A press that conforms to these standards of accuracy may be classified as being in "good" condition. A press whose accuracy is such as to halve these tolerances may be considered to be in "excellent" condition. A press whose accuracy deviates slightly below the minimum standards described as "good" may be classified as "fair". These classifications are used in tabulating the survey data which are shown in Table 17.

Table 17 is a summary of operating, design, and performance factors encountered in an examination of 40 carbide dies operating successfully under varying conditions. Only

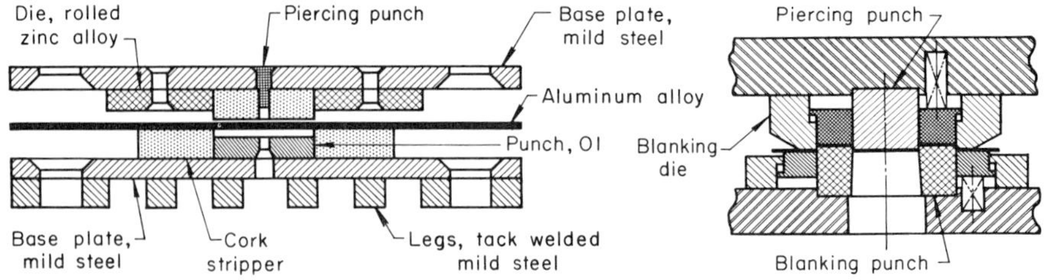

Fig. 7. Tooling for making parts 2 and 3 (illustrated in Fig. 6). Tools at left above are for short-run production with zinc alloy die and O1 steel punch; those at right are for all-steel construction.

17% reported their press equipment in "excellent" condition. No reports of "poor" presses were received. Reports on press tonnage reveal that the average user holds press load at about 50% of capacity. Figures on stripper construction indicate the argument between spring and box stripper adherents remains about even. In terms of performance, no appreciable difference was noted because of stripper type.

Sharpening Equipment. Assuming the availability of a surface grinder of sufficient capacity to sharpen steel dies, additional equipment needed for carbide is: a diamond wheel of adequate size, a wheel mount, and a simple diamond dresser. The wheel is rather costly but has an appreciable service life. Wheel consumption is approximately $1 per hour of use. An important requirement is that an operator be properly trained in the procedures of sharpening with a diamond wheel. Figures on the cost of such training are not available.

Costs of Design Changes. Most changes in high-production piece parts develop as minor modifications over a period of time, requiring simple die revisions. Drastic changes in part design are fairly uncommon. The cost of resectioning or modifying a carbide die follows about the same ratio as the initial steel versus carbide die cost (carbide 2¼ to 3 times as much as steel). Electrical discharge machining makes possible various die revisions with a minimum of original section replacement.

Cost and Production Data. Based on information supplied by carbide users, considerable data on both die costs and the performance of dies in production have been made available. Figure 8 compares die wear for carbide and D2 tool steel dies based on millions of hits per grind for nine different blank designs. The designs include both simple and complex shapes. For each set of dies compared (carbide versus tool steel), the sheet material, sheet thickness, and part size and shape are identical. In all cases, the performance of properly designed carbide dies is superior, providing from about 7 to 12 times more life between grinds than their tool steel counterparts. Also, although carbide wear is affected by the relative complexity of the blank design, carbide appears to be less sensitive than tool steel to the type of material blanked. The nine blank designs compared in Fig. 8 are shown below the chart, together with two other designs (parts 20 and 21) that are mentioned in connection with Fig. 9 and 10.

Table 18 is another summary of production results obtained with seven carbide dies operating side-by-side with seven tool steel dies. Again, parts and operating conditions were identical. Although the maximum allowable burr was de-

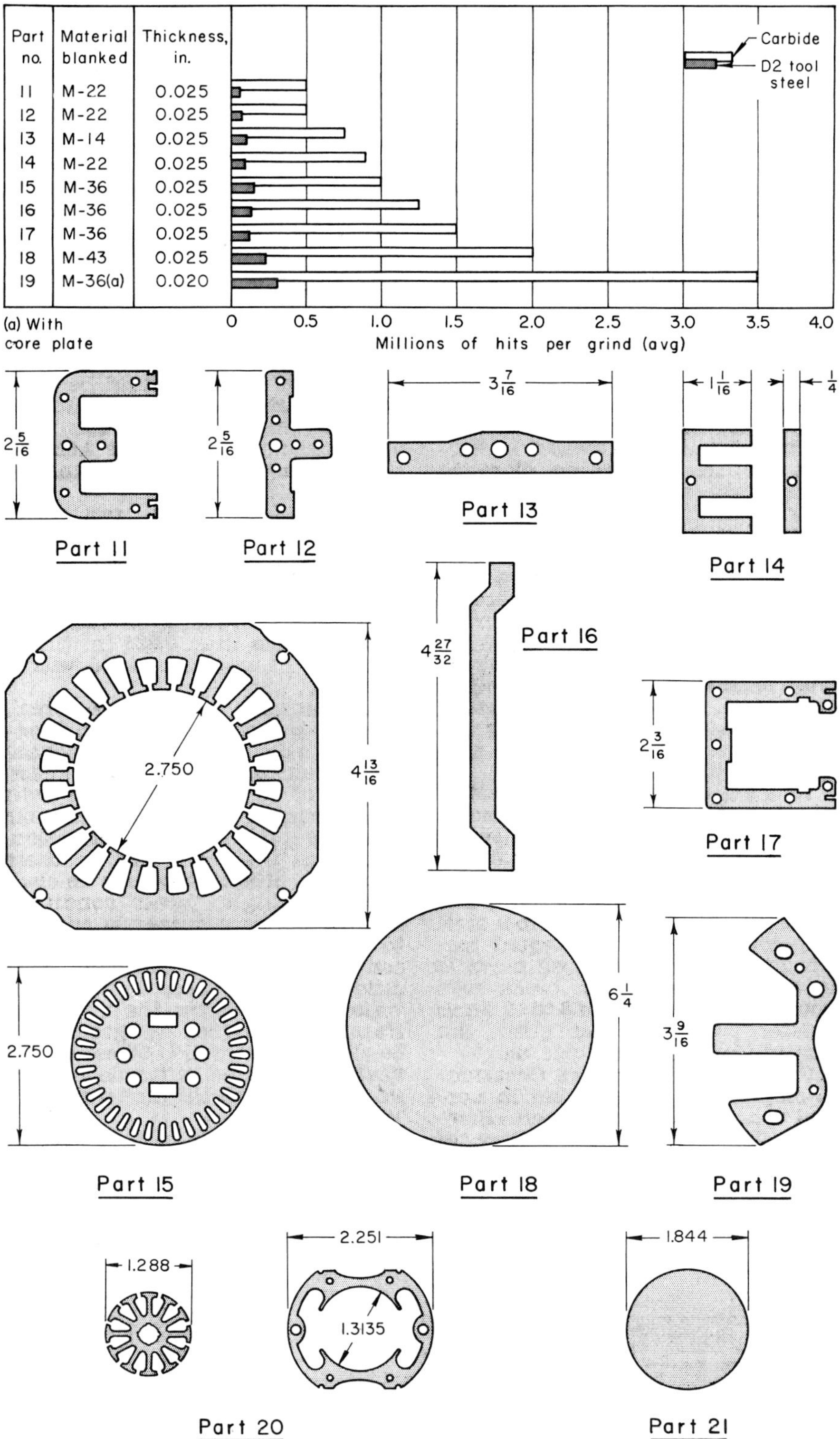

Fig. 8. Press performance of carbide and D2 tool steel dies; actual production figures comparing die life for nine different blank designs, based on millions of hits per grind in blanking electrical sheets (designated by standard AISI M-numbers). For each comparison, the blank material, sheet thickness and part design are identical. (Parts 20 and 21, shown at bottom, are not included in the above comparisons but are dealt with in the comparisons charted in Fig. 9 and 10, on the next page.)

Table 17. Survey of Operating Conditions for 40 Carbide Dies in Various Plants

Condition	Value
Press condition(a)	
Excellent	17%
Good	48%
Fair	35%
Press tonnage utilized, % of rated capacity (avg)	47%
Feed, in. per min (avg)	474
Carbide removed per grind, in. (avg)	
Punch	0.006
Die	0.006
Depth of carbide chipping, in. (avg)	0.012
Stripper construction preferred	
Spring	45%
Solid	41%
Combination	14%
Hits per grind (avg) (silicon steels only)	976,000

(a) See text for definitions of "excellent", "good" and "fair" press conditions.

Part no.	Material blanked	Thickness, in.	% Si	Number of dies required: Carbide	D2
17	M-36	0.025	1.40	0.70	10.00
13	M-14	0.025	4.50	2.50	8.33
15	M-36	0.025	1.40	0.60	4.60
16	M-36	0.025	1.40	0.50	3.50
19	M-36(a)	0.020	1.40	0.14	2.00
20	M-50(a)	0.0185	0.40	0.33	1.66
21	Tin plate	0.008	(b)	0.06	0.66

(a) With core plate
(b) 0.30% or less

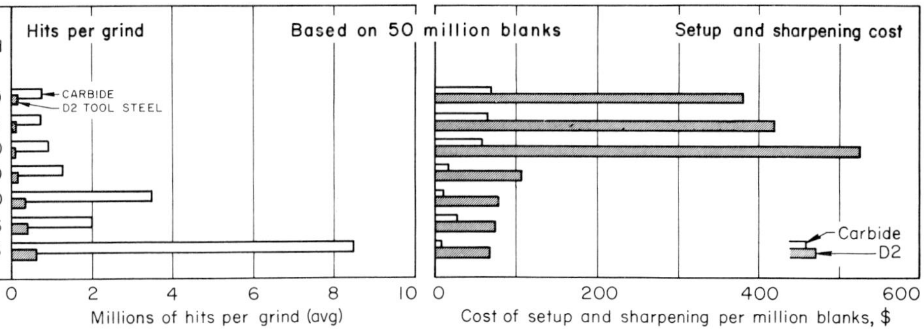

Fig. 9. Production results and operating costs for carbide dies and D2 tool steel dies used to make seven different parts. Parts (shown by number in Fig. 8) and operating conditions were identical for both carbide and tool steel dies. A portion of the data contained in Table 18 is summarized above. Setup and sharpening costs are derived from data in Table 19, comparing the number of dies required to produce 50 million blanks. It was indicated by users of these dies that all of the carbide dies were resharpened before the maximum allowable burr height was reached. Had they been run until they produced a burr comparable to that from their tool steel counterparts, the ratio of hits per grind would have been higher.

creased for only two parts, it was indicated by the users that all of the carbide dies were being pulled before they reached the maximum allowable burr. Had they been run until they produced a burr equal to that formed by their tool steel counterparts, the ratio of hits per grind would have been considerably higher than indicated. Part of the data contained in Table 18 is summarized graphically in Fig. 9. The electrical steels used are indicated by their standard AISI designations (M-14, M-36, etc), and the effects of silicon content and core plate on die performance are evident.

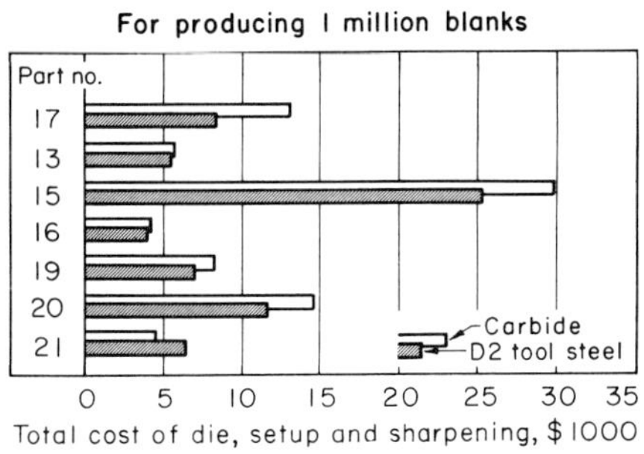

Fig. 10. Comparison of total cost of die, setup and sharpening for producing one million blanks of each of seven shapes, which are shown in Fig. 8.

Table 19 compares cost, maintenance, and setup time on identical carbide and tool steel dies, for a run of 50 million pieces. Only one part required more than a single carbide die to produce this quantity. The savings realized by the use of carbide dies are based here on sharpening and setup costs only. Total savings are not recorded, but would be considerably higher, since they would include such factors as reduction in press downtime and initial die costs.

Setup and sharpening costs plotted in Fig. 9 are derived from data contained in Table 19. Figure 9 com-

Table 18. Production Results on Carbide and D2 Tool Steel Dies

Part No. (Fig. 8)	Material: Die	Material: Blank	Max burr allowed, in.	Clearance: Punch and die, in. per side	Clearance: Angular, in./in. per side	Average grind, in., on: Punch	Average grind, in., on: Die	Average grinding time, hr	Average hits per grind	Estimated total hits per die	Usable die life, in.
17	Carbide	M-14	0.003	0.001	0.002	0.002	0.002	5	750,000	70,000,000	3/16
17	D2	M-14	0.003	0.001	0.002	0.010	0.005	5	125,000	5,000,000	3/16
13	Carbide	M-36	0.003	0.0012	0.0017	0.009	0.007	4	750,000	20,000,000	5/16
13	D2	M-36	0.003	0.0012	0.0013	0.013	0.010	3	100,000	6,000,000	½
15	Carbide	M-36	0.003	0.0009	0.0015	0.005	0.005	6	900,000	80,000,000	¾
15	D2	M-36	0.003	0.0009	0.0015	0.008	0.008	5	80,000	11,000,000	1¼
16	Carbide	M-50(a)	0.0025	0.0007	0.002	0.004	0.004	5	2,000,000	150,000,000	⅝
16	D2	M-50(a)	0.004	0.0007	0.002	0.010	0.010	2	400,000	30,000,000	¾
19	Carbide	M-36	0.001	0.001	Cutoff	0.005	0.005	2½	1,250,000	140,000,000	½
19	D2	M-36	0.002	0.001	Cutoff	0.009	0.009	2	140,000	14,000,000	1¼
20	Carbide	M-36(a)	0.003	0.0008	0.0025	0.006	0.006	4	3,500,000	350,000,000	⅝
20	D2	M-36(a)	0.003	0.0008	0.0025	0.010	0.010	3	300,000	25,000,000	¾
21	Carbide	Tin plate	0.0007	0.0004	0.0015	0.004	0.004	8	8,500,000	875,000,000	¾
21	D2	Tin plate	0.0007	Sheared in	0.0015	0.007	0.007	5	600,000	72,000,000	¾

(a) With core plate

Table 19. Operating Costs of Carbide Versus D2 Tool Steel Dies Producing 50 Million Identical Parts

Part No. (Fig. 8)	Die material	No. of dies needed	Die cost differential (a), %	Setup time, hr	Sharpening time, hr	Cost: Setup (b)	Cost: Sharpening (b)	Cost: Total (b)	Cost: Per 1,000,000 blanks	Cost: Per 1,000 blanks	Savings per 50 million blanks(c), $
17	Carbide	0.7	....	200	333	$ 1,200.00	$ 2,331.00	$ 3,531.00	$ 70.62	$ 0.0706	$ 15,669.00
17	D2	10	+700	1200	2000	7,200.00	12,000.00	19,200.00	380.40	0.3804	....
13	Carbide	2½	....	264	264	1,584.00	1,848.00	3,432.00	68.64	0.0686	17,568.00
13	D2	8⅓	+128	2000	1500	12,000.00	9,000.00	21,000.00	420.00	0.4200	....
15	Carbide	0.6	....	111	333	666.00	2,331.00	2,997.00	59.94	0.0599	25,253.00
15	D2	4.6	+383	1250	3125	7,500.00	18,750.00	26,250.00	525.00	0.5250	....
16	Carbide	⅓	....	75	125	450.00	875.00	1,325.00	26.50	0.0265	2,425.00
16	D2	1⅔	+150	375	250	2,250.00	1,500.00	3,750.00	75.00	0.0750	....
19	Carbide	½	....	20	100	120.00	700.00	820.00	16.40	0.0164	4,532.00
19	D2	3½	+350	178	714	1,068.00	4,284.00	5,352.00	107.40	0.1074	....
20	Carbide	1/7	....	14	59	84.00	413.00	497.00	9.94	0.0099	3,497.00
20	D2	2	+700	166	498	996.00	2,988.00	3,984.00	79.68	0.0797	....
21	Carbide	1/17	....	12	48	72.00	336.00	408.00	8.16	0.0082	3,078.00
21	D2	⅔	+560	166	415	996.00	2,490.00	3,486.00	69.72	0.0697	....

(a) Assuming carbide die cost at twice that of D2 steel (high-carbon high-chromium) dies. (b) Based on an average setup cost of $6.00 per hour for either type of die; sharpening cost of $6.00 per hour on steel dies, $7.00 per hour on carbide dies. ($7.00 includes $1.00 per hour diamond wheel wear.) (c) Savings in setup and sharpening costs through the use of carbide.

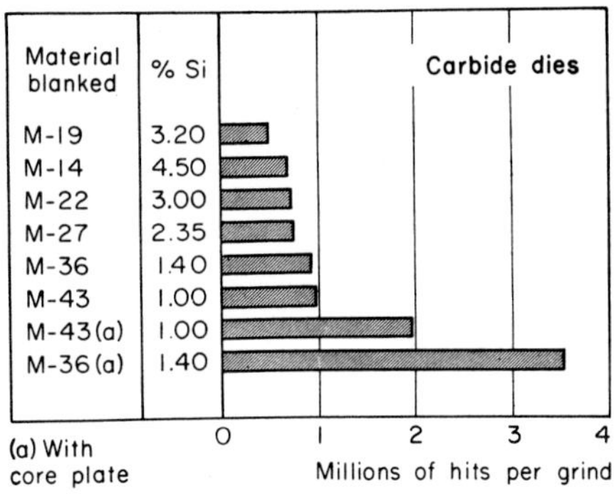

In general, die life is shortened as silicon content increases, but the beneficial effect of core plate on die life is far more pronounced than silicon content. At 1.00% Si, core plate extends die life by approximately 100%. M-numbers are standard AISI designations for electrical steel sheet.

Fig. 11. Effect of silicon content and core plate of electrical sheet steel on the service life of sintered carbide dies

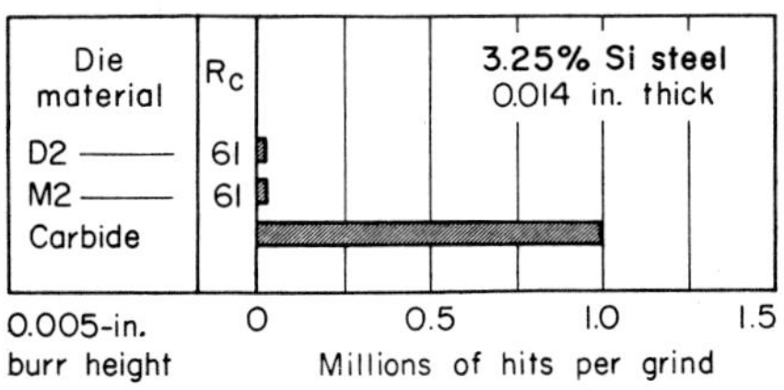

Fig. 12. Life of three die materials under the same operating conditions. The difference in wear between tool steels of different compositions is negligible, compared with the difference between tool steel and carbide.

pares the number of dies (carbide and tool steel) required to produce 50 million blanks with the cost of setup and sharpening in dollars per million blanks. Because only a relatively few operations are likely to require quantities this large, the original setup and sharpening cost data (Table 19) were combined with initial die costs to estimate total costs for an assumed production of one million pieces. These estimated costs are presented graphically in Fig. 10 for the same seven parts included in Fig. 9.

Despite certain operational advantages provided by carbide dies, it cannot be assumed that the conversion of any steel die to carbide will automatically increase production to the level of a million hits per grind. On the average, however, a carbide die can be expected to provide from 10 to 12 times the production normally achieved by its tool steel counterpart. If a tool steel produces about 50,000 hits, it is unlikely that a comparable carbide die will produce more than about 600,000.

Lamination Dies

As shown in Fig. 8 and 9, a tremendous gain is achieved through the use of core plate material. Although a gradual increase in hits per grind is experienced with decreasing silicon content (Fig. 11), other variables are known to influence blanking productivity. Factors such as steel reduction method, annealing procedure, lubrication during stamping, and punch and die clearance are also of considerable importance.

Die life is influenced by a number of variables, most of which pertain to the die material, die design, or the nature and condition of the material being blanked. In a study of die wear encountered in the blanking of 3.25% silicon steel (0.014 in. thick),* the measure of die life (for purposes of comparison) was set at the number of strokes before a 0.005-in. burr appeared on the stamped material. Figure 12 compares three die materials, indicating that die material is a major variable only if carbide is chosen to replace tool steel. The difference in wear of tool steels D2 and M2 is negligible. (Both steels were heat treated to Rockwell C 61.)

In addition to the factors discussed in the preceding section, consistency of production experience is sometimes an important consideration in the choice between tool steel and carbide. One company producing many millions of laminations discontinued the use of tool steel dies because of the wide variation in die life, as shown for five similar parts in Fig. 13. When carbide dies were used to make the same parts, die life was increased to 750,000 pieces per grind with a high degree of consistency in performance.

Surface Condition. The effect of the surface condition of silicon sheet on die wear is shown in Fig. 14. With tool steel dies, this effect is highly significant and stems primarily from the presence or absence of an abrasive scale on the blank material, and from the use of lubricants. Those treatments that increase the abrasiveness of the surface of the blank material (high-temperature annealing in the presence of MgO, and the use of a silica-containing varnish) affect die life adversely. Pickling removes abrasive surface materials and increases die life. Similarly, die life can be extended by the use of lubricants.

*"Improving Punchability of Silicon Steel" by C. S. Wukusick and R. S. Zeno, *The Tool Engineer*, Dec. 1958. Mr. Asa Kaplan's additions to this discussion are gratefully acknowledged.

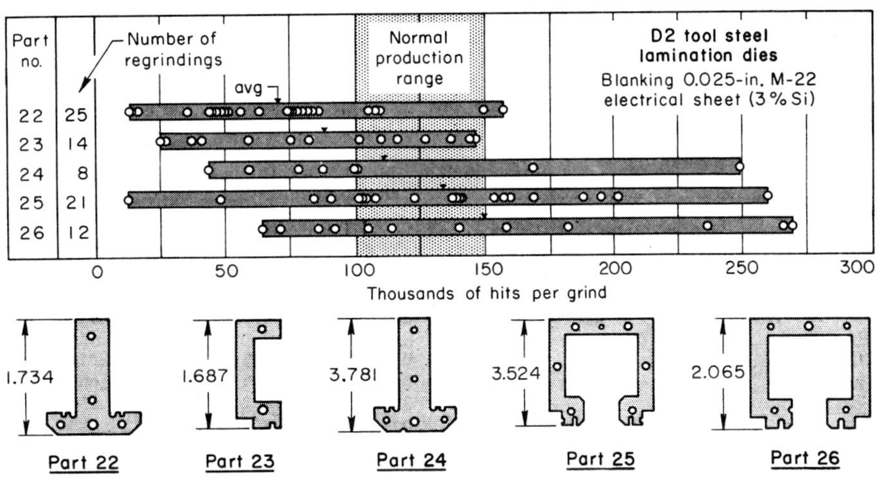

Fig. 13. Production experience with D2 tool steel lamination dies blanking the parts shown. All the dies were run on either a 60-ton or a 100-ton Henry & Wright dieing machine with a high degree of press accuracy. Criterion for regrinding for all dies was a maximum permissible burr of 0.004 in.

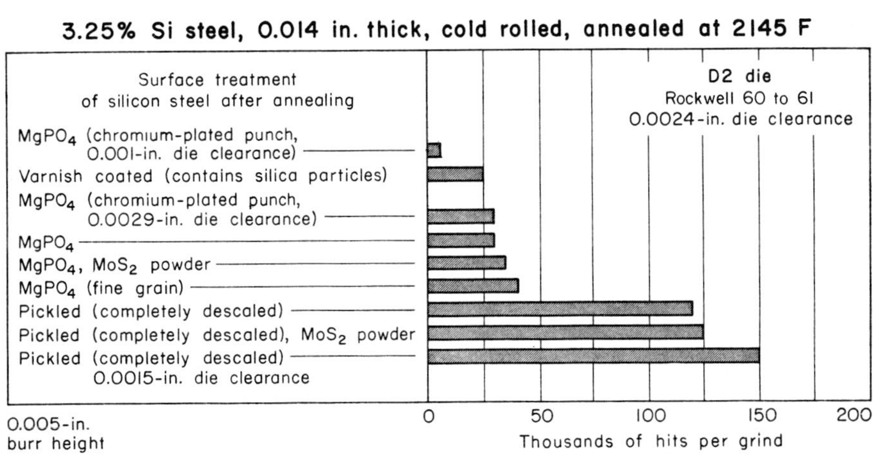

Fig. 14. Effect of surface condition of strip on the life of D2 tool steel dies blanking 3.25% Si electrical sheet. With tool steel dies, the effect of surface condition derives mainly from the presence or absence of an abrasive scale on the blank material and the use of lubricants. The presence of MgO and silica-containing varnish (both abrasive) affects die life adversely. On the other hand, pickling and the use of lubricants both serve to extend die life.

Section VI: Forming and Drawing Dies

The Selection of Material for Press Forming Dies

By the ASM Committee on Press Forming Dies

THIS ARTICLE is intended to help the reader select die materials for use in the press forming of sheet metal, specifically in those forming operations that involve bending or moderate stretching, or both. Separately considered, the materials used to make dies for deep drawn and cup-shaped parts are discussed in the next article in this Handbook.

In this article, the criterion of die material selection is the performance of that material in production. Thus, the article summarizes the production experience encountered in more than 500 specific forming applications. Based on this broad range of experience in materials selection, the reader may gain an understanding of die materials that can be applied to many analogous selection problems.

The useful performance of a forming die is measured in terms of its wear. Total wear is affected primarily by the length of the production run and the severity of the forming operation. Hence, quantity is the principal variable shown in the accompanying selection tables, and the importance of severity is recognized by the inclusion of a separate table for each of six parts of different size or shape.

The metal being formed, its thickness, and the finish and dimensional tolerances required in the part are also influential factors and are included as additional variables in the tables.

The amount of wear on a given die material during forming is proportional to the distance the sheet metal slides over the die at a given pressure between the surfaces in contact. Thin, soft or weak sheet metals exert the least pressure and thus cause the least wear; strong, thick metals cause the most rapid wear. However, the rate of wear for each combination of metals may be considerably different, depending on their surface characteristics, the speed of forming and type of lubrication. In the forming of parts with dies that produce wrinkles, high localized pressures are developed on the tools, and attempts to iron out such wrinkles almost always produce prohibitively high rates of wear and galling.

The recommendations made in this article are based on observed performance of a wide range of die materials in producing a great variety of parts. These recommendations are listed in the selection tables for the six parts shown in Fig. 1, ranging from the simplest, such as parts 1 and 2, to moderately severe, such as parts 5 and 6.

Recommended die materials include the range from plastics for low production of simple to moderate parts, up to the most wear-resistant tool steels, surface hardened by nitriding, for making severely formed parts. Parts of even greater severity, or those run in quantities larger than one million, may require dies or inserts of cemented carbide.

Where the tables show more than one material for the same conditions of tooling, the materials are listed in the order of preference with regard to the expected cost per piece; the ultimate choice will often depend on availability, rather than on the small differences in performance or cost of the materials for a given set of conditions.

The tools for parts 1 and 2 consist of a punch and an upper and lower die. None of these die components would wear much, because there is little deformation of the sheet metal during forming of such extremely simple parts, and therefore there is little or no sliding of the sheet over the lower die and little movement over the punch. Thus, large quantities of such parts can be formed by tools made of any die material available in a convenient size.

Tooling for part 3 consists of a punch and lower die. In operation, the punch pushes the blank through the lower die, which results in wear on the lower die. The metal closely envelops the punch, with little sliding. Thus, the punch generally will produce ten times as many parts with the same wear as the lower die made of the same material. However, wear and, possibly, galling will occur at the areas of moderate shrinkage of this part, particularly when the part is formed on these single-action dies. For this

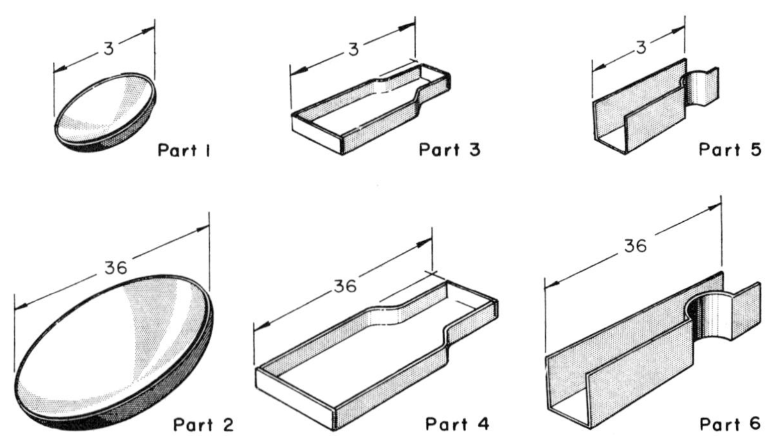

Fig. 1. Parts for which die materials are recommended in the selection tables

small die, type D2 tool steel may be used for production quantities as low as 10,000, since the cost of steel is of minor importance for a small die. If galling occurs with the high-carbon high-chromium tool steel, the tool can be nitrided after trial, for effective prevention of galling.

Tooling for part 4 consists of a punch and an upper and lower die. Without the upper die, excessive wrinkling would be expected at the shrink flanges. As in part 3, less wear-resistant material is required for the punch and upper die than for the lower die. Under conditions for which tool steel is recommended for making part 4, the tool steel should be in the form of inserts in a cast iron lower die, and the punch should be made of a cast tool steel such as D2. For example, for 10,000 to 100,000 pieces a cast iron die might be used with A2 or D2 inserts at points of most wear. For quantities greater than 100,000 pieces, or when this part must have close tolerances, the dies would have inserts of type D2 tool steel at all surfaces subject to wear. For quantities less than 100,000, the entire lower die would be made of the material indicated in the selection table, without inserts. The punch would be made of a less wear-resistant material, usually as indicated in the first column to the left of the quantity being considered.

The constriction in parts 5 and 6 is produced by stretching, rather than by shrinking of the metal along the constricted region, and the tooling therefore consists of a punch and lower die with no upper die required. The metal envelops the punch with only a minor amount of sliding, but produces about ten times more wear on the lower die than on the punch. Cast iron dies with inserts of type A2 or D2 tool steel are needed for production of 10,000 parts or more. However, the same material is indicated for both the punch and lower die for part 5 in the selection table, because of the small size and minor material cost.

In the selection table for part 6, the recommendations refer to the material to be used for the wearing edges of the lower die. The body of the lower die would be made of cast iron with wearing edges of tool steel in those instances where tool steel is recommended in the table. The punch for making part 6 could be of a material about one tenth as wear resistant as tool steel—for example, alloy cast iron. For a quantity large enough to require a punch material of greater wear resistance than alloy cast iron or cast steel, a tool steel insert should be used at the constricted section.

Sheet Thickness

Thick sheets of any metal will exert greater pressure on the dies than thin sheets of the same metal. The selection tables list recommendations only for sheets 0.050 in.

Selection Table for Part 1. Recommended Punch Material for Forming a Small Part of Minimum Severity

Part (made of 0.050-in. sheet) — Metal being formed	Requirements — Finish	Toler-ance, in.	Lubrica-tion(b)	Recommended punch material(a) — Total quantity of parts to be formed — 100	1,000	10,000	100,000	1,000,000
1100 aluminum, brass, copper(c)	None	None	Yes	Epoxy-metal, mild steel	Epoxy-metal, polyester-metal, mild steel	Epoxy(d), polyester (d), mild steel	Polyester-glass (d), mild steel	O1
1100 aluminum, brass, copper(c)	Best	None	No	Epoxy-metal, mild steel	Epoxy-metal, polyester-metal, mild steel	Epoxy(d), mild steel	Polyester-glass (d), mild steel	O1
Magnesium or titanium(e)	Best	None	Yes	Mild steel	Mild steel	Mild steel	Mild steel	A2
Low-carbon steel, to ¼ hard	None	None	Yes	Epoxy-metal, polyester-metal, mild steel	Epoxy-glass-metal, polyester-glass-metal, mild steel	Polyester-glass(d), mild steel	Mild steel	O1, A2
Type 300 stainless, to ¼ hard	None	None	Yes	Epoxy-metal, polyester-metal, mild steel	Epoxy-glass-metal, polyester-glass-metal, mild steel	Polyester-glass(d), mild steel	Mild steel	O1, A2
Low-carbon steel, to ¼ hard	Best	None	Yes	Epoxy-metal, polyester-metal, mild steel	Epoxy-glass-metal, polyester-glass-metal, mild steel	Mild steel	Mild steel	O1, A2
High-strength aluminum or copper alloys	Best	None	No	Epoxy-metal, polyester-metal, mild steel	Epoxy-glass-metal, polyester-glass-metal, mild steel	Mild steel	Mild steel	O1, A2
Heat-resisting alloys	Best	None	No	Epoxy-metal, polyester-metal, mild steel	Polyester-glass-metal, mild steel	Mild steel	Mild steel	O1, A2
Type 300 stainless, to ¼ hard	Best	None	No	Epoxy-metal, polyester-metal, mild steel	Epoxy-glass-metal, polyester-glass-metal, mild steel	Mild steel	Mild steel	O1, A2

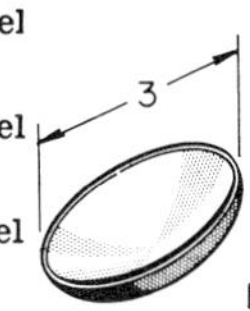

Part 1

Selection Table for Part 2. Recommended Punch Material for Forming a Large Part of Minimum Severity

Metal being formed	Finish	Toler-ance, in.	Lubrica-tion(b)	100	1,000	10,000	100,000	1,000,000
1100 aluminum, brass, copper(c)	None	None	Yes	Epoxy-metal, zinc alloy	Epoxy-metal, polyester-metal, zinc alloy	Epoxy(d), polyester (d), zinc alloy	Polyester-glass (d), zinc alloy	Cast iron
1100 aluminum, brass, copper(c)	Best	None	No	Epoxy-metal, zinc alloy	Epoxy-metal, polyester-metal, zinc alloy	Epoxy(d), polyester (d), zinc alloy	Zinc alloy	Cast iron
Magnesium or titanium(e)	Best	None	Yes	Zinc alloy	Cast iron	Cast iron	Cast iron	Alloy cast iron
Low-carbon steel, to ¼ hard	None	None	Yes	Epoxy-metal, polyester-metal, zinc alloy	Epoxy-glass-metal, polyester-glass-metal, zinc alloy	Polyester-glass(d), zinc alloy	Cast iron	Alloy cast iron
Type 300 stainless, to ¼ hard	None	None	Yes	Epoxy-metal, polyester-metal, zinc alloy	Epoxy-glass-metal, polyester-glass-metal, zinc alloy	Polyester-glass(d), zinc alloy	Cast iron	Alloy cast iron
Low-carbon steel, to ¼ hard	Best	None	Yes	Epoxy-metal, polyester-metal, zinc alloy	Epoxy-glass-metal, polyester-glass-metal, zinc alloy	Polyester-glass(d), zinc alloy	Cast iron	Alloy cast iron
High-strength aluminum or copper alloys	Best	None	No	Epoxy-metal, polyester-metal, zinc alloy	Epoxy-glass-metal, polyester-glass-metal, zinc alloy	Polyester-glass(d), zinc alloy	Cast iron	Cast iron
Heat-resisting alloys	Best	None	No	Epoxy-metal, polyester-metal, zinc alloy	Epoxy-glass-metal, polyester-glass-metal, zinc alloy	Cast iron	Alloy cast iron	Alloy cast iron
Type 300 stainless, to ¼ hard	Best	None	No	Epoxy-metal, polyester-metal, zinc alloy	Epoxy-glass-metal, polyester-glass-metal, zinc alloy	Cast iron	Alloy cast iron	Alloy cast iron

NOTE: Where mild steel is recommended for forming fewer than 10,000 pieces, the dies are not heat treated. For forming 10,000 pieces and more, such dies should be carburized and hardened. For information concerning the effect of sheet thickness on selection of die materials, see Table 1. For information on the cost of tooling, see Table 4.

(a) Description of materials is given in Table 3. When more than one material for the same conditions of tooling is given, the materials are listed in order of cost preference with the least expensive being first; however, the final choice often will depend on availability rather than on small differences in cost or performance of the materials. (b) Refers to specially applied lubrication rather than mill oil. (c) Soft. (d) With inserts. (e) Heated sheet.

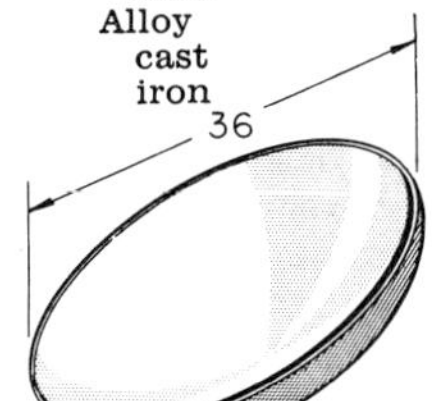

Part 2

thick. Table 1 recommends materials for dies to make parts 3 and 4 in three sizes, from sheet of four different thicknesses. The selections in Table 1 illustrate the increasing effect of galling and wear as parts are made from thicker sheets—especially parts with shrink flanges.

Table 1 deals only with die materials for forming steel parts. The forming pressure depends not only on the thickness but also on the strength of the sheet being formed. Wear and galling are less severe with any thickness of soft metal, such as aluminum and copper alloys, than with low-carbon steel, but more severe with high-strength metals such as stainless steels and heat-resisting alloys.

Size of Part

For small stampings, such as parts 1, 3 and 5, cast or plastic dies are uneconomical unless made from a model already available and with only minor finishing operations required on the dies. When the cost of patternmaking is included, cast or plastic dies will usually be more expensive than dies machined from other materials. The cost of the steel is usually a small fraction of the cost of dies for a small part, and the availability of material in a size that will minimize machining on the dies will usually be a greater factor in cost than any other.

Selection Table for Part 3. Recommended Lower Die Material for Forming a Small Part of Mild Severity

Part (made of 0.050-in. sheet)	Requirements			Recommended lower die material(a) — Total quantity of parts to be formed				
Metal being formed	Finish	Tolerance, in.	Lubrication(b)	100	1,000	10,000	100,000	1,000,000
1100 aluminum, brass, copper(c)	None	None	Yes	Epoxy-metal, mild steel	Polyester-metal, mild and 4140 steel	Polyester-glass(d), mild and 4140 steel	O1, 4140	A2, D2
1100 aluminum, brass, copper(c)	None	±0.005 R	Yes	Epoxy-metal, mild and 4140 steel	Polyester-metal, mild and 4140 steel	Polyester-glass(d), mild and 4140 steel	4140, O1, A2, D2	A2, D2
1100 aluminum, brass, copper(c)	Best	±0.005 R	Yes	Epoxy-metal, mild steel	Polyester-metal, mild and 4140 steel	Polyester-glass(d), mild and 4140 steel	4140, O1, A2	A2, D2
Magnesium or titanium(e)	Best	±0.005 R	Yes	Mild steel	Mild and 4140 steel	A2	A2	A2, D2
Low-carbon steel, to ¼ hard	None	None	Yes	Mild and 4140 steel	Mild and 4140 steel	4140, mild steel chromium plated, D2	A2	D2
Type 300 stainless, to ¼ hard	None	None	Yes	Mild and 4140 steel	Mild and 4140 steel	Mild and 4140 steel	A2, D2	D2
Low-carbon steel	Best	±0.005 R	Yes	Mild and 4140 steel	Mild and 4140 steel	Mild and 4140 steel	A2, D2, nitrided D2	D2, nitrided D2
High-strength aluminum or copper alloys	Best	±0.005 R	No(f)	Mild and 4140 steel	Mild and 4140 steel	Mild steel chromium plated and 4140	Cr plated O1; A2	D2, nitrided D2
Type 300 stainless, to ¼ hard	None	±0.005 R	Yes	Mild and 4140 steel	Mild and 4140 steel	Mild steel and 4140	Cr plated O1; A2	D2
Type 300 stainless, to ¼ hard	Best	±0.005 R	Yes	Mild and 4140 steel	Mild and 4140 steel	Mild steel chromium plated, D2	D2, nitrided D2	D2, nitrided D2
Heat-resisting alloys	Best	±0.005 R	Yes	Mild and 4140 steel	Mild and 4140 steel	Mild steel chromium plated, D2	D2, nitrided D2	D2, nitrided D2
Low-carbon steel	Good	±0.005 R	No(f)	Mild and 4140 steel	Mild and 4140 steel	Mild steel chromium plated	D2, nitrided D2	D2, nitrided D2

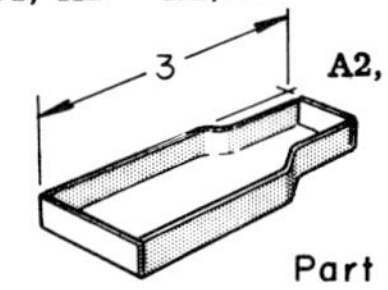

Part 3

Selection Table for Part 4. Recommended Lower Die Material for Forming a Large Part of Mild Severity

1100 aluminum, brass, copper(c)	None	None	Yes	Epoxy-metal, polyester-metal, zinc alloy	Polyester-metal, zinc alloy	Epoxy or polyester-glass(d), zinc alloy	Alloy cast iron	Cast iron or A2(g)
1100 aluminum, brass, copper(c)	None	±0.005 R	Yes	Epoxy-metal, polyester-metal, zinc alloy	Polyester-metal, zinc alloy	Alloy cast iron	Alloy cast iron	Alloy cast iron
1100 aluminum, brass, copper(c)	Best	±0.005 R	Yes	Epoxy-metal, polyester-metal, zinc alloy	Polyester-metal, zinc alloy	Alloy cast iron	Alloy cast iron	Alloy cast iron, A2(g)
Magnesium or titanium(e)	Best	±0.005 R	Yes	Cast iron, zinc alloy	Cast iron, zinc alloy	Cast iron	Alloy cast iron	Alloy cast iron, A2(g)
Low-carbon steel, to ¼ hard	None	None	Yes	Epoxy-metal, polyester-metal, zinc alloy	Epoxy-glass, polyester-glass, zinc alloy	Epoxy or polyester-glass(d), cast iron	Alloy cast iron	
Type 300 stainless, to ¼ hard	None	None	Yes	Epoxy-metal, polyester-metal, zinc alloy	Epoxy-glass, polyester-glass, zinc alloy	Epoxy or polyester-glass(d), alloy cast iron	A2(g)	D2(g)
Low-carbon steel	Best	±0.005 R	Yes	Zinc alloy	Epoxy-glass, polyester-glass, zinc alloy	Alloy cast iron	D2; nitrided A2(g)	D2, nitrided D2(g)
High-strength aluminum or copper alloys	Best	±0.005 R	No(f)	Zinc alloy	Polyester-glass, zinc alloy	Alloy cast iron	Alloy cast iron	Nitrided A2(g), nitrided D2(g)
Type 300 stainless, to ¼ hard	None	±0.005 R	Yes	Zinc alloy	Zinc alloy	Alloy cast iron	D2; nitrided A2(g)	D2(g), nitrided D2(g)
Type 300 stainless, to ¼ hard	Best	±0.005 R	Yes	Zinc alloy	Zinc alloy	Alloy cast iron	Nitrided D2	Nitrided D2(g)
Heat-resisting alloys	Best	±0.005 R	Yes	Zinc alloy	Zinc alloy	Alloy cast iron	Nitrided D2	Nitrided D2(g)
Low-carbon steel	Good	±0.005 R	No(f)	Zinc alloy	Zinc alloy	Alloy cast iron	Nitrided D2	Nitrided D2(g)

NOTE: Where mild steel is recommended for forming fewer than 10,000 pieces, the dies are not heat treated. For forming 10,000 pieces and more, such dies should be carburized and hardened. Where 4140 is recommended for fewer than 10,000 pieces, it should be pretreated to a hardness of Rockwell C 28 to 32. Flame hardening of high wear areas is recommended for quantities greater than 10,000 pieces. For the effect of sheet thickness on selection of die materials for making parts 3 and 4, see Table 1. For information on the cost of tooling to make these two parts, see Table 4.

(a) Description of materials is given in Table 3. When more than one material for the same conditions of tooling is given, the materials are listed in order of increasing cost; however, the final choice often will depend on availability rather than on small differences in cost or performance. Under conditions for which tool steel is recommended for making part 4, the lower die should be of cast iron with inserts of wrought tool steel and the punch of a cast tool steel such as D2. For example, for 10,000 to 100,000 pieces a cast iron die might be used with A2 or D2 inserts at the shrink flanges; over 100,000 pieces, dies would have D2 inserts at all wear surfaces. (b) Specially applied lubrication, rather than mill oil. (c) Soft. (d) With inserts. (e) Heated sheet. (f) Use lubrication to make 1 to 100 parts. (g) Use as inserts in cast iron body.

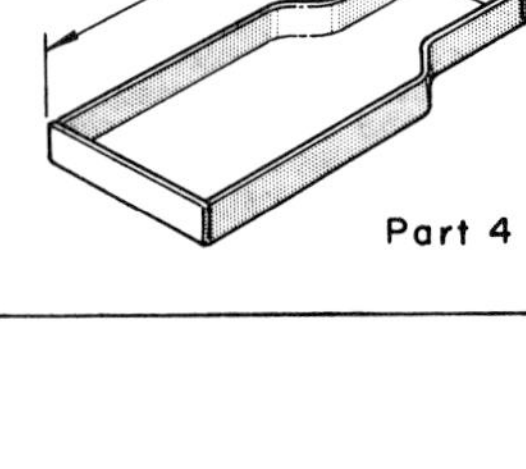

Part 4

As the size of the part increases, cost savings resulting from minimizing machining by the use of a casting close to final size will more than offset the cost of a pattern. However, inserts of tool steel or carbide must be used on high-production dies subject to severe wear and galling. The selection of both the material and the locations of the inserts should be on the conservative side in those instances where production must not be interrupted for alteration of the tooling. If tools can be taken out of production, cast iron dies may be used, with the wear surface flame hardened and inserts added later if needed.

Quantity

The number of parts to be produced is an important factor in large dies in which the cost of material is equal to or greater than the cost of machining. In dies smaller than about 6 in., the difference between the cost of the most expensive and cheapest steels is less important than the assurance of long life without the necessity for rebuilding tools if the quantity should be increased above original expectations, or if the die material should prove to have been underspecified. However, for large dies, both the choice of tool material and the design of the dies depend on the number of parts to be produced, particularly when quantities exceed 1000 parts.

On parts such as part 6, in which there is considerable movement of the sheet over die surfaces, greatly increased die wear is likely. With such parts, die inserts can be shimmed to allow for wear of the die, whereas a solid die cannot be adjusted readily for wear. Therefore, risks may be taken in the selection of steel for inserts on jobs where production can be interrupted for this operation. However, if the insert is small, the cost of material may be of much less consequence than the cost of diemaking and shimming, which may run from $10 to $100 per hr for lost time of the press, plus labor. Thus, cheap steel inserts may be uneconomical.

Adjustable inserts are often impractical for small dies. Therefore for high-production dies working under severe wear conditions and producing parts to close tolerances, it is often desirable to use a complete insert or to make the entire die of extremely wear-resistant material such as sintered carbide or nitrided D2 tool steel.

Tolerance Requirements

Tolerance requirements of the part may have an important effect on the choice of tool material when the part is to be finished without restriking. If the part is to be restruck, the material used in the restriking die is of less importance, because it usually will be subjected to less wear than the die that completes the primary operation. A major factor in the choice between a wear-resistant material and a less costly and less wear-resistant material is the necessity for maintenance during the production run.

Selection Table for Part 5. Recommended Lower Die Material for Forming a Small Part of Moderate Severity

Part (made of 0.050-in. sheet)				Recommended lower die material(a)				
	Requirements			Total quantity of parts to be formed				
Metal being formed	Finish	Tolerance, in.	Lubrication(b)	100	1,000	10,000	100,000	1,000,000
1100 aluminum, brass, copper(c)	None	None	Yes	Mild steel	Mild steel	Mild steel	O1, A2	A2
1100 aluminum, brass, copper(c)	Best	±0.020 R	Yes	Mild steel	Mild steel	Mild steel	O1, A2	A2
Magnesium or titanium(d)	Best	±0.020 R	Yes	Mild steel	Mild steel	A2	D2	D2
Low-carbon steel, to ¼ hard	Best	None	Yes	Mild steel, O1	Mild steel, O1	Mild steel, A2	Mild steel, chromium plated, A2	D2
Type 300 stainless, to ¼ hard	None	None	Yes	Mild steel, O1	Mild steel, O1, A2 tool steel	Mild steel chromium plated, A2	Mild steel, chromium plated, D2	D2
Low-carbon steel	Best	±0.020 R	Yes	Mild steel, O1 or nitrided A2	Mild steel, O1 or nitrided A2	Mild steel chromium plated, D2	Mild steel, chromium plated, nitrided D2	Nitrided D2
High-strength aluminum or copper alloys	Best	±0.020 R	No(e)	Mild steel, O1	Mild steel, O1	Mild steel chromium plated	A2	D2
Type 300 stainless, to ¼ hard	None	±0.020 R	Yes	Mild steel, O1	Mild steel	Mild steel chromium plated, A2	D2	D2
Type 300 stainless, to ¼ hard	Best	±0.020 R	Yes	Mild steel, O1	Mild steel, O1	Mild steel chromium plated	Nitrided D2, nitrided A2	Nitrided D2
Low-carbon steel	Good	±0.020 R	No(e)	Mild steel, O1	Mild steel, O1	Mild steel, A2	A2	D2

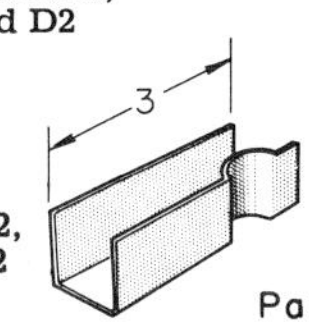

Part 5

Selection Table for Part 6. Recommended Lower Die Material for Forming a Large Part of Moderate Severity

Metal being formed	Finish	Tolerance, in.	Lubrication(b)	100	1,000	10,000	100,000	1,000,000
1100 aluminum, brass, copper(c)	None	None	Yes	Zinc alloy	Zinc alloy	Cast iron	Alloy cast iron	D2, A2
1100 aluminum, brass, copper(c)	Best	±0.031 R	Yes	Zinc alloy	Zinc alloy	Cast iron	Alloy cast iron	A2, D2
Magnesium or titanium(d)	Best	±0.031 R	Yes	Cast iron	Cast iron	Alloy cast iron	A2	D2
Low-carbon steel, to ¼ hard	None	None	Yes	Zinc alloy	Cast iron	Cast iron	Alloy cast iron	D2, A2
Type 300 stainless, to ¼ hard	None	None	Yes	Zinc alloy	Cast iron	Alloy cast iron	D2, A2	D2, nitrided D2
Low-carbon steel	Best	±0.031 R	Yes	Zinc alloy	Cast iron	Alloy cast iron	Alloy cast iron	Nitrided D2
High-strength aluminum or copper alloys	Best	±0.031 R	Yes	Zinc alloy	Zinc alloy	Alloy cast iron	Alloy cast iron	D2, A2
Type 300 stainless, to ¼ hard	None	±0.031 R	Yes	Zinc alloy	Zinc alloy	A2	A2	D2
Type 300 stainless, to ¼ hard	Best	±0.031 R	Yes	Zinc alloy	Alloy cast iron	D2; nitrided A2	Nitrided D2	Nitrided D2

NOTE: Where mild steel is recommended for forming fewer than 10,000 pieces, the dies are not heat treated. For forming 10,000 pieces and more, such dies should be carburized and hardened. For information concerning the effect of sheet thickness on selection of die materials, see Table 1. For information on the cost of tooling to make parts of different sizes, see Table 4.

(a) Description of material is given in Table 3. When more than one material for the same conditions of tooling is given, the materials are listed in order of increasing cost; however, the final choice often will depend on availability rather than on the small differences in cost or performance. Cast iron dies with inserts of type A2 or D2 tool steel would be needed for making 10,000 parts; however, the same material is recommended for punch and lower die for part 5 because of the small size and minor cost of material. Recommendations for making part 6 are the materials for the wearing edges of the lower die; body of the lower die would be of cast iron with wearing edges of the tool steel recommended; the punch could be of a material about one tenth as wear-resistant as tool steel. (b) Refers to specially applied lubrication rather than mill oil. (c) Soft. (d) Heated sheet. (e) Lubrication should be used to make 1 to 1000 parts.

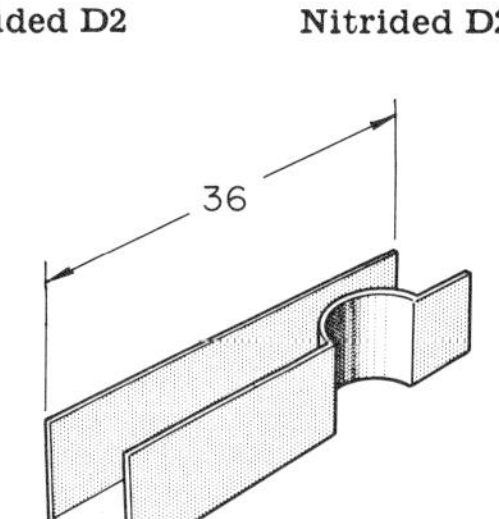

Part 6

Work Metal

The sheet metals of higher hardness usually will wear dies more rapidly, but other factors, such as the presence of scale on the surface of hot rolled, unpickled steels, will cause two to five times more wear. However, scaled surfaces cause less galling, which, on tool materials, may be an even more serious condition than wear, because galling or "pickup" on a die will cause repeated interruptions for reconditioning the die.

Soft brass and aluminum cause less wear and galling than carbon steel, while stainless steels and heat-resisting alloys cause more wear and galling. Where galling is anticipated, it is desirable to use materials such as D2 tool steel that can be treated subsequently, if necessary, to eliminate the difficulty. Possible treatments include chromium plating of any hardened steel, the hardening of alloy cast iron, and the nitriding of tool steels such as A2 and D2, which are preferable for nitriding because of the presence of nitride formers such as chromium and molybdenum in the steels.

Lubrication

In making parts at low and medium production (up to 10,000 pieces), it is often economical to use lubricants. With zinc dies, it is necessary. However, the most effective lubricants are difficult to apply and remove, and they add significantly to the cost. Efficient application of lubricants is particularly difficult in high-production operations where presses are being fed automatically. In such operations, it is usually economical to use die metals that are more costly but more resistant to galling in combination with the usually less effective lubricants that can be applied automatically. Examples of these materials are aluminum bronze, nitrided D2 tool steel and sintered carbides, which can be used for forming low-carbon steel with only mill-oil lubrication.

Prevention of Galling

Attempting to stretch sheet metal beyond practical limits, poor tool fitting, and rough finish on the surface of tools are common causes of galling. For short and medium runs, surface hardened hot rolled steel dies will produce parts equal to those produced from most tool steel dies. Exceptions may be encountered in severe reductions or in forming metals that show a greater tendency to gall, such as austenitic stainless steel. However, tool steel dies may also gall under these conditions.

When galling is encountered, the tool fit and the thickness of the metal being formed should be checked first, to determine whether clearance is adequate. Attempts to iron out wrinkles will often cause galling. Whenever possible, wrinkles should be prevented by the design of the tools.

Nitriding minimizes or prevents the galling of dies made from alloy steel or alloy tool steel such as A2 or D2. It is not recommended for steels that contain no nitride-forming elements such as chromium or molybdenum. Nitrided surfaces on such steels may spall off on radii smaller than about 1/8 in.; and plated surfaces, on radii less than 1/4 in., thus increasing the risk of failure in these locations.

Hard chromium plating will usually eliminate galling of mild steel, alloy steel and tool steel dies, and it is often used for severe duty. For operations involving high local pressures, hardened steels such as alloy or tool steels will be less likely to yield plastically and to cause cracking of the hard chromium plate.

Dirt, grit and shot fragments on the sheet cause greater damage to nitrided and chromium-plated tools than to hardened tool steels; hard particles may cause minute spalling and small pits, especially at radii in areas of high forming pressure. When these pits in the die cause scratches on the formed sheet, the pits must be stoned out if high finish is required on the part. After such repairs have been made several times in the same area, the soft underlying metal will be exposed and renewal of the nitrided case or the chromium plate will be necessary.

Galling is less likely if the die ma-

Table 1. Recommended Materials for Lower Die for Forming Steel Parts of Three Sizes, Each in Four Thicknesses

(Assuming no finish or tolerance requirements)

Sheet thickness, in.	Recommended lower die material(a) — Total quantity to be produced: 100	1,000	10,000	100,000(b)	1,000,000(b)
For Part 3, Maximum Dimension of 3 In.					
0.031......	Mild steel and 4140	Mild steel and 4140	Mild steel and 4140	4140, mild steel chromium plated, D2	D2
0.062......	Mild steel and 4140	Mild steel and 4140	Mild steel and 4140	4140, mild steel chromium plated, D2	D2
0.125......	Mild steel and 4140	Mild steel and 4140	Mild steel, 4140 and A2	D2	D2
0.250......	Mild steel and 4140	Mild steel and 4140	A2 and D2	D2	D2
For a 12-In. Part Similar to Parts 3 and 4					
0.031......	Mild steel, 4140 and zinc alloy	Mild steel, 4140 and zinc alloy	Mild steel,4140, zinc alloy and alloy cast iron	Alloy cast iron(c), mild steel chromium plated, 4140	D2
0.062......	Mild steel, 4140 and zinc alloy	Mild steel, 4140 and zinc alloy	Mild steel, 4140, zinc alloy and alloy cast iron	Alloy cast iron(c) and A2	D2
0.125......	Mild steel, 4140 and zinc alloy	Mild steel, 4140 and cast iron	Mild(c) and A2 steels	A2 and D2	D2
0.250......	Mild steel, 4140 and cast iron	Mild steel and alloy cast iron	Mild(c) and A2 steels	A2 and D2	D2
For Part 4, Maximum Dimension of 36 In.					
0.031......	Epoxy-metal, polyester-metal, zinc alloy	Epoxy-glass, polyester-glass, zinc alloy	Epoxy or polyester-glass(d), alloy cast iron, zinc alloy(d)	Alloy cast iron(c)	D2
0.062......	Epoxy-metal, polyester-metal, zinc alloy	Epoxy-glass, polyester-glass, zinc alloy	Epoxy or polyester-glass(d), alloy cast iron, zinc alloy(d)	A2	D2
0 125......	Zinc alloy	Zinc alloy, cast iron	Alloy cast iron(c)	A2 and D2	D2
0.250......	Zinc alloy, cast iron	Alloy cast iron	Alloy cast iron(c)	A2 and D2	D2

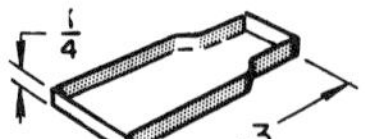

Part 3

Part 3. For a sketch of the die for this part, see Fig. 4

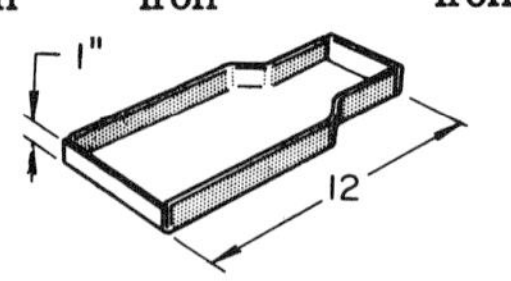

A 12-in. part similar to parts 3 and 4. For a sketch of the die for this part, see Fig. 4. The cost of tooling is given in Table 4.

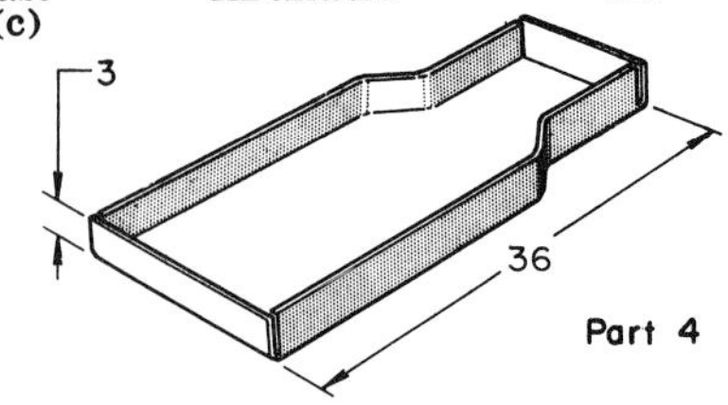

Part 4. Tooling for this part is shown in the sketch of Fig. 5.

(a) See Table 3 for description of materials. Materials recommended in above table are for forming parts from low-carbon steel; wear and galling will be less with aluminum and copper alloys, and more with stainless and heat-resisting alloys, than with low-carbon steel. (b) All A2 or D2 tool steels nitrided after tryout if required. (c) Inserts of A2 tool steel are recommended for wear edges for the shrink flange. (d) At the quantity level indicated, use of inserts with these materials is required.

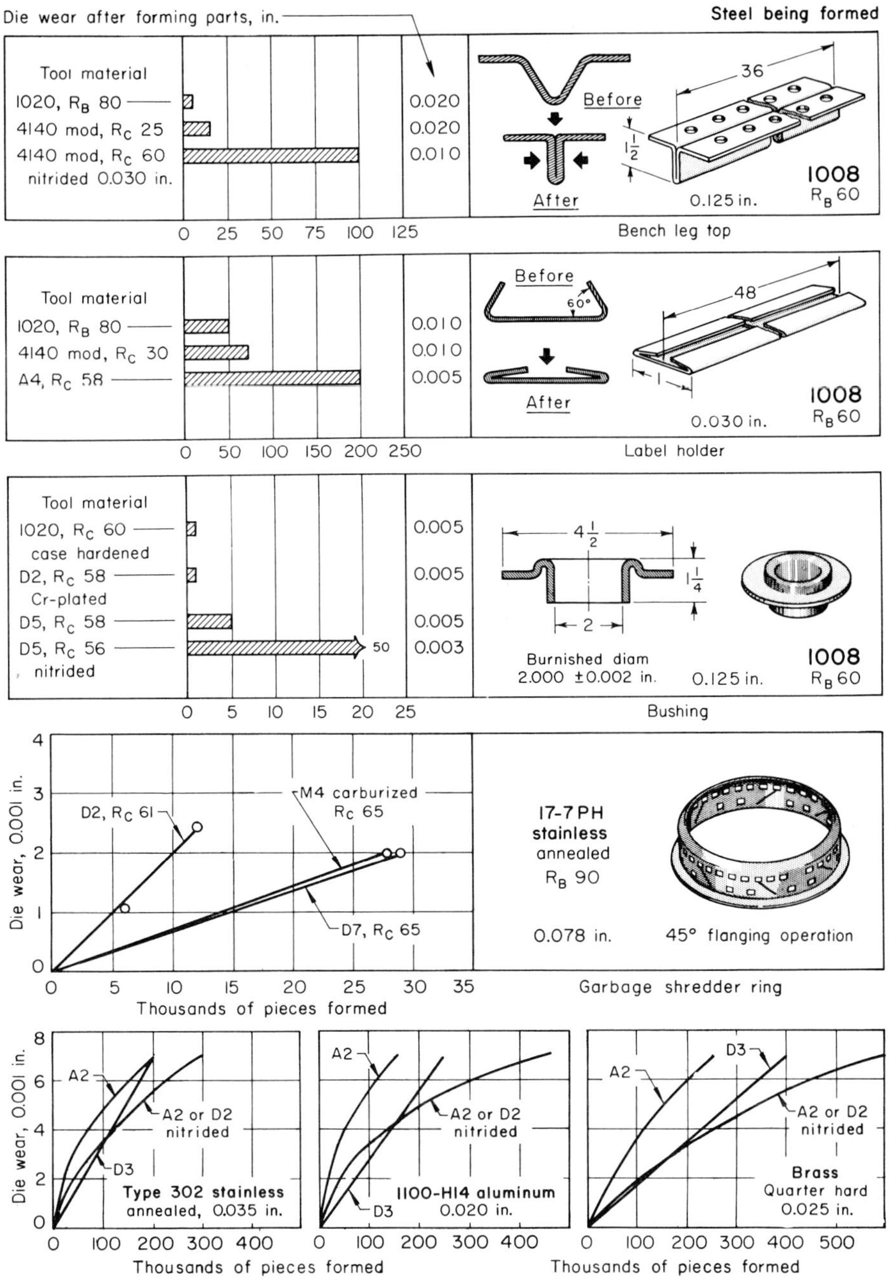

Fig. 2. Data on die wear and die life. The three graphs in the bottom row relate to small instrument parts having a maximum area of 3 sq in.

terials and the metal being formed are dissimilar in hardness, chemical composition and surface characteristics. For instance, effective combinations are: (*a*) aluminum bronze tools for forming carbon steel and stainless steel; (*b*) tool steel for forming aluminum and copper alloys; and (*c*) carbide tools for carbon steel, stainless steel and aluminum parts.

Die Finish. Despite the fact that, in many instances, galling can be minimized or eliminated by careful selection of die materials, die finish is an important factor affecting susceptibility to galling. The probability of galling increases as the severity of forming increases.

The influence of die finish, with other factors constant, is illustrated in the forming of a hat section where the walls of the hat were nearly perpendicular to the original blank, thus representing fairly drastic severity. The material being formed was annealed 1010 steel. The punch portion of the die was made from A6 tool steel at Rockwell C 60 with a finish of 30 to 35 micro-in. The die ring was made from nitrided medium-carbon alloy steel having a hardness of Rockwell C 63 and was also finished to 30 to 35 micro-in.

Both components were copiously lubricated, but after ten hits there were indications of metal pickup, and after 30 hits the tools were removed because of excessive galling. Both die components were refinished to about 20 micro-in., after which 300 parts were produced before galling was excessive. The tools were then refinished to 8 to 10 micro-in., after which 3000 pieces were produced with no sign of galling. However, producing this high finish all over dies is costly and is not always warranted, but it should be considered when galling is experienced in severe forming. Sometimes the use of lubricants can be minimized by improving the die finish.

Die Wear and Die Life

Composition and hardness of die material and part, and thickness, shape and quantity of the parts to be formed are of primary importance in die wear and life (the total useful life of the die).

Effects of these factors are shown graphically in Fig. 2. Type 4140 steel shows the effect of die hardness by increasing about 800% in die life when nitrided (Fig. 2, upper graph). Differences in tool material composition and hardness are also seen in forming the label holder, and in comparing D5 with and without nitriding in forming the bushing. M4 and D7 at Rockwell C 65 gave more than twice the die life of D2 at Rockwell C 61, in flanging the shredder ring of 17-7 PH stainless steel.

The lower row of graphs in Fig. 2 shows the influence on die wear and life of the composition of the material being formed into parts of similar shape. Rate of wear gradually decreases for the nitrided steels, resulting from the wearing away of the superficial surface that is partially composed of softer "white layer".

Effect on die life of composition and increased resistance to deformation by the metal being formed is shown in Fig. 3. Also shown are comparisons of different tool materials forming the same work metal. Table 2 shows additional die life data from several specific operations.

Tool Steels

Table 3 lists nominal compositions of tool steels for which recommendations or performance data are given.

Water-hardening tool steels are the least costly and the best in machinability and weldability, but distort badly during heat treatment. They have better resistance to galling and wear than O1, but not as good as A2 or D2. They may be used for such operations as forming of aluminum with dies that would not distort much during heat treatment, or for dies that can be selectively hardened, as for the dies for making part 1 of aluminum. Carbon tool steel can be used in the annealed condition, hard faced with bronze or other materials that have greater resistance to galling and wear.

Steel O1 overcomes most of the heat treating problems of water-hardening tool steel, and has almost as good machinability. However, it is only slightly less costly than a more wear-resistant steel such as A2, has poor weldability, and is less resistant to galling and wear than the water-hardening steels. Steel O1 is suitable for moderate conditions and where distortion in heat treatment precludes water-hardening steels.

A2, only slightly more costly than O1, distorts in heat treatment as little as, or less than, any other tool steel and is superior to O1 in resistance to wear and galling, which can be improved still further by nitriding. Welding properties are also poor. A2 can be used for inserts or cast to shape at a slight impairment of performance in forming carbon steel and metals of similar strength.

D2 is more costly and less easily machined and welded than A2 but deforms very little in heat treatment. Its excellent resistance to galling and wear can be improved by nitriding.

D3, D5 and D7 are similar to D2, and can be used as solid dies, castings, or inserts for forming materials such as stainless steel in severe applications. D3 must be quenched in oil. D5 contains an addition of cobalt, costs more, is more difficult to machine and quench, and can be nitrided to increase its useful life (Fig. 3). D7 contains additions of molybdenum and vanadium. It is hardened in air and is the most costly and most difficult of all the high-carbon high-chromium tool steels to machine and grind. For these reasons it is rarely used except for extremely small solid dies, and for inserts in areas where larger dies have excessive wear.

The high-vanadium high speed steels such as M4 can be used for forming dies and may be desirable in specific applications — for instance, carburized M4 was used in making the shredder ring in Fig. 2. However, these steels cost more and are less readily available than the high-carbon high-chromium types. Carburized high speed steels have been used successfully.

Shock-resisting tool steels such as S1 are used primarily for die components that are fragile or subject to impact in service. S1 has good machinability, distorts little in heat treatment, and when carburized has good resistance to galling and wear.

Cast Iron

Cast iron is a useful die material for forming parts larger than about 12 in. (Table 1). Its performance makes it suitable for use in medium production runs or in short runs of large parts. When cast iron is used with inserts, it will produce greater quantities, as shown in the tables.

Both types of cast iron listed in Table 3 should have predominantly fine pearlite with no massive carbides and a minimum of ferrite. Graphite should be of AFS-ASTM type A distribution (shown in Fig. 4, page 371, in the article on Gray Iron) with a preferred flake size of 4 to 5.

Alloy cast iron should be flame hardenable to the equivalent of 450 Bhn with air cooling. The alloy iron listed in Table 3 is sufficiently hardenable for oil quenching of small dies and local flame hardening with air cooling for dies too large to be quenched. When no heat treating is anticipated, the unalloyed iron is usually preferable, because of lower cost and better machinability. However, alloy iron will usually wear less, even without flame hardening or other heat treatment.

Zinc Alloy

Zinc alloy has high pattern fidelity when cast, but requires very accurate shrink patterns to minimize hand labor in finishing. Zinc alloy is economical only for die components at least 12 in. long, and it is most economical in dies about 36 in. long, as is evident in the selection tables. One of the principal values of zinc alloy is for complicated dies that would require intricate machining and hand tooling if made of cast iron, but which can be cast more closely to shape in zinc alloy. In tensile strength, compressive strength and hardness, zinc tool alloys are inferior to other metals used in die construction.

If compound curves on the face of a stamping have no sharp creases or embossments with sharp corners, it is not unusual for zinc alloy dies to give the length of service indicated in the selection table for part 2. Zinc alloy tools are particularly sensitive to surface roughness, thickness and hardness of the sheet metal being formed. The increase of any one of these multiplies unit pressures over a small area of the die face.

In production use, zinc alloy drawing dies in double-action presses have been found to give the longest life. Contrary to conventional diemaking practice, zinc alloy

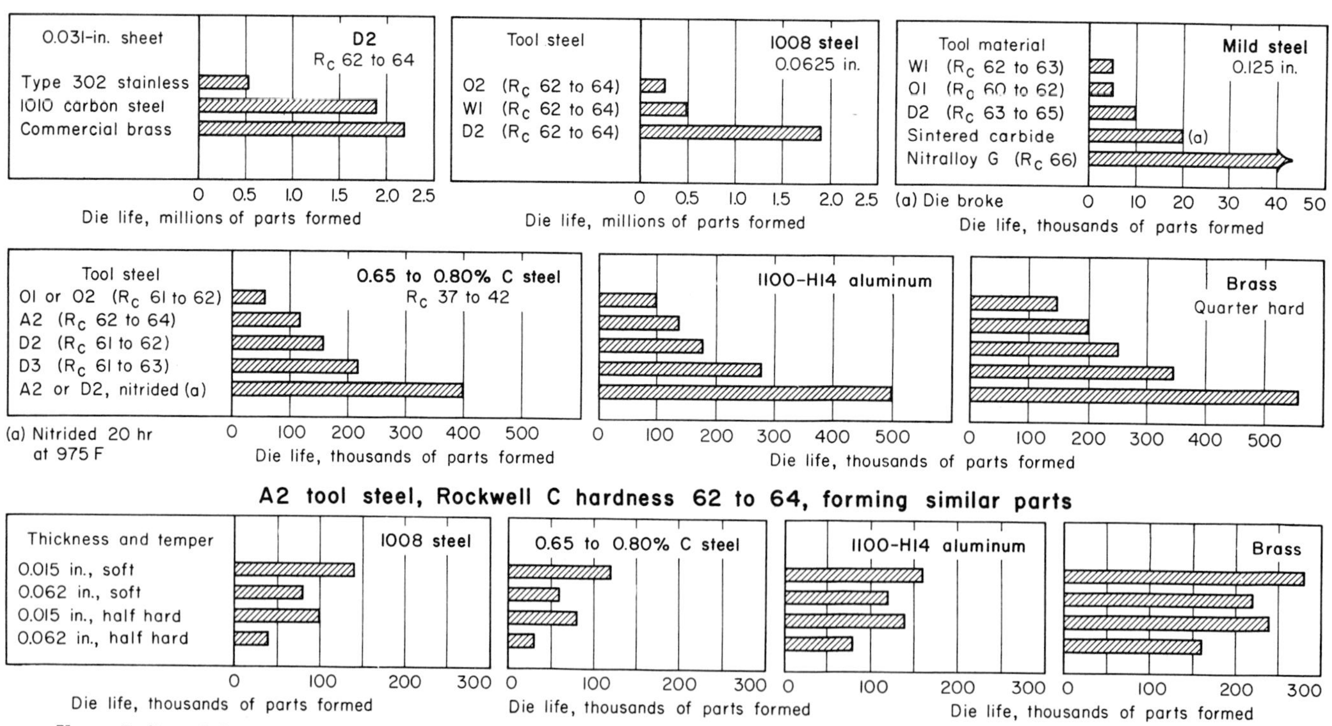

Upper Left and Center. Effect of material being formed on life of D2 dies, and comparison of die life for O2, W1 and D2. These parts have a maximum area of approximately 4 sq in., and the forming is largely medium severity (upper left) and simple 90° bend (upper center).

Upper Right. Severe forming of drawing quality 1008 hot rolled strip. The blank is curled to form a full round bushing.

Center Three Graphs. Comparison of different die materials and effect on die life of the material being formed. Charts indicate summary results for dies forming a variety of small instrument parts having a maximum area of 3 sq in. and thickness of 0.062 in.

Lower Four Graphs. Effect of temper, thickness and composition of material being formed on the life of A2 dies. The formed parts to which these data refer are identical to those parts dealt with in the center graphs above.

Fig. 3. Die life data for various press forming operations

Table 2. Typical Life of Some Press Forming Dies

Tool material and hardness	Part material and hardness	Blank size, in.	Tool life of individual dies, thousands of pieces	Remarks
4140 nitrided, Rc 62..	6140 annealed, RB 90	0.078 by 17 diam	10	(a)
O1 punch, Rc 58 to 59	Type 302, annealed, R15T 88	0.008 by 1/2	100 to 150	(b)
M2 punch, Rc 60	10 karat gold on nickel, R15T 82	0.009 by 5/8	150 to 200	(c)
Carbide die	10 karat gold on nickel, R15T 82	0.009 by 5/8	750 to 1000	(d)
W1 punch, Rc 58 to 59	10 karat gold on nickel, R15T 82	0.008 by 7/8	50 to 150	(e)
O1 punch, Rc 58 to 59	10 karat gold on nickel, R15T 82	0.008 by 3/4	50 to 70	(e)
S1	1008 steel, HR	0.219 thick	200	(f)
A2	Type 430 strip	0.032 by 3/8	28	(g)
D2, Rc 60	....	0.040 by 4 3/16 wide	100	(h)
D7, Rc 65	....	0.040 by 4 3/16 wide	2000	(h)
D2, Rc 56 to 60	1008 strip steel, CR, CQ, RB 60 max	0.0329 by 5 3/4	49	(i)
D2, Rc 56 to 60	1008 strip steel, CR, CQ, RB 60 max	0.0329 by 6 1/8	33, 44, 48	(j)
D2, Rc 56 to 60	1008 strip steel, CR, CQ, RB 60 max	0.0329 by 6 1/8	107, 108	(k)
D2, Rc 56 to 60	1008 strip steel, CR, CQ, RB 60 max	0.0425 by 6 1/8	45	(l)
D2	....	0.187 by 7 by 12	500	(m)
D2	Aluminum	0.036 by 15 by 24	750	(n)
D2	1008 steel HR, P and O	0.095 thick	...	(o)
D2 die	1008 steel, P and O	0.070 thick	Die, 155	(p)
D2, Rc 58	1008 steel, RB 50	0.080 thick	40, 250	(q)
D2	1010 HR coil, P and O	0.255 thick	500	(r)
D2	1045 steel, spheroidize annealed	0.040 thick	2000	(s)
D2, Rc 57	1025 steel, RB 77	1/4 thick, 2 1/4 wide	250	(t)
D2, Rc 59	1010 steel, RB 60	3/16 thick, 12 wide	135	(u)
D2, Rc 60	1010 steel, RB 55	0.090 thick, 6 wide	1000	(v)
Cast iron	1008 sheet steel, CR	0.036 by 32 by 56	250 to 300	(w)

CR = cold rolled. HR = hot rolled. CQ = commercial quality. P and O = pickled and oiled.

(a) Part is a reel head with a ¼-in. radius rim, formed in two operations. Die was reground after 0.002 in. wear. (b) Part is a slide of ladies' bracelet being formed by bending 90°; 100 pieces per minute using mineral oil lubrication; 0.004 in. wear to regrind. (c) Same forming operation as in (b); 100 pieces per minute using mineral oil lubrication; 0.004 in. wear to regrind. (d) Mating die for (c); 0.002 in. wear to regrind. (e) Part is a link for men's bracelet, bent 90°; 100 pieces per minute using mineral oil lubrication; 0.002 in. wear to regrind. (f) Die wear, 0.001 in. to regrind. (g) Part is a coil clip. (h) Part is a 20-mm ammunition link; 0.001 in. die wear to regrind.

(i) Part is an end top plate for a toaster; bending operation with forming to 1-in. depth; 800 pieces per hour with sulfur-chlorine-base oil lubrication. (j) Part is a center top plate for a toaster; bending operation with forming to ¾-in. depth; 940 pieces per hour with sulfur-chlorine-base oil lubrication. (k) Part is a bread rack for toaster being formed to ½-in. depth; 538 pieces per hour with sulfur-chlorine-base oil lubrication. (l) Part is a side frame for toaster formed by bending to 45°; 840 pieces per hour using sulfur-chlorine-base oil lubrication. (m) Part is a refrigerator compressor case; die galls at 500,000 pieces and must be polished; after 950,000 pieces, die must be recut; die life is 4,000,000 pieces at time of replacement. (n) Part is a rotary shelf for a refrigerator; the shelf has voids punched out and the remaining strips are formed to ridges. (o) Part is a drum shell drawn to a depth of ¾ in.

(p) Part is a flywheel spider produced at a press speed of 45 strokes per minute; die is reground after 0.002 in. wear. (q) Forming consists of 90° bending at a press speed of 65 strokes per minute. Measured wear is 0.002 in. at 40,000 pieces and 0.010 in. at 250,000 pieces formed. (r) Part is a trailer rim lug which has been formed to a contour of 1¼-in. depth by ⅝ in. wide. A 0.09-in. radius classifies this part as having severe formability requirements. Press speed is 65 strokes per minute. The 500,000 pieces cause 0.015 in. wear at the radius before regrind is necessary. The dies occasionally break. (s) Wear is 0.005 in. (t) Auto hinge arm with inserts of same material used at high wear areas; lubricated with zinc oxide base and paraffin oil; press speed, 14 strokes per minute. (u) Trailer coupler; lubricated with zinc oxide base and paraffin oil; press speed, 5 strokes per minute. (v) Auto lock component; lubricated with zinc oxide base die lubricant; press speed, 100 strokes per minute. (w) Outer door for refrigerator, formed to a depth of 3 in.

Table 3. Nominal Compositions of Tool Materials for Press Forming Dies

Material	Composition
Tool Steels	
W1, water-hardening	0.60 to 1.40 C
S1, oil-hardening, shock-resisting ...	0.50 C; 1.50 Cr; 2.50 W
O1, oil-hardening	0.90 C; 1.00 Mn; 0.50 Cr; 0.50 Mo
A2, air-hardening, medium-alloy	1.00 C; 5.00 Cr; 1.00 Mo
A4, low-temperature air-hardening .	1.00 C; 2.00 Mn; 1.00 Cr; 1.00 Mo
D2, high-carbon high-chromium	1.50 C; 12.00 Cr; 1.00 Mo
D3, high-carbon high-chromium	2.25 C; 12.00 Cr
D5, high-carbon high-chromium	1.50 C; 12.00 Cr; 1.00 Mo; 3.00 Co
D7, high-carbon high-chromium	2.35 C; 12.00 Cr; 1.00 Mo; 4.00 V
M2, molybdenum high speed	0.80 C; 4.00 Cr; 5.00 Mo; 6.00 W; 2.00 V
M4, molybdenum high speed	1.30 C; 4.00 Cr; 4.50 Mo; 5.50 W; 4.00 V
Other Ferrous Alloys	
Hot rolled mild steel	Steels 1010 to 1018
Unalloyed cast iron, 185 to 225 Bhn..	3% total C; 1.6 Si; 0.7 Mn (or equivalent)
Alloy cast iron, 200 to 250 Bhn	3% total C; 0.7% combined C; 1.6 Si; 0.4 Cr; 0.4 Mo (or equivalent)
Cast carbon steel, 185 to 225 Bhn....	0.75 C
Cast alloy steel, 200 to 235 Bhn	0.45 C; 1.10 Cr; 0.40 Mo
4140 alloy steel	0.40 C; 0.60 Mn; 0.30 Si; 1 Cr; 0.20 Mo
Nitralloy	0.35 C; 1.20 Cr; 0.20 Mo; 1.00 Al
Nonferrous Alloys	
Zinc alloy	4 Al; 3 Cu; 0.03 Mg; rem, Zn
Aluminum bronze, 270 to 300 Bhn....	13 Al; 4 Fe; rem, Cu
Plastics	
Polyester-glass	50% polyester plastic, 50% glass in the form of cloth, strand or chopped fibers
Epoxy-glass	50% epoxy plastic, 50% glass as above
Polyester-metal	Polyester plastic reinforced with metal powder
Epoxy-metal	Epoxy plastic reinforced with metal powder
Nylon-metal	Polyamide plastic reinforced with metal powder
Polyester or epoxy-glass-metal	Polyester or epoxy plastic with both glass and metal as above

dies, cast to shape, are made of as few components as possible. This not only results in a lower cost because fasteners are largely eliminated, but it adds to the simplicity and durability of the tool. Otherwise, zinc alloy tools for conventional press production are designed according to the usual practice.

Virtually all the zinc alloy in a tool can be salvaged for reuse simply by melting. For greatest economy, the user must have melting and molding facilities available; the commercial scrap value of the alloy is about 65% of the original cost, while remelting costs the user about as much as melting purchased metal. These facilities, plus the necessary auxiliaries, cost a minimum of $50,000 to install.

Cycle time or press speed has a great effect on the life of zinc alloy tools. A die component moving more than 25 in. per min during its working cycle is highly destructive to zinc alloy. The best results have been obtained on double-action presses that average three strokes per minute and involve drawing to a depth of more than 6 in. For this work, hydraulic presses are more desirable than mechanical presses. Therefore, tools made from zinc alloys have a minimum size limit. This is true because even the slowest speed of small presses is greater than is permissible with zinc alloy tooling, and the cost is excessive if a small die is run on a large press merely to suit the speed requirements. Zinc alloy dies cannot be used with unlubricated stock because wear will be too rapid.

A zinc alloy containing 4 Al, 3.25 Cu, 0.1 Mg, 0.7 Ni and 0.2 Ti has been in general use in a few large plants for more than five years. Based on this experience, it has provided die life up to three times that indicated for the standard zinc alloy in the selection tables.

Hot Rolled Mild Steel

Where a limited variety of tool materials are available, hot rolled mild steel plate with carbon content from 0.10 to 0.20% is in major use as a die material, and is also widely used elsewhere for short-run forming of small parts. This type of steel is not recommended without surface hardening except for quantities of less than 10,000 parts. Dies are made by torch cutting, then normalizing and grinding the stock. The dies are easily machined and generally are carburized, quenched and tempered after finish machining. No grinding should be done after carburizing. For a production quantity of 100,000 parts such as part 5, the die is chromium plated after hardening.

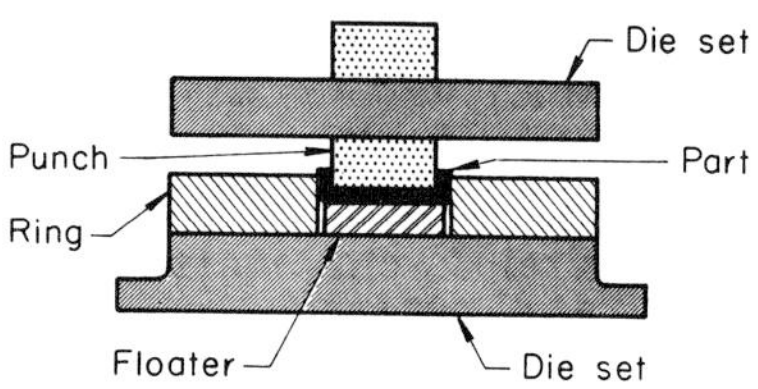

Fig. 4. Die for Tables 1 and 4, to make 3-in. and 12-in. parts

Hot rolled mild steel can be used for dies only where straightening facilities are available for correcting distortion induced by heat treatment. Straightening is necessary for all except symmetrical and thick shapes such as a large round punch.

Medium-Carbon Alloy Steels

Several medium-carbon alloy steels such as 4140 are available in plate form and are useful for some types of forming dies. For example, 4140 (or a slightly modified composition sold as a proprietary steel) is available prehardened to about Rockwell C 30. Dies machined from

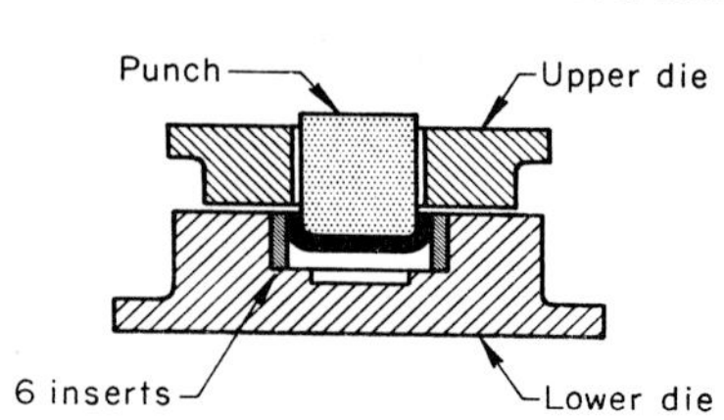

Fig. 5. Die for Tables 1 and 4, to make 36-in. part

Table 4. Relative Cost of Tooling to Form Parts 3 and 4
(Tools are shown in Fig. 4 and 5, in column at left.)

Item	Cost of die components(a), $: Die set	Punch	Ring	Floater	All others	Templet	Pattern	Totals
Tooling for Part 3, Maximum Dimension of 3 In. (See Fig. 4)								
Solid D2 tool steel								
Material	$ 34	$ 4	$ 18	$ 3	$ 2	...	....	$ 61
Labor	60	120	200	50	20	...	....	450
Total	94	124	218	53	22	...	....	511
Solid A2 tool steel								
Material	34	2	12	2	2	...	....	52
Labor	60	100	160	50	20	...	....	390
Total	94	102	172	52	22	...	....	442
Unalloyed gray cast iron								
Material	34	1	3	1	1	...	....	40
Labor	60	70	100	50	20	...	....	300
Total	94	71	103	51	21	...	....	340
Hot rolled mild steel, carburized								
Material	34	1	1	1	1	...	....	38
Labor	60	80	120	50	20	...	....	330
Total	94	81	121	51	21	...	....	368
4140 alloy steel								
Material	34	1	2	1	1	...	....	39
Labor	60	80	120	50	20	...	....	330
Total	94	81	122	51	21	...	....	369
Epoxy-metal(b)								
Material	34	....	3	....	1	...	$ 2	40
Labor	60	....	160	....	20	...	80	320
Total	94	....	163	....	21	...	82	360
Tooling for a 12-In. Part Similar to Parts 3 and 4 (See Fig. 4)								
Cast iron with inserts of D2 tool steel								
Material	$ 127	$ 9	$107	$ 7	$ 5	...	....	$ 255
Labor	80	200	620(c)	100	40	...	....	1040
Total	207	209	727	107	45	...	....	1295
Cast iron with inserts of A2 tool steel								
Material	127	5	72	7	5	...	....	216
Labor	80	160	580(d)	100	40	...	....	960
Total	207	165	652	107	45	...	....	1176
Unalloyed gray cast iron								
Material	127	9	18	7	2	...	....	163
Labor	80	130	240	100	40	...	....	590
Total	207	139	258	107	42	...	....	753
Hot rolled mild steel, carburized								
Material	127	6	13	5	2	...	....	153
Labor	80	140	260	100	40	...	....	620
Total	207	146	273	105	42	...	....	773
4140 alloy steel								
Material	127	18	32	14	2	...	....	193
Labor	80	140	260	100	40	...	....	620
Total	207	158	292	114	42	...	....	813
Alloy die iron, 225 Bhn								
Material	127	9	18	7	2	...	....	163
Labor	80	130	240	100	40	...	....	590
Total	207	139	258	107	42	...	....	753
Alloy die iron with inserts of aluminum bronze, 285 Bhn								
Material	127	9	53	7	2	...	....	198
Labor	80	130	800	100	40	...	....	1150
Total	207	139	853	107	42	...	....	1348
Plastic cast (epoxy, iron filled)								
Material	127	20	15	6	....	...	....	168
Labor	80	240	120	100	....	...	....	540
Total	207	260	135	106	....	...	....	708
Epoxy-metal(b)								
Material	50	15	....	....	2	...	$ 5	72
Labor	40	120	....	....	15	...	60	235
Total	90	135	....	....	17	...	65	307

Item	Cost of die components(a), $: Punch	Upper die blank-holder	Lower die: Blank-holder	Lower die: Inserts	All others	Templet	Pattern	Totals
Tooling for Part 4, Maximum Dimension of 36 In. (See Fig. 5)								
Cast iron with inserts of D2 tool steel								
Material	$ 224	$ 88	$172	$233	$ 8	...	$ 12	$ 737
Labor	480	460	480	680	160	$80	560	2900
Total	704	548	652	913	168	80	572	3637
Cast iron with inserts of A2 tool steel								
Material	224	88	172	151	8	...	12	655
Labor	480	460	480	560	160	80	560	2780
Total	704	548	652	711	168	80	572	3435
Unalloyed gray cast iron, 210 Bhn								
Material	100	51	110	....	5	...	12	278
Labor	400	440	620	....	160	80	560	2260
Total	500	491	730	....	165	80	572	2538
Zinc alloy (Zn-Al-Cu-Mg)								
Material	Pattern,	$325	Metal, 3450 lb at $0.2525 per lb,				$871	1196(e)
Labor	Foundry,	$660	Machining, $400; spotting, fitting,				$800	1860
Total		....			....			3056(e)
Alloy die iron, 225 Bhn								
Material	120	61	131	....	5	...	12	329
Labor	400	440	620	....	160	80	560	2260
Total	520	501	751	....	165	80	572	2589
Alloy die iron with inserts of aluminum bronze, 285 Bhn								
Material	120	61	131	100	5	...	12	429
Labor	400	440	670	620	160	80	560	2930
Total	520	501	801	720	165	80	572	3359
Combination of cast and laminated construction (epoxy, iron filled)								
Material	157	146	240	....	....	...	....	543
Labor	1010	480	640	....	....	...	....	2130
Total	1167	626	880	....	....	...	....	2673

(a) The cost of labor includes overhead and was computed at $10 per hr. (b) Costs listed under punch for epoxy-metal include ring and floater also. (c) The labor cost for making the inserts is $360. (d) The labor cost for making the inserts is $320. (e) Allowing 65% scrap value for the metal reduces total cost from $3056 to $2490.

this material are used without heat treatment for moderate forming or short runs — for example, the label holder die of Fig. 2. For longer runs or more severe forming, dies made from this prehardened steel can be nitrided — for instance, the die for a bench leg top, Fig. 2. These steels are also suitable for flame hardening and may be flame hardened by a hand torch for smoothly contoured parts similar to part 3 where the flame hardening operation could progress continuously along the wearing surfaces.

The 4140 steel die to form part 3 is a simple five-element tool consisting of a punch, a die mounted to a pair of die shoes, and a spring or rubber-backed pad to eject the part from the lower die. A spring or rubber-backed stripper is used to strip the part from the upper punch. Steel 4140 is selected for the punch and die because a component of this shape and size lends itself to low-cost design and toolmaking techniques. The punch and die can be cut from plate with a band saw so that only a small amount of filing and fitting is necessary. The edges of both the punch and die are then flame hardened, which eliminates furnace heat treatment. After grinding and polishing out the edges as necessary, the punch and die are fitted to mating die shoes, and the pad and stripper are added, along with the necessary gages for locating the blanks.

When properly hardened and finished, such a die can produce 100,000 pieces or more. Some tools of this kind have produced more than 200,000 pieces from 0.060-in. cold rolled 1010 steel.

Steels 8640 and 5140 are suitable alternates for 4140 and cost about the same.

Several medium-carbon free-cutting grades sold as proprietary steels are also used for flame hardened dies. Typical compositions are: 0.45 C, 1.10 Mn, 0.25 S; and 0.50 C, 1.10 Mn, 0.08 S, and 0.15 Mo. These steels, which are not recommended for nitriding, cost less than 4140 but usually require water quenching to develop full hardness after flame heating, whereas 4140, because of its higher hardenability, does not.

Cast Steel

Cast steel is used for forming parts larger than 12 in. long where the shape of the part makes a casting more feasible than wrought stock. Plain carbon steel (0.65 to 0.85 C) and alloy cast steel (0.40 to 0.50 C, 0.90 to 1.25 Mn, 0.90 to 1.25 Cr, 0.35 to 0.50 Mo) are two commonly used compositions. Cast steel is more costly and more difficult to machine than cast iron and is more likely to gall. However, it is tougher and more resistant to abrasive wear than cast iron. Both the alloy and plain carbon grades of cast steel can be flame hardened; the plain carbon grade must be water quenched, whereas the alloy grade will harden in air.

One important advantage of cast steel, particularly the plain carbon grade, is its weldability. It can be rewelded with steel or hard faced with tool steel, aluminum bronze or other hard facing material. Because of its poorer resistance to galling, it is less desirable for dies for forming carbon and stainless steels than for aluminum and copper alloys. Cast steel is also useful for restrike, flanging and other types of dies that are less likely to gall or pick up metal from the sheet.

Aluminum Bronzes

Aluminum bronzes, which are alloys principally of copper and aluminum, are available in a wide range of hardness from 120 to 340 Bhn. These alloys have excellent resistance to galling and are desirable for dies where best finish is required on carbon and stainless steel parts. The softer grades (120 to 270 Bhn) wear rapidly, particularly where wrinkling is likely to occur on the formed part. The harder grades (270 to 340 Bhn) wear less rapidly, but are difficult to machine, drill and tap. Where elimination of scratches is extremely important, aluminum bronzes should be considered, but for medium to high production (10,000 to 100,000), replacement inserts should be available to permit reconditioning of worn tools.

Plastics

The raw materials used in plastics formulations are readily available, as are the prepared tool plastics. While satisfactory dies may be produced from plastics, even in initial experiments, quality and economy improve with experience.

Requiring no outlay for special equipment, plastic tooling may be produced on the usual toolmaking facilities. Patterns, prototype parts or models must be available for casting the resin (if no reinforcement is to be used) or for placing the resin-impregnated glass reinforcement on the face of dies that are built with reinforcement throughout. When a pattern of some kind is not available, there will be little or no cost advantage in using plastic dies rather than metal dies.

Such terms as nylon-metal, epoxy-metal, polyester-metal, which appear in the selection tables, refer to plastics reinforced by the addition of metal powder to the uncured resin. The reference to inserts in footnote (d) of the selection tables for parts 1, 2, 3 and 4 indicates that plastic dies selected to produce the quantities shown are reinforced at the areas of highest pressure and wear with inserts of tool steel, cast iron or alloy steel bolted or doweled in place in the die component.

The strength (10,000 to 40,000 psi) and, in particular, the hardness of plastics are inferior to those of metallic die materials. In some instances, the problem of obtaining the required strength in a plastic die must be solved by extra work in designing or developing; but in many dies the regions of high pressure are confined to local areas, which, if anticipated, are strengthened with metallic inserts.

When plastic tooling is selected for a part that is subsequently changed in design, the necessary alterations of the tooling will be costly. In addition, if maintenance is required on a plastic die, the bonding of repair plastic to worn areas may be inadequate because of the previous absorption of oil by the plastic.

The cost of plastic per unit volume, without reinforcing material, is from one to two times that of cast iron but less than that of primary zinc. However, unlike cast iron or zinc, it has no scrap value. The molding material costs from 45 to 90¢ per lb, and glass reinforcing material costs as much as $1 per sq yd.

Except for very short runs, plastics should not be selected for blankholder material where burred edges of the blank slide over the plastic surface. Since plastic materials cannot withstand such abrasion, these areas will wear severely on plastic dies.

Cost of Tooling

Table 4 shows the relative cost of dies for forming parts such as parts 3 and 4 in sizes of 3, 12 and 36 in. The cost covers only the basic die components indicated; accessories, such as gages, knockouts, wipers, guards, blankholder plates and the like, have been omitted because they do not affect the comparison of cost. For the 3 and 12-in. parts, the die consists of punch, ring and floater only, as shown in Fig. 4; for the 36-in. part, the die consists of a punch with upper and lower dies, as shown in Fig. 5.

In one set of dies shown in Table 4, the lower die had inserts of A2 tool steel; in another, D2 tool steel inserts were used; in both sets the punch was made of cast steel without inserts.

Cost of a pattern was included for the 3-in. plastic die. If this pattern were already available from some other operation and could be used without added expense for making the die, the cost would be reduced from $360 to $278.

For the 36-in. part, costs are not shown for hot rolled mild steel, because distortion in heat treatment makes this size impractical. Alloy steel 4140 was omitted because it is not generally available in a size this large.

The Selection of Material for Deep Drawing Dies

*By the ASM Committee on Deep Drawing Dies**

THIS ARTICLE deals with the selection of material for dies to draw round and square cup-shaped metal parts in a press, primarily by conventional drawing in which each reduction is made in the same direction. Types of dies and other tooling components are illustrated in Fig. 1.

Deep drawing operations are associated with stretching and bending of the sheet over the punch nose, together with a gathering and compression of the sheet metal as it is drawn through the die. For economy of manufacture, the drawn part is always produced in the fewest steps and the least number of dies possible. Ironing is used almost universally in multi-operation drawing, to produce deep draws in the fewest operations and to obtain uniform wall thickness. Each operation is designed for maximum practical reduction of the metal being drawn. Accordingly, the examples given in this article involve nearly maximum reductions of about 35%.

The selection of material for a drawing die (or any other kind of die) is aimed at production of the desired quality and quantity of parts with the least possible tooling cost per part. In small dies (for instance, up to a part size of about 3 in.), performance is the predominant consideration. Cost is a minor factor, since the cost of tool steel in a die this small ranges from about 5% of the total cost of the die for the lower-cost steels, such as W1, O1, and A2, to about 10% for the higher-alloy steels, such as D2.

In dies for parts larger than about 8 in., the tool material becomes a major factor in cost, and in a die for a 12-in. part, it may amount to about 55% when an A2 forged ring insert is used in a flame hardened alloy cast iron die.

The performance of a drawing die is determined by the amount of wear or galling during a production run. Wear of a given die material is determined largely by the kind and thickness of the sheet metal, sharpness of die radii, lubrication, die construction and finish. The amount of wear on die radii can vary by a factor of 20 between the sharpest and most liberal radii. Use of the most liberal radii is the best practice, and this practice is assumed in the accompanying selection tables.

In drawing square cups, the formation of wrinkles at the corners, accompanied by high localized pressures, may produce prohibitively high rates of wear. The square cups dealt with in this article have liberal corner radii consistent with favorable die life. Small radii will produce greater wear and shorter life of the dies.

Quantity is the principal variable in the selection tables. The kind of material to be drawn is also recognized as a variable because it determines the relative importance of wear and galling.

High localized pressures caused by inaccurate fitting of dies during tryout, rough surface on the drawn sheet or on the working surfaces of the die, inadequate lubrication, and poor maintenance (stoning) of dies are typical of uneconomical practice. This article makes no attempt to deal with improvement of die life under poor production practices, since such mistakes should never be alleviated by selection of die material.

Selection Tables

In the many instances where the tables list more than one recommended material for the same conditions of tooling, the materials are listed in the order of increasing cost per piece, but the ultimate choice will usually depend on availability, rather than on the small differences in either the performance or the cost of the tool materials.

Drawing quality sheets of aluminum and copper alloys, steel and stainless steel are considered in the tables. The maximum hardness for aluminum and copper alloys is 58 Bhn, Rockwell F 61, and Rockwell 30T 28. Maximum hardness of steel sheets dealt with is Rockwell B 70, and for stainless steel the maximum is Rockwell B 95, with Rockwell B 85 preferred.

The operations shown in the tables involve reductions of about 35% max along with ironing of the sides to a uniform thickness; if shallow draws of 15 to 20% were to be considered, the die life would be about twice as much for the indicated materials.

Table 1 shows the selection of draw ring or draw die material for cups up to about 3 in. in diameter or square. For such mild service, selection of a low-cost die material might be expected. However, the ratio of the cost of tool material to total tool cost is so small in small dies that negligible savings would

*S. R. PRANCE, *Chairman,* Chief Metallurgist, Inland Manufacturing Div., General Motors Corp.; EMIL C. BLOCKS, Research Metallurgist, Ekco Products Co.; VICTOR J. BOLL, General Manager, Tool and Die Div., Sheffield Corp.; JOHN J. CARROLL, Manufacturing Engineering, Grumman Aircraft Engineering Corp.; GEORGE L. CHASE, Works Manager, General Manufacturing Div., Scovill Mfg. Co.; L. L. JAFFE, Supervisor, Manufacturing Research Laboratory, Frigidaire Div., General Motors Corp.; ERVIN M. SYLVESTER, Chief Die Engineer, Stamping Div., Chrysler Corp.; J. H. WOLFHART, Chief Tool Engineer, Kaiser Metal Products, Inc.; CARL H. GERLACH, *Secretary,* Project Engineer, Publication Development Div., American Society for Metals.

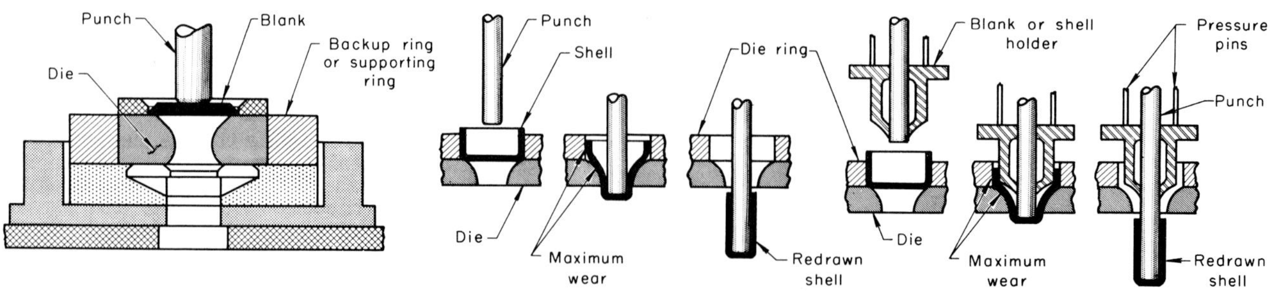

Fig. 1. Types of dies and other tooling components used in the first draw and in conventional redrawing operations

result. In fact, the selection problem for such dies is easy because there is such a slight cost factor involved in using even the most expensive tool steels.

If only 50 or 100 small cups were to be made of aluminum and if there were no ironing involved, materials such as zinc, plastics, hot rolled soft steel or masonite might be economical, provided experience in constructing dies from such lower-cost materials were sufficient to accelerate tool making.

The materials recommended for making the square cup (part 3) involve conservative radii at punch nose, punch nose corners, and punch and die corners where the sheet material gathers and may wrinkle in proportion to the sharpness of the radius. More liberal radii than those shown would not greatly increase die life nor influence material selection. Sharper radii would increase the scrap rate in drawing and might increase die wear as much as 20 times unless tool material selections were upgraded.

This square cup (part 3) is more difficult to draw and, for equal die life, requires better draw ring material at the corners than is needed in the draw ring for part 2. The fact that the recommended materials are the same for all parts in Table 1 indicates that the recommendations for the two cylindrical cups are more conservative.

Table 2 recommends materials for the wearing edge of the die for drawing large cups, like parts 4 and 5, of 12 and 24-in. diam, respectively, and 18 in. deep. Under conditions for which tool steel or aluminum bronze is indicated in the table, the material is used as an inserted ring, preferably made from a forging but otherwise, with slightly less die life, from a casting. A sectional ring could be used, but this is an inferior alternative and would not perform with the die life indicated in the tables, because die surfaces and corners would deteriorate along the joints between sections. As noted in footnote (b) of Table 2, inserts are usually pressed into a die made of flame hardened alloy cast iron. However, the alloy iron could be left unhardened, or soft gray iron could be used, to make 1000 cups from any sheet metal shown in Table 2.

While the recommendations in Table 2 apply particularly to parts of 12 to 24-in. diam, there is no adequate material of less cost than alloy cast iron into which inserts can be placed satisfactorily. Larger dies, therefore, would be made of the same materials as in Table 2. However, for deeper draws, the life of the indicated draw ring materials would be less.

Table 3 deals with the materials for corner inserts at the point of greatest wear of the draw die for a large square cup such as part 6, or for any similar part more than 8 in. square of equal depth and with equally conservative corner radii. For drawing aluminum and steel, the inserts extend 2 in. past the radius at the corner and are fitted to blend into the adjoining flame hardened draw die of alloy cast iron. Drawing of this part from type 300 stainless steel requires inserts completely around the wearing surfaces of the flame hardened alloy cast iron die if the best finish is to be maintained. The selection of nitridable steels and the more expensive aluminum bronze for drawing stainless steel indicates that galling is a more serious problem than wear. Nitrided A2, D2 and D3, and aluminum bronze have excellent resistance to galling; the nitrided steels are superior in wear resistance also.

Short-run quantities of 1000 pieces made of aluminum or copper alloys or of carbon steel can be made satisfactorily from dies of flame hardened alloy cast iron without corner inserts. One thousand pieces of stainless steel can be run using a die of flame hardened alloy cast iron (preferably not over 400 Bhn) or with aluminum bronze corner inserts, where the best possible and most uniform surface appearance is necessary.

Table 4 gives the recommended tool materials for the wearing edge to draw a shallow 54 by 36-in. pan shaped like part 7. This part is shallower and of thinner sheet (0.031-in.) than part 6 and requires less wear-resistant die materials. About 1000 of these parts could be drawn from aluminum or steel sheet with an unhardened alloy iron die, but for a stainless steel part of good finish, aluminum bronze inserts would be needed in the corners. When cost is a minor consideration, aluminum bronze inserts can be used for all wear surfaces in order to obtain the best finish.

Table 1. Recommended Draw Ring Materials for Drawing and Ironing Cups up to About 3 In. in Diameter or Square from 0.062-In. Sheet (Parts 1, 2 and 3)

Metal to be drawn	Total quantity of parts to be drawn: 10,000	100,000	1,000,000
Drawing quality aluminum and copper alloys	W1, O1	O1, A2	A2, D2
Drawing quality steel	W1, O1	O1, A2	A2, D2
Series 300 stainless steel	W1 chromium plated, aluminum bronze	Nitrided A2, aluminum bronze	Nitrided D2 or D3, sintered carbide

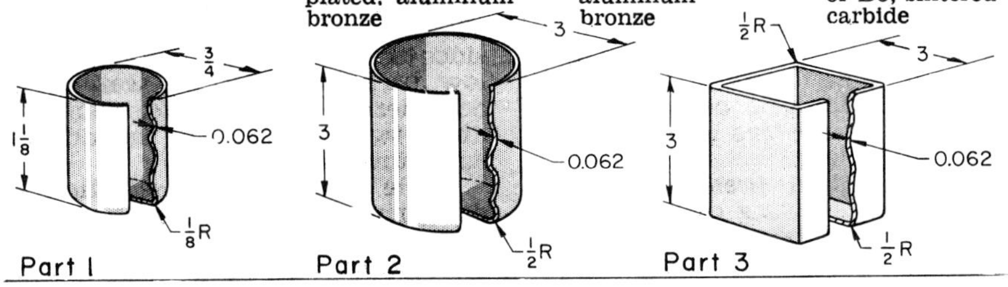

Table 2. Recommended Draw Ring Material for Drawing and Ironing Large Round Cups 12 In. or More in Diameter from 0.062-In. Sheet (Parts 4 and 5)

Metal to be drawn	Total quantity of parts to be drawn: 10,000	100,000	1,000,000
Drawing quality aluminum and copper alloys	Alloy cast iron (a)	Alloy cast iron (a) or A2 inserts (b)	A2 or D2 inserts (b)
Drawing quality steel	Alloy cast iron (a)	Alloy cast iron (c) or A2 inserts (b)	A2 or D2 inserts (b)
Series 300 stainless steel	Alloy cast iron (d) or aluminum bronze inserts (b)	A2 or aluminum bronze inserts (b)	A2 or D2 nitrided inserts (b)

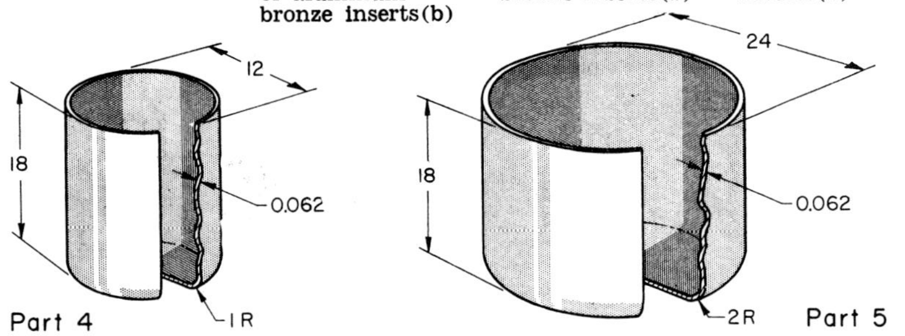

(a) Wearing surfaces flame hardened. (b) In flame hardened alloy cast iron. (c) Quenched and tempered for part 4; flame hardened for part 5. (d) Flame hardened on wearing surfaces to not over 420 Bhn.

Thickness of Sheet Metal

In drawing thick sheets of a given metal, the pressure on the dies increases in proportion to the square of the sheet thickness. The pressure involved is concentrated on the draw radius, and increasing sheet thickness will localize the wear in this high-pressure area without a similar effect on other surfaces of the die.

Thick stock will wrinkle less than thin stock. As a result the pressures required to prevent wrinkling are no greater, and may even be less, than those required for thin stock. In fact, heavy sheets are often drawn without a pressure pad. As a result, the wear-resistance requirements for pressure pads used on heavy stock are usually no greater than for thin material.

Table 5 recommends tool materials for drawing and ironing sheet steel and plate ranging in thickness from 0.015 to 0.500 in. Where alloy cast iron is recommended, for the long runs and thick sheets, it is intended to be flame hardened; for small quantities and thin sheets, it may be used as cast; for the mildest applications, even unalloyed gray iron at 200 to 250 Bhn may be used.

The O1, A2 and D2 tool steels recommended in the tables are preferably forged rings, but may be cast rings inserted into flame hardened dies.

Punches, Blankholders and Other Components

Table 6 lists materials for punches and blankholder pressure pads or sleeves for drawing and ironing steel parts 2, 4 and 5. The same die materials are satisfactory for all operations in the sequence that makes the part, provided all reductions are no greater than 35%.

Punches, except those of cast iron, may be chromium plated to facilitate removal of the part. As indicated in the table, large punches to produce more than 10,000 parts have a nose insert of tool steel.

Punch steel selection depends on the size of punch and the drawing pressure required to make the cup. Increases in sheet thickness produce a disproportionate increase in drawing pressure for a fixed punch diameter. As the resulting punch pressure increases from low to maximum values, the selection of punch material for short runs changes in the following order: alloy cast iron, 4140, W1, W2, O1, and carburized S1. For long runs, the selection as pressure increases is as follows: W1, W2, O1, S1 (slightly carburized), A2, D2 and M2.

Usually, in drawing without ironing, there is slight metal movement over the punch, which is subjected to relatively little wear regardless of the thickness of metal drawn. When the piece is ironed while being drawn, the pressure and movement of metal over the punch become considerable, and a tool material of better wear resistance is required.

The selection of a punch material to resist wear is the same regardless of the alloy being drawn. For example, in a cup like part 2, if the last diameter reduction is 35% and if reduction is combined with ironing, the punch material would be the same to make identical quantities, whether the sheet being drawn is aluminum, steel or stainless steel, without a finish requirement. The selection, therefore, would be based on Table 6. Frequently, tool steel ironing punches are chromium plated 0.0002 to 0.0008 in., to facilitate removal of the part from the punch. Plating also improves punch life but usually is less effective than selecting the next better tool steel.

The materials shown in the column headed "10,000" are the lowest in cost that can be recommended for punches where ironing is combined with drawing, even for quantities considerably less than 10,000.

The blankholder pressure pad or sleeve to make part 2 is subject to little wear, as indicated by the recommendation of W1 or O1 as the tool material for 1,000,000 parts in Table 6. In fact, unhardened mild steel may be used for making 10,000 such parts whenever galling is not objectionable on the inside of the part. Table 7 is a continuation of Table 6, showing recommended materials for drawing the square cups, parts 3, 6 and 7.

Inserts are not used in blankholder pressure pads except possibly on the first operation for square parts, or for maintaining finish inside the cup.

Pressure sleeves (also called

Table 3. Recommended Insert Materials for Draw Rings for Drawing and Ironing Large Square Cups Similar to Part 6 from 0.062-In. Sheet

Metal to be drawn	Total quantity of parts to be drawn: 10,000	100,000	1,000,000
Drawing quality aluminum and copper alloys	W1	O1, A2	A2, D2
Drawing quality steel	W1	O1, A2	A2, D2, nitrided A2 or D2
Series 300 stainless steel	W1 or aluminum bronze	Nitrided A2, aluminum bronze	Nitrided A2 or D2

For drawing aluminum, copper and steel, the tool material would be used as corner inserts. For drawing stainless steel parts, the inserts would be used for all wear surfaces. Recommended selections apply to both the push-through and inverted-draw methods of tooling.

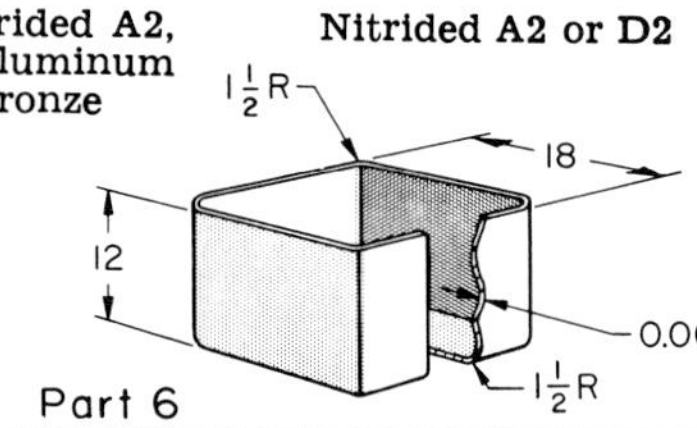

Part 6

Table 4. Recommended Draw Ring Material for Drawing and Ironing a Large Pan Similar to Part 7 from 0.031-In. Sheet

Metal to be drawn	Total quantity of parts to be drawn: 10,000	100,000	1,000,000
Drawing quality aluminum and copper alloys	Alloy cast iron(a)	Alloy cast iron(a) or A2 corner inserts(b)	Nitrided A2 or D2 inserts(b)
Drawing quality steel	Alloy cast iron(a)	Alloy cast iron(a) or A2 corner inserts(b)	Nitrided A2 or D2 inserts(b)
Series 300 stainless steel	Alloy cast iron(c) or aluminum bronze inserts(b)	Nitrided A2 or aluminum bronze inserts(b)	Nitrided A2 or D2 inserts(b)

(a) Wearing surfaces flame hardened. (b) In flame hardened alloy cast iron. (c) Flame hardened on wearing surfaces to not over 420 Bhn.

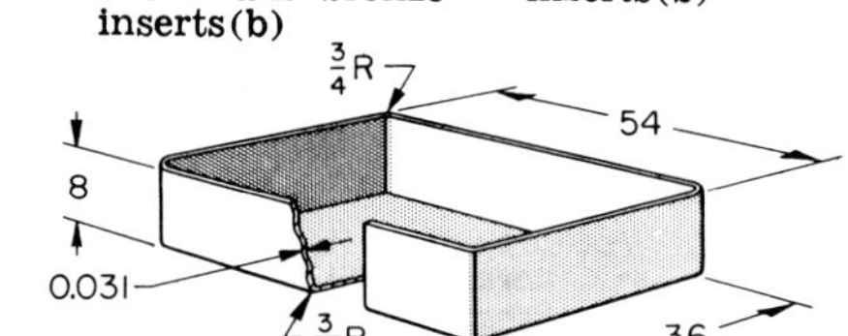

Part 7

Table 5. Recommended Draw Ring Material for Drawing and Ironing Part 4 from Flat Rolled Steel of Six Different Thicknesses

Thickness of steel, in.	Total quantity of parts to be drawn: 1,000	10,000	100,000	1,000,000
0.015	Alloy cast iron(a)	Alloy cast iron(a)	Alloy cast iron(a)	Alloy cast iron(b), O1, A2
0.031	Alloy cast iron(a)	Alloy cast iron(a)	Alloy cast iron(b)	A2, D2
0.062	Alloy cast iron(a)	Alloy cast iron(b)	Alloy cast iron(b), A2	A2, D2
0.125	Alloy cast iron(b)	Alloy cast iron(b)	A2, D2	D2
0.250	A2	A2	D2	D2
0.500	A2	A2	D2	D2

(a) Flame hardening is not necessary.
(b) Wearing surfaces are flame hardened.

Tool steels are inserted into flame hardened alloy cast iron. In drawing the 0.500-in. plate with A2 or D2 inserts, press speed would be slower than for the thinner stock and the plate would be phosphate coated.

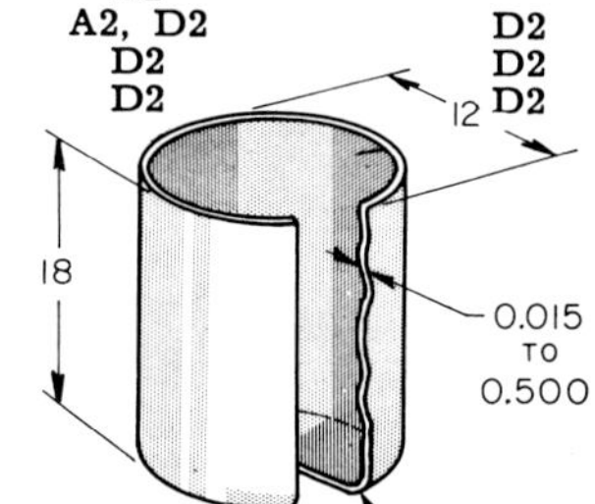

Part 4

pressure pads) are used where minimum pressure is required close to the punch. The contact area is generally less than for the pressure pads that are used where greater blank holddown surface is required.

An example would be a pressure pad for part 7 (Table 4) and a pressure sleeve for the operation illustrated in Fig. 7.

Materials recommended for either pad or sleeve depend on the size of the blank and the number of parts to be drawn. As the blank size increases or the number of parts decreases, softer and less expensive materials are used.

Holding, shrink or backup rings for holding the pressed-in draw ring insert, usually about three times the radial width of the draw ring, are made of heat treated medium-carbon steel and, for assembly, are heated to about 1000 F to obtain expansion and to allow pressing-in of the inserts.

Draw beads, frequently needed to control the flow of metal in irregular drawn shapes, are made of a material the same or nearly the same as the draw die. Beads applied by rivets or other fasteners are made of hardened W1 bead stock for medium-size dies or of hardened W2 bead stock for large dies. Inserted locking beads are usually made of hardened W1 or W2.

Ironing Tools

Table 8 shows the effect of various sidewall reductions on the selection of punches and dies for ironing. Blankholder pressure pad material is independent of ironing reduction and can be selected from Tables 6 and 7. Cast iron is not used for ironing punches, because it sinks and breaks. Sidewall reduction is measured at the top, at a point where the drawn cup is to be trimmed to a specific length.

Table 6. Recommended Materials for Punch and Blankholder Pressure Pad or Sleeve for Drawing and Ironing Round Steel Cups Like Parts 2, 4 and 5

Die component	Total quantity of parts to be drawn: 10,000	100,000	1,000,000
	Part 2		
Punch(a)	4140 carburized, W1	W1; S1 carburized	A2, D2
Blankholder pressure pad or sleeve	W1, O1	W1, O1	W1, O1
	Parts 4 and 5		
Punch(a)	Alloy cast iron(b)	O1(c)	A2(b), D2(b)
Blankholder pressure pad or sleeve	Alloy cast iron(b)	Alloy cast iron(d)	O1, A2

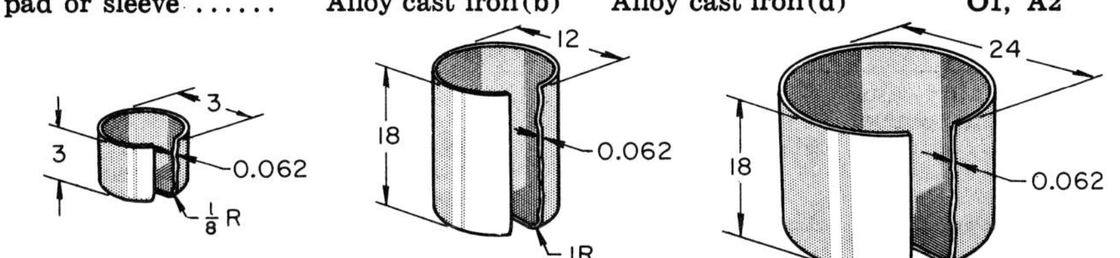

Part 2 Part 4 Part 5

(a) Chromium plating is optional on these punches, to reduce friction between the part and the punch for removal of the part. Cast iron, however, should not be plated. (b) Flame hardening not necessary. (c) The punch holder is flame hardened alloy cast iron with a nose insert of the indicated tool steel. (d) For part 4, this blankholder is quenched and tempered; for part 5, it is flame hardened.

Table 7. Recommended Materials for Punch and Blankholder Pressure Pad or Sleeve for Drawing and Ironing Square Steel Cups Like Parts 3, 6 and 7

Die component	Total quantity of parts to be drawn: 10,000	100,000	1,000,000
	Part 3		
Punch(a)	4140 carburized, W1	W1; S1 carburized	A2, D2
Blankholder pressure pad or sleeve	W1, O1	W1, O1	W1, O1
	Parts 6 and 7		
Punch(a)	4140 carburized(b)	W1, O1(c)	Nitrided A2; D2(c)
Blankholder pressure pad or sleeve	Alloy cast iron(d)	W1, O1	O1, A2

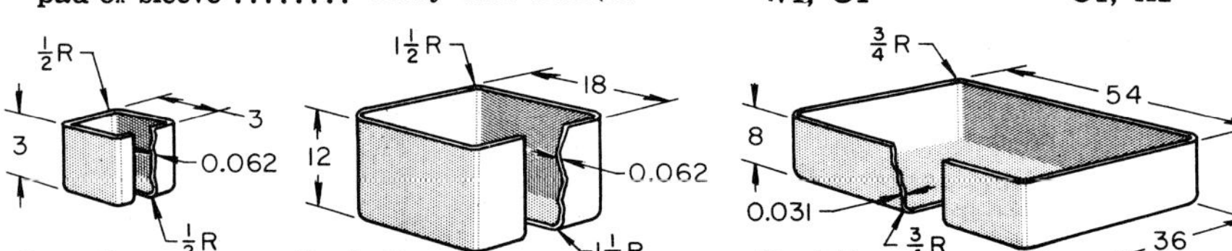

Part 3 Part 6 Part 7

(a) Chromium plating is optional on these punches to reduce friction between the part and the punch for removal of the part. (b) The punch holder is alloy cast iron with a nose insert of the indicated steel. (c) The punch holder is flame hardened alloy cast iron with nose insert of indicated tool steel. (d) Flame hardening unnecessary.

Reverse Redrawing Tools

Occasionally, specifications for redrawn shells may require a wrinkle-free sidewall of uniform thickness, or may require a section in the bottom of the cup to be sharply raised and formed in two operations. Such operations are difficult, impossible or uneconomical in the conventional single-action drawing method but are easy in reverse drawing. Figure 2 shows terminology of tooling for reverse redrawing of thin-wall shells.

Table 9 recommends punch and die materials for the reverse redrawing of steel cups of various sizes and wall thicknesses. Pressure-ring materials may be the same as die materials.

The draw ring to produce quantities of about 100 small, thick-wall parts can be made of 1018, 4140, 8620, 6150 or any similar steel, either soft, or carburized and hardened, provided distortion in heat treating is not a problem. In large parts where die material cost is an important factor, steels such as 4140 or 6150, when carburized and finely polished, can be expected to make 10,000 parts; any carbon or alloy steel will make 1000 parts when heat treated but only a few hundred when left soft.

To redraw 1000 small, thick-wall parts the recommended punch material is a hardened alloy steel such as 4140 or 6150. For all types of parts, when quantities are as low as 100, the punch may be made of the same steel, unhardened.

Combined Operations

Combined operations, including a combination of drawing and coining in one operation and a combination of successive or tandem operations called double drawing or double ironing, have made great progress since the second World War. This advance has matched developments such as better selection of drawing steels, improvements in engineering and construction of tools, and especially in surface treatments such as those using zinc phosphate with emulsified soap.

While these operations have increased production by doubling reductions and decreasing the number of operations at the expense of larger press requirements, they have also required tool steels of greater resistance to compression and heat.

For instance, in deep cold drawing, a blank of 0.30% C steel ½ in. thick and 4-in. diam can be cupped with a 35% sidewall reduction in a 250-ton press. A blank of the same structure and composition but with a thickness of 7⁄16 in. and approximate diameter of 3⅞ in., when cupped by the combined process, can be given a 50% coining reduction and an average sidewall reduction of 60% with a press of 1500 tons. The combined deep cold drawing operation saves raw material and time, and produces a more accurate cup that aids in obtaining closer tolerances in a subsequent operation.

Selection of tool steels for combined drawing and coining with reductions over 40% has little or no relation to the quantity to be produced, up to 500,000 pieces. The minimum steel to be used for both punches and dies is undecarburized D2. However, D2 should not be used for long, small-diameter punches because misalignment or accidents may cause hazardous shattering. Instead, such punches should be made of honed, slightly carburized, chromium-plated S1. A reduction in punch life should be expected in such applications.

Sidewall reductions of 60% or more, combined with quantities of 500,000 to 1,000,000 pieces, require carbide draw rings with a 0.010-in. shrink fit into the backup ring. Several S1 tool steel punches will be required to produce these larger quantities.

Double drawing or double ironing operations are successive operations in one tooling setup, with two dies placed in tandem so that a punch forces the cup through one die and then directly through the second die while the cup is still hot. Punches are longer and, because of their slenderness, S1 tool steel is preferred. Die materials are much the same as in single operations except that when temperatures are high in the second operation, selection is confined to tool steels such as A2 and D2. These more temper-resistant steels can better withstand the effects of the higher temperatures.

Work Metal Classification

Table 10 classifies various sheet metals into groups requiring the same die materials for equivalent tooling circumstances. Although aluminum and copper are listed separately from steel in Table 10, nonferrous alloys to quarter hard or even half hard cause almost as much die wear as drawing quality steel, as shown in the various selection tables. Ferritic stainless steels are similar to carbon and alloy steels of the same hardness in their rate of wear on dies.

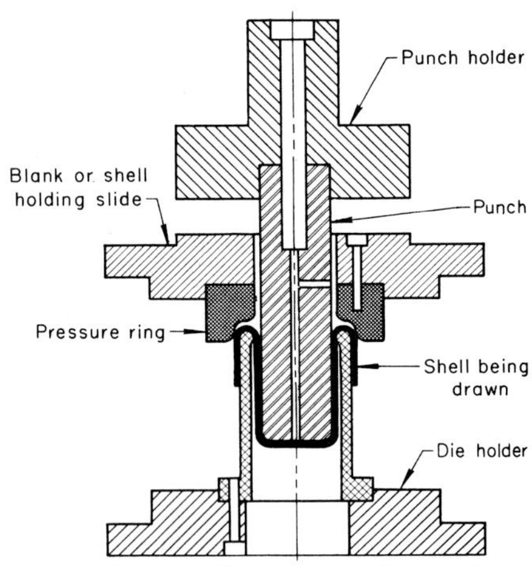

Fig. 2. Terminology for the tooling components used in reverse redrawing of thin-wall shells

Die Life

Table 11 shows performance differences in punches and dies made of various tool and constructional steels. The effect of chromium plating of punches is evident from the results of the tests made in production of the part illustrated in the table. The life, or end point, of the punch or die was taken as that production quantity where the tool no longer was capable of maintaining dimensional tolerances of ±0.003 in. for the 0.062 and 1⁵⁄₁₆-in. dimensions and ±0.015 in. for the 2⅝-in. dimension.

Production Quantity

The runs shown in the selection tables should be achieved with one set of dies under the conditions outlined in this article. For small dies, in which material cost is a small percentage of total cost, whenever the quantities are likely to be increased above the original order, overspecification of die material is justified, or a nitridable tool steel, such as A2 or D2, should be selected.

Large dies that cannot be removed from the production line obviously must be made of material certain to last for the required production period, as given in the selection tables. In drawing rectangular cups such as part 6, there is considerable crowding of the sheet in the corners with considerable probability that wrinkles will form at the corners, and greatly increased die wear is possible. Where inserts are used and the dies may be shimmed to allow for wear, risks may be taken in selecting tool material. However, if the insert is small, the cost of material may be inconsequential, compared with the cost of diemaking and shimming, which may run from $10 to $100 per hr for lost time of the press, plus labor. Thus, it is usually uneconomical to underspecify the steel for inserts.

Table 8. Recommended Steels for Punches and Dies to Iron Various Reductions of Steel Sheet up to Rockwell B 75, or Softer Metals

Ironing reduction, %	Total quantity of shells to be ironed: 1,000	10,000	100,000	1,000,000
Ironing Punches(a)				
Up to 25	W1	O1	A2	A2; S1 carburized
25 to 35	W1	A2	A2; S1 carburized	D2
35 to 50	A2	A2; S1 carburized	D2	D2
Over 50	D2	D2	D2	D2
Ironing Dies				
Up to 25	W1(b)	O1	O1	D2
25 to 35(c)	W1(b)	O1	D2	D2
35 to 50(c)	O1	D2	D2	D2
Over 50(c)	D2	D2	D2	D2

(a) All tool steel punches should be chromium plated 0.0002 to 0.0004 in., for easier removal of the part from the punch. (b) W1 is quenched on the inside and tempered to Rockwell C 60 min for these applications. (c) Draw rings must be inserted in shrink rings to iron reductions greater than 25% and for quantities of more than 10,000.

Table 9. Recommended Punch and Die Materials for the Reverse Redrawing of Steel Cups of Various Sizes and Wall Thicknesses

(For parts having no specific finish or tolerance requirements)

Die component	Total quantity of parts to be redrawn: 1,000	10,000	100,000	1,000,000
Small Thick-Wall Cups				
Die and pressure ring	O1	O1(a)	A2(b)	D2(b)
Punch(c)	4140, 6150	O1, A2	D2	D3
Medium and Large Thin-Wall Cups				
Die and pressure ring	1018(d), 4140	4140(e), O1	A2(b)	D2(b)
Punch(c)	W1	A2	D2	D2, D3

(a) The die is polished and chromium plated. (b) A2 and D2 should be nitrided. (c) All punches to make more than 1000 pieces should be heat treated to Rockwell C 60 to 62, polished and chromium plated. (d) Carburized, hardened, and polished to a fine finish. (e) 4140 or 6150 may be used when carburized and highly polished.

Table 10. Classification of Sheet Metals That Require Similar Drawing Die Materials

Sheet metals listed in Tables 1 to 9	Maximum hardness	Metals that require similar drawing die materials
Drawing quality aluminum and copper alloys	58 Bhn (500 kg) Rockwell F 61 Rockwell 30T 28	All aluminum and clad aluminum alloy sheet, copper and alloys, zinc and alloys, silver, pewter and Monel
Drawing quality steel	Rockwell B 70 Rockwell B 75	1008 to 1020 1021 to 1030
Stainless steel	Rockwell B 95	301, 302, 304, 305, 308, 310, 316, 317, 410 and 430; carbon steels clad with stainless; copper clad with stainless; magnesium drawn at 400 to 600 F with no ironing of sides; 17-4 PH, 17-7 PH and 15-7 PH stainless steels

Short-Run Tooling

Where a cup is to be ironed, selection of draw ring material for production quantities of 10,000 pieces is the same as for quantities smaller than 10,000 pieces. Lower-cost die materials can be recommended only in a few instances where less than 10,000 pieces of small parts are to be drawn and ironed. For production quantities of 1000 of parts 1 and 2, any carburized steel will make the run but with a lower-quality finish. However, the difference in tooling cost compared with W1 is insignificant, and distortion in heat treating the carburized steel would be expected to exceed that of W1.

If 1000 pieces were to be run of parts 4 and 5 made of aluminum or copper alloy or carbon steel, a draw ring material of cast iron, either alloyed or unalloyed, with a hardness of 200 to 250 Bhn would be recommended. Dies made of flame hardened alloy cast iron without corner inserts will satisfactorily produce 1000 of parts 6 and 7 made of carbon steel or aluminum.

When low production quantities of 10 to 100 cups such as parts 4 and 5 are to be made without finish requirements and without ironing, the selection of tooling material is not critical. The draw ring may be made of either hardenable or unhardenable steel. If the die is not designed to iron the cup, plastic, masonite and zinc alloys are used to produce up to 1000 cups like those shown in Fig. 3 and 4 from thin sheets of easily drawn alloys. Such materials are especially well adapted to fast tooling-up, particularly when the user can make his own dies.

Figure 3 shows a 4 by 5-in. part made of 0.031-in. aluminum alloy 2014 drawn 2½ in. deep over a 3⁄16-in. radius using a masonite punch with a 1-in. nose radius, a masonite blankholder pressure pad, and a phenolic die held in a masonite backup. This setup produces 1000 parts without rework of tools.

Table 11. Service Life of Nine Tool Materials in Drawing a Steel Cup

Punch and die material	Punches: Honed and not plated	Punches: Honed and chromium plated(a)	Dies smoothly polished
1040(b)	34,000	42,000	....
2340(b)	45,000	55,000	....
W2	46,000	59,000	52,000
3140(b)	50,000	66,000	....
4140(b)	62,000	87,000	....
O1	67,000	79,000	78,000
O6	68,000	....	78,000
A2	83,000	104,000	82,000
D5	107,000	125,000	87,000

(a) Chromium plating was 0.0002 to 0.0005 in. thick. (b) Punch material with carburized case of 0.010 in. min after grinding.

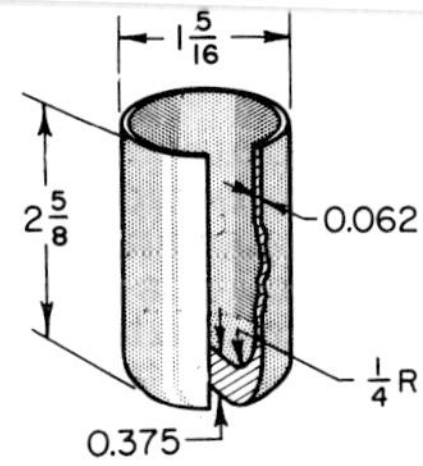

The production cup at right was made of 1020 hot rolled and annealed steel with a hardness of Rockwell B 65 to 70 before the draw, and lubricated with a zinc phosphate and soap coating. Reductions were 40% of the wall and 7% of the diameter at 20 strokes per minute in a 100-ton mechanical press with a 10-in. stroke.

Figure 4 illustrates an aluminum pan 25 in. wide, 35 in. long and 6 in. deep, drawn on a double-action press from 2014 aluminum alloy sheet 0.051 in. thick. Five hundred pieces were made with tooling consisting of a punch nose insert of epoxy resin impregnated with fiber glass, with ½-in. punch nose radius, and a zinc alloy punch holder. The die was cast phenolic resin, and the blankholder pressure pad was cast iron. The punch nose insert was reworked once to run the 500 parts.

Galling

Common causes of galling are: attempting to stretch sheet metal beyond practical limits, poor tool fitting with poor alignment or insufficient die clearance for the sheet thickness, wrinkles, the use of galling-susceptible tool steel without improving the lubricants, and rough finish on the surface of tools. For short runs, carburized hot rolled steel or hardened alloy steel dies will, in many instances, produce parts equal to those made from most tool steel dies. Exceptions may be encountered in severe reductions or in drawing metals that show a greater tendency to gall, such as austenitic stainless steel. However, tool steel dies may also gall under these conditions. The longest life can be expected if die surfaces are finished to the least roughness, with final surface scratches parallel to the direction of drawing.

Die materials to resist galling can be selected on the following basis: For parts drawn of carbon steel or nonferrous alloy sheet, the die material may be selected without regard to galling and then, as a finishing operation, the punch and die should be either nitrided or chromium plated. If tool steels such as A2, D2, D3 or D4, which contain chromium and molybdenum, have been selected, the smoothly ground tools should be nitrided, and polished or buffed after nitriding. Otherwise, the tools should be hardened to at least Rockwell C 60, smoothly finished on the wearing surfaces, hard chromium plated, and the plating polished or buffed. Punches should be plated 0.0002 to 0.0004 in. To prevent spalling or flaking of the plate, dies should not be plated more than 0.0002 in. thick.

When the parts are drawn from stainless steel or from high-nickel alloy steel, the draw ring material with best resistance to galling is aluminum bronze. The second choice is D2, D3 or D4, smoothly ground, nitrided and polished. The third choice is alloy cast iron, quenched and tempered to 400 to 420 Bhn.

Punch material is selected without regard to galling and is then chromium plated as in the previous example, unless it is cast iron.

Sintered carbide, which is only as efficient as the lubricants used, has proved economical for nonferrous alloys, carbon steel and stainless steel in many long, continuous runs, but galling will occur as soon as lubrication becomes faulty.

Lubrication

Correct lubrication of the parts being drawn is essential to reduce friction, wear and galling. In fact, deep drawing is impossible without lubrication. In actual practice, die materials are selected after trials employing the production lubricants. If excessive wear or galling occurs, a better lubricant is usually applied. For extremely difficult draws, the best lubricants are usually applied at the outset.

Tool material recommendations given in Tables 1 to 9 are predicated on the concurrent selection of suitable lubricants.

Table 12 recommends basic lubricants in accordance with composi-

Table 12. Recommended Basic Lubricants for Deep Drawing

Material drawn	Severity of draw: 10% or less	Severity of draw: 25% average	Severity of draw: 50% or more
Aluminum and its alloys	Straight mineral oil, 100-sec viscosity(a) Mineral oil with approx 10% lard oil	Straight mineral oil, 200 to 250-sec viscosity(a) Mineral oil with approx 15% lard oil	Mineral oil with EP additives—sulfur and others Coating of soap or wax dried on blanks or draws prior to draw or redraw
Copper and its alloys	5% soap solution Lard and soap emulsion	10% soap solution with stearic or oleic acid Lard oil and mineral oil with stearic acid	Lard oil blended with 50% mineral oil Coating of soap or wax dried on blanks or draws prior to draw or redraw
Carbon steel ...	Mineral oil, 250 to 350-sec viscosity(a) 5% soap solution	Emulsions of lard oil, mineral oil and sulfonated oils	Phosphate coating impregnated with dried soap or wax
Stainless steel	Castor oil and soap emulsion	Castor oil with fillers such as mica or zinc oxide	Boiled linseed oil with mica or lithopone Phosphoric acid etch with dried soap or wax film

NOTE: When more than one lubricant is recommended, they are listed in the order of increasing effectiveness.
(a) Saybolt viscosity at 100 F

tion and severity of the parts being drawn. Lubricants are marketed under proprietary names, but any supplier of lubricants can select a commercial compound from the information given in Table 12.

Chromium Plating

Chromium plating is used on tool steel draw rings to improve life. On punches, its primary function is to reduce the frictional forces and facilitate removal of the parts from the punch after the sidewalls have been ironed tight to the punch. Usually chromium plating improves punch life somewhat less than would the next best tool steel.

For successful performance of tools, chromium plating must always be deposited on a surface harder than Rockwell C 50, and the thickness of the coating should preferably be 0.0002 to 0.0004 in., and never less than 0.0001 in. This gives the required hardness and reduction of friction without excessive spalling or chipping at corners. Chromium-plated dies should be heated to 300 to 400 F for a minimum of 3 hr immediately after plating to minimize the possibility of hydrogen embrittlement.

Tool Steels

Table 13 lists the nominal chemical compositions of the die materials recommended in the selection tables.

W1 water-hardening tool steel has the advantage of low cost and good machinability. Wear resistance is fair to good, increasing with carbon content. W1 is usually limited to dies having balanced sections that are least likely to distort or crack during quenching. The success of the water-hardening grade W1 recommended in the tables for draw rings is attributed to hole hardening or inside-diameter hardening in a brine quench, followed by shrink fitting into heated (1000 F) medium-carbon steel backup rings approximately three times the width of the draw ring. When through hardened instead of hole hardened, W1 may fail by cracking after a short service life. Rockwell C 60 to 62 is the usual working hardness for W1 punches and dies.

W2 may be used where shallow hardening is acceptable and an increase in toughness is desirable. W6 may be used for deeper hardening requirements.

O1 oil-hardening tool steel costs almost twice as much as the lowest-cost water-hardening grade but only a little more than the most expensive grades. However, with oil-hardening steels, distortion resulting from heat treatment is greatly reduced. Wear resistance is slightly better than for the water-hardening grades, and machinability is good to fair.

Among the oil-hardening grades, O1 is the most readily available in almost every warehousing district in the United States. Where it is available, O2 is equivalent to O1, and the graphitic tool steel O6 has greater wear resistance than O1. Rockwell C 57 to 62 is the usual working hardness for O1.

A2 air-hardening tool steel is only slightly more costly than the oil-hardening grades O1, O2 and O6, but it is more difficult to machine. Its advantages include the best nondeforming properties and much greater resistance to wear and galling than O1. It can be nitrided to greatly improve resistance to wear and galling. A2 is machined more easily than D2; both steels can be obtained as castings, as well as in wrought form. Punches made from A2 are usually hardened to Rockwell C 58 to 60, and dies to Rockwell C 60 to 62.

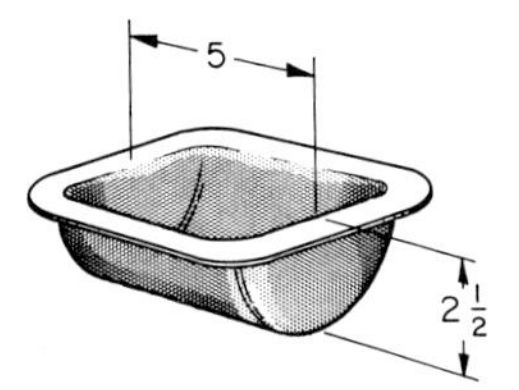

Fig. 3. Example of a part made on masonite and plastic tooling from 2014 aluminum alloy, 0.031 in. thick

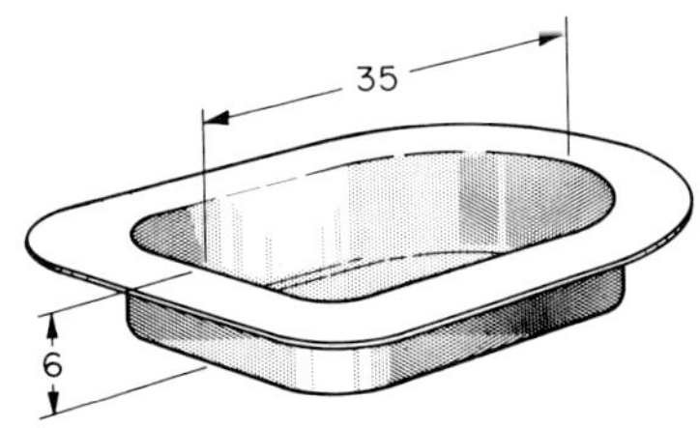

Fig. 4. Example of a part made on glass-reinforced plastic tooling from 2014 aluminum alloy, 0.051 in. thick

Table 13. Nominal Chemical Compositions of Tool Materials Recommended in Tables 1 to 9

Material	Compositions
Tool Steels	
W1	0.95 to 1.10 C
O1	0.90 C, 1.00 Mn, 0.50 Cr, 0.50 W
A2	1.00 C, 5.00 Cr, 1.00 Mo
S1	0.50 C, 1.50 Cr, 2.50 W
D2	1.50 C, 12.00 Cr, 1.00 Mo
D3	2.25 C, 12.00 Cr
Other Ferrous Alloys	
Alloy cast iron	3% total C, 0.7% combined C, 1.6 Si, 0.4 Cr, 0.4 Mo (or equivalent)
4140	0.40 C, 0.85 Mn, 0.30 Si, 1 Cr, 0.20 Mo
6150	0.50 C, 0.80 Mn, 0.30 Si, 0.95 Cr, 0.20 V
Nonferrous Alloys and Sintered Carbide	
Aluminum bronze	10 to 12 Al, 3 to 4 Fe, rem Cu
Sintered carbide(a)	5.6 C, 85.4 W, 9.0 Co

(a) The sintered carbide referred to has a hardness of Rockwell A 90, a density of 14.60 g per cu cm and a grain size of 1 micron.

D2 high-carbon high-chromium air-hardening tool steel is the most readily available of its type. D3 is at least equal to D2 in resistance to wear and galling, especially for drawing stainless steel; however, it distorts more in hardening. D4 and D5 are equal to, or better than, D2 and D3 in wear resistance and may have longer life in drawing cups like part 2 from sheet steel 1/8 in. or more thick. Resistance to wear and galling is greatly improved by nitriding. Long slender punches made of the D steels may fail by shattering. The recommended hardness of D2 and other D steels for both punches and dies is Rockwell C 57 to 59.

S1 shock-resisting steel, often used where susceptibility to breakage of punches is high, does not wear well unless the surface is slightly carburized during heat treatment, to obtain 0.70 to 0.75% C to a depth of 0.010 to 0.020 in. S2 is tougher, less costly and less wear resistant than S1. S1 at Rockwell C 53 to 58 has excellent resistance to breaking under high compressive loads when used as long slender punches for either short or medium runs, except where much heat is generated.

All of the above-mentioned tool steels are frequently used for wear inserts.

Tool steel is obtainable in hot rolled bars, hammered bars, forgings and castings. Small dies are usually produced from hot rolled bars up to 8 in. in diameter. Larger die sections require the use of hammered bars, forgings or castings. For small as well as large dies, cast-to-shape tool steel is often used. Many dies that could be made from bar stock or forgings only in sections can be produced from castings in a single piece.

Tool steel bars, forgings and castings must be ordered oversize to allow for the removal of surface defects and decarburization. Failure to remove decarburization and surface defects may result in excessive distortion and breakage during heat treating. The wear characteristics of any die will also be adversely affected if decarburization is not removed. It is recommended that an equal amount of stock should be removed from all sides when making a die.

Standard Alloy Steels

SAE 4140 alloy steel is suitable for short runs, and die life can be prolonged by carburizing or other surface hardening treatments. Steels of this type are seldom selected for long production runs. Similar alloy steels that will give comparable results are 6140 and 8640. For small dies, material costs are a small part of the total cost. Therefore, tool steels are usually a wiser selection, even for short runs, especially when production requirements are not immediately known.

Alloy Cast Iron

One of the commonly used die materials for large deep drawing dies is alloy cast iron. In designing dies for parts 8 in. in diameter or larger, alloy cast iron should be considered as a die material.

If properly made, alloy iron has reasonably good wear resistance in the as-cast condition and can be used for short to moderate runs, depending on the severity of the draw. If it is of the proper composition and microstructure, it will respond to flame hardening, and this will increase its wear resistance. For long runs and heavy pressures, it becomes necessary to use hardened tool steel inserts at the points of wear in the die sections. These hardened tool steel inserts must be large enough to cover the entire area of heavy wear; otherwise, because of the different rates of wear of the two materials, a step-off will develop at the interface where they meet, resulting in broken parts or unsatisfactory shapes.

Alloy cast iron flame hardened to the maximum value (above Rockwell C 50) is not recommended for dies used for stainless steel, because of the high probability of galling. Unhardened cast iron is compatible with stainless steel but wears rapidly so that it is unsuitable for long runs. Alloy iron heat treated to about 400 Bhn will give longer die life without the high probability of galling when drawing stainless steel. This hardness level can be obtained by oil quenching the dies from between 1500 and 1600 F and then tempering them at 600 F.

In selecting an alloy cast iron for deep drawing dies, the two most important properties required are a suitable microstructure and sufficient hardenability; composition is secondary, provided these two requirements are met. Good foundry practice is imperative in obtaining the above two properties; therefore, when cast iron sections are being purchased, foundries should be selected that have had experience and success with these types of castings.

A typical composition of a cast iron suitable for die sections 3 to 12 in. thick is as follows:

Total carbon	2.80 to 3.50%
Combined carbon	0.60 min
Manganese	0.60 to 0.90
Phosphorus	0.12 max
Sulfur	0.12 max
Silicon	1.50 to 2.25
Molybdenum	0.35 to 0.50
Chromium	0.35 to 0.50

The silicon content should be adjusted to produce the hardness specified.

While the composition given above is that of an iron that has been found suitable for die sections, it is not the only satisfactory iron (Table 13). Other compositions that will produce the properties listed below will be equally satisfactory:

Hardness (after stress relieving at 1000 F)	207 to 255 Bhn
Microstructure	Predominantly fine pearlite; minimum ferrite; no massive carbides
Graphite distribution	Type A
Preferred graphite flake size (AFS-ASTM)	3 to 5
Hardness after hardening (see below)	Rockwell C 50, minimum

A test coupon 3 by 3 by 6 in. is cast with the die. One 6-in. and one 3-in. side at right angles to each other are machined to remove 0.250 in. of metal, flame heated above the critical temperature, and air cooled.

Aluminum Bronze

Aluminum bronzes are available in a wide range of hardness from 120 to 340 Bhn. These alloys have excellent resistance to galling and are desirable for dies where best finish is mandatory on carbon steel and stainless steel parts, and sometimes to avoid iron pickup on the surface of stainless steel parts that might later give the parts a rusty appearance. The softer grades (120 to 270 Bhn) wear rapidly, particularly where wrinkling is likely to occur. The harder grades (270 to 340 Bhn) wear less rapidly but are difficult to mill, drill and tap. Where elimination of scratches on the part is extremely important, aluminum bronzes should be considered, but for medium to high production (10,000 to 100,000 parts) replacement inserts should be available to permit reconditioning of worn tools.

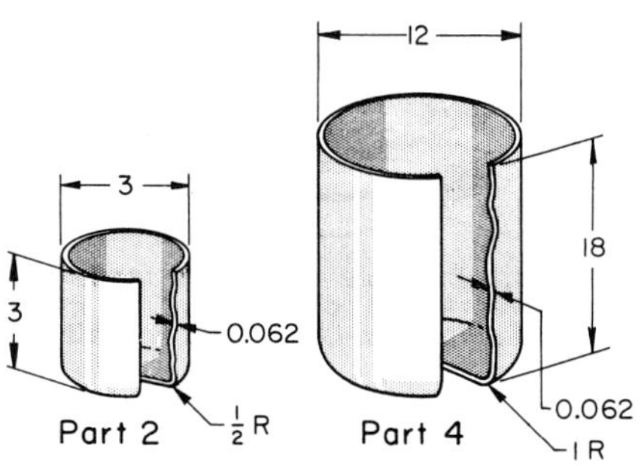

Sintered Carbide

For long runs, inserts of sintered carbide are now widely used in deep drawing dies. In dies of about 8 in. or less for continuous production of over 1,000,000 pieces, carbide has in many instances proved to be the most economical die material. Such dies have maintained size in drawing 60% reductions of more than 500,000 pieces and have made as many as 1,000,000 parts with reductions greater than 40% when the steel to be drawn was surface treated with zinc phosphate and soap. However, carbide is not superior to tool steels such as D2 in complex deep drawing operations, such as those with reductions greater than 40% which combine drawing with coining or forming.

Plastics

Even more than with other tooling materials, the quality and economy of plastic tooling improves with shop experience and volume. Polyester, epoxy, phenolic resin and nylon have been used.

The plastic dies of longest life are constructed so that the wearing surface is faced with glass cloth that has had most of the plastic material forced out under pressure before and during curing. Plastics are most economical when a model is available as a pattern for the lay-up of the plastic-impregnated glass cloth facing, which is backed up with chopped glass fibers impregnated with 50% resin.

Except for very short runs, plastics should not be selected for blankholder material where burred edges of the blank slide over the plastic

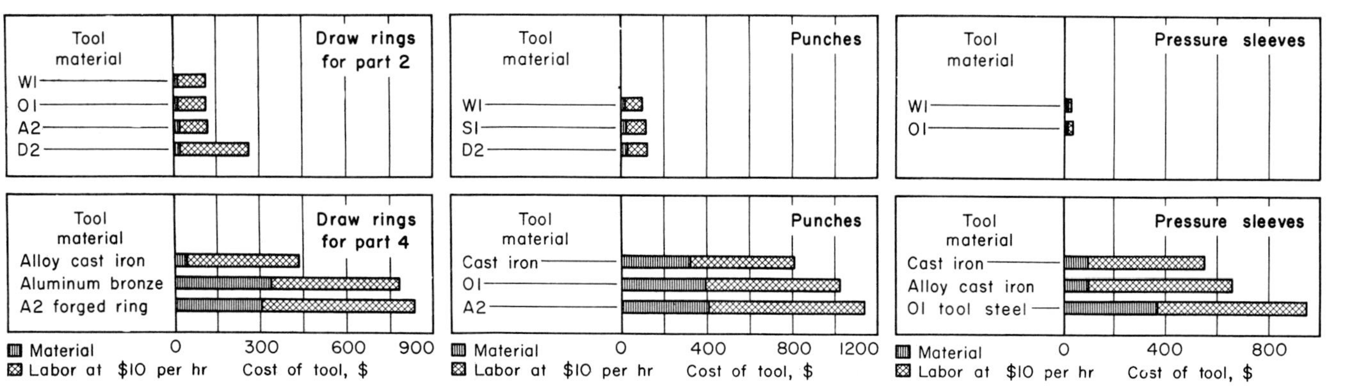

Fig. 5. Relative cost of tools to produce parts 2 and 4

surface. Such areas will wear severely on plastic dies. Two examples of plastic dies are included in the paragraphs on short-run tooling in this article.

Zinc Alloy

Zinc alloy is economical only for die components 12 in. or more in size, and it is most economical in dies for short-run quantities of cups such as parts 4 and 5. In tensile strength, compressive strength and hardness, zinc tool alloys are inferior to other metals used in die construction. Therefore, dies that both draw and iron the part should not be made of zinc.

Zinc alloy tools have application only in drawing small quantities of thin-wall parts where no wrinkling is likely to occur during the drawing cycle, and only where lubricated stock is being drawn. Further pertinent information about cost and processing with zinc dies is included in the article on press forming dies, on page 705.

Cost of Tooling

Figure 5 indicates the cost of tools to make parts 2 and 4. Tooling is shown in Fig. 6 and 7. These costs are computed for a single set of tools, excluding overhead charges, profit allowance, tool design, die tryout and development, none of which influence relative tooling cost. Costs are given for a flame hardened alloy cast iron draw ring. An unhardened ring would cost $60 less and would not be recommended for making more than 1000 parts.

Dies requiring section sizes of 8 in. or larger are most economically made as castings with inserts, because rolled sections of this size are not available. Therefore, dies 8 in. and larger require patterns and castings for the holding or backup ring and, if the draw ring insert is cast rather than forged, also require a pattern for the insert. Figure 5 includes the cost of backup ring casting but not the pattern cost. Insert costs are for forged rings.

The costs given in Fig. 5 are for die components used in the last draw of parts 2 and 4, which requires the smallest die of the three operations to make this part. Dies for prior operations would be larger and more costly. For example, the flame hardened alloy cast iron draw ring for part 4 costs $422 for the final (third) operation and $476 for the first operation. Moreover, a flame hardened alloy cast iron pressure sleeve costs $660 for the third operation, and the pressure pad for the first operation would cost $1000 using the same material.

Analysis of this table with respect to cost of large and small dies shows that material cost in small dies is a minor percentage of die cost. In large dies, however, material cost increases appreciably and may approach half the total cost.

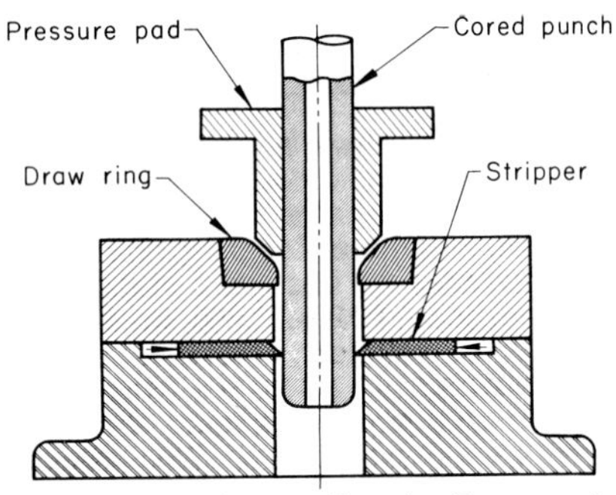

Fig. 6. Cast iron tooling for the second and final draws of parts 2 and 4

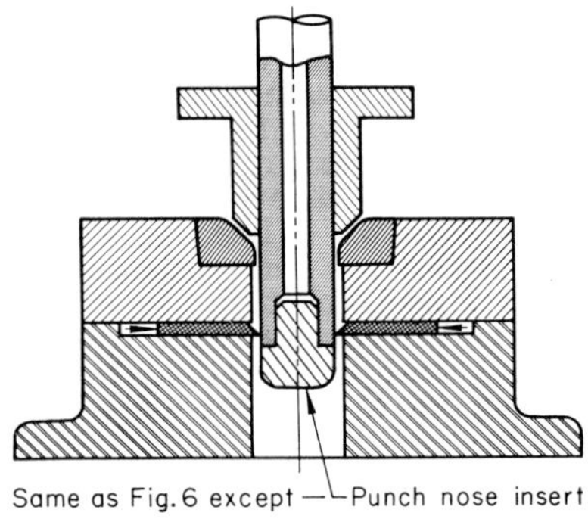

Fig. 7. Tool steel tooling for the second and final draws of parts 2 and 4

Section VII: Lubricants

Selection and Use of Lubricants in Forming of Sheet Metal

LUBRICATION is of two main types: fluid (hydrodynamic) and boundary. (Extreme-pressure lubrication is a special type of boundary lubrication.) These types are illustrated schematically in Fig. 1.

Fluid lubrication is typified by metal surfaces separated by a continuous film of lubricant having a thickness considerably greater than the height of the surface asperities of the metal (Fig. 1a). This type of lubrication, common in bearing systems, is seldom observed in sheet-metal forming.

Boundary lubrication is typified by metal surfaces separated by a lubricant film only a few molecules thick, with considerable metal-to-metal contact between asperities of the two surfaces (Fig. 1b). In most metal-forming operations, because of the high pressures and low speeds involved, lubrication is of the boundary type.

Nonpolar lubricants (such as mineral oils) that are used for fluid lubrication have insufficient adherence to metal surfaces for effective boundary lubrication. For this reason, polar lubricants, such as fatty oils, fatty acids, soaps and waxes, are used as lubricants in sheet-metal forming. The main function of a boundary lubricant in press forming is to interpose between the work metal and the tools a film that minimizes metallic contact and that is itself easily sheared. In deep drawing, this film also is of major importance in controlling friction of the work metal against the drawing die and the blankholder or pressure ring.

The molecules of polar substances in boundary lubricants have long carbon chains with one or more chemically active (polar) groups at one end. The polar groups adhere strongly to the work metal and the tool metal, causing the long carbon chain to be oriented at approximately a right angle to the metal surfaces. These oriented films of polar substances (sometimes called oiliness agents), which are several molecules thick, minimize metal-to-metal contact and friction. They also retard the plastic flow of asperities and thus minimize galling or welding.

The classification of lubricants according to severity of press operations, as presented in this article, was developed by the editorial staff of this Handbook from an analysis of examples in this volume and from severity relations tabulated in the articles that begin on pages 319 and 330 in Volume 1 of this Handbook.

Much of the information on characteristics and compositions of lubricants, as well as on their suitability for press forming of steel, was supplied by Leon Salz, of Fiske Brothers Refining Co., Newark, N. J. Other sources of information concerning lubricants are cited in the list of references at the end of this article.

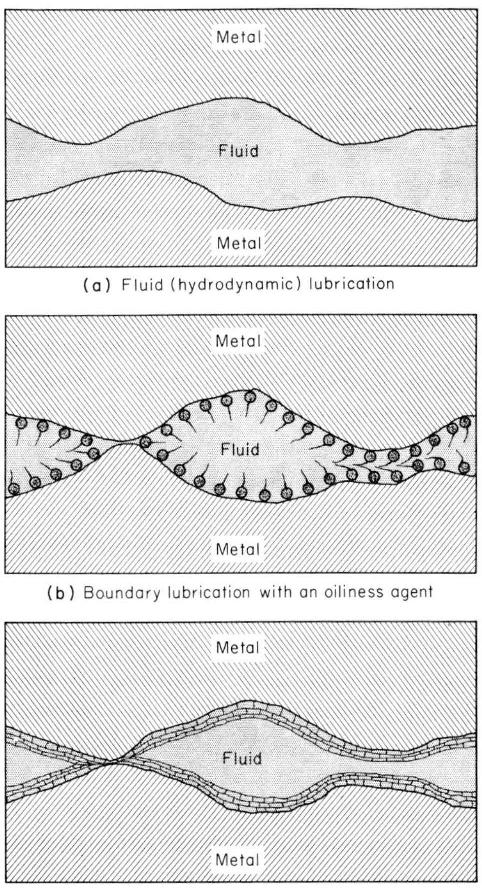

(a) Fluid (hydrodynamic) lubrication

(b) Boundary lubrication with an oiliness agent

(c) Extreme-pressure boundary lubrication with a pigment (mechanical EP agent)

Fig. 1. Schematic representation of the types of lubrication

Extreme-Pressure (EP) Lubrication. Substances that have boundary-lubrication properties in an optimum combination are called extreme-pressure (EP) lubricants. One type contains finely divided inorganic solids (pigments, or mechanical EP agents) that physically separate workpieces from the tools. The use of extreme-pressure boundary lubrication with a pigment is illustrated in Fig. 1(c). When subjected to sufficiently high pressure, the layer of solid particles can be squeezed out from between some of the asperities, permitting metallic contact at those points (as at the left in Fig. 1c). To avoid metal-to-metal contact at asperities where particles of pigment can be squeezed out in forming, phosphate coatings are sometimes applied to the work metal. Besides providing an integrally bonded solid barrier layer, the porous phosphate coating improves the retention of the lubricant.

A second type of EP lubricant contains chemical EP agents. These are chlorinated, sulfurized or phosphated organic materials that react with the metal surface to form firmly bonded chemical compounds of chlorine, sulfur or phosphorus. Lubrication with chemical EP agents can be represented in a form similar to that shown in Fig. 1(b) but with denser packing of the boundary-film molecules.

Components of Lubricants

A lubricant for use in press forming operations may contain:

1 Vehicles or diluents
2 Oiliness agents
3 Pigments (mechanical EP agents)
4 Chemical EP agents
5 Emulsifiers
6 Special-purpose additives.

Vehicles or diluents usually are low-cost materials, such as water or mineral oil, that act as carriers or extenders for the more costly friction-reducing and antigalling components of a lubricant. They also help to dissipate the heat generated in forming; in some applications, this is their most important function. Water is the most effective coolant in common use, being about twice as effective in the removal of heat as mineral oil.

One of the main technical advantages of mineral oil as a vehicle for lubricants is that it is available in a number of viscosity ranges and chemical types. This permits selection of the most suitable grade for a particular forming application.

Oiliness agents are polar organic materials such as fatty oils, fatty acids, soaps, waxes and certain synthetic oils. These polar materials provide boundary lubrication.

Fatty oils include liquid oils and oils that are solid at room temperature. (Solid oils are also called fats.) The fatty oils commonly used include lard oil, castor oil, palm oil, and tallow.

The grades of fatty oils used in lubricants for press operations contain free fatty acids as an impurity, usually in amounts of 2 to 5%, but sometimes as much as 15%. In these amounts, fatty acids slightly increase the lubricating effectiveness of the oil, but when present in amounts of 10% or more, they also slightly increase the rusting of steel workpieces.

Fatty acids are sometimes added to lubricants for press forming of steel, particularly to those based on mineral or fatty oils, or to sulfochlorinated oil blends.

Saturated fatty acids such as stearic and palmitic acids, which are less likely to form varnishlike residues on severely formed parts, are preferred to unsaturated acids such as oleic acid. A free acid content greater than about 5% often results in the formation of a residue of iron soaps that are difficult to remove if allowed to remain on the workpiece for two or three days after application.

Most soaps used in lubricants (usually the fatty acid soaps of sodium or potassium) are soluble in water.

Wax may be a natural type such as beeswax or carnauba wax, or a petroleum derivative, and may be chemically modified to make it emulsifiable.

Synthetic oils, such as polyalkylene glycols and similar compounds, are expensive, but usually are more stable than natural oils at high temperature and are selected for the physical properties desired.

The response of a polar substance to changes in temperature is important in lubrication, not only with regard to suitability for warm or hot forming, but also because forming pressures increase the temperature of the metal surface and the lubricating film. The effectiveness of a polar substance in a boundary lubricant decreases sharply above its melting point, at which temperature the film loses its orientation. As the temperature is increased still further, the effectiveness continues to decrease, until the film is completely desorbed. An effective boundary-lubricant film should have: (*a*) a high melting point, (*b*) strong lateral attraction between molecular chains, (*c*) low shear strength at forming pressures, and (*d*) strong polarity.

Pressure also influences the behavior of lubricants in metal forming. Provided that temperature does not become excessive, liquid lubricants undergo a great increase in viscosity at the high pressures existing at regions of metal-to-metal contact in severe forming, whereas solid lubricants increase greatly in shear strength. These phenomena may explain why mineral oil and some pigmented lubricants provide more effective lubrication in press forming than would be expected from their properties at atmospheric pressure.

Pigments (mechanical EP agents) are chemically inactive solids that physically separate the work metal from the tool surfaces and prevent welding under conditions of extreme pressure. The most widely used of these pigments, or "spacing agents", are chalk, zinc oxide, lithopone, mica, talc, graphite, and molybdenum disulfide. Chalk, zinc oxide, and lithopone are the most commonly used of a number of inexpensive light-colored pigments that are readily available in a finely divided and relatively nonscratching form. Mica, talc, graphite, and molybdenum disulfide have a weak lamellar, or platelike, structure that allows the particles to slide readily over one another.

Pigments may be suspended in either water-base or oil-base diluents. Graphite and molybdenum disulfide are the most suitable pigments for high-temperature forming operations.

In deep drawing, the pigment is compacted by pressure between one side of the work metal and the die, and between the other side of the work metal and the blankholder or pressure ring. In addition to minimizing metal-to-metal contact, the compacted pigment equalizes the pressure exerted by the die and the pressure ring on the work metal. This equalization of pressure over the confined area of the work metal hinders preferential movement of metal in some regions and prevents or minimizes wrinkling.

Chemical EP agents react with the metal surface to form a solid film that is firmly bonded with the metal and that resists welding of the work metal to the punch or die under extreme pressure. The active element usually is sulfur or chlorine. To a lesser extent, phosphorus compounds are used as EP agents.

The most commonly used chemical EP agents are concentrated chlorinated oils and sulfurized oils. The former are chlorinated mineral oils, paraffin oils, fatty oils, or waxes. They are usually liquids, may differ greatly in viscosity, and contain 30 to 50% chlorine.

Sulfurized oils are made by chemically reacting either mineral oil or a fatty oil with sulfur at an elevated temperature. Sulfurized mineral oils, which are light-colored and transparent, contain an approximate maximum of 1.5% sulfur. Sulfurized fatty oils, which are dark brown in color, contain about 12% sulfur ("inactive" grades of oil) or about 16% sulfur ("active" grades). Elemental sulfur ("flowers of sulfur") can be suspended in oils or in water, but it is too corrosive for use in most press operations on steel. To minimize or prevent chemical attack on the work metal and on dies (particularly, dies made from ordinary grades of hardened steel), inactive grades of sulfurized fatty oil are preferred.

Sulfurized and chlorinated oils can be blended to make what are called sulfochlorinated oils. These blends are also used as chemical EP agents.

Chlorinated and sulfurized oils react with steel workpieces to form boundary films of iron chlorides and iron sulfides, respectively. The rate of formation of these boundary films depends on the local temperature, the chemical activity of the metal surface, and the tightness of bonding of chlorine or sulfur in the lubricant molecules. During the forming of metal, the boundary film is reformed almost instantaneously where it is broken, because the sites of damage to the film are also regions where the temperature rises and where a fresh, active metal surface becomes exposed.

The compounds formed by the reaction of chemical EP agents with steel retain their effectiveness as boundary lubricants up to temperatures approaching their melting points (iron chlorides, to about 1100 F; iron sulfides, to about 1800 F), as contrasted with fatty acid soaps, which lose most of their effectiveness above about 210 F.

Emulsifiers are chemicals (such as soaps, organic sulfonates, esters of fatty acids, polyoxyethylene acids, and alcohols) that modify mineral oils, fatty oils and waxes to permit them to form a stable emulsion readily in water. The modified materials are called soluble oils or emulsifiable oils or waxes. A combination of soaps and sulfonates or other emulsifiers is used for rapid dispersion in water, rust resistance, and stability in hard water.

Special-purpose additives are included in lubricants in low concentrations to modify either the chemical or the physical properties of the lubricant. Depending on the type of lubricant and on work-handling procedures, a die lubricant may contain corrosion inhibitors, antifoaming agents or antimicrobial agents.

Corrosion inhibitors may be needed in water-base lubricants to prevent corrosion of work metal and die surfaces, lubricating equipment, and storage containers for lubricants. They are not intended to provide long-term protection of the work metal, but rather to permit the storage of lubricated workpieces for periods ranging from several hours to several weeks without significant damage.

Emulsifiers such as mineral oil sulfonates form sodium salts that function as corrosion inhibitors in emulsions, and sodium or potassium soaps are effective in protecting against corrosion. Sodium nitrite or an organic amine may be used as a corrosion inhibitor in such water-base lubricants as emulsions and diluted soap-fat pastes.

Corrosion inhibitors are not generally used in oil-base lubricants, except to reduce the corrosiveness of concentrated chlorinated oils or blends of chlorinated oils.

Antifoaming agents are sometimes used to prevent excessive buildup of stable foam in water-base lubricants that contain soaps. Soluble-oil emulsions, which ordinarily are the only lubricants used in recirculating systems present no foaming problem.

Antimicrobial agents (bactericides and fungicides) are used more frequently in water-base than in oil-base lubricants. They prevent the growth of bacteria and fungi, and thus inhibit the production of insoluble matter, frothing, discoloration, odor formation, corrosion and demulsification. Substances commonly used as antimicrobial agents include tarry acids, chlorine-containing compounds, and certain alcohols.

Selection Factors

Table 1 summarizes the characteristics of water-base and oil-base lubricants used on low-carbon sheet steel in the common press forming operations. The formulations shown may be modified by the addition of other lubricants or of component materials. For example, low percentages of chemical EP agents are sometimes added to the water-base or oil-base lubricants that are used in low-severity forming. Also, fatty additives may be introduced into EP lubricants, and occasionally are applied over dry soap or wax films at points of exceptionally severe deformation. The reason for balancing formulations or using more than one lubricant is that polar agents are more effective than EP agents in reducing friction, whereas EP agents are much more effective in producing and maintaining a boundary film between work metal and tools under conditions of severe deformation.

Viscosity has an important influence on the effectiveness and suitability of a lubricant for a specific operation. In general, lubricants of high viscosity are used on thick work metal and for severe deformation of the work metal, and lubricants of low viscosity are used on thin work metal and for mild deformation. Lubricants of a given chemical type are generally available in a range of viscosities, or the viscosity can be modified by blending two or more lubricant materials.

In deep drawing, the adjustment of lubricant viscosity to help obtain an optimum amount of friction between the blank and the restraining surfaces of the die and the blankholder may be as important as the chemical composition of the lubricant.

Viscosity must also be considered in relation to the method of lubricant application, and (together with tackiness or the presence of pigments) in terms of effects on the handling of blanks and formed parts — particularly in automated operations.

Method of Application. When selected in a suitable viscosity range, most lubricants can be applied by spray, roller, drip feed, dip, flood, swab or brush. Pigmented lubricants cannot be sprayed, and are difficult to apply by any method except brushing. Dry soap or wax films are almost always produced by dipping the workpiece in a hot aqueous solution. Recirculation and flooding are seldom used in press operations on sheet metal; these methods are useful primarily for applying soluble oils in high-speed operations on small parts.

Toxicity, Flammability, Odor and Stability. Toxicity and skin-irritation effects should be checked before introducing a lubricant with which the user has no experience. Forced ventilation and other safety precautions may be needed in using lubricants that contain flammable or volatile substances. Some lubricants have objectionable odors.

Fatty oils (including fats) can become rancid on standing, producing foul odors, acidity and gummy deposits.

Table 1. Lubricants Commonly Used in the Press Forming of Low-Carbon Sheet Steel(a)

Type or composition of lubricant	Ease of removal by: Water-base cleaners	Degreasers or solvents	Protection against rusting
Water-Base Lubricants			
Low-Severity Press Operations			
1 Water emulsion of 5 to 20% general-purpose soluble oil or wax	Very good	Good	Fair
Moderate-Severity Press Operations			
2 Water solution of 5 to 20% soap	Very good	Very poor	Fair
3 Water emulsion of heavy-duty soluble oil (contains sulfurized or chlorinated additives)	Very good	Good	Fair
High-Severity Press Operations			
4 Soap-fat paste, diluted with water (may contain wax)	Fair	Poor	Fair
5 Water emulsion of heavy-duty soluble oil (contains a high concentration of sulfurized or chlorinated additives)	Very good	Good	Fair to poor
Maximum-Severity Press Operations			
6 Pigmented soap-fat paste, diluted with water	Poor	Very poor	Good
7 Dry soap or wax (applied from water solution or dispersion); may contain soluble filler such as borax	Good	Very poor	Good
Oil-Base Lubricants			
Low-Severity Press Operations			
8 Mill oil, residual (usually about 100 SUS at 100 F)	Good	Very good	Fair
9 Mineral oil (40 to 300 SUS at 100 F)	Good	Very good	Fair
10 Vanishing oil	Removal not required		...
Moderate-Severity Press Operations			
11 Mineral oil (100 to 300 SUS at 100 F) plus 10 to 30% fatty oil	Good	Very good	Fair
12 Mineral oil plus 2 to 20% sulfurized or chlorinated oil (EP oil)	Good to fair	Good	Fair to poor
High-Severity Press Operations			
13 Fatty oil	Fair	Fair	Fair
14 Mineral oil (100 to 300 SUS at 100 F) plus 5 to 50% of:			
(a) Nonemulsifiable chlorinated oil	Poor	Good	Very poor
(b) Emulsifiable chlorinated oil	Good	Good	Very poor
15 Concentrated phosphated oil	Fair	Fair	Fair
Maximum-Severity Press Operations			
16 Blend of pigmented soap-fat paste with mineral oil	Poor	Poor	Fair
17 Concentrated sulfochlorinated oil (may contain some fatty oil):			
(a) Nonemulsifiable	Very poor	Fair	Poor
(b) Emulsifiable	Good	Fair	Poor
18 Concentrated chlorinated oil:			
(a) Nonemulsifiable	Very poor	Fair	Very poor
(b) Emulsifiable	Good	Fair	Very poor

(a) See text for discussion and for explanation of severity classifications.

Water-base lubricants are susceptible to the growth of bacteria and fungi.

Deterioration is much less troublesome with lubricants for press forming than with the cutting fluids used in machining, because the interval between application and removal of forming lubricants is ordinarily only a few hours to a few weeks. It can be a serious problem in recirculating lubricant systems, but these are used in only a small fraction of press forming applications. Lubricants containing suitable stabilizers and other additives are used in these applications, and the lubricant is changed and the system cleaned thoroughly at regular intervals.

Cost and Final Selection. In general, the cost of lubricants increases with the order of listing in each half of Table 1, or with ability to lubricate effectively under conditions of severe deformation. However, the cost of the lubricant itself is often small in comparison with the effect of lubricant selection on over-all manufacturing cost and capability.

The choice between oil-base and water-base lubricants, and the final selection of one lubricant composition for a given forming application may require consideration of the effect of lubrication on the manufacturing sequence as a whole. Factors to be considered include the condition of the workpiece before forming, the condition required after forming, storage time and environment at various stages of processing, and the type of equipment available for applying and removing the lubricant and for handling the workpiece before, during and after forming.

Sometimes it is necessary to use different lubricants in successive press operations on a single workpiece, as in the application described in the following example from the article on Deep Drawing, in this volume:

Example 265. Multiple-operation drawing and ironing of a shell 8 in. in inside diameter by 62 in. deep from low-carbon steel blanks 0.220 in. thick, with the following lubrication steps:

1 Lubricate with sulfur-containing oil before cold drawing the flat blank to a cup 20 in. in diameter by 15 in. deep in a 1500-ton press.
2 Apply oil plus graphite for hot redrawing in three stages to a shell 8¼ in. in diameter by 29 in. deep in 500-ton presses.

3 Pickle and apply a dry soap film before cold redrawing and ironing to a shell 8 1/16 in. in diameter by 45 in. deep in a 500-ton press.
4 Pickle and apply a dry soap film before cold redrawing and ironing to a shell 8 in. in diameter by 62 in. deep in a 500-ton press.

Final selection of a lubricant for a specific application is done largely by full-scale trial.

Laboratory tests to assess lubricant performance may sometimes be misleading, but are used in some plants and are of value at least for screening purposes. One technique that has been used is the Parkinson cup test (see E. L. H. Bastian, "Metalworking Lubricants", McGraw-Hill, 1951).

Lubricant stability can be observed readily by prolonged exposure of the lubricant, and corrosion behavior can be assessed by exposure of lubricant-coated test panels under shop conditions or in a humidity chamber.

Severity Classification

The primary factor to be considered in selecting a lubricant is the severity of the press operation, because the forces at the region of contact between work metal and tools are the primary cause of galling or welding, and hence of damage to the workpiece and wear of the tools.

In Table 1, lubricants are arranged according to their suitability for press operations of four categories of severity: low, moderate, high and maximum.

In the four sections of this article that follow, the four severity categories are described in terms of typical blanking, piercing, bending, forming, drawing and deep drawing operations. Types of lubricants suitable for operations of each severity category are given in Table 1, along with ratings of removability and corrosion behavior, and are discussed in the sections that follow. In general, a lubricant can also be used for press forming operations of lower severity than the operations for which it is shown in the table.

Water-base lubricants (No. 1 to 7) are listed in the top half of Table 1, and oil-base lubricants (No. 8 to 18) are listed in the bottom half. In each half of the table, the lubricants are shown in order of increasing suitability for severe press operations, but differences in effectiveness within a severity group are generally small and may vary with the concentration of the active components.

Selection within each of the four severity groups, as discussed in the following sections, often depends primarily on the removability of the lubricant and the protection it provides against rusting or its corrosiveness to steel (see Table 1). Selection also is influenced by the other factors discussed above (viscosity, facilities for application, toxicity, flammability, odor, stability, cost, and effect on other steps in the manufacturing sequence).

Lubricants for Press Operations of Low Severity

Low-severity operations are defined here as those of severity approximately equal to the following, on low-carbon commercial quality sheet steel:

Blanking and piercing of sheet 0.105 in. thick with a punch-to-die clearance of 10% of sheet thickness
Bending 90° around a radius equal to the sheet thickness (*t*), with a flange width of 25*t* or greater
Forming with maximum localized stretch of 10% in a 1-in. length
Drawing a cup with 10% reduction in diameter.

For such operations, the following four lubricants (numbered as in Table 1) are commonly used:

1 Water emulsion of 5 to 20% general-purpose soluble oil or wax
8 Residual mill oil (usually about 100 SUS at 100 F)
9 Mineral oil (40 to 300 SUS at 100 F)
10 Vanishing oil.

These lubricants are essentially nonpolar in their lubricating behavior. None of them presents any problem in removability or corrosion behavior, as shown by the ratings in the last three columns of Table 1.

Water Emulsion (Lubricant 1). Because of their excellent cooling characteristics, water-base emulsions are well suited for high-speed press operations when applied to work metal and dies by flooding in a recirculating system. They are applied also by drip-feed and spray methods, and are used for a wide variety of operations, including the forming of large parts. The versatility of these emulsions is illustrated by five examples that are presented in this volume:

Example 151. Forming a large, shallow rectangular pan from 0.075-in.-thick sheet.
Example 153. Forming a large dish-shape part from 0.062-in.-thick sheet.
Example 159. Four-operation forming of a 9-in.-diam center disk from 0.028-in.-thick stock.
Example 186. Flattening and crimping 0.028-in.-thick tabs.
Example 187. Forming a small ball socket and staking small projections, using 0.125-in.-thick stock in a progressive die run at about 1000 pieces per hour. The strip was sprayed with lubricant as it entered the die.

Mill Oil (Lubricant 8). Cold rolled steel usually is supplied with a coating of mill oil on the surface, and hot rolled steel is often ordered for delivery in the pickled and oiled condition. Mill oil is ordinarily a mineral oil of low viscosity (about 100 SUS at 100 F), and is intended primarily to provide the steel with resistance against rusting in transit and in storage.

In many applications, residual mill oil provides adequate lubrication for mild forming. For instance, see the following examples in this volume:

Example 149. Forming 0.218 and 0.250-in.-thick backing plates for an automotive brake drum from pickled and oiled stock.
Example 177. Multiple-operation fabrication (notch, trim, cut off and form) of a motor cradle from 0.120-in.-thick strip in a progressive die.
Example 179. In making a deflector in separate dies, residual mill oil was a satisfactory lubricant, but light mineral oil was used also when progressive dies (pierce, notch, form and cut off) were adopted to meet increased demand.

Mineral oil (Lubricant 9) was used in the applications described in the following examples:

Example 51. Blanking and piercing of 0.025-in.-thick laminations. Mineral oil was applied by spraying and by dipping.
Example 135. Press-brake forming (blanking, notching, piercing, forming and embossing comprised the complete production sequence) of a small bracket from 0.089-in.-thick hot rolled steel.
Example 160. Progressive-die forming of a complex lever from 0.075-in.-thick stock at 2400 pieces per hour.
Example 173. Progressive-die forming of an eggbeater side frame from 0.040-in.-thick stock at 360 pieces per hour.
Example 174. Progressive-die piercing, shaving and forming of a shaver-comb support from 0.042-in.-thick stock at 7200 pieces per hour. (Shaving dies lubricated with mineral oil had a short life of 5000 to 10,000 pieces between regrinds, compared with 50,000 pieces per regrind for piercing and forming dies with the same lubricant.)
Example 179. In making a deflector in separate dies, residual mill oil was a satisfactory lubricant, but light mineral oil was used also when progressive dies (pierce, notch, form and cut off) were adopted to meet an increase in production requirements.

Vanishing oil (Lubricant 10) is a low-viscosity, chemically inert, colorless and odorless organic liquid that evaporates completely in about an hour or less. It is a special-purpose lubricant for use when cleaning of the completed part would be impractical, or for use when it will result in a saving in cost or some other production advantage.

Vanishing oil has no effect on the surface of plated, lithographed, painted or plastic-coated metal, and is approved by the U.S. Food and Drug Administration for use on materials that will come into contact with food. In order to realize the advantages of vanishing oil, work metal and die surfaces must be kept free of oil, grease or other contaminants.

Because vanishing oils are usually flammable, they should be handled and used with adequate ventilation, and the customary precautions for flammable liquids should be observed.

Lubricants for Press Operations of Moderate Severity

Moderate-severity operations are defined here as those of severity approximately equal to the following, on low-carbon commercial quality sheet steel:

Blanking and piercing of sheet thicker than 0.125 in.
Bending 90° around a radius less than sheet thickness (*t*) with a flange width of 15*t*
Forming with maximum localized stretch of 20% in a 1-in. length
Drawing a cylindrical shell with 20% reduction in diameter.

For such operations, the following four lubricants (numbered as in Table 1) are commonly used:

2 Water solution of 5 to 20% soap
3 Water emulsion of heavy-duty soluble oil (contains sulfurized or chlorinated additives)
11 Mineral oil (100 to 300 SUS at 100 F) plus 10 to 30% fatty oil
12 Mineral oil plus 2 to 20% sulfurized or chlorinated oil (EP oil).

In contrast to the lubricants for low-severity press operations, which are essentially nonpolar, the lubricants for moderate-severity use contain moderate to high concentrations of polar substances such as soap or fatty oil, or low concentrations of EP agents.

Soap Solution (Lubricant 2). Water solutions containing 5 to 20% of fatty acid soaps of sodium or potassium are commonly used for light-severity work. Potassium soaps have the advantage of being more easily dissolved in water than sodium soaps.

Synthetic soaps made by saponifying a fatty acid with an organic base such as triethanolamine are completely ashless, and need not be removed before annealing or other heat treating operations that are done at temperatures high enough to burn them off (usually, above 1000 F). Also, residues of such soaps often do not interfere with welding operations.

Heavy-Duty Emulsion (Lubricant 3). Water emulsions of heavy-duty soluble oil are preferred for blanking and piercing of thick sheet, particularly in high-speed production where their low foaming and superior performance as coolants are added advantages in recirculating systems.

Mineral Oil Plus Fatty Oil (Lubricant 11). Blends of 10 to 30% fatty oil in mineral oil show excellent friction-reducing characteristics and form polar films of sufficient strength for moderate-severity presswork. These blends are available in a wide range of viscosity, are easily removed in either water-base or solvent cleaning systems, and provide steel with fair protection against rusting.

Mineral Oil Plus EP Oil (Lubricant 12). Blends of 2 to 20% sulfurized or chlorinated EP oils with mineral oil have a wide range of usefulness in moderate-severity operations. Almost any viscosity desired can be obtained by selection of a suitable grade of mineral oil, and the polar or chemical activity can be altered to suit the severity of the operation by adjusting the content of EP additive.

Blends of EP oils in mineral oil are easily removed in solvents; their removability in water-base cleaners depends on the concentration of EP oil in the blend. The sulfur-containing types generally provide fair resistance against rusting; but blends of chlorinated oils have the disadvantage of being corrosive to steel (approximately in proportion to their EP activity), and usually must be removed within a few hours to a few days after application.

The use of oil-base lubricants of these types is illustrated by the following examples in this volume:

Example 19. Blanking, forming and notching of a brake-drum back from 0.097-in.-thick steel at 1050 pieces per hour, with blended chlorinated oil as lubricant.
Example 23. Blanking and piercing of a cam from 0.187-in.-thick stock at 3600 pieces per hour, with blended sulfurized oil as lubricant.
Example 34. Blanking and piercing of a 0.250-in.-thick flyweight at 300 pieces per hour, with roller-applied sulfurized oil as lubricant.

Mineral oils of extremely high viscosity contain significant concentrations of heterocyclic compounds and other polar constituents. Hence they have sufficient boundary-lubrication capability for some moderate-severity operations, even without the addition of oiliness agents or EP agents.

Lubricants for Press Operations of High Severity

High-severity operations are defined here as those of severity approximately equal to the following, on low-carbon drawing quality sheet steel:

Bending stock thicker than 0.125 in. 90°, with flange width of 10t or less
Forming with maximum localized stretch of 30% in a 1-in. length
Drawing a cylindrical shell with 30% reduction in diameter.

For such operations, the following five lubricants (numbered as in Table 1) are commonly used:

4 Soap-fat paste diluted with water (may contain wax)
5 Water emulsion of heavy-duty soluble oil (contains a high concentration of sulfurized or chlorinated additives)
13 Fatty oil
14 Mineral oil (100 to 300 SUS at 100 F) plus 5 to 50% of: (*a*) nonemulsifiable chlorinated oil, or (*b*) emulsifiable chlorinated oil
15 Concentrated phosphated oil.

To provide boundary films of sufficient strength and adherence for high-severity press operations, all of these lubricants contain high concentrations of oiliness agents or moderate to high concentrations of EP agents.

Diluted Soap-Fat Paste (Lubricant 4). The water dilutions of soap-fat pastes (nonpigmented) provide excellent lubrication for many high-severity press operations, particularly when these pastes also contain wax. They are frequently used in the deep drawing of products that will be porcelain enameled, such as tubs, sinks, and washer and refrigerator liners.

These materials can be removed by alkaline cleaning in soak tanks or spray washers; when completely saponified, they present no redeposition problems. They are more difficult to remove in a solvent system, especially if they have dried to a hard crust. They give fair protection against rusting.

On some deep drawn parts, residues from soap-fat drawing lubricants need not be removed before welding or annealing, since they yield only a low percentage of ash when burned.

Heavy-Duty Emulsion, Concentrated (Lubricant 5). Emulsions of heavy-duty soluble oils containing a high percentage of sulfur or chlorine are effective where a high rate of cooling is needed in high-speed press operations, in addition to EP activity. Emulsions containing chlorine must be removed shortly after use, to avoid excessive corrosion of the work metal, but those based on sulfur generally give fair resistance against rusting.

The effectiveness of a water emulsion of a chlorinated oil as a coolant is illustrated in the following example, which is presented in the article on Deep Drawing in this volume:

Example 239. Drawing of an oil-filter shell from 0.089-in.-thick steel, using three dies in a 550-ton mechanical press. A pigmented lubricant had been tried for this part, but the work heated up to 400 F during drawing and required handling with asbestos gloves. Changing to a water-soluble chlorinated oil kept the work cool enough to be handled without special gloves and also improved cleanability of the drawn shell.

Fatty Oil (Lubricant 13). Fatty oils, which usually contain from 2 to 5% free fatty acid as an impurity, are extremely effective in reducing friction. However, their removability in either water-base or solvent systems is only fair (unless the oil is modified to make it emulsifiable for easy removal in aqueous cleaners). In removing fatty oils in water-base cleaners, spray washers are preferred to soak tanks, in order to avoid flotation of oil and possible redeposition on the work.

An advantage of fatty oils is that (like organic-base synthetic soaps) they can be burned off in annealing operations where the temperature exceeds 1000 F, leaving no residue. Under these circumstances, the cleaning step usually needed before annealing can be eliminated.

Fatty oils are sometimes applied to blanks over dry soap films to give a further reduction in friction in maximum-severity press operations (see discussion of Dry Soap or Wax Films, Lubricant 7, on the next page).

Mineral Oil Plus Chlorinated Oil (Lubricant 14). Blends of mineral oil and chlorinated oil are readily removed in either water-base or solvent systems, except for nonemulsifiable grades, which are difficult to remove in aqueous cleaners. As with fatty oils, spray washers are preferred to soak tanks for removal with water-base cleaners, in order to avoid flotation of oil and possible redeposition on the work.

Chlorinated blends are corrosive to steel and require removal shortly after use, followed by the application of a rust-preventive film.

Phosphated Oil (Lubricant 15). Concentrated phosphated oils are used to a limited extent for high-severity operations on steel sheet. These lubricants, which contain ester phosphates as EP agents, are much less corrosive than concentrated chlorinated or sulfochlorinated oils, and are readily removable (except for nonemulsifiable grades in aqueous cleaners).

Lubricants for Press Operations of Maximum Severity

Maximum-severity operations are defined as those of severity approximately equal to the following, on low-carbon drawing quality sheet steel:

Forming with maximum localized stretch (about 40% in a 1-in. length)
Deep drawing a cylindrical shell with maximum reduction in diameter (reduction of 40% or more).

For such operations, the following five lubricants (numbered as in Table 1) are commonly used:

6 Pigmented soap-fat paste, diluted with water
7 Dry soap or wax (applied from water solution or dispersion), which may contain soluble filler such as borax
16 Blend of pigmented soap-fat paste with mineral oil
17 Concentrated sulfochlorinated oil (may contain some fatty oil), either nonemulsifiable or emulsifiable
18 Concentrated chlorinated oil, nonemulsifiable or emulsifiable

Each of these lubricants contains a high concentration of either pigments (mechanical EP agents) or chemical EP agents.

Diluted Pigmented Paste (Lubricant 6). Because of the cushioning effect of the pigment, water dilutions of pigmented soap-fat pastes give greater protection than unpigmented pastes against galling of work metal and dies in high-severity press operations. They permit deep draws with sharp radii, even on thick material, and are often used in cupping operations and for break-in of new dies. The extent of dilution is usually slight, and the pastes are sometimes used without dilution. The presence of a pigment in a water-base soap-fat lubricant can sometimes eliminate metal tearing in drawing steel of ductility that is borderline for the application.

A major disadvantage of the pigmented type of lubricant is the difficulty in removing the pigment after forming, especially by degreasers, and in solvent systems that are not supplemented by scrubbing or brushing. The pigment, or spacing agent, frequently remains on the metal after the soap and fat have been removed.

In addition, problems are encountered in keeping the pigment in suspension and avoiding the clogging of small openings. Thus, application by spray is not usually possible, and any type of recirculation is difficult. Also, these materials are likely to build up hard-to-remove accumulations on the tools and press equipment.

Dry soap or wax films (Lubricant 7) elongate with the metal during forming and prevent metal-to-metal contact almost completely, even under high pressure. These dry films are easily removed in water-base cleaners; some types are removable in plain hot water.

The six examples cited below, which appear in the article on Press Forming of Low-Carbon Steel in this volume, illustrate the use of dry soap for press forming operations that involve maximum stretch. In forming these irregularly shaped parts, highly effective cushioning action and uniform flow of metal were needed in the critical regions. As noted in the text of several of these examples, it was also necessary to apply fatty-oil compounds to critical regions or on certain lots of material for further reduction in friction.

Example 146. Severe forming of 0.096-in.-thick steel in a 500-ton double-action toggle press at 720 pieces per hour.

Example 157. Severe forming of an 0.089-in.-thick side rail for an automobile frame in an 800-ton column press at about 600 pieces per hour.

Example 161. Severe forming and edge bending of side rails up to 0.250 in. thick for automobile or truck frames in 500-ton to 4000-ton presses at up to 900 pieces per hour.

Example 165. Severe hole flanging in forming a 0.16-in.-thick automobile control arm at 200 to 325 pieces per hour.

Example 166. Severe forming of a 0.150-in.-thick automobile control arm at 350 to 575 pieces per hour.

Example 167. Severe forming of a 0.179-in.-thick truck frame member in an 1800-ton hydraulic press at 250 to 310 pieces per hour.

The principal use of dry soap or wax films is in drawing large parts such as refrigerator doors, bathtubs and gas cylinders. Illustrations of the use of dry soap films in maximum-severity deep drawing include the following examples, which appear in the article on Deep Drawing in this volume.

Example 241. Drawing a 36-in.-diam by 11½-in.-deep cup from 0.048-in.-thick low-carbon steel in a 700-ton double action mechanical press at 150 pieces per hour.

Example 244. Direct drawing and reverse redrawing in making a 21-in.-diam by 14-in. or 16-in.-deep washing-machine tub from 0.048-in.-thick low-carbon steel.

Example 245. Direct drawing and reverse redrawing in forming a large complex shape for a washing-machine base from 0.060-in.-thick low-carbon steel at a production rate of 220 pieces per hour.

Example 246. Forming a flanged shell 14½ in. in diameter by 24 in. deep from 0.103-in.-thick high-strength low-carbon steel in one step by reverse redrawing.

Example 264. Two-stage drawing of 14½-in.-diam by 24-in.-deep halves for a compressed-gas cylinder from 0.100-in.-thick high-strength low-alloy steel, applying a dry soap film before each draw.

Dry soap films are also used in conjunction with phosphate coatings in cupping and deep drawing of cartridge cases and other cylindrical cups with a high depth-to-diameter ratio (see the section "Special Dry Barrier-Type Films", which follows).

Disadvantages of the use of dry soap films include: (*a*) incompatibility with mill oil; and (*b*) the need to prepare solutions of dry powder in hot water, to apply the solution hot, to control its concentration, and to use ovens or heat lamps to dry the film on the blanks. These films are of little value in high-speed production of small drawn parts.

Pigmented Paste Plus Mineral Oil (Lubricant 16). Concentrated blends of pigmented soap-fat paste containing a small percentage of mineral oil (in addition to their normal water content) perform well in maximum-severity deep drawing and can be used where the corrosiveness of chlorinated or sulfochlorinated oils cannot be tolerated. Lubricants of this type generally resemble water dilutions of pigmented pastes (Lubricant 6) in removability, but the problems are less severe because the oil reduces hard crusting of residues.

Sulfochlorinated Oil (Lubricant 17). Concentrated sulfochlorinated EP oils are also used for maximum-severity press operations, often with additions of fatty oils or other fatty additives. The content of both sulfur and chlorine in these lubricants is high. At equal chlorine content, they would have the same corrosive action as straight chlorinated oils; substituting sulfur for some of the chlorine content reduces corrosivity somewhat.

In viscosity and removability, sulfochlorinated oils resemble the concentrated chlorinated oils (Lubricant 18), which are discussed next.

Chlorinated Oil (Lubricant 18). Concentrated chlorinated oils are unsurpassed in prolonging die life and in enabling maximum-severity press operations without scratching, galling or tearing of metal. They are widely used for maximum-severity deep drawing when their corrosiveness can be tolerated, and are well suited for automated operations.

These lubricants are applied just before forming and are removed within a day or two, after which a rust-preventive compound is applied if needed. In removing chlorinated oils, the use of spray cleaners is preferred to cleaning in soak tanks, in order to avoid redeposition on the work.

Although the original concentrated chlorinated oils were highly viscous and not suitable for pumping and spraying, chemically modified types or formulations that contain small percentages of mineral oil or fatty oil are available at the same level of chemical activity and in viscosities as low as that of SAE 10 motor oil.

Typical parts on which concentrated chlorinated oils have been used include bumper guards, deep drawn automotive parts, and ⅛-in.-thick cages for roller bearings. Other illustrations of the use of these lubricants for maximum-severity forming include the following examples, which are presented in the article on Press Forming of Low-Carbon Steel in this volume:

Example 138. Forming a channel-shape control arm with flanged holes from 0.089-in.-thick stock in a 1600-ton transfer press at 1110 pieces per hour.

Example 145. Forming a panel for a glove compartment door from 0.032-in.-thick stock in a transfer die at 500 pieces per hour.

Example 168. Forming a corner bracket from 0.043-in.-thick stock in a progressive die at 600 pieces per hour.

On each of these parts, maximum-severity deformation was encountered in the critical region. The selection of concentrated chlorinated oils for these three applications was based not only on their effectiveness as EP lubricants, but also on their efficient heat transfer and suitability for automated high-speed press operations.

Special Dry Barrier-Type Films

Phosphate-type chemical conversion coatings are used in conjunction with dry soap films or other lubricants to provide an impervious barrier film between work metal and tools and to improve retention of the lubricant in regions of high forming pressures.

In the following two examples, which appear in the article on Deep Drawing in this volume, the work metal was phosphated and then coated with lubricant before drawing:

Example 243. Drawing a shell 5.76 in. in inside diameter by 21¼ in. deep from 0.075 to 0.089-in.-thick 4130 steel in four stages, including two reverse redraws. The raw stock was lubricated for drawing by phosphating and then applying a dry soap film. The shell was lubricated after each of two anneals (before the third and fourth draws).

Example 249. Drawing a flanged cup 1⁷⁄₁₆ in. in diameter by 1¹¹⁄₁₆ in. deep from 0.047-in.-thick enameling iron in two draws and a restriking operation.

Example 259. Drawing an 8.872-in.-OD by 3.648/3.618-in.-deep cup with a reverse spherical bottom from 0.188-in.-thick low-carbon steel, in three stages, with drastic thinning of the walls only. The blanks were phosphated and coated with a pigmented lubricant before cupping.

In Volume 2 of this Handbook, a reduction in finishing cost for automobile bumpers was attributed to the use of phosphating before forming:

Example 23 (Vol. 2, page 545). Forming an automobile bumper from blanks that were phosphated before applying a dry soap film. Hand polishing of the bumpers, which was needed when the soap film alone was used, was eliminated when the phosphating operation was added to the cycle.

Widely different types of solid barrier-type films are also used in the press forming of steel.

Organic dry films such as cured vinyl, polyethylene, polypropylene and other resinous organic coatings are sometimes used in drawing highly finished steel sheet that must be protected from scuff marks or scratches. These films can be applied as solutions in an organic solvent or as dispersions in water.

Clear vinyl films are applied to the highly finished side of automobile-bumper stock, and the stock is coated on both sides with a supplementary conventional drawing lubricant, such as a water dilution of a soap-fat paste, before the deep drawing operation. Organic films and adhesive-bonded or loose sheets of paper are also used in drawing decorative chromium plated steel, as described in Examples 211 and 212; the plastic film used in Example 211 was a strippable coating that was removed after drawing.

Example 211. Forming a broiler top from 0.025-in.-thick decorative chromium preplated steel, using strippable plastic and paper in place of a conventional lubricant.

Example 212. Forming a hotplate top from 0.030-in.-thick decorative chromium preplated steel that was coated on the face side with low-tack adhesive paper, using no additional lubricant.

Permanent paint and plastic coatings on sheet steel also function as solid barrier-type lubricants in forming and drawing operations. In addition, electroplating or hot dip coating with a soft ductile metal such as tin, zinc, copper or aluminum provides lubrication in press operations on sheet steel, as described in the article on Forming of Coated Steel, which begins on page 137.

Examples in this volume that describe the forming or deep drawing of steel coated with a soft, ductile metal are:

Example 140. Forming a housing wrapper from unlubricated 0.039-in.-thick galvanized steel in six stages in compound and single-operation dies.

Example 152. Forming a flanged and contoured bracket from unlubricated 0.075-in.-thick galvanized steel in two operations at 2400 pieces per hour.

Example 207. Piercing, notching, cutting off and bending a flame-spreader flange from unlubricated 0.036-in.-thick hot dip aluminum-coated steel.

Example 208. Multiple-operation forming of a burner-body half from 0.037-in.-thick hot dip aluminum-coated steel strip, using an oil-base lubricant that was applied to the strip by roller. Forming was done more easily, and less work was needed to recondition the tools, than with steel that did not have the aluminum coating.

Example 209. Deep drawing a rocker-arm cover from 0.031-in.-thick electrolytic tin-coated steel. The thin, unfused electrolytic tin coating helped retain the dry drawing lubricant used.

Example 210. Forming flat 180° bends and sharp corners on folded-end baking pans made from unlubricated 0.022-in.-thick electrolytic tin plate with a 1.5-lb tin coating.

Lubricants for Stainless Steel

Lubrication needs are more critical with stainless steels than with low-carbon steels, because the higher friction, strength and thermal resistivity of stainless steels means more heat is generated at the surface being formed, and less is being conducted away, than in forming a comparable workpiece from low-carbon steel. Stainless steel is more likely to gall than low-carbon steel. The demand for surface finish is usually more exacting for stainless steel parts than for low-carbon steel parts. The following two examples from the article "Forming of Stainless Steel" demonstrate the typical care exercised in the lubrication of stainless in preparation for forming. In both examples an oil-base lubricant was used.

Example 457. A muffler header of type 321 annealed stainless steel 0.032 in. thick was formed nearly to the limits of formability of the material in two press operations. The dies were brushed with oil between pieces.

Example 458. A dome section was drawn in a blank that had to be held at an angle other than 90° to the motion of the punch in order to accommodate a re-entrant angle in the drawn dome. The blank was clamped on two flanges, an upper and a lower. Blankholder load was evenly divided between the two flanges, but since the upper flange had more area than the lower, material was likely to be drawn from the upper flange and cause wrinkling in the dome. A bead was added to the upper flange, and the lower flange was lubricated with fatty-acid-type, nonpigmented drawing compound, to encourage the drawing of dome material from the lower flange.

Soap solutions and ordinary emulsions of water-soluble oils are not effective in most forming operations on stainless steel. Oils that contain significant concentrations of sulfur are not ordinarily used on stainless steel, because they can permanently stain the metal surface; if used, they should be removed promptly and completely.

The chemical EP activity of chlorinated oils is effective in heavy-duty forming and drawing of stainless, as in the following four examples from the article "Forming of Stainless Steel".

Example 461. A shell was blanked and severely drawn in one operation. Material was 0.037-in.-thick type 302 stainless steel. The formed piece was restruck in the same die to sharpen the draw radius. The lubricant was an emulsified chlorinated concentrate.

Example 463. A small bracket was made of type 302 stainless steel 0.015 in. thick in a seven-station progressive die. A chlorinated, inhibited oil was used as the lubricant.

Example 465. A mounting frame was made of 0.022-in.-thick type 430 stainless in a nine-station progressive die. The lubricant was an emulsifiable chlorinated-oil concentrate.

Example 472. Fountain-pen caps were deep drawn in seven stages in different presses from blanks of 0.011-in.-thick type 302 stainless. The lubricant was sulfur-free chlorinated oil mixed 1 to 3 with inhibited hydraulic oil having a viscosity of 250 SUS. It was furnished to all presses through a central pumping station.

The ability to adjust the chemical EP activity and viscosity of a wide range of chlorinated lubricants and the ease with which they can be formulated for easy removal make them very useful in severe forming and deep drawing of stainless steel. The percentage of active chlorine is adjusted to control chemical EP activity: any viscosity from SAE 10 to liquids that are so thick they barely flow can be obtained by selection of EP concentrate and blending with mineral oil or fatty oils.

In the first of the two examples that follow, both high chlorine content and high viscosity were needed to draw acceptable parts; in the second example, a viscous mineral oil was replaced by a low-viscosity mineral oil containing a chlorinated wax.

Example 459. A wheel cover was drawn from type 302 stainless steel 0.028 in. thick. At first, a lightly chlorinated oil (10% Cl) of medium viscosity (1500 SUS at 100 F) was used. Although the draw was shallow, rejection rate was 12% for splits and scratches. A change was made to a much more highly chlorinated oil (36% Cl) with a much higher viscosity (4000 SUS at 100 F), and rejection rate dropped to less than 1%. After forming, the wheel covers were cleaned in a vapor degreaser.

Example 470. A coffeepot was deep drawn from a type 302 stainless steel blank 0.032 in. thick in two draws and one bulging operation. At first, the blanks were lubricated by brushing both sides with mineral oil having a viscosity of 6000 SUS at 100 F. The workpiece wrinkled in the first draw, and galling and tearing occurred in redrawing and bulging. The lubricant was replaced with a thinner mineral oil (viscosity, 500 SUS at 100 F) that was fortified with a chlorinated wax. The lubricant was brushed on, as before. The modified lubricant not only eliminated wrinkles in the first draw, but also prevented galling in the two other operations.

In some severe drawing applications, there is no substitute for the physical separation and equalization of pressure provided by pigments, as demonstrated in the following example:

Example 471. Chlorinated and sulfochlorinated oils with viscosities of 4000 to 20,000 SUS at 100 F failed to eliminate welding to the dies and splitting at the corners in deep drawing a rectangular shell from 0.035-in.-thick type 304 stainless steel. The oils had been chosen because they could be removed later in a vapor degreaser. The original lubricants were replaced by a highly pigmented water-miscible soap-fat paste diluted with two parts of water. The paste, which was applied by roller to both sides of the blank, eliminated welding and splitting, but it had to be removed later in a hot alkaline solution.

Lubrication of stainless steel for forming is discussed on page 354 in the article "Forming of Stainless Steel", in this volume. Table 3 in that article (page 355) lists lubricants generally used in forming of stainless steels, and indicates their suitability for use in various forming operations.

Lubricants for Heat-Resisting Alloys

Most mild forming of heat-resisting alloys can be done with the use of polar lubricants, such as fatty oils. However, for more severe forming operations, metallic soaps or EP lubricants such as chlorinated, sulfochlorinated or sulfurized oils are used.

Pigments (such as mica) are sometimes added to both these types of lubricant. Pigments containing lead, zinc or sulfur should not be used on workpieces that will be annealed, because of the difficulty of removing pigments and because of the harmful effects of these elements on heat-resisting alloys at high temperature.

In the following example from the article "Forming of Heat-Resisting Alloys" in this volume, a chlorinated oil was used as the lubricant in the press forming of alloy A-286.

Example 496. Aerospace end caps 9 in. in diameter were press formed in two stages from 0.045-in.-thick alloy A-286. The lubricant was a sulfur-free chlorinated oil.

Lubrication during the forming of heat-resisting alloys is discussed on page 373 in the article "Forming of Heat-Resisting Alloys".

Refractory Metals. Lubricants used in forming refractory metals are oils, extreme-pressure (EP) lubricants, soaps, waxes, silicones, graphite, molybdenum disulfide, copper plating, and enamels made by suspending powdered copper in acrylic resin. Columbium and tantalum are often formed with ordinary oils and greases, because they are generally formed at room temperature, as in the following example from the article "Forming of Refractory Metals":

Example 505. A deep drawn shell was produced from 0.010-in.-thick columbium sheet in four operations as follows: draw cup; bulge; draw smaller cup in bottom; pierce. A mixture of graphite and oil was used as lubricant.

For severe forming operations on columbium and tantalum, petrolatum is frequently used.

Solid lubricants and suspensions of suitable pigments such as molybdenum disulfide with or without colloidal graphite are used in hot forming of tungsten and molybdenum.

Lubricants for Aluminum Alloys

In selecting lubricants for forming of aluminum and aluminum alloys, special attention must be given to avoiding staining or corrosion. The following two examples, presented in the article "Forming of Aluminum Alloys", are typical of general practice in lubricating aluminum for severe press forming.

Example 524. A large rectangular box was made in two draws from 0.072-in.-thick aluminum alloy 5052-O. A special drawing lubricant was used: mineral oil containing 10% fatty oil and 10 to 15% EP additive (sulfurized fatty oil).

Example 525. A 22½-in.-deep rectangular bin was drawn in three stages from 0.102-in.-thick alloy 3003-O. A heavy oil-base drawing compound was used to prevent galling.

Table 1 on page 380 in the article on Forming of Aluminum lists ten lubricants used in forming of aluminum alloys, in approximate order of increasing effectiveness. Lubrication in specific processes applied to aluminum is discussed in the sections on Blanking and Piercing, Press-Brake Forming, Contour Roll Forming, Deep Drawing, Hot Drawing, Spinning, Stretch Forming, Drop Hammer Forming, and Bending of Bars and Tubing, in the same article (pages 379 to 402).

Lubricants for Copper Alloys

Avoidance of staining or corrosion is an important consideration in the selection of lubricants for forming of copper and copper alloys.

Because of their effectiveness as coolants, water-base lubricants are favored for high-speed press operations, but they are likely to cause staining and must be removed promptly. Soap-fat pastes have the widest range of usefulness of the water-base lubricants. Use of this type of lubricant for severe drawing is illustrated in the next three examples from the article "Forming of Copper and Copper Alloys". (Soluble-oil emulsions are suitable where forming pressures are less severe, as for redrawing in Examples 590 and 593.)

Example 590. A tapered ferrule was drawn and trimmed in eight stages from 0.021-in.-thick alloy 260 (cartridge brass, 70%) with a grain size of 0.030 to 0.050 mm. The workpiece was annealed after the first drawing operation (cupping). Water-base lubricants were used, in a 1-to-30 mixture with water: for cupping, a dilute soap-plus-fat paste; for redraws, an emulsion of soluble oil.

Example 592. A synchro-converter case was deep drawn from 0.016-in.-thick alloy 752 (nickel silver, 65-18) with a grain size of 0.025 to 0.050 mm. Eleven drawing operations were required for producing this complex, dual-diameter case. The lubricant was diluted soap-plus-fat paste.

Example 593. A rectangular relay body was deep drawn from alloy 735 (nickel silver, 72-18) in four operations. The stock was 0.022 in. thick and had a grain size of 0.020 to 0.040 mm. The body was 1¼ in. deep by ½ in. wide by 1¹⁄₁₆ in. long. Dilute soap-plus-fat paste was used for cupping, and an emulsion of soluble oil and water was used for redraws.

Pigmented soap-fat pastes are used in more severe drawing, as described in the next three examples:

Example 591. A sugar bowl was produced from alloy 230 (red brass, 85%) by deep drawing in one operation and closing in. The blank had a grain size of 0.015 to 0.020 mm. A pigmented soap-plus-fat paste, applied to the workpiece by dipping, was used as the lubricant in drawing.

Example 596. A cup was drawn from alloy 110 (ETP copper) in two stages. The stock was 0.026 in. thick. Reduction was 40% in the first operation and 24% in the second operation. A pigmented soap-plus-fat paste was used as the lubricant.

Example 598. A preform for a spun pillar was made by deep drawing a shell from 0.032-in.-thick alloy 220 (commercial bronze, 90%) with a grain size of 0.010 to 0.020 mm. A 50% reduction in diameter was produced by the draw. Lubricant was a pigmented soap-plus-fat paste.

Oil-base lubricants free from acidity are less likely than water-base lubricants to stain workpieces on prolonged contact. Stabilized chlorinated oils are effective for maximum-severity forming and high-speed operations, as illustrated in the following example:

Example 589. A small cylindrical housing was deep drawn from annealed alloy 230 (red brass, 85%), having a nominal grain size of 0.025 mm, in eight operations (including four redraws) in an eyelet machine. Lubricant was a stabilized chlorinated oil with 30 to 50% combined chlorine.

Information on lubricants used in the forming of copper is presented in the section "Lubrication" on pages 408 and 409 in the article "Forming of Copper and Copper Alloys", where a severity classification and guide for selection of lubricants is given in Table 5. Additional information on the selection of lubricants for the forming of copper and its alloys is given in the same article in the sections on Spinning, Contour Roll Forming, and Coining.

Lubricants for Magnesium Alloys

Lubrication is more important in hot forming than in cold forming of magnesium and magnesium alloys, because these alloys are more likely to gall in hot forming. Dry soap films are satisfactory at forming temperatures below 250 F; graphite and molybdenum disulfide are used at higher temperatures, but may be difficult to remove.

Following are two examples from the article "Forming of Magnesium Alloys" in this volume, in which graphite was used as the lubricant in the forming of heated magnesium alloy.

Example 606. Rectangular boxes 9⅜ in. long by 6⅛ in. wide by 9½ in. deep were drawn from 0.081-in.-thick alloy LA141A. Temperature of the pressure pad was 540 F, of the draw ring, 440 F, and of the punch, 300 F. Clamp pressure was 75 psi. Drawing speed was approximately 2 ft per min.

In the same press, boxes of the same size were drawn from the same thickness of alloy AZ31B or AZ31C. For these boxes, draw-ring and punch temperature was 750 F; clamp pressure was 100 psi, and drawing speed was 1 ft per min. A graphite lubricant was used for all three alloys.

Example 607. A dome-shape piece was drawn from a 0.100-in.-thick blank of alloy ZE10A-H24 that was 38 in. in diameter. Blanks were graphite coated and heated to 400 F. Tools were heated by gas to 400 F. Clamping force varied from 1500 to 2000 lb. The punch had a spherical radius of 29.087 in. It traveled 7⅜ in. to form the dome.

When lubricants cannot be tolerated, thin sheets of paper or fiber glass, depending on forming temperature, are placed between the work metal and the die. Lubrication used in forming of magnesium alloys is discussed in the section on pages 426 to 427 in the article "Forming of Magnesium Alloys".

Lubricants for Nickel Alloys

Nickel alloys, depending on their composition, may gall readily during forming. Fatty oils and blends of fatty oils with mineral oil provide satisfactory lubrication for many mild and moderate forming or drawing operations, but EP lubricants (chlorinated or sulfurized oils) are generally needed for severe forming of nickel alloys. The EP lubricant must be removed completely after forming, as sulfur will cause embrittlement on heating and chlorine causes pitting on long exposure.

Lubrication of nickel alloys during forming is discussed under "Lubricants" on page 432 in the article "Forming of Nickel Alloys", and in the sections in that article that deal with Shearing, Blanking and Piercing; Bending Tube and Pipe; Expanding Heat-Exchanger Tube; Cold Heading and Cold Extrusion; and Straightening.

Lubricants for Titanium Alloys

Lubricants for cold forming of titanium alloys are generally similar to those for severe forming of aluminum.

A dried soap film provided lubrication for forming of titanium in the next two examples from the article "Forming of Titanium Alloys", in this volume:

Example 611. A rectangular blank of commercially pure titanium 0.090 in. thick was formed in two operations to an irregular shape that included two oblong drawn cups that tapered to a depth of 3½ in. Preforming was done at 550 F in a 7000-ton press, using beryllium copper inserts for a draw ring; and the part was then sized at 1000 F in a drop hammer, using zinc alloy dies. Both sets of dies were lubricated with soap.

Example 612. Blanks of alloy Ti-8Al-1Mo-1V or Ti-6Al-4V, 0.050 in. thick, were reduced 50% on diameter to produce 8-in.-diam hemispheres in a rubber-diaphragm press. Maximum forming pressure was 14,000 psi. A disk of 1018 steel, ⅛ in. thick, was used as an overlay on the titanium to protect the diaphragm and prevent wrinkling of the work. The punch side of the blank was coated with wax, which was removed from the workpiece immediately after forming.

Temperature-resistant lubricants for hot forming contain graphite or molybdenum disulfide, and they are often applied over zinc phosphate conversion coatings.

Lubricants used in cold and hot forming of titanium alloys are listed in Table 3 on page 439 in the article "Forming of Titanium Alloys"; additional information on lubricants is presented in the sections "Power Spinning" (page 442) and "Bending of Tubing" (page 446) in that article.

Selected References

Richard L. Jentgen, Lubrication in Metal-Deformation Processes, DMIC Report 234, 10 June 1967, p 74-103.

W. Torkington, Lubricants as Related to Shearing, Blanking and Forming; Technical Paper 150 presented at a symposium sponsored by the Metal Fabricating Institute at Purdue University, Apr 1969.

D. A. Shenton, paper presented to a symposium on metalworking (drawing) lubricants, sponsored by ASTM Committee E-26, at Hamilton, Ontario, Canada, 23 Apr 1966.

Felix M. Giordano, A Critical Review of Die Lubricants in Pressworking, ASTME Technical Paper SP 63-21.

Ferdinand L. Ewald, How to Choose a Lubricant for Stamping Steel Parts, *Metal Progress*, Jan 1966, p 72-75.

E. L. H. Bastian, "Metalworking Lubricants", McGraw-Hill, 1951.

Selection of power presses

By V. J. MANKOWSKY, vice-president, Engineering Research and Development and Product Development – USI Clearing*

Proper selection of a press is essential if it is to be operated successfully and economically. This article outlines the basic factors relating to the design of conventional mechanical and hydraulic presses, together with the advantages associated with the various types of presses in these two categories. By evaluating more accurately the type of machine he needs, the potential press buyer is in a better position to discuss his particular requirements with press suppliers.

Un choix adéquat de presse est indispensable si elle doit être utilisée avec succès et de manière économique. Cet article décrit les facteurs de base se rapportant à la conception des presses classiques mécaniques et hydrauliques, ainsi que les avantages associés aux différents types de presses dans ces deux catégories. En évaluant avec plus de précision le type de machine dont il a besoin, l'acheteur éventuel se trouve mieux à même de discuter de ses besoins particuliers avec les fournisseurs de presse.

Die richtige Auswahl einer Press ist underlässlich, wenn diese erfolgreich und wirtschaftlich betrieben werden soll. Dieser Artikel beschreibt die grundsätzlichen Faktoren, die die Konstruktion von herkömmlichen mechanischen und hydraulischen Pressen bestimmen, zusammen mit den Vorteilen, die mit den verschiedenen Pressentypen in diesen beiden Kategorien verbunden sind. Der Käufer einer Presse kann seine speziellen Erfordernisse besser mit den Pressenlieferanten erörtern, wenn er den von Ihm benötigten Pressentyp vorher genauer bewertet.

La elección correcta de una presna es indispensable si tiene que funcionar con éxito y económicamente. En este artículo se describen en líneas generales los factores básicos relativos al diseño de prensas mecánicas e hidráulicas conventionales, así como las ventajas asociadas con los diversos tipos de prensas de estas dos categorías. Evaluando más exactamente el tipo de maquina que necesita, el posible comprador de prensas estará mejor situado para discutir con los proveedores lo que en particular le interesa.

PURCHASE OF a metal forming press represents a large capital investment. Also, the return on this investment will be determined by how well the press does the job it was meant to do. The best way to insure this return is to ensure that the press is suited to the operations that the dies will perform. Nobody benefits when a press is not capable of making a part or if it is in constant need of repair, regardless of where the fault lies. Therefore, it is as much to the benefit of the press manufacturer as to his customer that the press user be familiar with the fundamentals that determine press capability. The following is a review of the basic information to be considered when undertaking the selection of a metal forming press.

Since both mechanical and hydraulic metal forming presses are available, consideration is given to the selection of both types. Comparisons are drawn between the two which highlight the advantages of each.

Mechanical presses

Mechanical presses, so far as their function is concerned, can be classified in two basic categories—single-action and multiple-action (Figs 1 and 2). The single-action press has one slide. The slide may be driven by a crankshaft or an eccentric. It may also have one, two or four points of suspension. Each of these points

* USI Clearing products are marketed in this country and Europe by Clearing Europe, 197 Knightsbridge, London SW7 1RU.

Fig 1

Fig 2

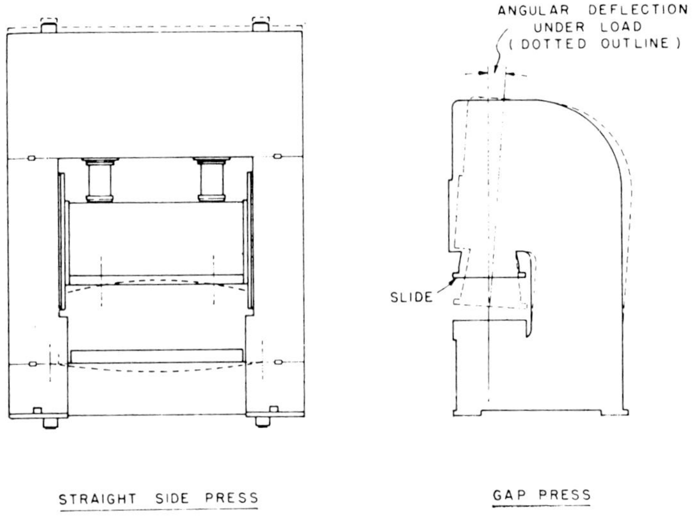

Fig 3

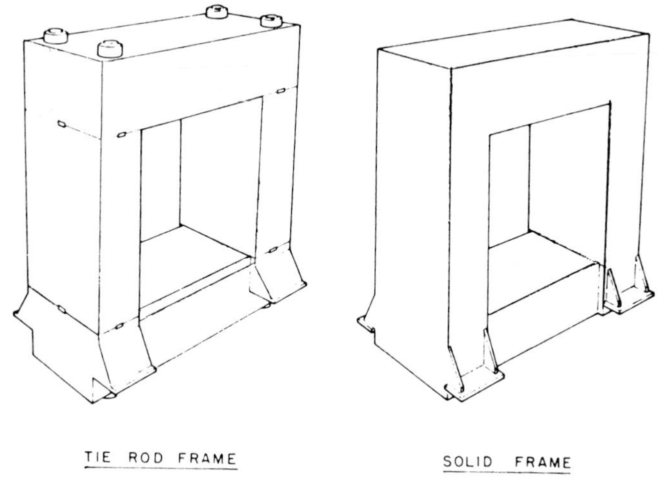

Fig 4

will be driven by a crank or eccentric. The single-action press can be used for many types of metal stamping operations such as blanking, embossing, coining and drawing. Drawing may sometimes require pneumatic die cushions for blankholding.

Multiple-action presses are commonly called toggle presses. A toggle press can be either a double or a triple-action machine. The double-action press has two slides; a blankholder slide and an inner or draw slide. The blankholder slide is a hollow rectangle, and the inner slide is a solid rectangle that moves up and down within the blankholder. The blankholder slide, as the name implies, holds the blank while the inner slide forms the part. The term toggle press is derived from the linkage used on all double-action presses to drive the blankholder slide and provide a dwell in its motion at the bottom of the stroke.

The double-action press is used for drawing. It is more suited to this than a single-action press. When making a draw on a single-action press, much of the flywheel energy is lost in depressing the air cushions. On the double-action press the blankholding pressure is maintained by the blankholder slide, and flywheel energy is expended only by the drawing action of the inner slide.

Types of press frames

Once the basic type of press, either single- or double-action, has been agreed upon, the next step is to decide what type of frame configuration should be used. Here again selection must be made between two categories:

1. Gap-frame presses
   This category is confined to single-action presses. It includes open-back-inclinable presses (OBI's), open-back-non-inclinable presses (OBS's), horn type and adjustable-bed presses.
2. Straight-side presses
   This category includes single- and multiple-action presses. They can have a solid one-piece frame or can be of the multiple-piece frame tie rod construction.

Gap-frame presses

Advantages of gap-frame presses are their relatively lower cost, versatility of application and material handling. The important point in considering a gap-type press is the amount of misalignment that can be tolerated by the dies. All gap frame presses will open up in the gap when subjected to a work load. This opening up or deflection causes a distortion of the press frame (Fig 3). As a result the face of the slide and the face of the bolster are no longer parallel as the punch goes through the work. The degree of misalignment is proportional to the force required to make the part. If punch and die alignment is critical, from the standpoint of either part accuracy or

Fig 5a

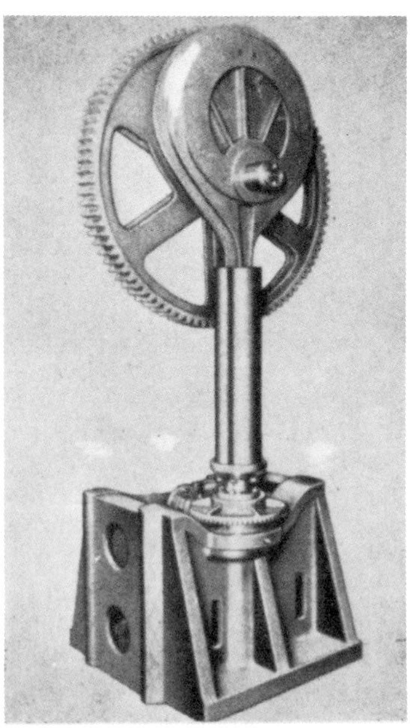

Fig 5b

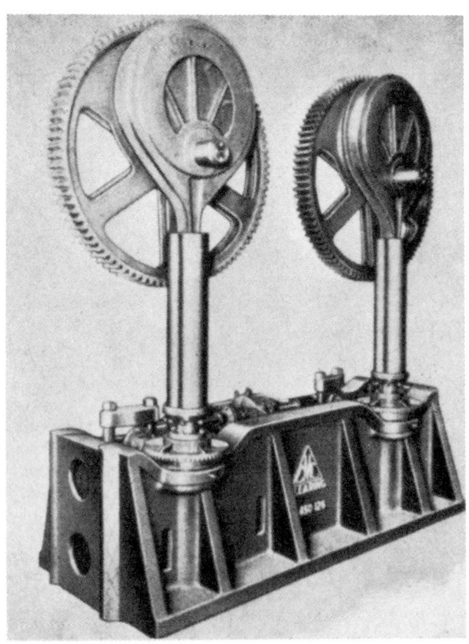

Fig 5c

excessive die wear, the gap type press would be the wrong selection.

Straight-side presses

The straight-side press provides the best conditions possible for maintaining die alignment. Frame distortion or deflection under load is straight and in a vertical direction. Deflection is in the same direction as the punch action and does not create alignment problems.

Solid-frame, straight side presses usually have a lower price tag than tie-rod presses but have limitations on the frame size because they must be transported in one piece. The physical limitations of railroad flat cars and truck drawn trailers impose this restriction. Machining facilities of the company manufacturing the press also will impose size restrictions. Another drawback to the solid-frame press is the problem created should the press become jammed. On a tie rod press, it is only necessary to heat the tie rods with a torch, causing the rods to expand, relieving the pre-load on the press frame. Heating the solid press frame in such a manner could cause distortion in the frame members which could not be corrected without an expensive repair bill. In the straight-side press with a tie rod frame, four structures, the crown, the bed and the two uprights are held together by four tie rods. The rods are 'shrunk in' by means of electrical or flame heating or by hydraulic jacks. Tie rods are pre-loaded to 200 per cent of rated press capacity (Fig 4).

Die space area usually presents no great problem when selecting a press. The dimensions of the die or dies to be used in the machine will establish this requirement. The number of points of slide suspension will be determined by the left-to-right and front-to-back dimensions of the slide face. As an example, a 200 ton press has a slide face dimension of 36in left-to-right by 36in front-to-back. This most certainly would be a single-point press. If the left-to-right dimension were increased to 60in it would be a two-point press. Increase both dimensions again to 108in left-to-right by 72in front-to-back and the press has four-point slide suspension. The foregoing applies to single-action presses and the inner slide on double-action presses. Unless the press user has a special requirement, the manufacturer will determine the number of points of slide suspension. The outer slide or blankholder always has four points (Figs 5a, 5b, 5c).

Small straight-side presses

A popular type of press used in metal stamping shops is the two-point, straight-side, crank press. The capacity usually ranges from 50 to 300tons. Such machines are offered in a number of different drive variations as follows:

1. Single-end flywheel drive.
2. Single-end, single-geared drive.
3. Twin-end, single-geared drive.
4. Twin-end, double-geared drive.

The cost advantage of the simplest drive available is readily apparent. However, it is important to know the limitations of each drive arrangement to be sure it meets the needs of the job the press is intended for (Fig 6).

The single-end flywheel drive is the simplest form of press drive. It is used for blanking and very shallow forming operations. Flywheel energy is limited and windup or torsional deflection in the crankshaft demands that the work be done very near to the bottom of the stroke. This type of drive is used mainly on high-speed presses. A single-end, single-geared press runs at slower speeds than a single-end drive flywheel press and has more flywheel energy available. This advantage is offset by the fact that, as before, torsional deflection in the crankshaft limits the distance up the stroke at which the work load can be applied. Deflection or torsional windup in a crankshaft can cause misalignment between the slide and bed. The longer the crank, the greater the effect. For this reason, press manufacturers generally limit the left-to-right dimension of the slide to 72in on a single-end press drive.

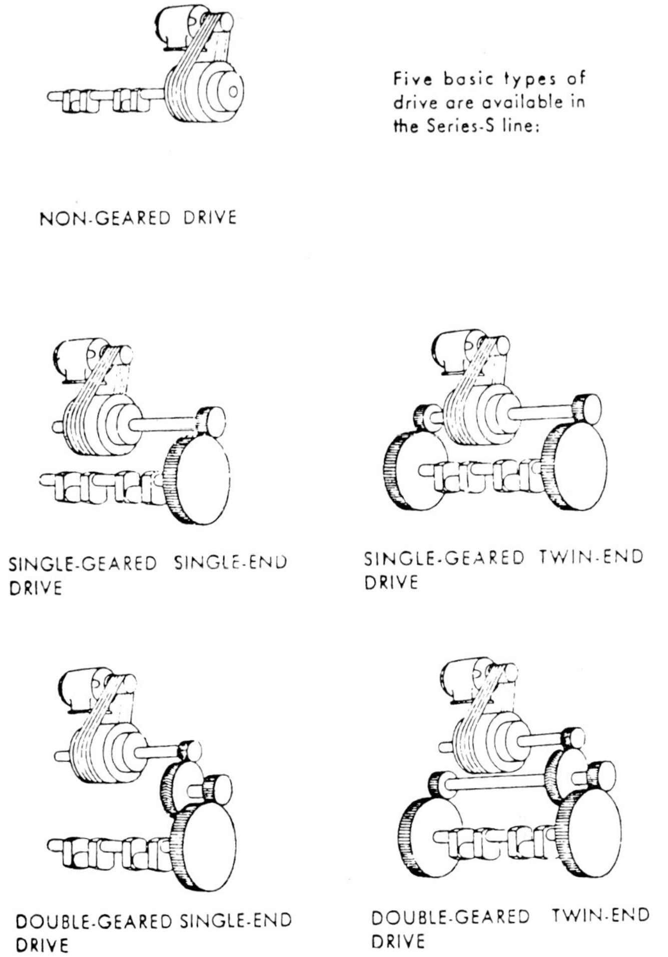

Fig 6

The problem of torsional deflection is overcome by the twin-end, single-geared press drive. Most of the crank presses sold are of this type. Longer press strokes are permissible without sacrificing slide and bed alignment. Also, the likelihood of a broken crankshaft is reduced. The single-geared press will run at fairly high speeds—up to 75 strokes/min and sometimes faster. Twin-end, double-geared press drives are used when the press speed is less than 20 strokes/min generally. Such presses usually have longer strokes and more flywheel energy is provided for making parts that require greater draw depth.

Fig 7

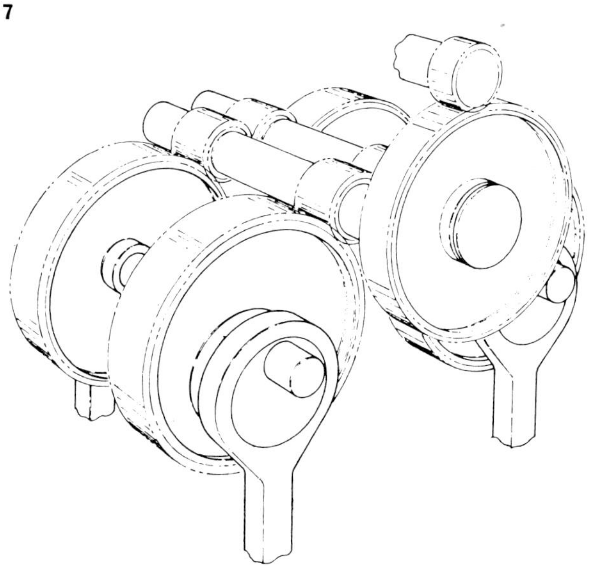

ECCENTRIC PRESS DRIVE

Large straight side presses

Large mechanical metal stamping presses range from 400 tons up. One press currently under construction will have 6000 tons capacity. Modern-day presses in this category, generally do not use crankshafts. Instead, eccentric gears, or where the stroke is exceptionally long, spool gears, take the place of cranks. Since each eccentric has its own driving gear the problems created by such things as crankshaft windup do not plague a press of this design. On presses with multiple-point slide suspension all the eccentric gears are locked in mesh with the gear train and the whole system moves in unison. This provides for a rigid drive that will move the press slide and punch through the work without incurring the misalignment that might tend to occur with adverse loading conditions (Fig 7).

Tonnage or capacity of a press is logically established by the die, requiring the greatest amount of force to make a part, that will be used in the machine. This tonnage may be obtained by calculating the forces required at each step of the forming process. In some cases, it may be the result of gauge readings taken when actual parts were made in a test set up in a hydraulic press. Quite often, tonnage will be established by virtue of previous experience where the same or a similar part has been made on an existing press. Undoubtedly, the second method would be ideal, but it is not always possible or practical. Under any circumstances, a calculation of the tonnage requirements for a part should be made before the press is selected.

Fig 8

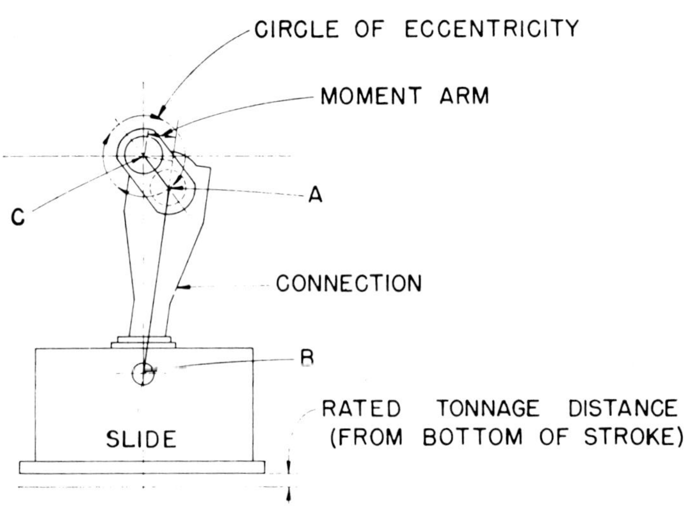

Fig 9

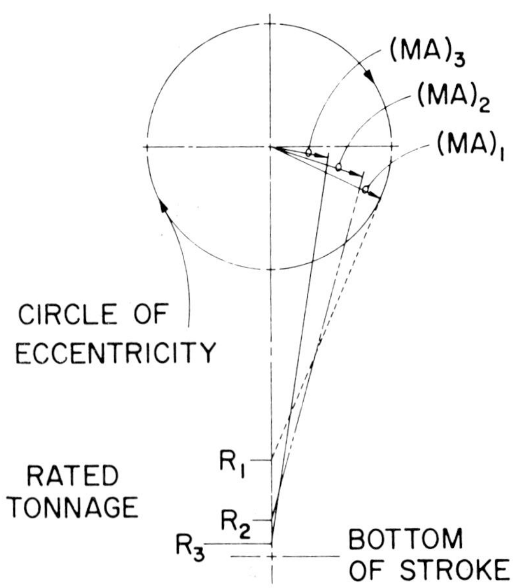

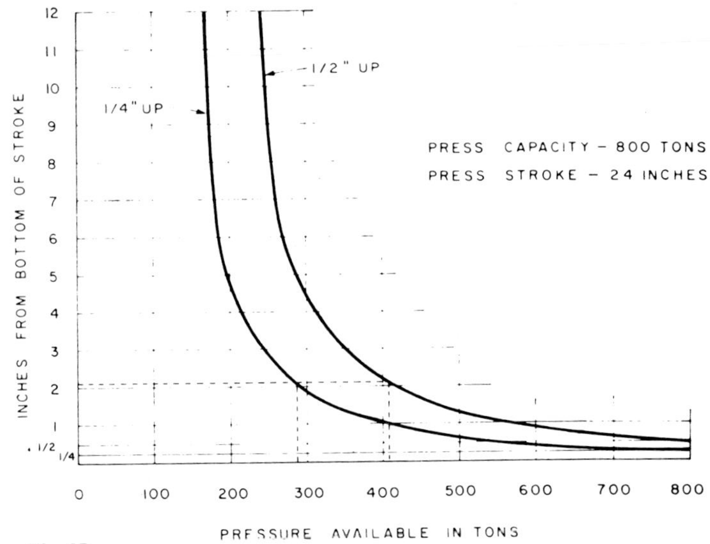

Fig 10

In order to compare the loading and working requirements with the capability of the press selected for the job, it is helpful to know the conditions for which the press was designed. All press manufacturers design their machines according to certain basic fundamentals. What they build into their machines to meet these requirements is what creates competition between press companies. The primary step in finding out if a press is suited to making a part is to check the capacity of the press drive. Basically, the determination of press drive capacity goes back to the crank and connection or eccentric and connection, as the case may be. All presses are rated on the tonnage that can safely be applied by the slide at a given distance up from the bottom of the stroke.

In Fig 8, this can be seen very graphically. As the crank revolves, the centre of the crank pin (point 'A') inscribes the circle of eccentricity and the slide moves up and down. At any point on the slide stroke, if the centre line of the connection is drawn from the centre of the wrist pin (point 'B' through 'A') and projected past the centre of the crankshaft (point 'C'), an imaginary line projected at a perpendicular from the centre line of the connection to meet point 'C' is the 'moment arm'. This theoretical 'moment arm' is used to calculate the torque capacity of the press drive. The value of this torque capacity is usually established with the slide ½in or ¼in up from the bottom dead centre. With the slide at any position higher up the stroke, the 'moment arm' gets longer, reducing the allowable force that can be applied by the slide. A typical pressure curve that has been plotted from values taken with the slide at various positions on the stroke is shown in Fig 9.

Using pressure curve

When choosing a press for making a part, the distance up on the stroke at which tonnage will be applied must be considered before it is possible to determine whether or not it has enough capacity. To take a case in point, the operation is a blank and draw in which the depth of the cup is 2in. Blanking will be done about 2⅛in from BDC. Calculations show that 360tons is required for blanking and that the draw requires 280tons. Suppose you have an 800ton press with a 24in stroke and the press drive is rated ½in up. Reference to Fig 10, shows that the 800ton press has a capacity of 390tons at 2⅛in from BDC. The tonnage required for the draw operation will not tax the drive because the capacity increases rapidly as the slide approaches the bottom of its stroke.

In addition to knowing that the press drive can provide adequate force, an important consideration is the press work capacity. This work capacity is dependent on flywheel energy. The flywheel must have enough energy to carry the slide through the work. In other words, to deliver enough force through the

Fig 11

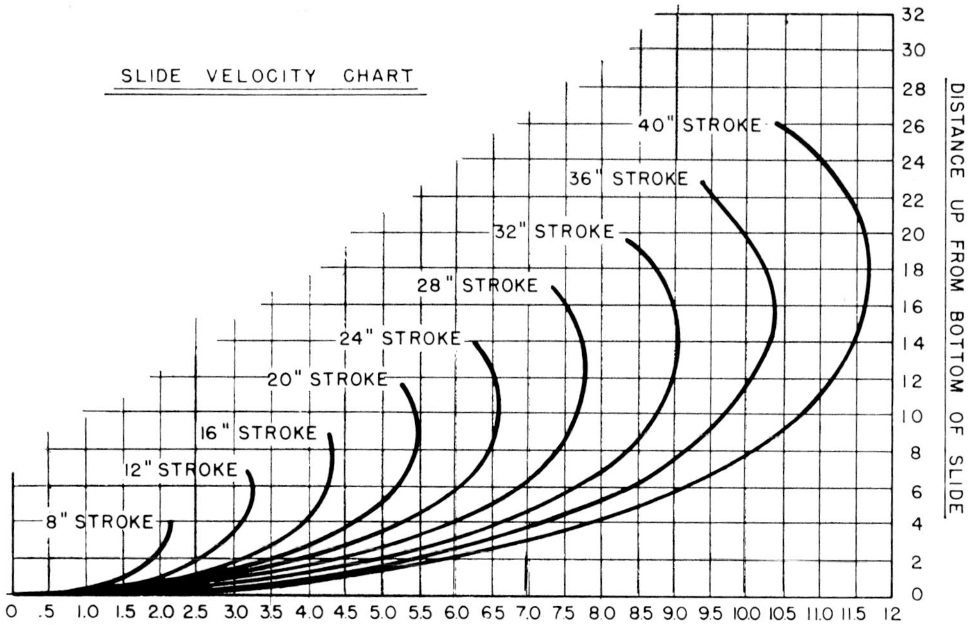

distance required to make the part. Considering the same part as before, assume that the 2in deep cup has a material thickness of $\frac{1}{8}$in and a cushion is required with 15 tons capacity. The energy for blanking would be 360 by 0.125 or 45in tons. The draw would require 280 by 2 or 560in tons. Depressing the air cushion takes 15 by 2.125 or approximately 32in tons. The total would be 637in tons. Since the 800ton press has 880in tons of flywheel energy available for work, the press has enough work capacity.

If this work capacity was not adequate, the flywheel slowdown during the working portion of the press stroke would be excessive. The result would be an overloaded main motor. If the flywheel slowdown is severe enough, the motor will not only overheat, but will cause the overload devices in the motor starter to open and stop the press.

The stroke of a press must be capable, at the very minimum, of forming the piece part and providing enough clearance when the die is open for the part to be removed. In some cases, the press user may consider buying a press with an extra long stroke to provide as much clearance or opening as possible for parts handling. One disadvantage of this practice is the fact that slide velocity increases proportionately with the press stroke as well as the press speed. The increase in slide velocity can be detrimental from the standpoint that increased draw speed can make it difficult to form parts with deeper draws. In addition, die life can decrease because of the higher rate of impact. It is important then that slide velocity be given some considerations when selecting the length of stroke for a press. Fig 11 can be used for making this comparison.

Load distribution in die area

Assuming that all capacity factors have been investigated and found to be satisfactory for the job in hand, consideration should be given to the load distribution in the slide area. Presses are designed on the basis of proportioning the tonnage capacity equally on each point of slide suspension. This means that a 400ton two-point press is capable of delivering 200tons at each eccentric, connection and adjusting screw. Care should be taken, particularly when multiple dies are used, to distribute the loading so that one point or the other carries its share but is not overloaded. Further, the press frame is designed on the basis of distributing the load over two-thirds of the left-to-right die space dimension. This loading condition approximates the best average condition for providing stiffness in the slide and bed to insure good tool life. The type of situation to avoid can best be typified by a case where one die, having a base dimension much smaller than the two-thirds of the die space previously mentioned, would be placed in the centre of the bed and slide. Excessive deflection would occur under load that could affect die performance both in wear and function (Fig 12).

There is no doubt that in some stamping shops the press that has a die space large enough to take the die gets the job. The fact that the press can make a part without stalling or breaking up confirms that it is right for the job. Perhaps this procedure works for some, but to do so is to court trouble. Remember, a mechanical press is a self-destructive type of machine. It can do work and destroy itself at the same time. The damage that results may not be apparent immediately. In some cases, it takes months or even years to develop to the point where a part failure occurs.

If the recommendations made here for mechanical press selection are followed, a decision may not be reached as quickly as by the cut-and-try method, but there will be fewer 'surprises'. A press that is not of the self-destructive type is the hydraulic

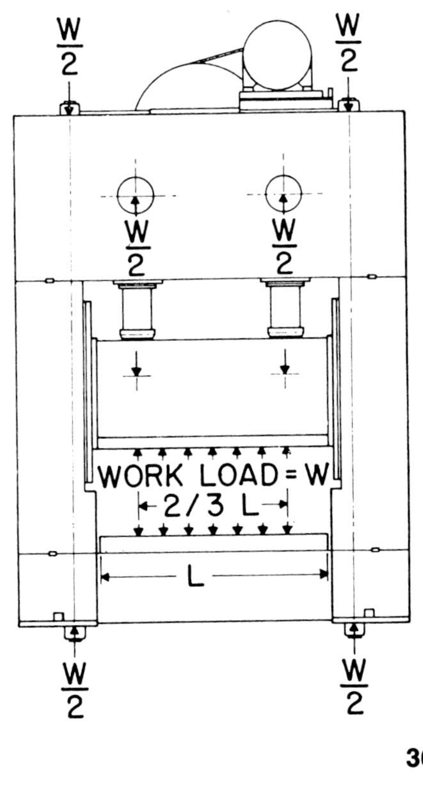

Fig 12

Fig 13 (above)

press. In the following, the basic fundamentals for comparing mechanical and hydraulic presses are presented.

Hydraulic press

Hydraulic presses, although not used as extensively as mechanical presses, have advantages to offer in many applications. In some cases, a hydraulic press is the only machine that will do a special job. For instance, hydraulic presses are the only type that can be built to deliver extremely high tonnages. During the US Air Force heavy press programme in the 1950's, several hydraulic presses were built with 50 000tons capacity. (One is now proposed that will deliver 200 000tons. Another example would be a hydraulic press for forging artillery projectiles. The stroke is 75in and the tonnage required is 800tons The drive for a mechanical press for this job would be of fantastic proportions.

Realizing that there will always be grey areas where one type of press, either mechanical or hydraulic, will do the job, the following is an attempt to classify the work best suited for each:

Mechanical

1. Shallow draw or forming
2. Metalworking—high production
3. Blanking
4. Piercing
5. Air forming
6. Transfer or progressive die operations

Hydraulic

1. Deep draws
2. Draws for irregular shaped parts
3. Straightening
4. Hobbing
5. Powdered metal briquetting
6. Die tryout
7. Bonding
8. Rubber pad forming
9. High tonnage forming
10. Multiple operation

Fig 14 (below)

Fig 15

Fig 16

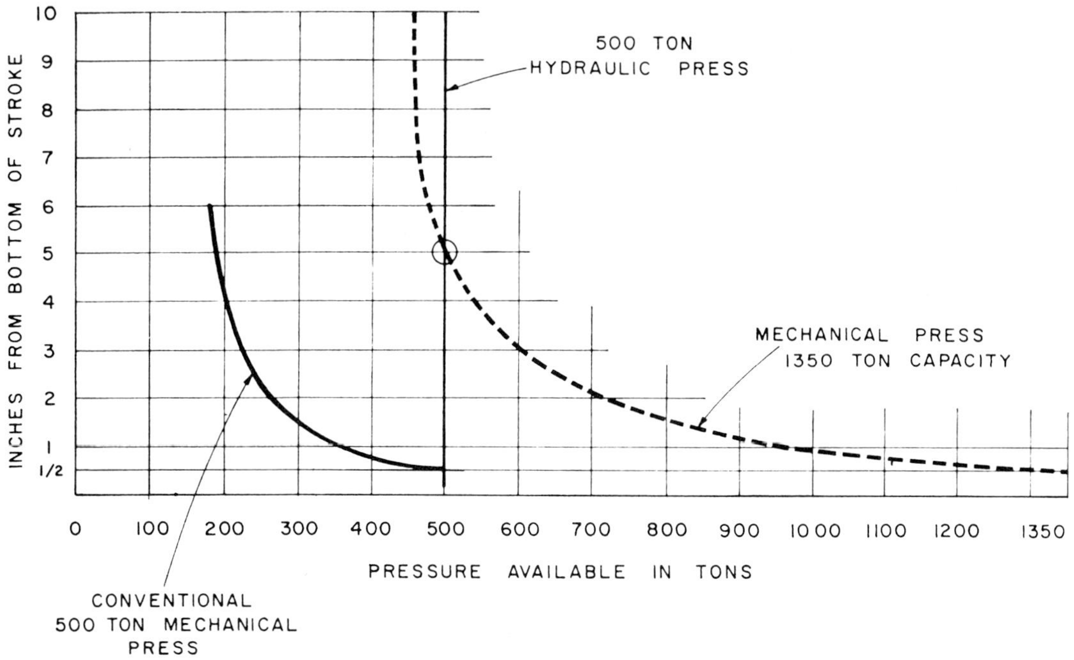

Mechanical or hydraulic

1. Medium draws
2. Sizing or coining
3. Trimming
4. Forging
5. Embossing
6. Automated operations

Hydraulic presses are available in a number of frame configurations. The most popular of these is the column type. This provides the simplest type of construction and is satisfactory for many kinds of applications. However, the straight side frame construction is used often when slide and bed alignment is more critical and the possibility of off-centre loading exists. The gap frame is also used. It is quite commonly seen on hydraulic straightening presses (Figs 13, 14, 15).

An hydraulic press, regardless of length of stroke, can have full tonnage available for the entire stroke. Due to the harmonic motion of a mechanical press drive, the full tonnage would be available only at a certain distance above bottom dead centre (usually ½in). The tonnage drops off at points higher in the stroke. For example, on the 500-ton press, only 175tons is available, 5in up from BDC. A mechanical press, with comparable tonnage to the hydraulic press at a point 5in up from BDC, would need a 1350 ton rating on the press slide. See Fig 16.

Controlled pressure

On a mechanical press, the pressure builds up at an increasing and uncontrolled rate after midstroke to a maximum at the end of the stroke. Since the tonnage is a function of the press drive, it cannot be adjusted for stock variations or die settings. Once in motion, the mechanical press slide cannot be reversed but must go through its stroke. If the resistance is great enough, the slide will stick on bottom or else break the die or the weakest point of the press.

The pressure on an hydraulic press is dependent on the setting of an adjustable relief valve and can be regulated automatically from an operator's console at floor level in a stepless range from a maximum setting down to approximately 20 per cent of maximum. The system pressure, through a complete work stroke, builds up only to that needed to overcome the resistance encountered in the work piece. Should the work piece demand excessive pressure due to stock thickness, double blanking or inaccurate die settings, the system will build up pressure only to the setting of the relief valve and no higher. It is almost impossible to overload an hydraulic press.

Exact pressure applied to the work at any point in the stroke is indicated on a direct reading pressure gauge. Thus, if a 200-ton job is to be done in a 500-ton press, the tonnage controls are set before hitting the run buttons with the assurance the pressure will build up to this setting and no higher, with the actual pressure used indicated by a gauge reading.

On an hydraulic press the length of stroke of the slide can be adjusted to suit any depth of draw. Limit switches, adjustable from floor level, determine the daylight from a maximum down to any desired minimum. Thus, a 60in stroke press can be set to do a 6in draw with just enough daylight to remove the part. Any desired stroke can be set in a matter of seconds. This is contrary to a mechanical press where the stroke cannot be altered. The full stroke of the crank must take place regardless of the actual depth of draw.

Ease of die setting

Die setting is accomplished on an hydraulic press with less skill and in less time than is required on mechanical presses. Since there is a fixed stroke and positive slide travel on a mechanical press, any variation in the shut height must be made carefully to avoid the possibility of excessive interference on closing of the

Fig 17

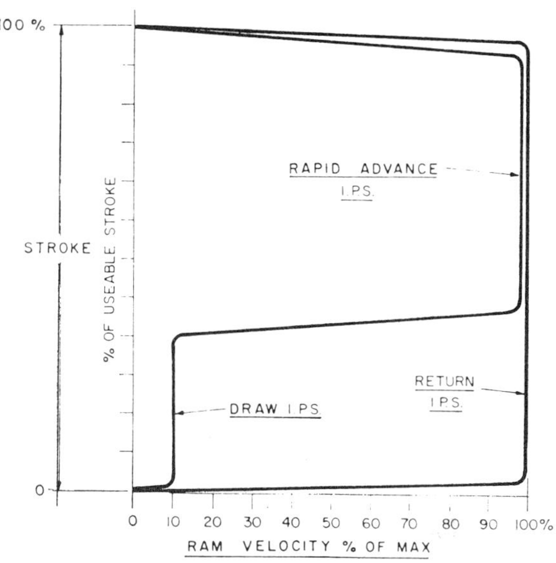

HYDRAULIC CALCULATIONS - SHEET

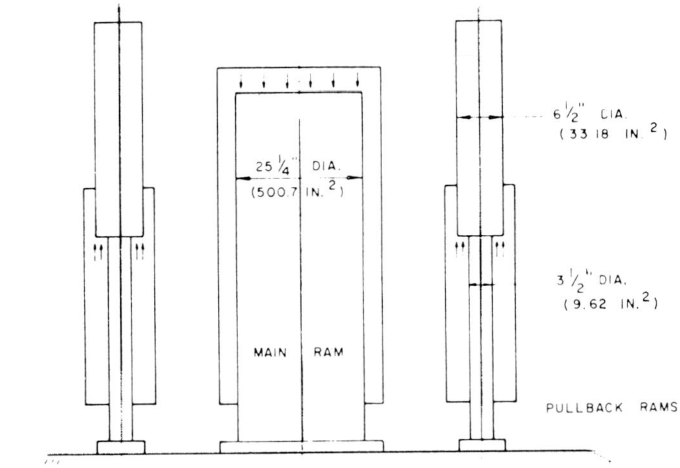

SLIDE

REQUIRED - 750 TON PRESSING
10% STRIPPING (APPROX.)
250 TON CUSHION
10 STROKES PER MINUTE - 5" DRAW

AREA - PRESSING - 500.7 IN.²
STRIPPING - 33.18 IN.²
− 9.62 IN.²
23.56 X 2 = 47.13 IN.²

H.P. REQUIRED (APPROX.) *: $\frac{\text{TONNAGE X PRESSING SPEED}}{250}$

$\frac{\text{750 TONS X 100 IPM}}{250}$ = 300 H.P.

PUMPS TO BE USED - (2) 108 GPM OILGEARS

SYSTEM PRESSURE - 3,000 P.S.I.

dies. The hydraulic press with no fixed stroke can adapt itself to the job. There is no fixed shut height. A uniform pressure is applied to the part regardless of the stock variation and without shimming the die.

This is particularly evident on double-action presses where the part is held by the blankholder during the draw. Each corner of the blankholder slide on an hydraulic press can be adjusted independently for pressure applied in holding the blank.

HYDRAULIC CALCULATIONS - SHEET 2

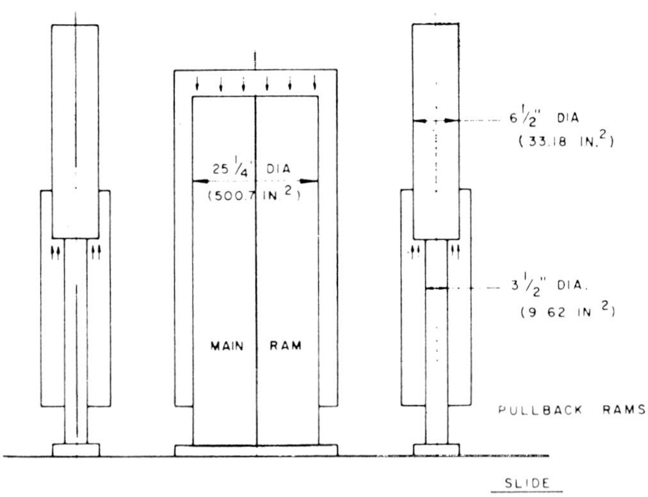

SLIDE

SYSTEM PRESSURE - $\frac{\text{750 TONS X 2000\#/TON}}{\text{500.7 IN.}^2}$ = 2950 P.S.I.

STRIPPING - $\frac{\text{70 TONS X 2000\#/TON}}{\text{47.13 IN.}^2}$ = 2960 P.S.I.

PRESSING SPEED - $\frac{\text{(2) X 108 GPM X 231 IN.}^3\text{/GAL.}}{\text{500.7 IN.}^2}$ = 100 I.P.M.

RETURN SPEED - $\frac{\text{(2) X 108 GPM X 231 IN.}^3\text{/GAL.}}{\text{47.13 IN.}^2}$ = 1060 I.P.M.

CLOSING SPEED - = 1060 I.P.M.

HYDRAULIC CALCULATIONS - SHEET 3

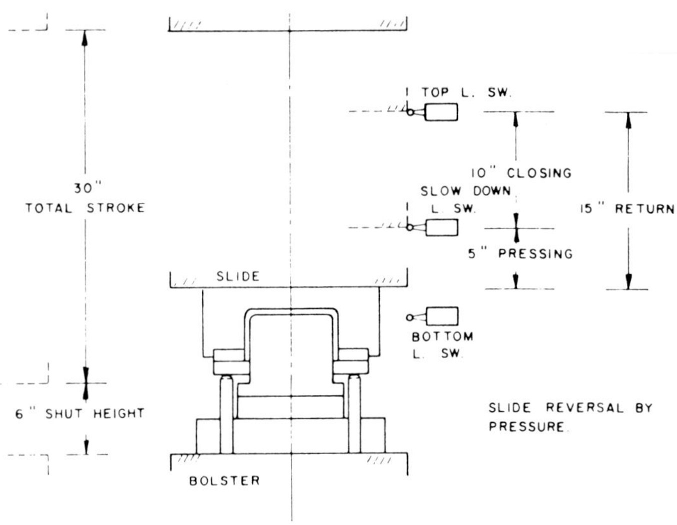

STROKE PER MINUTE - BASED ON A 5" DRAW

STROKES PER MINUTE WILL VARY DEPENDING ON DEPTH OF DRAW. TO FIGURE SPM SUBSTITUTE LENGTH OF CLOSING, PRESSING AND RETURN STROKES IN FORMULA.

10" CLOSING - $\frac{\text{10" CLOSING X 60 SEC./MIN.}}{\text{1060 IN./MIN.}}$ = .55 SECONDS

5" PRESS - $\frac{\text{5" PRESSING X 60 SEC./MIN.}}{\text{100 IN./MIN.}}$ = 3.00 SECONDS

15" RETURN - $\frac{\text{15" RETURN X 60 SEC./MIN.}}{\text{1060 IN./MIN.}}$ = .85 SECONDS

FACTOR FOR COMPRESSION OF OIL - PUMP SHIFT - DECOMPRESSION, ETC. = 1.50 SECONDS

TOTAL TIME PER CYCLE = 5.90 SECONDS

STROKES PER MINUTE = 10 S.P.M.

The controls are at floor level with readings visible on a gauge. If one corner is wrinkling or tearing, pressure can be adjusted as necessary before or while the slide is in motion. There is no shimming required.

Pressing speed variable

The slide on an hydraulic press is capable of slowing down from its rapid closing speed to pressing speed just prior to contacting the dies and work piece. The work is then done at a constant speed throughout the entire length of the draw. This eliminates the possibility of die damage and tearing of the part, so prevalent on mechanical presses where the work is started high up in the stroke. There are no problems of midstroke velocity or tearing due to impact on an hydraulic press (Fig 17).

On an hydraulic press the slide travel can be stopped at any point and reversed. In a work stroke the press slide can either be reversed by closing the dies and building to a desired pressure or on depth of draw in an open die. An hydraulic press is basically self-lubricating with added lubrication required only for the slide gib ways. Since the hydraulic force is developed and delivered through pumps and valves, additional controls can often be added to change the basic nature of the work being done.

Maintenance

An hydraulic press is usually more difficult to maintain since it requires a greater degree of skill and specialization in the personnel involved. Breakdowns on a mechanical press are visual and more easily detected although usually more severe. A thorough understanding of the electrical and hydraulic circuitry is required to determine the source of an hydraulic problem before repairs can be started. Once detected, actual repair on an hydraulic press is usually much simpler to accomplish.

Dirt, poor housekeeping in general and poor preventive maintenance are of far greater detriment to an hydraulic press operation than to a mechanical press operation. Due to the pipes, valves and controls involved, an hydraulic press can easily develop oil leaks if the circuit is not kept in adjustment and maintained.

To obtain a high stroke/min rating, an hydraulic press requires far more horsepower than a mechanical press although actual horsepower used is relatively the same. It is very difficult to obtain more than 20 strokes/min on an hydraulic press even with shallow draws. Blanking, piercing, trimming, etc., introduce a shock to either an hydraulic or a mechanical press. The effect of the shock, however, is more critical on the hydraulic press.

Summary

The criteria presented in the foregoing should enable a potential press buyer to make an evaluation of the type of machine he needs. In addition, by being familiar with the basic factors affecting press design, he will be able to discuss more knowledgeably his own requirements with the press supplier. Since we have only been concerned with the selection of conventional presses, no effort has been made to discuss special purpose machines.

If nothing else, having knowledge of the information contained herein should enable the press user to determine whether or not he needs a conventional or special purpose press. Under any circumstances, the press manufacturer will be more than glad to supply whatever information is needed to fill in the gaps and give the press buyer a complete picture of the press he should select.

Presses and Auxiliary Equipment for Forming of Sheet Metal

*By the ASM Committee on Forming of Sheet Metal in Presses**

PRESSES described in this article are mechanically or hydraulically powered machines used for producing parts from sheet metal.

Power presses can be classified according to the following characteristics: source of power, type of frame, method of actuation of slides, and number of slides in action. Presses in any of these classes are available in a range of capacities (tonnage or bed area), although the range is not necessarily the same for all types of presses. Characteristics of 18 types of presses are summarized in Table 1.

JIC Identification System. The Joint Industry Conference (JIC) system of identifying press characteristics is in general use. In a typical sample:

S4-750-96-72

the press is identified by the S as a single-action model (D is used for double-action, T for triple-action, and OBI for open-back inclinable); by the 4 as having four-point suspension; by the 750 as being rated at 750-ton capacity; and by the 96 and 72 as having a bed measuring 96 in. left-to-right and 72 in. front-to-back. Any other press can be so identified, by substitution of appropriate numerals for number of suspension points, tonnage rating, and bed dimensions. These characteristics are discussed later in this article.

The JIC also recommends that a metal tag be attached permanently to the press, stating the stroke length, shut height, kind and length of adjustment, strokes per minute, size and weight. If a die cushion is provided, an additional tag should describe it.

Source of Power

Power presses for sheet-metal work can be driven hydraulically or mechanically. The performance characteristics and other operational features of hydraulic and mechanical presses are compared in the following list.

1 Force is exerted constantly throughout the stroke of a hydraulic press. Force developed by a mechanical press varies with the position of the slide.
2 The length of stroke is easily adjusted and controlled in a hydraulic press. In a mechanical press, the stroke is fixed by the throw of the crank or eccentric.
3 The speed of a hydraulic press is adjustable over a wide range, whereas the speed of a mechanical press is limited by the type of drive.
4 A hydraulic press cannot be overloaded. It can deliver only a preset force, and slide motion stops when that force is reached. A mechanical press can be overloaded, resulting in damage to the press, if it is not equipped with overload protection.
5 Mechanical presses cycle faster and are better suited to high production than are hydraulic presses.
6 Because energy is stored in the flywheel, a mechanical press can use a smaller motor. For some applications, the size of the motor in a hydraulic press may be as much as

*H. L. ANTHONY III, *Chairman,* Staff Fellow, Mellon Institute, Carnegie-Mellon University; M. GLENN BARBER, Senior Manufacturing Research Engineer, Michigan Manufacturing & Engineering Div., Burroughs Corp.; PEHR O. BOLLING, Overland-Bolling Co.

M. L. CLARK, Producibility Analyst, Small Aircraft Engine Dept., General Electric Co.; WILLIAM B. FORSBERG, Product Engineer, Carr Fastener Div., United-Carr Fastener Corp.; R. HARNIST, Tool and Design Supervisor, F. H. Lawson Co.; A. A. HERRADA, Supervisor of Tool Design Unit, Manufacturing Engineering Dept., General Parts Div., Ford Motor Co.; ROBERT S. NEILSON, Project Engineer, Fabrication Engineering Dept., Lukens Steel Co.; JOHN E. PARKS, HPM Div., Koehring Co.

J. W. RATH, Supervisor of Tool Engineering, Delco Appliance Div., General Motors Corp.; G. D. REED, Production Engineer, West Bend Co.; MERRILL RIDGWAY, Application Engineer, Minster Machine Co.; STEPHEN K. SZUCK (deceased), formerly Chief Manufacturing Engineer, Ronson Corp. of Pennsylvania.

The assistance of HAROLD R. DANIELS, senior associate editor, *Metalworking* magazine, in preparing this article is gratefully acknowledged.

Table 1. Characteristics of 18 Types of Presses

Type of press	Type of frame: Open-back	Gap	Straight-side	Arch	Pillar	Solid	Tie rod	Position of frame: Vertical	Horizontal	Inclinable	Inclined	Action: Single	Double	Triple
Bench	X	X	..	..	..	X	..	X	..	X	X	X	..	..
Open-back inclinable	X	X	..	..	..	X	..	X	..	X	..	X	X	..
Gap-frame	X	X	..	..	..	X	X	X	X	X	X	X	X	..
Adjustable-bed horn	..	X	..	..	..	X	..	X	..	..	..	X	..	..
End-wheel	..	X	..	..	..	X	..	X	..	..	..	X	..	..
Arch-frame	..	..	X	X	..	X	..	X	..	..	X	X	..	..
Straight-side	..	..	X	X	..	X	X	X	X	..	X	X	X	X
Reducing	X	X	X	..	..	X	X	X	X	..	..	X	..	..
Knuckle-lever	..	..	X	..	..	X	X	X	..	..	X	X	X	..
Toggle-draw	..	..	X	..	..	X	X	X	..	..	..	..	X	X
Cam-drawing	X	X	X	..	..	X	X	X	..	X	..	..	X	..
Two-point single-action	..	X	X	..	..	X	X	X	..	..	X	X	..	..
High-production	..	..	X	..	..	X	X	X	..	..	X	X	..	..
Dieing machine	..	..	..	..	X	..	..	X	..	..	..	X	..	..
Transfer	..	X	X	..	..	X	X	X	..	..	..	X	..	..
Flat-edge trimming	..	..	..	X	..	X	..	X	..	..	..	..	X	..
Hydraulic	..	X	X	..	X	..	X	X	X	X	..	X	X	X
Press brake	X	X	..	..	..	X	..	X	..	..	..	X	..	..

Type of press	Method of actuation: Crank	Front-to-back crank	Eccentric	Toggle	Screw	Cam	Rack and pinion	Piston	Type of drive: Over direct	Geared, overdrive	Under direct	Geared, underdrive	Suspension: One-point	Two-point	Four-point	Ram: Single	Multiple	Bed: Solid	Open	Adjustable
Bench	X	..	X	..	X	..	X	X	X	..	..	..	X	..	..	X	..	X	X	X
Open-back inclinable	X	..	X	..	..	X	..	X	X	X	..	..	X	X	..	X	X	..	X	..
Gap-frame	X	X	X	X	..	..	..	X	X	X	..	..	X	X	X	X	X	X	X	..
Adjustable-bed horn	X	..	X	..	..	..	..	X	X	X	..	..	X	X	..	X	..	X	X	X
End-wheel	..	X	X	..	..	..	..	..	X	X	..	..	X	..	..	X	..	X	X	..
Arch-frame	X	..	..	..	..	..	..	..	X	X	..	..	X	..	..	X	..	..	X	..
Straight-side	X	X	X	..	..	..	X	X	X	X	X	X	X	X	X	X	X	X	X	..
Reducing	X	..	..	..	..	..	..	X	X	X	..	..	X	..	..	X	..	..	X	..
Knuckle-lever	..	..	..	X	..	..	..	..	X	X	..	..	X	X	..	X	X	X	X	..
Toggle-draw	X	..	X	X	..	..	..	..	..	X	..	X	X	X	X	..	X	..	X	..
Cam-drawing	X	..	..	..	..	X	..	..	X	X	..	..	X	..	..	..	X	..	X	..
Two-point single-action	X	..	X	..	..	..	..	..	..	X	..	X	..	X	..	X	..	X	X	..
High-production	X	..	X	..	..	..	..	..	X	X	..	..	X	X	..	X	..	..	X	..
Dieing machine	X	..	..	..	..	..	..	..	X	..	X	..	X	X	..	X	..	..	X	..
Transfer	X	X	X	X	..	..	..	..	X	X	..	..	..	X	..	X	..	..	X	..
Flat-edge trimming	..	..	..	..	..	X	..	..	..	X	..	..	X	..	..	X	..	..	..	..
Hydraulic	..	..	..	..	..	..	..	X	..	..	..	..	X	X	..	X	X	X	X	X
Press brake	X	..	..	..	..	..	..	X	..	X	..	..	..	X	X	X	..	X	..	..

2½ times as large as in an equivalent mechanical press.

7 Ram velocity in a mechanical press is higher, making this equipment more useful in operations such as blanking and piercing, in which a high-impact blow is needed. Blanking and piercing can be done in a hydraulic press, but the shock of the punch breaking through the metal can cause damage to the hydraulic system.

Mechanical Presses. In most mechanical presses, a flywheel is the major source of energy that is applied to the slides by cranks, gears, eccentrics or linkages during the working part of the stroke. During operation, the flywheel runs continuously and is engaged by the clutch only when a press stroke is needed. In some very large mechanical presses the drive motor is connected directly to the press shaft, thus eliminating the need for a flywheel and a clutch.

Two basic types of drive are used to transfer the rotational force of the flywheel to the main shaft of the press — the nongeared drive and the gear drive.

Nongeared Drive. In a nongeared drive (often known as a flywheel drive), the flywheel is on the main shaft (Fig. 1a), and its speed, in revolutions per minute, controls the slide speed. Press speeds with this type of drive usually are high, ranging from 60 to 1000 strokes per minute. The main shaft can have a crankshaft as shown in Fig. 1(a), or an eccentric.

Energy stored in the flywheel should be sufficient to insure that the reduction in the speed of the flywheel will be no greater than 10% per press stroke. If the energy in the flywheel is not sufficient to maintain this minimum in speed reduction, a press with a gear drive should be used.

Gear drives (Fig. 1b, c and d) have the flywheel on an auxiliary shaft that drives the main shaft through one or more gear reductions. Either single-reduction or multiple-reduction gear drives are used, depending on size and tonnage requirements. In gear-driven presses, there is more flywheel energy available for doing work than in the nongeared presses, because the speed of the flywheel is higher than that of the main shaft. The flywheel shaft of a gear-driven press often is connected to the main shaft at both ends, as shown in Fig. 1(c), which results in a more efficient drive.

A single-reduction gear drive develops speeds of 30 to 100 strokes per minute. Speed for a multiple-reduction twin-gear drive (Fig. 1d) usually is 10 to 30 strokes per minute, which provides exceptionally steady pressure.

Hydraulic Presses. Hydrostatic pressure against one or more pistons provides the power for a hydraulic press. Most hydraulic presses have a variable-volume, variable-pressure, concentric-piston pump to provide them with a fast slide opening and closing speed, as well as with a slow working speed at high forming pressure.

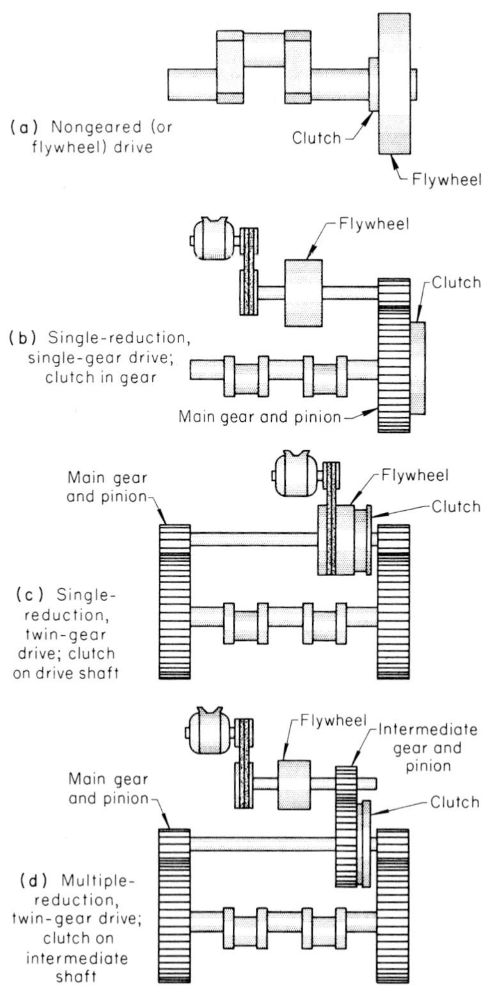

Fig. 1. Four types of drive-and-clutch arrangements for mechanical presses

Pumps, reservoirs and other components of the hydraulic system usually are housed in the frame and in the crown of the press, although in some presses these components are housed below the bed. If all components are readily accessible for maintenance, their location is unimportant. The principal components of a typical hydraulic press are shown in Fig. 2. A bolster plate is attached to the bed to support the dies and to guide the pressure pins between the die cushion and the pressure pad.

The tonnage that a hydraulic press can exert depends on the diameter of the hydraulic pistons and on the rated maximum hydraulic pressure, the latter being a function of the pump pressure and related mechanisms.

Hydraulic presses with capacities up to 50,000 tons have been built, but most have a capacity of less than 15,000 tons. The typical hydraulic press is rated at 100 to 1000 tons. Gap-frame presses are rated at 5 to 50 tons.

Because of their construction, hydraulic presses can be custom designed at relatively low cost. They can be designed with a number of slides and motions, or separate hydraulic circuits can be used for various independent actions.

Side action can be provided also within the frame of the press by separate cylinders. Such side action in a mechanical press usually is provided by cams and is complex and expensive. Most hydraulic presses are straight-side models, but small, fast, gap-type presses designed to compete with mechanical open-back inclinable (OBI) presses have been developed.

Hydraulic press slides, or platens, are actuated by numerous combinations of hydraulic drives. Hydraulic presses usually have a longer stroke than mechanical presses, and force can be constant throughout the stroke. Hydraulic presses have an adjustable stroke for one or more slides. Accumulators or large-volume pumps can provide fast motion for a slide to open and close. High-pressure pumps provide the working force at a slower speed.

Usually all slides are operated by one pumping system. The relation of each action to the others, interaction, and timing all depend on the controls.

Type of Press Frame

Presses are classified broadly according to the type of frame used in their construction, into two main groups — gap-frame presses and straight-side presses. Details of construction vary widely in each group.

Gap-frame presses are sometimes called C-frame presses because the frame resembles the letter "C" when viewed from the side. The gap makes the die area accessible from either side as well as from the front, for ease in die setting or for feeding stock. Coil stock often is supplied to gap-frame presses by feeders from stock reels and straighteners.

Workpieces usually are ejected through an opening in the press bed, or through the back of an open-back press. (A few gap-frame presses with solid backs are in service, but solid-back presses are in increasing disfavor because work cannot be ejected through the back of the press, and because the design in general is less convenient than the open-back type.)

Deep-throat presses are made, but the throat depth (distance from bolster centerline to frame) on standard presses varies from 6 in. in a 22-ton press to 17 in. in a 200-ton press.

Gap-frame construction has one disadvantage — the gap opens under load, thus causing angular deflection. (A straight vertical deflection would be less critical.) Gap deflection resulting from overload causes misalignment of punches and dies, which is a major cause of premature die wear. This condition can be avoided by using a larger press; however, gap-frame presses have sufficient rigidity if not overloaded.

Tie rods extending from the top of the frame to the front corners of the bed can be used to minimize deflection. Since the tie rods close the gap, the accessibility to the die area and the width of parts fed into the die are limited. Tie rods can be removed for die setup.

Gap-frame presses are made in four frame styles, based on frame position:

1. Vertical frame with a fixed or an adjustable-height bed
2. Inclined frame fixed at an angle of 20° to 30° (Inclination can be custom built to any angle.)
3. Inclinable frame that can be tilted to a maximum angle of 20° to 30°, in stepless increments or in two or three increments, depending on design
4. Horizontal frame (not a common design).

Gap-frame presses generally have less over-all height than straight-side presses of the same tonnage rating. This can be important when overhead clearances are low.

Open-back inclinable (OBI) presses are gap-frame presses that can be inclined so that gravity assists in ejection of the workpiece. The smaller OBI presses can be tilted as much as 45°, whereas the larger OBI presses are usually designed to tilt to a maximum of 20°.

The Joint Industry Conference (JIC) has developed standards for eight sizes of OBI presses ranging in capacity from 22 to 200 tons. Larger OBI presses have been built, but most are within this range. Machined surfaces are provided for mounting a die cushion in the press bed. On small presses, die cushions frequently are mounted on the lower side of the bolster plate.

JIC standards include bed dimensions, bolsters, T-slot and mounting-bolt spacing, and the location of the pressure-pin holes in the bolster; thus bolsters and dies can be interchanged between different makes of presses. The dimensions of slide faces, size and location of T-slots and die-anchorage holes and the upper-shoe stem hole, and minimum shut heights also have been standardized.

Inclinable gap-frame presses are made in tonnage ratings other than those standardized by the Joint Industry Conference.

End-wheel presses are gap-frame presses with a deep throat and, usually, a closed back. The crankshaft runs from front to back, with the flywheel and drive in the back of the press. End-wheel presses, sometimes referred to as deep-throat presses, are best suited to hole punching and other blanking or forming applications requiring a deep reach-in of tools from the edge of the workpiece.

Adjustable-bed presses serve much the same function as gap-frame presses, except that instead of being part of a solid "C", the bed is a separate component. The bed, supported by an adjusting screw, can be moved up or down to provide a greater range of shut heights; consequently, the use of die cushions in this press is impractical.

An advantage of the adjustable-bed press is that the bed usually is hinged and can be swung out of the way. A horn or mandrel can be inserted in a hole in the press frame, so that the press can be used as a horn press for cylindrical work. A typical application of an adjustable-bed press is the forming and insertion of flanges on oil drums.

Straight-side presses have a frame made up of a base, or bed, two columns, and a top member (crown). In most straight-side presses, steel tie rods hold the base and crown against the columns. Straight-side presses have either

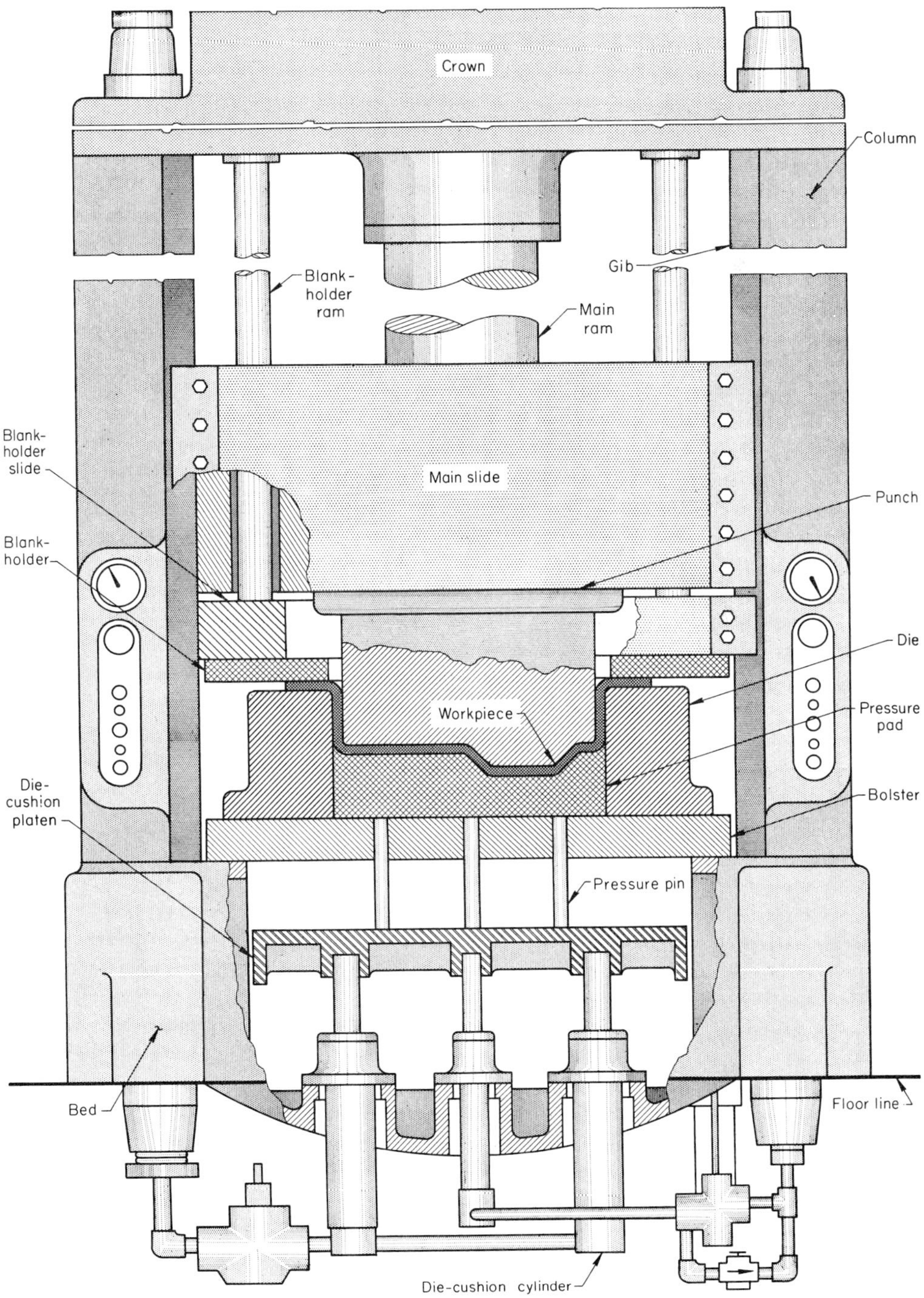

Fig. 2. Principal components of a double-action hydraulic press with a die cushion

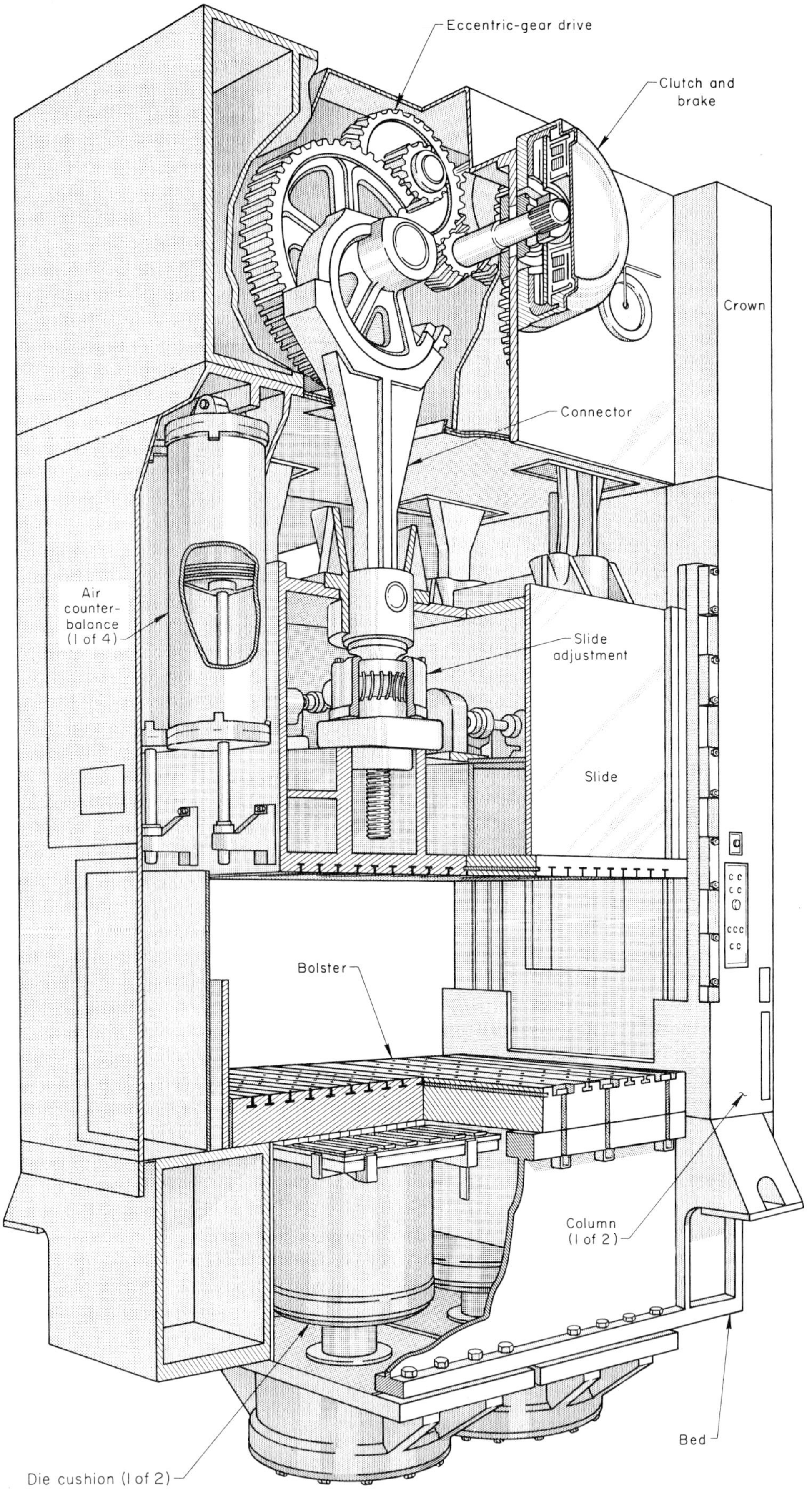

Large-bed, four-point-suspension, eccentric-drive press with slide counterbalance cylinders. Slide adjustment is motorized. Presses with smaller beds have one or two-point suspension and have nongeared (flywheel) or eccentric-shaft drives, but no slide counterbalance cylinders.

Fig. 3. Principal components of a single-action straight-side mechanical press

crankshaft, eccentric-shaft or eccentric-gear drives. (See the section "Slide Actuation in Mechanical Presses", which begins on the next page.)

A single-action straight-side press is shown in Fig. 3. The slide in this illustration is equipped with air counterbalances to assist the drive in lifting the weight of the slide and the upper die to the top of the stroke. Counterbalance cylinders provide a smooth press operation and easy slide adjustment. Die cushions are used in the bed for blankholding and for ejection of the work.

The straight-side design permits the use of an endless variety of bed and slide sizes. Presses range from 20-ton capacity and a bed of 20 by 15 in., to 4000-ton capacity with a bed as large as 360 in. left-to-right by 180 in. front-to-back. The size and shape of the slide usually determine the number of points of suspension, or connections between the main shaft and the slide, that are needed.

The straight-side design also can provide high pressures with minimum deflection. A straight-side press will deflect less under off-center loads than a gap-frame press.

Size and location of anchorage holes and slots, T-slots, and pressure pin holes for the press bed, bolster plate and ram were standardized by JIC.

Rail presses combine the long, narrow bed of a press brake with the construction features of a straight-side press. They are used for forming channels and for operations in progressive or multiple dies or utility tooling.

Tryout presses are straight-side presses used almost exclusively for die tryout. Press capacities range from 50 to 300 tons. The drive is in the bottom of the press. The upper part of the press is a pivoted head that rotates around a horizontal axis at the front of the press. The head, carrying the upper die, swings to the front of the press, so that the tooling can be finished or corrected without removing it from the press. Once corrected, the tooling is installed in a conventional press for production.

Notching presses are small, high-speed, short-stroke presses that have a single-column, closed-back frame like that of a horn press. They are high-production presses, used primarily for notching and piercing of motor laminations. (See the article "Blanking and Piercing of Magnetically Soft Materials", which begins on page 60 in this volume.)

Arch-frame presses (of straight-side design, with bowed-out columns to provide extra die area) are becoming obsolete.

Pillar presses are small, solid-frame, straight-side presses specially built for use with a subpress instead of a die set.

Underdrive presses have the drive mechanism below the bed and pull the slide down, instead of pushing it as in conventional design. They usually are built in capacities of 400 tons or more and have a minimum left-to-right dimension of 72 in. They are driven by rocker arms or eccentric gears and are of single, double or triple-action construction. Because of their large size, they have either double or triple-reduction gearing.

These presses should be installed in a plant with two levels. The upper level is confined to production work, the lower level to maintenance operations. Most of the press mechanism is at the lower level; therefore, underdrive presses require less ceiling height than top-drive presses. The presses are mounted on a network of I-beams, which are supported on the lower-level piers. Because the I-beams are readily accessible, it is fairly simple to relocate both the beams and the presses, when necessary to facilitate production.

Another type of underdrive press is the dieing machine (see the section "Presses for High Production", page 16 in this article). This press is mounted on the floor with the bed at eye level.

Slide Actuation in Mechanical Presses

Rotary motion of the motor shaft on a mechanical press is converted into reciprocating motion of the slides by one of the following:

- Crankshaft
- Eccentric shaft
- Eccentric-gear drive
- Knuckle-lever drive
- Rocker-arm drive
- Toggle mechanism.

Crankshafts. The most common mechanical drive for presses with capacities up to 300 tons is the crankshaft drive (Fig. 4). A crankshaft is used in both gap-frame and straight-side presses. The crankshaft drive is used most often in the single-suspension design, although some double-crank (two-point suspension) presses — particularly in the 100-to-200-ton range — also have crankshafts.

The crankshaft imparts a sine-curve speed relation to the press slide. The stroke of a crankshaft-actuated press can be as short as 1 in., in a small gap-frame press, to as much as 30 in., in a straight-side press. However, most mechanical presses with longer strokes are actuated by an eccentric gear, because it provides greater strength. Crankshaft drives usually are limited to strokes of 6 to 12 in.

The main advantage of a crankshaft-driven press is its lower cost, particularly when capacities do not exceed 300 tons.

Points of suspension refer to the number of connections between the slide and the actuating mechanism. Presses can have single-point, two-point, or four-point suspension, according to the number of points at which the slide is pushed or pulled. The simplest mechanical presses have a pitman that connects the eccentric shaft or the crankshaft to the slide at only one point.

Many wide mechanical presses are built with two-point suspension by connecting the slide to the crankshaft (or eccentric) with two pitmans instead of one, for better distribution of force on the slide.

The largest straight-side mechanical presses usually have four-point suspension, for more uniform loading of large slides. Four-point suspension usually is accomplished by two interconnected crankshafts or eccentrics and four pitmans; each pitman is connected near a corner of the slide.

Hydraulic presses also can have one, two, or more points of suspension, by operating the slide with as many rams as desired.

Eccentric shafts are similar to crankshafts. The eccentric completely fills the space between the supporting bearings of the press crown, thereby eliminating the deflection commonly caused by the unsupported portion where the crank cheeks normally would be. Eccentric drives (Fig. 4) are often used in high-speed short-stroke straight-side presses with progressive dies.

The height of the workpiece is the main limitation of the eccentric-shaft drive, because the stroke always is equal to twice the eccentricity. When eccentricity is increased, the space available in the press crown will determine the maximum stroke that can be used. In most presses of this type, the maximum stroke usually is limited to 6 in. A few presses have been built and used in high-speed operations in which strokes longer than 6 in. were needed. This is accomplished in the press by balancing the eccentric shaft to minimize vibration.

Eccentric-gear drives (Fig. 4) are used almost universally for large straight-side presses that operate at speeds of less than 50 strokes per minute. In place of a crankshaft, an eccentric is built as an integral part of the press drive gear. The eccentric gear permits strokes as long as 50 in.; however, with such long strokes, speeds usually are only 8 to 16 strokes per minute. With the eccentric as part of the gear, accuracy of alignment of the slide is determined by accuracy and alignment of the gears. In a two-point suspension, the parallel condition of the ram is determined by the alignment of the driving gears. The principal advantage of the eccentric gear is that it permits greater torque loads at points above the bottom of the stroke. The eccentric gear also permits multiple-point construction with greater versatility and range of stroke length than is possible with a crankshaft.

The chief limitation of the eccentric-gear design is that usually it requires an overhung flywheel. In addition, a single-gear eccentric press usually costs more than a crankshaft or eccentric-shaft press of equal capacity. A second limitation is that eccentric-gear presses are more likely to stick at the bottom of the stroke than crankshaft presses. Sticking is caused by greater friction in the connector, which is inherent in the large-diameter eccentrics needed for an equal press stroke. Sticking usually will occur during setup, if the press is moved slowly until the bottom of the stroke is reached. At this point, a skilled setup man usually can detect whether the press is likely to stick.

Knuckle-lever drives combine the motions of a crank and a knuckle lever to drive the press slide (Fig. 4). Their use is limited to operations such as coining or embossing, in which the work is done almost entirely at the bottom of a short stroke. The knuckle-lever mechanism permits large capacity in a relatively small press. High mechanical advantage is inherent at the bottom of the stroke. These presses are rated to deliver full tonnage at $\frac{1}{16}$ to $\frac{1}{4}$ in. above the bottom of the stroke. The very quick increase in force as the slide nears the bottom of the stroke is the reason its usefulness is limited to operations performed at the bottom of the stroke. Knuckle-lever presses usually have capacities of 150 to 1000 tons.

Rocker-arm drives apply crank or eccentric motion to a rocker arm that is connected to the press slide (Fig. 4). In this mechanism, the linkage is driven by an eccentric gear and a connecting rod. The rocker-arm drive is a variation of the knuckle-lever drive. However, a press with rocker-arm drive is not limited to coining operations, but can be used also for drawing or forming operations.

The rocker-arm drive is used mainly in large-bed underdrive presses. The linkage operates from below the press bed and pulls the slide into the work by a link running up through each of the press columns. In most rocker-arm

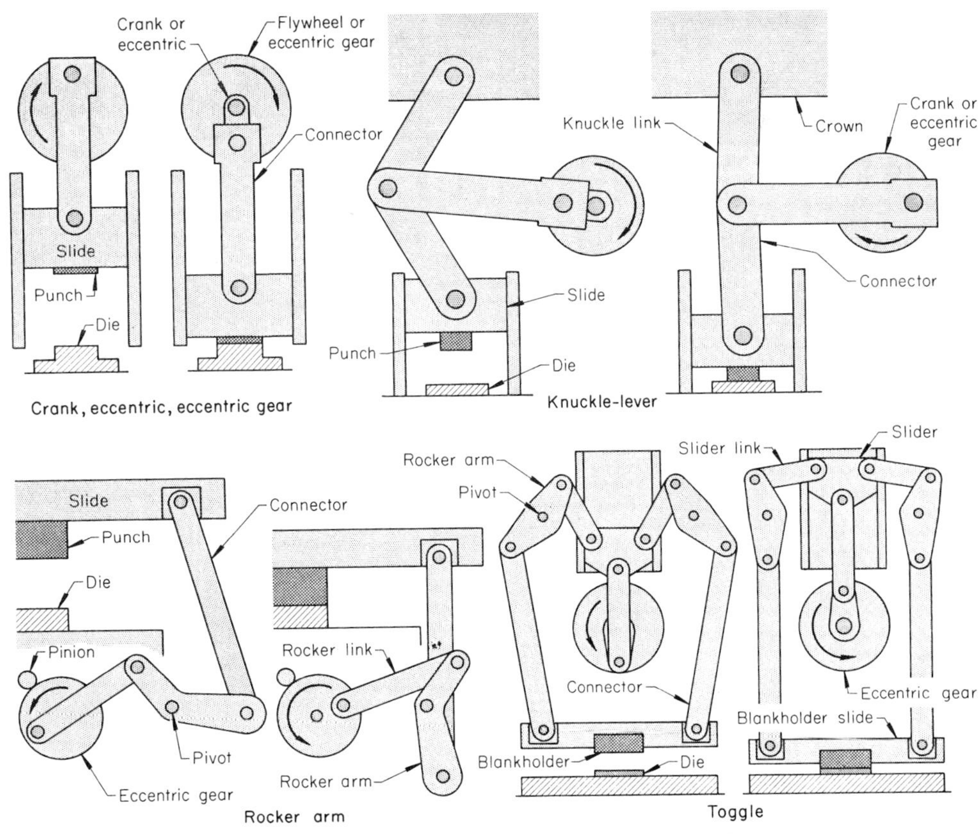

Fig. 4. Principles of operation of drives for mechanical presses

drives, the rocker pin and the connecting eccentric pin do not stop in a vertical plane; thus, the load on the eccentric shaft is relieved at the point where the maximum load on the slide is exerted, and sticking at the bottom of the stroke is prevented. In addition, a high press capacity is obtained because of the mechanical advantage.

Toggle mechanisms are the most widely used means of providing the second action in double-action mechanical presses. The toggles operate an outer slide, which clamps the blank against the die, while the punch, operated by the inner slide directly from the crankshaft, performs the draw operation. Principal components of a toggle mechanism are shown in Fig. 4.

The diagram in Fig. 5 shows relative positions on the stroke during one revolution of the crankshaft for the outer slide (blankholder) and the inner slide (punch holder) in a typical draw operation. As shown in Fig. 5, the blankholder dwells while the punch performs the draw operation. The shaded area of Fig. 5 represents the recommended working portion of the press stroke. By incorporating mechanical slow-down linkages or two-speed clutches to accelerate the slides before and after the actual working stroke, the number of press strokes per minute can be increased without affecting the draw speed. An accelerated-motion curve for one press stroke is shown in Fig. 6.

Number of Slides

Mechanical presses have one, two or three slides and are referred to as single, double or triple-action presses. Each slide can be moved in a separately controlled motion.

Single-action presses (see Fig. 3) have one moving slide that applies force to the workpiece. These presses are built with one, two or four points of suspension, depending on the bed size.

Single-point presses are of the crankshaft, eccentric-shaft, eccentric-gear or knuckle-lever construction. The same drives are used in two and four-point presses, except that a rocker arm is used instead of a knuckle lever. The two and four-point presses may be of top-drive or underdrive construction.

Single-action presses are used for all sheet-metal pressworking operations.

Double-action presses have two slides — an outer, or blankholder, slide, and an inner, or punch-holder, slide. This type of press is generally used for drawing and forming operations in which the outer slide carries the blankholder and the inner slide carries the punch. The outer slide, having a shorter stroke than the inner slide, dwells and holds the blank while the inner slide, which carries the punch, descends to perform the drawing operation. (Relative motions of the inner and outer slides are shown in Fig. 5 and 6.)

The outer slide also can carry one member of a blanking die for cutting blanks from strip or sheet. The blank is then held for the drawing operation.

Triple-action presses have three slides with motions properly synchronized for drawing, redrawing and forming operations. The blankholder and punch-holder slides are located in the crown, as in a double-action press, and a third slide is located in the press bed. The punch-holder slide has a small dwell at the bottom of its stroke. The blankholder slide usually has slightly more dwell than in a comparable double-action press. The lower slide usually is actuated by an eccentric or a crank. As a rule, the lower-slide mechanism develops a long stroke, but only during the top portion of the stroke is work

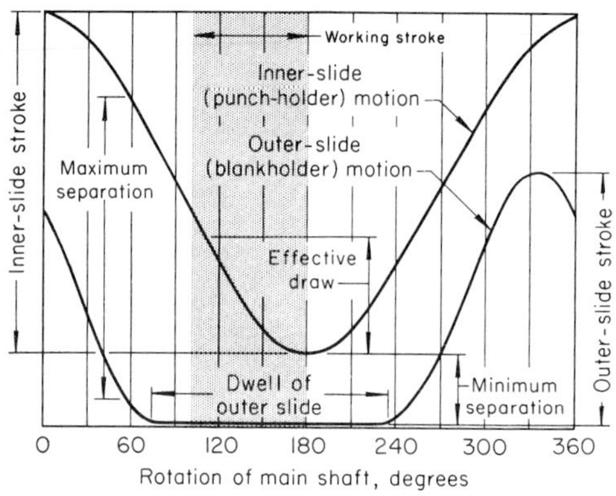

Fig. 5. Relative cycles of motion of inner slide (punch holder) and outer slide (blankholder) during one stroke of a double-action mechanical press

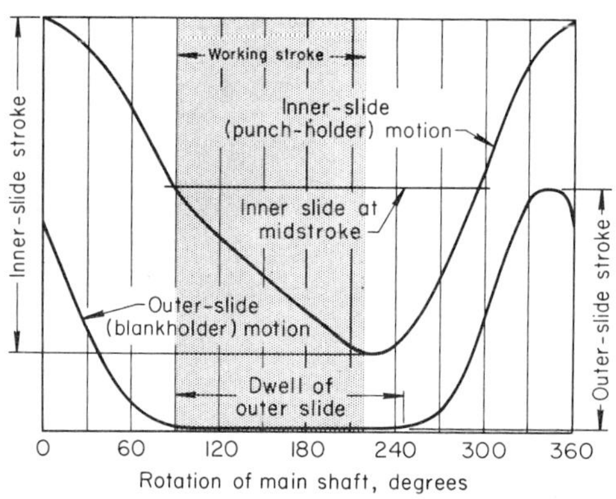

Fig. 6. Accelerated-motion curve for one stroke of a double-action mechanical press. Slides descend to work rapidly, slow down during draw cycle, and return at high speed.

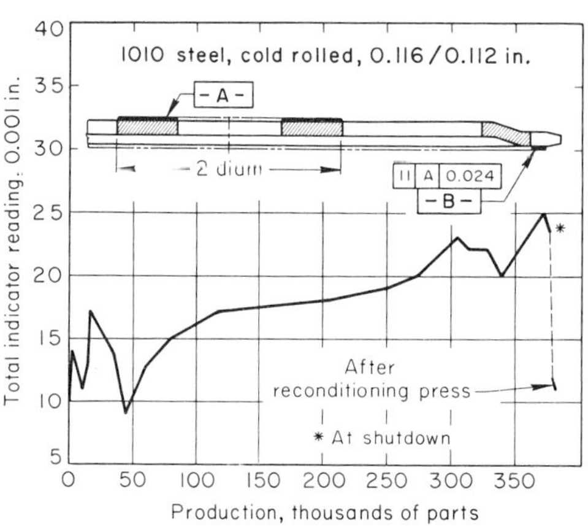

Item	Measurement, in. Original	At shutdown
Slide-to-bolster parallelism:		
Right to left, in. per ft ...	0.001	0.005
Front to back, in. per ft ...	0.001	0.004
Gib clearance, in.	0.002	0.008

Fig. 7. Variation in parallelism between two offset formed surfaces that was attributed to wear of gib surfaces (Example 1)

done by the slide. In many triple-action presses the lower-slide mechanisms are synchronized with the upper-slide motions, although separately driven lower motions electrically interlocked to the upper motions are used.

The third slide can be used for reverse drawing, forming, blanking or lancing operations. The lancing or blanking may be done before the drawing is completed, to provide extra metal. Drawing or forming is done when it is not possible to use a bottoming form block in the die.

Press Accuracy

Suggested criteria for the accuracy of a press are:

1 Maximum tolerances for parallelism between slide and bed; 0.001 in. per foot at the bottom of the stroke for all slides; 0.003 in. per foot at midstroke for punch slides; and 0.005 in. per foot at midstroke for blankholder slides.
2 Feed, if used, should be accurate within ±0.003 in. at 900 in. per minute.
3 Gib clearance should be set as close as required to do the job.

For many operations, such as blanking and piercing, the gibs are set with a minimum clearance. On small-bed presses this may be within 0.0015 in. When doing drawing operations with dies fitted with heel blocks, it is better to have the gibs set loose. This allows the heel blocks to function without interference from the gibs. For this reason, provision to adjust the gib surfaces of the inner slide of a double-action press is not commonly found. If minimum gib clearance is desired, the formula $c = ph/w$ can be used as a guide. Where c is the gib clearance; p, the total parallelism of slide to bed; h, the height of slide guides; and w, the left-to-right dimension of the slide. (All dimensions are in inches.)

A press that conforms to these standards is rated as being in good condition. If a press is accurate within half of these tolerances, it is rated in excellent condition. A press that meets only some of these tolerances is rated as being in fair condition.

A running log of the accuracy and condition of each press in a plant is helpful in scheduling maintenance and in routing work.

Press slides that are not parallel with the bed at the bottom of the stroke can result in uneven stock thickness when the punch bottoms against the die surface. (In Example 172 in the article on Press Forming, the parallelism of the slide and bolster was maintained within 0.003 in. per foot to maintain the accuracy specified for the part being formed.)

The gradual decrease in accuracy on formed parts that occurs as slide and bolster become increasingly out-of-parallel, and as gib clearance increases, with continued use of a press is described in the following example. The restoration of the original accuracy on the parts after reconditioning of the press also is discussed.

Example 1. Decrease in Dimensional Accuracy on Formed Parts Caused by Press Wear During Continued Use (Fig. 7)

A press-formed 46-tooth sprocket (a cross section of which is shown in Fig. 7) was not acceptable with out-of-parallelism greater than

0.024 in. TIR (total indicator reading) between offset surfaces A and B indicated on the cross section in Fig. 7. To attain this degree of accuracy, the press used for flattening was rebuilt to the slide-to-bolster parallelism and gib-clearance specifications given in the "Original" column in the table with Fig. 7. After flattening 375,000 pieces (approximate annual output), the press was shut down and measurements were taken of the slide-to-bolster parallelism and gib clearance. Both measurements are given in the table with Fig. 7.

Measurement of workpiece samples that were randomly selected during one year produced the data shown graphically in Fig. 7. Out-of-parallelism approached the permissible maximum at about 315,000 pieces (350 samples), and the frequency of measurement at this point was increased to 200 samples from 60,000 pieces.

At about 370,000 pieces, the rejection rate had increased substantially, but production was continued to a total of 375,000 pieces, salvaging the rejects by restriking.

Then the press was shut down and restored to its original accuracy. As shown in Fig. 7, parallelism on the pieces formed after reconditioning the press was equivalent to that obtained at the beginning of production.

Press Capacity

Capacity, or tonnage rating, of a press is the maximum force that the press can apply. Hydraulic presses can exert maximum force during the full press stroke. Mechanical presses exert maximum force at a specified distance above the bottom of the stroke (see Table 2), and the force decreases to a minimum at midstroke.

The tonnage rating of a press may have little relation to the bed area. This is especially true in the automotive and appliance industries, where presses have large bed area and large die space, but relatively low tonnage rating. Coining presses have small bed area and high tonnage rating.

Overloading of the press can cause damage to the die and the press. Several devices, based on the strain-gage principle, have been developed for accurately measuring the load on a mechanical press with a given die. Misfeeds or double blanks are common causes of press overloading. Detectors built into the die stop the press before overloading occurs.

The capacity of a mechanical press involves consideration of the frame capacity, drive capacity, flywheel energy, and motor size.

Frame Capacity. The press frame must be able to work at its rating without deflecting beyond predetermined standard limits. For general-purpose applications, the bed deflection should not exceed 0.002 in. per foot between tie-rod centers (or per foot of left-to-right bed dimension on presses without tie rods), when the rated load is evenly distributed over the middle 60% of the distance between tie-rod centers (or of left-to-right bed length). Slide deflection should not exceed 0.002 in. per foot between pitman centers, when the rated load is evenly distributed between pitmans. Both bending and shear deflections are considered. These specifications can be revised to suit more precise applications.

Drive capacity is the tonnage a mechanical press develops through the gear train and linkage. The tonnage can vary because of the mechanical advantage developed by different types of press linkage, and generally is expressed in distance above the bottom of the stroke, as shown in Table 2.

Generally, straight-side, twin-drive, geared presses of the crank or eccentric-gear type are rated at ½ in. above the bottom of the stroke, whereas gap-frame gear presses are rated at ¼ in. or less above the bottom of the stroke. The rating point above the bottom of the stroke depends on the press design. However, the values given in Table 2 are generally applicable.

Variation in the capacities of presses with eccentric-gear, crankshaft or eccentric-shaft drive at any point in the stroke is almost equal. However, the capacity decreases between the bottom of the stroke and midstroke. The capacity of both the knuckle-lever drive and the rocker-arm drive, however, is much less above the point of rating on the stroke than that of the crank drives, because the knuckle-lever drive loses mechanical advantage more rapidly than crank drives. In addition to loss in capacity, velocity of the slide in the knuckle-lever drive and the rocker-arm drive is considerably greater at points high above the bottom of the stroke, than in the eccentric-gear or crankshaft drives.

Table 2. Distance From Bottom of Stroke at Which Maximum Rated Tonnage Is Obtained in Mechanical Presses

Capacity of press, tons	Distance from bottom of stroke, in. — Flywheel drive (non-geared)	Single reduction — Single-gear drive	Single reduction — Twin-gear drive	Multiple reduction — Single-gear drive	Multiple reduction — Twin-gear drive
Gap-Frame, Open-Back Inclinable, or Straight-Side Presses					
Up to 32	1/32	1/8	..	..	..
45 to 200	1/16	1/4	1/2	1/4	1/2
Over 200	1/8	1/4	1/2	1/4	1/2
Knuckle-Lever Presses					
Up to 400	1/32	1/16	..	1/16	..
Over 400	1/16	1/8	..	1/8	..
Rocker-Arm Presses					
All sizes	..	..	1/2	..	1/2

Figure 8 is a nomograph showing the relation of press capacity and position on the stroke for standard double-gear presses in which tonnage capacity is rated at ½ in. above the bottom of the stroke. The chart may be used to determine the force exerted at specific positions on the stroke for various stroke lengths.

Sample Calculation Using Fig. 8. To find the force exerted 2 in. from the bottom of a 20-in. stroke, first find the point at which the curve for 2-in. working stroke intersects the vertical line for 20-in. stroke length. From this point, project horizontally (broken line) to the 0.5-in. curve (position of rating). From the 0.5-in. curve, project downward to the lower (horizontal) 20-in.-stroke line. As shown in Fig. 8, the intersection occurs at 53%, which means that the press can exert 53% of its rated force at a stroke position 2 in. above the bottom.

Most press manufacturers supply similar curves for determining press capacity at any point, or tables of such capacity data.

Flywheel energy for a given job may be insufficient, although the press frame and shaft may be adequately strong. For a greater working distance or for faster operation, more energy and power must be provided.

Blanking operations are completed in a brief portion of the press cycle. The flywheel instantly supplies practically all of the energy required by its resistance to deceleration. The motor may take the remainder of the press cycle to restore lost energy to the flywheel by bringing it back up to speed. Draw operations may take up to one fourth of the press cycle.

For intermittent operation, 20% is arbitrarily considered the maximum the flywheel may be slowed down when

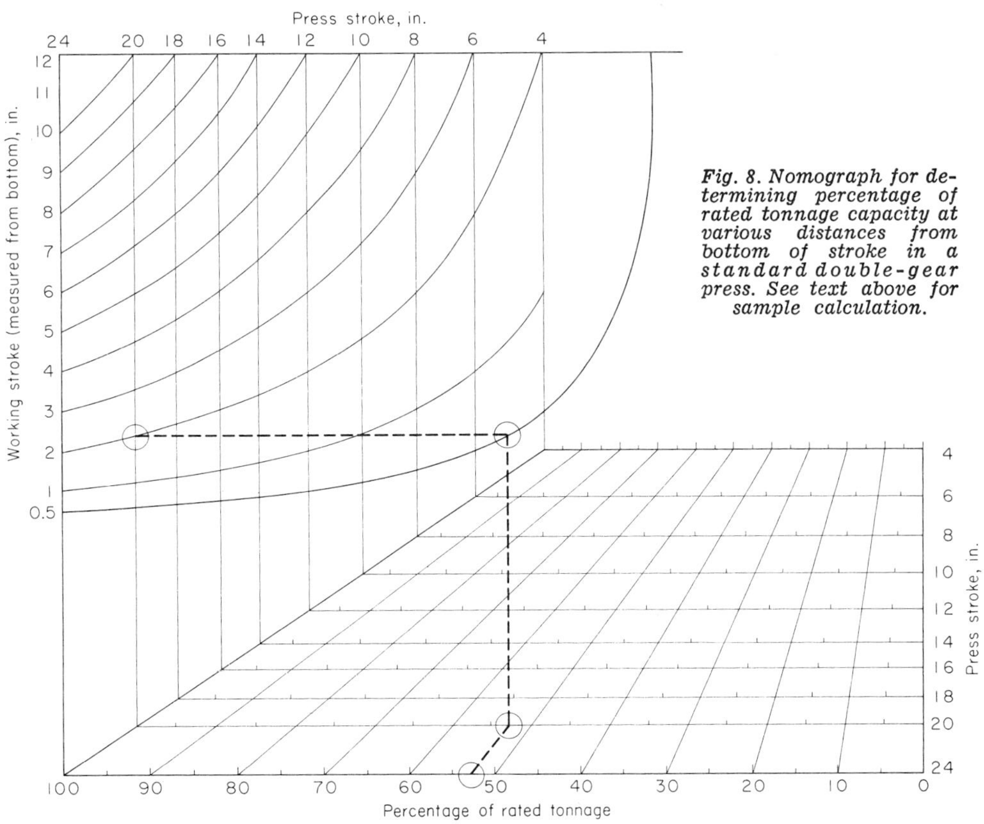

Fig. 8. Nomograph for determining percentage of rated tonnage capacity at various distances from bottom of stroke in a standard double-gear press. See text above for sample calculation.

using energy from it. For continuous operation, 10% is considered the limit, because of the short time available to restore lost energy. The low-speed torque characteristics of the press drive motor will greatly affect how much the flywheel can be safely slowed, because the ability of the motor to restore lost kinetic energy is a function of these characteristics.

The amount of energy available at 10% slowdown can be determined from the following formula:

$$E = N^2D^2\frac{W}{5.25 \times 10^9}$$

where E is energy, in inch-tons; N is number of flywheel revolutions per minute; D is the outside diameter of the flywheel, in inches; and W is the weight of the flywheel, in pounds. A constant includes necessary conversion factors.

If calculation indicates that the flywheel will not furnish the necessary energy, it may be necessary to increase the weight, diameter or speed of the flywheel, or to use a different type of drive or motor.

For safety with respect to centrifugal force, cast iron flywheels should not have a speed greater than 5500 sfm.

To change the press speed, speed-change devices must change the speed of the flywheel. Energy of the flywheel is directly proportional to the square of its speed of rotation. Therefore, the standard energy of a variable-speed press is calculated at its slowest speed. The intended operating speed should be used in checking the suitability of a press for a specific operation.

Motor sizes specified for punch presses frequently are high for blanking and coining and low for deep drawing. If the flywheel alone can deliver the total energy requirements, the motor need only return the wheel to speed, spreading its delivery of power over the entire cycle. The motor horsepower should be approximately $0.0048E \times$ spm, where E is the energy (in inch-tons) required per stroke, and spm the strokes per minute. Energy here is the force, in tons, required to do the operation, multiplied by the distance, in inches, through which the force acts.

Motors designed for punch presses can be slowed down 10 to 15% without serious overloading. In contrast, a general-purpose motor is limited to a slowdown of 3 to 5%.

In a drawing operation with a long working stroke, the flywheel may slow down as much as 20% in order to give up enough energy to do the job. A general-purpose motor on such a job would resist the slowdown and probably would take on a greater and greater proportion of the load. Unless the motor has considerable capacity, it will become overloaded. A punch-press motor will slow down and let the wheel do the work. Then the motor will return the wheel to speed during the comparatively long period between strokes.

Clutches and Brakes in Mechanical Presses

Clutches and brakes are essential to the operation of mechanical presses. No other part must work more perfectly if the press is to operate successfully. The clutch must deliver and control the surge of force that is required to shape the work metal. When the press runs continuously, the clutch transmits power from the flywheel to the main shaft. In a single-stroke press, the clutch must accelerate the rotating parts of the drive from stop to full speed at each stroke of the press. The brake must decelerate this moving mass in order to stop the slide at the end of each upstroke. The brake must be large and efficient enough to stop the press in an emergency, or during inching.

Clutches and brakes in presses that are stopped at the end of each stroke need more maintenance than those in presses operating continuously or stopped only a few times a day.

Positive clutches are used mainly on presses of less than 100-ton capacity. Positive clutches always are on the main shaft and use pins, keys or jaws to lock the shaft and flywheel together. They usually are engaged by a foot treadle or by an air cylinder.

Positive clutches can be engaged or disengaged only once during each press stroke. Usually a throwout cam disengages the keys, pins or jaws near the top of the stroke. These clutches can be arranged for one-stroke or continuous operation.

Positive clutches accelerate the slide very rapidly, because there is no slip. Because they are shaft mounted, they have a minimum mass to move.

Mechanical positive clutches cost less than other types and are compact and easy to operate, but are limited in many respects and usually require excessive maintenance. They are not recommended for one-stroke work, because wear on the clutch would be severe.

Pneumatic positive clutches are much more efficient than the mechanical types. A pneumatic positive clutch usually is a jaw clutch with 16 or more points of engagement. The jaws are engaged by an air cylinder or a diaphragm, and are disengaged by springs. No throwout cams are used. The clutches have electric controls like those used with friction clutches. A press with a positive pneumatic clutch can be used for a single stroke, operated continuously, jogged in either direction, and stopped for emergencies. The brake used with this type of clutch is usually spring-operated and air-released as a fail-safe measure.

Friction clutches are preferred to positive clutches for most press applications. They are mounted on either the crankshaft, eccentric shaft, intermediate gear shaft, or the drive shaft, as shown in Fig. 1. The location of the clutch-brake unit is determined by factors such as press size, press speed, type of clutch-brake unit, and proper attention to the inertia of the press drive.

Friction clutches use air pressure to force the friction surfaces together; brakes use springs. Brakes are spring-operated rather than air-operated, so the press will stop if the power or air fails. Acceleration and deceleration of the press drive parts is relatively shock-free; thus less maintenance is needed than with positive clutches.

There are three basic types of friction clutches: integral, combination and separate mounting. In the integral type of friction clutch, the clutch and brake are mounted at one point on the shaft, but are not built as a single unit on a common sleeve. In the combination type of friction clutch, the clutch and brake components are assembled on a common sleeve. Both the integral and the combination types use the same air chamber to actuate the clutch and release the brake simultaneously. The brake is actuated by springs as a fail-safe measure. An advantage of these clutches is that clutching and braking are mechanically synchronized. A disadvantage is that most of the clutch and brake parts must be started and stopped each time the press is single-stroked.

For faster single-stroke rates, a third type of friction clutch has been developed, in which the clutch and brake have separate mountings and separate controls. Separate air chambers engage the clutch and disengage the brake. The effect is to reduce the number of rotating parts and consequently to reduce inertia.

Another type of friction clutch uses air or hydraulic pressure to force together two bronze plates, one of which is a driving member. The plates are submerged in oil. This design improves heat dissipation — a problem always present in friction clutches used in high production.

The air chambers for friction clutches may be simple cylinders, diaphragms or tubes. Diaphragms and tubes have some operating advantages and usually are more easily serviced.

One of the greatest advantages of friction clutches is their compatibility with electric and electronic controls. In this respect, they are superior to the positive clutch, with the exception of the pneumatic type.

Some air friction clutches can be used as overload protection devices when mounted on the main shaft. The clutch air pressure is gradually reduced, while making stampings, until the clutch begins to slip. The air pressure is then increased about 5 psi above the slipping point. If a load greater than normal is encountered, the press will not carry through its stroke, because the clutch will slip and stall the press. The air-pressure adjustment must be repeated for every die setup.

Eddy-Current Clutches. A high degree of control over the press is the major advantage of eddy-current clutches, which are, in reality, press drives. Ram speed at any point on the stroke can be programmed.

Eddy-current drives consist of a constant-speed flywheel, a variable-speed clutch-and-brake rotor, and a stationary brake field assembly. The clutch-and-brake rotor is directly connected to the press drive shaft.

Clutching and braking are controlled by the current in the coils. In practice, the press drive usually is controlled automatically to speed up the motion of the slide during the idle portion of the stroke, and to slow it down just before the tooling contacts the work. Figure 6 shows accelerated-motion curves for slides operated in this manner.

Clutch Location. The location of the clutch-brake unit depends on the type and amount of work to be done, press size, slide velocity, and the type of unit. For maximum torque, the clutch should be mounted on the shaft that revolves at the highest speed. However, mounting a friction clutch on the slowest shaft will make it more efficient, in terms of heat generation, because fewer and slower operating parts must be stopped and started.

The torque requirement of the clutch is reduced in direct proportion to its increase in speed. The heat generated by a friction clutch increases as the square of the speed.

In nongeared presses, the clutch usually is in the flywheel. In single-gear drive presses, the clutch is more often in the housing of the main gear, as shown in Fig. 1(b).

In twin-gear drive presses, the clutch is usually in the flywheel, which is mounted on the pinion drive shaft, as shown in Fig. 1(c). In multiple-reduction presses, the clutch is mounted on the intermediate gear (see Fig. 1d).

When the friction clutch is on the pinion drive shaft, overload protection is lost, because the load cannot be sensed through the gear train. Thus, some other form of protection is necessary.

Press Accessories

The setup and operation of a mechanical press are made more versatile through the use of built-in accessories. Included are bolster plates, rolling bolster assemblies, speed-change mechanisms, shut-height adjusters, and slide counterbalances.

Bolster plates are used in most hydraulic and mechanical presses, between the bed and die. They provide a flat surface on which to mount the dies and can be remachined to remove nicks and worn areas. T-slots in the top surface facilitate clamping the die to the bolster plate. Clearance holes through the bolster plate provide for pressure pins extending from the die cushion to the die. Some press beds have a large hole through the top surface for drop-through of parts or for mounting of die cushions. For this reason, bolster plates are thick to minimize deflection and to support the die properly. When parts are ejected through the die, holes of the proper size and location are cut in the bolster.

The width, length and thickness dimensions of bolster plates have been standardized for each bed size, as have the size and location of T-slots, pressure pin holes, and holes for fastening to the bed. Standardization facilitates interchangeability of dies between presses. Filler plates can be used either above or below the bolster to reduce the shut height of the press. This is in addition to the normal slide adjustment.

Rolling bolster assemblies are made for some large straight-side presses, for fast tooling change. Dies are set up on an assembly outside the press. When a press run is finished, the punch and blankholder are unclamped from the slides and the assembly is moved out of the press. Then another assembly is moved into place, the punch and blankholder are clamped to the slides, adjustments are made, auxiliary equipment is set up, and the press is made ready to run.

Speed-change drives are used mostly to change the number of strokes per minute, but some drives also can be used to change speeds during the press stroke for fast approach to the work, slower working stroke, and quick return. Changing speed during the stroke permits an increase in production without increasing the working speed. Press builders commonly supply charts that show the slide speed at any point on the press stroke.

In simple blanking operations the speed of the working stroke is not critical. In drawing and some forming operations, the plastic flow characteristics of the material being formed impose specific speed limitations.

A two-speed drive is combined with clutching and braking in some presses. A clutch can have planetary gears for a two-speed drive. Some two-speed drives have two flywheels with a common brake. With a two-speed gearbox and two speeds from the clutch, a press can have four speeds.

Variable-speed drives may incorporate a speed-change belt with adjustable cone pulleys connecting the motor to the flywheel or a steplessly variable electric drive. The eddy-current drive, originally developed for inching of mechanical presses, also provides variable speed.

Shut-height adjustment is provided in mechanical presses for changing the distance between the slide and the bed to fit dies of different sizes.

Small, single-point presses have a screw arrangement to provide this adjustment. In heavier presses, a gear drive makes it easier for the operator to move the massive slide. As press size increases, this gear drive is motorized. Motorized slide adjustment is used also in many smaller presses. Air counterbalances on most large presses relieve the load of the slide and die from the adjusting mechanism.

Some presses have dials that indicate the shut height in thousandths of an inch. If the dies to be used in the press are similarly marked, die-setting time will be greatly reduced. Other presses have a motorized adjustment with a dial control. The operator sets the desired shut height, and the slide automatically positions itself.

Counterbalances in press slides provide smooth cycling and reduce backlash and gear wear, by:

1. Counteracting the moving weight of slides, components and die members attached to the slides
2. Reducing the load on the press brake, thus providing faster stopping
3. Taking up clearance on the main bearings, reducing the breakthrough shock for cutting operations
4. Reducing backlash in the drive gearing
5. Easing the adjustment of slides by reducing the load on the adjusting screws.

Excessive counterbalance pressure can prevent the normal breathing of bearings and consequently can prevent good lubrication of the bearings.

Most presses manufactured with slide counterbalances use pneumatic cylinders as a counterbalance means, although springs have been used. To prevent too great an increase of pressure through the full range of press stroke, a surge tank is used in conjunction with the cylinders. The tanks are of such size that the pressure does not increase more than 20 to 25%. A pressure control valve allows the counterbalance pressure to be adjusted to take care of variation in die weights.

The counterbalance cylinder is attached to the press frame, and the cylinder rod to the press slide. Usually the cylinders are attached either to the crown or the press uprights.

Die Cushions

A die cushion is a tooling component that provides a reasonably uniform blankholding force in a single-action press (Fig. 9a) or secondary blankholding force in a double-action press (Fig. 9b). Die cushions are used also in stripping finished parts from the punch or the die.

Blankholding and stripping action in blanking and piercing dies and shallow forming and drawing dies for single-action presses can be achieved by springs or rubber pads incorporated in the die design. The force exerted by compression springs increases as they are depressed. For a relatively small increase in force, extremely long springs are required. On deep draws, the increase in blankholding force and the decrease in flange area under the blankholder result in an increase in force per unit area (pressure) on the flange metal. The workpiece will fracture if the force pulling it into the die exceeds the tensile strength of the work metal.

Die cushions are often used in double-action presses to keep a stamping flat, to hold it to shape, or to prevent it from slipping and distorting during drawing. Die cushions are used also to actuate the ejector plates or pins that push the finished part from the die cavity or from the punch.

In a single-action press using a die cushion (Fig. 9a), the punch is on the lower shoe, and the die is on the upper shoe. Through pins, force is applied to the pressure ring by the die cushion to hold the outer portion of the blank against the die during drawing. The force increases a maximum of 20 to 25% throughout the stroke. (If springs were used instead of the die cushion, the force would increase much more during the working stroke.) The die cushion is used also to remove the part from the punch.

In a double-action press (Fig. 9b), the die is on the lower shoe, and the punch is on the inner slide. A blankholder carried by the outer slide holds the blank against the die during drawing. The die cushion is used to hold the blank firmly against the bottom of the punch and to eject the formed workpiece from the die.

The capacity of a die cushion generally is 15 to 20% of the rated capacity of the press; some cushions have a capacity about one-third that of the press. The stroke of the die cushion generally is one-third to one-half the length of the press stroke. The size of the press bed opening limits the size, type and capacity of the cushion.

Pneumatic die cushions are used for short, fast operating strokes and when

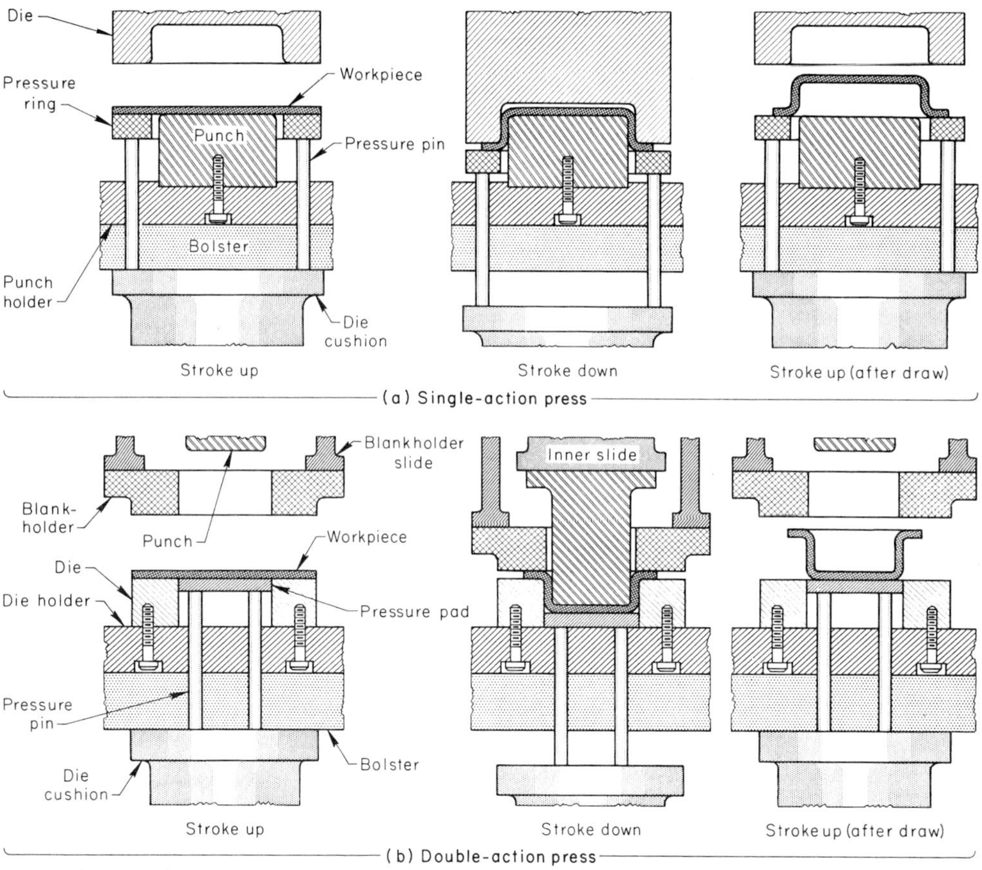

Fig. 9. Setups incorporating die cushions in single-action and double-action presses

the capacity does not need an operating air pressure of more than 100 psi. Hydropneumatic die cushions are recommended when the capacity is greater than can be obtained with 100 psi in a pneumatic cushion. A hydraulic die cushion can be used when hydraulic power is available. Hydropneumatic and hydraulic die cushions are slower acting than pneumatic cushions, and therefore generally are used in larger and slower presses.

A pneumatic die cushion consists basically of a cylinder, a piston and a surge tank. Either the piston or the cylinder is fixed. Downward movement of the blankholder, through pressure pins, forces the movable element against a cushion of air inside the cylinder and moves the air back into the surge tank. On the upstroke, the air in the surge tank forces the movable part back to its original position.

The air in the cushion cylinder and surge tank compresses on the downstroke of the press slide, and the force exerted by the cushion can increase as much as 20 to 25% over a full cushion stroke. This buildup in pressure is not enough to be detrimental to a drawing operation.

Hydropneumatic Die Cushions. Figure 10 shows the essential components of a hydropneumatic die cushion. Two individually controlled air lines are required — one to the operating valve of the cushion, and one to the top of the surge tank. The air pressure supplied to the air-piston assembly determines the tonnage capacity of the cushion on the downstroke. The air pressure in the surge tank determines the stripping pressure on the upstroke.

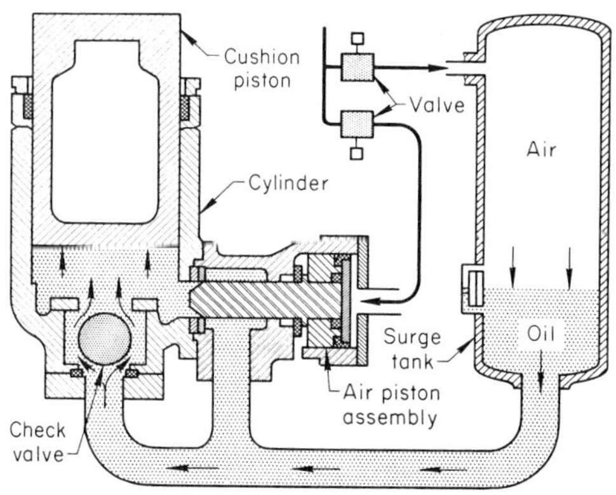

Fig. 10. Components and mechanics of a hydropneumatic die cushion

Auxiliary Equipment

Most primary press operations are automated, so that equipment for feeding and unloading is used even for fairly short runs. Hand feeding, with its attendant hazards, is often confined to second operations on partly completed workpieces.

Planning for automated operations should include the following goals:

1 Maximum safety to the operator and to the equipment
2 High or nearly continuous production
3 Improved quality of the product and minimum scrap
4 Reduction in cost of the finished parts.

The shape and position of the part before and after each operation must be carefully studied to determine whether design changes, such as providing tabs or extra stock on the blank, will facilitate handling.

Automatic handling equipment can be divided into the following categories: feeding equipment, unloading equipment, and equipment for transferring the work from one press operation to the next.

Straighteners are used to remove coil kinks and to flatten stock before feeding into a die. Roller levelers improve formability of the stock by plastic working. Coil-handling equipment moves coiled stock to the press area and uncoils it with a minimum of damage to the stock and danger to the tools and operator.

Feeding Equipment (Feeds)

Mechanical feeds are important for high production, for combined operations, and for operator and press safety. Some feeds supply the presses with stock from a strip or coil; others feed blanks or partly completed workpieces. Either kind, with or without auxiliary hand feeding, can be used with almost any kind of press. For progressive-die work, the feed length should be accurate and repeat within ±0.003 in. The stock must advance accurately so that the pilot pin can easily enter the piloting hole and position the strip. Too great a variation in feed length could result in distorted pilot holes and scrap parts.

Feeds for coil stock feed the work metal from a coil to the press. Choosing the optimum type of feed depends mainly on the type of press, strokes per minute, length of feed per stroke, accuracy needed, and the kind of strip — its width, thickness, stiffness and surface condition. The two most common kinds of feeds are slide and roll feeds.

Slide feeds are made in a variety of sizes and capacities. The basic principle of a slide feed is the use of a feed block that is moved between positive stops to advance the material the distance required at each stroke. Slide feeds are very accurate and are particularly suitable for use with coil stock. When strip stock is used, the ends of the strip must be hand fed into the press.

Some slide feeds are powered by the press through an eccentric mounted on the crankshaft extension. The eccentric can be a simple one-piece unit keyed to the crankshaft, or it can be adjustable to vary the feed in relation to the rotation of the crankshaft. When changes in feed length are frequent, the adjustable type is usually warranted.

The feed block is mounted on hardened slides and either has a feed-blade holder with an adjustable feed blade (usually carbide-tipped), or has a pair of eccentric gripping cylinders. The material is gripped during the feed stroke and released on the return stroke. Accurate control of feed length is obtained by the use of adjustable stops.

The direction of feed — left to right, right to left, or front to back — is governed by the location of the crankshaft extension on the press and the arrangement of the die. A mechanical slide feed, feeding from left to right on a press with a front-to-back crankshaft, can be provided by using an appropriate linkage.

Slide feeds also can be powered by air cylinders or cams on the press slide.

Air-actuated slide feeds are recommended for production jobs involving long feed lengths and small press tonnages. This feed is adaptable to hydraulic presses or other presses without accessible crankshafts, and the feed length can be easily set.

Cam-actuated slide feeds can feed stock in any direction and can be used on presses that do not have accessible crankshafts. The propelling force is generated by one or more springs that are compressed by the action of the cam as the punch slide descends. The springs feed the stock into the die on the upstroke of the press. The mechanical action of a cam feed is shown in Fig. 11.

Cam-actuated feeds are inexpensive devices, and are ordinarily mounted permanently on the dies with which they are used. They are well suited for fast blanking operations and for use with progressive dies. They can feed stock up to 8 in. wide and 0.031 in. thick. (One type can feed stock up to 0.050 in. thick, but only in narrow widths.)

Cam-actuated feeds use gripper blades or rolls to grip the stock during feeding. The feed shown in Fig. 11 has a pair of gripping rolls on the slide, which grip the stock during the feed stroke. A pair of stationary rolls prevents the stock from moving back when the slide is retracted. The rolls are eccentrically mounted so that they grip and release the stock automatically. This feed is not suited to a long feed length. Because of the cam action, the length of feed cannot exceed the approximate length of stroke of the press. The length of the springs also has an effect on feed length.

Another type of spring-powered feed has a rack slide in the body of the feed. With it, a feed length about four times as long as the press stroke is possible. This feed has a stock release that opens the pinch rolls, making the feed suitable for use with dies with pilots. Because only the edge of the material being fed is gripped, wide stock can be used. This feed is particularly suitable for use in draw operations.

Roll feeds are available in sizes suitable for use with almost any width and thickness of stock and are used in every type of presswork, from blanking to complex operations in progressive dies.

A roll feed consists essentially of a pair of rolls that can turn in one direction only. The rolls exert pressure on the stock by the use of springs or some other device. The rolls are rotated by the motion of the press crankshaft.

Roll feeds are suitable for extremely thin material and material with highly polished surfaces. If hard-chromium-plated rolls are substituted for standard ground-steel rolls, polished surfaces will not be scored or marked during feeding. Rubber-coated or plastic-coated rolls can be used on soft finished or prepainted stock.

There are two advantages in using roll feeds for feeding thin stock. With patterned rolls, a flange can be formed on a waste edge of the stock as a stiffener. With a single roll feed, the stock usually is pulled through the die.

The best method of feeding extra-thin stock is the use of double roll feeds (Fig. 12), in which roll feeds at each side of the die are set so that the stock between them is always under slight tension. Double roll feeds eliminate manual feeding of end sections when strip stock is processed, and are suitable only when a substantial scrap skeleton remains.

Rack-and-pinion-actuated roll feeds are available in almost all sizes but are used most in relatively heavy stamping and drawing operations. In larger presses it is common to use double roll feeds of the rack-and-pinion type that are attached to the press bolster.

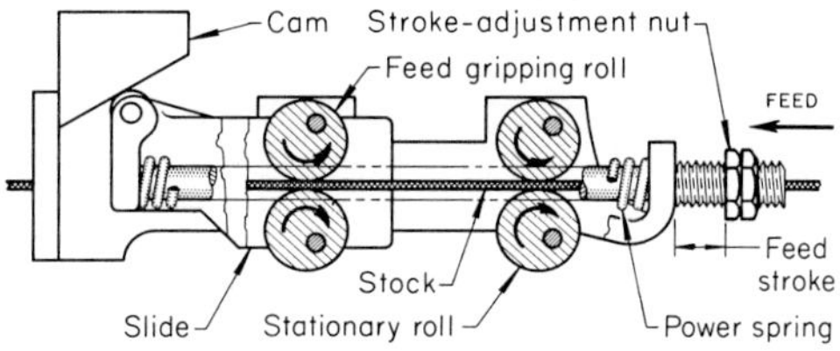

Fig. 11. Cam-actuated slide feed with pinch rolls that grip the stock for feeding. Arrows on the rolls indicate gripping direction. Reverse rotation of rolls releases the stock.

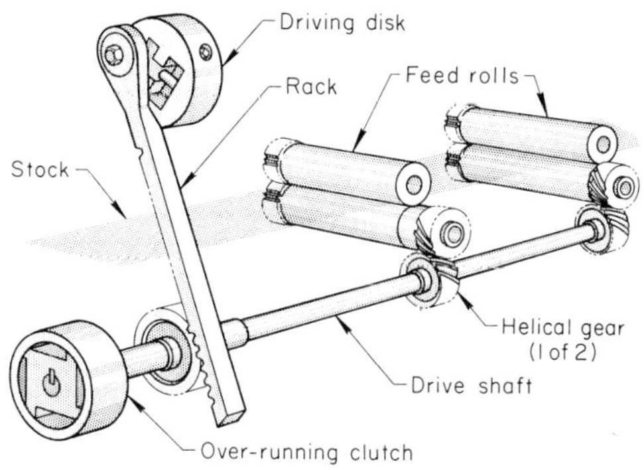

Fig. 12. Components of a double roll feed

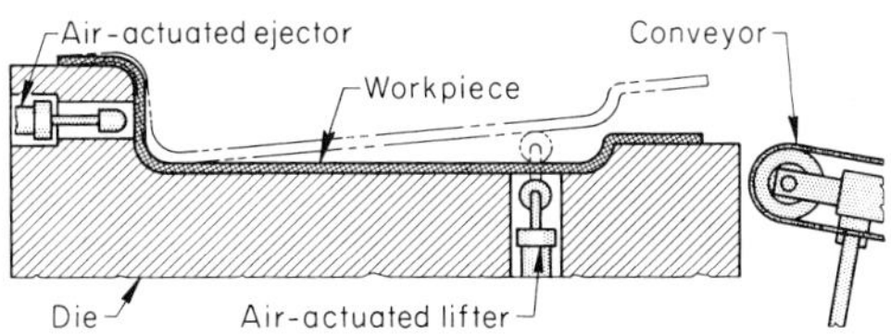

Fig. 13. Air-actuated device for lifting a workpiece from the die cavity and ejecting it onto a conveyor belt

Roll vs Slide Feeds. Roll feeds are the most used because: (*a*) they can feed stock as wide as 72 in. in increments as large as 90 in., (*b*) they are easy to set up and require little maintenance, and (*c*) they run well at high speed, with feeds of 1 to 3 in. per press stroke.

Disadvantages of roll feeds include: (*a*) lower accuracy than slide feeds, especially for longer feed lengths and faster speeds; (*b*) lack of a built-in measuring device; and (*c*) higher cost than slide feeds.

Disadvantages of slide feeds include: (*a*) less versatility than roll feeds, (*b*) greater difficulty in setting up than roll feeds, (*c*) slower feed than roll feeds, and (*d*) more marking of work metal.

Feeds for Blanks. Unlike slide and roll feeds, which are used solely with strip and coil stock, feeds for blanks are used with work metal that has already been blanked from the scrap skeleton, and for feeding partly completed workpieces into a press for secondary operations. The most common types are dial feeds, transfer feeds, and various hopper, chute and magazine feeds. Size and shape of the workpiece, production rate and quantity, and type of press are the principal factors in the choice of a feeding device.

Dial Feeds. Basically, a dial feed is a rotating table. It can either be fixed to the press bed, or be used as a separate piece of equipment. The dial is set up to rotate and index, as required, to deliver the workpiece from station to station. One type of dial takes the work to a series of dies or work stations; another carries the dies beneath a series of rams or punches. Workpieces sometimes are loaded into dial feeds by hand; more often, dial feeds are loaded from hoppers, chutes, vibrator feeders, or feed fingers.

Dial feeds are used for making slots, one at a time, in stator and rotor laminations.

Dial feeds can be applied to blanking, coining, staking, piercing, forming and drawing operations. Their use is limited only by the size of the dies and the workpiece. Dial feeds have been used on presses exceeding 150 tons in capacity.

Transfer feeds are used for multiple-operation stamping when it is impractical to move the blank from station to station by using a strip, as in a progressive die. Coil stock, flat blanks, or preforms are fed to the first station. Transfer of the work material is done by cam-actuated gripping fingers shaped to fit the workpiece. The fingers grip the part, carry it to the next station, return to their initial station, and repeat the cycle.

Feeds for Small Parts. Air-operated feeds for small parts can be mounted on the side of a gap-frame press. Parts are transferred to the die from a magazine or a preload station outside the die area. The feed unit cycles the press, and an air jet ejects the part from the die. Pickup heads are interchangeable and may be actuated mechanically or magnetically, or by vacuum.

Another type of feed for small parts consists of an arm, movable in three planes, with adjustable fingers. The unit has a memory system, and the arm is programmed by manually passing it along the desired path. Once programmed, it will continue to cycle along the same path.

Hopper feeds are widely used for feeding partly completed workpieces or, occasionally, small blanks to the next operation. The workpieces usually are dumped by hand into a hopper, where they are oriented by the feed. There are many kinds of hoppers and hopper feeds; selection depends on the size and shape of the workpiece. They may be driven by the press or by a separate power source. Although most hopper feeds are made up of standard components, some modification is usually needed to suit the specific job.

Chute feeds are simple, gravity-operated feeds that are used almost exclusively with blanks or fairly flat workpieces. They usually have a device at the bottom of the chute that allows the blanks to drop out one at a time.

Magazine feeds are a development of the chute feed. They are gravity fed but have a slide at the bottom with a blade of the same thickness as the workpiece to push the bottom piece out.

Stack feeds can feed blanks and lengths of sheet or strip from the top of the stack by use of vacuum pads or suction cups that grasp the work. A magnetic or air-operated sheet separator almost always is needed with this type of feed, especially when working with thin or oily material. Most stack feeds have an air or mechanically actuated transfer arm capable of simple motions.

Unloading Equipment

Stampings can be removed from dies by falling through the die or sliding down an incline. Those, however, that need special handling must sometimes be removed by hand or by mechanical devices.

A short blast of air can be used to unload small light parts. The air-blast control valve can be operated by a cam on the crankshaft. When the press cycles, the cam opens the valve to permit a jet of air to escape near the top of the stroke, or just before the press stops.

Air-actuated ejectors can be used to remove parts that are too heavy for air-blast removal, particularly if the press cannot be inclined. A simple air-actuated ejector can propel the part onto a conveyor or into a chute after the part has been lifted out of the die cavity, as in the setup shown in Fig. 13.

Mechanical Unloaders. Various linkages, arms, trays, pusher slides, and ram-operated devices are used in dies to unload finished parts automatically. Many of these devices are designed for a specific die and are difficult to adapt to another die.

Commercial unloaders are actuated by air or hydraulic cylinders and are synchronized with the press stroke. These devices have a swinging or horizontal motion to lift and carry the parts from the dies. The finished parts are raised out of the die cavity by lifter units built into the die. From this position, jaws grip the parts and carry them from the press to a conveyor or transfer fixture.

A swinging-arm unloading device is mounted on the press frame above the die space, as shown in Fig. 14. Two air cylinders operate the device — one for swinging the arm and one for opening and closing the jaws. The jaws are opened and closed by a wedge cam, moved by an air cylinder. The locking cam prevents the jaws from opening while carrying a load. In addition, they can close, without adjustment, on stock of any thickness within the span of the jaw opening. The cylinders are controlled by limit switches tripped by the press slide or by the rotary-cam switch of the press. The jaws are replaceable with standard jaws, or jaws designed to suit the part being moved.

The unloader shown in Fig. 14 is mounted on rails and can be easily moved aside to permit die change or die repair. Further, the unloader can be positioned to grip the part at the best point for removal.

The straight-line press-unloader is an air-operated device with adjustable stroke. It uses the same kind of jaws as the swinging-arm type. The unloader reciprocates in a straight line to remove the part, and can be mounted on a portable A frame, on conveyors, or on the press. Parts can be unloaded from positive-knockout dies, which eject at the top of the stroke, by replacing the jaws with a basket or shovel.

Equipment for Transfer Between Presses

Workpieces that require processing in two or more presses are transported from one press to another by various types of transfer units. Conveyor or shuttle-type machines receive the workpiece from an unloader and move it to the next press, where it is loaded into the die either manually or automatically. The type of transfer equipment used depends on the material, weight, size and shape of the workpieces, on the orientation of the work required at the next station, and on the importance of preventing scratches.

Adjustable flat-belt conveyors can be used for transferring stampings in press-line operations. The conveyor can be lengthened or shortened quickly. The entire conveyor, or one end at a time, can be raised or lowered to suit the die.

Turnover devices (Fig. 15) are used to turn the workpiece over as it moves from one press to another. The workpiece is placed above the arm of the

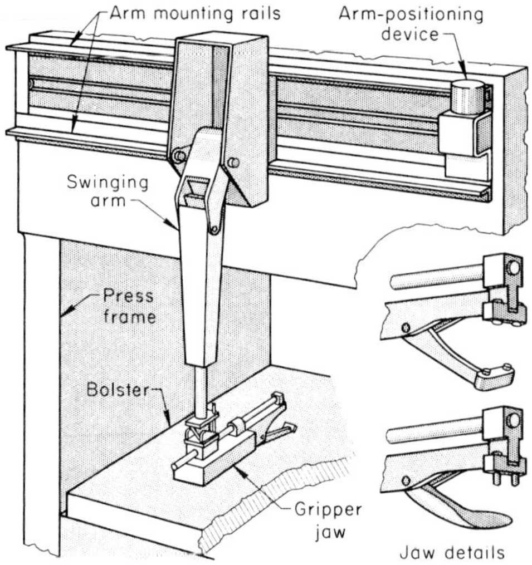

Fig. 14. Rail-mounted swinging-arm press unloader with interchangeable gripper jaws

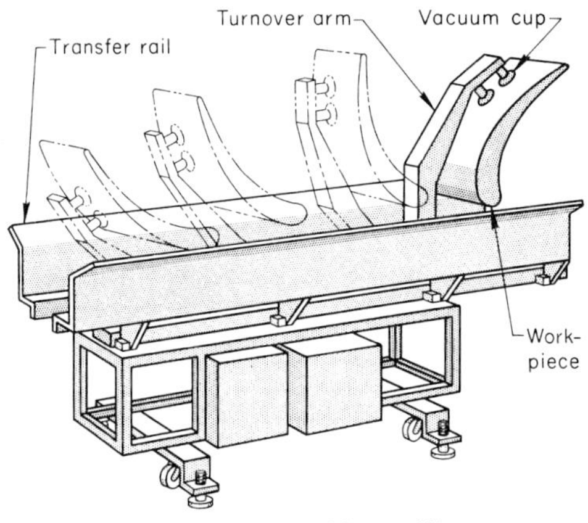

Fig. 15. Turnover machine with vacuum cups to hold workpiece during 180° rotation

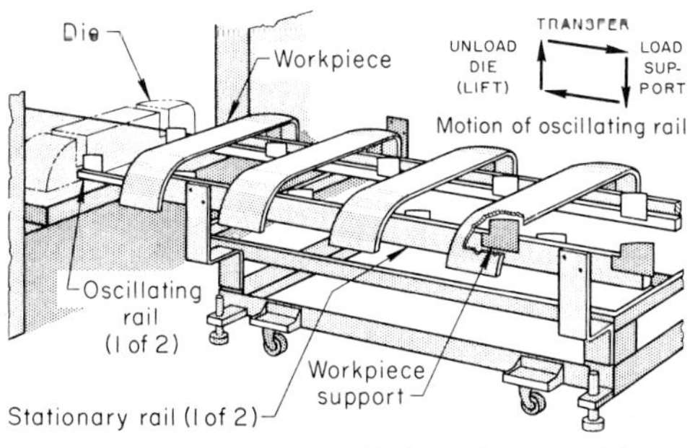

Fig. 16. Oscillating-rail transfer machine

turnover mechanism, the arm raises, the vacuum cups engage the workpiece, and the arm rotates approximately 180°, thus turning the workpiece over. The vacuum is released, and the arm returns and repeats the cycle. The workpiece is positioned above the arm by a press unloader. The arm can turn the part over from a fixed pivot point, or the arm can traverse between the presses to move the part from an unload to a load position. The turned-over workpiece is placed on a table ready for feeding into the next press.

Transfer devices of various designs move workpieces between presses. Some machines incorporate devices for unloading the die, turning the workpiece over, rotating, carrying the workpiece, and loading it into another press. Transfer is done with a shuttle table by reciprocating pusher dogs, or by chain or belt conveyors. The units are driven by pneumatic, hydraulic or electric power, and are synchronized with the presses.

A shuttle transfer, used between presses, is shown in Fig. 16. The oscillating rail removes the part from the die and carries it along the stationary rail to the next press. The transfer machine is supported on casters and can be moved aside for die repair or die change. Adjustable feet make it easy to set the rails to suit the die.

Applicators for Lubricants

In blanking or in forming, a lubricant usually is applied to metal that is fed into the press from coils. The lubricant can be swabbed or brushed on the metal as it leaves the reel, but this is inefficient and wasteful and produces inconsistent results. An automatic applicator improves efficiency and uniformity. The type of applicator used depends on whether the lubricant is a powder or a liquid and, if a liquid, on its viscosity and flow characteristics.

Liquid Lubricants. The simplest automatic applicator is a dropper suspended above the stock as close to the feed as possible. Usually these are manifold-type applicators so that orifices spaced along the stock width can be opened to control the flow of the lubricant. Disadvantages of this method are that only one side of the metal is coated, and that some lubricants for deep drawing are too thick to flow freely.

Wicks made of felt or other absorbent material can be used to apply the lubricant. The wicks are kept wetted with lubricant, and they wipe one or both sides of the metal as it passes between them. Wicks are inefficient with some viscous lubricants, such as chlorinated mineral oils, and the amount of lubricant delivered is limited.

Roller-coating devices, similar to those used for painting, can be used with most lubricants. They are installed just ahead of the feed, and do not waste lubricant. Sometimes the lubricator is placed between the die and feeder so that the lubricant will not affect the accuracy of the feed length.

Dry lubricants, in the form of powder or dried soap solutions, are used where liquid lubricants are unsuitable. An efficient arrangement for applying dry lubricants is described and illustrated on page 68 in the article "Blanking and Piercing of Magnetically Soft Materials", in this volume.

Straighteners

Straighteners have upper and lower rolls alternately mounted in a staggered position. The minimum number of rolls that can be used is three; however, five-roll or seven-roll straighteners are most common for the usual range of stock thickness. Straightening of stock less than 0.020 in. thick requires more rolls; as many as 17 have been used for some thin stock.

When three-roll straighteners are used, frequent adjustment of the center roll is necessary to compensate for the change in curvature of the metal from the outside to the inside of the coil. The metal unrolled from a new coil has a large radius of curvature, and only a small amount of offset between the rolls is necessary to remove the curvature. As the coil is unwound, the roll offset must be increased, because the radius of curvature of the coil has decreased. If the offset is too great, a reverse curvature is put in the stock.

Straighteners with five or more rolls use this principle by bending the stock more than is necessary to remove the smallest radius of curvature. The following set (or sets) of rolls have a smaller offset and straighten the metal, because the induced curvature is the same whether the outer or inner layers of the coils are being straightened. The arrangement of straightener rolls is the same as that of the leveling rolls illustrated in Fig. 17(a).

Some straighteners have a separate screw adjustment for each of the upper rolls; others have one adjustment for the entire series of upper rolls. A straightener should not be overloaded. When stiff, thick metal is passed through a straightener designed for thin metal, it may deflect the rolls permanently or break their shafts. Stiff, thick stock requires larger, stronger rolls spaced well apart. Thin metal requires more straightening rolls than thick metal. These rolls usually are smaller in diameter and more closely spaced.

Stock straighteners are available in a wide range of capacities and speeds, with powered or nonpowered straightening rolls. Either the upper or lower set of rolls, or sometimes both sets, are powered. Nonpowered rolls can be used when there is enough pulling or pushing force to get the stock through the rolls. Powered pinch rolls are used to push or to pull the stock through the straightener.

Thin stock requires more working to straighten than thick stock. For this reason two sets of pinch rolls are used and all straightening rolls are driven.

The speed of powered straighteners can be adjusted so that the material is delivered by the rolls at the rate it is fed into the press, plus 10%. The ideal condition is to have the stock run through the rolls continuously, so that there are no breaks or bends in the stock when it is stopped on the rolls. When straighteners are operated intermittently, breaks or bends occur in the stock and are almost impossible to remove.

The size of the slack loop between the straightener and press can be controlled by several devices. A paddle or roller above or below the stock, and connected to an arm, controls the size of the loop by providing a signal to operate a clutch. A photoelectric cell or proximity switch regulates the loop size without contacting the metal surface. This is an advantage when using a very soft metal, or stock with a polished surface, such as stainless steel, or preplated or prepainted stock. All of these surfaces can be damaged by a rolling or sliding control device.

Straighteners often are combined with reels or cradles, or with feeds, to conserve floor space. Separate units are easily moved from press to press.

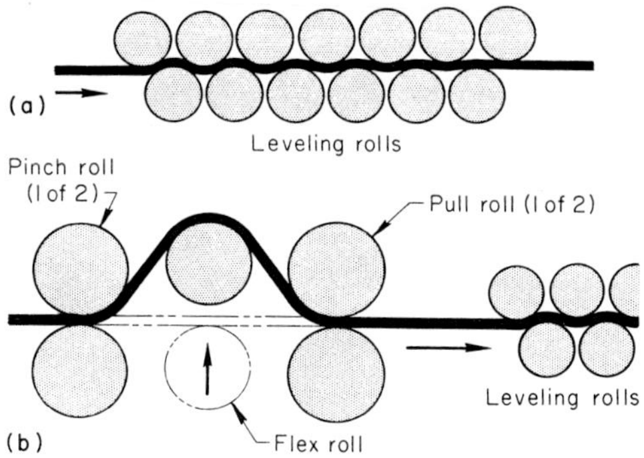

Fig. 17. Arrangement of rolls for roller leveling of sheet or strip with and without flex rolling (see text for discussion)

Roller Levelers

Roller levelers, like straighteners, have staggered pairs of meshing rolls, as shown in Fig. 17(a), but the rolls are smaller and more closely spaced, and as many as 39 rolls are used. All of the rolls are powered and some of the upper and lower working rolls have backup rolls. Levelers with backup rolls can impose strains on the metal to remove slack edges or a crowned center.

To secure good leveling, the material must be worked beyond its yield point. This is done by adjusting the leveling rolls at the entrance end so that the metal is worked beyond the yield point. The exit end of the rolls is set open an amount equal to stock thickness. Very slight adjustment of the exit-end rolls will turn the strip up or down, depending on the roll arrangement. The tops of the lower rolls usually are level or parallel with the material. The leveling rolls in the upper bank are adjusted to achieve the proper cold working.

A more effective type of roller leveling includes a break roll, or flex roll. In this type, the stock is gripped by a pair of pinch rolls and a pair of pull rolls (Fig. 17b). After the sheet is gripped by these rolls, the flex roll moves upward, flexing the sheet as shown in Fig. 17(b). The flex roll is usually actuated by hydraulic or mechanical means and can be adjusted to any height within the limits of the machine. After flex rolling, the metal passes through a series of staggered rolls, as in conventional roller leveling.

Roller levelers can be equipped with washing facilities to provide uniformly clean blanks for feeding into the dies. Brush rolls and high-volume solvent sprays remove dirt, steel slivers, and mill lubricants from the surface of the stock.

Flex rolling is frequently done as a continuous operation in a blanking or shearing line as the metal is uncoiled. As in straightening, the rolls run at a speed slightly greater than the rate at which the metal is fed into the press. This eliminates the need to start and stop the flex roller, prevents bends or breaks in the metal, and assures flexing of the entire strip. Because of the age-hardening phenomenon, scheduling of stock to the drawing operation dictates to what extent in-line flex rolling or roller leveling can be utilized. Timing is important, because the metal will age harden within 24 hr after flex rolling.

In the event of die or press breakdown, a sheet flex roll processor may be needed to reroll the stock before it can be formed. Rerolling of cut blanks — that is, turning 180° and flex rolling again — is done on a blank for a critical part where stretcher strains could appear close to the leading and trailing ends of the blank. This is not needed for blanks cut from coil stock that was flex rolled in the shearing line.

The tension type of roller leveler is used to flatten metals too thin (0.020 to 0.005 in.) to be processed on ordinary roller levelers. Flatness is achieved by elongation of the strip, which results from the combination of bending stress and applied tension. The leveler has pre-flex stands and a roller-leveling station. Pre-flex rolls work the metal beyond its elastic limit; elongation, precisely controlled to a preset percentage, takes place, producing a flat, stress-equalized strip. Control of strip elongation is maintained through bridles that are interconnected by a mechanical draw drive.

The function of roller leveling is discussed in the article "Press Forming of Low-Carbon Steel", page 112.

Coil-Handling Equipment

Coil cradles, reels, uncoilers, re-coilers and other types of coil-handling equipment are important to the successful operation of a press.

Coil cradles may be either nonpowered or powered. In the nonpowered type, the stock is pulled from the coil by a powered feed, straightener or pinch rolls, or by the equipment being fed. A powered cradle is preferable for coils that weigh more than 2000 lb or when stock is going directly from the reel to the press feed.

In a powered cradle, the coil is supported by chain-driven or gear-driven rolls, or by a driven steel-slat conveyor belt. The drive should be automatically self-equalizing, to prevent skidding of the coil.

Coil cradles should have motors that can stand frequent starting of inertia loads. A slack loop is created between the coil and straightener or feed devices by starting and stopping the motor intermittently on signal from a dancer roll, paddle or other control. This intermittent operation may cause a standard motor to fail prematurely. With a variable time delay (electronic or adjustable-cam), the motor can overrun to a controllable extent after the control has commanded it to stop.

A variable-speed drive reduces the number of starts and stops, prolongs the life of the motor and drive, and often makes it possible to match the speed of the cradle to that of the machine being fed. A clutch can be used so that the motor will run continuously and the clutch is engaged to drive the slat conveyor or rolls only when stock is required.

Guide plates should extend between the support rolls and should be easily adjustable from either side for coils of different widths. Rotary guide plates are recommended only for tilting cradles, which are generally used to feed OBI presses, and for some heavy-duty shaft-driven cradles.

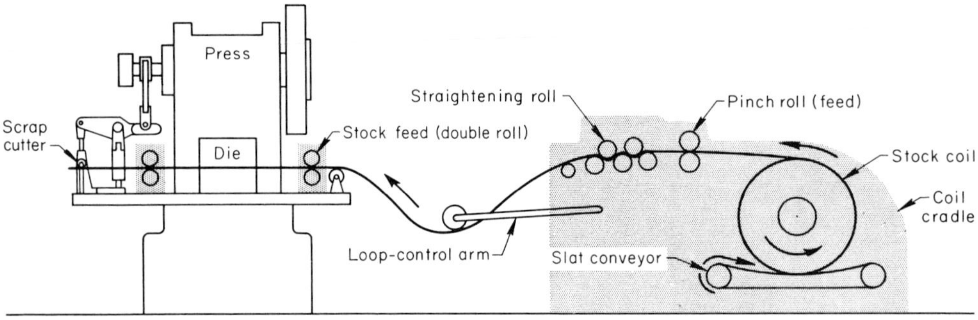

Fig. 18. Power-operated coil cradle, with straightening rolls and a loop-control arm, supplying coil stock to a double roll feed in a gap-frame press

Some cradles have rollers that reduce friction between the side of the rotating coil and the guide plates. These rollers protect the edges of some kinds of material.

Pinch rolls, or take-out rolls, add extra pull to the action of the support rolls, and should be used in cradles for feeding extra-thin or extra-heavy coil, for oily material, for spring-temper or lamination stock, for coils with slitting camber, and for feeding from irregular coils. Some pinch rolls swing out from the cradle to make loading easier. Pinch rolls are provided in some cradles, but they can also be bought as separate items.

Coil catchers, or bumpers, cushion the weight of the coil and lower it gently into position on the support rolls, or conveyor, to prevent shock to the equipment and damage to the coils.

Coil cradles often are combined with straightening heads. This results in a more compact unit requiring less floor space, reduces capital investment, promotes faster loading and adjustment, and reduces power requirements and maintenance. The addition of a straightener requires hold-down rolls to prevent the outer wrap of the coil from unwinding (hold-down rolls are useful even with a plain cradle), coil breakers, and peelers. These devices serve primarily to straighten the leading and trailing edges of the stock for easier feeding into the straightener, or for use with an automatic threader. A combination cradle-and-straightener supplying coil stock to a double roll feed is shown in Fig. 18.

Cradle Limitations. For some applications, a reel or other type of uncoiler is more suitable than a cradle. Ordinarily, cradles are not recommended for use with prefinished stock, thin material, telescoped coils, or soft material.

Prefinished Stock. The scuffing that inevitably occurs when stock is uncoiled from a cradle (even from a cradle with a cleaning and oiling attachment) will mar plated, polished, vinyl-coated and otherwise prefinished material.

Thin Material. When a heavy coil of thin material is unwound from a cradle, the stock may elongate unevenly between the support rolls and drift toward the side guides in a wobbling motion. Some producers of equipment do not recommend a cradle for use with stock less than 0.050 in. thick, but the thickness-to-weight ratio, rather than the thickness alone, is the critical factor.

When thin material is not strong enough to resist the corrective action of the side guides, a flange may be produced that cannot be removed by the straightener. Even slight flanging makes it impossible to feed the material into a die without corrective work.

Telescoped coils rub heavily against the side guides of a coil cradle and may be impossible to control. Such coils can usually be unwound more efficiently from a reel or other type of uncoiler.

Soft material, very loosely wound material, and egg-shaped coils also present problems when unwound from a cradle.

Despite the limitations mentioned, coil cradles and cradle-and-straightener combinations have many uses. If they are selected for the work expected of them, they are a highly efficient means of handling coil stock, especially in the heavier thicknesses. The main advantage of a coil cradle is that it unwinds the stock at a constant rate regardless of the outside diameter of the coil.

Stock Reels and Uncoilers. Commercial stock reels can accept coils weighing as much as 50,000 lb. There are reels of the proper size and type for almost any pressworking application.

Selection of a reel should be based on the maximum coil weight and the widths of stock to be unwound. In selecting a reel, it is better to overestimate future requirements than to underestimate and find that reel capacity limits improvement in equipment and production methods.

Plain or nonpowered reels usually are adequate when the press feed or stock straightener has pinch rolls with enough gripping power to pull the stock from the reel. When stock is going directly from reel to press feed, the reel should be powered so that the feed does not have the job of both feeding the press and unwinding the coil. If the stock becomes taut between the reel and the feed, the feed may start to advance and the stock slip, resulting in a short feed length. If a straightener is used between the press feed and the reel, a plain reel can be used. However, materials with low tensile strength and lightweight materials should be unwound from a powered reel; otherwise, they might be stretched between the reel and the feeding device.

Powered reels with variable speed and a loop control are preferred for smooth operation. Noncontact sensor units, such as photoelectric cells or proximity switches, on the loop control should be used for soft metals, polished surfaces, and prepainted stock. These prevent damage inherent with contact-type (rolling or sliding) sensor units. Without powered reels or loop control a sudden pull can cause the stock feed to slip and mark the work metal. A motor-driven reel with loop control and automatic centering is shown in Fig. 19. The arms expand to grip the inside of the coils, regardless of their inside diameter. Unreeling is smooth, and the coil is always centered.

Automatic centering by adjustable arms is available in plain and powered reels and on both single-head and double-head models. It is recommended for all but the smallest coil-processing applications. (Double-head reels have two reels on a common base. While one side of the reel is feeding the process equipment, the empty side is being loaded, thus minimizing downtime for changing reels.)

Reels for light duty are available in a wide variety of plate and arm reels. A horizontal pan reel feeds stock from either the inside or the outside of the coil. This type requires no centering of the coil, nor does it require a brake. It can be used for stock up to 8 in. wide and coil weights to 300 lb. Its best application is for narrow widths of springy material.

Plate reels, in which the stock is confined between two disks, are suitable for coils with a small inside diameter and a relatively small outside diameter.

Arm reels are used when the inside and outside diameters of the coils and stock widths vary considerably. Arm reels have manual adjustments for these variations. Most arm reels use quick-clamping forks for rapid loading and unloading of the coil. Some can be adjusted for height and can be tilted 90°, primarily for use with OBI presses.

Manual braking is controlled by a drag that can be adjusted to keep the stock from overrunning.

Automatic braking is preferable for pressworking of thin material. Usually, the stock loop between the press feed or stock straightener and the coil contacts an arm that engages and disengages the brake. When stock is being fed into the press, the length of the loop is reduced, causing the loop control to disengage the brake on the stock reel. The brake remains disengaged as long as the stock loop is not too large. When feeding stops, the stock loops down, dropping the loop-control arm, causing the brake to engage and prevent overrunning of the stock.

Medium and Heavy-Duty Reels. Reels with capacities of 2,500 to 10,000 lb are rated as medium-duty. Reels with capacities of 10,000 to 50,000 lb are rated as heavy-duty. Both types are available in plain and powered models. Automatic centering is essential with heavier coils.

Because they must support great weights from one side only, these reels must be massive and must have ample bearing capacity to support overhanging weight. Sometimes these reels are used with outboard supports to help counteract deflection.

Cone-Type and Arbor Reels. The cone-type uncoiler has opposed, steep-angle cones that enter the coil at each end. The greater the angle of the cone, the greater the range of inside diameters that can be handled. Minimum inside diameter is a limitation of cone-type uncoilers. Another limitation is that they are likely to deform the inner layers of the coil. This type of equipment cannot be used for stock less than 9 in. wide.

Stub-arbor uncoilers, which have two opposed stub mandrels, can uncoil stock as narrow as 4 in. in width. Screw-driven mechanical or hydraulic expansion of the jaws does not deform the stock. Stub-arbor uncoilers are loaded and unloaded rapidly.

Other equipment useful for handling coil stock includes re-coilers, turnstiles, downlayers, coil cars, coil grabs, and coil ramps.

Re-coilers are used for winding coil stock after slitting and for winding the scrap skeleton after pressworking.

Turnstiles (or horns) are two-arm or three-arm devices used to store coils temporarily before processing. In function, a turnstile resembles a coil ramp. Turnstiles may be equipped with hydraulic push-off devices, which add to their speed and efficiency.

Downlayers, sometimes called up-enders, are turnover devices for rotating the coil from horizontal to vertical position.

Coil cars are used for transporting heavy coils to the press service area.

Coil grabs, for use with cranes, are devices that can handle stock in the horizontal or vertical position. Some similar devices are available for use with fork-lift trucks. Other devices will pick up a coil and change the position from horizontal to vertical.

Coil ramps are inclined storage units for use with reels or cradles. Most coil ramps operate by gravity.

For additional information on material handling in pressrooms, see "Metalworking Automation", by G. H. De Groat (McGraw-Hill, 1962), and "Techniques of Pressworking Sheet Metal", by D. F. Eary and E. A. Reed (Prentice-Hall, 1958).

Operator Safety and Protection

The safest press is one operating continuously with a stock feeder and part unloader. This type of machine does not require the full attention of an operator, and there is no need for him to reach into the danger area. Flywheels, gears and other moving parts likely to catch an operator or passerby usually are covered.

For hand feeding, shields should be interlocked with press controls, so that the press will not run unless the shields are in place. Best practice is to make the guard or shield a part of the die, so that protection is automatically in place when the setup is made or installed by the setup man. Shields also can be attached to the press frame and adjusted for various kinds of work. These guards should suit all the work done in the press, should be easy for the setup man to adjust, and should give the operator an unobstructed view.

Usually, it is more difficult to guard hand-fed second operations, because the workpiece requires special handling. However, if production rate and quantity warrant the expenditure, standard or special devices can replace hand feeding of presses.

Available safeguards include (*a*) barriers or interlocking guards that keep the operator away from danger, (*b*) sweep and pulling devices that push the operator's hands away, and (*c*) devices that require both hands to trip the press. All safeguards should be inspected and adjusted before and after every press run.

Important considerations in choosing safety devices are: number of operators at the press, size and type of press, size and shape of workpieces, length of press stroke, and number of strokes per minute.

Protective devices cannot do the job by themselves; they should be used with a well-planned and strictly enforced safety program.

Rod or Basket Guards. For primary, strip-feed operations and for use with dies provided with feeds, adjustable barrier guards are used. These are usually attached to the press frame and have adjustable front and side sections that can be made to fit many dies. Some guards can be swung clear of the die area. These are interlocked with the press so that it cannot be tripped when the guard is out of position. The guards can also be attached to the die for operator protection. Operators must be trained not to remove or adjust them under any circumstances; adjustment should be the diesetter's responsibility.

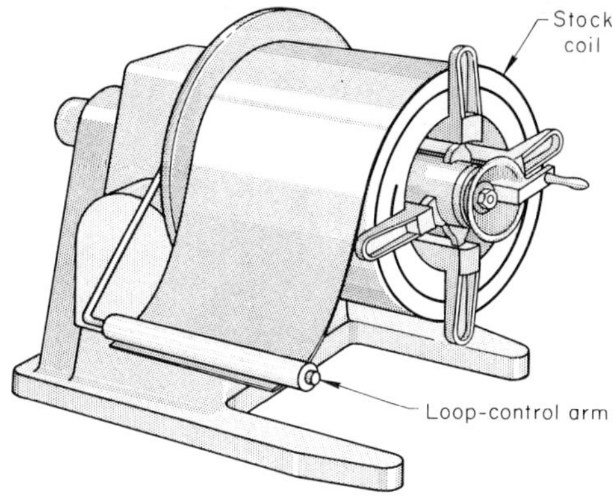

Fig. 19. Motor-driven automatic-centering reel with loop control

Drop gates usually have a barrier or gate that drops when the foot treadle is depressed. If the gate is blocked by the operator's hand or any other object, the press will not cycle. Operators should be trained not to reach around the barrier or tamper with it.

Drop gates can also be used as fixed barriers for strip operations. A deficiency of drop gates that are connected to the trip mechanism is that they do not guard against repeats.

Sweep guards have a single or a double arm that passes across the front of the press between the die and the operator. They are actuated by the press ram and are intended to brush the operator's hands out of the danger area. To be effective, they must sweep the front of the die area during the beginning of the press stroke. Sometimes the whipping action can injure the operator's hands. If an operator has a hand against a guide pin or other obstruction, the sweep guard can hold or trap the operator.

Sweep guards have been most effective on small presses. During slow, long-stroke operations using large dies, the operator may be able to reach over a sweep guard before the ram is at the bottom and be injured.

Pullbacks, attached to the operator's wrists or arms, pull his hands away from the danger area as the ram descends. Properly adjusted, they guard against repeats, major malfunctions that cause the ram to drop, and operator error.

Pullbacks must be adjusted after every change of dies or operator. They should also be checked regularly for wear. The cable must be attached to the operator in such a manner that it cannot wrap around a guide post or other obstruction in front of him; otherwise, his hand will be pulled into the danger area instead of away from it.

Two-hand controls are based on the principle that the operator must engage a control with each hand to operate the press and therefore cannot have either of his hands in the die area.

Three types are in general use: (*a*) mechanical controls that trip the press through a linkage system, (*b*) electric two-hand controls that actuate a solenoid or a solenoid air valve, and (*c*) air-operated two-hand devices that control an air cylinder that actuates the press. A special two-hand system, and one of the safest, has dual valves, pneumatically in series and electrically in parallel, with single-stroke operation and complete monitoring.

Although the principle of two-hand control is sound, many injuries have been caused by presses equipped with these devices. Almost invariably, the cause is inferior components or faulty installation. Merely wiring two switches in series or hooking up two air valves does not make a safe circuit. Nor is the device safe if the operator can tie down one of the switches and still operate, or if he can use his knee or elbow to depress one switch. Also, two-hand controls do not provide foolproof protection on presses with positive clutches at speeds below 200 strokes per minute.

Because of the many protective features that must be included in a truly safe two-hand control circuit, only reliable components should be used. High-reliability valves are made specifically for press control. Generally the design and installation of control circuitry are beyond the capability of the average shop electrician.

Die Protection

Electrical, mechanical or optical sensors can be incorporated in tooling or mounted on the press to detect stock buckling, misfeeds, doubles, failure of the workpiece to eject, and other malfunctions that can damage dies.

Electrical sensors include spring sensors, wire barriers, and proximity switches. A spring sensor basically is a piece of wire attached to an insulated rod. When a grounded metal part (for instance, a piece of buckled stock) touches it, it provides an electrical signal that can be amplified and used to stop the press.

Spring sensors are the least expensive protective devices, but they can be set off by oil droplets and sometimes are difficult to incorporate in progressive tooling.

Wire barriers consist of a grid of fine wires. Every other wire has ground potential. Ejected parts, striking the wires, make a bridge between the ground and the sensor wires and produce a signal. Usually the wires are arranged so that lack of a signal, indicating that a part was not ejected, will stop the press. They are not suitable for small flat work that can fly between the wires without touching them. Another disadvantage is that a part can hit the wires and fall back into the die if the barrier is improperly set up.

Proximity switches function in much the same way as spring sensors. They are easier to install but cost more.

Mechanical sensors include sensitive switches and transducer pickup plates. A sensitive switch functions when a part contacts and deflects a lever arm that closes a circuit. A transducer functions like a phonograph pickup — when an ejected part strikes a transducer control, a minute signal is generated. This signal can be detected by a sensitive circuit and amplified. Transducers are not suited to heavy parts, because of wear of the sensor, nor to press speeds above 250 strokes per minute.

Optical sensors consist of a photoelectric cell and a light source. Usually the light source makes a curtain of light (sometimes infrared) through which the part must pass. In doing so, it makes a signal, which is read by the photocell and is amplified.

Optical sensors are the most expensive. They have a high degree of reliability and flexibility, however, and their usefulness increases directly with the speed of the press.

Press Protection

Mechanical presses should be protected against overloading. (Hydraulic presses are inherently protected against overloading.) A small error on the part of the diesetter in adjusting the slide can result in an overload that can fracture the frame members or the crown. Even when the damage is less severe — sometimes the crankshaft or eccentric shaft will break — bearing damage almost always occurs.

There are two types of overload devices — permanent and replaceable. Both protect against overloads occurring near the bottom of the stroke. (Torque overloads occurring near the top of the stroke are far less severe a problem.) Overloads can be caused by double blanks, misfeeds, or improper die setting, or by trying to do a job in too small a press.

Replaceable overload devices are of either the stretch-link or the shear-washer type. The former, used only on underdrive presses, consists of a link that will stretch if a given load is exceeded. It is more an overload indicator than a protective device.

A shear washer (or collar) is mounted on the slide adjustment. It is designed to shear or fracture when overloaded, leaving sufficient displacement on the adjustment nut to relieve the overload, if it is within ½ in. of the bottom of the stroke (where most overloads occur).

Disadvantages of shear washers are: (*a*) they must be replaced after each overload; (*b*) they will fail under less than rated load after long use; and (*c*) because shear washers on a multiple-point-suspension press do not fail simultaneously, there can be severe side loading, which can damage the gibs.

Permanent overload devices do not require replacement after an overload. In most of these devices, a hydraulic cylinder is mounted in the slide adjustment. A preset pressure is maintained by a hydraulic pump, and a piston in the cylinder trips a limit switch if the preset load is exceeded. Another type uses air cylinders and a linkage attached to the tie rods of the press. It will allow the press crown to lift slightly if the preset pressure is exceeded.

These devices can be used to relieve a stuck press, and then can be quickly reset to the proper tension on the tie rods.

Friction clutches can be used for overload protection by regulating the air pressure or by automatically reducing air pressure near the bottom of the stroke with relief valves. Full air pressure is restored on the upstroke.

Slip couplings can be used with either positive or friction clutches on crankshaft presses. The couplings have a friction disk built into the flywheel and arranged so that the drive will slip under severe overload.

Presses for High Production

Mass-produced parts are often formed in presses that are made especially for high-production operation. High speed, or the highest number of strokes per minute, is not the only factor in a high production rate. The capability of a press to run continuously for several hours without full operator attention and with a minimum of wear and vibration contributes more to high productivity than running at high speed for a short period and then stopping for reconditioning of dies.

The more common types of high-production presses are dieing machines, multiple-slide machines, two principal types of transfer presses, and "high-productivity" presses.

Dieing machines are set up with conventional progressive dies for long-run operation. These machines are used extensively for the blanking of laminations; however, drawing and forming can be done. The height of the bed above the floor makes it easy to install stacking chutes for laminations and other parts.

Dieing machines are single-action underdrive presses. The drive mechanism for a dieing machine is located beneath the press bed. Four guide rods from a guided lower crosshead pass up through bronze bushings in the bed and are fastened to a platen to which the upper die half is attached. The lower crosshead is reciprocated by a crankshaft through connecting rods. By this action, the die halves are pulled together, rather than pushed together as in a conventional press.

The size of the guide rods and bushings results in excellent die alignment and long die life. The underdrive construction keeps the center of gravity of the press low. The progressive dies mounted in the machine are near eye level, and there are no columns or side members to obstruct the operator's view. Ejection chutes for finished parts and scrap are comparatively high above the floor so that containers are easy to position.

Pneumatic cushions, fastened to the top of the platen for better accessibility for service and adjustment, are used as strippers and blankholders.

Stock is fed through the guide rods with either single or double roll feeders. A scrap cutter can be mounted on the end of the machine. Both devices are operated by the upper platen or by a power takeoff on the end of the crankshaft.

Multiple-slide machines are fully automatic machines for mass production of small parts from metal strip or wire in coil form. Detailed information is given in the article on Forming of Steel Strip in Multiple-Slide Machines, page 154 in this volume. Tooling for parts made of wire is discussed in the article on Forming of Wire, page 347.

Transfer Presses. Two types of machines are called transfer presses. One type has single stations in which straight punches, die inserts that fit standard holders, transfer fingers and simple strippers are mounted. This type is the successor to, and often is still called, the eyelet machine. The second type is basically a wide-bed straight-side press with a horizontal slide for transfer-type operations. Individual single-operation or compound dies are used in these presses.

Eyelet-type transfer presses use many of the principles of the eyelet machine, but can form larger parts of heavier material and can perform many secondary operations with appropriate die design or attachments. Such operations include bottom stamping and marking, threading, beading, knurling and side lettering, side forming and piercing, and side slotting. Idle stations may be used (*a*) to provide space for mechanisms serving adjacent stations, (*b*) to position parts for special side marking, (*c*) to heat a part by an induction coil, or (*d*) to avoid moving an attachment already installed, powered and adjusted.

The tooling is mounted in each separate station and is actuated either by individual cam-actuated plungers for each station, or by a single cam-actuated ram acting against individually adjusted plungers. The single-station principle makes servicing and maintenance easy. The tools can be removed for repair or replacement without affecting the rest of the setup or its relationship, because the tooling elements are more nearly independent of other elements within the tool than with progressive dies.

The presses are used primarily for draw work. The first few stations involve blanking, an initial draw, and several redraws. Thus, elements developed for one job are common to many other jobs. In general, tool costs are about 50% of the cost of comparable progressive-die tooling.

The part is blanked in the first station from coil stock running at right angles to the tooling progression. The blank is moved from station to station with cam-actuated slides carrying transfer fingers. Because no stock allowance for pilots or carrying tabs is needed, as much as 30% less material is required than for comparable operations in a progressive die.

Multiple rows of blanks may be cut by using a re-coiling reel at the back of the press. After the stock has been run through the press once, it is transferred to the front and run again. A reversible feed system for multiple-row blanking is available. Another practice is to feed the stock through the first station, then loop the partial skeleton over the press and back through a second set of tools.

Setup time for a previously run job is less than one hour for each station. New jobs require more time, because the transfer fingers must be fitted to each individual station and the punches and dies may need alteration to eliminate wrinkles and draw marks. Tools from other jobs may be used in the initial stations, thus reducing toolmaking and setup time.

The production run needed to justify an eyelet-type transfer press compared to other methods is about 50,000 parts. On complex parts with fairly deep draws, the figure may be less than 25,000 parts. Runs of as few as 5000 parts have been made when anticipated production is not high enough to amortize other tooling, particularly a progressive die. Estimates of break-even point must be made to determine the economical length of run more closely.

Eyelet-type transfer presses have from 7 to 15 stations with a capacity of 15 to 75 tons, depending on the press design. Production rate ranges from 30 to 275 strokes per minute. This can be doubled by using a dual-feed arrangement. The maximum blank diameter ranges from 1½ to 3⅞ in., with a maximum draw depth of 13/16 to 3⅛ in., respectively. The maximum press stroke is approximately twice the maximum draw depth plus ¼ in.

Straight-side transfer presses are wide-bed straight-side presses in which a series of dies are set up to produce a finished stamping. Mechanical handling equipment takes a blank from the blanking die in the first station, or from a stack of precut blanks, and simultaneously transfers all stampings from one die to the next until the part is completed.

A straight-side transfer press setup is used when a part cannot be successfully produced on a progressive die and the additional cost above a series of separate dies is justified. Operations that sometimes become difficult or impossible to perform in progressive dies — such as deep drawing, restriking, cam piercing, and trimming — can be done in a transfer press. The transfer press eliminates the need for conveyor systems between presses or storage areas between operations.

The feed or transfer motion is basically the same on all straight-side transfer presses. It can be either press-driven or powered separately by electric motors, hydraulic motors, air or hydraulic cylinders, or a combination of these methods. Most transfer mechanisms are designed so that stampings are fed while the press is running through the top portion of its stroke. As the press closes through the midstroke position, the mechanical fingers are retracted. At the bottom of the working stroke the fingers return unclamped to the pickup position. The fingers clamp on the workpiece and the transfer cycle starts to repeat as the press opens through the midstroke position. All transfer dies in a set are designed to feed at the same level. In each station, the stamping must be elevated to the feed-line height of the die to permit transfer to the next die without interference.

Two basic types of stock feed are used — roll feed and stack feed. The choice of feed type is dictated by the following considerations:

1 Shape of the workpiece
2 Efficient use of roll stock
3 Availability of a blanking press
4 Effect of coil loading and its frequency on production rate
5 Effect of blanking die, if used, on other operations.

The roll feed brings coil stock into a blanking die. A stack feed is used for precut blanks, in which a blanking die is set up in a separate blanking press.

High-productivity presses are precision machines designed to operate continuously at speeds equal to or greater than conventional presses and are equipped with precision automatic feeding devices. They are available in gap-frame design, either fixed or inclinable, for relatively light operations. For work above about a 60-ton rating, straight-side presses are recommended because of their rigidity.

Gibs and slides, designed so that clearance and deflection are minimized, are machined to close tolerances and parallelism, because the alignment of the slide and bed is critical. Crankshafts, connections and bearings have been carefully designed to minimize vibration and cumulative bearing clearances, as both have an adverse effect on die life.

The advantage of carbide progressive dies can be fully utilized, because these presses are precision machines.

Definitions of Terms Relating to Presses

bed. The stationary, lower part of the press that serves as a table or support for the die or bolster plate. The bed in an inclinable press tilts back as the press is inclined, but otherwise the bed of a press is usually horizontal. The die may be bolted to the press bed, but usually is supported by a bolster that is secured to the bed. One or more fillers or risers can be used on the bed to decrease the shut height, putting the bolster atop the fillers, to raise the height of the stationary parts. The bed is sometimes called the lower platen, as in some hydraulic presses. When moving bolsters are used, the bolster, bolster-handling device, and supports replace the conventional press bed.

blankholder slide. The outer or upper slide of a double-action press that moves parallel to the main, or punch-holder, slide. It usually provides force to clamp the outer portion of the workpiece (blank) during drawing, to control metal flow. It may be used for blanking before drawing.

bolster. A plate or structure that holds the die, in some presses. Instead of being bolted to the bed, the die can be fastened to the bolster, thus raising the effective level of the press by the thickness or height of the bolster. JIC has standardized bolster dimensions and size and location of T-slots, pressure-pin holes and anchor holes. Some large, straight-side presses have bolsters that slide or roll out the side of the press for ease of die setup.

crown. The upper part of the press frame or structure, sometimes called the head, or the dome. Most presses have the press drive in the crown.

daylight. See *shut height of a press.*

die. A press tool, a pair of mating tools, or a combination of pairs of mating tools. Sometimes only the female (stationary) tool is called the die, and the mating male (moving) tool is called the punch. In an inverted die assembly, the punch is stationary and the die is movable.

die set. A unit consisting of an upper shoe, a lower shoe, and guide pins and posts. The die elements are attached to the shoes. Alignment and position are maintained with guide pins and bushings.

ejector. The mechanism for ejecting the workpiece from the die. The ejector may be actuated by a crossbar through a slot in the slide, a bell crank worked by a cam, a die cushion, springs or a rubber pad.

frame. The main structure of a press.

knockout. See *ejector.*

pitman. A connecting rod that carries force from a rotating or oscillating member to the slide. This member contains a means for adjusting the slide.

press slide. The part of the press that moves, carrying the punch, or upper part of the die, toward the mating tool part. On a hydraulic press it may be called the platen, or the upper platen.

pressure pad. A general term used for the part of a die that delivers holding pressure to the metal being worked. It can also be used as an ejector.

punch. The male part of a die; usually, the upper and moving part.

punch slide. The inner slide in a double-action press, which usually carries the punch. See also *press slide.*

ram. See *press slide.*

shut height of a press. Distance from the top of the bed to the bottom of the slide with the stroke down and adjustment up. In general, the shut height of a press is the maximum die height that can be accommodated for normal operation, taking the bolster thickness and any fillers into consideration.

slide. See *blankholder slide, press slide* and *punch slide.*

slide adjustment. The distance that the slide of a mechanical press can be moved to lessen the die space, decreasing the shut height. Adjustment may be made by hand or by power.

stroke. Distance the press slide moves from one end of its motion to the other.

strokes per minute. Rating of a press, in cycles of operation, for continuous running. When single-stroking a press (stopping after each stroke), strokes per minute may depend on the type of clutch and on the skill and dexterity of the operator in loading and unloading the die.

working stroke. That portion of the stroke that performs the work or operation (see Fig. 5 and 6). It cannot be greater than half the press stroke. In forming or drawing operations, the relation of the working stroke to the press stroke should be such that the slide velocity at the start of work does not exceed the drawing speed of the work metal.

Selected References

General Sources

Frank W. Wilson (editor in chief), "Die Design Handbook", prepared for the American Society of Tool and Manufacturing Engineers, Second Edition. McGraw-Hill, New York, 1965. [Section 23 (28 p) gives a detailed classification of the available types of press with concise descriptions of their principal features, and discusses the principles of press selection. The standards of the Joint Industry Conference of press users and manufacturers (JIC) for dimensions and markings of the principal types are reproduced. Section 22 (15 p) is an illustrated account of limit stops and safety guards for protecting presses and operators, and Section 19 (26 p) deals with most types of press feeds, with unloading devices, and with transfer equipment between presses.]

Frank W. Wilson (editor in chief), "Tool Engineers Handbook", prepared for the American Society of Tool Engineers, Second Edition. McGraw-Hill, New York, 1959. [Section 56 (24 p) gives a concise, illustrated classification of power presses (distinguishing clearly between knuckle-joint and toggle presses), summarizes the main features of clutches, brakes, controls, die cushions, feeding and unloading devices, and illustrates some special-feature presses. The JIC press standards are summarized, and there are notes on safety provisions.]

Heinrich Mäkelt, "Die mechanischen Pressen", Carl Hanser Verlag, Munich (W. Germany), 1966; 266 p. [A comprehensive textbook dealing with the design, construction, and uses of all types of mechanical presses, fully illustrated with diagrams and photographs. Includes sections on multiple-slide and transfer presses and feeding devices; 188 references. In German.]

Gerhard Oehler, "Die hydraulischen Pressen", Carl Hanser Verlag, Munich (W. Germany), 1962; 294 p. [A systematic account of the construction and characteristics of hydraulic presses of all types; 135 references. In German.]

L. Scherer, "Les Machines à Travailler les Métaux en Feuilles. Tome VI, Emboutissage", La Société des Publications Mécaniques, Paris (France), 1954; 106 p. [An illustrated catalog of presses of French manufacture, with dimensions and capacities. French safety requirements are listed. In French.]

Aldo Berruti, "Stampi e Presse", S. Lattes, Turin (Italy), 1951; 481 p. [The construction and operation of the basic types of press are described, illustrated with clear sketches, and accompanied by a systematic account of the functions of all press and die parts in blanking, bending, and forming operations. In Italian.]

A. Geleji, "Forge Equipment, Rolling Mills, and Accessories" (translated by J. Sivó), Akadémiai Kiadó, Budapest (Hungary), 1967. [A textbook on press design. Chapter 4 (47 p) considers calculations of forces and power requirements in cutting, bending and deep drawing with friction, eccentric, and crank-operated presses.]

C. W. Hinman, "Press Working of Metals", Second Edition, McGraw-Hill, New York, 1950. [Contains chapters outlining press types and their selection with simple computations of pressure and tonnage ratings; die construction; press accessories; the design of feeding equipment; formulas and reference tables with details of methods for developing shells and blanks.]

E. Molloy (editor), "Power Presses", George Newnes Ltd., London (England), 1945. [Includes chapters giving an illustrated survey of power press types (55 p), a brief survey of feed mechanisms (9 p), a useful account of fixed, automatic and interlock guards (21 p), and a discussion of the principles of lubrication (9 p).]

J. Dudley Jevons, "The Metallurgy of Deep Drawing and Pressing", Second Edition, Chapman and Hall, Ltd., London (England), 1941. [Includes chapters describing and illustrating the basic types of press (22 p), surveying press tool design and tool materials (70 p), and discussing in some detail the principles of lubrication (44 p).]

George Sachs, "Principles and Methods of Sheet Metal Fabricating", Second Edition revised and enlarged by Henry E. Voegeli, Reinhold, New York, 1966. [Chapter 16 (19 p) outlines the principles of press and tool construction and of lubrication practice.]

E. V. Crane, "Plastic Working of Metals and Non-Metallic Materials in Presses", Third Edition, John Wiley and Sons Inc., New York, 1944. [Basic press characteristics and some special-purpose modifications are described in Chapter 13 (42 p).]

J. A. Grainger, "Presswork and Presses", Second Edition, Machinery Publishing Co. Ltd., London (England), and The Industrial Press, New York, 1952. [Includes chapters describing and illustrating typical examples of presses (19 p), cushions for single-action presses (7 p), and the design of safety guards (13 p).]

John W. Langton, "Sheet Metal Working Machinery", Sir Isaac Pitman and Sons Ltd., London (England), 1963. [Chapter 4 (57 p) is a simple descriptive introduction to the main features of power presses.]

B. Kurseman, "Economics of Press Selection", ASTME Paper SP 63-68, 15 p, 1963. [Proper choice of press can reduce operating costs. Types considered include gap-frame, OBI (open-back inclinable), straight-side, high-productivity, and transfer presses.]

J. A. Grainger, Modern Trends in Power Press Design, *Sheet Metal Industries,* **39,** Oct 1962, 721-728. [Trend is seen to smaller, not overelaborate presses of greater power. High-speed friction clutches and high-energy-rate forming machines are described as under development.]

"Entwicklungstendenzen bei Stufenpressen", *Blech,* **12,** Aug 1965, 396, 398-400. [Compares design and performance of two 25-ton presses built in 1956 and 1964. Improvements include faster tool changes, higher speed of working, and more sophisticated control systems. Illustrated. In German.]

"Metalworking's Mechanical Press Handbook", Metalworking Publishing Co., Boston, 1960. [Describes the design, construction and use of mechanical presses. Discusses applications of straight-side, gap-frame, transfer and high-productivity presses.]

Special-Purpose Presses

J. H. Hayes, Automatic Presswork on Multi-Slide Machines, *Sheet Metal Industries,* **37,** Apr 1960, 267-276; discussion, 276-278. [Describes the design, operation, tooling, and application of multiple-slide machines, and discusses savings possible by their use.]

D. W. Bonnar, Press Automation with Short Run Flexibility, ASTME Paper 720, 11 p, 1965. [Discusses the use of moving-bolster presses integrated with automatic feeding equipment to give large outputs with quick change-overs.]

R. H. Eshelman, Change Tooling in Minutes on Sliding Bolster Press, *Iron Age,* **182,** Dec 25, 1958, 47-49. [Time of change-over can be reduced from 4-8 hr to 15-30 min.]

L. Schuler and E. E. Michaelis, The Speed of C-Frame Power Presses, *Sheet Metal Industries,* **35,** Sept 1958, 705-709. [A survey of the Schuler range of presses, using fast electromagnetic friction clutches to minimize tool wear at very high operating speeds.]

Feeding Devices

H. F. Hawkins and R. J. L. Lloyd, Development of Roller-Feed Mechanisms for Coiled Strip, *Sheet Metal Industries,* **39,** Feb 1962, 101-123; discussion, 123-126, 135. [Describes research leading to the development of improved mechanisms for positively controlled roller-transport feeds. Methods of combating the ill effects of drag, snatch, lack of flatness, and the presence of oil and dirt on accurate registration are examined.]

W. W. Schug, Dial Feeds Can Boost Your Press Production, *Tooling and Production,* **26,** Oct 1960, 65-68. [Recommendations for efficient use are given.]

A. P. J. Soepnel, Feeding Band and Strip Material to Automatic Presses, *Sheet Metal Industries,* **38,** Aug 1961, 588-594. [Discusses accuracy obtainable by roller and clamp transport and describes a twin-slide feeder.]

Safety

"Safety Code for Power Presses and Hand and Foot Presses", United States Standard USA B11.1-1960, United States of America Standards Institute, New York.

Frank E. McElroy (editor), "Accident Prevention Manual for Industrial Operations", Fifth Edition, National Safety Council, Chicago, 1964. [Chapter 22 (30 p) gives principles, types, methods and materials for guarding, with advantages and limitations of guard method for many machines and auxiliary transmission equipment; relates plant layout to safe practice. Chapter 23 (44 p) lists power press definitions, discusses purpose of operation guarding, auxiliary devices, feeding and ejecting devices for various presses, and handling, setting up, and removing heavy and light dies.]

A. P. J. Soepnel, Safety Devices for Eccentric Power Presses; Some Further Comments on Design of Eccentric Presses, *Sheet Metal Industries,* **35,** Oct 1958, 737-743; **36,** Oct 1959, 647-652, 656. [Constructional details of some safety features and devices are outlined.]

S. Hoffer, Electronic Die Protection Techniques, ASTME Paper SP 65-84, 18 p, Jan 1965. [Outlines the available systems and discusses their applications.]

E. E. Michaelis, Development in Photo-electric Guarding, *Sheet Metal Industries,* **40,** Dec 1963, 894-896, 904. [Describes circuits and guards developed by Institution of Automation, Munich.

Metric Conversion Data

Common Units for Converting From the English to the Metric (SI) System

The International System of Units (SI for short) is a modernized version of the metric system. It is built upon seven base units and two supplementary units. Derived units are related to base and supplementary units by formulas in the right-hand column. Symbols for units with specific names are given in parentheses. The information in this Data Sheet, adapted from the revised "Metric Practice Guide," Standard E380 ASTM, includes a selected list of factors for converting U. S. customary units to SI units.

Metric Units and Conversion Factors

Quantity	Unit	Formula
	Base Units	
length	metre (m)	
mass	kilogram (kg)	
time	second (s)	
electric current	ampere (A)	
thermodynamic temperature	kelvin (K)	
amount of substance	mole (mol)	
luminous intensity	candela (cd)	
	Supplementary Units	
plane angle	radian (rad)	
solid angle	steradian (sr)	
	Derived Units	
acceleration	metre per second squared	m/s^2
activity (of a radioactive source)	disintegration per second	(disintegration)/s
angular acceleration	radian per second squared	rad/s^2
angular velocity	radian per second	rad/s
area	square metre	m^2
density	kilogram per cubic metre	kg/m^3
electric capacitance	farad (F)	A • s/V
electric conductance	siemens (S)	A/V
electric field strength	volt per metre	V/m
electric inductance	henry (H)	V • s/A
electric potential difference	volt (V)	W/A
electric resistance	ohm (Ω)	V/A
electromotive force	volt (V)	W/A
energy	joule (J)	N • m
entropy	joule per kelvin	J/K
force	newton (N)	kg • m/s^2
frequency	hertz (Hz)	(cycle)/s
illuminance	lux (lx)	lm/m^2
luminance	candela per square metre	cd/m^2
luminous flux	lumen (lm)	cd • sr
magnetic field strength	ampere per metre	A/m
magnetic flux	weber (Wb)	V • s
magnetic flux density	tesla (T)	Wb/m^2
magnetomotive force	ampere (A)	—
power	watt (W)	J/s
pressure	pascal (Pa)	N/m^2
quantity of electricity	coulomb (C)	A • s
quantity of heat	joule (J)	N • m
radiant intensity	watt per steradian	W/sr
specific heat	joule per kilogram-kelvin	J/kg • K
stress	pascal (Pa)	N/m^2
thermal conductivity	watt per metre-kelvin	W/m • K
velocity	metre per second	m/s
viscosity, dynamic	pascal-second	Pa • s
viscosity, kinematic	square metre per second	m^2/s
voltage	volt (V)	W/A
volume	cubic metre	m^3
wavenumber	reciprocal metre	(wave)/m
work	joule (J)	N • m

Metric Conversion Factors

To convert from	To	Multiply by
atmosphere (760 mm Hg)	Pa	$1.013\,25 \times 10^{5}$
Btu (International Table)	J	$1.055\,056 \times 10^{3}$
Btu (International Table)/hour	W	$2.930\,711 \times 10^{-1}$
calorie (International Table)	J	4.186 800*
centipoise	Pa • s	$1.000\,000^{*} \times 10^{-3}$
centistoke	m^2/s	$1.000\,000^{*} \times 10^{-6}$
circular mil	m^2	$5.067\,075 \times 10^{-10}$
degree Fahrenheit	°C	$t_C = (t_F - 32)/1.8$
foot	m	$3.048\,000^{*} \times 10^{-1}$
$foot^2$	m^2	$9.290\,304^{*} \times 10^{-2}$
$foot^3$	m^3	$2.831\,685 \times 10^{-2}$
foot-pound-force	J	1.355 818
foot-pound-force/minute	W	$2.259\,697 \times 10^{-2}$
$foot/second^2$	m/s^2	$3.048\,000^{*} \times 10^{-1}$
gallon (U.S. liquid)	m^3	$3.785\,412 \times 10^{-3}$
horsepower (electric)	W	$7.460\,000^{*} \times 10^{2}$
inch	m	$2.540\,000^{*} \times 10^{-2}$
$inch^2$	m^2	$6.451\,600^{*} \times 10^{-4}$
$inch^3$	m^3	$1.638\,706 \times 10^{-5}$
inch of mercury (60 F)	Pa	$3.376\,85 \times 10^{3}$
inch of water (60 F)	Pa	$2.488\,4 \times 10^{2}$
$kilogram\text{-}force/centimetre^2$	Pa	$9.806\,650^{*} \times 10^{4}$
kip (1000 lbf)	N	$4.448\,222 \times 10^{3}$
$kip/inch^2$ (ksi)	Pa	$6.894\,757 \times 10^{6}$
ounce (U.S. fluid)	m^3	$2.957\,353 \times 10^{-5}$
ounce-force (avoirdupois)	N	$2.780\,139 \times 10^{-1}$
ounce-mass (avoirdupois)	kg	$2.834\,952 \times 10^{-2}$
$ounce\text{-}mass/ft^2$	kg/m^2	$3.051\,52 \times 10^{-1}$
$ounce\text{-}mass/yard^2$	kg/m^2	$3.390\,575 \times 10^{-2}$
pint (U.S. liquid)	m^3	$4.731\,765 \times 10^{-4}$
pound-force (lbf avoirdupois)	N	4.448 222
pound-mass (lbm avoirdupois)	kg	$4.535\,924 \times 10^{-1}$
$pound\text{-}force/inch^2$ (psi)	Pa	$6.894\,757 \times 10^{3}$
$pound\text{-}mass/inch^3$	kg/m^3	$2.767\,990 \times 10^{4}$
$pound\text{-}mass/foot^3$	kg/m^3	$1.601\,846 \times 10$
quart (U.S. liquid)	m^3	$9.463\,529 \times 10^{-4}$
ton (short, 2000 lbm)	kg	$9.071\,847 \times 10^{2}$
torr (mm-Hg)	Pa	$1.333\,22 \times 10^{2}$
watt-hour	J	$3.600\,000^{*} \times 10^{3}$
yard	m	$9.144\,000^{*} \times 10^{-1}$
$yard^2$	m^2	$8.361\,274 \times 10^{-1}$
$yard^3$	m^3	$7.645\,549 \times 10^{-1}$

*Exact

Multiplication Factors	Prefix	SI Symbol
$1\,000\,000\,000\,000 = 10^{12}$	tera	T
$1\,000\,000\,000 = 10^{9}$	giga	G
$1\,000\,000 = 10^{6}$	mega	M
$1\,000 = 10^{3}$	kilo	k
$100 = 10^{2}$	hecto*	h
$10 = 10^{1}$	deka*	da
$0.1 = 10^{-1}$	deci*	d
$0.01 = 10^{-2}$	centi*	c
$0.001 = 10^{-3}$	milli	m
$0.000\,001 = 10^{-6}$	micro	μ
$0.000\,000\,001 = 10^{-9}$	nano	n
$0.000\,000\,000\,001 = 10^{-12}$	pico	p
$0.000\,000\,000\,000\,001 = 10^{-15}$	femto	f
$0.000\,000\,000\,000\,000\,001 = 10^{-18}$	atto	a

*To be avoided where possible

English/Metric (SI) Stress Conversion Factors

Look up stress to be converted in boldface column. If in ksi (1,000 psi), read kg/sq mm and MPa in righthand column. If in kg/sq mm, read ksi in lefthand column.

Note: 1 MPa (megapascal)=1 MN per sq m (meganewton per sq m).

0 to 100

ksi		kg/sq mm	MPa	ksi		kg/sq mm	MPa
—	**0**	—	—	71.10	**50**	35.16	344.7
1.42	**1**	0.70	6.89	72.52	**51**	35.86	351.6
2.84	**2**	1.41	13.79	73.94	**52**	36.57	358.5
4.27	**3**	2.11	20.68	75.37	**53**	37.27	365.4
5.69	**4**	2.81	27.57	76.79	**54**	37.97	372.3
7.11	**5**	3.52	34.47	78.21	**55**	38.68	379.2
8.53	**6**	4.22	41.37	79.63	**56**	39.38	386.1
9.95	**7**	4.92	48.26	81.05	**57**	40.08	393.0
11.38	**8**	5.63	55.16	82.48	**58**	40.79	399.9
12.80	**9**	6.33	62.05	83.90	**59**	41.49	406.8
14.22	**10**	7.03	68.95	85.32	**60**	42.19	413.7
15.64	**11**	7.74	75.84	86.74	**61**	42.90	420.6
17.06	**12**	8.44	82.74	88.16	**62**	43.60	427.5
18.49	**13**	9.14	89.63	89.59	**63**	44.30	434.4
19.91	**14**	9.85	96.53	91.01	**64**	45.01	441.3
21.33	**15**	10.55	103.4	92.43	**65**	45.71	448.2
22.75	**16**	11.25	110.3	93.85	**66**	46.41	455.1
24.17	**17**	11.95	117.2	95.27	**67**	47.12	462.0
25.60	**18**	12.66	124.1	96.70	**68**	47.82	468.8
27.02	**19**	13.36	131.0	98.12	**69**	48.52	475.7
28.44	**20**	14.06	137.9	99.54	**70**	49.23	482.6
29.86	**21**	14.77	144.8	100.96	**71**	49.93	489.5
31.28	**22**	15.47	151.7	102.38	**72**	50.63	496.4
32.71	**23**	16.17	158.6	103.81	**73**	51.34	503.3
34.13	**24**	16.88	165.5	105.23	**74**	52.04	510.2
35.55	**25**	17.58	172.4	106.65	**75**	52.74	517.1
36.97	**26**	18.28	179.3	108.07	**76**	53.45	524.0
38.39	**27**	18.99	186.2	109.49	**77**	54.15	530.9
39.82	**28**	19.69	193.1	110.92	**78**	54.85	537.8
41.24	**29**	20.39	199.9	112.34	**79**	55.56	544.7
42.66	**30**	21.10	206.8	113.76	**80**	56.26	551.6
44.08	**31**	21.80	213.7	115.18	**81**	56.96	558.5
45.50	**32**	22.50	220.6	116.60	**82**	57.67	565.4
46.93	**33**	23.21	227.5	118.03	**83**	58.37	572.3
48.35	**34**	23.91	234.4	119.45	**84**	59.07	579.2
49.77	**35**	24.61	241.3	120.87	**85**	59.77	586.1
51.19	**36**	25.32	248.2	122.29	**86**	60.48	593.0
52.61	**37**	26.02	255.1	123.71	**87**	61.18	599.8
54.04	**38**	26.72	262.0	125.14	**88**	61.88	606.7
55.46	**39**	27.43	268.9	126.55	**89**	62.59	613.6
56.88	**40**	28.13	275.8	127.98	**90**	63.29	620.5
58.30	**41**	28.83	282.7	129.40	**91**	63.99	627.4
59.72	**42**	29.54	289.6	130.82	**92**	64.70	634.3
61.15	**43**	30.24	296.5	132.25	**93**	65.40	641.2
62.57	**44**	30.94	303.4	133.67	**94**	66.10	648.1
63.99	**45**	31.65	310.3	135.09	**95**	66.81	655.0
65.41	**46**	32.35	317.2	136.51	**96**	67.51	661.9
66.83	**47**	33.05	324.1	137.93	**97**	68.21	668.8
68.26	**48**	33.76	331.0	139.36	**98**	68.92	675.7
69.68	**49**	34.46	337.8	140.78	**99**	69.62	682.6
71.10	**50**	35.16	344.7	142.20	**100**	70.32	689.5

100 to 200

ksi		kg/sq mm	MPa	ksi		kg/sq mm	MPa
142.20	**100**	70.32	689.5	213.30	**150**	105.49	1,034
143.62	**101**	71.03	696.4	214.72	**151**	106.19	1,041
145.04	**102**	71.73	703.3	216.14	**152**	106.89	1,048
146.47	**103**	72.43	710.2	217.57	**153**	107.59	1,054
147.89	**104**	73.14	717.1	218.99	**154**	108.30	1,062
149.31	**105**	73.84	724.0	220.41	**155**	109.00	1,069
150.73	**106**	74.54	730.8	221.83	**156**	109.70	1,076
152.15	**107**	75.25	737.7	223.25	**157**	110.41	1,082
153.58	**108**	75.95	744.6	224.68	**158**	111.11	1,089
155.00	**109**	76.65	751.5	226.10	**159**	111.81	1,096
156.42	**110**	77.36	758.4	227.52	**160**	112.52	1,103
157.84	**111**	78.06	765.3	228.94	**161**	113.22	1,110
159.27	**112**	78.76	772.2	230.36	**162**	113.92	1,117
160.69	**113**	79.47	779.1	231.79	**163**	114.63	1,124
162.11	**114**	80.17	786.0	233.21	**164**	115.33	1,131
163.53	**115**	80.87	792.9	234.63	**165**	116.03	1,138
164.95	**116**	81.58	799.8	236.05	**166**	116.74	1,145
166.38	**117**	82.28	806.7	237.47	**167**	117.44	1,151
167.80	**118**	82.98	813.6	238.90	**168**	118.14	1,158
169.22	**119**	83.68	820.5	240.32	**169**	118.85	1,165
170.64	**120**	84.39	827.4	241.74	**170**	119.55	1,172
172.06	**121**	85.09	834.3	243.16	**171**	120.25	1,179
173.48	**122**	85.79	841.2	244.58	**172**	120.96	1,186
174.91	**123**	86.50	848.1	246.01	**173**	121.66	1,193
176.33	**124**	87.20	855.0	247.43	**174**	122.36	1,200
177.75	**125**	87.90	861.8	248.85	**175**	123.07	1,207
179.17	**126**	88.61	868.7	250.27	**176**	123.77	1,213
180.60	**127**	89.31	875.6	251.69	**177**	124.47	1,220
182.02	**128**	90.01	882.5	253.12	**178**	125.18	1,227
183.44	**129**	90.72	889.4	254.54	**179**	125.88	1,234
184.86	**130**	91.42	896.3	255.96	**180**	126.58	1,241
186.28	**131**	92.12	903.2	257.38	**181**	127.29	1,248
187.70	**132**	92.83	910.1	258.80	**182**	127.99	1,255
189.13	**133**	93.53	917.0	260.23	**183**	128.69	1,262
190.55	**134**	94.23	923.9	261.65	**184**	129.39	1,269
191.97	**135**	94.94	930.8	263.07	**185**	130.10	1,276
193.39	**136**	95.64	937.7	264.49	**186**	130.80	1,282
194.81	**137**	96.34	944.6	265.91	**187**	131.50	1,289
196.24	**138**	97.05	951.5	267.34	**188**	132.21	1,296
197.66	**139**	97.75	958.4	268.76	**189**	132.91	1,303
199.08	**140**	98.45	965.3	270.18	**190**	133.61	1,310
200.50	**141**	99.16	972.2	271.60	**191**	134.32	1,317
201.92	**142**	99.86	979.1	273.02	**192**	135.02	1,324
203.35	**143**	100.56	986.0	274.45	**193**	135.72	1,331
204.77	**144**	101.27	992.9	275.87	**194**	136.43	1,338
206.19	**145**	101.97	999.7	277.29	**195**	137.13	1,344
207.61	**146**	102.67	1,007	278.71	**196**	137.83	1,351
209.03	**147**	103.38	1,014	280.13	**197**	138.54	1,358
210.46	**148**	104.08	1,020	281.56	**198**	139.24	1,365
211.88	**149**	104.78	1,027	282.98	**199**	139.94	1,372
213.30	**150**	105.49	1,034	284.40	**200**	140.65	1,379

Look up stress to be converted in boldface column. If in ksi (1,000 psi), read kg/sq mm and MPa in righthand column. If in kg/sq mm, read ksi in lefthand column.

Note: 1 MPa (megapascal) = 1 MN per sq mm (meganewton per sq m).

200 to 300

ksi		kg/sq mm	MPa	ksi		kg/sq mm	MPa
284.40	**200**	140.65	1,379	355.50	**250**	175.81	1,724
285.82	**201**	141.35	1,386	356.92	**251**	176.51	1,731
287.24	**202**	142.05	1,393	358.34	**252**	177.22	1,737
288.67	**203**	142.76	1,400	359.77	**253**	177.92	1,744
290.09	**204**	143.46	1,407	361.19	**254**	178.62	1,751
291.51	**205**	144.16	1,413	362.61	**255**	179.32	1,758
292.93	**206**	144.87	1,420	364.03	**256**	180.03	1,765
294.35	**207**	145.57	1,427	365.45	**257**	180.73	1,772
295.78	**208**	146.27	1,434	366.87	**258**	181.43	1,779
297.20	**209**	146.98	1,441	368.30	**259**	182.14	1,786
298.62	**210**	147.68	1,448	369.72	**260**	182.84	1,793
300.04	**211**	148.38	1,455	371.14	**261**	183.54	1,800
301.46	**212**	149.09	1,462	372.56	**262**	184.25	1,806
302.89	**213**	149.79	1,469	373.98	**263**	184.95	1,813
304.31	**214**	150.49	1,475	375.41	**264**	185.65	1,820
305.73	**215**	151.20	1,482	376.83	**265**	186.36	1,827
307.15	**216**	151.90	1,489	377.25	**266**	187.06	1,834
308.57	**217**	152.60	1,496	379.67	**267**	187.76	1,841
310.00	**218**	153.30	1,503	381.09	**268**	188.47	1,848
311.42	**219**	154.01	1,510	382.52	**269**	189.17	1,855
312.84	**220**	154.71	1,517	383.94	**270**	189.87	1,862
314.26	**221**	155.41	1,524	385.36	**271**	190.58	1,868
315.68	**222**	156.12	1,531	386.78	**272**	191.28	1,875
317.11	**223**	156.82	1,538	388.20	**273**	191.98	1,882
318.53	**224**	157.52	1,544	389.63	**274**	192.69	1,889
319.95	**225**	158.23	1,551	391.05	**275**	193.39	1,896
321.37	**226**	158.93	1,558	392.47	**276**	194.09	1,903
322.79	**227**	159.63	1,565	393.89	**277**	194.80	1,910
324.22	**228**	160.34	1,572	395.31	**278**	195.50	1,917
325.64	**229**	161.04	1,579	396.74	**279**	196.20	1,924
327.06	**230**	161.74	1,586	398.16	**280**	196.91	1,931
328.48	**231**	162.45	1,593	399.58	**281**	197.61	1,937
329.90	**232**	163.15	1,600	401.00	**282**	198.31	1,944
331.33	**233**	163.85	1,606	402.42	**283**	199.02	1,951
332.75	**234**	164.56	1,613	403.85	**284**	199.72	1,958
334.17	**235**	165.26	1,620	405.27	**285**	200.42	1,965
335.59	**236**	165.96	1,627	406.69	**286**	201.12	1,972
337.01	**237**	166.67	1,634	408.11	**287**	201.83	1,979
338.44	**238**	167.37	1,641	409.53	**288**	202.53	1,986
339.86	**239**	168.07	1,648	410.96	**289**	203.23	1,993
341.28	**240**	168.78	1,655	412.38	**290**	203.94	1,999
342.70	**241**	169.48	1,662	413.80	**291**	204.64	2,006
344.12	**242**	170.18	1,669	415.22	**292**	205.34	2,013
345.55	**243**	170.89	1,675	416.64	**293**	206.05	2,020
346.97	**244**	171.59	1,682	418.07	**294**	206.75	2,027
348.39	**245**	172.29	1,689	419.49	**295**	207.45	2,034
349.81	**246**	173.00	1,696	420.91	**296**	208.16	2,041
351.23	**247**	173.70	1,703	422.33	**297**	208.86	2,048
352.66	**248**	174.40	1,710	423.75	**298**	209.56	2,055
354.08	**249**	175.11	1,717	425.18	**299**	210.27	2,062
355.50	**250**	175.81	1,724	426.60	**300**	210.97	2,068

300 to 400

ksi		kg/sq mm	MPa	ksi		kg/sq mm	MPa
426.60	**300**	210.97	2,068	497.70	**350**	246.13	2,413
428.02	**301**	211.67	2,075	499.12	**351**	246.83	2,420
429.44	**302**	212.38	2,082	500.54	**352**	247.54	2,427
430.86	**303**	213.08	2,089	501.96	**353**	248.24	2,434
432.29	**304**	213.78	2,096	503.39	**354**	248.94	2,441
433.71	**305**	214.49	2,103	504.81	**355**	249.65	2,448
435.13	**306**	215.19	2,110	506.23	**356**	250.35	2,455
436.55	**307**	215.89	2,117	507.65	**357**	251.05	2,461
437.97	**308**	216.60	2,124	509.07	**358**	251.76	2,468
439.40	**309**	217.30	2,130	510.50	**359**	252.46	2,475
440.82	**310**	218.00	2,137	511.92	**360**	253.16	2,482
442.24	**311**	218.71	2,144	513.34	**361**	253.87	2,489
443.66	**312**	219.41	2,151	514.76	**362**	254.57	2,496
445.08	**313**	220.11	2,158	516.18	**363**	255.27	2,503
446.51	**314**	220.82	2,165	517.61	**364**	255.98	2,510
447.93	**315**	221.52	2,172	519.03	**365**	256.68	2,517
449.35	**316**	222.22	2,179	520.45	**366**	257.38	2,523
450.77	**317**	222.93	2,186	521.87	**367**	258.09	2,530
452.19	**318**	223.63	2,193	523.29	**368**	258.79	2,537
453.62	**319**	224.33	2,199	524.72	**369**	259.49	2,544
455.04	**320**	225.03	2,206	526.14	**370**	260.20	2,551
456.46	**321**	225.74	2,213	527.56	**371**	260.90	2,558
457.88	**322**	226.44	2,220	528.98	**372**	261.60	2,565
459.30	**323**	227.14	2,227	530.40	**373**	262.31	2,572
460.73	**324**	227.85	2,234	531.83	**374**	263.01	2,579
462.15	**325**	228.55	2,241	533.25	**375**	263.71	2,585
463.57	**326**	229.25	2,248	534.67	**376**	264.42	2,592
464.99	**327**	229.96	2,255	536.09	**377**	265.12	2,599
466.41	**328**	230.66	2,261	537.51	**378**	265.82	2,606
467.84	**329**	231.36	2,268	538.94	**379**	266.53	2,613
469.26	**330**	232.07	2,275	540.36	**380**	267.23	2,620
470.68	**331**	232.77	2,282	541.78	**381**	267.93	2,627
472.10	**332**	233.47	2,289	543.20	**382**	268.64	2,634
473.52	**333**	234.18	2,296	544.62	**383**	269.34	2,641
474.95	**334**	234.88	2,303	546.05	**384**	270.04	2,648
476.37	**335**	235.58	2,310	547.47	**385**	270.75	2,654
477.79	**336**	236.29	2,317	548.89	**386**	271.45	2,661
479.21	**337**	236.99	2,324	550.31	**387**	272.15	2,668
480.63	**338**	237.69	2,330	551.73	**388**	272.86	2,675
482.06	**339**	238.40	2,337	553.16	**389**	273.56	2,682
483.48	**340**	239.10	2,344	554.58	**390**	274.26	2,689
484.90	**341**	239.80	2,351	556.00	**391**	274.97	2,696
486.32	**342**	240.51	2,358	557.42	**392**	275.67	2,703
487.74	**343**	241.21	2,365	558.84	**393**	276.37	2,710
489.17	**344**	241.91	2,372	560.27	**394**	277.08	2,717
490.59	**345**	242.62	2,379	561.69	**395**	277.78	2,723
492.01	**346**	243.32	2,386	563.11	**396**	278.48	2,730
493.43	**347**	244.02	2,392	564.53	**397**	279.18	2,737
494.85	**348**	244.73	2,399	565.95	**398**	279.89	2,744
496.28	**349**	245.43	2,406	567.38	**399**	280.59	2,751
497.70	**350**	246.13	2,413	568.80	**400**	281.29	2,758

Source: Based on Nickel Alloy Steels Data Book, Sec. 8, Appendix B. International Nickel Co.

Index